EXPLORATION OF THE PLANETARY SYSTEM

INTERNATIONAL ASTRONOMICAL UNION
UNION ASTRONOMIQUE INTERNATIONALE

SYMPOSIUM No. 65
(COPERNICUS SYMPOSIUM IV)
HELD IN TORUŃ, POLAND, 5–8 SEPTEMBER, 1973

EXPLORATION OF THE PLANETARY SYSTEM

EDITED BY

A. WOSZCZYK AND C. IWANISZEWSKA

Astronomical Observatory of Nicolaus Copernicus University, Toruń, Poland

D. REIDEL PUBLISHING COMPANY

DORDRECHT-HOLLAND / BOSTON-U.S.A.

1974

Published on behalf of
the International Astronomical Union
by
D. Reidel Publishing Company, P.O. Box 17, Dordrecht, Holland

Sold and distributed in the U.S.A., Canada, and Mexico
by D. Reidel Publishing Company, Inc.
306 Dartmouth Street, Boston,
Mass. 02116, U.S.A.

Library of Congress Catalog Card Number 73–94458

ISBN-13:978-90-277-0450-4   e-ISBN-13:978-94-010-2206-4
DOI: 10.1007/978-94-010-2206-4

500th ANNIVERSARY

OF THE BIRTH OF NICOLAUS COPERNICUS

1473–1973

TABLE OF CONTENTS

PART III / OUTER PLANETS AND THEIR SATELLITES

PART IV / FUTURE EXPLORATIONS OF THE SOLAR SYSTEM

THE SCIENTIFIC ORGANIZING COMMITTEE

On Behalf of the IAU

P. Swings, *Chairman*, Liège University, Belgium

A. Woszczyk, *Secretary*, Toruń University, Poland

A. Dollfus, Paris Observatory, France

F. D. Drake, Cornell University, Ithaca, U.S.A.

I. Koval, Main Astrophysical Observatory, Kiev, U.S.S.R.

V. I. Moroz, Sternberg Astronomical Institute, Moscow, U.S.S.R.

T. Owen, State University of New York, Stony Brook, U.S.A.

H. C. van de Hulst, Leiden Observatory, The Netherlands

COSPAR Representatives

A. D. Kuzmin, Academy of Sciences, Moscow, U.S.S.R.

C. Sagan, Cornell University, Ithaca, U.S.A.

PREFACE

The idea of a symposium devoted to the contemporary knowledge of the world of Copernicus – the planetary system – to commemorate the 500th anniversary of his birth, came during the XIV General Assembly of IAU in Brighton. The Executive Committee has approved it in the program of the Extraordinary (Copernicus) General Assembly of IAU in Poland in 1973.

The IAU Symposium No 65 (Copernicus Symposium IV) on the 'Exploration of the Planetary System' was held in Copernicus' native town – Toruń, Poland, from 5th to 8th September, 1973 under the auspices of Commissions 16 (Physical Study of Planets and Satellites) and 40 (Radio-astronomy) and the co-sponsorship of COSPAR. There were about 140 invited participants from 29 countries and about the same number of other participants to the Extraordinary General Assembly of IAU who came to Toruń to attend the sessions of this symposium. Special funds of the Polish Academy of Sciences made possible the participation of several young astronomers in this meeting.

We are very grateful to Professor P. Swings, the Director of the Astrophysical Institute of the University of Liège, Belgium, for accepting the task of chairing this symposium. His expert and enthusiastic guidance helped us constantly in the preparation. The efforts of the Members of the Scientific Organizing Committee are also very much appreciated. Special thanks are due to Professors A. Dollfus and T. Owen.

A number of review and introductory papers were invited but the participants were also allowed to submit the papers of their own choice. The final program was organized in 12 sessions devoted to particular scientific subjects. The sessions were chaired by P. Swings, A. Woszczyk, W. A. Baum, H. C. van de Hulst, W. Brunk, A. Dollfus, I. K. Koval, J. H. Smith, D. Menzel, T. Gehrels, T. Owen and G. S. Golitsyn. We regret that several colleagues were prevented from coming to Toruń at the very last minute.

For all Polish Copernicus symposia and the Extraordinary General Assembly of IAU one Local Organizing Committee under the chairmanship of Prof. J. Smak was formed. I am very grateful to my colleagues Dr A. Stawikowski and Dr A. Czacharowski as well as to Mrs U. Kucza and Mr M. Kaczmarek, and to other members of the University staff, who helped me in the local arrangements in Toruń, and to Dr C. Iwaniszewska and my colleagues from the Toruń Observatory for their help in the smooth running of the Symposium. The financial assistance of the Polish Academy of Sciences in the local arrangements is very much appreciated. We met constant encouragement from Professor W. Iwanowska, the Director of Toruń Astronomical Observatory and from the academic authorities of Copernicus University. The local city and provincial authorities made a great effort helping us in the preparation of the accompanying cultural and social programs. We wish to thank them for the hospitality to the participants.

We also wish to thank Drs Louise and Andrew Young who acted as referees for many papers.

Finally, I address my warmest words of thanks to my wife Irene who very efficiently assisted in all stages of the preparation of the symposium and of the editing of this volume.

ANDRZEJ WOSZCZYK

INAUGURAL ADDRESS

WITOLD ŁUKASZEWICZ

Rector Magnificus of Nicolaus Copernicus University

Monsieur le Président, Mesdames, Messieurs,

J'ai le grand honneur de souhaiter très chaleureusement la bienvenue, dans les murs de notre Almae Matris Copernicanae Thorunensis, aux éminents savants venus du monde entier pour participer aux Symposia de l'Assemblée Générale Extraordinaire de l'Union Astronomique Internationale. Bienvenue en mon nom propre, celui du Sénat, des chercheurs et de la jeunesse académique, c'est-à-dire au nom de toute notre famille universitaire, en ce 500e anniversaire du Patron de notre Université et du Grand Fils de Toruń – Nicolas Copernic, un des personnages les plus illustres de la science polonaise et de la science mondiale, ardent patriote et défenseur de la Poméranie Polonaise.

Nicolas Copernic et l'oeuvre de sa vie sont la propriété de toute la civilisation humaine. C'est pourquoi les célébrations coperniciennes de 1973 se déroulent dans le monde entier et sont placées sous le Haut-Patronage de l'UNESCO. Le but des célébrations coperniciennes n'est pas seulement de rendre hommage à un des plus grands génies, mais surtout de créer une atmosphère propice à l'élargissement et à l'approfondissement de la coopération internationale dans les domaines des sciences, de l'instruction publique et de la culture, et à la consolidation de la paix dans le monde. Ces symposia et les Colloquia Copernicana, dont l'intérêt des sujets et leur actualité sont grands, et auxquels participent quelques centaines de savants éminents, polonais et étrangers, contribueront certainement à la réalisation de ces fins élevées.

Cet anniversaire exceptionnel ne donne pas uniquement l'occasion de multiplier nos expressions d'hommage à Nicolas Copernic, savant génial, chercheur universel et illustre représentant de l'époque de la Renaissance, mais donne avant tout l'occasion de développer les idées coperniciennes qui, malgré les dogmes et les préjugés, frayèrent un chemin à la pensée humanistique, unissant, de façon harmonieuse, le progrès scientifique et technique aux valeurs culturelles et sociales, à la pensée d'où naissent toutes les impulsions nobles et élevées de l'action humaine.

Je puis vous assurer que, parallèlement à l'hommage qu'il rend aux mérites et à la grandeur de son Patron Nicolas Copernic, à l'occasion du 500e anniversaire de sa naissance, le milieu universitaire de Toruń s'efforce également de développer le plus possible et de graver la pensée copernicienne dans les jeunes esprits.

Je vous souhaite, Mesdames, Messieurs, des discussions très fructueuses et un séjour agréable à Toruń – ville natale de Copernic.

Quod felix faustum fortunatumque sit!

PART I

ORIGIN AND GENERAL PHYSICS OF THE PLANETARY SYSTEM

PLANETARY FORMATION

I. P. WILLIAMS

Queen Mary College, University of London, England

Abstract. Existing theories of planetary formation (based on a continuous solar nebula or on discrete objects) are briefly reviewed.

1. Introduction

The formation of the planetary system has been a problem which has stimulated the thought of man since time immemorial and the number of different theories cannot be far short of the number of human beings that have existed. From time to time reviews of the most promising of these theories have been published (for example, Williams and Cremin, 1968; Woolfson, 1969; ter Haar and Cameron, 1963; McCrea, 1972). Even though a number of theories have been formulated since the last of these reviews were published I do not propose to give here another review which simply brings these other reviews up to date. Instead, I wish to discuss the current ideas regarding possible formation of planets and try to follow the pattern in the development of such ideas. In this way, I hope that most recent work will be mentioned while at the same time the development towards possible complete theories will become more evident than by giving a chronological summary of recent papers. References are by no means complete and are intended only as a guide to the principal work that has been carried out.

2. General Outline of the Problem

In its simplest form, the problem we are discussing is 'how can a planetary system like our own form from some acceptable initial state.' Even in this simple statement of the problem there are at least two points on which theorists can argue, even before a theory has been formed. Starting from the wrong end, the first concerns the 'acceptable starting conditions'. To any theorist his own initial conditions are obviously acceptable. I shall to some extent avoid this issue stating simply that any situation where the material is less organized than in the present system and where such a state might reasonably be expected to have existed is acceptable. In practice the argument centres around the question of the existence or non existence of the Sun and possibly other stars when the planets started to form. The second is the statement 'a system like our own'. How much like our own is meant, is it taken to imply that we are trying to form a theory to explain our own solar system, or a theory for forming a system which has the major characteristics of our system. Most theorists, including myself, mean the latter, we are looking for a general mechanism for forming planets which is capable of generating both our own systems and systems similar to it. In compiling a list of the properties to be explained one should therefore take account of all known planetary systems. Since our system is the only one known in any detail, the list in-

Woszczyk and Iwaniszewska (eds.), Exploration of the Planetary System, 3–12. All Rights Reserved

evitably consists only of the properties of our solar system and the question therefore is whether all the properties of our system are general properties of any system. There are at least five stars that have companions of planetary mass and Barnard's star appears to have a number of such companions (see Williams, 1973; van de Kamp, 1971; for more information), but the only possible property that can be deduced is that the companions of Barnard's star appear to be non-coplanar (Black and Suffolk, 1973) though recently doubts have been cast onto Barnard's star observations. A list of properties which a general planetary system should have is therefore:

(1) There should be many planets. (Solar System 9, Barnard's 2 +, no information on any of the others.)

(2) These planets should, in general, move on coplanar prograde orbits, though there may be exceptions. (Solar System very well aligned, Barnard's star has angle of 50° between companions.)

(3) There is a central star, much more massive than the sum of the planets, which nevertheless possesses little angular momentum.

(4) There appear to be three groups of planets distinct from one another in mass, chemical composition and position, the terrestrial, major and outer planets, the relevant information being (Allen, 1963; Ramsey, 1967).

TABLE I

Main characteristics of planetary groups

Type	Mass (g)	Basic composition	Interstellar (by mass)
Terrestrial	5–6×10^{27}	Refractory, i.e. Si, Fe, Mg + compounds	0.3%
Major	1–2×10^{30}	H, He + impurities	100%
Outer	9×10^{28}	C, N, O + compounds Some H, He	5%

(Solar System, no information on the others.)

(5) There are a number of other miscellaneous objects, notably satellites, asteroids and comets in the system. (Solar system.)

These properties have to be produced by a mechanism to be described in our theories. Point (3) may or may not need an explanation as it is presumably allowable for a theorist to attempt to form a planetary system about an already formed star. It has also been demonstrated that the solar wind is capable of removing angular momentum from the Sun so that the slow rotation of the Sun need not have anything to do with the formation of the planets. However, it is clear that, irrespective of the type of theory that is postulated, the following general stages must have occurred in order to form a planetary system.

(a) Sufficient material (at least 10^{31} g) must be acquired from which to form the planets and sufficient angular momentum given to it.

(b) This material must be arranged, so that in general the final outcome is a near coplanar system.

(c) There must be a redistribution (either by accumulation or fragmentation) into about 10 objects.

(d) There must have occurred a stage of chemical segregation so that the terrestrial planets are composed of refractory material and so on.

Note that the listing given above need not be in chronological order and that in fact it is quite likely that some stages can occur simultaneously.

There are, of course, more restrictions one can place on the theories, some arising from the cosmochemical work of Anders (1972) and others. However, as I see it, we should first find what types of theories have been generate to account for stages (a) to (d) above, thus satisfying the crude requirements of the system. We should then investigate whether any of them can be rejected when it comes to accounting for the finer points in our knowledge.

3. Possible Theories

The first process to consider is the acquiring of the material for the planets. In practice this can either occur after the Sun has formed or as the tail end of the process of the formation of the Sun. In many reviews in the past this point has been one of crucial division between theories (see for example Williams and Cremin, 1968), possibly because at the time the slow rotation of the Sun was thought to be of crucial importance. I do not propose to make such a division here but rather concentrate on a division between the different forms which the captured material can have, or to be more precise the form in which the material first plays a part in the evolution of the system. Here the division is obvious, the material can either appear in the form of a continuous distribution of dust and gas, the traditional solar nebula, as has been postulated by Hoyle (1946, 1960), ter Haar (1950), Cameron (1962), Schatzman (1967), Pendred and Williams (1967), and Alfvén and Arrhenius (1970a, b) amongst others, or alternatively the matter appears as a collection of discrete objects. These discrete objects can be proto-planets themselves, as is the case in the tidal theories, made famous by Jeans (1916), the most modern theory of this type being due to Woolfson (1964), and modified by Dormond and Woolfson (1971). Alternatively they can appear as smaller objects but nevertheless distinct, as has been suggested by McCrea (1960) and Urey (1966).

Of course, there are many differences between theories, in either of the two classes mentioned, for example Cameron's nebula is much more massive than that postulated by Hoyle. There are also considerable variations in the way the planetary material acquires its angular momentum. In the theories of Cameron (1962) and Schatzman (1967) for example, the angular momentum is that originally in the gas cloud from which the Sun formed and this planetary material does not fall into the Sun because of its high angular momentum, the same being true of the theories of McCrea (1960) and Urey (1966), theories in the other category. In Hoyle's (1960) theory the angular momentum is transferred from the Sun by means of magnetic coupling. Alfvén (1954) had also shown that the solar magnetic field could transfer angular momentum to a plasma and also segregate out the grains. Sarvajna (1970) has also considered magnetic effects as a mechanism for transferring angular momentum as indeed had Birkland (1912) much earlier.

In spite of these differences, I believe that it gives a simple way of subdividing theories. There are now, of course, different difficulties facing the two streams of theories. In the case of the continuous nebula type of theory, one of the main problems is the accumulation of small grains and gas into planets, which causes no particular worries in the discreet accumulation theories. On the other hand, the prograde co-planar motion has to be explained in the discreet accumulation theories while it is an obvious consequence of the continuous nebula type of theory. It is therefore better to discuss the two streams of thought separately.

3.1. Continuous solar nebula

In this stream the next stage in the development of the system is that of chemical segregation. That some segregation comes about is obvious; since the Sun is at the centre of the system there is a temperature gradient through the nebula with the obvious result that near to the Sun only the refractory material is able to condense into grains while further out the CNO group also can condense. Throughout hydrogen and helium remains as a gas, but it is possible that since these are the lightest elements they will actually evaporate from the solar nebula near its outer edges. The obvious thought emerged as early as 1944 (Schmidt, 1944) that the terrestrial planets accumulated from this dust condensation while the major planets accumulated both dust and gas. The outer planets accumulated dust and CNO grains but little gas since most of it had evaporated. The work of Alfvén (1954) may also be important here, where the hydrogen plasma has a high charge to mass ratio and so is contained far from the Sun, the grains on the other hand have a smaller charge to mass ratio and fall nearer the Sun. There therefore exist almost two nebula segregated chemically and positionwise as required. In recent years much work has been done in investigating further the growth of planets in such a segregated nebula. In most investigations, grains are thought to adhere so that small accumulations form. In time the gravitational field of the growing accumulation becomes important which leads to a more rapid growth rate. One particular detailed investigation of this process is by Safronov (1969). Dole (1970) has simulated such a process on a computer and obtains good agreement with the existing solar system. Alfvén and Arrhenius (1970b) introduced the idea of the focusing of particles on a Keplerian orbit into jet streams with increasingly similar orbital elements and velocities of individual particles. The growth of embryos and subsequent development into planets then becomes easier. One of the points which these theorists have found difficult to explain is the actual distance and mass of the planets; that is to reproduce the Titus-Bode law. For a comparison of theories and reality see Nieto (1973). Hills (1970) has investigated tidal interaction in this context and found it to be important, though Dermott (1972) tends to doubt these conclusions.

Recently, however, a new idea has been introduced, following a suggestion first made by Lyttleton (1968). This is that the dust grains of refractory material, instead of directly accumulating into planets, first settle into the invariable plane of the rotation and form a disc of no more than a few centimetres thickness. Accumulation into planets in this disc becomes much easier since the density is now higher which means that the disruptive tidal effect of the Sun is much less important. Such ideas have been discussed more fully by Cameron (1972) and Lyttleton (1972), while a variation in which the grains grow as they settle as suggested by McCrea and Williams (1965) has been investigated by Schatzman (1971) and Williams and Handbury (1973). The conclusion of all these investigations is that it is possible to form such a thin disc in a comparatively short period of time and that fairly large objects are easily formed. So far, however, it has not been shown that the correct number of planets will be formed at the correct distances inside this disc.

The major planets form by accumulating hydrogen and helium which give the higher mass while in the region of the outer planets the disc also contains CNO grains but evaporation of hydrogen and helium has occurred. Most of the basic features have therefore been explained by this type of theory.

3.2. DISCRETE ACCUMULATION

In theories of this nature, such as suggested by McCrea (1960) and Urey (1966) the first problem is to generate a coplanar prograde system. (Note that in the tidal theories (e.g. Woolfson, 1964) this is automatic but it is then coincident that the Sun rotates about a near parallel axis.) It is suggested that the random collisions between the globules (for lack of a better word) will achieve both these ends. This contention has been subject to a number of investigations, primarily by Williams and Galley (1971), Aust and Woolfson (1971), and Williams and Donnison (1973). The conclusion of the latest of these, being a computer simulation of the process by Williams and Donnison (1973), shows that there is certainly a tendency for a settling and for removing retrograde orbits but that very many collisions are required to generate a system that is as coplanar as the existing planetary system, but it is not clear whether or not there will be sufficient collisions occurring in practice. As a result of these collisions and accumulations a system of basic object is built up. In the case of McCrea's theory and the tidal theories, these basic objects are of mass similar to the major planets while in Urey's theory they are somewhat smaller. It is now required to segregate the chemical elements and the obvious mechanism is by the settling of the grains to the centre of these objects, as first suggested by McCrea and Williams (1965). A detailed investigation by Williams and Crampin (1971) shows that a heavy core can form in a relatively short period of time, providing the falling grains accumulate all other grains that they collide with. The terrestrial planets then lose their outer hydrogen envelopes because of the proximity of the Sun while the major planets do not. There exists more of a problem with the outer planets. However, I make a suggestion which as far as I know has not been investigated anywhere. The outer protoplanets will be cooler and so the CNO compounds will also be falling as grains. Now Williams and Crampin (1971) showed that the normal refractory material on settling to the centre released potential energy sufficient only to disperse about 7% of the protoplanets. If, however, the CNO compounds also fall then much more energy is released which may well disperse the gaseous (HHe) envelope of the outer planets.

Urey with smaller original globules forms moons rather than planets from the core of his globes and so has a second accumulation process for planets. Again, however, the main features are explained. The difference between the two types of theories are shown in the following

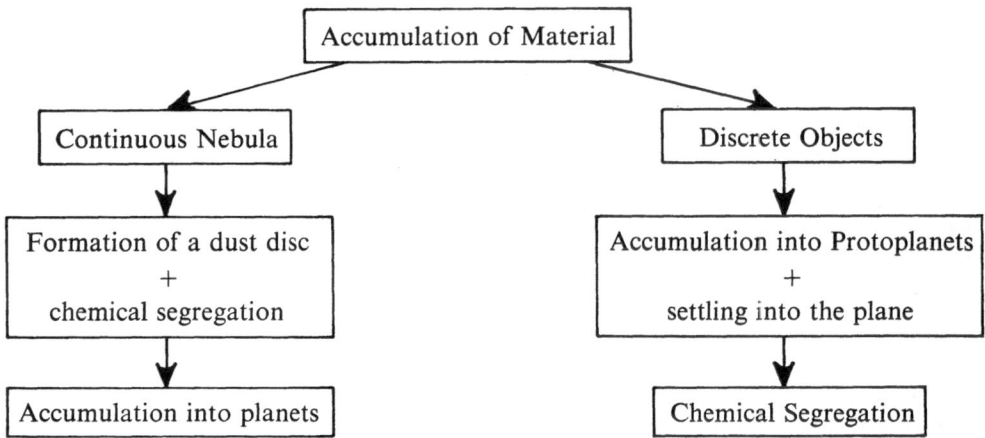

This then gives some indication of most of the general ideas currently prevalent amongst cosmogonists.

We can, as we have indicated already, see a few substreams amongst the two main streams and a very crude summary is given below.

3.2.1. *Continuous Nebula*

3.2.1.1. *Large nebula as part of star formation.* The solar nebula is formed because of the excess angular momentum in the gas cloud from which the Sun forms, the nebula is never part of the star but is left behind by the contracting star. The solar wind, magnetic fields etc. slows the Sun down to an acceptable level. Dust settles in the plane and near the Sun the terrestrial planets form from an accumulation of this nonvolatile material. Further away hydrogen and helium are also accumulated. Some of the names associated with this kind of theory are Cameron (1973), Schatzmann (1967), and Pendred and Williams (1967).

3.2.1.2. *Small nebula as part of star formation.* Here a protostar is formed from the interstellar gas but because of angular momentum difficulties, during the contraction of a star material is left behind in the equatorial plane as a solar nebula, this nebula is much smaller than that in case 3.2.1.1 and transfer of angular momentum is necessary between star and nebula. Again, the terrestrial planets form in a dust disc close to the Sun while further out gases are also accumulated. Typical names associated with work on this type of nebula are Hoyle (1960) and Lyttleton (1972).

3.2.1.3. *Nebula resulting from the accretion of material.* A nebula is acquired by the passage of the already formed Sun through an interstellar cloud and, following Schmidt (1944), the development is then similar in general to that in case 3.2.1.1. Alfvén (1954) suggested that condensation of the infalling matter would occur resulting in the magnetic fields segregating grains (close to the Sun) from the gas. Another new idea introduced by Alfvén is the concept of jet streaming leading more easily to the formation of planets.

3.2.2. *Discrete Objects*

3.2.2.1. *Tidal theories.* In the tidal theories material is drawn out of a passing star and captured by the Sun. A computer simulation by Woolfson (1964) shows that this is possible. Break up of this filament forms protoplanets and the terrestrial planets are formed by the settling of grains to form a core.

3.2.2.2. *Others.* The two main authors are McCrea (1960) and Urey (1966). Material is captured in random orbits. Subsequent collision and amalgamation leads to a degree of settling in the plane for the protoplanets. The terrestrial planets are formed by settling of solid grains and the subsequent loss of gas.

We now discuss some of the differences and possible developments.

4. General Differences and Difficulties

In the continuous nebula type of theory the alignment into the plane of the material is very efficient and one would therefore expect almost complete alignment. In the discreet object type of theory the alignment comes about as a statistical effect and one would not expect the end effect to be perfectly aligned. One can therefore ask whether the existing solar system is too well aligned for the discrete object theory or not well enough aligned for the continuous nebula type of theory. Unfortunately the solar system itself is not clear cut either way and is within the 'error bars' of both types of theory. It is here that Barnard's star and other similar stars may be important. If the analysis of Black and Suffolk (1973) is correct that of the planets detected, moves at a considerable (50°) angle to the other then the continuous nebula theory has difficulties in explaining it. Whipple *et al.* (1972) has also drawn attention to the problem of the alignment (or lack of it) of Pallas and underlined the problem of explaining it in the context of the continuous nebula theory. Observations of Barnard's star (and others) from an extra terrestrial environment offering a longer base line for parallax work may be of use here.

Another possible strange result arising from the study of other solar systems is that all the planets so far discovered are all very close in mass to the major planets of the solar system. In a continuous nebula theory one would expect there to be some dependence on the mass of the parent star in the mass of the planet while the mass in the discreet object theory is more likely to be dependent on outside influences only. Of course, there is a very considerable selectional effect in the data in that planets much smaller than Jupiter cannot be detected while any found much more massive are dismissed as possible stellar companions. Nevertheless, in theory this could prove an interesting test. Though this point has been introduced as a possible strength of the discreet object theories, in fact it also contains their main weakness. It is the question of the origin of the discreet globules which accumulate, they virtually arrive on the scene as thinly disguised postulates, though it could be mentioned that Walker (1972) believes that young stars do show evidence of accreting matter in a discrete form.

One other problem with the discrete theories is the well known under abundance of xenon (Suess, 1949) and other rare gases, the problem basically being that evaporation is probably the main way in which the terrestrial planets lost their hydrogen and helium, but xenon is too heavy to evaporate and so it should be over abundant. The only obvious solution is to assume that in some way the xenon succeeded in finding its way into the interior of the planets.

The major difficulty in the continuous nebula type of theory is to explain why, from the original dust carpet, planets grew to their existing sizes and existing positions. In particular why are all the planets (after compensating for composition) all of very similar mass and why is the spacing of the planets as found, with the peculiar gap where the asteroids are. There are many other difficulties, primarily found by cosmochemists concerning the abundancies of elements and isotopes. Unfortunately, in general these have been couched in the framework of particular theories and some problems vanish with a simple change of model. I will not go into any details here since my main concern was to give a general account of the state of theories today as I see it. In conclusion, we are not at present able to decide which of the main streams of thought are likely to be correct, however, I believe that in the near future this may become more likely for the following reasons.

(1) More observations of Barnard's star and similar systems may give an answer to the question of how well aligned planetary systems should be.

(2) Exploration of other planets may lead to useful information regarding the abundance of xenon and other elements there.

(3) More use of computer models will lead to a better understanding of some of the more complex physical problems involved, for example the collision and accumulation of discreet objects or the growth of planets in a dust carpet.

(4) Laboratory experiments on the adhesiveness of grains in collision are continuing (e.g. Kerridge and Vedder, 1972). More information may become available as the possibility of experimentation in outer space becomes practicable. Some evidence is already available from a study of moon rocks.

(5) Further studies of young stars may indicate whether a *continuous* nebula exists around them (as opposed to a circumstellar dust cloud) and give an estimate of its dimensions.

We may therefore look forward with pleasure and anticipation to the next few years of exploration and computation.

References

Alfvén, H.: 1954, *Origin of the Solar System*, Oxford U.P.
Alfvén, H. and Arrhenius, G.: 1970a, *Astrophys. Space Sci.* **8**, 338.
Alfvén, H. and Arrhenius, G.: 1970b, *Astrophys. Space Sci.* **9**, 3.
Allen, C. W.: 1963, *Astrophysical Quantities*, Athlone.
Anders, E.: 1972, *Proceedings Nice Symposium on the Origin of the Solar System*, 179.
Aust, C. and Woolfson, M. M.: 1971, *Monthly Notices Roy. Astron. Soc.* **153**, 21P.
Birkland, K.: 1912, *Compt. Rend. Hebd. Seanc. Sci.* **155**, 892.
Black, D. C. and Suffolk, G. C. J.: 1973, *Icarus* **19**, 353.
Cameron, A. G. W.: 1962, *Icarus* **1**, 18.
Cameron, A. G. W.: 1972, *Proceedings Nice Symposium on the Origin of the Solar System*, p. 56.
Cameron, A. G. W.: 1973, *Icarus* **18**, 407.
Dermott, S. F.: 1972, *Proceedings Nice Symposium on the Origin of the Solar System*, p. 320.
Dole, S. H.: 1970, *Icarus* **13**, 494.
Dormand, J. R. and Woolfson, M. M.: 1971, *Monthly Notices Roy. Astron. Soc.* **151**, 307.
Hills, J. G.: 1970, *Nature* **225**, 840.
Hoyle, F.: 1946, *Monthly Notices Roy. Astron. Soc.* **106**, 406.
Hoyle, F.: 1960, *Quart. J. Roy. Astron. Soc.* **1**, 28.
Jeans, J. H.: 1916, *Mem. Roy. Astron. Soc.* **77**, 84.
Kerridge, J. F. and Vedder, J. F.: 1972, *Proceedings Nice Symposium on the Origin of the Solar System*, p. 282.
Lyttleton, R. A.: 1968, *Mysteries of the Solar System*, Clarendon Press.
Lyttleton, R. A.: 1972, *Monthly Notices Roy. Astron. Soc.* **158**, 463.
McCrea, W. H.: 1960, *Proc. Roy. Soc.* **A256**, 245.
McCrea, W. H.: 1972, *Proceedings Nice Symposium on the Origin of the Solar System*, p. 2.
McCrea, W. H. and Williams, I. P.: 1965, *Proc. Roy. Soc.* **A287**, 143.
Nieto, M.: 1973, *Titus-Bode Law of Planetary Distance*, Pergamon.
Pendred, B. W. and Williams, I. P.: 1967, *Icarus* **8**, 129.
Ramsey, W. H.: 1967, *Planetary Space Sci.* **15**, 1609.
Safronov, V. S.: 1969, *Evolution of the Preplanetary Cloud and the Formation of the Earth and Planets*, Moscow.
Sarvajna, D. K.: 1970, *Astrophys. Space Sci.* **6**, 258.
Schatzman, E.: 1967, *Ann. Astrophys.* **30**, 963.
Schatzman, E.: 1971, *Physics of the Solar System*, Goddard Institute X630-71-380.
Schmidt, O. Y.: 1944, *Dokl. Akad. Nauk SSSR* **45**, 245.
Suess, H. E.: 1949, *J. Geol.* **57**, 600.
Ter Haar, D.: 1950, *Astrophys. J.* **111**, 179.
Ter Haar, D. and Cameron, A. G. W.: 1963, in R. Jastrow and A. G. W. Cameron (eds.), *Origin of the Solar System*, Academic Press, p. 1.
Urey, H. C.: 1966, *Monthly Notices Roy. Astron. Soc.* **131**, 199.
Van de Kamp, P.: 1971, *Ann. Rev. Astron. Astrophys.* **9**, 103.
Walker, H. F.: 1972, private communication.
Whipple, F. L., Lecar, M., and Franklin, F. A.: 1972, *Proceedings Nice Symposium on the Origin of the Solar System*, p. 312.
Williams, I. P.: 1973, *Hermes* **20**, 57.
Williams, I. P. and Crampin, D. J.: 1971, *Monthly Notices Roy. Astron. Soc.* **152**, 261.
Williams, I. P. and Cremin, A. W.: 1968, *Quart. J. Roy. Astron. Soc.* **9**, 40.
Williams, I. P. and Donnison, J. R.: 1973, *Monthly Notices Roy. Astron. Soc.* **165**, 293.
Williams, I. P. and Galley, G. J.: 1971, *Monthly Notices Roy. Astron. Soc.* **151**, 207.
Williams, I. P. and Handburry, M. J.: 1973, under preparation.
Woolfson, M. M.: 1964, *Proc. Roy. Soc.* **A282**, 485.
Woolfson, M. M.: 1969, *Progr. Phys.* **32**, 135.

DISCUSSION

Icke: You considered the most massive bodies as the most important. In this the natural thing to do? I would think that specific angular momentum is a more relevant parameter in a contracting rotating nebula, so that comets and meteorites should be considered.

Williams: If you do not consider mass then you have no planetary system to explain.

Vsehsvyatsky: The hypothesis of Williams seems a backward one compared with the classical. It does not explain the basic peculiarities of the solar system, namely the planets activity in the history of the solar system, and the eruptive formation of comets and other small bodies.

Poss: The observations of solar systems would be crucial for distinguishing among various general theories. Could you comment further on the reliability of the observational data on Barnard's star, the different interpretations (i.e. the existence of one, two, three or more planets) and whatever observations might be used.

Williams: Observations are only by van de Kamp, but there is a suggestion that a systematic error exists in his observations.

Owen: The observations of Barnard's star have been repeated in the sense that a different series of plates was analysed and no periodic oscillation was found.

STATUS OF MOLECULAR OPACITIES OF INTEREST
IN THE MODELING OF A PROTO-SOLAR NEBULA

W. F. HUEBNER and L. W. FULLERTON

Los Alamos Scientific Laboratory, Los Alamos, N.M., U.S.A.

Abstract. The report centers on general procedures applicable to the calculation of constitutive properties (equation of state and opacity) of media that serve as models for the solar nebula during planet
formation and for the atmospheres of some planets. Specifically considered are the equilibrium compositions of a mixture of atoms, molecules, and their ionic species in the gaseous phase, condensation
into grains with refractory cores and mantles of volatile compounds, and the 'optical' properties of
the grain-gas medium. A summary of available and still needed basic (input) data and some currently
available results are presented.

1. Motivation for Opacity Calculations

In atmospheres energy is transported by radiation and convection. In the radiative
mode, transfer is determined by the coefficients for emission, absorption, and scattering; these, in turn, depend on the atomic and molecular structure and equation of
state of the medium. General methods are reviewed to predict these properties at
temperatures, densities, and compositions suitable for modeling conditions to be expected in a proto-solar nebula before and during the time of planet formation, in
proto-planets, and in sun-lit atmospheres throughout the history of planets. Although
local thermodynamic equilibrium (LTE) does not always apply it has been adopted
as a simplifying assumption.

The energy transfer in the proto-solar nebula determines the rate of its cooling and
thus affects its chemical composition and the chemical makeup of the planets (Larimer,
1967; Larimer and Anders, 1967; Studier *et al.*, 1968; Hayatsu *et al.*, 1968; Anders,
1968, 1971). The diffusion approximation, using an extinction coefficient averaged
over all photon energies (i.e., the Rosseland opacity), can be applied to optically thick
atmospheres. For optically thin atmospheres photon-group mean opacities (Rosseland
or Planck) are required to solve the more complicated radiation transfer equation.

2. Earlier Calculations

Some of the earlier predictions of opacities used in models for different stages of
evolution during planet formation were carried out by Hayatsu and Nakano (1965),
Gaustad (1963), Auman and Bodenheimer (1967), and Cox and Stewart (1965). Each
of these applied only to a particular phase: grains, molecules, ions, or a composite
synthesized artificially and very approximately from these subresults.

A more satisfactory calculational procedure is one in which all stages are treated in
one program permitting continuous transformations from one phase to the next with
simultaneous coexistence of several phases, e.g., molecular absorption must still be

Woszczyk and Iwaniszewska (eds.), Exploration of the Planetary System, 13–20. All Rights Reserved

considered after grains have formed and atomic absorption still contributes after molecules have formed.

3. Continuity of Opacity During Phase Transformations

The Rosseland mean opacity is defined by

$$\kappa_R \equiv \frac{\int\limits_0^\infty \partial B_\nu(T)/\partial T \, d\nu}{\int\limits_0^\infty \kappa_\nu^{-1} \, \partial B_\nu(T)/\partial T \, d\nu}, \tag{1}$$

and the Planck mean opacity is

$$\kappa_P \equiv \frac{\int\limits_0^\infty \kappa_\nu^{(a)'} B_\nu(T) \, d\nu}{\int\limits_0^\infty B_\nu(T) \, d\nu}, \tag{2}$$

where the mass extinction coefficient depending on frequency, ν, is

$$\kappa_\nu = \kappa_\nu^{(a)'} + \kappa_\nu^{(s)}, \tag{3}$$

$\kappa_\nu^{(a)'}$ is the mass absorption coefficient corrected for stimulated emission, $\kappa_\nu^{(s)}$ the mass scattering coefficient, and $B_\nu(T)$ is the Planck function at temperature T. The major differences between these mean opacities are: (1) in the Planck mean the large values of $\kappa_\nu^{(a)'}$, i.e., the strong absorption lines, add most heavily, while in the Rosseland mean the low values of κ_ν, i.e., the continuum extinction in the windows between lines contributes most; (2) the individual absorption processes, e.g., atomic and molecular line and continuum absorption and absorption by grains, are additive in the Planck mean, they do not add in the Rosseland mean; (3) the weighting functions for the Planck mean peaks at $u = h\nu/kT \approx 2.8$, while that for the Rosseland mean peaks at $u \approx 3.8$; (4) the Rosseland opacity includes scattering, the Planck opacity does not. One frequently overlooked, but very important property of the Rosseland mean at low temperatures is the required behavior of the absorption coefficient in the long wavelength limit. Dividing the range of integration in the denominator of Equation (1) into two regions: from 0 to ν_1 and from ν_1 to ∞ with the constraint that $h\nu_1 < kT$ and expanding the exponentials occurring in the derivative of the Planck function in a polynomial, one obtains for the integral in the first region

$$\int\limits_0^{\nu_1} \frac{1}{\kappa_\nu} \frac{\partial B_\nu(T)}{\partial T} \, d\nu = \frac{2k^4 T^3}{c^2 h^3} \int\limits_0^{u_1} \frac{u^2}{\kappa_u} \left(1 - \frac{u^2}{12} + \cdots \right) du. \tag{4}$$

The usual procedure of writing the integral in terms of $u = h\nu/kT$ has been adopted (the functional form of κ_u is the same as for κ_ν) and $u_1 < 1$. If the leading term of a polynomial expansion of κ_u is const $\times u^\alpha$ then α must be less than 3. For $\alpha \geqslant 3$ Equation (4) will diverge and the Rosseland opacity is zero independent of the behavior of the extinction coefficient for $u > 0$. For Rayleigh scattering by atoms, molecules, and grains the cross section is proportional to λ^{-4}, i.e., $\kappa_u \sim u^4$. Hence Rayleigh scattering alone is insufficient to yield a non-zero Rosseland opacity and other absorption processes *must* be included even if they appear to have negligibly small cross sections. At temperatures sufficiently high for free electrons to exist, inverse bremsstrahlung provides the needed additional absorption. At low temperatures it can be provided by electrically conducting grains (non-zero imaginary part of the refractive index) or molecular absorption, i.e., by the vibration-rotation or pure rotational spectrum of molecules still in the gaseous phase.

Figure 1 illustrates the relationship of the weighting function of Equation (1) (the

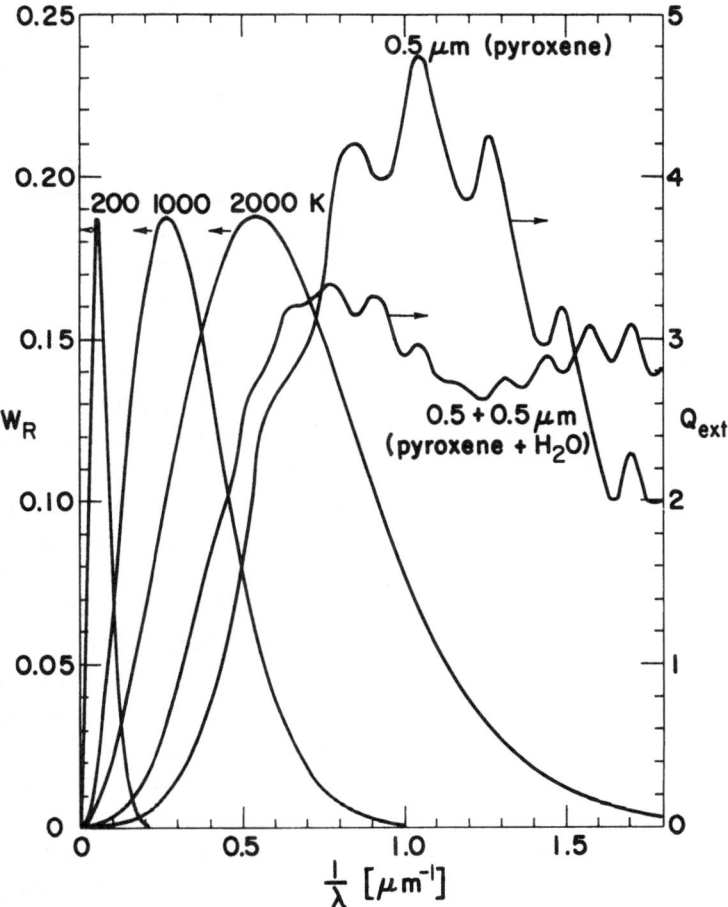

Fig. 1. Extinction efficiency, Q_{ext}, for pyroxene grain ($r = 0.5\ \mu m$) and pyroxene with H_2O mantle (scale on right) vs $1/\lambda$ compared to Rosseland weighting functions, W_R, for three different temperatures (scale on left).

Rosseland weighting function) to the extinction efficiency Q_{ext} for a spherical ($r = 0.5$ μm) pyroxene, $(Mg, Fe)SiO_3$, grain and a pyroxene grain covered with a 0.5 μm thick ice-mantle. The refractive index for the ice was assumed constant and real, with $m = 1.33$; for pyroxene data from Huffman and Stapp (1971) for frequency dependent real and imaginary parts were used. Their measurements indicate, however, that the imaginary part of the refractive index goes to zero as $1/\lambda$ becomes small. Thus pyroxene and ice-covered pyroxene cross sections exhibit pure Rayleigh scattering at long wavelength. Another feature illustrated in Figure 1 is that the large extinction features of submicron grains are in the high photon energy tail of the Rosseland weighting function. For the Planck mean the weighting function is still smaller in the region where submicron grain absorption is large. Only small refractory grains will exist at temperatures of about 1000 to 2000 K, above several thousand degrees all grains will have vaporized.

The vibration-rotation transitions in the infrared and the pure rotational transitions in the microwave region will be very important contributors to opacity – for the Rosseland mean because they prevent it from going to zero, for the Planck mean because of their relatively high absorption cross sections. There are about 600 known microwave transitions originating from 34 simple, volatile molecules composed of H−C−N−O−S atoms (Wacker and Pratto, 1964). Most of these lie in the range $1/\lambda = 10^{-1}$ to 10^1 cm^{-1} and a few below 10^{-1} cm^{-1}. At very low temperatures more of the molecules will condense on grains and their ice mantles will grow. Absorption from some of the largest grains will approach the peak of the weighting functions, but the microwave transitions of the few remaining volatile molecules will still be important to the Rosseland opacity.

At higher temperatures (up to about 20 000 K) electronic transitions in the molecules dominate the opacity. Much of the basic work on these transitions has been carried out by many investigators, but the application to high temperature molecular opacities was pioneered by Gilmore (1965, 1967), Armstrong and Nicholls (1967), Avilova et al. (1969a, b), Johnston et al. (1972), and Generosa and Harris (1973). The theories and calculations of all of these apply to air, however, they take into account the continuity and coexistence of molecules, atoms, and ions. Although some astrophysical molecular opacities had been calculated earlier (e.g., Tsuji, 1964, 1971; Auman and Bodenheimer, 1967; Linsky, 1969; Carbon et al., 1969; Alexander et al., 1971), efforts to preserve continuity and coexistence of atoms and molecules in astrophysical opacities are just getting started (Merts and Magee, 1973). The main reason for the delays in molecular opacity calculations are the lack of basic data and the difficulties encountered in obtaining them. Data for absolute measurements exist only for room temperature absorption bands, i.e., for transitions involving the vibrational level $v'' = 0$ in the electronic ground state. Most absorption data for excited states is deduced from emission measurements or from calculations. Unfortunately emission usually involves transitions between many closely spaced initial and final states, making analysis difficult and often impossible. Calculated results often suffer because a multiconfiguration approximation may be required but is too complex and therefore not used. The N_2 Birge-Hopfield band system is a typical example of the complexities encountered in relatively simple molecules. Figure 2 illustrates how, in the energy range from about 12.5 to 14 eV above the ground state of N_2, the unperturbed potential curves belonging to pairs of states with the same symmetry character intersect. Near the intersections the superposition of these states is the cause for strong perturbations altering the transition probabilities and destroying the otherwise regular spacing of the vibrational levels. One pair of intersecting states has been labeled $b^1\Pi_u$ and $c^1\Pi_u$, the other one $b'^1\Sigma_u^+$ and $c'^1\Sigma_u^+$. The Birge-Hopfield band systems 1 and 2 are associated with the valence states b' and b, respectively. States c' and c are Rydberg states. The proper identification of these levels succeeded only relatively recently

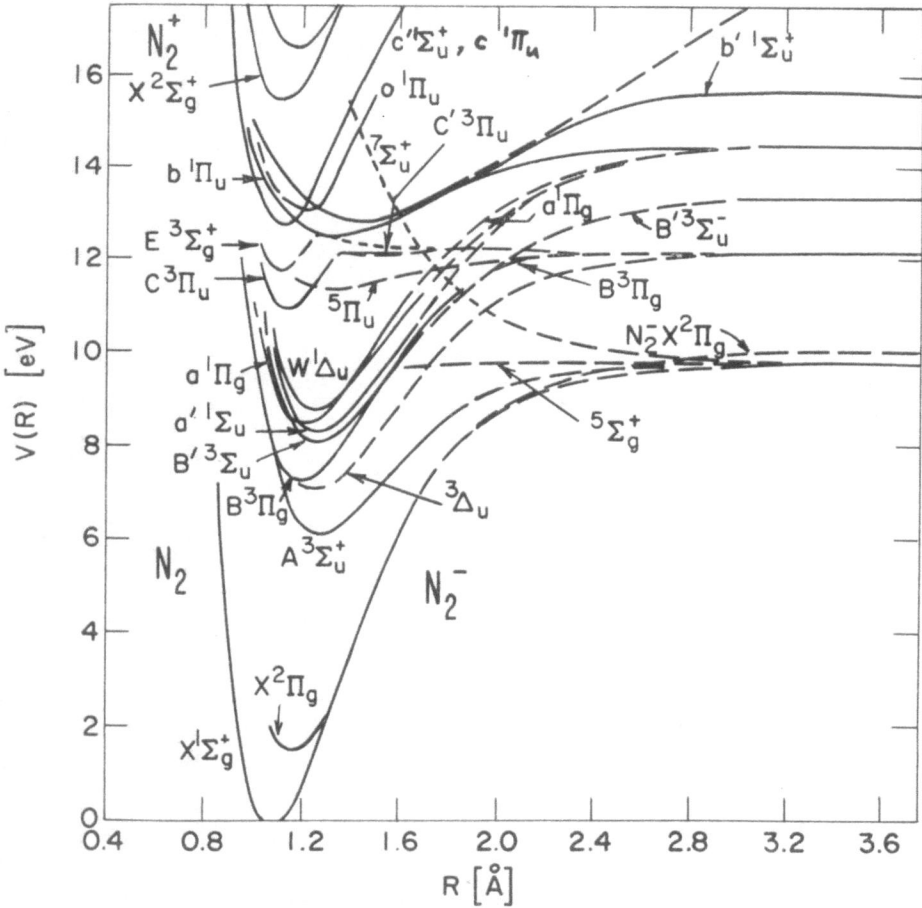

Fig. 2. Potential energy curves for nitrogen (taken from Gilmore, 1967) showing the intersection of b, b', c, and c'. Curves c and c' are drawn as one curve. The separated atom states at large R from top to bottom correspond to $N(^2D^\circ)+N(^2P^\circ)$, $N(^2D^\circ)+N(^2D^\circ)$, $N(^4S^\circ)+N(^2P^\circ)$, $N(^4S^\circ+N(^2D^\circ)$, $N(^4S^\circ)+N^-(^3P)$, and $N(^4S^\circ)+N(^4S^\circ)$. (1 eV $\equiv 0.8066$ μm^{-1}).

(Dressler, 1969) and is a necessary prerequisite for determining Franck-Condon factors and for making theoretical predictions of the band absorption. Similar difficulties exist with many other simple molecules.

4. Status of Opacity Calculations

Two basic sets of information are needed for opacity calculations: the equation of state of the medium and the radiative cross sections.

The equation of state is determined by the equilibrium between ions (and electrons), atoms, molecules, and molecular ions (all in the gaseous phase), and condensed matter.

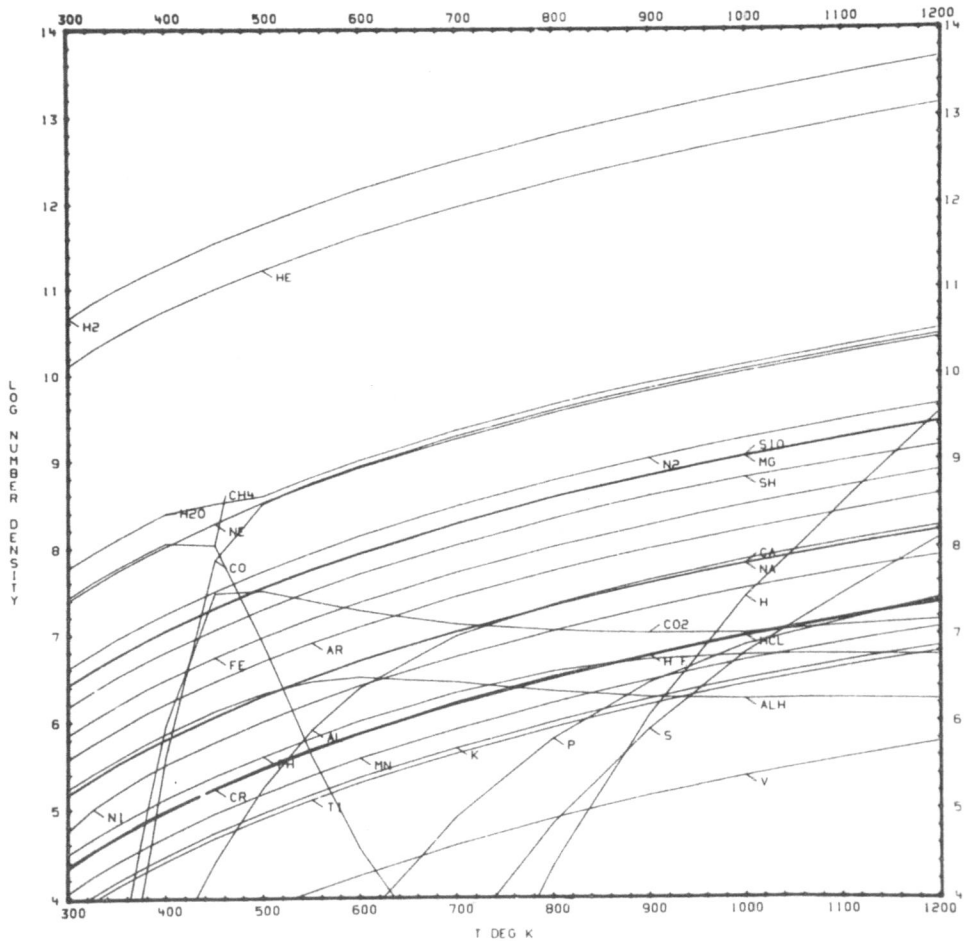

Fig. 3. Chemical abundances (without condensation) in proto-solar nebula assuming $\varrho = (T\mathrm{K}/10^5)^5$ g cm^{-3}. For clarity only a limited temperature range is shown. NH$_3$ increases very rapidly just below 300 K. Above 12000 K a few more diatomics (primarily OH and MgH) contribute significantly. Standard astrophysical abundances were assumed for the elements.

Programs to compute the equilibrium in the gaseous phase alone – i.e., the chemistry, see, e.g., Figure 3, – have been available for some time. Only very limited efforts have been made to include the condensed phase of matter in the equilibrium (Duff, 1962; Tsuji, 1966).

The collective absorption and scattering by particles (grains as well as liquid drops) also requires knowledge of the particle size distributions. Usually Gaussian or power-law distributions with cutoffs are adopted.

The physics for commencement of condensation in the microscopic realm is insufficiently understood. It is not clear whether a better understanding of this process would shed more light on particle distributions. The absorption and scattering of radiation by 'grains' is well developed in the Mie theory. Programs to compute the cross sections for 2-component refractory core-ice mantle spheres using complex and wavelength-dependent refractive indices are available (e.g., see Figure 1). Similar programs for spheroidal particles could be developed. Some parameters for the geometric asymmetry could be obtained, e.g., from observed polarization of zodiacal light (Greenberg, 1970).

Data for microwave transitions and infrared rotation-vibration absorption is available, but more data will be needed. Data for electronic transitions involving highly excited states of molecules are considerably more sparse and unreliable. There are no difficulties with the atomic-ionic cross sections of astrophysical interest that could not be resolved with presently available methods and programs.

Acknowledgement

Our gratitude goes to R. K. M. Landshoff, who provided insight and valuable discussions on molecular transitions.

References

Alexander, D., Collins, J., Fay, T., and Johnson, H. R.: 1971, *Bull. Am. Astron. Soc.* 3, 380.
Anders, E.: 1968, *Accounts Chem. Res.* 1, 289.
Anders, E.: 1971, *Ann. Rev. Astron. Astrophys.* 9, 1.
Armstrong, B. H. and Nicholls, R. W.: 1967, 'Thermal Radiation Phenomena, Vol. 2, The Equilibrium Radiative Properties of Air-Theory', Lockheed Missiles and Space Company report LMSC-3-27-67-1.
Auman, J. A. and Bodenheimer, P.: 1967, *Astrophys. J.* 149, 641.
Avilova, I. V., Biberman. L. M., Vorobjev, V. S., Zamalin, V. M., Kobzev, G. A., Lagar'kov, A. N., Mnatsakanian, A. Ch., and Norman, G. E.: 1969a, *J. Quant. Spectrosc. Radiat. Transfer* 9, 89.
Avilova, I. V., Biberman. L. M., Vorobjev, V. S., Zamalin, V. M., Kobzev, G. A., Lagar'kov, A. N., Mnatsakanian, A. Ch., and Norman, G. E.: 1969b, *J. Quant. Spectrosc. Radiat. Transfer* 9, 113.
Carbon, D., Gingerich, O. J., and Latham, D. W.: 1969, in S. S. Kumar (ed.), *Low Luminosity Stars*, Gordon and Breach Science Publishers, New York, p. 435.
Cox, A. N. and Stewart, J. N.: 1965, *Astrophys. J. Suppl.* 11, 22.
Dressler, K.: 1969, *Can. J. Phys.* 47, 547.
Duff, R. E. and Bauer, S. H.: 1962, *J. Chem. Phys.* 36, 1754.
Gaustad, J. E.: 1963, *Astrophys. J.* 138, 1050.
Generosa, J. I. and Harris, R. A.: 1973, Air Force Weapons Laboratory, Kirtland AFB, N.M., private communication.

Gilmore, F. R.: 1965, *J. Quant. Spectrosc. Radiat. Transfer* **5**, 125.

Gilmore, F. R.: 1967, 'Thermal Radiation Phenomena, Vol. 1, The Equilibrium Radiative Properties of Air', Lockheed Missiles and Space Company report LMSC-3-27-67-1.

Greenberg, J. M.: 1970, in T. M. Donahue, P. A. Smith, and L. Thomas (eds.), *Space Research* **X**, North-Holland Publishing Company-Holland, p. 225.

Hayashi, C. and Nakano, T.: 1965, *Progr. Theor. Phys.* **34**, 764.

Hayatsu, R., Studier, M. H., Oda, A., Fuse, K., and Anders, E.: 1968, *Geochim. Cosmochim. Acta* **32**, 175.

Huffman, D. R. and Stapp, J. L.: 1971, *Nature Phys. Sci.* **229**, 45.

Johnston, R. R., Landshoff, R. K. M., and Platas, O. R.: 1972, 'Radiative Properties of High Temperature Air', Lockheed Missiles and Space Company report LMSC D 267205.

Larimer, J. W.: 1967, *Geochim. Cosmochim. Acta* **31**, 1215.

Larimer, J. W. and Anders, E.: 1967, *Geochim. Cosmochim. Acta* **31**, 1239.

Linsky, J. L.: 1969, *Astrophys. J.* **156**, 989.

Merts, A. L. and Magee, N. H.: 1973, *Bull. Am. Astron. Soc.* **5**, 338.

Studier, M. H., Hayatsu, R., and Anders, E.: 1968, *Geochim. Cosmochim. Acta* **32** 151.

Tsuji, T.: 1964, *Proc. Japan Acad.* **40**, 99.

Tsuji, T.: 1966, *Publ. Astron. Soc. Japan* **18**, 127; corrected: 1967, *Publ. Astron. Soc. Japan* **19**, 119.

Tsuji, T.: 1971, *Publ. Astron. Soc. Japan* **23**, 553.

Wacker, P. F. and Pratto, M. R.: 1964, 'Spectral Tables, Line Strength of Asymmetric Rotors', NBS Monograph 70 – Vol. II, U.S. Government Printing Office, Washington, D. C.

ON THE GROWTH MECHANISM OF GRAINS IN A
PRIMORDIAL STAGE OF THE SOLAR NEBULA

A. CARUSI, A. CORADINI, and C. FEDERICO

Istituto di Geologia e Paleontologia, Università di Roma, Italy

and

M. FULCHIGNONI and G. MAGNI

Laboratorio di Astrofisica Spaziale, CNR, Frascati, Italy

Abstract. Grain accretion processes in a protoplanetary nebula have been studied regarding: (a) the distribution function of grain velocities; (b) electrostatic and electromagnetic mechanisms between grains. The velocity distribution function has been investigated for grains embedded in a turbulent gaseous medium. Results have been obtained for protoplanetary nebula densities ranging from 10^{-19} to 10^{-10} g cm^{-3}. Considering interactions between two grains, photoelectrically charged by galactic ultraviolet flux and by charged-particle capture, and solid-solid interactions (dipole fluctuation effect), the authors estimate the physical cross section $\sigma(v)$ with respect to the geometric one σ_0. Then a statistical approach for an assembly of grains gives the accretion or destruction rates for these small particles. Therefore, according to their characteristic velocities, the following processes have been studied: rupture, fusion, vaporization.

Introduction

The problem of formation of a planetary body may be investigated by studying the accretion processes of dust grains and gas inside a normal interstellar nebula. Many authors have developed this problem, and have obtained analytical solutions for some cases (Cameron, 1972; Safronov, 1972).

Our aim is to investigate grain-accretion processes and to define in some detail the more realistic physical adhesion mechanisms between particles.

The processes examined are:

(a) electrostatic and electromagnetic adhesion;

(b) adhesion by shock-melted material;

(c) destructive mechanisms: rupture or fragmentation by shock melting and vaporization.

These processes are supposed to occur in a turbulent medium: the turbulence may arise in a primordial stage of the nebular collapse (Cameron, 1972).

1. The adopted initial parameter set is consistent with the most recent experimental data (Lequeux, 1972), and also seems to be consistent with the initial theoretical parameters of a collapsing cloud (Larson, 1969, 1972).

For the succeeding calculations, different parameters reflecting the subsequent stages of a collapse have been chosen, in order to evaluate perturbations on the postulated mechanisms of grain growth.

It is difficult to investigate these growth mechanisms in great detail because the chemical composition and the morphological surface structures of grains are poorly

Woszczyk and Iwaniszewska (eds.), Exploration of the Planetary System, 21–35. All Rights Reserved
Copyright © 1974 by the IAU

TABLE I

Set of adopted parameters

$\varrho = 10^{-19}$ g cm^{-3}
$T = 10$ K
$g = 10^{-10}$ cm s^{-2}
$Z = 1.8 \times 10^{-2}$ (heavy atoms fraction)

known, as are their physical interactions. Therefore, a homogeneous chemical composition (silica) is adopted for the grains because the cosmical abundances of elements and the interstellar optical extinction seem to agree with a grain chemical composition mixture of silica, graphite, ice, SiC, iron, etc. (Spitzer, 1968; Watson, 1972; Woolf and Ney, 1969; Gillett et al., 1971).

TABLE II

Grain parameters

$\varrho_s = 2.3$ g cm^{-3}	density
$\tau_s = 300$ dyn cm^{-1}	surface tension
$k_B = 4 \times 10^{11}$ g cm^{-1} s^{-2}	bulk modulus
$Q = 6.75 \times 10^9$ dyn cm^{-2}	stress
$\bar{k} = 1.43 \times 10^5$ erg s^{-1} cm^{-1} K^{-1}	thermal conductivity
$E_F = 8.5 \times 10^9$ erg g^{-1}	melting energy
$E_V = 5 \times 10^{11}$ erg g^{-1}	vaporization energy
$T_F = 1103$ K	melting temperature
$T_V = 2473$ K	vaporization temperature
$c_p = 1.5 \times 10^7$ erg g^{-1} K^{-1}	mean specific heat at $p = $ const.
$\varepsilon_0 = 3.75$	static dielectric const.
$T_D = 375$ K	Debye temperature
$\alpha_{12} = 0.1$ ⎞	shock wave coefficients
$\beta_{12} = 2 \times 10^{-14}$ g erg^{-1} ⎠	

This chemical composition has been chosen because the grains are probably not mainly metallic, owing to the low cosmic abundance of metals in comparison with the mass needed for interstellar extinction (Watson, 1972).

Experimental data show a fluffy and amorphous surface structure caused by cosmic-ray bombardment (Watson, 1972; Greenberg, 1972; Maurette and Bibring, 1972).

In this paper grains are nearly spherical, but corrected by a suitable factor, due to structure roughness.

2. To obtain the different cross sections a Maxwellian distribution of grain velocities has been considered:

$$f(v) = \frac{4}{\pi^{1/2}} \frac{v^2}{\langle v^2 \rangle^{3/2}} \exp(-v^2/\langle v^2 \rangle), \tag{1}$$

where $\langle v^2 \rangle = \frac{2}{3}w^2$, and w^2 is the variance of the distribution.

This choice may be justified on the ground of its physical reality and mathematical simplicity (Saffman and Turner, 1956). The physical content of the distribution, however, is summarized in its variance, evaluated in two different ways.

The first model consists of equalating the relaxation length of grains to the size of the smallest eddy, when the relative grain-eddy velocity is comparable to the velocity difference across the smallest eddy (Cameron, 1972; Landau-Lifchitz, 1971).

$$\langle v^2 \rangle_I^{1/2} = 2.27 \, (\varrho_s S)^{1/2} \left(\frac{T}{\varrho} \right)^{1/4}. \tag{2}$$

The second model for the variance calculation accounts for the following dynamical effects acting on a grain: turbulence and gravitational field. This theory assumes the following conditions:

(a) grain size smaller than the smallest eddies.

(b) turbulence is isotropic.

The root mean square velocity $\langle v^2 \rangle^{1/2}$ (deduced by other authors in the past) has been modified, taking into account that when the gas density is low, the drag exerted on the grain is not given by Stokes' law (Cameron, 1972; Epstein, 1924). The time scale of the grain-grain impact phenomenon has been defined by supposing that the relaxation time is equal to the eddies' decay time. In this way it is possible to calculate the length scale of the eddy that governs the impact phenomenon. So the distribution variance is obtained:

$$w_{II}^2 = \tfrac{3}{2} \langle v^2 \rangle_{II} = \tfrac{1}{3} (s_1 + s_2)^2 \frac{\varrho \varepsilon}{\eta} + \left(1 - \frac{\varrho}{\varrho_s} \right)^2 \times$$
$$\times [g^2 (\tau_1 - \tau_2)^2 + 4.6 \times 10^4 |(\tau_1 - \tau_2)| \, T \varrho^{1/2}], \tag{3}$$

where

$$\tau_{1,2} = \left(\tfrac{2}{9} \pi s_{1,2}^2 \frac{\varrho_s}{\varrho} \right) / v_{1,2}^*;$$
$$v_{1,2}^* = \eta^*/\varrho = 2.16 \times 10^3 \, T^{1/2} s_{1,2}$$
$$\varepsilon \quad = \text{energy dissipation } g^{-1} s^{-1}.$$

η^* is a coefficient derived by Epstein's relationship, $s_{1,2}$ is the radius of the particle, and $\eta = 2.38 \times 10^{-6} \, T^{2/3}$ (Cameron, 1972) is the normal Stokes viscosity for the assumed chemical composition.

3. Now some possible physical grain-grain interactions are examined. The first one is related to the possible electric charge of grains. Many authors have considered detailed balance between negative and positive charge, due to the interstellar radiation field and plasma (Feuerbacher et al., 1973; Watson, 1972).

In the UV radiation field the grain potential is calculated by balancing the photoelectron flux and the charge due to bombardment by plasma particles. We assume

$$F_0 \approx 1.8 \times 10^{-7} \, \text{erg cm}^{-2}\text{s}^{-1} \, \text{Å}^{-1} \, \text{sr}^{-1},$$

where F_0 is the UV flux (Hayakawa et al., 1969).

As regards the charge flux due to ion bombardment, two relationships must be considered.

For negative grain potentials,

$$\dot{N}_{\text{stick}} = \frac{2n_0\sigma_0 a}{\pi^{1/2}} \left\{ \left(\frac{2kT}{m_i}\right)^{1/2} \left[1 - \left(\frac{m_i}{m_e}\right)^{1/2} \exp\left(eV/300\,kT\right)\right] + \right.$$
$$\left. - \frac{eV}{300}\left(\frac{2}{kTm_i}\right)^{1/2}\right\} \qquad (4a)$$

and for positive grain potentials,

$$\dot{N}_{\text{stick}} = \frac{2n_0\sigma_0 a}{\pi^{1/2}} \left\{ \left(\frac{2kT}{m_i}\right)^{1/2} \left[\exp\left(\frac{-eV}{300\,kT}\right) - \left(\frac{m_i}{m_e}\right)^{1/2}\right] + \right.$$
$$\left. - \frac{eV}{300}\left(\frac{2}{kTm_i}\right)^{1/2}\right\}. \qquad (4b)$$

For the photoelectric effect, the following holds:

$$\dot{N}_{\text{photo}} = \sigma_0 Y_0 Q_0 F_0 G(V), \qquad (4c)$$

where

$$\begin{cases} G(V) = \dfrac{1}{(E_0 + V)^2} - \dfrac{1}{13.6^2} & 0 < V < 13.6 - E_0 \\[2mm] G(V) = \dfrac{1}{E_0^2} - \dfrac{1}{13.6^2} & V < 0 \\[2mm] G(V) = 0 & V > 13.6 - E_0. \end{cases}$$

In Equations (4), we have

n_0 = concentration of plasma ions
a = capture probability
σ_0 = geometrical cross section
m_i, m_e = ion and electron masses
T = plasma temperature
Y_0 = photoemission yield
Q_0 = photon absorption coefficient
E_0 = work function
V = potential

Now two grains with radii s_1 and s_2, masses m_1 and m_2, and electric potential V_1 and V_2 attract each other both in the trivial case of oppositely charged grains and in the case of similarly charged grains. The latter occurs when $s_1/s_2 \gg 1$; in fact there is an inductive force produced by the smaller on the bigger particle, able to overcome the normal repulsive effect.

One obtains:

$$U_{\text{tot}}^{\text{es}}(r) = \frac{V_1 V_2}{r} s_1 s_2 - \frac{1}{8}\frac{\varepsilon_0 - 1}{\varepsilon_0}\frac{V_1^2 s_1^2}{r - s_1}, \qquad (5)$$

where r = distance between the centers of grains. The sticking condition between grains is then

$$\tfrac{1}{2}\mu v^2 \leqslant U^{es}_{tot}(r_0) - U^{es}_{tot}(s_1 + s_2) = \Delta U^{es}_{tot}, \tag{6}$$

where $\mu = m_1 m_2 / (m_1 + m_2)$, and v is the relative velocity.

The maximum value of $U^{es}_{tot}(r)$ is found to correspond to r_0:

$$r_0 = s_1 \left\{ 1 + \left[2^{3/2} \left(\frac{\varepsilon_0}{\varepsilon_0 - 1} \right)^{1/2} \left(\frac{V_2}{V_1} \right)^{1/2} \left(\frac{s_2}{s_1} \right)^{1/2} - 1 \right]^{-1} \right\}. \tag{7}$$

When $r_0 > s_1 + s_2$, the electromagnetic interaction is attractive (see Figure 1).

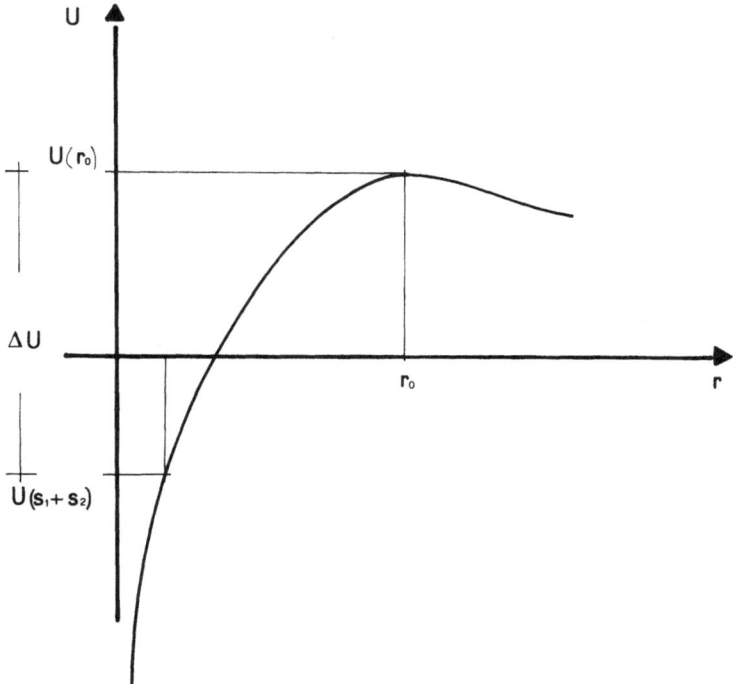

Fig. 1.   Electrostatic potential vs. radial distance. In this case the value r_0 is greater than $s_1 + s_2$, so it is possible to have an attractive potential.

Another effect may be significant when the surfaces of two grains are very close. By applying electromagnetic fluctuation theory, the efficiency of this interaction can be evaluated.

The electromagnetic potential is now critically dependent on the distance: when $r > \lambda_0$ (characteristic wavelength of the solid) the attractive potential becomes (for plane surfaces):

$$U^{em}_{tot}(r) = - \frac{A}{r^3}, \tag{8}$$

where

$$A = \frac{hc\pi^2}{720} \left(\frac{\varepsilon_0 - 1}{\varepsilon_0 + 1}\right)^2 \varphi(\varepsilon_0)$$

$\varphi(\varepsilon_0) \simeq 0.4$ (Landau-Lifchitz, 1969).

In the opposite case $(r < \lambda_0)$, the macroscopic relationship for a spherical grain is (see Appendix)

$$U_{tot}^{em}(r) = -\frac{\hbar}{8r^2} \pi^2 v_0 \alpha^2 n^2, \tag{9}$$

where

$$v_0 = \frac{kT_D}{h}$$

$$\alpha = \frac{1}{4\pi} \frac{\varepsilon_0 - 1}{n}$$

n = particle concentration.

In the case of two grains, the total potential becomes (see Appendix):

$$\Delta U_{tot}^{em} = 3\pi A P_4(r_0, \lambda_0, s_2) + 2\pi B P_3(r_0, \lambda_0, s_2), \tag{10}$$

where

$$A = \frac{2hc}{720} \left(\frac{\varepsilon_0 - 1}{\varepsilon_0 + 1}\right)^2 \varphi(\varepsilon_0); \qquad B = \frac{h\alpha^2 n^2 v_0}{4}$$

and r_0 = minimum approach distance between the grain surfaces.

4. When the two grain strike each other, the collision may be partially inelastic, if the impact velocity is sufficiently high (sound velocity in the grain). Then an amount of energy may be irreversibly trapped.

The interpolation of the results of Carusi *et al.* (1971) and Coradini (1970) relative to shock coefficients gives the following expression for the amount of trapped energy ΔE when the target and projectile are siliceous:

$$\Delta E = E\left[\alpha(m_1, m_2) + \frac{\beta}{m_2}(m_1, m_2) E\right] \quad (m_1 > m_2)$$

$$E = \tfrac{1}{2}\mu v^2. \tag{11}$$

It is assumed two particles stick when ΔE is enough to melt at least 10% of the mass of the smaller one.

That is achieved when:

(a) the radiating time of the trapped heat is less than the characteristic shock time:

$$\tau_v > \tau_1 = \gamma\mu E_F \left[\frac{\bar{k}\left(\frac{4}{3}\pi s_2^3\right)^{2/3}}{s_2} + \sigma_B T_f^4 (4\pi)^{1/3} \left(\frac{\mu}{3\varrho_s}\right)^{2/3}\right]^{-1}, \tag{12}$$

where σ_B is the Stefan-Boltzmann constant and γ is the fraction of molten material.

It is also possible to obtain another characteristic velocity:

$$v_1 = \frac{\bar{k}\left(\frac{4}{3}\pi s_2^3\right)^{2/3} + \sigma_B T_f^4 (4\pi)^{1/3} \left(\frac{\mu}{3\varrho_s}\right)^{2/3}}{\gamma \mu E_F}$$

(b) ΔE is not enough to vaporize the smallest particle or a fraction γ' of it (at least 10%):

$$\Delta E < \mu \left[E_F + c_p T_V + \gamma' E_V \right] \tag{13}$$

(c) if condition (b) is fulfilled, but not condition (a), the smallest particle completely melts, and adhesion is obtained if the configuration with joined particles is more stable than the disjoined one:

$$E\left(1 - \alpha_{12} - \frac{\beta_{12}}{m_2} E\right) = \tau_s (n_p S_n - S) \tag{14}$$

with

n_p = number of molten fragments, deriving from the original drop;
S_n = surface of the nth fragment;
S = surface of the original drop.

(d) obviously, the shock must not be able to cause the break-up of the smaller grain. The limit condition results:

$$E_R\left(1 - \alpha_{12} - \frac{\beta_{12}}{m_2} E_R\right) = \frac{(QS\tau_c)^2}{2\mu}, \tag{15}$$

where E_R is the breaking energy, and

$$\tau_c = 2\pi \left[\mu^{2/3} \left(\frac{4}{3}\pi\varrho_s\right)^{1/3} / k_B \right]$$

is the free oscillation time of the solid.

5. It is possible now to define for every mechanism of interaction a characteristic velocity: thus the conditions for constructive and destructive impact may be found.

Electrostatic and electromagnetic interactions cause adhesion between particles when the residual kinetic energy after impact (a fraction of the energy is spent in the shock wave produced) is less than the potential wall produced by these interactions:

$$E\left(1 - \alpha_{12} - \frac{\beta_{12}}{m_2} E\right) = \Delta U_{tot}^{es} + \Delta U_{tot}^{em}; \tag{16}$$

if $\Delta U_{tot}^{es} + \Delta U_{tot}^{em} > 0$, the characteristic velocity v_0 is obtained.

Moreover, several characteristic velocities are related to the adhesion of grains partially melted by impact (Section 4), which permits one to define the following four characteristic velocities:

(a) $$E\left(\alpha_{12} + \frac{\beta_{12}}{m_2} E\right) = \mu\left(\gamma E_F + c_p T_F\right) \Rightarrow v_1$$

(b) $$E\left(\alpha_{12} + \frac{\beta_{12}}{m_2} E\right) = \mu\left(E_F + c_p T_V + \gamma' E_V\right) \Rightarrow v_2$$

(c) $$E\left(1 - \alpha_{12} - \frac{\beta_{12}}{m_2} E\right) = \tau_s\left(n_p S_n - S\right) \Rightarrow v_3$$

(d) $$E\left(1 - \alpha_{12} - \frac{\beta_{12}}{m_2} E\right) = (Q n_p S \tau_c)^2/2\mu \Rightarrow v_4,$$

where $n_p = 2$ (breaking of the smaller particle in two equal parts).

TABLE III

Constructive cross-section
$\sigma_c(v) > 0$ for:

A	B	C
$v < v_0$	$v > v_0$	$v > v_0$
$v < v_4$	$v > v_1$	$v > v_I$
	$v < v_I$	$v > v_1$
	$v < v_4$	$v < v_2$
	$v < v_2$	$v < v_3$
		$v < v_4$

6. The collision of grains, if the impact velocity is great enough, causes their fragmentation. From (14) and (15) it is possible to obtain critical velocities related to fragmentation in the solid (rupture into n_p fragments) and liquid states (distribution of matter melted in the shock in n_p drops) of grains.

A significant destructive cross section $\sigma_D(v)$ is related to the impact-velocity conditions given in Table IV.

It is thus possible to obtain cross sections for constructive and destructive processes, respectively:

$$\sigma_c(v) = \sigma_0 C(v)$$
$$\sigma_D(v) = \sigma_0 D(v),$$

(17)

TABLE IV

Destructive cross-section
$\sigma_D(v) > 0$ for:

I	·II
$v_4 \geqslant v_3$	$v_4 < v_3$
$v > v_3$	$v > v_4$

where

$$\sigma_0 = \pi (s_1 + s_2)^2 \left[1 + (\Delta U_{tot}^{es} + \Delta U_{tot}^{em})/\tfrac{1}{2}\mu \langle v^2 \rangle \right]$$

is the geometrical cross section corrected for interaction processes between grains, and $D(v)=1$ and $C(v)=1$ if the conditions in Table III and Table IV are satisfied; otherwise $C(v)=D(v)=0$.

The collision frequency for the larger particle is

$$\dot\nu = N_g \int\limits_0^\infty \sigma(v)\, vf(v)\, dv, \qquad (18)$$

where N_g is the number of the smaller grains per cm^3.

7. The probabilities of impact between particles have been evaluated for several values of their masses and of the nebular density. A collapsing one solar mass nebula of uniform density has been assumed; cross sections are evaluated at the boundary of the nebula where the gravity is a function of density only.

Results have also been obtained for both models of characteristic turbulent velocity (Section 2).

One can see from the rms turbulent velocity vs nebular density (Figures 2, 3, 4) that the general trend of velocity is a decrement with increasing density. The velocity in the first model is, moreover, systematically lower than in the second model, depending on the different characteristic scale length of the turbulence; this effect is particularly evident at low density. The turbulent velocity depends considerably on the mass of the larger particle. In the first model this velocity ranges from 600 to 60 cm s^{-1} (largest-mass grain equal to 1 g) and from 60 to 6 cm s^{-1} (largest-mass grain equal to 10^{-12} g). For the second model in the same mass ranges the velocities run from 10^6 to 3×10^3 cm s^{-1} and from 5×10^3 to 60 cm s^{-1}.

Figures 2, 3, 4 also illustrate computed values for the frequencies of constructive and destructive collisions, $\dot\nu_c$ and $\dot\nu_D$, between particles of masses m_1 and m_2 with $m_1 > m_2$. The cross sections are computed for the larger particles when about 0.015 by mass of the nebular medium is in the smaller grains (Cameron, 1972). The geometrical cross section frequency of collision $\dot\nu_0$ is also plotted; so $\dot\nu_c/\dot\nu_0$ and $\dot\nu_D/\dot\nu_0$ show the efficiency of constructive and destructive processes.

The only efficient constructive mechanism is adhesion by interaction between particles, which is governed by velocities v_0 and w (Sections 2 and 3). For grains of silica, adhesion of molten grains is possible only for extremely low masses; only with a large thermal conductivity (metallic grains) adhesion does become possible in a reasonable mass range (Orowan, 1969). Destructive processes on the contrary are driven by v_4, because $v_4 < v_1$, and breaking precedes melting in the case of silica.

The characteristic velocity v_0, if $m_1 \gg m_2$, is essentially driven, for low densities, by electrostatic interaction; while at higher densities photoelectric ejection and ionic capture tend to balance, and electromagnetic interactions become more efficient. For

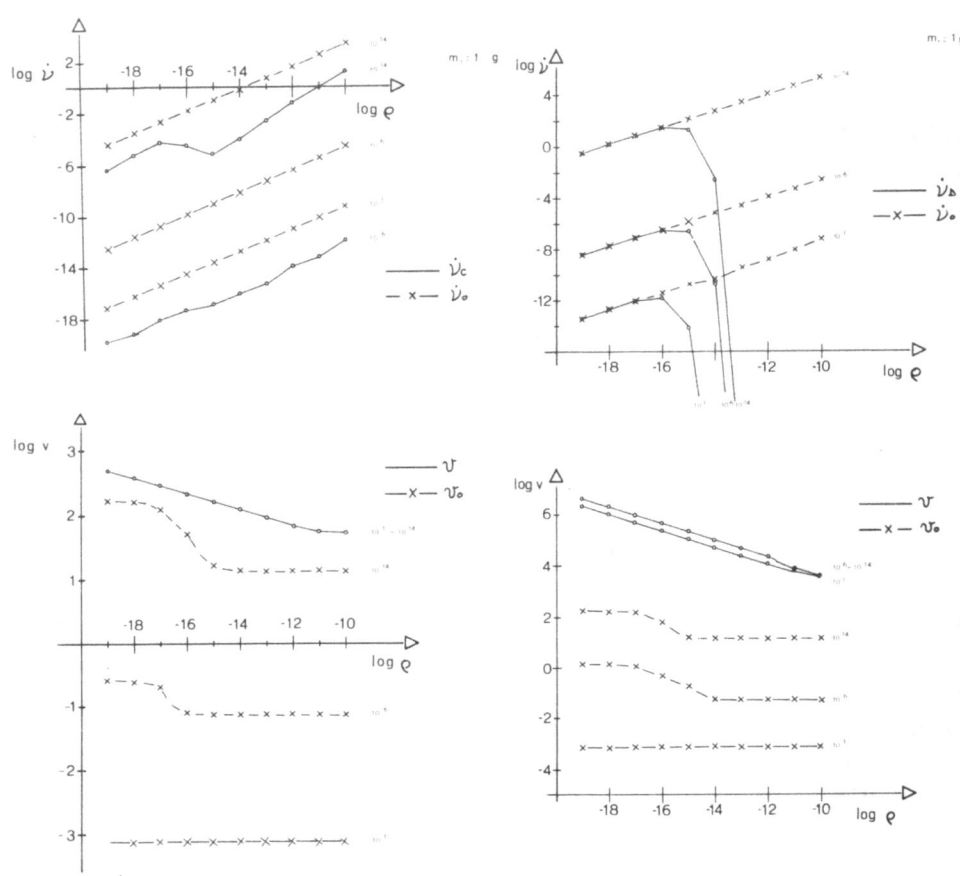

Fig. 2. The left side contains data for model I, the right – for model II. In the upper part is shown
the collision frequencies vs nebular density for models I and II, when the mass of the larger grain is
1 g. For model I, the solid line represents the constructive and the dashed line the geometrical collision frequency. For model II, the solid line is the destructive collision frequency and the dashed line
is again the geometrical collision frequency. The lower part represents the turbulent and critical
velocities in the same conditions: for both models the solid line represents the turbulent velocity for
grains of 10^{-1}, 10^{-6}, and 10^{-14} g, the dashed line represents the characteristic adhesion velocity.

grains of similar masses, the electrostatic interaction is only repulsive; the sole efficient interaction is the electromagnetic one.

A gobal analysis of Figures 2–5 gives some features:

(a) The constructive and destructive processes are essentially governed respectively
by the ratios w/v_0 and w/v_4; so the collision rates are very different, due to the adopted
model for the characteristic velocity of turbulence. Indeed, model I shows constructive
collision frequencies higher than model II because of a lower mean impact velocity,
while compared with model II it gives small values for destructive collision frequencies.

(b) The collision rates strongly depend on the masses of the colliding particles. In
general, a constructive collision is more probable for small colliding particles, or for

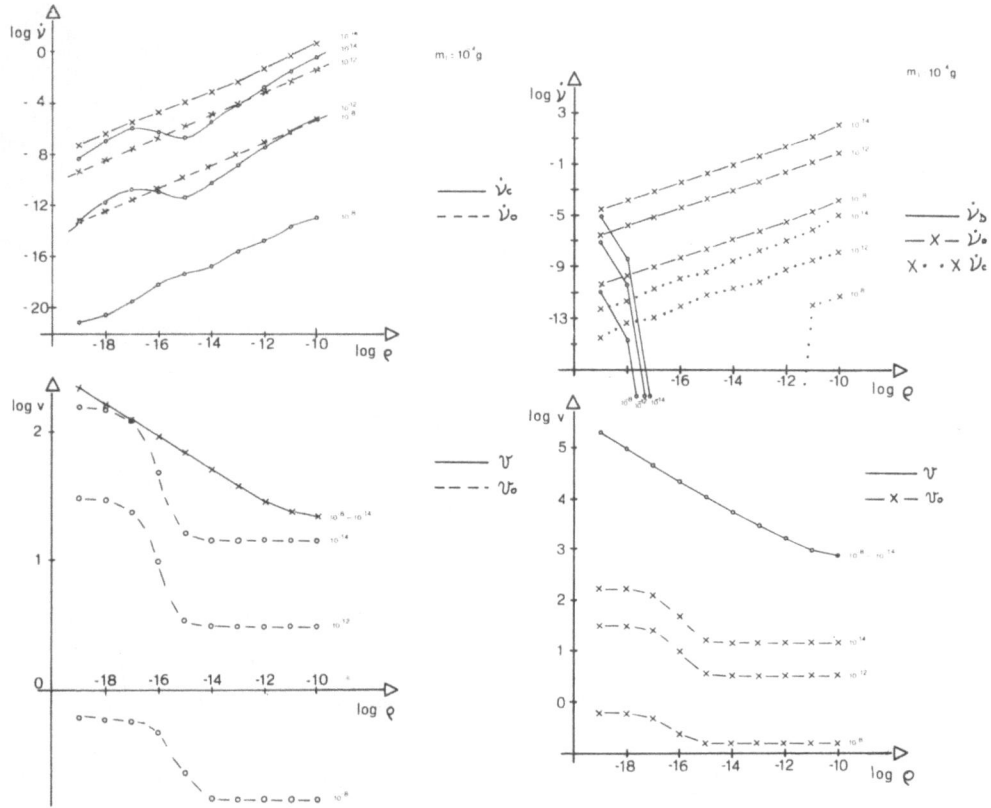

Fig. 3. The left side contains data for model I, the right – for model II. In the upper part is shown the collision frequency vs nebular density for a mass of the larger particle of 10^{-4} g, and masses of impacting particles ranging from 10^{-8} to 10^{-14} g. For model I, the solid line represents the constructive collision frequency, the dashed line the geometrical one. For model II the solid line represents the destructive collision frequency, the dashed line the geometrical, and the dotted line the constructive one. In the lower part for the same mass ranges, the turbulent velocity and characteristic adhesion velocity are represented.

particles with a great mass ratio. This is because the mean impact velocity increases with increasing particle mass, and because the interactions considered decrease with increasing mass. In both models, therefore, the frequencies of constructive velocity are negligible for masses of about 1 g.

Efficiency for lower masses can approach 100% (e.g. the impact of masses of 10^{-12}–10^{-13} g is more efficient than 10% for $\varrho > 10^{-13}$ g cm^{-3}).

(c) For model II it is possible to have a destructive interaction that can also reach an efficiency of 100% if the mass of the larger grain is high enough. The processes' efficiency rapidly decays with increasing density, and becomes less efficient as the particle mass is reduced. It must be taken into account that the collision frequencies refer to the boundary of a one solar mass nebula. Compared to the first model, the

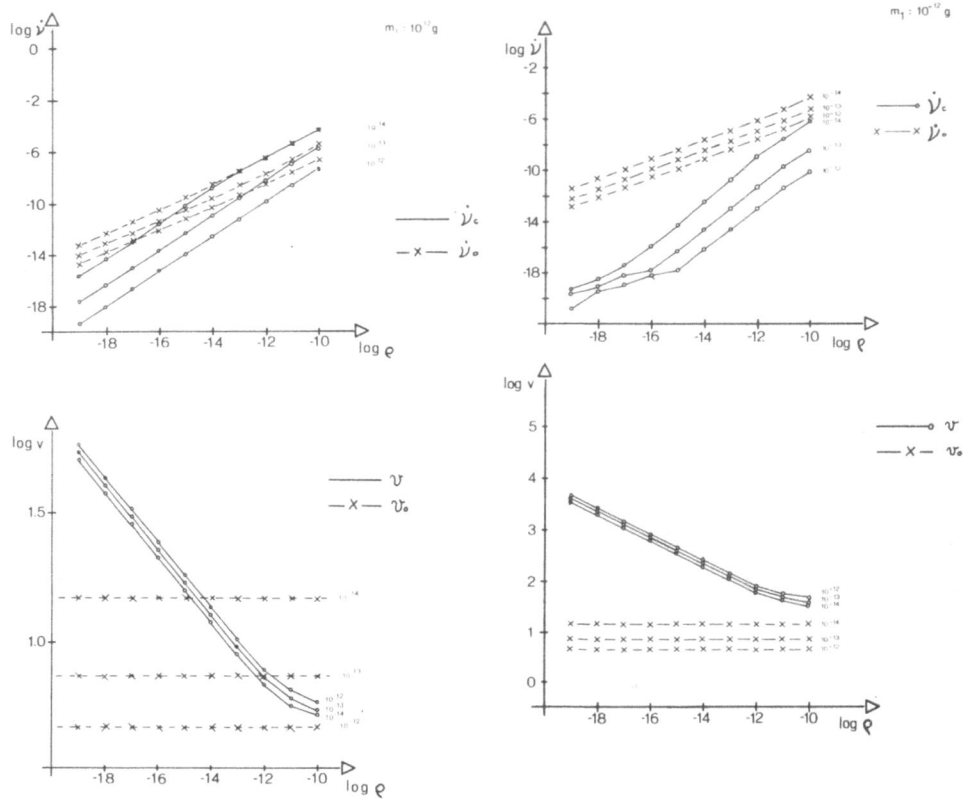

Fig. 4.   The left side contains data for model I, the right – for model II. In the upper part is shown the collision frequency vs nebular density for masses of the larger particles of 10^{-12} g, the masses of impacting particles ranging from 10^{-14} to 10^{-12} g. For both models the dashed line represents the geometrical frequency distribution, the solid line the constructive one. In the lower part the turbulent velocity (solid line) and characteristic adhesion velocity (dashed line) for the two models are represented.

lower gravity implies a smaller characteristic turbulent velocity and, consequently, a lower constructive-collision efficiency.

Conclusions

A model based on the adhesion and fragmentation mechanism contains several uncertainties, connected with:

(a) the difficulty of a correct definition of grain interaction processes; and

(b) the uncertainty of physical parameters and the structure and dynamical process of the nebula such as turbulence, radiative balance and opacity.

However, the model can produce many results that, if valid, would indicate that aggregation processes between grains can work on masses less than 10^{-4} g with sufficient efficiency during several thousand years.

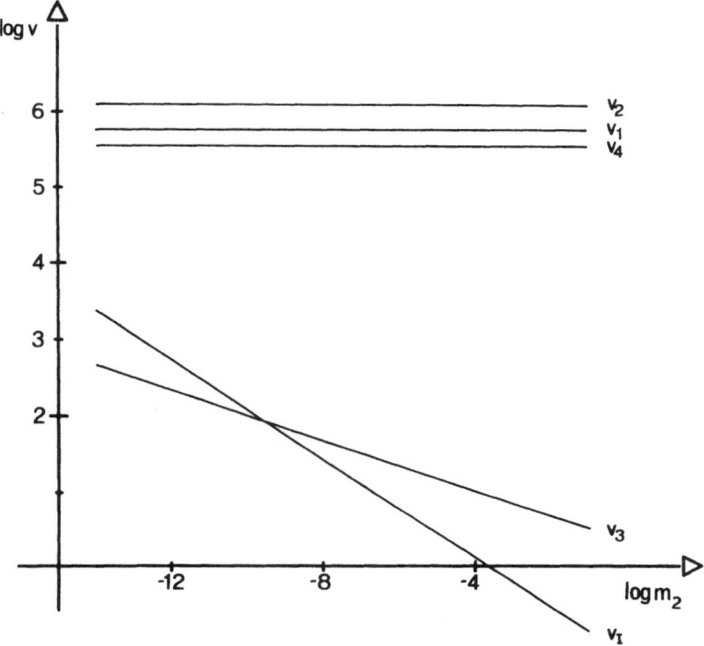

Fig. 5.   Critical velocities: v_1, v_I, v_2, v_3 and v_4 vs the mass of the smaller considered particles are represented.

A final answer on the probability of grain accretion requires a detailed study of the time evolution of grain masses in a collapsing nebula. The authors are completing a numerical method for defining the time evolution of a fixed-mass distribution function of grains in conditions of gravitational collapse, making use of processes studied in this paper, and also considering accretion of nebular gas captured by grains.

Acknowledgements

We wish to thank Prof. B. Accordi, who made this work possible; L. Spinozzi, M. Salomone, and O. Fanucci took care of the pictures and graphs.

Appendix

The electromagnetic interaction between two grains of radii s_1 and s_2 may be considered, if $s_1 \gg s_2$, as an interaction between a sphere of radius s_2 and semispace (Figure 6). The layer of depth dr contributes to the energy of interaction by

$$dU(r, \theta) = -\left[\frac{3A}{4} u(r - \lambda_0) + \frac{2B}{r^3} u(\lambda_0 - r)\right] (\pi s_2^2 \sin^2 \theta) s_2 \sin \theta \, d\theta,$$

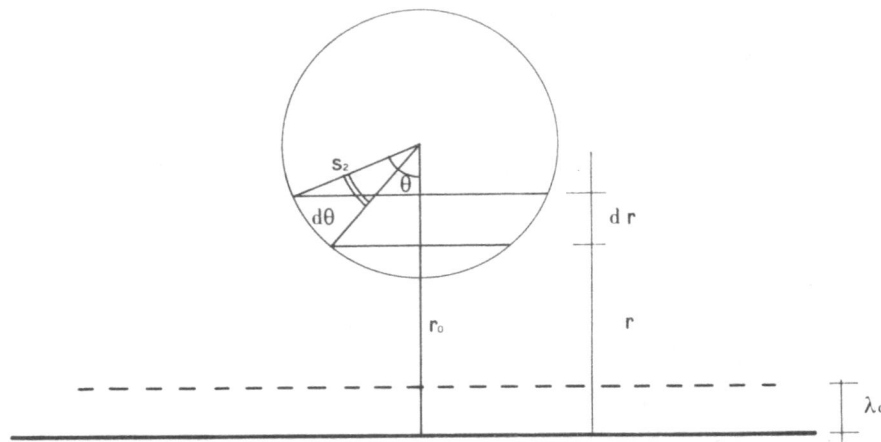

Fig. 6.   Illustration of the model considered in the appendix.

where

$$u(r - \lambda_0) = 0 \qquad r < \lambda_0$$
$$u(r - \lambda_0) = 1 \qquad r \geqslant \lambda_0.$$

The total energy is

$$U_{tot} = \int_0^\pi dU(r, \theta) = 3A\pi s_2^3 P_4(r_0, \lambda_0, s_2) + 2\pi s_2^3 B P_3(r_0, \lambda_0, s_2),$$

where if

(1) $\lambda_0 \leqslant r_0 \Rightarrow P_4 = \int_0^\pi \dfrac{\sin^3 \theta \, d\theta}{[r_0 + s_2 - s_2 \cos \theta]^4} = \int_0^\pi F_4(\theta, r_0, s_2) \, d\theta$

$P_3 = 0$

(2) $r_0 < \lambda_0 \leqslant 2s_2 + r_0 \Rightarrow P_4 = \int_{\theta_0}^\pi F_4(\theta, r_0, s_2) \, d\theta$

$P_3 = \int_0^{\theta_0} \dfrac{\sin^3 \theta \, d\theta}{[r_0 + s_2 - s_2 \cos \theta]^3} = \int_0^{\theta_0} F_3(\theta, r_0, s_2) \, d\theta$

$\theta_0 = \arccos\left[1 + \dfrac{r_0 - \lambda_0}{s_2}\right]$

(3) $\lambda_0 > r_0 + 2s_2$

$P_4 = 0; \qquad P_3 = \int_0^\pi F_3(\theta, r_0, s_2) \, d\theta$

and where the integrals have an easy analytical resolution.

References

Cameron, A. G. W.: 1972, *Symposium on the Origin of the Solar System*, Nice, p. 52.

Carusi, A., Coradini, A., Fulchignoni, M., and Magni, G.:1972, in S. K. Runcorn and H. C. Urey (eds.), 'The Moon', *IAU Symp.* **47**, 180.

Chapman, S. and Cowling, T. G.: 1961, *Mathematical Theory of Non Uniform Gases*, Cambridge University Press.

Coradini, A.: 1970, Laurea Thesis, University of Roma.

Epstein, P. S.: 1924, *Phys. Rev.* **23**, 710.

Feuerbacher, B., Willis, R. F., and Fitton, B.: 1973, *Astrophys. J.* **181**, 101.

Gillet, F. C., Merrill, K. M., and Stein, W. A.: 1972, *Astrophys. J.* **164**, 83.

Greenberg, M. J.: 1972, *Symposium on the Origin of the Solar System*, Nice, p. 135.

Hayakawa, S., Yamashita, K., and Yoshioka, S.: 1969, *Astrophys. Space Sci.* **5**, 493.

Landau, L. and Lifchitz, E.: 1969, *Electrodynamique des milieux continus*, MIR.

Landau, L. and Lifchitz, E.: 1971, *Mécanique des fluides*, MIR.

Larson, B. R.: 1969, *Monthly Notices Roy. Astron. Soc.* **145**, 271.

Larson, B. R.: 1972, *Monthly Notices Roy. Astron. Soc.* **156**, 437.

Larson, B. R.: 1972, *Symposium on the Origin of the Solar System*, Nice, p. 142.

Lequeux, J.: 1972, *Symposium on the Origin of the Solar System*, Nice, p. 118.

Maurette, M. and Bibring J. P.: 1972, *Symposium on the Origin of the Solar System*, Nice, p. 284.

Orowan, E.: 1969, *Nature* **222**, 867.

Saffman, P. G. and Turner J. S.: 1956, *J. Fluid. Mech.* **1**, 16.

Safronov, V. S.: 1972, *Symposium on the Origin of the Solar System*, Nice, p. 89.

Spitzer, L., Jr. and Tomasko, M. G.: 1968, *Astrophys. J.* **152**, 971.

Watson, W. D.: 1972, *Astrophys. J.* **176**, 103.

Wickramasinghe, N. C.: 1967, *Interstellar Grains*, Chapman & Hall.

Woolf, N. J. and Ney, E. P.: 1969, *Astrophys. J.* **155**, L181.

THE ESCAPE OF PLANETARY ATMOSPHERES

DONALD H. MENZEL

Harvard College Observatory, Cambridge, Mass., U.S.A.

Abstract. The problem of escape of atmospheres from the Moon and planets has roots deep in ancient history. Many of the great philosophers of the past regarded the Earth's atmosphere as a medium extending to infinity, with a stationary Earth imbedded at the center. Indeed, it was this concept that led Ptolemy, among many others, to conclude that the Earth could not be moving, for otherwise it would be subject to a gale-force wind caused by its own motion. This idea fostered many of the early stories of interplanetary visitations. Lucian, for example, writing in the second century A.D., has his Icarome nippus fly to the Moon and beyond by means of wings attached to his body.

Not until Copernicus, did this concept of air as an universal medium completely die out, though men had by then become aware of the fact that atmospheric density is less over mountains than over the lowlands. The question then began to take form whether or not the Moon possessed an atmosphere. And this idea was gradually extended to other planets.

The lunar observers of the late eighteenth century set out to ascertain whether the Moon indeed showed visible traces of an atmosphere. In 1792, Schröter recorded what he interpreted as traces of a lunar twilight, which indicated the presence of a thin atmosphere some 29 times more tenuous than our own.

Bessel expressed doubts about this observation, basing his argument on the well-known instantaneous disappearance of a star occulted by the Moon. This argument, which I heard repeatedly in my youth, is not fully correct. A thin atmosphere, if it existed, would cause a delay in the disappearance, by refractive displacement of the star's position. But it would not cause the star to disappear gradually as proponents of the theory argued.

The lunar air battle, which endured for more than a century, involved dozens of famous scientists. The estimates of allowable surface density decreased. Bessel set the figure as 1/500 that of the Earth, in good agreement with the figure set by Newcomb of 1/400. Sir John Herschel urged a much lower figure, of 1/2000, a figure that Airy claimed was in accord with his occultation measures, which indicated the slight retardation of an occultation. Comstock, on similar grounds, reduced the figure to 1/5000.

Paul and Prosper Henry believed they had detected traces of twilight, caused by a thin lunar atmosphere. And W. H. Pickering, observing an occultation of Jupiter by the Moon, claimed to have detected traces of a distortion of the planet, which indicated a density of 1/4000. Yet even this minute amount, as Pickering typically pointed out, required the presence of 'hundreds of tons of atmosphere per square mile of lunar surface.' Pickering also accorded changes in the visibility of certain lunar fea-

Woszczyk and Iwaniszewska (eds.), Exploration of the Planetary System, 37–39. All Rights Reserved

tures, which he ascribed in part to presence of lunar fog or clouds as well as to possible vegetation on the surface of the Moon.

In the midst of such arguments, Johnstone Stoney quietly called attention (in 1898) to the fact that the kinetic theory of gases implies that some fraction of molecules would acquire velocities in excess of the velocity of escape and that atmospheres of the Moon and planets would thus gradually evaporate into space. The speed of disappearance would depend on the body's gravitational potential and upon the boundary temperature of the atmosphere. Stoney elaborated his theories in a series of papers that appeared in the *Transactions of the Royal Society of Dublin* between 1898 and 1904.

In the modern era, Dollfus from measurements of polarization and brightness of the Moon, near quadrature, set the limiting density first at 10^{-6} and, later, 10^{-9} that of the Earth. Elsmore, from radio occultations of the Crab Nebula by the Moon set an even smaller limit.

Such scientists as Sir George Darwin, Cook, Bryan, Emden, and others also considered the problem of atmospheric escape. But it was Sir James Jeans who brought the studies into focus, in his book *Dynamical Theory of Gases*, published in 1916.

Jeans considered the escape of an isothermal atmosphere from a spherical planet. The calculation was a model of simplicity. He calculated the total amount of original atmosphere, figured the rate of escape, and then, dividing the first figure by the second, derived a mean time for atmospheric decay.

Nowhere does Jeans mention the well-known fact – of which he must have been aware – that the total mass of an isothermal atmosphere is infinite. He does not correct for the gravitational pull of the atmosphere upon itself. Indeed, he implicitly assumes that the atmospheric mass is zero. Jeans' formula also predicts an infinite rate of escape of such an atmosphere. The mean decay time, thus derived, is the result of dividing infinity by infinity, always a precarious procedure.

That Jeans was aware of this defect in his analysis I have no doubt. He writes paragraphs about the escaping atmosphere completely detached from the planet, apparently in an effort to justify his loose mathematical treatment of the subject. I have made a careful study of Jeans' argument and am forced to conclude that his times of decay have no physical significance whatever.

William H. Pickering noted that the value of gravity at the surface of a red giant star, such as α Orionis, would be far less than that of the Moon and hence such stars would be expected to dissipate rapidly into space. E. A. Milne pointed out that the rate of dissipation depends on the *gravitational potential* not on the *gravitational acceleration* at the surface. He then develops a modern theory of atmospheric escape, 'with special reference to the boundary of a gaseous star.' The analysis was performed with the 'elegance' characteristic of Milne.

Milne's analysis is applied specifically to a star, whose internal temperature varies according to the law of an Emden 'polytrope'. Milne applied his results to the Earth, but they seem to have largely escaped the attention of astrophysicists.

Spitzer made an important contribution, in analyzing the physical properties of the

Earth's *exosphere*. And various other scientists, especially Chamberlain, Öpik, Singer, and Lifshitz, have extended and improved earlier results.

The high-speed computer has attracted analysts. However, a computer is limited by the way in which the problem is given to it.

I have tried to improve on previous studies, by taking more properly into account the effects of atmospheric temperature gradient. The escape rates prove to be extremely sensitive to the model assumed. I have made some improvement, I think, by inventing a 'collision depth', analogous to 'optical depth' in the problem of radiative transfer.

I conclude, however, that no existing model properly allows for the known physical properties of the exosphere, including the ionosphere. The effect of the variable solar wind needs to be taken into account. Extrapolation backwards in time, from present physical conditions of a moon or planet, is also dangerous. I can construct models of the Moon, for example, which can retain an atmosphere – including water – for some thousands of millions of years. I do not insist that this proves anything, because I can find other models where the decay constant is thousands of times smaller. If we are to draw a conclusion from such studies, it would consist – I think – of a warning not to take too seriously the results from any given model. The Moon, planet, or the universe can readily find some method of evading the predictions made on almost any model.

DISCUSSION

McCrea: I have two questions: (a) It is widely believed that the Earth lost its original atmosphere; could this have happened in accordance with your theory? (b) Many years ago, E. A. Milne developed an extensive theory of the escape of atmospheres; is your work similar to Milne's?

Menzel: Milne's approach is similar to mine. But he was concerned with evaporation from a star and hence did not give a figure for lifetime of an atmosphere above a planet with a definite surface. I see no problem about escaping of the entire atmosphere and later reconstitution. There was probably a time when the Earth was quite hot.

Khodak: Comment to Dr Menzel's paper; The evolutionary and comparative approach is very important for studying planetary problems. For example, the terrestrial planets form a sequence, from the smaller, more primitive, and unevolved bodies like Mercury and the Moon (which lack atmospheres), through the intermediate level represented by Mars (with a thin CO_2 atmosphere and a frozen or liquid hydrosphere), to the largest and most complex planets, Venus and Earth. The dense CO_2 atmosphere of Venus represents a qualitatively new stage in planetary development after Mars; while the Earth, with an oxygen-nitrogen atmosphere, a global hydrosphere, and a biosphere, represents a still higher stage. Thus the evolution of a dense CO_2 atmosphere is to be sought between the levels of Mars and Venus, while the evolution of our present atmosphere lies in the stages between Venus and Earth, buried in the Earth's past. This problem is, simultaneously, relevant to the evolution of a biosphere.

An atmosphere, and the closely-related glacial or liquid hydrosphere (on Mars, Earth, and also, slightly, on Venus), should be treated *fundamentally* as results of endogenous (internal) processes of evolution. That is, they are results of geological outgassing of solid bodies; these atmospheres are secondary, by contrast with the primary atmospheres of the Jovian planets. Thus the straightforward approach needed for the study of planets and satellites is the historical, comparative study of all planetary data with the methods of structural geology and petrochemistry: the 'geology' of planets, planetology (see Yu. A. Khodak, 'Geography and Geology of Planets – Planetology', a course of lectures, Pedagogical Institute Press, Moscow, 1972).

MODERN RESEARCH IN HIGH RESOLUTION INFRARED
SPECTROSCOPY AND ITS IMPORTANCE
TO PLANETARY PHYSICS

K. NARAHARI RAO

Department of Physics, The Ohio State University, Columbus, Ohio, U.S.A.

Abstract. The scope of research in the field of high resolution infrared spectroscopy has expanded in numerous directions in recent years. Investigations of high resolution molecular spectra performed under controlled conditions in the laboratory are of significance to planetary studies because of the high resolution which one can achieve at the present time in the spectra of planets. In making use of laboratory data it is important to develop a proper perspective; due to technological advances the quality of experimental results has improved continuously during the past two decades, and, therefore, one has to exercise proper judgement when combining information from different laboratories. Somewhat detailed discussions are presented for the spectra of CO and CO_2 to illustrate what modern measurements are capable of accomplishing. In the case of CO_2, the extensive measurements of Courtoy and Herzberg done in the early 50's are not summarized in this paper because they are all consolidated in three previous publications. On the other hand, more recent measurements on CO_2 are scattered in several publications and, therefore, an attempt has been made to collect them together in this article. This is not a comprehensive review paper but it is intended to provide only a glimpse into the expanding frontiers of research in high resolution infrared spectroscopy.

1. Introduction

During the past decade many notable advances have been made in the field of high resolution infrared spectroscopy as a result of which the scope and applicability of the field has expanded by leaps and bounds.

From a spectroscopist's point of view, by studying high resolution infrared spectra of molecules we are primarily contributing to the understanding of the structures of molecules in their ground electronic states. By this we mean, determination of the geometrical arrangement of the nuclei in molecules, obtaining precise values for internuclear distances and bond angles and evaluation of force fields applicable to molecular vibrations. For achieving these goals, it is necessary to obtain precision in spectral positions and intensities and develop procedures for correlating experimental results with theory.

The sophisticated technology (Rao and Mathews, 1972) that has become available to us in recent years has been enabling us not only to achieve the above objectives better than ever before but has contributed to the expansion of the capabilities of the field in numerous other directions. This is illustrated in the following summary which gives an overall view of many of the current interests in high resolution infrared spectroscopy.

Structures of Molecules: {Molecular Geometry

Internuclear Distances

Force Fields

Planetary and Stellar Spectra: Fourier Spectroscopy

(since 1962)

Woszczyk and Iwaniszewska (eds.), Exploration of the Planetary System, 41–56. All Rights Reserved
Copyright © 1974 by the IAU

Molecular Laser Emissions:
Transitions between levels not attained ordinarily by absorption spectroscopy
Earth's Atmosphere:
Photochemistry of stratosphere (in particular)
NO_2, HNO_3, (identified 1968)
Concentrations of O_3, H_2O, NO_2, CH_4, N_2O, HNO_3, CO, NO ... (requiring measurement)

First, the high resolution attained in the infrared spectra of cosmic sources by employing techniques in Fourier transform spectroscopy has provided a new dimension to the type of information available to astrophysicists. Second, the molecular laser emissions are playing a significant role in spectroscopic research because a large number of the energy levels involved in the transitions of the laser lines are unusual in that they are not easily reached by conventional absorption and emission techniques in infrared spectroscopy. Finally, in view of future large scale supersonic transport operations, the need to understand more about our own stratosphere has become important and for this purpose the experimental studies of stratospheric constituents conducted in the laboratory under controlled conditions have assumed vital significance.

There is much unity in all these aspects of research in high resolution infrared spectroscopy and a few examples will be discussed in this paper in order to emphasize that scientists involved in any one of these areas should develop more than a passing interest in all the other areas.

2. Improved Planetary Spectra in the Infrared

Use of Fourier transform spectroscopy (FTS) in the recording of planetary spectra in the infrared has provided a gain in the resolving power in the spectrum of Venus by a factor of at least 100 as compared to the best spectra recorded with a scanning grating spectrometer. In addition to high resolution capabilities, the techniques in Fourier spectroscopy enable the determination of spectral positions with high accuracy and, therefore, comparison of planetary data with laboratory measurements is easier to make. As a result of such superior quality data obtained for planetary spectra, Connes and associates made important discoveries. For instance, they were able to identify for the first time the presence of CO, HCl and HF in the Venus atmosphere (Connes et al., 1967, 1968) and of CO in Mars (Kaplan et al., 1969).

3. Large Gratings and Fourier Transform Spectrometers

By high resolution infrared spectra we refer to the ones in which the measured full widths at half heights in unblended spectral lines are at least of the order of 0.02 cm^{-1}. At the present time, it is possible to achieve this kind of spectral resolution in day-to-day observations obtained with spectrographs equipped with large modern plane gratings (with dimensions 40 cm × 20 cm) (Rao and Mathews, 1972). In the context of the present paper, spectrographs equipped with such large gratings and Fourier transform spectrometers are considered to be the two kinds of instruments which are important for obtaining good high resolution infrared spectra. These instruments have complementary advantages. If employed properly, Fourier transform spectroscopy leads to high spectral resolution and with it one can determine wavelengths relative to the primary standard of length. Moreover, the use of high quality gratings for recording spectra is a well-established and well-understood technique; not only is it possible to achieve precision in the determination of spectral positions with grating instruments but it is very necessary to use them in research work involving the determination of intensities of spectral lines. This is because we want to assure ourselves that information is not lost as might happen in the mathematical operations used in obtaining Fourier transforms.

At the present time, for specific investigations, tunable lasers are helpful in attaining very high resolving power and, with this technique there will undoubtedly be attempts made to achieve accuracy in spectral positions as well as ability to scan over large extents of the infrared spectrum. To be sure, the vibration rotation bands of molecules occur in different parts of the near infrared between 1–30 μ, and for many spectroscopic studies it is important to have available as much information as feasible for as many bands as possible.

4. Aspects of the Infrared Spectrum of CO

The infrared spectrum of the CO molecule continues to provide fascinating possibilities for research. The interdisciplinary nature of infrared spectroscopic research can be recognized when we consider a few of the numerous types of programs undertaken involving the spectra of this molecule. Many of these programs are the result of recent research. Mention has already been made of the occurrence of CO in the atmospheres of Venus and Mars. In 1970, the CO molecule was identified (Wilson *et al.*, 1970) as one of the more abundant molecules in interstellar space. In the laboratory measurement of high resolution infrared spectra, the infrared CO lines have been useful as secondary standards of wavelength (Rao, 1950). It is well known that studies of the abundance of this molecular species in the Earth's atmosphere have been (Migeotte, 1959; Migeotte and Neven, 1950; Locke and Herzberg, 1953) and continue to be of interest to atmospheric scientists. Finally, studies of CO laser spectra (Mantz *et al.*, 1970) have enabled a critical examination (Mantz *et al.*, 1971; Fleming and Rao, 1972; Kirschner and Watson, 1973) of the methods of developing the potential functions of diatomic molecules. This last aspect will be amplified a little more because of its impact in interpreting recent observations of solar spectra.

First, a few comments about the absorption spectrum of CO in the infrared: The vibration-rotation fundamental of CO at 4.7 μ is easy to study by absorption techniques. It requires only an absorption path of about 2 cm-atm to observe. The spectrum presents a simple appearance and the spacings between the successive lines in its rotational structure (3–4 cm^{-1}) is such that, as mentioned earlier, CO lines have been useful in calibrating infrared spectra. The third overtone band of CO (4–0 band) occurs in the photographic infrared and Herzberg and Rao (1949) used an absorption path of nearly 800 meters at a pressure of 1 atm to observe it.

Figure 1 shows a sketch of the potential energy curve for CO. It may be noted that absorption studies as indicated above allow us to obtain information for only a very small portion of the potential energy curve. Attempts to evaluate the potential energy function for the ground electronic state of CO have met with some measure of success only during recent years. The energy levels sketched in Figure 1 up to $v=37$ can all be studied experimentally from studies of the infrared spectra of a CO laser (Roh and Rao, 1974). The vibrational and rotational term value expression used to fit the observational data including all the laser transitions is given below:

$$
\begin{aligned}
T(v, J) = {} & G(v) + F_v(J) \\
= {} & \omega_e(v + \tfrac{1}{2}) - \omega_e x_e(v + \tfrac{1}{2})^2 + \omega_e y_e(v + \tfrac{1}{2})^3 - \omega_e z_e(v + \tfrac{1}{2})^4 + \\
& + \omega_e a_e(v + \tfrac{1}{2})^5 - \omega_e b_e(v + \tfrac{1}{2})^6 + \cdots + \\
& + J(J + 1)\,[B_e - \alpha_e(v + \tfrac{1}{2}) + \gamma_e(v + \tfrac{1}{2})^2 - \cdots] - \\
& - J^2(J + 1)^2\,[D_e - \beta_e(v + \tfrac{1}{2}) + \varepsilon_e(v + \tfrac{1}{2})^2 - \cdots] + \\
& + J^3(J + 1)^3\,[H_e - \delta_e(v + \tfrac{1}{2}) + \cdots] - \cdots.
\end{aligned}
\tag{1}
$$

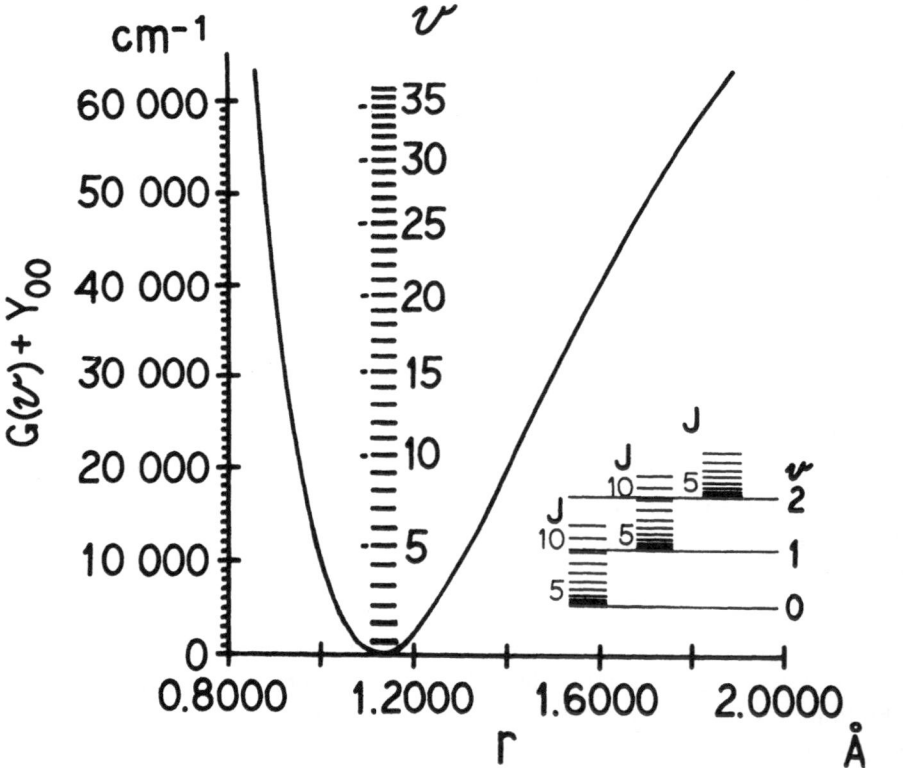

Fig. 1.   Vibrational energy levels observed in a CO laser.

The numerical values for the constants obtained from experimental data for $^{12}C^{16}O$ are presented in Equation (2) expressed in cm^{-1}.

$$
\begin{aligned}
T(v, J) = & (2169.817346 \pm 0.000670)\,(v + \tfrac{1}{2}) - \\
& - (13.289545 \pm 0.000280)\,(v + \tfrac{1}{2})^2 + \\
& + (1.06521 \times 10^{-2} \pm 0.00350 \times 10^{-2}(v + \tfrac{1}{2})^3 + \\
& + (5.037 \times 10^{-5} \pm 0.197 \times 10^{-5})\,(v + \tfrac{1}{2})^4 + \\
& + (1.1486 \times 10^{-6} \pm 0.0510 \times 10^{-6})\,(v + \tfrac{1}{2})^5 - \\
& - (3.3124 \times 10^{-8} \pm 0.0484 \times 10^{-8})\,(v + \tfrac{1}{2})^6 + \cdots + \\
& + J(J + 1)\,[(1.931283656 \pm 484 \times 10^{-9}) - \\
& - (1.751003 \times 10^{-2} \pm 0.000110 \times 10^{-2})\,(v + \tfrac{1}{2}) + \\
& + (1.7214 \times 10^{-6} \pm 0.0260 \times 10^{-6})\,(v + \tfrac{1}{2})^2 - \cdots] - \\
& - J^2(J + 1)^2\,[(6.125781 \times 10^{-6} \pm 0.000780 \times 10^{-6}) - \\
& - (2.927 \times 10^{-9} \pm 0.437 \times 10^{-9})\,(v + \tfrac{1}{2}) + \\
& + (4.313 \times 10^{-10} \pm 0.763 \times 10^{-10})\,(v + \tfrac{1}{2})^2 - \cdots] + \\
& + J^3(J + 1)^3\,[(7.723 \times 10^{-12} \pm 0.749 \times 10^{-12}) - \\
& - (3.364 \times 10^{-13} \pm 0.772 \times 10^{-13})\,(v + \tfrac{1}{2}) + \cdots] + \cdots.
\end{aligned}
\tag{2}
$$

The error limits quoted in the above equation for each constant are the standard deviations. In evaluating these molecular constants, by least squares techniques, other measurements in the infrared (Rao et al., 1966; Rank et al., 1957, 1965; Weinberg et al., 1965) and in the microwave regions (Rosenbloom et al., 1958; Cowan and Gordy, 1957; Helminger et al., 1970) have been included by weighting the data proportional to $(\Delta v)^{-2}$, where Δv is the estimated uncertainty of measurements.

Now, we come to the importance of all this work to astrophysics. In the June 1973 issue of the *Astrophysical Journal*, Hall (1973) reported some solar data in the fundamental region of the CO molecule at 4.7 μ in the infrared. This is a continuation of his work in the first overtone region at 2.3 μ. He has a much richer CO spectrum at 4.7 μ and in the ordinary variety of carbon monoxide (viz., $^{12}C^{16}O$) he reported transitions involving high J ($J > 100$) between vibrational levels up to $v=7$. He also announced the detection of the isotopic varieties $^{13}C^{16}O$, $^{12}C^{18}O$ and $^{12}C^{17}O$ in the solar spectra. These identifications have been made by calculating the CO spectrum employing the molecular constants involving laser data up to $v=28$ (Mantz et al., 1970). For evaluating the spectra of the isotopic varieties, Hall employed the theoretical relations given by Dunham (1932). All these CO studies demonstrate the aspect of relationship between the different areas of interest in high resolution infrared spectroscopy mentioned earlier in this paper. Incidentally, the results of Hall are consistent with the hypothesis that in the solar atmosphere all the three isotopes, viz., carbon-13, oxygen-18 and oxygen-17 are within their terrestrial abundances.

Before closing the discussion of the CO molecule, it should be mentioned that the solar data were not included in the least squares program mentioned above for determining molecular constants. The constants are sensitive to the weighting of the observational data which, as stated earlier, is proportional to $(\Delta v)^{-2}$. There was no clear indication about the observational uncertainties (Δv) of the solar data. When that aspect is clarified it would indeed be highly desirable to include the solar data for the types of computations of interest to molecular spectroscopists. In fact, as may be seen below, the values calculated for some of the solar lines from Equation (2) do not quite agree with the measurements (Hall, 1973).

TABLE I

Wave numbers (expressed in vac. cm⁻¹) for the 7–6 band of $^{12}C^{16}O$

R(J)	Solar data	Calc.[a]	Dif.
R 65	2140.930	2140.913	0.017
66			
67	2142.811	2142.796	0.015
68	2143.688	2143.669	0.019
69			
70			
71	2146.056	2146.018	0.038
R101	2146.361	2146.119	0.242
102			
103	2144.929	2144.552	0.377
104	2144.156	2143.692	0.464
105	2143.336	2142.781	0.555
106			
107			
108	2140.603	2139.738	0.865

[a] The author wishes to thank Dr Won B. Roh for supplying this information.

This suggests two possibilities: a more careful study of the solar spectrum should be made especially because the (7–6) band forms a head and the lines R (101)–R (108) form the returning branch. On the other hand, it is entirely possible that further refinements should be made in the CO constants by including more experimental data especially those involving high J values.

5. CO_2

The infrared spectra of this molecule provide an example of how by combining the laboratory data with properly measured planetary infrared spectra it has been possible to obtain valuable information both for the structure of this molecule as well as for the understanding of the planetary atmospheres. Good high resolution data are available for several infrared bands of the ordinary variety of carbon dioxide, namely, $^{12}C^{16}O_2$. With somewhat less precision are the $^{13}C^{16}O_2$ bands known. In view of the importance of this molecule it would be useful to summarize the available data for these and other isotopic varieties of carbon dioxide and that has been done in Tables II–VIII. The ground state molecular constants available for the different isotopic species of carbon dioxide are given in Table IX. Combining the results obtained

TABLE II

Molecular constants (expressed in cm⁻¹) of $^{12}C^{16}O_2$ bands
Reference numbers refer to the references typed as a footnote to this table and they are not the same as references in the text)

Upper	Lower	ν_0	$(B'-B'')\times 10^5$	$(D'-D'')\times 10^8$	Ref.	Other references
03^10	10^00	544.26	58	2.75	3	
$03^{1c}0$	$02^{2c}0$	597.337 ± 0.008	− 90.1		8	3,7
$03^{1d}0$	$02^{2d}0$		1.8			
02^00	$01^{1c}0$	618.033 ± 0.005	− 15.9	2.3	8	3, 7, 21
$01^{1c}0$	00^00	667.380 ± 0.005	42.4 ± 0.7	0.3 ± 0.3	19	2, 3, 5, 7, 8, 20, 21
$01^{1d}0$	00^00		103.7 ± 0.9	0.3 ± 0.3		
$02^{2c}0$	$01^{1c}0$	667.750 ± 0.005		0.1	8	3, 7
$02^{2d}0$	$01^{1d}0$		41.2	0.0		
10^00	$01^{1c}0$	720.808 ± 0.005	− 43.4	−1.0	8	3, 7, 21
$11^{1c}0$	$02^{2c}0$	741.730 ± 0.008	− 128.5		8	3,7
$11^{1d}0$	$02^{2d}0$		− 32.1			
12^20	03^30	757.47	− 84	−0.8	3	
11^10	02^00	791.43	− 7	−2.75	3	
$01^{1c}1$	$11^{1c}0$	927.17	− 303		12	11
$01^{1d}1$	$11^{1d}0$					
00^01	10^00	960.95857	− 304.739	1.837	1	4, 15
00^01	02^00	1063.73457	− 334.076	− 2.382	1	4, 15
02^00	00^00	1285.40	19		22	
10^00	00^00	1388.15	− 3		22	
$03^{1c}0$	00^00	1932.477 ± 0.004	51.3 ± 0.2	1.1 ± 0.2	13	3, 7
$03^{1d}0$	00^00		146.7 ± 0.4			
$11^{1c}0$	00^00	2076.890 ± 0.003	16.7 ± 0.2	− 0.7 ± 0.1	13	3
$11^{1d}0$	00^00		110.1 ± 0.3			
$12^{2c}0$	$01^{1c}0$	2093.356 ± 0.004	89.2 ± 0.9		13	3
$12^{2d}0$	$01^{1d}0$		28.2 ± 0.9			
20^00	$01^{1c}0$	2129.769 ± 0.005	− 3.3 ± 0.5		13	3
00^09	00^08	2150.65 ± 0.02	− 304 ± 7	1 ± 6	19	
00^08	00^07	2175.268 ± 0.007	− 305.1 ± 1.8	0.4 ± 0.9	19	
00^07	00^06	2199.953 ± 0.005	− 306.6 ± 1.5	− 0.4 ± 0.8	19	
00^06	00^05	2224.697 ± 0.004	− 307.2 ± 1.3	− 0.42 ± 0.49	19	
00^05	00^04	2249.492 ± 0.003	− 305.6 ± 0.6	0.34 ± 0.31	19	
00^04	00^03	2274.355 ± 0.002	− 307.3 ± 0.4	− 0.19 ± 0.15	19	
00^03	00^02	2299.255 ± 0.001	− 307.35 ± 0.19	0.00 ± 0.08	19	10
$02^{2c}1$	$02^{2c}0$	2324.142 ± 0.005	− 302.3 ± 1.0	0.6 ± 0.6	16	10
$02^{2d}1$	$02^{2d}0$		− 302.3 ± 1.0	0.6 ± 0.6		
00^02	00^01	2324.188 ± 0.001	− 307.67 ± 0.15	− 0.04 ± 0.06	19	10
10^01	10^00	2326.600 ± 0.006	− 312.8 ± 1.0		16	10
02^01	02^00	2327.433 ± 0.005	− 297.9 ± 1.0	0.08 ± 0.1	10	
$01^{1c}1$	$01^{1c}0$	2236.633 ± 0.003	− 304.4 ± 0.5	0.00 ± 0.1	16	10
$01^{1d}1$	$01^{1d}0$		− 306.4 ± 0.5	0.00 ± 0.1		
00^01	00^00	2349.147 ± 0.001	− 307.88 ± 0.07	− 0.05 ± 0.02	19	9, 10, 14, 15, 16, 17
10^01	02^00	2429.37	− 341	− 3.5	15	
$01^{1c}1$	00^00	3004.016 ± 0.005	− 262.3 ± 0.7	− 0.3 ± 0.3	19	
$01^{1d}1$	00^00		− 202.9 ± 0.9	− 0.3 ± 0.3		
05^10	00^00	3181.45	80		18	
13^10	00^00	3339.34	− 19	0.0	18	
05^31	03^30	3528.25	− 277		18	

Table II (Continued)

Levels		ν_0	$(B' - B'') \times 10^5$	$(D' - D'') \times 10^8$	Ref.	Other references
Upper	Lower					
04^21	02^20	3552.872 ± 0.010	-273 ± 2		9	27
04^01	02^00	3568.218 ± 0.008	-229.7 ± 1.0			9
$03^{1c}1$	$01^{1c}0$	3580.327 ± 0.003	-285.8 ± 0.07	1.3 ± 0.3	9	6, 15
$03^{1d}1$	$01^{1d}0$		-255.3 ± 0.07	1.8 ± 0.3		
12^01	10^00	3589.646 ± 0.005	-365.5 ± 0.8		9	
02^01	00^00	3612.843 ± 0.003	-271.1 ± 0.5	2.6 ± 0.2	19	6, 7, 9, 15, 21
12^01	02^00	3692.416 ± 0.005	-394.1 ± 0.7		9	
20^01	10^00	3711.475 ± 0.010	-269.4 ± 1.0	-1.8 ± 2.0	9	
10^01	00^00	3714.782 ± 0.002	-315.5 ± 0.3	-1.9 ± 0.1	19	6, 7, 9, 21
$11^{1c}1$	$01^{1c}0$	3723.250 ± 0.003	-327.4 ± 0.7	-1.1 ± 0.5	9	6
$11^{1d}1$	$01^{1d}0$		-302.5 ± 0.7	-1.8 ± 0.3		
$12^{2c}1$	$02^{2c}0$	3726.647 ± 0.005	-313.9 ± 0.7	-0.8 ± 1.0	9	
$12^{2d}1$	$02^{2d}0$		-313.5 ± 0.7	-0.8 ± 0.6		
00^03	00^00	6972.49	-922.5 ± 0.5		5, 9	
02^05	00^00	12672.274	-1456.1	3.9	23	
10^05	00^00	12774.727	-1567.8	-2.4	23	

Important Note

In addition to the above data, ν_0, ΔB and ΔD values for several other bands of $^{12}CO_2$ can be obtained from the following three classic publications. The constants quoted in these papers have been obtained by graphical techniques. Since the basic measurements are also given in these papers they can be reprocessed should that become necessary.

I. G. Herzberg and L. Herzberg, 'Rotation-Vibration Spectra of Diatomic and Simple Polyatomic Molecules with Long Absorbing Paths. XI: The Spectrum of Carbon Dioxide (CO_2) Below 1.25 μ', *J. Opt. Soc. Amer.* **43**, 1037 (1953).

II. C. P. Courtoy, 'Spectres de vibration-rotation de molécules simples diatomiques ou polyatomiques avec long parcours d'absorption. XII: Le spectre de $^{12}C^{16}O_2$ entre 3500 et 8000 cm^{-1} et les constantes moléculaires de cette molécule', *Can. J. Phys.* **35**, 608 (1957).

III. C. P. Courtoy, 'Spectre infrarouge à grande dispersion et constantes moléculaires du CO_2', *Ann. Soc. Sci. Bruxelles, Sér I: Math. Astron. Phys.* **73**, 5 (1959).

References for $^{12}C^{16}O_2$ data

1. K. M. Baird, H. D. Riccius, and K. J. Siemson, *Opt. Commun.* **6**, 91 (1972); see also T. Y. Chang, *Opt. Commun.* **2**, 77 (1970).
2. Joseph J. Barrett and Alfons Weber, *J. Opt. Soc. Am.* **60**, 70 (1970).
3. W. S. Benedict, *Mem. Soc. Roy. Sci. Liège* **2**, 18 (1957). [M. Migeotte, L. Neven, and J. Swensson, 'The Solar Spectrum from 2.8 to 23.7 Microns. II: Measures and Identifications.]
4. T. J. Bridges and T. Y. Chang, *Phys. Rev. Letters* **22**, 811 (1969).
5. Charles P. Courtoy, *Ann. Soc. Sci. Bruxelles, Ser. 1: Math. Astron. Phys.* **73**, 5 (1959).
6. W. L. France and F. P. Dickey, *J. Chem. Phys.* **23**, 471 (1955).
7. Howard R. Gordon, Ph. D. Dissertation, The Pennsylvania State University (1965).
8. Howard R. Gordon and T. K. McCubbin, Jr., *J. Mol. Spectrosc.* **18**, 73 (1965).
9. Howard R. Gordon and T. K. McCubbin, Jr., *J. Mol. Spectrosc.* **19**, 137 (1966).
10. Yu Hak Hahn, Ph. D. Dissertation, The Pennsylvania State University (1967).
11. B. Hartmann and B. Kleman, *Can. J. Phys.* **44**, 1609 (1966).
12. John A. Howe and R. A. McFarlane, *J. Mol. Spectrosc.* **19**, 224 (1966).
13. Arthur G. Maki, Earle K. Plyler, and Robert J. Thibault, *J. Res. Nat. Bur. Stand.* **67A**, 219 (1963)
14. T. K. McCubbin, Jr., and Yu Hak Hahn, *J. Opt. Soc. Am.* **57**, 1373 (1967).
15. M. Migeotte, L. Neven, and J. Swensson, *Mem. Soc. Roy. Sci. Liège* **2**, 5 (1957).
16. Ralph Oberly, K. Narahari Rao, Y. H. Hahn, and T. K. McCubbin, Jr., *J. Mol. Spectrosc.* **25**, 138 (1968).

17. Earle K. Plyler, Lamdin R. Blaine, and Eugene D. Tidwell, *J. Res. Nat. Bur. Stand.* **55**, 183 (1955).
18. Earle K. Plyler, Eugene D. Tidwell, and W. S. Benedict, *J. Opt. Soc. Am.* **52**, 1017 (1962).
19. Robert Earl Pulfrey, Ph. D. Dissertation, The Pennsylvania State University (1972).
20. Kurt Rossmann, K. Narahari Rao, and Harald H. Nielsen, *J. Chem. Phys.* **24**, 103 (1956).
21. Kurt Rossmann, W. L. France, K. Narahari Rao, and H. H. Nielsen, *J. Chem. Phys.* **24**, 1007 (1956).
22. B. P. Stoicheff, *Can. J. Phys.* **36**, 218 (1958).
23. L. D. Gray Young, A. T. Young, and R. A. Schorn, *J. Quant. Spectrosc. Radiat. Transfer* **10**, 1291 (1970).

TABLE III

Molecular constants (expressed in cm^{-1}) of $^{13}C^{16}O_2$

Levels Upper	Levels Lower	ν_0	$(B' - B'') \times 10^5$	$(D' - D'') \times 10^8$
$11^{1c}0$	00^00 [a]	2037.093 ± 0.006	-210.0 ± 2.0	1.40 ± 0.70
$01^{1c}1$	$01^{1c}0$ [b]	2271.764 ± 0.005	-292.8 ± 1.0	
$01^{1d}1$	$01^{1d}0$ [b]		-293.6 ± 1.0	
00^01	00^00 [b]	2283.490 ± 0.004	-296.3 ± 1.0	
$04^{2c}1$	$02^{2c}0$	3473.673 ± 0.003	-235.1 ± 0.5	
$04^{2d}1$	$02^{2d}0$		-235.3 ± 0.5	
04^01	02^00	3482.202 ± 0.002	-209.7 ± 0.4	
$03^{1d}1$	01^10	3498.718 ± 0.002	-209.8 ± 0.4	1.96 ± 0.14
12^01	10^00	3517.285 ± 0.002	-289.6 ± 0.3	
02^01	00^00	3527.706 ± 0.001	-220.9 ± 0.2	2.20 ± 0.03
20^01	10^00	3621.260 ± 0.005	-299.1 ± 0.7	
12^01	02^00	3621.531 ± 0.006	-407.8 ± 0.8	
10^01	00^00 [c]	3632.917 ± 0.005	-352.4 ± 0.7	-1.70 ± 0.30
10^01	00^00	3632.879 ± 0.002	-351.5 ± 0.3	-1.46 ± 0.08
$11^{1c}1$	$01^{1c}0$	3639.182 ± 0.2	-345.8 ± 0.1	
$11^{1d}1$	$01^{1d}0$		-327.8 ± 0.1	
$12^{2c}1$	$02^{2c}0$	3641.543 ± 0.004	-332.9 ± 0.7	
$12^{2d}1$	$02^{2d}0$		-333.3 ± 0.8	
$06^{2c}1$	$02^{2c}0$	4673.657 ± 0.005	-179.9 ± 1.5	
$06^{2d}1$	$02^{2d}0$		-171.0 ± 1.3	
06^01	02^00	4685.641 ± 0.011	-211.0 ± 1.6	
$05^{1c}1$	$01^{1c}0$	4708.475 ± 0.3	-199.1 ± 0.6	1.77 ± 0.26
$05^{1d}1$	$01^{1d}0$		-111.8 ± 0.5	4.54 ± 0.21
04^01	00^00	4748.016 ± 0.002	-140.1 ± 0.3	4.68 ± 0.08
14^01	02^00	4853.761 ± 0.005	-339.1 ± 0.8	
14^21	02^20	4858.143 ± 0.017	-323.8 ± 0.5	
$13^{1c}1$	$01^{1c}0$	4871.403 ± 0.002	-352.3 ± 0.1	
$13^{1d}1$	$01^{1d}0$		-302.2 ± 0.1	
22^01	10^00	4871.857 ± 0.004	-384.9 ± 0.6	
12^01	00^00	4887.343 ± 0.002	-338.6 ± 0.2	1.52 ± 0.05
22^01	02^00	4976.085 ± 0.009	-499.9 ± 2.2	
20^01	00^00	$4991.304 + 0.002$	$-352.9 + 0.3$	$-3.79 + 0.06$
30^01	10^00	4993.503 ± 0.004	-270.0 ± 1.1	
$21^{1c}1$	$01^{1c}0$	5013.733 ± 0.003	-373.7 ± 0.5	-2.14 ± 0.31
$21^{1d}1$	$01^{1d}0$		-328.4 ± 0.5	-2.69 ± 0.20
$22^{2c}1$	$02^{2c}0$	5028.728 ± 0.003	-352.9 ± 0.6	
$22^{2d}1$	$02^{2d}0$		-350.6 ± 0.8	
06^01	00^00	5951.532 ± 0.003	-58.2 ± 1.0	10.77 ± 0.67

Table III (Continued)

Levels		ν_0	$(B' - B'') \times 10^5$	$(D' - D'') \times 10^8$
Upper	Lower			
$15^{1c}1$	$01^{1c}0$	6088.160 ± 0.003	-328.6 ± 0.5	
$15^{1d}1$	$01^{1d}0$		-235.7 ± 0.6	
14^01	00^00	6119.556 ± 0.002	-264.5 ± 0.3	4.63 ± 0.09
22^01	00^00	6241.932 ± 0.002	-438.0 ± 0.3	-2.37 ± 0.10
$23^{1c}1$	$01^{1c}0$	6243.505 ± 0.011	-419.0 ± 3.0	
$23^{1d}1$	$01^{1d}0$		-365.1 ± 0.5	
30^01	00^00	6363.580 ± 0.002	-320.4 ± 0.5	-4.21 ± 0.21
$31^{1c}1$	$01^{1c}0$	$6397.512 + 0.003$	-375.5 ± 0.5	
$31^{1d}1$	$01^{1d}1$		-297.7 ± 0.7	
$01^{1c}3$	$01^{1c}0$	6745.047 ± 0.002	-879.3 ± 0.3	
$01^{1d}3$	$01^{1d}0$		-884.9 ± 0.3	
00^03	00^00	6780.140 ± 0.002	-887.2 ± 0.2	
34^01	00^00	7481.367 ± 0.003	-489.3 ± 1.1	-15.69 ± 0.78
22^01	00^00	7599.935 ± 0.002	-431.8 ± 0.4	
02^03	00^00	7981.175 ± 0.002	-799.1 ± 0.3	
$11^{1c}3$	$01^{1c}0$		-927.6 ± 1.1	
$11^{1d}3$	$01^{1d}0$	8070.892 ± 0.006	-926.4 ± 1.6	
10^03	00^00	8085.007 ± 0.003	-948.5 ± 0.5	

[a] Arthur G. Maki, Earle K. Plyler, and Robert J. Thibault, *J. Res. Nat. Bur. Stand.*, *Sect. A*, **67**, 219 (1963).

[b] Ralph Oberly, K. Narahari Rao, Y. H. Hahn, and T. K. McCubbin, Jr., *J. Mol. Spectrosc.* **25**, 138 (1968).

[c] Howard R. Gordon and T. K. McCubbin, Jr., *J. Mol. Spectrosc.* **19**, 137 (1966). *Note*: The molecular constants for the data other than those pertaining to the above references have been obtained from R. E. Oberly, Ph. D. Dissertation, The Ohio State University (1970). In determining these constants, Oberly used the basic data given by Charles P. Courtoy, Ann. Soc. Sci. Bruxelles **73**, 5–230 (1959).

from the Venus spectra (Connes *et al.*, 1969) we have a fairly good set of data* available for the following isotopic varieties of carbon dioxide:

$$^{12}C^{16}O_2 \qquad ^{16}O^{12}C^{18}O$$
$$^{13}C^{16}O_2 \qquad ^{16}O^{13}C^{18}O$$
$$^{12}C^{18}O_2 \qquad ^{17}O^{12}C^{18}O$$
$$^{13}C^{18}O_2$$
$$^{14}C^{16}O_2$$

Using all this information, Cihla and Chedin (1971) determined the potential energy

* In view of the strong Fermi resonance between levels v_1, v_2^l, v_3 and $(v_1 - 1)$, $(v_2 + 2)^l$, v_3 theory requires that the designation of the levels cannot be made by past conventional procedures. The alternate solutions are somewhat complex to adopt and assimilate. Also, one quickly loses an intuitive appreciation of what bands we are referring to when we use the newer designations. This is an area which requires careful discussion and reasonable solutions. The designation of the levels given in Tables II–VIII has followed the notation used for the spectra of triatomic molecules. Many experimental results were reported in the literature employing these designations. Anyone interested in evaluating vibrational constants for CO_2 should consult with the papers of Cihla and Chedin (1971) and Chedin and Cihla (1973).

function for the carbon dioxide molecule. By extrapolating calculations to transitions involving higher v values Chedin and Cihla (1973) predicted to good accuracy the emission lines involving levels 00°9, 00°8, ..., 00°5 measured subsequently. More experimental work is required to be done in the laboratory on the carbon dioxide

TABLE IV

Molecular constants (expressed in cm^{-1}) of $^{12}C^{18}O_2$ bands

Levels		ν_0	$(B' - B'') \times 10^5$	$(D' - D'') \times 10^8$
Upper	Lower			
$01^{1c}1$	$01^{1c}0$	2301.798 ± 0.004	-269.7 ± 0.7	35.2 ± 8.5
$01^{1d}1$	$01^{1d}0$		-270.9 ± 0.4	
00^01	00^00	2314.052 ± 0.002	-272.6 ± 0.2	
04^01	02^00	3491.597 ± 0.002	-250.1 ± 0.7	3.1 ± 0.5
$03^{1c}1$	$01^{1c}0$	3493.894 ± 0.001	-287.8 ± 0.2	0.7 ± 0.1
$03^{1d}1$	$01^{1d}0$		-270.5 ± 0.3	1.7 ± 0.1
02^01	00^00	3525.205 ± 0.001	-295.9 ± 0.1	1.20 ± 0.02
04^21	02^20	3549.388 ± 0.002	-293.7 ± 1.2	1.4 ± 1.0
12^01	02^00	3602.934 ± 0.002	-258.6 ± 0.6	-2.9 ± 0.4
10^01	00^00	3638.067 ± 0.001	-216.3 ± 0.1	-1.33 ± 0.03
$11^{1c}1$	$01^{1c}0$	3646.043 ± 0.001	-246.5 ± 0.2	-0.7 ± 0.1
$11^{1d}1$	$01^{1d}0$		-216.8 ± 0.2	-1.0 ± 0.1
20^01	10^00	3642.143 ± 0.002	-212.3 ± 0.4	
12^21	02^20	3649.538 ± 0.002	-238.7 ± 0.6	
$05^{1c}1$	$01^{1c}0$	4674.256 ± 0.002	-286.5 ± 0.6	2.1 ± 0.3
$05^{1d}1$	$01^{1d}0$		-247.5 ± 0.5	3.0 ± 0.2
04^01	00^00	4721.920 ± 0.001	-280.0 ± 0.2	3.2 ± 0.1
$13^{1c}1$	$01^{1c}0$	4827.253 ± 0.002	-295.8 ± 0.2	
$13^{1d}1$	$01^{1d}0$		-254.0 ± 0.2	
16^01	00^00	4833.262 ± 0.001	-288.3 ± 0.3	-2.0 ± 0.1
20^01	00^00	4989.234 ± 0.002	-156.2 ± 0.3	-3.0 ± 0.1
$21^{1c}1$	$01^{1c}0$	5009.413 ± 0.002	-213.0 ± 0.3	
$21^{1d}1$	$01^{1d}0$		-149.9 ± 0.3	

R. Oberly, K. Narahari Rao, L. H. Jones, and M. Goldblatt, *J. Mol. Spectrosc.* **40**, 356 (1971).

TABLE V

Molecular constants (expressed in cm^{-1}) of $^{13}C^{18}O_2$ bands

Levels		ν_0	$(B' - B'') \times 10^5$	$(D' - D'') \times 10^8$
Upper	Lower			
$01^{1c}1$	$01^{1c}0$ [a]	2235.78	-259	
$01^{1d}1$	$01^{1d}0$		-262	
00^01	00^00 [a]	2247.24	-261	
10^01	00^00 [b]	3545.029 ± 0.002	-248.0 ± 0.4	-1.6 ± 0.2

[a] C. Chackerian, Jr. and D. F. Eggers, Jr., *J. Mol. Spectrosc.* **27**, 59 (1968).
[b] R. Oberly, K. Narahari Rao, L. H. Jones, and M. Goldblatt, *J. Mol. Spectrosc.* **40**, 356 (1971).

TABLE VI

Molecular constants (expressed in cm^{-1}) of $^{12}C^{16}O^{18}O$ bands

Levels		ν_0	$(B' - B'') \times 10^5$	$(D' - D'') \times 10^8$
Upper	Lower			
03^10	00^00 [a]	2049.68	48	1.0
00^01	00^00 [c]	2332.15 $\pm$ 0.006	-291.0 ± 1.0	
20^00	00^00 [a]	2500.73	25	0.0
12^00	00^00 [a]	2614.20	-43	0.0
04^00	00^00 [a]	2757.14	90	0.0
02^01	00^00 [d]	3571.140 $\pm$ 0.001	-287.8 ± 0.2	1.7 ± 0.1
10^01	00^00 [d]	3675.139 $\pm$ 0.001	-262.3 ± 0.3	-2.2 ± 0.1
00^02	00^00 [b]	4639.530 $\pm$ 0.003	-590.2 ± 0.8	0
02^02	00^00 [e]	5858.022	-567.4	-1.47

[a] C. V. Berney and D. F. Eggers, Jr., *J. Chem. Phys.* **40**, 990 (1964).
[b] F. A. Diaz, *J. Opt. Soc. Am.* **53**, 203 (1963).
[c] R. Oberly, K. Narahari Rao, Y. H. Hahn, and T. K. McCubbin, Jr., *J. Mol. Spectrosc.* **25**, 138 (1968).
[d] R. Oberly, K. Narahari Rao, L. H. Jones, and M. Goldblatt, *J. Mol. Spectrosc.* **40**, 356 (1971).
[e] L. G. Young, *Icarus* **13**, 270 (1970).

TABLE VII

Molecular constants (expressed in cm^{-1}) of $^{13}C^{16}O^{18}O$ bands

Levels		ν_0	$(B' - B'') \times 10^5$	$(D' - D'') \times 10^8$
Upper	Lower			
00^01	00^00 [a]	2265.973 $\pm$ 0.010	-275.8 ± 0.7	
00^02	00^00 [b]	4508.748	-557.82	0

[a] R. Oberly, K. Narahari Rao, Y. H. Hahn, and T. K. McCubbin, Jr., *J. Mol. Spectrosc.* **25**, 138 (1968).
[b] L. G. Young, *Icarus* **11**, 66 (1969).

TABLE VIII

Molecular constants (expressed in cm^{-1}) of $^{12}C^{17}O^{18}O$ bands

Levels		ν_0	$(B' - B'') \times 10^5$	$(D' - D'') \times 10^8$
Upper	Lower			
10^01	00^00 [a]	3655.130 $\pm$ 0.002	-238.1 ± 0.8	-2.7 ± 0.6

[a] R. Oberly, K. Narahari Rao, L. H. Jones, and M. Goldblatt, *J. Mol. Spectrosc.* **40**, 356 (1971).

TABLE IX

Molecular constants for the ground states of the carbon dioxide molecule and its isotopic varieties

Isotopic species	B_{00^00} (cm^{-1})	D_{00^00} (cm^{-1})	Reference
$^{12}C^{16}O_2$	0.3902191 ± 0.0000032	(13.31 ± 0.12) × 10^{-8}	1
$^{12}C^{16}O^{18}O$	0.36818 ± 0.00005	(11.6 ± 5.0) × 10^{-8}	2
$^{12}C^{17}O^{18}O$	0.35702 ± 0.00005	(14.7 ± 2.7) × 10^{-8}	2
$^{12}C^{18}O_2$	0.346799 ± 0.00013	(10.4 ± 2) × 10^{-8}	2
$^{13}C^{16}O_2$	0.39027	13.70 × 10^{-8}	3
$^{13}C^{18}O_2$	0.34679	(9.3 ± 1.9) × 10^{-8}	2
$^{13}C^{16}O^{18}O$	0.36812		3
$^{14}C^{16}O_2$	0.39021	11 × 10^{-8}	4

1. T. K. McCubbin, Jr., Josef Pliva, Robert Pulfry, William Telfair, and Terry Todd, 'The Emission Spectrum of $^{12}C^{16}O_2$ from 4.2 to 4.7 Microns', *J. Mol. Spectrosc.* **49**, 136 (1974). The authors give aslo $H_{00^00} = (-1.26 ± 1.33) × 10^{-13}$ cm^{-1}.
2. R. Oberly, K. Narahari Rao, L. H. Jones, and M. Goldblatt, *J. Mol. Spectrosc.* **40**, 356 (1971).
3. R. E. Oberly, Ph. D. Dissertation, The Ohio State University (1970). The basic data given in Reference 4 have been used to recompute the constants.
4. Charles P. Courtoy, Infrared spectrum at high dispersion and molecular constants of CO₂, *Ann. Soc. Sci. Bruxelles* **73**, 5–230 (1959). (In French.) [G. R. Wilkinson values, unpublished, are quoted on p. 150.]

molecule and its isotopic varieties including measurements at high temperatures, long paths, and high pressures.

6. NO₂, HNO₃ and NO

The interest in the infrared spectra of NO₂ and HNO₃ have been prompted as a result of the identification of these molecular species in the Earth's stratosphere (Goldman *et al.*, 1970; Ackerman and Muller, 1972; Murcray *et al.*, 1968, 1969). The infrared spectra of NO₂ are complex and their interpretation has progressed significantly during the past year (Hurlock *et al.*, 1973a, b). The question of the occurrence and abundance of NO in the stratosphere is being vigorously pursued at the present time and references to two most recent articles (Girard *et al.*, 1973; Ridley *et al.*, 1973) are given to illustrate the type of experiments being undertaken with balloon borne samplers and by measurements made during flights of a supersonic transport. In addition, more recently a letter on detection of nitric oxide in the lower atmosphere has been published (Toth *et al.*, 1973). Needless to mention that the laboratory data for the spectral positions and intensities of some of the infrared bands used in these pursuits are very significant undertakings.

7. A Few Comments on Intensity Studies in Infrared Spectra

It should be emphasized that even if we are interested in intensities, it is important not to lose sight of the value of having accurate spectral positions because they are vitally needed for the assignments of the bands studies. This includes the assignments of the rotational structure observed in these bands recorded under high resolution. Such assignments assure the identifications of the bands to specific molecular species.

At the present time, excellent intensity work is being conducted in different laboratories employing mainly medium resolution infrared spectrographs. In the future, higher resolution instruments will be used and should provide valuable information.

An examination of the published literature reveals avoidable confusion especially in connection with the nomenclature adopted by spectroscopists reporting intensity data. The following sample shows that there exist numerous preferences for the units in which the strengths of the lines are being reported:

Strengths of the lines, S

$cm^{-1} (cm\ atm)^{-1}$
$cm^{-1} (cm\ atm)^{-1}_{STP}$
$cm^{-1} (m\ atm)^{-1}$
$cm^{-1} (km\ atm)^{-1}$
$cm^{-1} (km\ amagat)^{-1}$
$cm^{-1} \left(\dfrac{molecules}{cm^2}\right)^{-1}$ or $cm^{-1} \left(\dfrac{molecules}{cm^3} \times cm\right)^{-1}$
$cm^{-1} \left(\dfrac{grams}{cm^2}\right)^{-1}$ or $cm^{-1} \left(\dfrac{grams}{cm^3} \times cm\right)^{-1}$

Wave numbers per molecule per cm^2 at $295\,°C$

The last usage in the above table is an unfortunate consequence of a trend to use the word 'wave number' as synonymous to cm^{-1}. The word 'wave number' means reciprocal of the word 'wavelength' and the units for it depend upon the units used for wavelength. Commission 14 of the IAU has done yeoman service in the past in providing leadership with respect to wavelength standards. Perhaps they can consider the various aspects of spectroscopic nomenclature and units, in general, so that all scientists interested in the field may communicate with each other without creating sources for errors.

Acknowledgements

I would like to express my gratitude to Dr Tobias Owen for presenting this paper on my behalf at the Copernicus Symposium IV held in Torun, Poland, during September 5–8, 1973. Some of the problems discussed here have been the result of an examination of the problems of interest to the Department of Transportation, Climatic Impact Assessment Program. Finally, my thanks are also due to Dr Lawrence Weller for his help in preparing the manuscript.

Note added in proof. In a very recent communication, S. T. Ridgway (*Astrophys. J.* **187** (1974), L41), reported the identification of ethane (C_2H_6) and acetylene (C_2H_2) in the atmosphere of Jupiter. This was done by using a Fourier transform spectrometer at a resolution of $1.3\ cm^{-1}$ in the range $750-1300\ cm^{-1}$ (10 micron window).

This identification points to important possibilities for the future. In fact, Ridgway concluded:

The spectra described here were obtained in 3 h with a prototype instrument. Detectors 10 times more sensitive and telescopes of 3 times greater aperture are available. Within a few years much higher spectral resolution in the 10-μ window will be available. The ethane and acetylene emission features will be useful for future studies of the temperature inversion. We wish to emphazise that at this time the analysis is limited by the scarcity of molecular data, especially line intensities, for relatively simple organic molecules.

References

Ackerman, M. and Muller, C.: 1972, *Nature* **240**, 300.

Cihla, Z. and Chedin, A.: 1971, *J. Mol. Spectrosc.* **40**, 337.

Chedin, A. and Cihla, Z.: 1973, *J. Mol. Spectrosc.* **47**, 554.

Connes, P., Connes, J., Benedict, W. S., and Kaplan, L. D.: 1967, *Astrophys. J.* **147**, 1230.

Connes, P., Connes, J., Kaplan, L. D., and Benedict, W. S.: 1968, *Astrophys. J.* **152**, 731.

Connes, J., Connes, P., and Maillard, J.: 1969, *Near Infrared Spectra of Venus, Mars, Jupiter and Saturn Atlas*, Vol. 1, C.N.R.S., Paris.

Cowan, M. and Gordy, W.: 1957, *Bull. Am. Phys. Soc.* **2**, 212.

Dunham, J. L.: 1932, *Phys. Rev.* **41**, 721.

Fleming, H. E. and Rao, K. Narahari: 1972, *J. Mol. Spectrosc.* **44**, 189.

Girard, A., Fontanella, J. C., and Gramont, L.: 1973, *Compt. Rend. Acad. Sci. Paris* **276**, Ser. B, 845.

Goldman, A., Murcray, D. G., Murcray, F. H., Williams, W. J., and Bonomo, F. S.: 1970, *Nature* **225**, 443.

Hall, D. N. B.: 1973, *Astrophys. J.* **182**, 977.

Helminger, P., DeLucia, F. C., and Gordy, W.: 1970, *Phys. Rev. Letters* **25**, 1397.

Herzberg, G. and Rao, K. Narahari: 1949, *J. Chem. Phys.* **17**, 1099.

Hurlock, S. C., Rao, K. Narahari, Weller, L. A., and Yin, P. K. L.: 1973a, *J. Mol. Spectrosc.* **48**, 372.

Hurlock, S. C., Lafferty, W. J., and Rao, K. Narahari: 1974, *J. Mol. Spectrosc.* **50**, 246.

Kaplan, L. D., Connes, J., and Connes, P.: 1969, *Astrophys. J.* **157**, L187.

Kirschner, S. M. and Watson, J. K. G.: 1973, *J. Mol. Spectrosc.* **47**, 234.

Locke, J. L. and Herzberg, L.: 1953, *Can. J. Phys.* **31**, 504.

Mantz, A. W., Nichols, E. R. Alpert, B. D., and Rao, K. Narahari: 1970, *J. Mol. Spectrosc.* **35**, 325.

Mantz, A. W., Watson, J. K. G., Rao, K. Narahari, Albritton, D. L., Schmeltekopf, A. L., and Zare, R. N.: 1971, *J. Mol. Spectrosc.* **39**, 180.

Migeotte, M.: 1959, *Phys. Rev.* **75**, 1108.

Migeotte, M. and Neven, L.: 1950, *Physica* **16**, 423.

Murcray, D. G., Kyle, T. G., Murcray, F. H., and Williams, W. J.: 1968, *Nature* **218**, 78.

Murcray, D. G., Kyle, T. G., Murcray, F. H., and Williams, W. J.: 1969, *J. Opt. Soc. Am.* **59**, 1131.

Rank, D. H., Guenther, A. H., Saksena, G. D., Shearer, J. N., and Wiggins, T. A.: 1957, *J. Opt. Soc. Am.* **47**, 686.

Rank, D. H., Pierre, A. G. St., and Wiggins, T. A.: 1965, *J. Mol. Spectrosc.* **18**, 418.

Rao, K. Narahari: 1950, *J. Chem. Phys.* **18**, 213.

Rao, K. Narahari, Humphreys, C. J., and Rank, D. H.: 1966, *Wavelengths Standards in the Infrared*, Academic Press, New York.

Rao, K. Narahari and Mathews, C. W. (eds.): 1972, *Molecular Spectroscopy: Modern Research*, Academic Press, New York, Ch. 7.

Ridley, B. A., Schiff, H. I., Shaw, A. W., Bates, L., Howlett, C., LeVaux, H., Megill, L. R., and Aschenfelter, T. E.: 1973, *Nature*, **245**, 310.

Roh, W. B. and Rao, K. Narahari: 1974, *J. Mol. Spectrosc.* **49**, 317.

Rosenbloom, B., Nethercot, A. H., Jr., and Townes, C. H.: 1958, *Phys. Rev.* **109**, 400.

Toth, R. A., Farmer, C. B., Schindler, R. A., Raper, O. F., and Schaper, P. W.: 1973, *Nature* **244**, 7.

Weinberg, J. M., Fishburne, E. S., and Rao, K. Narahari: 1965, *J. Mol. Spectrosc.* **18**, 428.

Wilson, R. W., Jefferts, K. B., and Penzias, A. A.: 1970, *Astrophys. J.* **161**, L43.

DISCUSSION

Feigelson: What are the upper limits of pressure and temperature for which we have good laboratory spectroscopic data for CO_2? We have not such data for real Venus conditions.

Owen and Young: It has not yet been possible to match the spectral resolution achieved by Dr Connes in Venus spectra with long path lengths in the laboratory in the region $1.2\,\mu$. At shorter wavelengths Gerhard Herzberg has published results in the early 1950's, and D. Rank in mid-1960's. Herzberg's absorption tube was capable of measuring a 1 km path of gas at pressures of 1–2 atm. Individual rotational lines were resolved, but at a resolution somewhat less than that which could be obtained for Venus. However Herzberg's absorption tube could not be heated or cooled, nor could it sustain pressures much larger than atmospheric pressure. Long paths at high pressure and temperature have also not received study in this region of the spectrum.

OPTICAL PROPERTIES OF PARTICLES
IN PLANETARY ATMOSPHERES: LABORATORY STUDIES

DUDLEY WILLIAMS

Kansas State University, Manhattan, Kan., U.S.A.

Abstract. The laboratory techniques employed for the determination of optical constants are discussed briefly. The values obtained for the real n and imaginary k (or n_i) parts of the refractive indices of water, ice, and liquid ammonia are presented graphically. Work currently in progress on solid ammonia and on sulphuric acid is discussed. Other approaches to studies of the properties of aerosols in planetary atmospheres are presented briefly.

1. Introduction

Most of our information concerning the nature of planetary atmospheres and planetary surfaces has been derived from studies of the reflection and scattering of incident solar radiation by the planets. With the development of infrared methods, the thermal emission spectra of planets have provided additional information. Any quantitative interpretation of the spectral reflection or spectral emission by the planets depends upon not only a detailed knowledge of the spectral characteristics of their atmospheric gases but also a knowledge of the real $n(v)$ and imaginary $k(v)$ or $n_i(v)$ parts of the refractive indices of the particles of cloud or dust in their atmospheres. If these latter optical properties are known, this theory can be applied to observed spectra to give estimates of the number density, the range of sizes, and the spatial distribution of the particles in planetary atmospheres; in cases of dense cloud cover, this knowledge can, in turn, provide important information regarding planetary heat balances.

Although much spectroscopic work has been done on many of the substances that may be present in planetary aerosols, a survey by Irvine and Pollack (1968) revealed that even in the case of the most abundant telluric aerosols, water and ice, published spectra did not form the basis for determining accurate values of $n(v)$ and $k(v)$ in much of the infrared. Since the time of the Irvine-Pollack survey, extensive quantitative infrared work has been done on water, ice, the condensed phases of ammonia, and on sulphuric acid, all of which exist as planetary aerosols.

2. Methods of Determining Optical Constants

The general methods of determining optical constants $n(v)$ and $k(v)$ have been treated definitively by Humphrey-Owen (1961). They may be determined from measurements of the reflectance of polarized radiation at two angles of incidence (Querry *et al.*, 1969) and subsequent solution of the generalized Fresnel equations (Querry, 1969). They can also be determined from measurements of fractional spectral transmittance $T(v)$ of a layer of thickness x which give the Lambert absorption coefficient $\alpha(v)$ from

Woszczyk and Iwaniszewska (eds.), Exploration of the Planetary System, 57–65. All Rights Reserved
Copyright © 1974 by the IAU

the relation

$$T(v) = \exp[-\alpha(v)x] \tag{1}$$

together with measurements of the reflectance of polarized radiation at a single angle of incidence or with measurements of the reflectance of unpolarized radiation at near-normal incidence. In the latter case $k(v) = \alpha(v)/4\pi v$, where v is expressed in cm^{-1}, and $n(v)$ can be subsequently determined from the simple relation

$$R(v) = \frac{[n(v) - 1]^2 + k(v)^2}{[n(v) + 1]^2 + k(v)^2}. \tag{2}$$

where $R(v)$ is the fractional reflectance at near-normal incidence.

Alternatively, Kramers-Kronig theorems can be applied in cases where $\alpha(v)$ has been measured over a large spectral interval from the relation

$$n(v) = 1 + (1/2\pi^2) P \int_0^\infty \frac{\alpha(v') - \alpha(v)}{v'^2 - v^2} \, dv'. \tag{3}$$

In using Equation (3) the values of $n(v)$ can be determined in excellent approximation inside the spectral range of measurement provided one makes reasonable assumptions regarding $\alpha(v)$ outside the range of measurement (Robertson *et al.*, 1973); Equation (3) provides exact values of $n(v)$ only if $\alpha(v)$ is known for all frequencies. In other cases where $R(v)$ has been measured over a large spectral interval, both $n(v)$ and $k(v)$ can be obtained within the spectral range of measurement by use of Kramers-Kronig phase-shift analysis (Robertson *et al.*, 1973). According to this Kramers-Kronig theorem, if the modulus $[R(v)]^{1/2}$ of the complex reflectivity $[R(v)]^{1/2}e^{i\phi(v)}$ is known for all frequencies, the phase shift $\phi(v)$ at frequency v is given by

$$\phi(v) = \frac{2v_0}{\pi} P \int_0^\infty \frac{\ln[R(v)]^{1/2}}{v^2 - v'^2} \, dv' ; \tag{4}$$

$n(v)$ and $k(v)$ can then be determined from $\phi(v)$. Even though $R(v)$ is not known for *all* frequencies, Equation (4) gives values of $\phi(v)$ within the range of measurement (Hale *et al.*, 1972) that are not strongly influenced by assumed values of $R(v)$ outside the range of measurement. In the case of water, there is excellent agreement between values of optical constants determined by the methods just described (Querry, 1969; Robertson *et al.*, 1973; Hale *et al.*, 1972; Rusk *et al.*, 1971; Pontier and Dechambenoy, 1965–1966; Zolatarev *et al.*, 1969).

In using any of the computation methods just described, it is necessary that extreme care be taken in the experimental determination of $\alpha(v)$ and $R(v)$. In determining $\alpha(v)$ for a liquid in the infrared, measurements of the intensity $I_0(v)$ of the incident radiation and the intensity $I(v)$ of the radiation transmitted through an absorption cell containing a liquid layer of thickness x lead to the relation

$$T(v) = [1 - \Re(v)] [1 - \mathfrak{A}(v)] \exp[-\alpha(v)x], \tag{5}$$

where $\Re(v)$ is the fraction of the incident radiation reflected at the inner and outer surfaces of the cell windows and $\mathfrak{A}(v)$ is the fraction of the incident radiation absorbed by the cell windows; $\Re(v)$ and $\mathfrak{A}(v)$ cannot be easily measured. Further difficulties are introduced by the required values of x as illustrated for the case of water by the curves in Figure 1, in which the main curve represents the fractional spectral transmittance $T(v)$ for a layer of liquid only 10 μm thick; it will be noted that a water layer of this thickness is essentially opaque in the 3400 cm^{-1} region. In order to deter-

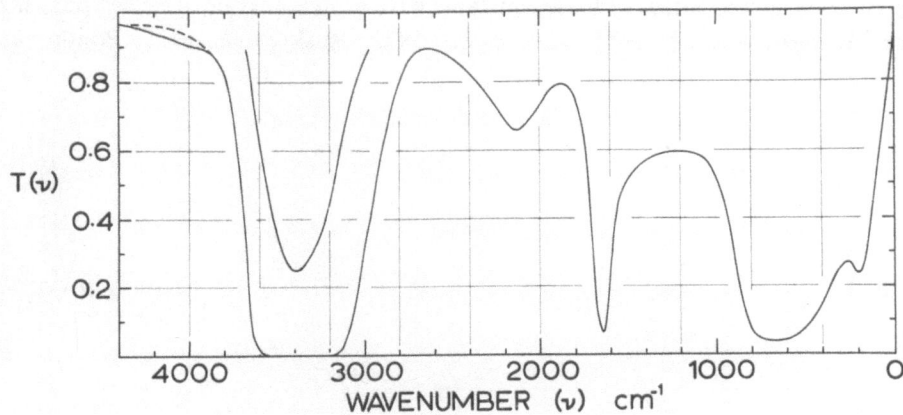

Fig. 1.   The spectral transmittance $T(v)$ of a 10 μm layer of water in the infrared; the upper curve gives $T(v)$ for a 1.125 μm layer in the vicinity of 3400 cm^{-1}.

mine $\alpha(v)$ in the 3400 cm^{-1} region, it would be necessary to prepare a uniform layer of thickness 1.125 μm, which would give the spectral transmittance given by the upper curve in the 3400 cm^{-1} region. Fortunately, Robertson and Williams (1971) have developed a technique involving a wedge-cell of thickness ranging from optical contact at one edge to 20 μm at the other edge; by passing such a cell laterally through the beam of radiation and measuring values of $T(v)$ at various positions corresponding to various thicknesses of the absorbing layer, one can eliminate the effects of $\Re(v)$ and $\mathfrak{A}(v)$ by taking ratios of $T(v)$ for different thicknesses; the thicknesses of the wedge-shaped layer can be measured and monitored by interferometric techniques.

   In the determination of absolute values of $R(v)$ we have adopted a technique of first measuring the nominal reflectance of a surface relative to a front surface reference mirror. The absolute spectral reflectance of the reference mirror is carefully determined by an independent measurement involving a reflectometer of a type developed by Strong (1938). From the two sets of experimental measurements, absolute values of $R(v)$ can be obtained.

   Since the values of $R(v)$ in most spectral regions are small, care must be taken to

see that detector-amplifier-recorder systems are strictly linear in giving chart deflec-
tions that are strictly proportional to the radiant flux reaching the detector. Rather
than relying on changes in amplifier gain settings, we have made use of optical attenua-
tors consisting of rapidly rotating calibrated sector wheels inserted in the beam when
the reference mirror is being used. Under these conditions we have found that thermo-
couples used with conventional amplifiers with fixed gain controls give satisfactorily
linear response.

3. Optical Constants

The most abundant particulate constituent of the Earth's atmosphere is liquid water
in the form of cloud droplets. The optical constants of water at 25°C are shown in
Figures 2 and 3; the values of these constants have been tabulated in earlier published
work (Robertson *et al.*, 1973; Hale *et al.*, 1972; Rusk *et al.*, 1971; Pontier and

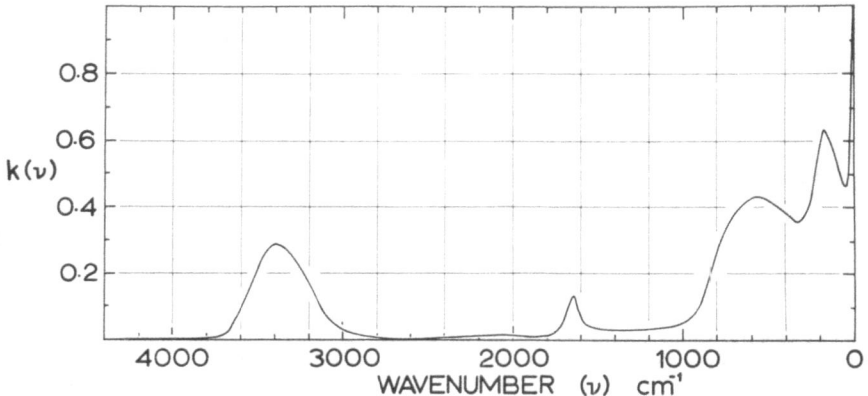

Fig. 2.   The imaginary part $k(v)$ of the complex index of refraction $N(v) = n(v) + ik(v)$
of water at 25°C.

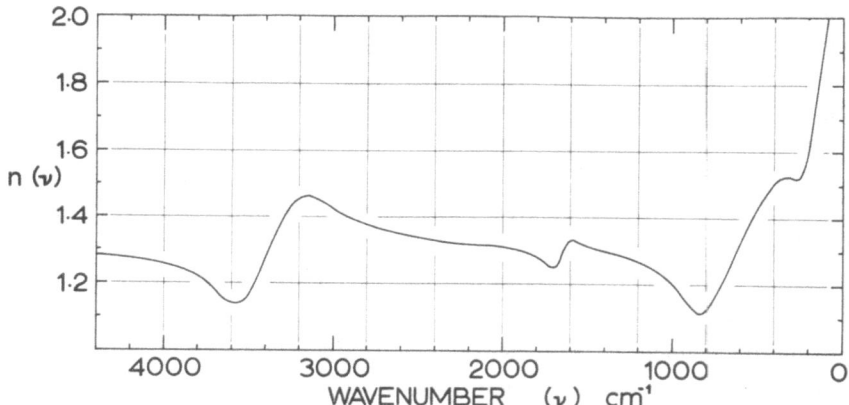

Fig. 3.   The real part $n(r)$ of the complex index of refraction $N(v) = n(r) + ik(v)$ of water at 25°C.

Dechambenoy, 1965–1966; Zolatarev *et al.*, 1969). The influence of temperature on the optical constants of water has been investigated by Hale *et al.* (1972). Optical constants in the submillimeter region have been determined by Davis *et al.* (1970) and by Zafar *et al.* (1973). Also present as an abundant aerosol in the Earth's atmosphere are ice crystals; the reflection spectrum of ice has recently been investigated by Schaaf and Williams (1973), whose values of the optical constants based on Kramers-Kronig phaseshift analysis are plotted in Figures 4 and 5. Ice particles have recently been detected in the Martian spectrum in the course of the recent Mariner flight.

Ammonia in condensed states is probably an important particulate constituent of the cloud cover of the major planets. The reflection spectrum of liquid ammonia has been measured in the visible and in the infrared by Robertson and Williams (1973a, b), who have also measured the transmission spectrum in the near infrared; because of the lack of suitable cell-window materials, these workers were unable to obtain transmission data in the far-infrared. Optical constants based on these studies of liquid ammonia are plotted in Figures 6 and 7. Robertson and Williams (1973) have also attempted to

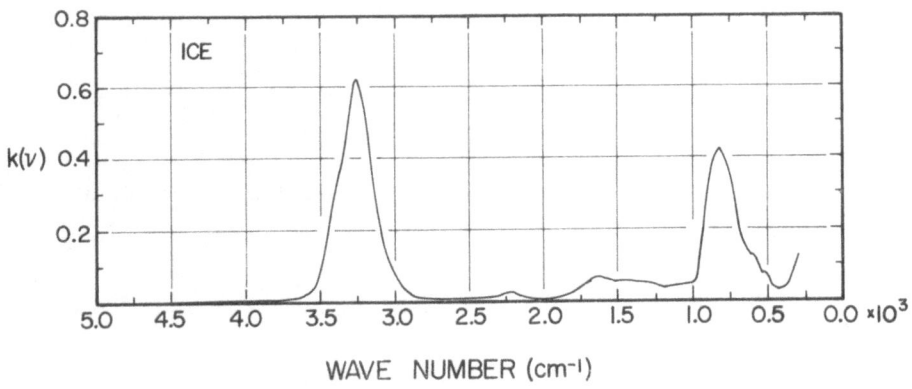

Fig. 4.   The imaginary part $k(v)$ of the complex refractive index of ice at $-7\,°C$.

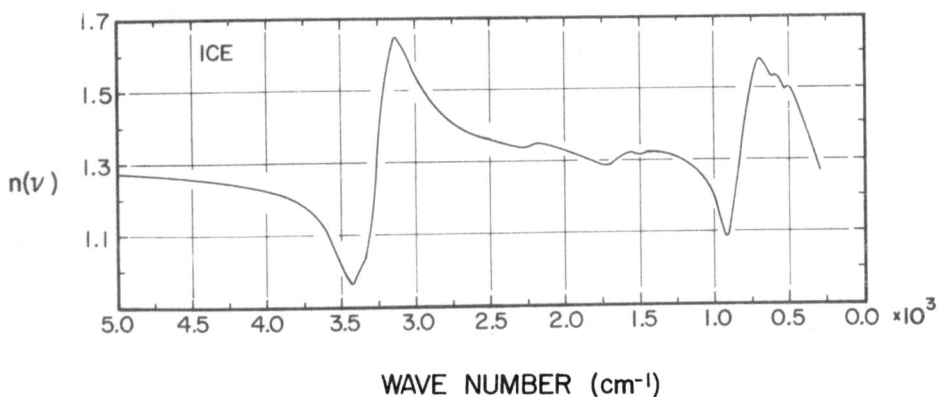

Fig. 5.   The real part $n(v)$ of the complex refractive index of ice at $-7\,°C$.

determine the optical constants of solid ammonia in the infrared at a temperature slightly below the melting point; they have succeeded in obtaining satisfactory transmission spectra in the 5000–900 cm^{-1} region in the near infrared. However, because of the high vapor pressure of solid ammonia, they have thus far been unsuccessful in obtaining acceptable reflection spectra; a good reflecting surface can be maintained for only brief periods. Other techniques for studying solid ammonia are being investigated.

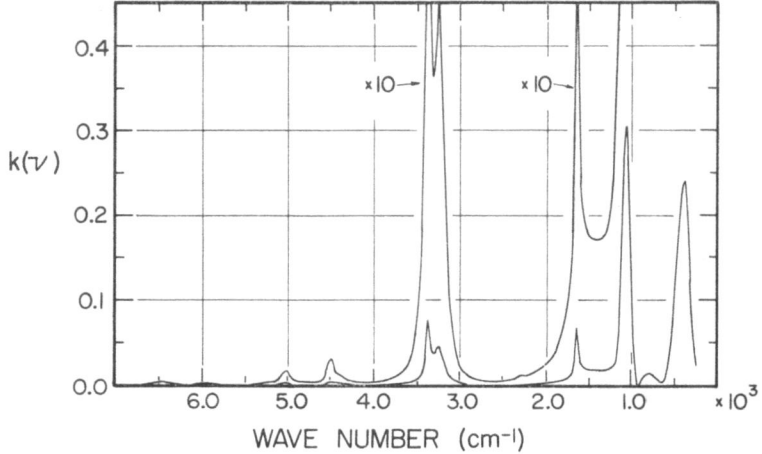

Fig. 6.   The imaginary part $k(\nu)$ of the complex refractive index of liquid ammonia at $-45\,°C$.

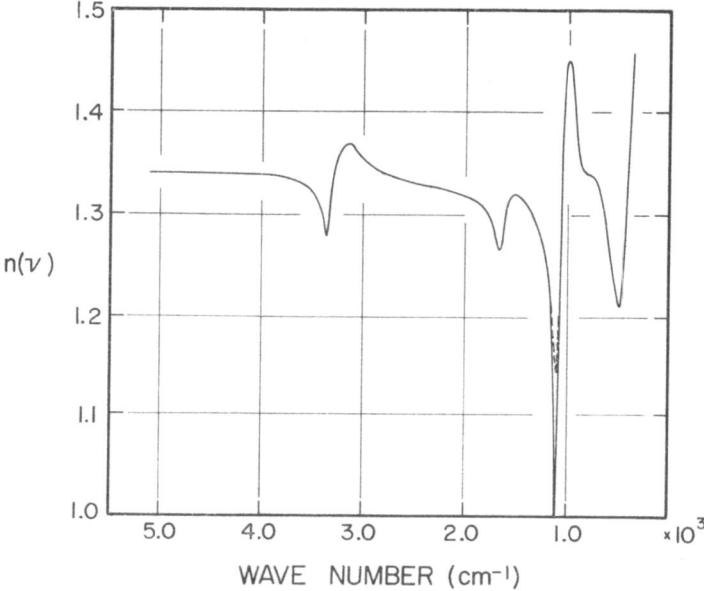

Fig. 7.   The real part $n(\nu)$ of the complex refractive index of liquid ammonia at $-45\,°C$.

The most abundant aerosol in the Earth's stratosphere consists of sulphuric acid droplets; it has also been suggested that sulphuric acid in liquid form is the dominant constituent of the clouds of Venus (Young, 1973). Palmer (1973) and Williams have measured the spectral reflectance of sulphuric acid in aqueous solution at a number of concentrations in the visible and intermediate infrared; transmission measurements are in progress in the near-ultraviolet-visible-near-infrared-region. The results of this study should provide values that will be useful in detailed studies of the clouds of Venus. Preliminary values of the optical constants for sulphuric acid solutions in the infrared, based on Kramers-Kronig phase-shift analysis of reflectance measurements, have been obtained; improved values based on our study will soon be available. Other reflection data providing values of $n(v)$ and $k(v)$ have been obtained by Querry and his colleagues (1974) for a 3–M solution of sulphuric acid. Remsberg (1973) has used attenuated-total-reflection (ATR) techniques to obtain the optical constants for 95 and 75% solutions of sulphuric acid in the 770–1250 cm^{-1} region, which covers an important window in the Earth's atmosphere. On the basis of a preliminary analysis of spectra taken from high-altitude aircraft, Pollack (1973) indicates that the spectrum in the 2–5 μm region of the Venus clouds resemble spectra reported for 75% sulphuric acid.

4. Actual Particles in Planetary Atmospheres

We recognize that planetary cloud covers contain numerous types of particles that are not composed of pure chemical substances of the type we have investigated; for example, in the case of the Earth's atmosphere, many of the water droplets over the sea are formed on salt nuclei, and many other types of aerosols are known to be present. In the case of the Earth and even more so in the case of Mars, dust storms can occasionally contribute to the particulate content of the atmosphere. In the case of the major planets, compounds other than pure ammonia are believed to contribute to the cloud cover. Thus, optical studies of pure substances can at best contribute only rough guides to an interpretation of astronomical observations.

Since the number of possible mixtures of materials in planetary cloud covers is infinite, it would appear pointless to attempt laboratory measurements of the 'effective optical constants' of various mixtures that have been suggested for the aerosol content of the atmospheres of various planets. Two courses appear possible; the first of these would involve mathematical combinations of the optical constants of pure substances to obtain values for 'effective optical constants' for various mixtures; the second course would be the careful selection of a limited set of materials believed to be present and to make careful laboratory studies of these selected materials.

An important example of this second course is the study (Pollack *et al.*, 1973) of a set of carefully selected terrestrial rocks that might be present in abundance in planetary surfaces and therefore in the atmospheric dust of planets. Rocks and most minerals in the Earth's crust vary in composition with locality; thus careful selection of specimens is also of great importance. Another example of the second course is the work of Volz (1972, 1973) who has obtained optical constants for aerosols of types actually collected in the Earth's atmosphere; this method will become possible for other planets only after particles have been collected and brought back to Earth. However, it may ultimately become of great importance in detailed studies of planetary heat balances. Meanwhile, we hope that our laboratory studies of pure materials will be of some use in the interpretation of observations.

Acknowledgements

We should like to express our appreciation to the U.S. National Aeronautics and Space Administration and to the U.S. Office of Naval Research, which have supported our work. Our thanks also go to Dr J. B. Pollack of the NASA Ames Research Center for assistance in our work on sulphuric acid and to Dr Andrew T. Young of the Jet Propulsion Laboratory for helpful discussions.

References

Davies, M., Pardoe, G. W. F., Chamberlain, J., and Gebbie, H. A.: 1970, *Trans. Faraday Soc.* **66**, 273.
Hale, G. M., Querry, M. R., Rusk, A. N., and Williams, D.: 1972, *J. Opt. Soc. Am.* **62**, 1103.
Humphrey-Owens, S. P. F.: 1961, *Proc. Roy. Soc. London* **77**, 949.
Irvine, W. H. and Pollack, J. B.: 1968, *Icarus* **8**, 324.
Palmer, K.: 1973, *Proc. Symposium on Molecular Spectra*, Ohio State Univ., p. 168.
Pollack, J. B.: 1973, private communication.
Pollack, J. B., Toon, O. B., and Khare, B. N.: 1973, *Icarus* **19**, 373.
Pontier, L. and Dechambenoy, C.: 1965, *Ann. Geophys.* **21**, 462.
Pontier, L. and Dechambenoy, C.: 1966, *Ann. Geophys.* **22**, 633.
Querry, M. R.: 1969, *J. Opt. Soc. Am.* **59**, 876.
Querry, M. R., Curnutte, B., and Williams, D.: 1969, *J. Opt. Soc. Am.* **59**, 1299.
Querry, M. R., Waring, R. C., Holland, W. E., Earls, L. M., Herman, M. D., Nijm, W. P., and Hale, G. M.: 1974, *J. Opt. Soc. Am.* **44**, 39.
Remsberg, E. E.: 1973, *J. Geophys. Res.* **78**, 1401.
Robertson, C. W.: 1973, *Proc. Symposium on Molecular Spectra*, Ohio State Univ., p. 167.
Robertson, C. W., Curnutte, B., and Williams, D.: 1973, *Mol. Phys.* **66**, 183.
Robertson, C. W. and Williams, D.: 1971, *J. Opt. Soc. Am.* **61**, 1316.
Robertson, C. W. and Williams, D.: 1973a, *J. Opt. Soc. Am.* **63**, 763.
Robertson, C. W. and Williams, D.: 1973b, *J. Opt. Soc. Am.* **63**, 188.
Rusk, A. N., Querry, M. R., and Williams, D.: 1971, *J. Opt. Soc. Am.* **61**, 895.
Schaaf, J. W. and Williams, D.: 1973, *J. Opt. Soc. Am.* **63**, 726.
Strong, J.: 1938, *Procedures in Experimental Physics*, Prentice-Hall, New York, p. 376.
Volz, F. E.: 1972, *Appl. Opt.* **11**, 755.
Volz, F. E.: 1973, *Appl. Opt.* **12**, 564.
Young, A. T.: 1973, *Icarus* **18**, 564.
Zafar, M. S., Hasted, J. B., and Chamberlain, J.: 1973, *Nature Phys. Sci.* **243**, 106.
Zolatarev, V. M., Mikhailov, B. A., Aperovich, L. I., and Popov, S. I.: 1969, *Opt. Spectrosk.* **27**, 790.

DISCUSSION

Irvine: How temperature dependent are the optical constants of NH_3? Would your results be applicable to Jovian conditions?

Williams: The present data could probably be extrapolated to Jovian conditions.

Owen: The upper clouds on Jupiter are most likely solid ammonia; but to study liquid droplets, it would be more appropriate to study NH_4OH, NH_4SH or $(NH_4)_2S$, since the lower, liquid droplet clouds at $T = 225$ K will not be pure NH_3.

Williams: The optical qualities of NH_4OH are at present under study in our laboratory.

Huebner: Do you have, or plan to obtain, frequency-dependent data of the complex refractive index for the refractory compounds like enstatite? These are important for interplanetary dust and meteoric matter.

Gulkis: A previous comment from the floor stated that the reflecting cloud temperature is about 225 K. The ammonia cloud and haze regions, however, form at 150 K. At higher temperatures one expects to find a solution or droplets of water and ammonia rather than ammonia ice.

FIRST RESULTS OF THE SOVIET-POLISH
SPACE EXPERIMENT 'INTERCOSMOS-KOPERNIK 500'

J. HANASZ

Polish Academy of Sciences, Institute of Astronomy, Astrophysical Laboratory, Toruń, Poland

V. I. AKSENOV

Institute of Radio Engineering and Electronics, Academy of Sciences U.S.S.R., Moscow, U.S.S.R.,

and

G. P. KOMRAKOV

Research Institute of Radiophysics, Gorky, U.S.S.R.

Abstract. The report contains a short description of the first Soviet-Polish space experiment, which has been accomplished aboard the 'Intercosmos-Kopernik 500' satellite, launched to commemorate the 500 anniversary of Copernicus birth. The purpose of the experiment is to observe sporadic radio emissions generated in the coronal and interplanetary medium at hectometric wave-length range and to measure parameters of the ionospheric plasma by means of the low- and high-frequency impedance probes. The idea of the experiments is presented together with the brief description of the measuring equipment and data recorded. First results of the space observations are shown. The correlation between electron density of the ambient plasma, ion-sheath capacitance of the aerial, and the upper hybrid frequency can be seen.

1. Introduction

The Soviet-Polish satellite, called 'Intercosmos-Kopernik 500' to commemorate 500th anniversary of Copernicus birth, was launched on the 19th of April 1973 from the territory of the Soviet Union to the orbit of 1551 km at apogee, 202 km at perigee, inclination of 48°43' and period of revolution 102.2 min.

The programme of experiments has been divided into 3 parts:

(1) Polish experiments which are aimed to investigate low-frequency radio-bursts travelling out through the corona and interplanetary space, and to measure the ionospheric resonances, both the experiments being made with the aid of the 4-channel radio-spectrograph, swept over the frequency range of 0.6 to 6.0 MHz with the period of repetition of 12 s and frequency resolution of 30 kHz; (Astrophysical Laboratory in Toruń of the Institute of Astronomy, Polish Acad. of Sci.);

(2) A Soviet experiment aimed to determine the characteristics of the ion-sheath of the aerial immersed in the ionospheric plasma and to determine the parameters like electron density and temperature in the ionosphere, the experiment being carried out with the aid of the low-frequency impedance probe (Institute of Radio Engineering and Electronics, Moscow);

(3) A Soviet experiment aimed to measure the electron density and electron density irregularities in the ionosphere with the aid of the high-frequency impedance probe (Research Institute of Radiophysics, Gorky).

Woszczyk and Iwaniszewska (eds.), Exploration of the Planetary System, 67–73. All Rights Reserved

2. Aerial System and Telemetry

The radiospectrograph uses a crossed electric 15-m dipole system, and the l.-f. and h.-f. impedance probes use 5-m and 70-cm cylindrical aerials, respectively. It should be mentioned that the measurements made with the h.-f. and l.-f. impedance probes enable the continuous calibration of the radiospectrograph aerial system.

Tape-recorded data over a whole revolution and resolution of 12 readouts per second are available for 7 to 8 hours daily. For the fast transmission of the radiospectrograph operational data to the ground station at the Ondrejov Observatory near Prague the special telemetry transmitter developped in Czechoslovakia is working aboard.

3. Bursts of Solar Radio Emission

During the first few months of the flight a great variety of results has been obtained, which at the moment are in course of reduction. Therefore only preliminary results can be presented here. The revue of satellite observations of solar radio bursts is given by Fainberg and Stone (1973).

The sample of radio-spectrograph data contains good examples of fast drifting type III bursts, of which one is shown in Figure 1. As it can be seen from this figure these bursts are characterized by their drift from upper to lower frequencies and by sudden rise and exponential decay. The burst shown in the figure had been observed while the satellite was deep in the ionosphere, and the ionospheric cut-off frequency limited the useful data in this case to the frequency range between 3 and 6 MHz. Coinciding type III bursts were observed at higher frequency range of 45 to 90 MHz by the ground station at IZMIRAN near Moscow (Institute of Terrestrial Magnetism, Ionosphere and Radio Wave Propagation of the Acad. of Sci. USSR). These bursts are commonly believed to be excited by fast electrons that travel along open-field lines out through the corona into interplanetary space (Hartz, 1964; Fainberg and Stone, 1970; Alvarez et al., 1972; Lin et al., 1973; Frank and Gurnett, 1972).

4. Aerial Capacitance Measurement

The low-frequency impedance probe makes it possible to measure reactive and active components of the impedance of an electric aerial immersed in the ionosphere at frequency of 50 kHz. At this frequency the aerial impedance for the ionospheric plasma is practically capacitive (Hugill, 1965; Gurnett et al., 1969). Its value is determined by an ion-sheath which developes in the aerial vicinity (Aksenov et al., 1970; Aksenov et al., 1971). For the cylindrical aerial the capacitance of the ion-sheath is given by

$$C = \frac{2\pi\varepsilon_0 l}{\ln\dfrac{a+s}{a}}, \tag{1}$$

where ε_0 is the free-space permittivity, l – length of the aerial, a – its radius, s – the

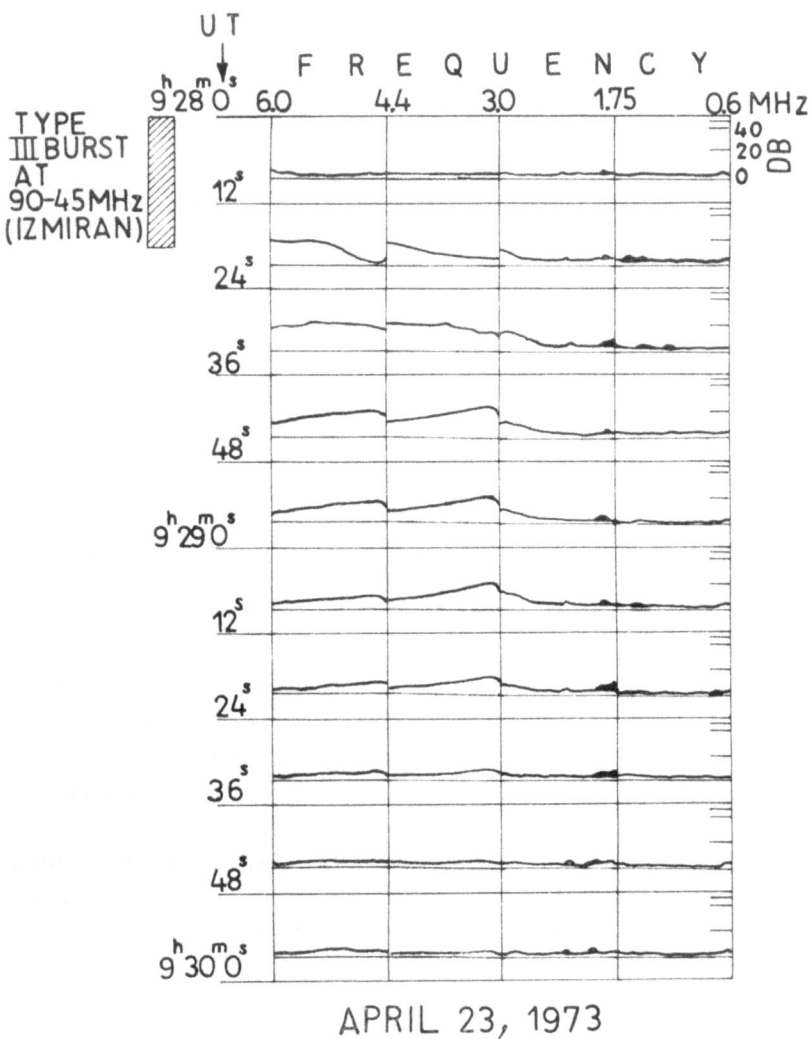

APRIL 23, 1973

Fig. 1. An example of the dynamic spectrum of the type III burst as recorded with the radio-spectrograph installed aboard 'Intercosmos-Kopernik 500' on April 23, 1973 at 9ʰ28ᵐ UT. The amplitude is in decibel scale. The shaded area shows the duration of the coinciding burst, which was also recorded at IZMIRAN, at 45 to 90 MHz frequency range.

sheath thickness. The value of the sheath thickness is near to the Debye's radius (Gurnett *et al.*, 1969):

$$s = 6.9 (T/N)^{1/2}, \tag{2}$$

where T – is the temperature of the charged particles, N – electron density.

Figure 2 shows an example of the results for the aerial capacitance measurement for the revolution No. 53. It can be seen that the altitude dependance of the capacitance corresponds to the electron density distribution in the ionosphere.

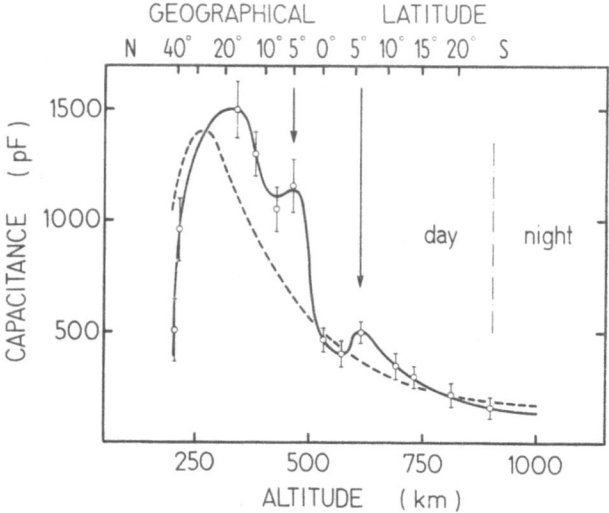

Fig. 2. Variations of the aerial capacitance in the ionosphere as measured with the low-frequency impedance probe during the 53rd revolution of 'Intercosmos-Kopernik 500' (April 23, 1973). Full line – experimental data, dashed line – results of calculation.

Additional maxima at the experimental curve (to north and south of the equator) are due to the latitudinal anomalies of the electron density in the ionosphere. For comparison this figure shows also the results of calculation for the ion-sheath capacitance with assumption that the sheath thickness is equal to one Debye's radius. Calculation has been carried out for the model of the ionosphere at the minimum solar activity. (The latitudinal anomalies have not been taken into account in the theoretical model). Satisfactory agreement between experimental and theoretical results proves that the l.-f. impedance probe can be used for the determination of the ionospheric plasma characteristics.

5. Electron Density Determinations

The determinations of electron density and its irregularities for the ambient ionospheric plasma, by means of the high-frequency impedance probe, are based on the measurements of the capacitance of the electrically short aerial (70 cm) at two frequencies 3.1 and 15 MHz, which is purely capacitive (13 pF in free space). The aerial capacitance for the collisionless ionosphere without magnetic field is proportional to the dielectric permittivity of the ambient plasma, which is equal to:

$$\varepsilon = 1 - \frac{4\pi e^2 N}{m\omega^2},\tag{3}$$

where e – electron charge, N – electron density, m – mass of the electron, $\omega = 2\pi f$ – frequency of observation.

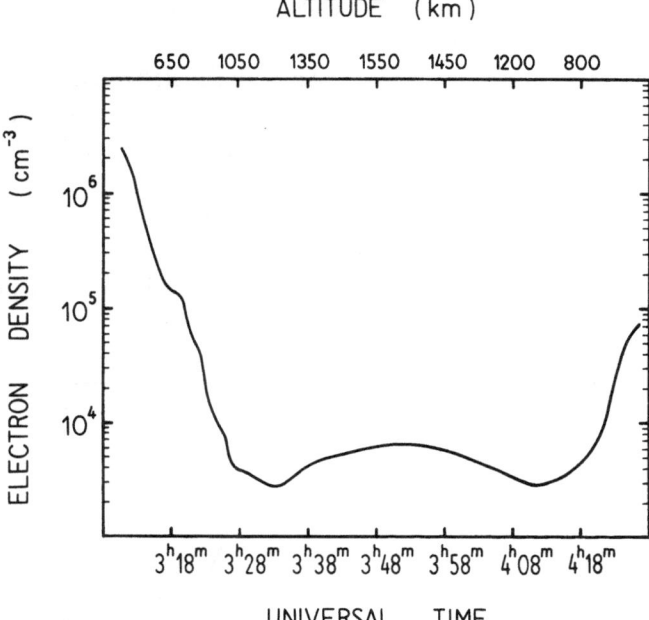

Fig. 3. Variations of the electron density in the ionosphere as determined with the high-frequency impedance probe during the 53rd revolution of 'Intercosmos-Kopernik 500' (April 23, 1973).

Variations of the aerial capacitance in the ionospheric plasma cause proportional frequency variations of the input oscillatory circuit, which after discrimination give proportional fluctuation of the telemetred d.c. voltage. To measure electron density irregularities a 0.05 to 20 Hz band-pass filter is applied to select slow variations of the d.c. output voltage. In this way the determinations of the electron density in the range of 10^3 to 1.6×10^6 electrons per cubic cm, and electron density irregularities of the size of 0.5 to 100 km and amplitude greater than 100 electrons per cubic cm are possible.

Finally, electron density of the ionosphere determined from the formula (3) can be corrected for the Earth magnetic field and the ion-sheath around the aerial and the satellite. The result of the electron density determination for the 53rd revolution is shown in Figure 3.

6. Radio-Noise at Upper Hybrid Frequency

Some data on the ionospheric plasma have also been obtained by means of the radio-spectrograph aboard. Bands of radio-noise at the frequency of upper hybrid resonance:

$$f_{uh}^2 = f_p^2 + f_H^2, \tag{4}$$

where f_p and f_H are plasma- and gyro-frequencies of the ambient plasma, have been recorded.

The main characteristics of this resonance, which have been concluded from the first inspection of the recordings are:

(1) a correlation of the resonance frequency with the electron density of the ambient plasma as measured by means of the h.-f. impedance probe, and with the capacitance of the ion-sheath of the aerial as measured by means of the l.-f. impedance probe, which can be seen in Figure 4;

(2) modulation of the amplitude of the resonance by the satellite rotation in the Earth magnetic field, which can be seen in Figure 5;

(3) variations of the band-width of the resonance, which is usually wider than 100 kHz;

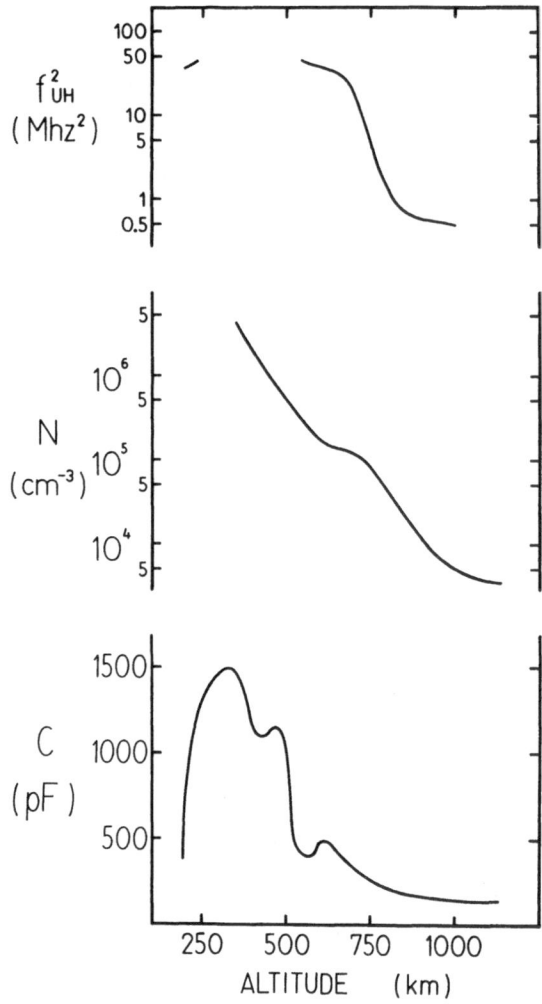

Fig. 4. Comparison of the capacitance of the aerial (C), electron density obtained with the high-frequency probe (N), and upper hybrid frequency (f_{uh}), as measured during the 53rd revolution aboard 'Intercosmos-Kopernik 500' (April 23, 1973).

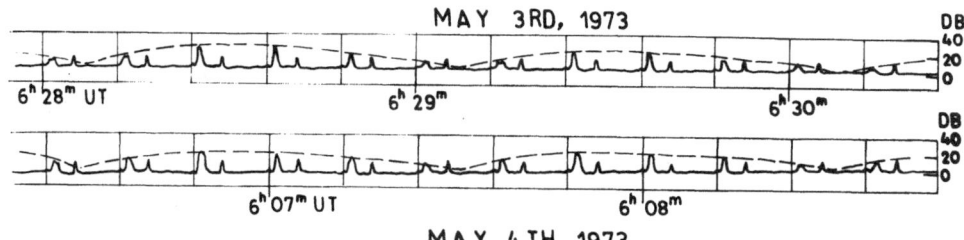

Fig. 5.   Modulation of the amplitude of the upper hybrid frequency resonance in the ionosphere
due to the satellite rotation as recorded with the radio-spectrograph
aboard 'Intercosmos-Kopernik 500'.

(4) variable asymmetry of the frequency profile.

The results presented here do not comprise all the data obtained during the fllight.
The 'Intercosmos-Kopernik 500' satellite is still orbiting and the experiments aboard
are continued.

The experiment reported here is a part of the 'Intercosmos' international space
research programme.

References

Aksenov, V. I. and Lishin, I. V.: 1970, *Radiotekh. Elektron.* **15**, 677 (in Russian).

Aksenov, V. I., Lishin, I. V., and Tikhomirnov, S. I.: 1971, *Proc. International Symp. on Electromagnetic Wave Theory*, Nauka, Moscow, p. 310 (preprint).

Alvarez, H., Haddock, F. T., and Lin, R. P.: 1972, *Solar Phys.* **26**, 468.

Fainberg, J. and Stone, R. G.: 1970, *Solar Phys.* **15**, 222.

Fainberg, J. and Stone, R. G.: 1974, submitted to *Space Sci. Rev.*; Preprint X-693-73-346, Goddard
Space Flight Center, Greenbelt.

Frank, L. A. and Gurnett, D. A.: 1972, *Solar Phys.* **27**, 446.

Gurnett, D. A., Pfeiffer, G. W., Anderson, R. R., Mosier, S. R., and Cauffman, D. P.: 1969, *J.
Geophys. Res.* **74**, 4631.

Hartz, T. R.: 1964, *Ann. Astrophys.* **27**, 831.

Hugill, J.: 1965, *Ann. Astrophys.* **28**, 255.

Lin, R. P., Evans, L. G., and Fainberg, J.: 1973, *Astrophys. Letters* **14**, 191.

DISCUSSION

Fox: Where and when will the data on your 0.6–6 MHz radio emission experiment be published?

Hanasz: This is the first report on the experiment as the satellite is still orbiting, and these data have
not been published as yet.

Smith: Is the 'Intercosmos-Copernicus' low-frequency receiver sensitive enough to receive Jupiter
radio noise, and if so have you detected it?

Hanasz: The sensivity of the receiver is just enough to receive galactic radiation. However, we have
not distinguished until now any signals which we could suspect originate from Jupiter.

Barrow: In reply to Dr Smith's question; Jupiter activity goes in an 11-yr cycle and has only
recently passed the minimum. Perhaps for this reason there has been little opportunity for you to
observe Jupiter so far.

PART II

TERRESTRIAL PLANETS

INFRARED SPECTRA OF VENUS

LOUISE GRAY YOUNG

Jet Propulsion Laboratory, California Institute of Technology, Pasadena, Calif., U.S.A.

Abstract. A historical account of observations of Venus and their interpretation is given. The major constituent of the atmosphere on Venus (CO_2) was detected spectroscopically forty years ago, and minor constituents (CO, HF, HCl) have been found more recently. The infrared spectra also provide a means of studying the motions of her cloudy atmosphere. The composition of the clouds has been sought in the reflection spectrum of Venus, and some of the evidence for their nature is discussed.

1. Early Observations of Venus

Venus has been known to possess an atmosphere since 1761. In that year, the Russian astronomer M. W. Lomonosov observed the transit of Venus across the face of the Sun and stated "the planet Venus is surrounded by a considerable atmosphere equal to, if not greater than, that which envelops our earthly sphere." The next reported observation of Venus' atmosphere was made by Schröter (1792) who noted that the horns of the crescent Venus extend beyond a semicircle. When Venus is near inferior conjunction (i.e. when the phase angle *i*, the angle between the Earth and the Sun as viewed from Venus, is nearly 180°) the horns of her crescent can be seen to extend around the arcumference of the planet. Herschel (1793) also remarked on the extension of the horns and wrote "the atmosphere of Venus is probably very considerable ..." Mädler (1849) saw 240° of the circumference of Venus illuminated, while Lyman (1866, 1874) succeeded in observing Venus when she appeared as a luminous ring.

Thus by the mid-nineteenth century Venus was known to have an atmosphere, but its composition remained unknown. Since the Earth and Venus were known to have a similar size and mass, it was logical to expect their atmospheres to be of a similar composition and extent. This notion was not completely dispelled until the Russian spacecraft Venera 4 entered the atmosphere of Venus, and made measurements of its physical and chemical properties.

With the development of the spectroscope, astronomers hoped to find more definite results, and the search for oxygen and water vapor began. The first attempt we know about was made by the astrophysicist Sir William Huggins, and Dr W. A. Miller, a chemist. Huggins (1864) reported that "the light of Venus gives a spectrum of great beauty" but their spectrum failed to reveal any lines not present in the corresponding solar spectrum. He suggested that this was because "the light is chiefly reflected, not from the planetary surface, but from masses of cloud in the upper strata of the atmosphere".

More extensive spectra were observed by Vogel (1874) who reported that telluric absorption features were enhanced in the spectrum of Venus. Tacchini and Ricco (1882) and Young (1885) also reported observations which indicated the presence of

Woszczyk and Iwaniszewska (eds.), Exploration of the Planetary System, 77–160. All Rights Reserved

water vapor in the atmosphere of Venus. There was one pitfall that these early observers failed to avoid: the variations in local humidity at their observing site and the different telluric air masses traversed by spectra of the Sun and Venus. On the basis of quite flimsy spectroscopic evidence, Scheiner concluded that "There can therefore be no doubt that the atmosphere of Venus exerts an absorption similar to our own, and hence the nature of the two atmospheres must be similar."

The faintness of the atmospheric lines of Venus indicates that the atmosphere is very thin, or else that the sunlight can penetrate only a short distance into it, being thus reflected from its upper strata. The latter explanation agrees well with other astronomical observations which show a thick envelope of clouds prevents a view of the true surface of the planet. This layer of condensed vapors would be naturally supposed to be situated at a considerable altitude in the atmosphere. Since Janssen's investigations show that the telluric lines are chiefly due to aqueous vapor, we may safely assume that the clouds of Venus consists of condensed aqueous vapor, thus again resembling those of the Earth." Arrhenius (1918) was even bolder: "The humidity is probably about six times the average of that on Earth and three times that in the Congo where the average temperature is 26 °C. The atmosphere of Venus holds about as much water vapor 5 km *above* the surface as does the atmosphere of the Earth *at* the surface. We must therefore conclude that everything on Venus is dripping wet. The vegetative processes are greatly accelerated by the high temperature. Therefore the lifetime of organisms is probably short."

The early observations of the spectrum of Venus have been made visually. A spectrum of Venus was compared with a spectrum of the sky (or of sunlight reflected from the Moon) and the relative intensities of the telluric absorption bands on the two spectra were estimated. Much more accurate measurements can be made at higher resolution when the well-known Doppler effect can be used to separate the absorption lines due to Venus' atmosphere from those of the Earth's atmosphere. Lowell (1905) suggested measuring oxygen and water vapor lines to see whether they were affected by the Doppler shift. V. M. Slipher photographed the spectrum of Venus and the sky, with a dispersion of 50 Å mm^{-1}, and Lowell had this to say: "Here again eye estimates by the writer subscribed to a shift in the α band (of oxygen), the water lines, very faint, concurring; ... As regarded differences in density, none was perceptible between $\cdots$ the solar and Venusian, either in the oxygen α band or the water vapor lines near (sodium) D. Water vapor is probably non-existent on the illuminated side of Venus. As for oxygen the results above show that the spectroscopic method is hardly a delicate enough in this respect to decide the question." Slipher (1908) added, "Although this attempt has failed to detect aqueous vapor in Mars and Venus, the conclusion should not be drawn that it does not exist in their atmospheres, nor that it will always remain impossible to discover it spectrographically." Slipher continued to search for atmospheric absorption in the spectrum of Venus with negative results. Slipher (1921) wrote, "The high albedo and telescopic appearance of Venus... seem to me to imply that our view of her is mainly a super-surface one, which may not be appreciably affected by light returned from her surface. And when the marked concentration of moisture in our air in its lowest stratum is considered, I become cautious about concluding that such results are proof of the absence of water on Venus ... There is telescopic evidence of incomplete light penetration of the atmosphere of Venus, hence we should expect that, because of the probable low-lying position of the moisture on her, its detection would be difficult. The depth of light penetration is obviously important."

An extensive series of measurements of the spectrum of Venus, at high dispersion (3 Å mm^{-1}), were made at Mt. Wilson Observatory between 1919 and 1921. Initially the spectra were taken to 'check Evershed's results' (St. John and Nicholson, 1920). Evershed (1918, 1919a–c) had made "a long series of measures of Venus and Fe arc spectra, and control plates of sunlight and Fe arc, with the very remarkable result, already indicated in previous work, that the integrated light of the Sun reflected by Venus differs from ordinary sunlight in the mean wave-length of the iron lines being quite appreciably smaller when the angle Venus-Sun-Earth exceeds 90°." Evershed was "reluctant to accept the Venus result since they seem to prove a recessive motion of solar gases controlled by the Earth."

St. John and Nicholson (1921) found "the main factors producing the displacements (of the solar lines reflected by Venus) are those depending upon the low altitude of Venus at the time of observation. When this effect is eliminated, the remaining residuals, which seem to vary with the relative position of Earth, Venus and Sun are more reasonably correlated with the varying diameter of the planet than with the angle VSE." "The correlation with the altitude at the time of observation points to (atmospheric) refraction as the controlling factor (producing systematic displacements of the solar lines in spectra of Venus) rather than to a repulsive 'Earth effect' acting on the solar vapors as suggested by Evershed ... the observed displacements are caused by unsymmetrical illumination of the slit due to the separation of the visual and photographic images by atmospheric refraction, and ... the unequal illumination is a function of the diameter of the image and the width of the slit."

Having dispelled the mythical Evershed Earth-repulsive effect, St. John and Nicholson then proceeded to look for oxygen and water vapor in their spectra of Venus. Five spectrograms of Venus were taken in the region of the α band of oxygen ($\lambda 6278$) when the relative velocity of Venus and Earth was -12.8 km s^{-1}. This corresponded to a Doppler shift of 0.268 Å to the violet, an amount sufficient to separate completely the terrestrial components. St. John and Nicholson (1922) reported "no lines are observable where they should appear if produced by oxygen in Venus' atmosphere." They also looked at the oxygen B band, $\lambda 6867$. This band is produced by a much smaller amount of O_2 than the α band and thus furnishes a more sensitive test for O_2 on Venus. King (1922) had shown that 39.4 m of air (or 8 m of O_2) at 72 cm pressure produces faint lines in the B band. Spectrograms of Venus taken when the relative velocity was -10.68 km s^{-1} (or a Doppler shift of -0.245 Å) and when it was $+11.36$ km s^{-1} (or a Doppler shift of $+0.286$ Å) showed no oxygen absorption in the atmosphere of Venus. The light path through the atmosphere of Venus was estimated to be "7.5 times the radial depth of the layer." Thus if an oxygen layer equivalent to 1 m atm$_{stp}$ were present in the atmosphere of Venus, detectable O_2 lines should have been observed. Hence an upper limit of 1 m atm$_{stp}$ could be set to the amount of O$_2$ observable on Venus.

Eight water vapor lines (λ5886–λ5932) were examined when the relative velocity of Venus and the Earth was -12.8 km s^{-1} (or a Doppler shift of -0.252 Å). Again St. John and Nicholson found "no traces of lines due to the planet's atmosphere are discernable on the spectrograms ... there must have been less than 1 mm of water in the layer of the planet's atmosphere traversed by the solar beam ..." They conclude: "These observations indicate that the previous spectroscopic evidence for oxygen and water vapor in the atmosphere of Venus, depending upon visual observations of a change in line intensity, is not reliable, that in fact there is no spectroscopic evidence of the presence of either. On the other hand, they do not show the complete absence of water vapor and oxygen from the planet's atmosphere, but that, to the depth penetrated by the solar beam, they are not present beyond a definite low limit."

A question which bothered St. John and Nicholson, as well as other observers, was how deep into the atmosphere of Venus the sunlight had penetrated before it was reflected back to observers on Earth. Before Russell's investigation (1899) the accepted view was that Venus' atmosphere was much denser than that of the Earth. Russell had shown that the prolongation of the cusps of Venus was mainly due to diffuse reflection of light in the planets atmosphere. He concluded that there is no satifactory evidence that the atmosphere of Venus at the apparent surface (cloud top) is more than one-third as dense as the Earth's at sea level. Russell thought the entire height of the atmosphere above the apparent surface might be 50 km and that the pressure there could be one-tenth that of the Earth's atmosphere at sea level.

On the other hand, Claydon (1909) argued for a dense atmosphere at least as extensive as that of the Earth. He assumed a high and heavy layer of pillared cumuli, which would account for the observed features on Venus, with a filmy veil of cirrus above it, which produces the prolongation of the cusps. This model indicated that the reflected sunlight could not have reached any closer to the real surface of Venus than the 43 km level, if her atmosphere was similar to that of the Earth. For in the case of Earth, one only finds 1 m atm of O$_2$ above the 43 km level. St. John and Nicholson's (1922) failure to detect water vapor did not appear to agree with part of Clayden's model of the Venus atmosphere: "If the dusky markings ... are a transient thinning of a cloudy envelope, it is probable that we there see down to levels at which humidity would be high in an atmosphere so heavily moisture laden that the planet is enveloped in a blanket of clouds ... If, however, the reflecting surface consists of a permanent layer of cirro-stratus, the quantity of water vapor traversed by the reflected beam would be small, as cirro-strati are formed in the upper troposphere where the temperatures are very low ... Reflection from a layer of cirri gives the shortest possible path of light in the planet's atmosphere. The water vapor above the cirrus level may be insufficient for detection by observations on the lines in the rain band."

St. John and Nicholson considered the following alternatives to water vapor clouds: "It is possible that a very small quantity of water vapor would produce an impenetrable haze-bank if the atmosphere of Venus contained minute hygroscopic centers of condensation capable of producing cloud particles in an atmosphere where the humidity is much below that which otherwise would be essential to cloud formation." "... it is conceivable that violent atmospheric circulation would cause clouds of dust to be permanent features of the planet's atmosphere ..." They conclude, "It has been too easily assumed, perhaps, that the atmospheric conditions on our nearest planetary neighbors are similar to those on Earth ... It was long ago suggested by Koene (1856) of Brussels, that all free oxygen may have been formed from carbonic acid in the air. Arrhenius (1908) says that probably all the oxygen of the air owes its existence to plant life. That a similar production of oxygen has apparently not taken place on Venus suggests that some conditions are wanting ... it may be that the exacting conditions for the origin of life have not been satisfied so that the existing atmosphere may consist of other permanent or semi permanent gases such as nitrogen or carbon dioxide."

Finally, we should mention that St. John and Nicholson assumed that the Doppler shifts due to the rotation of Venus "were assumed to be negligible, as would be the case for the long rotation period shown by Slipher's (spectroscopic) observations."

1.1. EARLY OBSERVATIONS OF THE ROTATION OF VENUS

We currently believe that 'motions' observed in the clouds of Venus are not directly related to the rotation of the planet, but for many years observers attempted to determine her solid-body rotation by observing cloud features. Spectroscopic studies can, however, reveal the velocity (or wind speed) of the cloud tops. The rotation period of Venus has been determined by radar observations of surface features to be 242.98 ± 0.04 days retrograde (Carpenter, 1970). This corresponds to ~ 117 Earth days for one Venus day (i.e. the time interval between noontimes on the surface of Venus is 117 times as long as between noontimes on Earth).

Barnard (1897) wrote "no other object has caused more controversy and produced more varied testimony in the determination of its rotation period than the planet Venus. This rotation controversy has raged for upwards of two centuries, with fitful periods of quiescence – after some observer more combative than the rest had definitely 'settled the question' only to break out again with renewed virulence when a new champion for rotational honors entered the field.

"The periods assigned to the planet vary all the way from 23 or 24 hours to 225 days. One of the short period men has gone so far as to produce a period, derived from drawings made a few days apart, with a decimal running to the ten-thousandth of a second, which ought certainly to be convincing enough, as a smaller subdivision of time would be an insensible quantity and ought never to be stickled for in determining the duration of a planetary day.

"These discrepancies are due in the main to the difficulty – from various causes – of seeing markings which really exist on the surface of Venus."

Among the observers active at the time Barnard was writing were G. V. Schiaparelli and C. Flammarion. Schiaparelli's (1890) opinion was "the planet makes one rotation in 224.7 days – that is to say, in a period exactly equal to the duration of its sidereal revolution about the Sun ... The rapid variations in the aspect of the planet (and espe-

cially in the horns of its crescent), which have been frequently noticed to repeat themselves in a period of about 24 hours cannot be adduced to support the hypothesis of a
rotation of about one day. Such variations arise from atmospheric causes, which tend
to repeat themselves in daily period." Flammarion (1894) had observed white features
at the poles of Venus remarking that "the general tone of the disc of Venus is a bright
yellow." He argued, "If these whitened poles are not an illusion, and if they represent
snow or cloud, their existence would be inexplicable in the hypothesis where the planet
rotates in the same time as it revolves. In fact, in the case where it always presents the
same face to the Sun, the maximum cold would be in the hemisphere opposed to the
Sun, and the maximum temperature towards the center of the illuminated hemisphere;
all the circumference of the illuminated hemisphere would be in the same condition
of climate ... The observation of spots are insufficient to determine the period of rotation, but they seem to indicate that it is not far from 24 hours."

Belopolsky (1900) observed Venus spectroscopically and found an equatorial
velocity of $v = 600 \pm 300$ m s^{-1} indicating a rotation period of less than one day.
Slipher (1903) observed Venus at superior conjunction and found velocities ranging
from 90 m s^{-1} (direct) to 120 m s^{-1} (retrograde) with an average velocity of 19 ± 12 m
s^{-1} (direct). He concluded "that there is no evidence that Venus has a short period of
rotation." Fifty years passed before another attempt was made to determine the rotation of Venus spectroscopically. Richardson's (1958) results, based on 102 measurements, indicated a mean velocity of 32 ± 33 m s^{-1} (retrograde). His measured velocities ranged from 900 m s^{-1} (direct) to 900 m s^{-1} (retrograde). Richardson found
values in the range 100–200 m s^{-1} (retrograde) occurred with the greatest frequency
and the measurements appeared to follow a normal error distribution.

1.2. Photographs of Venus

Wright (1927) took a number of photographs of Venus at both infrared (7600 Å) and
violet wavelengths. Venus was featureless in the infrared photographs but showed
bright streaks or bands in the violet. He remarked "It appears reasonable to assume
that the violet photographs represent an upper level of the atmosphere of Venus and
that the hazy markings ... are therefore atmospheric phenomena. In that case they are
probably variable in form, and, as they are the only planetary markings found at the
time, it seems likely that markings reported on occasion by observers with the telescope are of this nature. This would explain the numerous discrepancies in the observations of these observers." Ross (1928, 1931) reported similar results: features showed
up in photographs he made through an ultraviolet filter but photographs of Venus
made through blue, red and infrared filters gave no trace of the markings. Ross (1928)
remarked that "there is nearly always a complete change of cloud formation from
day to day" in the ultraviolet photographs of Venus. On the interpretation of his
photographs, Ross had this to say:

"Granted that Venus has an atmosphere which hides the surface from visual observation, it is normally to be expected that photographs taken in infrared light, which
would perhaps penetrate the atmosphere, might disclose surface features, just as

during the great war photographers were able to take pictures of the Earth's surface from points several miles above it by the use of infrared light ... But in photographing Venus, this does not appear to hold, the short waves now being the useful ones." Ross suggests the following explanation: ... "the apparent white surface which we see is imagined to be a uniform shell of light cirrus clouds overlying a dense yellow atmosphere. On occasions of violent atmospheric disturbance the uniform cloud-covering is broken up, and we see the underlying yellow atmosphere to which are due the dark markings seen visually and photographically. Or the cirri may in certain regions be piled up in heavy masses, again seen visually and photographically as brilliant white clouds.

"The observed visual albedo of Venus is not out of harmony with the hypothesis of a covering of light cirrus clouds. The coefficient of reflection of dense white clouds is 78%. According to Russell (1916) the coefficient of reflection of the outercloud-mantle of Venus is 49% ... The clouds on Venus thus have considerably less reflectivity than white clouds on Earth, a fact in harmony with the hypothesis of a scarcity of water vapor ... "

Ross was the first person to make systematic observations of Venus in ultraviolet light, which clearly revealed the day-to-day variations in her atmosphere. Twenty years went by before professional astronomers again looked for the dark markings in ultraviolet photographs. Observations were made by Dollfus (1953, 1955a, b) in the 1940's, Kuiper (1954) in 1950 and Richardson (1955) in 1954. In 1957 Boyer began a long series of photographic observations of Venus. He noticed that a dark cloud feature recurred on photographs taken 4 days apart. This suggested the possibility of a retrograde rotation of the upper clouds with a 4 day period (Boyer and Camichel, 1961). Boyer and Guérin (1966) then obtained sequences of ultraviolet photographs, at intervals varying from 2 to 6 hours, and the dark cloud markings were observed to move. Measurements of the photographs indicated they moved with an equatorial mean velocity of -105 ± 11 m s^{-1} (Boyer and Guérin, 1969); the measured velocities of cloud features varied from -68 m s^{-1} to -229 m s^{-1}. Smith (1967) confirmed the rapid cloud motions but he commented, "... investigation of a large number of our own photographs taken since 1959 has failed to reveal any *well-defined* repetitive patterns, although somewhat similar (cloud) formations often reoccur at 3-to-5 day intervals. Because of the uncertainties in the lifetimes of Venus cloud formations, there is no assurance that similar patterns reappearing after 4 or 5 days are in fact the original clouds. ... Indeed, some of our 1967 plates show strikingly similar cloud patterns at intervals of only 2 days although individual cloud displacements during several hours on these same dates clearly exhibit motions corresponding to a period of 5 days."

Scott and Reese (1972) reported, "The ultraviolet markings appear to be randomly distributed and quite ephemeral in nature, rarely enduring in a recognizable pattern for more than 20 days and usually much less." They found cloud velocities which varied from -66 to -127 m s^{-1} at the equator, with a mean equatorial velocity of -97.7 ± 6.4 m s^{-1}, a mean sidereal rotation period of 4.57 ± 0.30 days retrograde, with individual periods ranging from 3.5 to 6.8 days. Caldwell's (1972) observations indicated a sidereal rotation period of 4.50±0.02 days, while Boyer and Guérin's study indicated a period of 4.067 days. The period for cloud features, and hence the wind velocities in the atmosphere above the clouds, may vary from day-to-day, which may account for the apparent lack of agreement between the Venus observations made at different observatories.

1.3. RADIOMETRIC OBSERVATIONS

Pettit and Nicholson (1923, 1924, 1927, 1930, 1936), Menzel (1923), Menzel *et al.* (1926), and Coblentz and Lampland (1924, 1925a, b, 1927) were among the first to measure the thermal radiation emitted by Venus. Pettit and Nicholson (1927) reported "Two methods of estimating the temperature of emission of the radiation from the dark side agree in assigning a value near 250 K. This is consistent with what might be expected from radiation emitted by a cirrus cloud covered atmosphere." They found the same temperature on the bright side of the planet which indicated the temperature was uniform over the planet. Later radiometric observations of Venus were made by Sinton and Strong (1960) who reported a brightness temperature of 234 K. They also found very little (2 K) difference in temperatures between the bright and the dark side of the disc. Chase *et al.* (1963) reported an average temperature for Venus of 225±2 K as measured by the Mariner II radiometer; they had difficulties in the calibration of their instrument and this value is not too reliable as a result. Murray *et al.* (1963a) also experienced difficulties in obtaining a good absolute calibration of their brightness temperature of Venus measured in the 8 to 14 μ wavelength region. Their measurements indicated a temperature of 208±2 K at the center of the disc. The differential accuracy of brightness temperatures measured by Murray *et al.* was about $\frac{1}{3}$ K, however, and their high resolution maps of the temperature distribution over the planet also show the absence of a strong night-to-day effect.

Sinton (1964a, b) had made a series of radiometric observations of Venus from 1958 to 1962 and concluded that "Venus does not exhibit an appreciable periodic variation in its infrared temperature (of 232 K) and that its atmosphere is uniformly opaque from 3 to 30 μ."

1.4. OBSERVATIONS OF MICROWAVE RADIATION FROM VENUS

The investigation of thermal radiation from planets, at radio wavelengths, began comparatively recently. Microwave emission from Venus was first observed in 1956 by Mayer *et al.* (1958). They made observations, at a wavelength of 3.15 cm, over a period of nearly two months. Mayer *et al.* reported an apparent blackbody temperature which varied from 630±110 K (m.e.) to about 560±73 K (m.e.) near inferior conjunction. "Two single observations at 9.4 cm wavelength suggest the bulk of the radiation follows a thermal spectrum, but, the accuracy of these measurements is low." Since that time many more measurements have been made, at wavelengths from 0.3 to 40 cm. These measurements have been summarized by Barrett and Staelin (1964), Dickel (1967), Kuzmin (1967), and Pollack and Morrison (1970).

The microwave brightness temperature was much higher than had been anticipated. The simplest explanation of the measurements is to assume that the atmosphere is almost transparent at these wavelengths and that the temperatures refer to the surface and lowest levels of the atmosphere. (Sagan 1960; Barrett, 1961; Barrett and Staelin, 1964). This explanation was unpopular since there was a general belief that Venus was not drastically different from the Earth. With the exception of Wildt's (1940a, b) suggestion, that the surface temperature of Venus might be as high as $366\,\mathrm{K} < T_s < 408\,\mathrm{K}$, there had been little observational evidence to indicate that the surface of Venus might

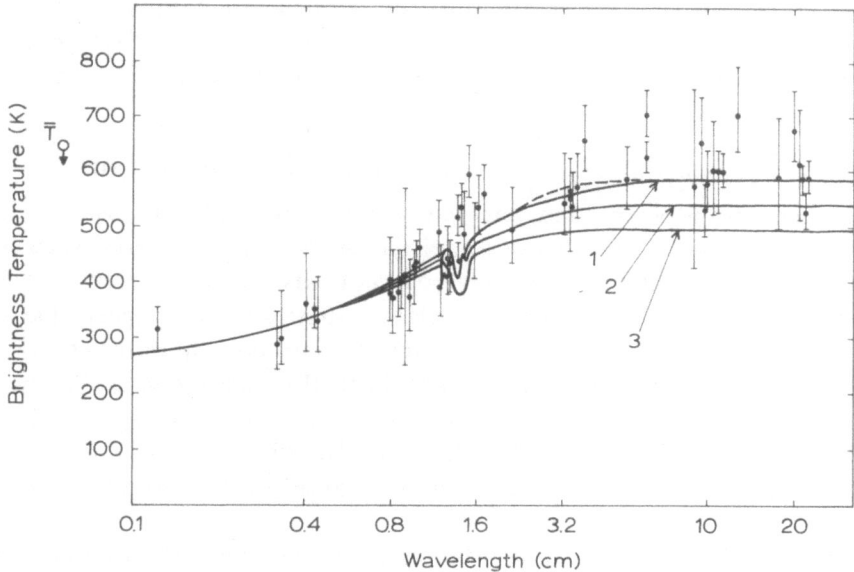

Fig. 1.   Brightness temperature of Venus in the microwave region. The solid curves were computed for surface temperatures of 650 K (1), 600 K (2), and 550 K (3) assuming a surface pressure of 20 atm; the dashed curve is for a surface pressure of 70 atm. The curves were computed for a mixing ratio of 0.4% H_2O in the atmosphere and show the 1.38 cm absorption line of water. (From Kuzmin and Vetukhnovskaya, 1968, *J. Atmospheric Sci.* **25**, 546.)

not be suitable for man to explore. Since the obvious interpretation of the data was unpalatable, there arose a number of fascinating hypotheses to 'explain' the observations (Jones, 1961; Priester *et al.*, 1962; Tolbert and Straiton, 1962; Danilov and Yatsenko, 1963; Danilov, 1964; Kuzmin, 1964; Vakhnin and Lobedinskii, 1966; Plummer and Strong, 1966, 1967; Drake, 1967). Unfortunately, none of the nonthermal 'explanations' were capable of agreeing with *all* the observational data. On the other hand, the original assumption of Mayer *et al.* (1958) that the microwave brightness temperature was due to thermal emission has proven to be correct.

Model atmosphere calculations assuming a high surface temperature had been made by Sagan (1960, 1967a, b), Barrett and Staelin (1964), Ho *et al.* (1966), and Young and Gray (1968) at a time when the reason for the high microwave temperature was disputed. The calculations all agreed that the surface pressure had to be $p_s > 10$ atm. Sagan (1967a, b, 1968) had shown that it was possible to deduce a high surface temperature ($T_s \approx 700$ K), [based on the difference between the optical and radar diameters of Venus, the known temperature at the cloud top and an assumed adiabatic temperature gradient between the clouds and the surface] independent of the microwave measurements. It was only after the *in situ* measurements of Venera 4 and Mariner 5 were reported (Avduevsky *et al.*, 1968; Kliore *et al.*, 1967; Kliore and Cain, 1968; Wood *et al.*, 1968; Eshleman, 1968) that the idea of a surface temperature for Venus $T_s \sim 500$ K was widely believed. The microwave measurements of brightness temperature were finally taken at face value, ten years after the first observations had been reported. The measurements from Venera 5, 6, and 7 confirmed the high surface temperature. Venera 7 measurements indicated a surface temperature of $T_s = 747 \pm 20$ K with a surface pressure, $p_s = 90 \pm 15$ bars (Marov, 1972).

1.5. OBSERVATIONS OF THE POLARIZATION OF SUNLIGHT REFLECTED BY VENUS

Measurements of the polarization of reflected sunlight are another means of studying a planetary atmosphere. They provide a reliable method of determining whether an atmosphere is clear, hazy or covered by an optically dense cloud cover. Furthermore, one can determine whether the cloud particles are spherical or of irregular shape, an estimate of the particle size distribution and information about the refractive index of the cloud particles. Regardless of the particle shape, the polarization can be used to separate the contribution of the gas molecules in the atmosphere (which follows a Rayleigh scattering law with optical depth, τ, varying with wavelength λ as λ^{-4}) from the contribution of the cloud particles (which is much less strongly wavelength dependent, according to Mie theory).

The first accurate measurements of the polarization of sunlight reflected from Venus were made in 1922 by Lyot (1929). Lyot also made laboratory measurements and noted that the primary effect of multiple scattering was to reduce the amount of polarization without changing the general shape of the polarization curve. The amount of polarization observed by Lyot was small, but his measurements were very accurate. Lyot (1929) found that his visual observations of Venus were in reasonably good qualitative agreement with clouds composed of water (refractive index $n_r \sim 1.33$) drops with a radius $r \sim 1.25 \mu$.

The comparatively low temperatures measured radiometrically for Venus suggested the idea that the clouds of Venus could be composed of water and still the atmosphere might contain very little water vapor above the clouds. This notion still has its proponents.

Like so many other techniques for observing Venus, polarization measurements were neglected for many years. Dollfus (1955a, b) resumed observations with the Lyot visual polarimeter and noted that the polarization varied from one point to another on the disc of the planet. Dollfus also noted that the polarization varied from day-to-day, over a particular location on the planet, on a time scale similar to the variation in the cloud features observed in ultraviolet photographs. In addition to his visual observations, Dollfus (1958, 1963a, b, 1966) also measured the polarization photoelectrically in the infrared. Subsequent observations of the polarization on Venus were made using colored filters which spanned the wavelength range from 9900 Å in the infrared to 3590 Å in the ultraviolet. (Gehrels and Samuelson, 1961; Coffeen, 1968; Coffeen and Gehrels, 1969; Dollfus and Coffeen, 1970). Coffeen, (1969) compared the observa-

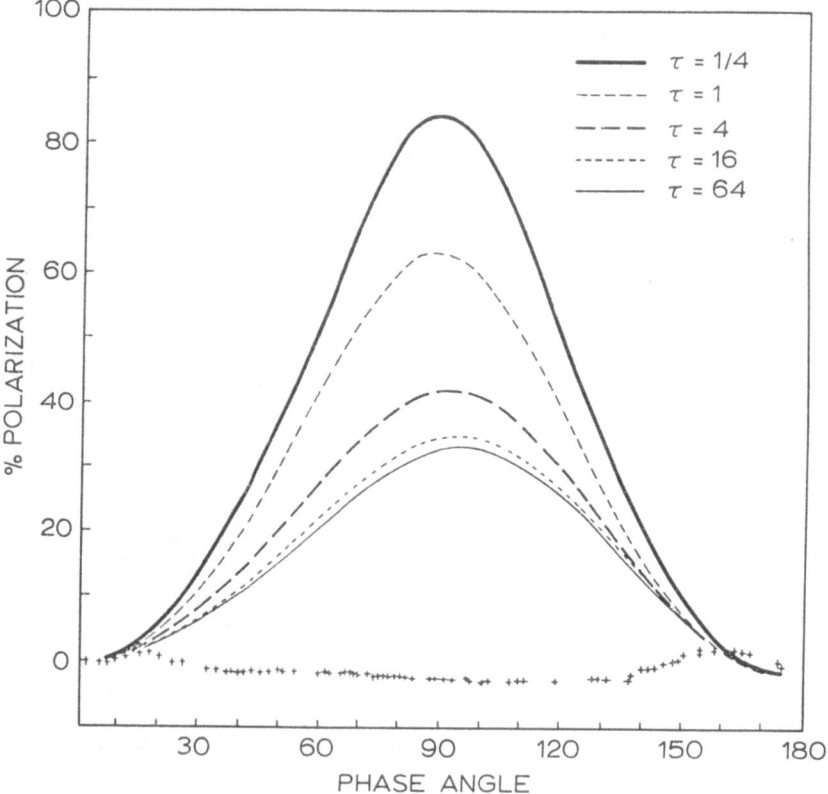

Fig. 2. Theoretical curves of the polarization calculated for Rayleigh scattering atmospheres as a function of the total scattering optical depth. The crosses represent Lyot's measurements of the polarization of visible light reflected from Venus. The lack of agreement between the observations and a Rayleigh scattering atmosphere is apparent. (From Hansen and Arking, 1971, *Science* **171**, 669).

tions of Coffeen and Gehrels (1969) with calculated values of the polarization due to the single-scattering of light by spherical particles. Coffeen concluded that the particles had an index of refraction $1.43 \leqslant n_r \leqslant 1.55$ and a particle radius, $r \sim 1.25\,\mu$.

Russian spacecraft which have entered the atmosphere of Venus have shown that her atmosphere is considerably more massive than that of the Earth. The pressure at the surface of Venus is 90 ± 15 bars and the temperature is 747 ± 20 K (Marov, 1972; Avduevsky *et al.*, 1970). This means that the temperatures of the clouds measured radiometrically (and spectroscopically) refer to some high level in the atmosphere. Similarly, it means that the polarization measurements refer primarily to photons which have been multiply scattered within the atmosphere rather than being reflected from the surface. As a result, the complete theoretical interpretation of the observa-

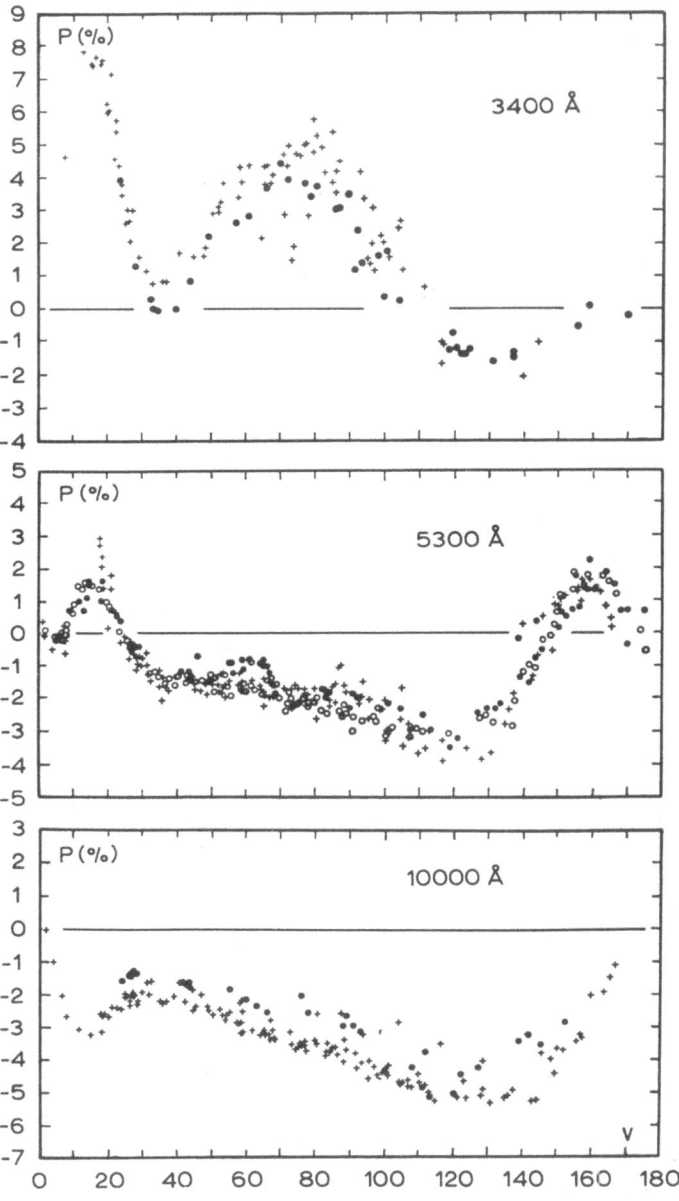

Fig. 3.   Observations of the polarization of sunlight reflected from Venus at various wavelengths as a function of phase angle. (From Dollfus and Coffeen, 1970, *Astron. Astrophys.* **8**, 251.)

tions must be based on solutions of the radiative transfer equation. Exact solutions are available for a Rayleigh scattering atmosphere (Chandrasekhar, 1950), i.e. when the radius of the scatterers, r, is much smaller than the wavelength of the light being scattered. However pure Rayleigh scattering produces much more polarization at visible wavelengths than is observed for Venus.

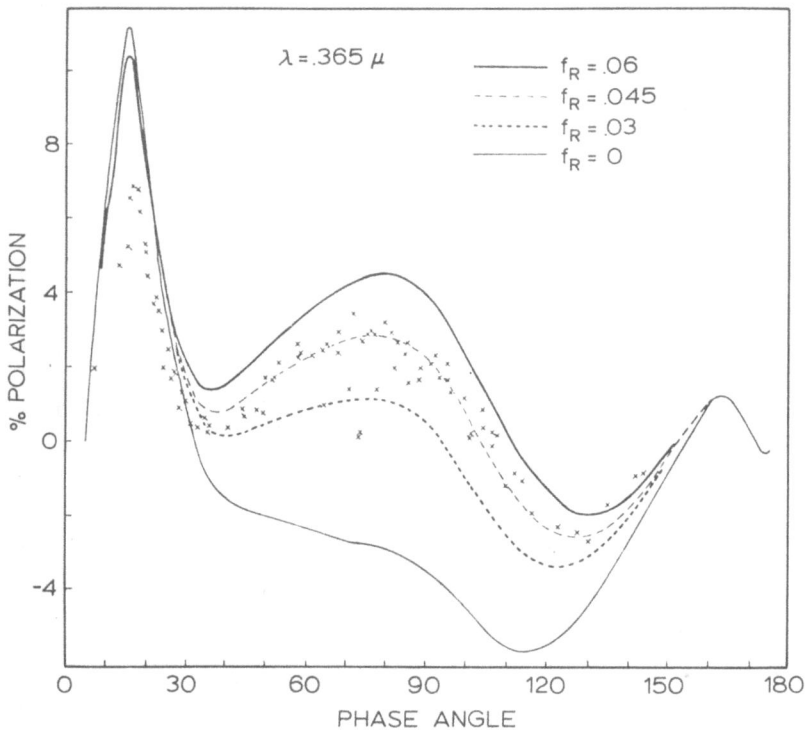

Fig. 4. Theoretical curves of the polarization in the ultraviolet for particles with an index of refraction $n_r = 1.46$ and a radius of 1.1 μ for different fractions of Rayleigh scattering, f_R. The crosses represent measurements of the polarization of Venus made by Coffeen and Gehrels. A value of $f_R = 0.045$ implies that the pressure in the clouds where the optical depth is approximately unity is about 50 mb. (From Hansen and Arking, 1971, *Science* **171**, 669.)

Coffeen (1969) found an upper limit to the Rayleigh scattering optical depth of $\tau \leqslant 0.072$ (in the ultraviolet). For a carbon dioxide atmosphere this meant the upper limit to the amount of gas above the clouds was 0.4 km atm or the cloud-top pressure was $p_c \leqslant 55$ mb. Subsequent calculations by Hansen and Arking (1971) indicated that the fraction of Rayleigh scatterers was $f = 0.045$ (or that the pressure at the cloud tops, $\tau = 1$, is $p_c \leqslant 35$ mb). Observations indicate the polarization of Venus in the ultraviolet is variable, and the cloud tops appear to occur at pressure levels between 30 and 60 mb based on this data (see also Sagan and Pollack, 1969). The recent analysis of the polarization data by Hansen and Arking indicated that most of the cloud particles "must be spherical" and "the mean particle radius is $r = 1.1 \pm 0.1 \; \mu$." They found that "the dispersion in particle sizes is amazingly small; this result is unexpected for dust, but it is not unusual for a liquid ... The particle shape and the disperison of particle sizes, taken together, strongly suggests that the cloud particles are liquid." Hansen

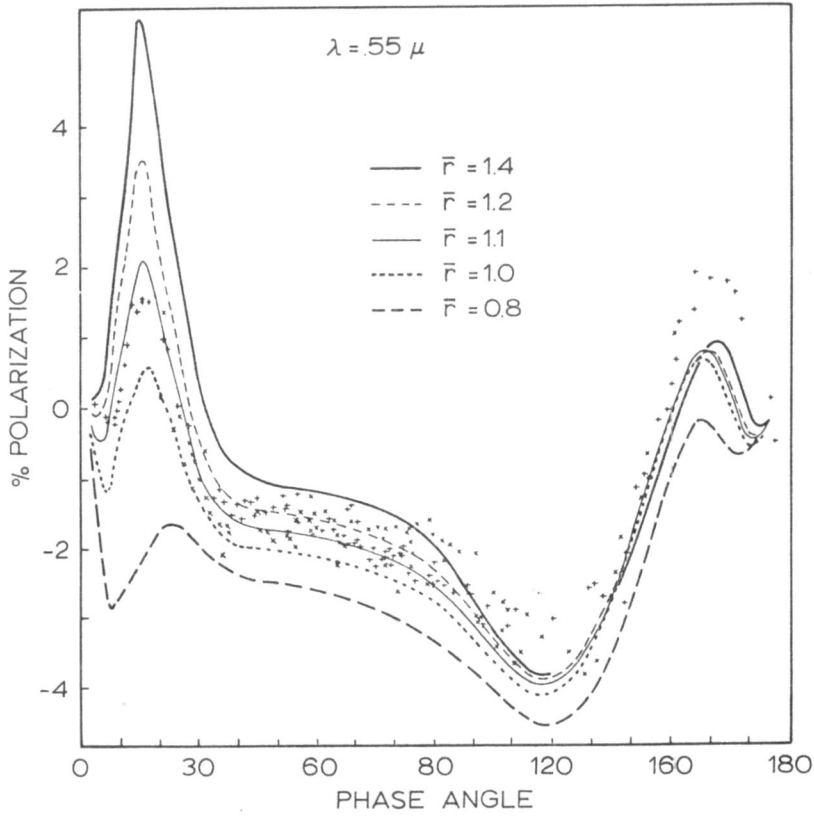

Fig. 5.   Theoretical curves of the polarization in the visible for particles with an index of refraction $n_r = 1.45$ and a fraction of Rayleigh scatterers of $f_R = 0.045$ for several values of the mean scatterng radius, $\bar{r}$. The crosses represent Lyot's measurements of the polarization of Venus. (From Hansen and Arking, 1971, *Science* **171**, 669.)

and Arking reported, "... the best fit to the observations occurs with a refractive index which decreases from $n_r \sim 1.46$ in the ultraviolet to $n_r \sim 1.43$ at $\lambda = 0.99 \; \mu$; the uncertainty in n_r is 0.02 at each wavelength."

These results for the refractive index rule out the possibility that the upper clouds of Venus can be composed of pure liquid water or ice. Since the visible clouds are composed of particles with a single index of refraction (as can be seen from the presence of a sharp rainbow in the ultraviolet data) they can not be composed of a mixture such as dust and H_2O. (If there were two or more cloud layers with different refractive indices, or different particle sizes, then features due to each type of particle would be present in the polarization data). Hansen and Arking's results stringently narrow the list of possible material composing the visible clouds of Venus. They also rule out many of the materials that have been proposed for the composition of the clouds. Table I lists the refractive index of some of the materials which have been, or could be, suggested for the cloud particles.

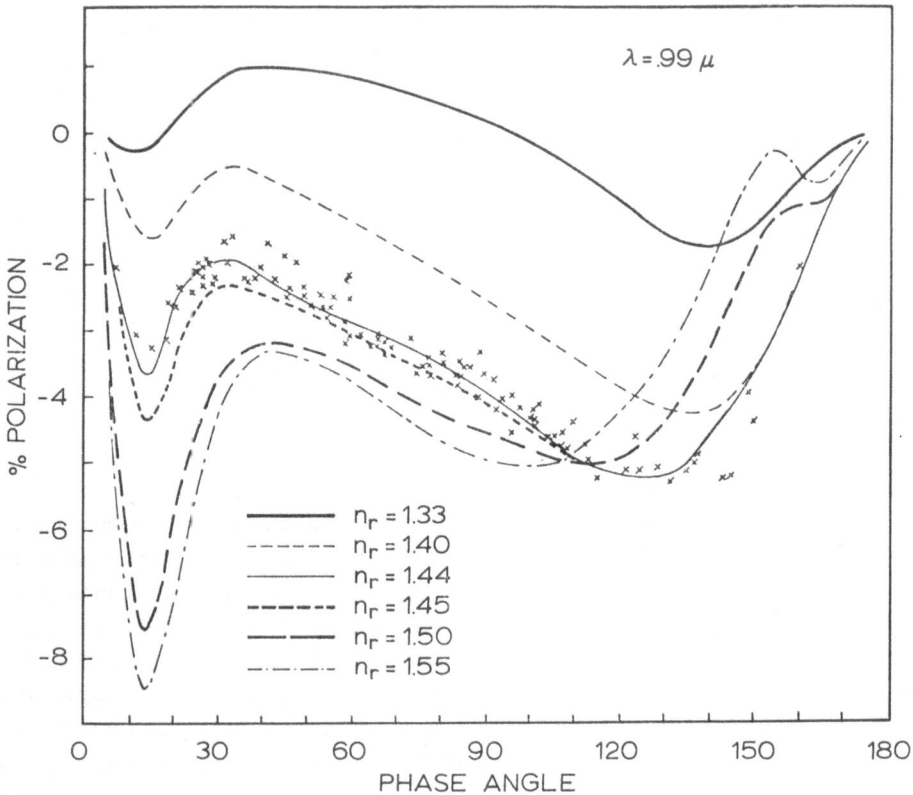

Fig. 6. Theoretical curves of polarization in the near infrared for spherical particles with different indices of refraction. The crosses represent the measurements of the polarization of Venus made by Coffeen and Gehrels. (From Hansen and Arking, 1971, *Science* **171**, 669.)

1.6. COMPOSITION OF THE CLOUDS OF VENUS

The early observations of Venus had given rise to the notion that the clouds of Venus were aqueous. Then it became fashionable to suppose they were composed of dust. Wildt (1940a, b) was the first to propose a more exotic chemical composition for the clouds: polymerized formaldehyde. Wildt noted "that the partial pressure of H_2O in Venus' atmosphere must be far below that required for saturation and condensation.

TABLE I

Refractive indices

H₂O(s)	1.31	Na₂SO₄	1.48
H_2O (l)	1.33	NH_4ClO_4	1.48
D_2O (l)	1.33	$FeSO_4.7H_2O$	1.48
NaF	1.34	Na_2SO_4	1.48
KF	1.36	KCl	1.49
C_2H_5OH (l)	1.36	K_2SO_4	1.49
HNO_3 (l)	1.40	C_6H_6 (l)	1.50
HCl (l)	1.41	$CaCl_2$	1.52
SO_2 (l)	1.41	$(NH_4)_2SO_4$	1.52
H_2SO_4 (l).2H₂O	1.41	$FeSO_4.5H_2O$	1.53
H_2O_2	1.41	NaCl	1.54
CaF_2	1.43	SiO_2 (quartz)	1.55
$Na_3PO_4.12H_2O$	1.43	$MgSO_4$	1.56
$MgSO_4.7H_2O$	1.43	$FeCl_2$	1.57
H_2SO_4 (l)	1.44	$Mg(OH)_2$	1.57
$ClSO_3H$	1.44	$CaSiO_3$	1.62
C_3O_2 (l)	1.45	CS_2 (l)	1.63
$Na_2B_4O_7.10H_2O$	1.46	NH_4Cl	1.64
$NaH_2PO_4.H_2O$	1.46	$CaCO_3$	1.65
$Al_2O_3SO_3.9H_2O$	1.46	$AgNO_3$	1.73
CCl_4 (l)	1.46	MgO	1.74
$Al_2(SO_4)_3$	1.47	CaO	1.84
$ZnSO_4.7H_2O$	1.47	$HgCl_2$	1.86
SiO_2 (tridymite)	1.47	Hg_2Cl_2	1.97

Consequently, the concept that the visible surface of Venus consists of a layer of clouds of the same nature as terrestrial ones has to be abandoned, and it appears necessary then to account for the high albedo of the planet in a different way. Gerasimovič (1937) demonstrated conclusively that the only model of Venus' (visible) surface which admits of an explanation both of the absolute brightness of the planet and of its phase curve is a stratum of large scattering particles, which he believes to be products of condensation, without further specifying their character." Wildt continued, "the extreme paucity, if not complete absence, of free oxygen is obviously related to the lack of water on Venus ... The absence of O_2 would deprive Venus of the protection this gas exerts against ultraviolet sunlight. Carbon dioxide and water vapor would be decomposed photochemically, and formation of formaldehyde would follow." Wildt searched for spectroscopic evidence of CH_2O on Venus, and saw none; he found an upper limit of 0.3 cm atm. "This result would seem to refute the alleged accumulation of formaldehyde at Venus' surface. However, gaseous formaldehyde is extremely susceptible to polymerization and precipitation in solid form. Therefore the spectroscopic test refers only to the small amount of gaseous CH_2O in equilibrium with the solid phase." Wildt (1942) later learned that "the vapors of linear formaldehyde polymers are monomeric formaldehyde" and a large fraction (92%) of the polymer clouds would be CH_2O. Thus Wildt's hypothetical plastic clouds on Venus were ruled out on the basis of his own observations.

Van de Hulst (1952) suggested that Lyot's polarimetric observations might be fit as well by small particles of SiO_2 as by H_2O droplets, but the index of refraction of quartz (1.55) is much larger than that found by Hansen and Arking (1971). The suggestions made by Suess (Kuiper, 1952) that the clouds could be composed of salts (NaCl, $MgCl_2$) also is ruled out by their high refractive index and the requirement that the cloud particles be liquid droplets.

Menzel and Whipple (1955) suggested that "oceans completely covering the surface of Venus could produce H_2O clouds with the general and detailed characteristics observed in the Venusian atmosphere." They based their arguments on Lyot's polarization measurements, the radiometric observations of Sinton, and the fact "that $-39\,°C\,(234\,K)$ is precisely the temperature assumed by many high-level cumulus clouds on the Earth, because this temperature is the lowest at which water can still exist in the liquid state. The droplets are super cooled but none the less liquid." Menzel and Whipple argued that the comparatively high surface temperatures suggested by Wildt (1940) of "366 K to 408 K appears to be denied by Sinton's measurements." We now know, from the Venera Spacecraft measurements, that the surface temperature of Venus is close to 740 K, much hotter than Wildt had suggested. However the idea that the planet Venus is significantly different from the Earth, and the idea that the radiometric temperatures measured in the infrared was not too much different from the surface temperature, were not to be given up easily.

Öpik (1955) disputed the hypothesis of water clouds. He argued that if the clouds were H_2O, the sunlit half of Venus should be "covered with cumulo-nimbus clouds of varying contours, producing contrasting surface markings which have not been observed. The monotony of the sunlit half of Venus would be more in accord with an indifferent haze or smoke of mineral origin, than with water clouds carrying up almost explosive amounts of latent heat." Öpik suggested that "there may be various kinds of dust, some of which may reveal a similar polarization effect to that caused by water clouds, even if they have not the same chemical composition." "It should be emphasized, in view of the negative outcome of the direct spectroscopic proof, the *approximate* agreement of the reflecting power and polarization curve of Venus with that for clouds of water droplets can not be attributed much weight" (Öpik, 1956).

To paraphrase Barnard, nothing has aroused more controversy and produced more varied testimony than the composition of the clouds of Venus. The controversy has raged with fitful periods of quiescence – only to break out again with renewed virulence. As we appear to be writing this in a quiescent period, it may be useful to recall the legions who have rallied to the defense of aqueous clouds on Venus and those who have opposed them. It is not the intent to give here a blow-by-blow account of all the arguments, pro and con, for water clouds. Table II lists some of the participants involved in the debate. The arguments in support of water clouds rely on the comparatively large amounts of H_2O detected spectroscopically by Bottema *et al.* (1964b), the measurements of H_2O by the spacecraft Venera 4 (Vinogradov *et al.*, 1970a, b), the similarity of the spectrum and polarization data to telluric water clouds, and the need of an additional substance to provide enough of a greenhouse effect to

TABLE II

Aqueous clouds on Venus?

Pro	Con
Vogel (1874)	Slipher and Edson (1939)
Tacchino and Ricco (1882)	Wildt (1940b)
Claydon (1909)	Kuiper (1952)
Arrhenius (1918)	Kuiper (1954)
Ross (1928)	Hoyle (1955)
Lyot (1929)	Öpik (1956)
Menzel and Whipple (1955)	Kuiper (1962)
Sagan (1960)	Moroz (1963)
Dollfus (1963a, b)	Sagan and Kellogg (1963)
Deirmendjian (1964)	Chamberlain (1965)
Bottema et al. (1964a, b, 1965a, b)	Espinola and Blau (1965)
Strong (1965)	Kuiper (1967)
Plummer and Strong (1965)	Kuiper and Forbes (1967)
Ohring (1966)	Rea and O'Leary (1968)
O'Leary (1966)	Coffeen (1968)
Pollack and Wood (1968)	Belton et al. (1968)
Pollack and Sagan (1968)	Lewis (1968a, b 1969)
Hansen and Cheyney (1968)	Hunten and Goody (1969)
Sagan and Pollack (1969)	Kuiper (1969)
Plummer (1969)	Rasool (1970)
Ohring (1969)	Schorn and Young (1971)
Pollack (1969)	Veverka (1971)
Plummer (1970)	Lewis (1971a, b)
O'Leary (1970)	Hansen and Arking (1971)
	Fink et al. (1972)
	Regas et al. (1972)

maintain the high surface temperature. The arguments against water clouds rely on the comparatively small amounts of H_2O detected spectroscopically by other observers, the fact that the refractive index of the cloud particles is significantly larger than that of liquid water or ice, and the contention that no additional substance is required to produce a greenhouse effect (Danielson and Solomon, 1966). It appears that the question of aqueous clouds will only be definitely settled by *in situ* measurements on the clouds of Venus. Some of the more imaginative suggestions for the cloud composition can be dealt with in a less ambiguous way.

Hoyle (1955) suggested that the negative results of the spectroscopic search for water vapor could be explained if a great excess of hydrocarbons existed on the primitive planet Venus. He surmised that the surface is now covered with the remainder of the hydrocarbons, and that the cloud layer is composed of smog. This hypothesis about the nature of the clouds of Venus was not exactly new. Velikovsky (1950) had made a similar prediction based on his interpretation of ancient myths and documents of varying antiquity. Velikovsky's version of cosmogony differs radically from that of physical scientists (Newton's Law of Gravitation doesn't hold, angular momentum isn't conserved, etc.) and the accuracy of his reserach has been questioned (Payne-Gaposchkin, 1950; Margolis, 1964). Mintz (1961), Kaplan (1963), and Robbins

(1965) also postulated hydrocarbon clouds. Kaplan said, "An attractive possibility is that the clouds are mostly composed of organic compounds, many of which have sufficiently high condensation or polymerization temperature: about 1 m atm_{stp} of almost any gaseous compound containing a C—H bond would be sufficient to close a spectral window around 3.5 μ, which would otherwise allow more radiation to escape from the surface than can possibly be compensated by incoming sunlight." These postulated hydrocarbons would explain two observational results: the low albedo of Venus in the 3–4 μ spectral region and the high surface temperature which is believed to be maintained by a greenhouse effect. One difficulty with the suggestion of hydrocarbons is the absence of an absorption feature near 2.4 μ in the spectrum of Venus. Plummer (1969) has pointed out that "Each hydrocarbon (from methane through the hydrocarbon waxes and tars) absorbs infrared radiation in a band of wavelengths centered between 2.3 and 2.5 μ, the position varying somewhat with molecular structure." Earlier searches for hydrocarbons in the spectrum of Venus had also been unsuccessful. Kuiper (1952) estimated there could be no more than 0.5 cm atm_{stp} of C_2H_6, 1.5 cm atm_{stp} of C_2H_4 or 10 cm atm_{stp} of CH_4 above the clouds of Venus and remain undetected in his low resolution spectra of Venus. The high resolution spectra of Connes et al. (1967) failed to reveal any detectable lines of CH_4, CH_3Cl, CH_3F, C_2H_6 or HCN; they estimated that there could be no more than 0.3 cm atm of any of these gases above the clouds. Thus all of the spectroscopic data are unfavorable to clouds composed of hydrocarbons.

Mueller (1964), Dayhoff et al. (1967), and Lewis (1968a, b) have placed stringent theoretical upper limits on various hydrocarbons through considerations of chemical equilibrium. To quote Lewis (1969), "Perhaps the most effective criticisms of these suggestions are based on (1) the complete absence of hydrocarbons and organic matter as derived from spectroscopic studies of Venus and (2) the thermodynamic instability of organic compounds particularly in a mildly oxidizing atmosphere."

The question of whether the clouds of Venus are 'smog' or not depends upon how one defines that complicated mixture of substances (which can include ozone, oxides of carbon, nitrogen and sulfur as well as hydrocarbons). The most popular source of 'smog' on Venus is volcanoes (Davidson and Anderson, 1967).

Another interesting suggestion for the composition of the clouds is C_3O_2. It was proposed by Sinton (1953), Kuiper (1957), and Harteck et al. (1963). Sinton and Strong (1960) note that "Carbon suboxide is an unstable molecule and readily polymerizes, on standing to a reddish or whitish mass. This polymerization may explain the clouds of Venus." Sagan (1961) raised the objection that carbon suboxide polymers have too low a reflectivity to be the main constituent of the clouds.

Plummer and Carson (1970) have presented convincing evidence that the clouds can not be either C_3O_2 or C_3O_2 polymers, "although very slight traces of low polymer might be present and visible as a yellow stain (in ice clouds, for example) without being otherwise detectable." Although gaseous C_3O_2 has never been observed in any spectra of Venus [upper limits of 5–10 cm atm, 0.5 cm atm, 0.2 cm atm and 0.1 cm atm have been given by Owen (1968a, b), Jenkins et al. (1969), Kuiper (1969), and

Plummer and Carson (1970) respectively] it was argued that this was to be expected if only polymers composed the clouds. In their experiments on C_3O_2, Plummer and Carson found, "Each polymer will continue to polymerize further as long as the ultraviolet light is present. By the time the monomer is exhausted, many of the molecules present are quite heavy, with a deep brown color. We have observed this process. Although polymerization could occur under solar ultraviolet irradiation in the Venus

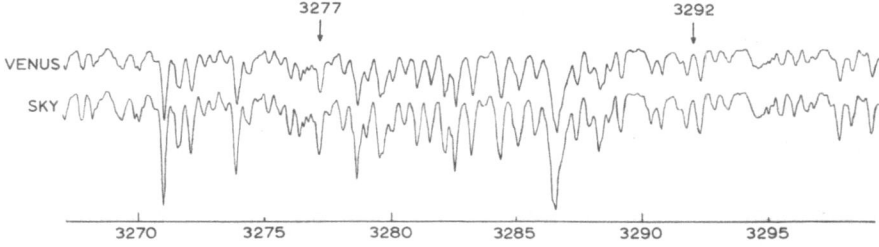

Fig. 7.   Spectra of Venus and the sky in the region of carbon suboxide absorption features at 3277 Å and 3292 Å. (From Owen, 1968, *J. Atmospheric Sci.* **25**, 583.)

clouds, we question whether the monomer could be entirely removed in this way without forming more than the slightest trace of heavier molecules which would show intense yellow, red, and brown coloration. Certainly the polymerization would proceed indefinitely, and a continual supply of monomer would have to be invoked to make the clouds perpetually white. We do not believe that the newly formed monomer could escape detection at 2.27 μ." The index of refraction of C_3O_2 is in agreement with that deduced from the polarization data. But the cloud particles would have to have a radius $r < 0.05 \mu$ (in contrast with a radius $r = 1.1 \pm 0.1 \mu$ required by the polarization data) in order to have a reflection spectrum like that observed for Venus.

The detection of HCl in the atmosphere of Venus (Connes *et al.*, 1967) led to the suggestions that the cloud particles could be made of NH_4Cl (Lewis, 1968a, b), aqueous HCl solutions (Lewis, 1968a, b), Hg_2Cl_2 and other compound of Mercury (Lewis, 1969) and incompletely hydrated $FeCl_2$ (Kuiper, 1969). Lewis (1968a, b) showed HCl solutions would evaporate at the temperatures measured for the clouds ($T_c = 235$ K from radiometric observations and $T_c = 250$ K from spectroscopic observations). In order for the clouds to be made of aqueous HCl solutions, they would have to contain about 25% by weight HCl and be at a temperature of 216 K. The discrepancy with the observed temperature led Lewis to conclude "It appears that NH_4Cl must be considered a possible cloud constituent on Venus. Aqueous HCl and solid HCl hydrate clouds, if anywhere present, must be extremely tenuous; pure water and ice seem impossible to reconcile with the best spectroscopic observations." From Table I we see that the refractive index of NH_4Cl is much larger than that required by the polarization measurements, so that it is ruled out as a possible cloud constituent. Lewis (1969) came to a similar conclusion also on the basis of chemical arguments: "Lewis' suggested NH_4Cl should probably be discounted in light of the fact the proposed condensate would almost certainly lie too deep in the atmosphere to account for the observed temperature of the cloud layer."

Lewis' suggestion (1969) that "The 'visible cloud deck' on Venus, at a temperature near 240 K, is composed of a thin haze of Hg_2Cl_2 overlying a deep cloud of liquid Hg droplets" is ruled out by the refractive index data: for Hg_2Cl_2, $n = 1.97$. Similarly, partially hydrated $FeCl_2$, $n \simeq 1.55$ does not agree with the polarization measurements.

Although hydrochloric acid clouds are incompatible with the observed temperature at the top of the clouds on Venus, they have been a popular contender for the composition of her clouds. For example, in a review article on the composition of the upper clouds of Venus, Rea (1972) says, "It is concluded that the leading candidate for the uppermost clouds is liquid drops of $HCl-H_2O$, that there is no recommended candidate for the second cloud deck, and that H_2O ice is at most a minor component of these cloud systems." ... "Lewis (1971a, b) has estimated theoretically that the refractive index of such a solution is between 1.42 and 1.43, in reasonable agreement with the polarization results. Moreover such a solution will absorb strongly in the 3–4 μ region because of strong broad bands of the H_3O^+ ion."

The fact that strongly ionized acids all show the absorption feature near 3 μ characteristic of the H_3O^+ led Young and Young (1973) to examine the spectra of HCl and other acids. While their survey showed the spectrum of Venus was incompatible with the spectrum of aqueous HCl solutions, the spectrum of sulfuric acid was in qualitative agreement. Clouds composed of sulfuric acid have an absorption feature at 11.2 μ which is observed in the spectrum of Venus (Sinton and Strong, 1960; Gillett et al., 1968; Hanel et al., 1968) and has not been identified. Young (1973) has presented some of the arguments in favor of sulfuric acid clouds: the refractive index agrees with the polarization measurements, sulfuric acid is produced photochemically in the Earth's atmosphere, etc. Sill (1973) had postulated sulfuric acid clouds on Venus at an earlier date, employing somewhat different chemical arguments. On the other hand, Lewis (1971a, b) believes that the failure to detect H_2S or COS in the spectrum of Venus argues against the possibility of the clouds being composed of a sulfur compound.

The question of the composition of the clouds does not appear to be closed. Their light yellow color and low albedo in the ultraviolet have not been explained by any of the postulated cloud substances other than hydrated ferrous chloride, which has been ruled out for other reasons.

2. Modern Spectroscopic Observations of Venus

2.1. DISCOVERY OF CARBON DIOXIDE

Although some of the observational methods employed today are very similar to those used by St. John and Nicholson in the 1920's, we have chosen to begin the 'modern' era with the discovery of absorption features in the spectrum of Venus.

Adams and Dunham (1932) set out to search once again for the elusive atmospheric O_2 and H_2O in the spectrum of Venus. "Recent progress at the Research Laboratory

of the Eastman Kodak Company in sensitizing photographic plates to the infrared
has made it possible to extend this investigation to the region of longer wavelengths
where the A band of oxygen at $\lambda 7594$ and the group of strong water-vapor lines in the
interval $\lambda 8150$–$\lambda 8300$ afford excellent material for a sensitive test of molecules of these
gases in the atmosphere of Venus." The photographic plates used by St. John and
Nicholson (1922) were not sensitive this far out in the infrared and they had been
restricted to a part of the spectrum where neither oxygen or water absorbed very
strongly. With the possibility of observing Venus in the infrared where the absorption
was known to be stronger, Adams and Dunham (1932) stood a much better chance of
detecting these molecules in the atmosphere of Venus. They found "no lines of measurable intensity due either to oxygen or to water vapor are present in the spectrum
of Venus," but they discovered three bands not appearing in the solar spectrum with
heads at $\lambda 7820.2$, $\lambda 7882.9$ and $\lambda 8688.7$. From measurements of the spacing of the
lines in these bands and theoretical knowledge of band structure, Adams and Dunham
presumed the bands to be due to carbon dioxide. This claim was substantiated by the
theoretical work of Adel and Dennison (1933). The carbon dioxide band at 8689 Å
was measured in an absorption cell by Adel and Slipher (1934). They reported, "The
lower limit of an estimate on the CO_2 content of the absorbing strata of Venus is apparently two mile atmospheres (3 km atm) whereas the amount actually present in
these layers is very probably several times greater... and this is, presumably, just a
very small fraction of the total CO_2 content of the entire atmosphere. In the upper
strata alone, Venus possesses 10^4 times as much CO_2 as is present in the entire atmosphere of the Earth."

2.2. Observations in the 'far' infrared

It is relatively recently that spectroscopic observations of the planets began to be made
in the 'far' infrared region of the spectrum. Infrared spectra of various gases have been
measured in the laboratory since the 19th century. For example, Paschen (1894)
recorded the spectra of carbon dioxide and water vapor in the wavelength region 1 to
5 μ. Rubens and Aschkinass (1898) measured the spectra of these gases out to a wavelength of 20 μ. Ångström (1890) discussed the influence of CO_2 and H_2O absorption
features in the solar spectrum on the spectral distribution of solar energy measured at
sea level. An extensive catalog of the infrared spectra of various substances was
prepared by Coblentz (1905). It must be admitted that these early measurements of
infrared spectra were made at comparatively low resolution because the detectors
were none too sensitive. For laboratory measurements, it was possible to obtain bright
light sources and there was little restriction on the length of time available to make the
measurement of the spectrum. The lack of sensitive detectors prevented many astronomical observations from being made in the 'far', i.e. non-photographic, infrared for
a number of years.

As one might expect, the Sun was the first astronomical object to have its spectrum
measured. Most of the strong absorption features in the infrared solar spectrum are
due to molecules in the Earth's atmosphere. Lemansky (1871) used a thermopile to

measure the energy in the solar prismatic spectrum and attributed the three absorption features which he recorded to the Earth's atmosphere. The first systematic analysis of the near infrared solar spectrum was begun by Langley in 1881 and completed in 1900 (Langley and Abbot, 1900). The solar spectrum was measured between 0.8 and 5 μ. Adel *et al.* (1935) extended the spectral coverage from 5 to 21 μ.

Kuiper (1947) was the first to measure the spectrum of Venus in the 'far' infrared. Kuiper was able to make measurements beyond 2 microns and his early spectra revealed 'nine strong CO_2 bands'. The McDonald Observatory report (Struve, 1948) notes that "Herzberg is engaged in the study of CO_2 with the long absorption tube. The bands found by Adams and Dunham as well as the new bands reported last year, in Venus, have been reproduced in the laboratory with appropriate pressures. The amount of CO_2 above the reflecting layer of Venus corresponds to at least 2200 m at atmospheric pressure. Since the intensity distribution within the bands, as photographed in the tube is similar to that observed in Venus, the appropriate layer in the atmosphere of Venus has about room temperature."

Kuiper (1949) continued his pioneering observations of Venus in the infrared, at McDonald Observatory, and found many more carbon dioxide bands in his spectra. He discovered the systematic variation in the CO_2 absorption band at 8689 Å; it is weakest near inferior conjunction ($i = 180°$) and increases in strength by almost a factor of 10 near superior conjunction ($i = 0°$). Kuiper (1952 also reported "that day-to-day fluctuations of considerable magnitude occur and that the observed distribution of CO_2 is often remarkably patchy over the disc. The patches are of such a size that they may correspond to the cloud features shown on ultraviolet photographs."

"Evidently, the Venus cloud layer is in violent motion ... It is recalled that the radiometric measures by Pettit and Nicholson have shown the dark side of Venus to emit nearly as much infrared radiation as the sunlit side, which also points to a vigorous convection in the Venus atmosphere."

2.3. The search for minor constituents in the atmosphere

Kozyrev (1954a, b, 1969) reported on two emission features which showed up in the violet region of the spectrum of Venus and also in the spectrum of smoke from volcanic eruptions. Warner (1960) suggested that some of the features noticed by Kozyrev occurred at the same wavelengths as known bands of N_2, N_2^+, O and O^+ and could be due to these species. However Newkirk (1959), Heyden *et al.* (1959), Richardson (1960), Spinrad (1962a–e), and Owen (1968a, b) searched for the emission features in their spectra of Venus and failed to find them. Spectroscopic searches for O_2 and oxides of nitrogen on Venus have proven fruitless which suggests that Warner's proposed identification of the Kozyrev bands may have little to do with the atmosphere of Venus. The bands are certainly seldom present in the spectrum of Venus and are possibly spurious.

Kuiper (1952) reported an upper limit of 100 cm atm for N_2O based on his infrared spectra. The NO_2 molecule absorbs in the visible region of the spectrum but these bands (Herzberg, 1966) have not been observed in the spectrum of Venus.

2.4. SCATTERING IN THE ATMOSPHERE OF VENUS

The next report on the spectrum of Venus (Chamberlain and Kuiper, 1956) had to do with the determination of the rotational temperature of a carbon dioxide band. Earlier estimates of the 'cloud-top' temperature had been based on visual comparisons of the Venus spectrum with laboratory spectra of carbon dioxide (Adel, 1937; Herzberg, 1951). Dunham (1949) had measured the intensities of the CO_2 lines in the Venus spectra photographed at Mt. Wilson Observatory and had estimated that the temperature was $T = 300 \pm 50$ K. Chamberlain and Kuiper (1956) say, "In these earlier estimates it was assumed in all cases that the relative intensities of the lines were proportional to the relative populations, N_J, of the lower levels of the transition. Van de Hulst (1952, p. 102) has pointed out, however, that in an optically thick atmosphere, where scattering and absorption are both important, this simple relation would not be expected to hold true. In particular, for an optically thick planetary atmosphere scattering light isotropically, the line absorption for weak lines is proportional to $(N_J)^{1/2}$." This was the first suggestion that the interpretation of the CO_2 absorption bands in the spectrum of Venus, could be much more complicated than their interpretation in laboratory spectra. Of course, the presence of clouds on Venus was well known, but here-to-fore it had generally been assumed that the measured absorption referrred to the amount of gas above the clouds; the clouds on Venus had been regarded as quite dense, so that most of the absorption occurred above the cloud deck and the infrared radiation did not penetrate very far into the clouds themselves.

On the basis of their square-root absorption law, and a slightly faulty relationship between the equivalent width and the intensity of rotational lines, Chamberlain and Kuiper (1956) found a mean temperature of $T_{rot} = 285 \pm 9$ K (p.e.). They remarked that "The most important systematic errors in T_{rot} probably arise from our neglect of the variation of oscillator strength with the different lines in a band and from departures of the absorption law from the asymptotic relation" $W(J) = \text{const} (N_J)^{1/2}$. Chamberlain (1965) subsequently estimated that use of the correct expression for the oscillator strength (or line intensity) "would lower all 'radiative transfer' values of T_{rot} by about 7% or 20 K. The correction to the 'simple reflection' values of T_{rot} would be about twice as great."

It was assumed, at that time, that any interpretation of a spectrum which postulated the clouds acted as a reflecting layer would have the absorption linearly proportional to the number density of absorbing molecules. That is, the equivalent width of a line, $W(J)$, was equal to the product of the line intensity $S(J)$ and the amount of absorbing gas, w:

$$W(J) = S(J) w. \tag{1}$$

It was well known from laboratory studies that this relation only held in certain situations. If the pressure, for line formation, is high enough that Doppler broadening can be neglected, then the lines are broadened by intermolecular collisions and have a dispersion line contour or Lorentz line shape. The spectral absorption coefficient of the line, $\varkappa_\omega(J)$, is given by

$$\varkappa_\omega(J) = [S(J)\,\gamma(J)/\pi]\,[(\omega - \omega_J)^2 + \gamma(J)^2]^{-1}, \tag{2}$$

where ω_J is the frequency (or wavenumber) of the center and $\gamma(J)$ is the halfwidth of the line at half of its maximum depth. Ladenburg and Reiche (1913) showed that the equivalent width of a Lorentz line is given by two asymptotic relations:

$$W(J) \simeq S(J)\,w \quad \text{for} \quad x(J) < 2/\pi \tag{3a}$$

and

$$W(J) \simeq 2[S(J)\,\gamma(J)w]^{1/2} \quad \text{for} \quad x(J) > 2/\pi, \tag{3b}$$

where

$$x(J) = [S(J)w/2\pi\,\gamma(J)].$$

The line intensity $S(J)$ is directly proportional to the number of molecules per unit volume, N_J, times the oscillator strength, and the line halfwidth $\gamma(J)$ is directly proportional to the total pressure. Depending on the pressure for line formation, absorption lines can follow either a linear or a square root absorption law *in the absence of scattering.*

This fact was largely ignored in the early interpretation of Venus spectra and has caused a certain amount of confusion in the literature. The idea that scattering particles would increase the effective absorption path in the atmosphere of Venus was sufficiently novel that some people assumed that the old laws of gaseous absorption no longer applied. A new brand of physics was needed to interpret the observations. Radiative transfer theory was called into play. Without going into all the mathematical details (which are sufficiently complicated that the physical processes involved are sometimes forgotten!) we will briefly summarize the situation. The transfer equation involves I_ω the radiant flux in the frequency interval between ω and $\omega + d\omega$ per unit projected area per unit solid angle; I_ω is termed the (spectral) intensity of radiation. (Note that this is a different use of the word intensity than is meant by the intensity, or strength, of a spectral line). The change in radiation intensity in the direction S is given by

$$\frac{dI\omega}{ds} = -(\varkappa_\omega + \sigma_\omega)\,I_\omega + \varkappa_\omega B_\omega + \sigma_\omega \int I_\omega \frac{d\Omega}{4\pi}. \tag{4}$$

Here σ_ω is the (spectral) scattering coefficient, B_ω is the Planck function, and Ω is the solid angle. For a plane-parallel atmosphere it is customary to use the geometrical depth measured from the top of the atmosphere,

$$dz = -ds\cos\theta,$$

where θ is the direction of radiation measured from the outward normal to the atmosphere, and to introduce the variables $\mu = \cos\theta$, the optical depth τ,

$$d\tau = (\varkappa_\omega + \tau_\omega)\,dz$$

and the dimensionless quantity $\tilde{\omega} = \sigma\omega(\varkappa_\omega + \sigma_\omega)^{-1}$. Then (4) can be written as

$$\mu \frac{dI_\omega}{d\tau} = I_\omega + (\tilde{\omega} - 1) R_\omega - \tilde{\omega} \int I_\omega \frac{d\Omega}{4\pi}. \tag{5}$$

For isotropic scattering, the flux integral in (5) reduces to

$$\int I_\omega \frac{d\Omega}{4\pi} = \tfrac{1}{2} \int I_\omega \, d\mu, \tag{6}$$

(but what real cloud behaves as an isotropic scatterer?). The basic idea of the Milne-Eddington approximation (Kourganoff, 1952) is to assume that the angular dependence of the intensity, I_ω, can be expressed in terms of a series of Legendre polynomials, $P_l(\mu)$. These form a complete set of orthogonal functions in the interval $(-1, 1)$, which is just the interval through which μ varies. The reason for doing this is to obtain closed-form solutions to the equation of transfer in terms of functions that are well known. There is a problem with this approach, however. The function I_ω is discontinuous and trying to represent it by a *finite* sum of Legendre polynomials is clearly impossible. Chandrasekhar (1950) tried to get away from the difficulty by introducing the method of discrete ordinates. He chose to fix the determination of $I_\omega(\tau, \mu)$, at the optical depth τ, to $2n$ points corresponding to $2n$ discrete values of μ_i which are more or less regularly distributed in the interval $-1 < \mu < 1$. This results in $2n$ linear differential equations for the $2n$ unknown functions $I_\omega(\tau, \mu_i)$. These can be integrated and the constants of integration are found from the boundary conditions.

   The observational data of Kuiper had shown that the equivalent width of absorption lines decreased as the Venus phase angle increased. In order to give a theoretical explanation to this effect, which is contrary to what would be expected for a reflecting cloud layer, Chamberlain and Kuiper (1956) utilized a solution to the equation of transfer obtained by Chandrasekhar (1950). This solution referred to a homogeneous planeparallel atmosphere which scattered radiation isotropically and had an albedo for single scattering $\tilde{\omega} < 1$. The intensity for diffuse reflection was given by (Chandrasekhar, 1950; p. 85) as

$$I(0, \mu) = \tfrac{1}{4}\varpi F \frac{\mu_0}{\mu + \mu_0} H(\mu) H(0), \tag{7}$$

where F is related to the net flux of radiation by

$$\pi F = \int I \cos\theta \, d\Omega.$$

The H-functions had been studied by Chandrasekhar. They are continuous functions which increase monotonically from $H(\tilde{\omega}, 0) = 1$ to $H(\tilde{\omega}, 1) \approx 3$. For isotropic scattering,

$$\int_0^1 H(\tilde{\omega}, \mu) \, d\mu = \frac{2}{\tilde{\omega}} [1 - (1 - \tilde{\omega})^{1/2}],$$

and for $\tilde{\omega} \sim 1$ there is an asymptotic relation (van de Hulst, 1952, p. 102)

$$H(\tilde{\omega}, \mu) \approx \frac{H(1, \mu)}{1 + \mu[3(1 - \tilde{\omega})]^{1/2}}. \tag{8}$$

Chamberlain and Kuiper used (7) and (8) to obtain the following relation for the absorption of a spectral line relative to the continuum ($\tilde{\omega}_c = 1$):

$$A = \frac{I_c - I}{I_c} = \left[1 - \tilde{\omega} \frac{H(\tilde{\omega}, \mu) H(\omega, \mu_0)}{H(1, \mu) H(1, \mu_0)} \right] \tag{9a}$$

or

$$A \simeq (\mu + \mu_0)[3(1 - \tilde{\omega})]^{1/2} = (\mu + \mu_0)[3\varkappa/\sigma]^{1/2}. \tag{9b}$$

This is how they arrived at the relation that the absorption in a scattering atmosphere should depend on the square root of the line intensity (or absorption coefficient). Chamberlain (1965) then considered the case where the continuum albedo differed from unity. This resulted in the following expression for the line absorption:

$$A \simeq 3^{1/2}(\mu + \mu_0)[(1 - \tilde{\omega})^{1/2} - (1 - \tilde{\omega}_c)^{1/2}]. \tag{9c}$$

He noted that ".. so long as the continuous absorption ($\varkappa_c$) is less than that due to the gas alone ($\varkappa$) (i.e. when $1 - \tilde{\omega}_c < \tilde{\omega}_c - \tilde{\omega}$), we may neglect $1 - \tilde{\omega}_c$ with little error. But when the inequality becomes reversed far out in the wings (of a line), the term in square brackets in (9c) rapidly diminishes." McClatchey (1967) showed that when the line absorption coefficient $\varkappa$ is much smaller than the continuous absorption $\varkappa_c$ the line absorption is given by

$$A \simeq 3^{1/2}(\mu + \mu_0) \frac{\varkappa}{2\varkappa_c} \left(\frac{\varkappa_c}{\sigma} \right)^{1/2}. \tag{9d}$$

McClatchey remarked, "Equation (9d) indicates that the absorption is a linear function of the absorption cross-section when $\varkappa \ll \varkappa_c$. This inequality means that the absorption in the line is much less than that due to the continuous absorption – a situation always realizable in practice for the weakest lines." The inequality of (9b) where $\varkappa_c \ll \varkappa$ "on the other hand can only be valid in the case of weak absorption lines when $\varkappa_c$ is itself a very small number. In the limit of weak lines (i.e. $\varkappa \to 0$) we must have $\varkappa_c \to 0$. But $\varkappa_c \to 0$ means that $\tilde{\omega}_c \to 1$." McClatchey concluded: ... "in the physically realistic case of $\tilde{\omega}_c < 1$, the square-root dependence on the absorption cross-section is not an asymptotic limit but rather a cloudy transition region... even though the absorption at the line center might correspond to a portion of the curve where it deviates from linearity (say $A = 0.30$), the equivalent widths of the lines would be expected to depart only very slightly from a linear law because most of the area under the absorption curve would result from absorption in the line wings where the monochromatic absorption does follow the linear law. Under this circumstance, it would seem most accidental if the equivalent widths of weak absorption lines in a scattering and absorbing atmosphere fell on the square root portion of a curve of growth."

The discussion of the absorption law (or laws) for line formation in cloudy atmospheres has continued up to the present date and we shall return to it later.

2.5. MORE OBSERVATIONS IN THE 'FAR' INFRARED

Reports of spectroscopic studies of Venus continued to be rare in the astrophysical literature. Four years after the publication of Kuiper's observations in the photographic infrared (Chamberlain and Kuiper, 1956), Sinton and Strong (1960) presented the results of their spectroscopic study, made in 1953 and 1954, of Venus at longer

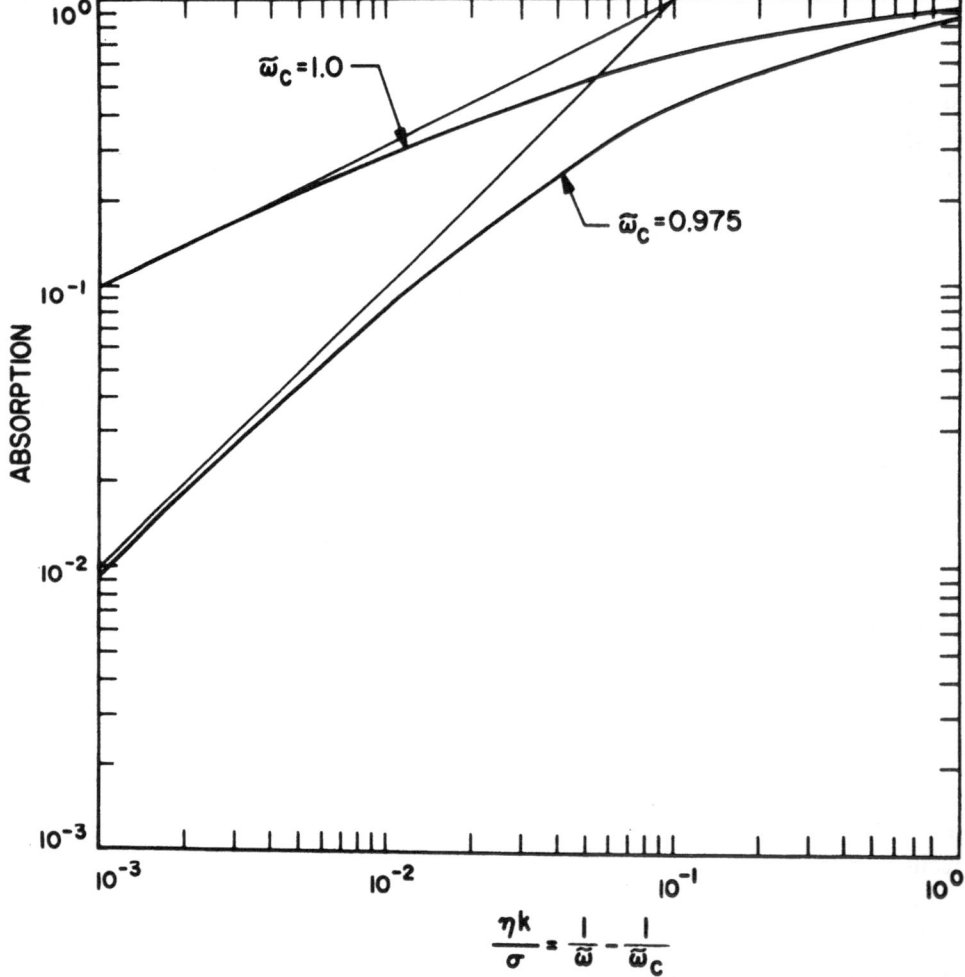

Fig. 8. Curves of growth for monochromatic absorption in a scattering atmosphere. The square-root asymptotic limit is shown for the continuum albedo equal to 1.0; the linear asymptotic limit is shown for the continuum albedo equal to 0.975. (From McClatchey, 1967, *Astrophys. J.* **148**, L93.)

wavelengths, between 8 and 13 μ. Sinton and Strong remarked that "Prior to obtaining these spectra it was expected that we would find the absorption bands of carbon dioxide at 9.4, 10.4 and 12.6 μ very intense in the spectrum of Venus. ... The spectrum shows no evidence for the 9.4 and 12.6 μ bands and the depression at 11 μ extends to too long a wavelength to be associated entirely with the band at 10.4 μ. ... A diffuse band at 11.2 μ was found in the spectrum of Venus in addition to a carbon dioxide

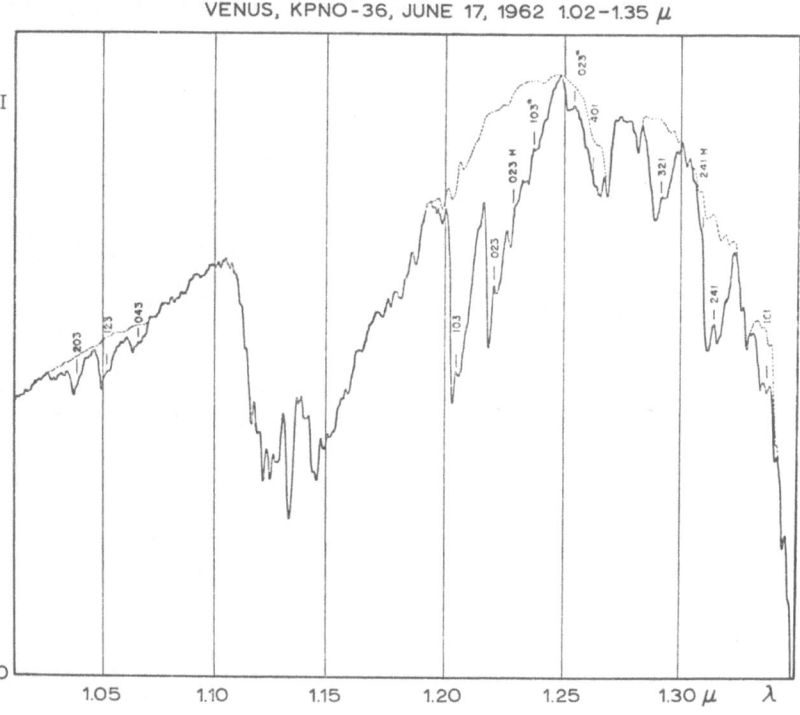

Fig. 9. Spectrum of Venus between 1.02 and 1.35 μ. The dashed upper curve shows the solar comparison spectrum, where it is different from the Venus spectrum. Carbon dioxide bands are identified by the quantum numbers of the upper state; an H indicates a 'hot' band; the C^{13} isotopic bands are indicated by an asterisk. (From Kuiper, 1962, *Comm. Lunar Planet. Lab.*, No. 15.)

band at 10.4 μ. The 10.4 band was found much weaker than expected." Their failure to observe some of the expected CO_2 bands let Sinton and Strong to postulate "That the radiation is being absorbed by a gray material above most of the CO_2. The low radiometric temperature compared with the higher temperature corresponding to the top of the cloud level obtained by Chamberlain and Kuiper (1956) and earlier by Adel (1937), also indicates that the radiating level is above the cloud top. ... The absorption may not really be gray but may be due to gases that just fill in the absorption around the CO_2 bands and perhaps have somewhat stronger absorption at 11.2 μ."

The first comprehensive study of the infrared spectrum of Venus was published by Kuiper (1962). It covered the spectral region 1.0–2.5 μ. Kuiper noted that "The Venus spectra give information on (a) the CO_2 abundance on Venus; (b) the $^{13}C/^{12}C$ ratio; (c) the $^{18}O/^{16}O$ ratio; (d) the hot bands." Kuiper found an abundance of "about 2 km atm of CO_2. ... However, since the strength of the CO_2 absorption on Venus is variable, both with the phase and from day-to-day, and even region to region on the planet (Kuiper, 1952), the amount of CO_2 found only applies to the date of observation." The isotope ratios $^{13}C/^{12}C$ and $^{18}O/^{16}O$ were "equal to that on the Earth

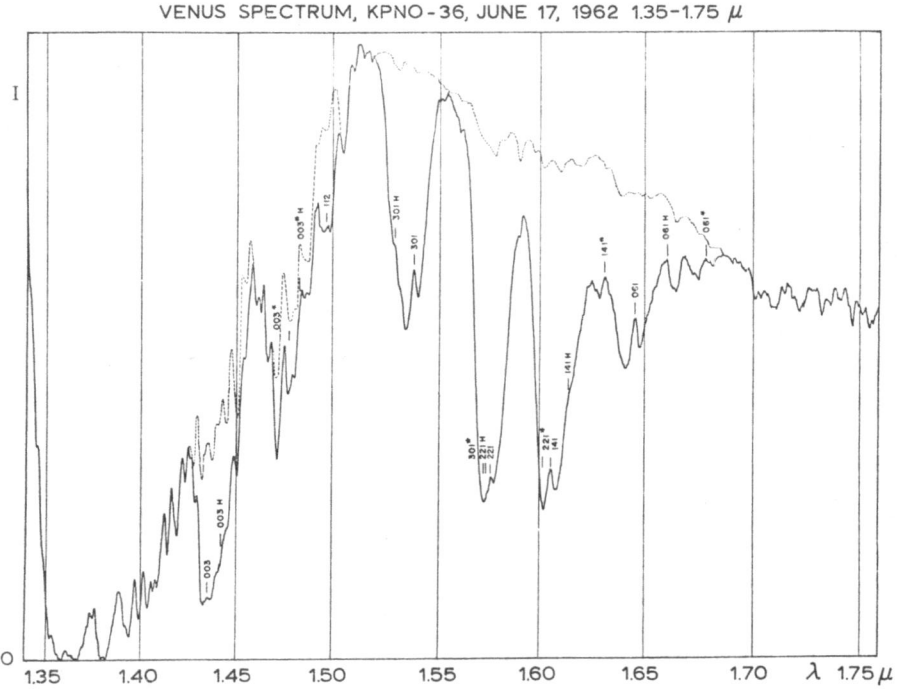

Fig. 10. Spectrum of Venus between 1.35 and 1.76 μ. (From Kuiper, 1962, *Comm. Lunar Planet. Lab.*, No. 15.)

within the error of measurement" which Kuiper estimated to have a precision of $\pm 10\%$. So-called 'hot' bands are those whose lower vibrational state is an excited one, rather than the ground state. As the temperature in the laboratory is increased above room temperature, the population of the excited vibrational states is increased relative to the ground state. This causes the intensity of the 'hot' bands to be enhanced and the intensity of the ground state bands to be diminished. As a result, the hotter the gas becomes, the more the absorption due to the hot bands increases and they appear much more prominently in the spectrum than they do at room temperature. Kuiper's (1962) measurements of the CO_2 hot bands suggested "that the Venus temperature is considerably higher than the laboratory temperature." Subsequent measurements made

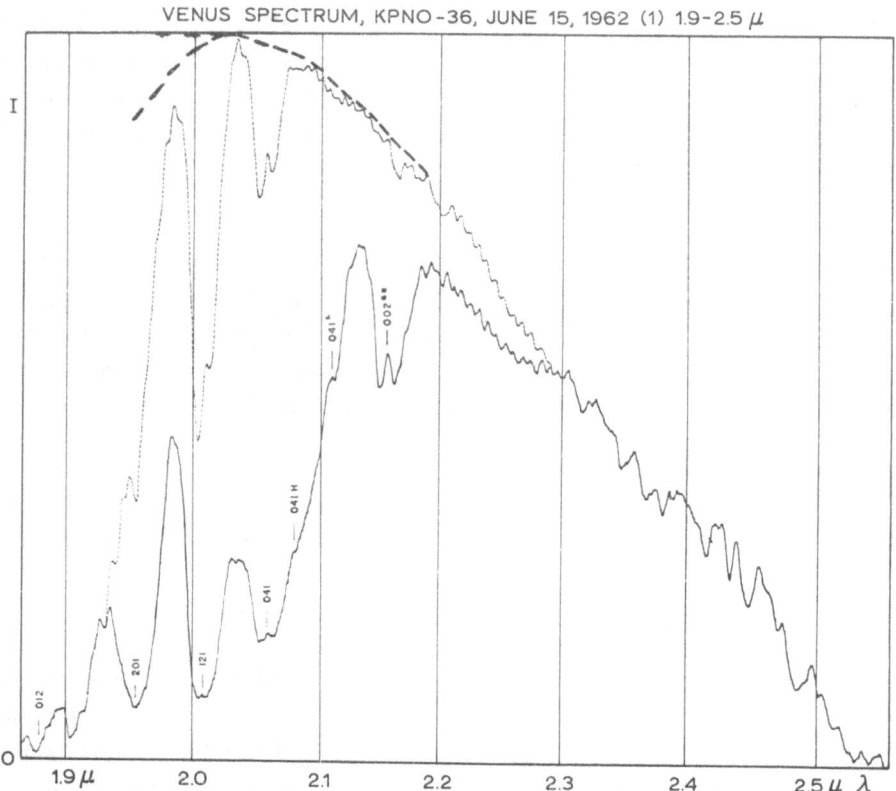

Fig. 11.   Spectrum of Venus between 1.9 and 2.6 μ. The solar comparison spectrum is indicated and
an estimate of the solar continuum is also shown. (From Kuiper, 1962,
Comm. Lunar Planet. Lab., No. 15.)

at higher resolution (Connes *et al.*, 1967) failed to confirm this suggestion when they
revealed that the hot bands were in many cases blended with isotopic CO_2 bands. The
blending results in an over-estimate of the equivalent width of the hot bands and
hence an over-estimate of the temperature. Kuiper was unable to detect the carbon
monoxide band at $\lambda 2.35$ μ and reported an "upper limit of 10 cm atm$_{stp}$, implying an
approximate upper limit of one-fourth of this amount for a vertical column in the
Venus atmosphere" (or a mixing ratio of less than 50 parts per million).

2.6. SIMPLE INTERPRETATIONS OF THE SPECTRUM OF VENUS

There was one peculiarity about the spectrum of Venus that was particularly trouble-
some. Herzberg (1951) had reported that the strong bands near 1.6 μ are roughly
matched by 88 m atm of CO_2 at a pressure of 1 atm in the laboratory but more than
1400 m atm of CO_2 were needed to match the weaker bands near 1.05 μ. Kaplan
(1961) believed this discrepancy was "mostly due to the fact that the laboratory
measurements were made at 1 atm, which is an order of magnitude higher than the

mean pressures we will later derive for the atmosphere above the reflecting layer."
Kaplan assumed that the individual lines in the 1.6 μ absorption band followed a
square-root absorption law and that the overlapping of adjacent lines was small. This
meant that the absorption by the entire band should follow a square-root absorption
law. Kaplan found that the laboratory measurements of Howard *et al.* (1951) of the
CO_2 band at 1.6 μ could be fit by

$$W = 8.4 [(wp)(1 + 0.3f)]^{1/2}, \tag{10a}$$

where W is the equivalent width of the band, in cm^{-1}; w is the amount of CO_2, in
m atm; p is the pressure, in atm, and f is the fraction of CO_2 by volume. For a uniformly mixed gas in a planetary atmosphere the appropriate pressure is one half that
at the reflecting layer ($p = p_r/2$) and the absorption path length $w = 2u \sec\theta$, where u
is the amount in a vertical column. Thus (10a) became for Venus

$$W = 8.4 [up_r \sec\theta (1 + 0.3f)]^{1/2}. \tag{10b}$$

Kuiper's (1947) observations indicated that the equivalent width of the 1.6 μ band
was at least as large as $W = 150 \ cm^{-1}$. Assuming $\sec\theta = 2$, Kaplan (1961) found

$$up_r = \frac{(150)^2}{2(8.4)^2 (1 + 0.3f)} = \frac{159}{1 + 0.3f}.$$

Assuming the atmosphere of Venus was composed of pure CO_2 ($f = 1$), then the
amount of CO_2 can be related to the cloud top pressure by $u = 5.93 \times 10^3 \ p_r$ meters.
This gave Kaplan a lower limit to the reflecting layer pressure of $p_r \geqslant 0.14$ atm (or
140 mb and $u \leqslant 830$ m). Kaplan argued that the temperature of 235 K measured by
Sinton and Strong (1960) is "a reasonable value for the cloud-top temperature." He
used the temperature of 285 K reported by Chamberlain and Kuiper (1956) as the
"temperature at a level having half the pressure of the effective reflecting surface."
Assuming an adiabatic lapse rate between the reflecting surface and the cloud top,
Kaplan concluded "the pressure at the cloud top will be about one-fourth of p_r."
Kaplan then went on to find the mixing ratio of CO_2 and the surface pressure.
He used the then current microwave measurements of 585 K for the surface temperature and de Vaucouleurs and Menzel (1960) measurements of the occultation of
Regulus. These data plus 'a reasonable guess' led Kaplan to conclude that the cloud
top pressure is "about 90 mb", the CO_2 concentration is "about 15% by volume",
"the surface pressure is of the order of two atmospheres and the total CO_2 (is) of the
order of 2 km atm$_{stp}$." While Kaplan's conclusions based on other data have subsequently proved to be faulty, the results of his interpretation of the spectra are not
all that different from current results.

   The first attempt to seek observational evidence as to whether the clouds of Venus
behaved as a scattering haze or as a reflecting layer was made by Spinrad (1962a). His
investigation "was originally motivated by interest in a possible variation of the CO_2
rotational temperature with Venus phase." To do this, Spinrad used "the ten best old
Mt. Wilson 100-in. coudé spectrograms of Venus" taken by Adams and Dunham in

the 1930's; the plates covered Venus phase angles from $51°$ to $113°$. Spinrad found rotational temperatures for the CO_2 band at 7820 Å in two ways: first, he used "the ordinary Boltzmann equation method with the assumption of W being directly proportional to $N(J)$, (which is) obviously true for weak lines"; second, he used "a radiative transfer scheme of Chamberlain and Kuiper where $W \sim [N(J)]^{1/2}$." Spinrad also made the faulty assumption of "equal oscillator strengths for all the J lines." For the linear absorption law, Spinrad obtained rotational temperatures ranging from 214 K to 445 K. For the square-root absorption law, he found temperatures ranging from 142 K to 433 K. Spinrad noted that, "In general the rotational temperatures given by the usual application of the Boltzmann equation are somewhat higher than those derived with the radiative transfer modification. ... If a choice of (models for) rotational temperatures is to be made at this time, the decision must be made on rather arbitrary empirical grounds. We shall reject the method which gives rotational temperatures far different from any other temperatures found for Venus by other means." Since the highest temperatures Spinrad had found from both of the two absorption laws were considerably colder than the microwave results for the 'surface temperature' (of ~ 600 K), he chose to reject the absorption law which yielded temperatures lower than the coldest reported 'measurement' of temperature. Menzel and de Vaucouleurs (1960) had derived a temperature of $T = 210$ K from their occultation data, based on the assumption that the atmosphere of Venus consisted solely of N_2. Spinrad argued that "The radiation transfer rotational temperatures... fall well below 210 K on two occasions, so on these grounds of incompatibility we shall rule against the Chamberlain-Kuiper method... ." Spinrad estimated that "both systematic and internal errors in the rotational temperatures... are probably smaller than ± 50 K... ."

He remarked, "One more very crucial point remains to be mentioned. The Venus CO_2 lines are lines whose total absorption is integrated over (an atmospheric) region with a rather large temperature gradient... Each line will be partially weighted by the high, cool carbon dioxide gas and also by the lower, hotter layers of the Venus atmosphere... The effect is to make the rotational temperature an average quantity which applies directly to some unknown, but intermediate level in the CO_2 absorbing region."

Spinrad next attempted to determine the pressure where the lines were formed. Making the usual kinetic theory assumption about the dependence of the line half-width γ on temperature and pressure,

$$\gamma = \gamma°(p/p_0)(T_0/T)^{1/2}, \tag{11}$$

Spinrad used the value of $\gamma°$ appropriate for nitrogen-broadened CO_2 lines. This was because "Kaplan (1961) finds the CO_2/N_2 mixing ratio to be about 0.2." Spinrad measured the apparent halfwidths of the CO_2 lines on the ten Mt. Wilson spectrograms. He made corrections "for the finite slit-widths of the spectrograph and microphotometer... The largest corrections to the halfwidths were about 15%." Young and Young (1972) have suggested that Spirad's failure to also correct for the point-spread function of the photographic plates led him to believe that the apparent halfwidths were quite similar to the true halfwidths.

Spinrad's familiarity with stellar spectra had caused him to make two assumptions which are valid for stellar spectra and very dangerous for planetary spectra: Stellar lines tend to be fairly broad and have a comparatively large halfwidth, hence the observed line profile is fairly close to the true line profile; in the planetary case, the lines are usually much narrower than the spectral slit width and the observed line profile is more likely to resemble the slit function than the true line shape. Secondly, a stellar line with an equivalent width of 25–100 mÅ will have an absorption that varies linearly with the amount of gas over the entire line profile; a planetary line of the same equivalent width may be strongly saturated (i.e. the absorption may increase with the amount of gas at a much slower rate than a linear one), near the center of the line profile. As a result, the halfwidth of the actual absorption line profile $A_\omega = 1 - \exp(-\varkappa_\omega \omega)$ (measured as the width of the line at half the maximum absorption depth) may be much larger than the halfwidth of the absorption coefficient, $\varkappa_\omega$.

Spinrad reported, "the pressure at the bottom of the Venus CO_2 path averages about 7 atm, but may occasionally reach 10 atm." He also found "Venus to have about 2 km atm of carbon dioxide" in the absorption path. "The *partial* pressure of 2 km atm of CO_2 is $\frac{1}{3}$ atm. The average *total* pressure down to the bottom of the CO_2 absorbing layer is about 7 atm. Thus the CO_2 abundance by mass, and presumably the CO_2/N_2 mixing ratio is about 5%. This estimate is uncertain by a factor of 2. Carbon dioxide appears to really be only a minor constituent of the Cytherian atmosphere."

Prior to the papers of Kaplan and Spinrad, carbon dioxide has appeared to be the major atmospheric constituent (since it was the only molecule that has been definitely

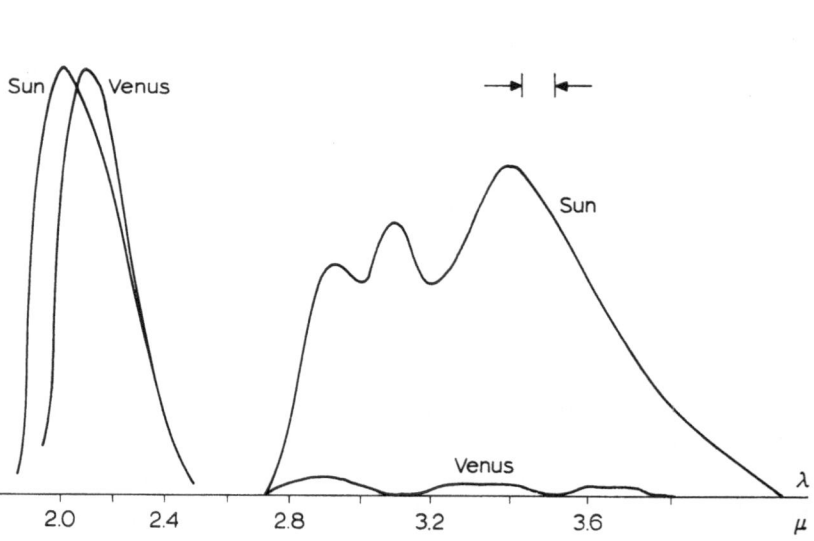

Fig. 12.   Prism spectrum of Venus between 2.0 and 3.8 μ. (From Moroz, 1964b, *Mem. Soc. Roy. Sci. Liège* **9**, 406.)

detected on Venus). When CO_2 was 'shown' to be a minor constituent, the possibility that the atmosphere of Venus was similar to that of the Earth was revived, if only O_2 or water vapor could be detected.

2.7. The search for H_2O and other constituents in the atmosphere of Venus

In another effort to detect H_2O lines on Venus by the Doppler shift method, Spinrad (1962b, c) selected "an excellent old high-dispersion spectrogram of Venus taken by Adams and Dunham at the 100-in. coudé focus;" the spectrogram was taken when the Venus lines would be Doppler shifted -0.37 Å from the telluric water lines, at 8180 Å. From his analysis of the CO_2 band at 7820 Å on this same plate, Spinrad found that it "refers to a region deep in the Venus atmosphere. This would tend to maximize the possibility of detecting any Cytherian H_2O lines. The rotational temperature of the λ7820 CO_2 band is $T_{rot} = 440$ K (Spinrad, 1962a) and the average pressure in the CO_2 absorbing layer is about 4 atm. ... Presumably, any very weak hypothetical Venus H_2O lines would also originate at a deep atmospheric level."

Adams and Dunham had taken the Venus plate on 23 March 1940. No Doppler-shifted H_2O line due to Venus could be detected. Spinrad estimated that the upper limit on any Cytherian H_2O component corresponded to only one part in fifty of the number of molecules producing the telluric H_2O line. Using "the Weather Bureau upper air data for their San Diego station indicates 7 mm precipitable H_2O above 5000 ft on that date. Since the Venus spectrogram was taken at $\overline{sec}\, z = 2$, we find the telluric H_2O lines on this plate were produced by 1.4 gm cm^{-2}." The old Mt. Wilson plate had been taken on a very wet day! Spinrad assumed "the solar radiation path equals about 4 thicknesses of the Venus atmosphere. Thus, the upper limit to the Venus water vapor content is really $(1/50) \times (1/4) = (1/200)$ that over Mt. Wilson or 7×10^{-3} gm cm^{-2} (70 μ)." Since Spinrad had found a base pressure of 8 atm or 8000 gm cm^{-2} from his analysis of the CO_2 lines, he estimated the mixing ratio of water might be less than 9×10^{-7} by mass. He concluded "water vapor is extremely difficult to detect in the Cytherian atmosphere at least down to a region where P is 8 atm."

Spinrad (1962b) also "made a careful search of the near ultraviolet and part of the blue region of the Cytherian spectrum ... Venus shows only the solar reflection spectrum ... The present philosophy of 'groping in the dark' for spectroscopic evidence of absorbers other than CO and CO_2 in the Venus atmosphere seems a painstaking but necessary task. The work will continue."

Spinrad's detailed examination of the Venus spectrograms in the Mt. Wilson Observatory plate files only revealed one new absorption feature: a carbon dioxide band with a head at 7158 Å (Spinrad, 1962c). Once again, the old Mt. Wilson plates indicated the presence only of CO_2 in the atmosphere of Venus.

Kaplan (1962) examined some of the microdensitometer tracings that Spinrad had made of the Venus plates taken by Adams and Dunham. He suggested that they showed two intensity maxima in the P branch: one corresponding to an effective temperature of 300 K and the other to an effective temperature of 700 K. "The obvious cause of the double maxima is a cloud layer at some level whose temperature is considerably in excess of 300 K, say at 350 to 400 K. ... The cloud, and the gas from which it presumably condenses, are of considerable importance in providing part of the green-house effect that is necessary to maintain surface temperatures in excess of 700 K." Supposedly, some of the sunlight was being reflected from the surface of Venus while the rest was reflected by her clouds. Young and Young (1972) showed that it was theoretically impossible to achieve a double maximum in the P branch of a CO_2 band for two *isothermal* layers of the gas unless the temperature ratio of the two layers exceeds ~9. In a planetary atmosphere, with a continuous variation in the tempera-ture between the planet's surface and the top of the atmosphere, the appearance of a double maximum would be very unlikely. Young and Young made statistical tests on the data for which Kaplan had reported two maxima and concluded, "The 'double maxi-mum' phenomenon appears to be exclusively associated with data having a relatively poor signal to noise ratio ... The effect reported by Kaplan was due to noise, and ... it is impossible to detect temperatures as high as 700 K in the atmosphere of Venus by means of ground-based observations of the 7820 Å CO_2 band."

Dollfus (1963a, b) reported an apparent detection of water vapor in the atmosphere of Venus on 21 January 1963, "Les comparaisons photométriques de la bande 1.4 μ de la vapeur d'eau sur Vénus et la Lune en haute montagne en hiver révèlent au moins 10^{-2} g cm^{-2} de vapeur d'eau au-dessus de la couche nuageuse éleveé de Vénus. Les voiles nuageux peuvent donc être de la glace ou de l'oeau". Kuiper and Forbes (1967) suggested that "The Dollfus result might be due to residual CO_2 absorption entering his 1.4 μ filter." Dollfus had found 280 μ of precipitable water in the total absorption path for Venus; using an air mass of 4, this corresponded to 70 μ in the vertical path.

Sinton (1962a, b, 1963, 1964a, b) reported "Spectra of Venus that were obtained at Lowell Observatory gave an indication that CO was present in this planet's atmosphere although the band was rather weak. This question is still not resolved because the shape of the observed band indicates that the CO is at a considerably lower temperature (80 K) than most people would assume is present – even high in the Venus atmosphere." The distortion in the shape of the CO band in the Venus spectra was due to unresolved blends in these spectra. The observations of Connes *et al.* (1968) at much higher reso-lution than Sinton's measurements revealed no peculiarities in the contour of the band; "Analysis of the lines of the principal isotopes shows that they are formed at cloud level, with an effective temperature of 240 K and a total effective CO path of 13 cm amagat. Comparison with CO_2 lines yields a CO/CO_2 ratio of 45 parts per million."

In 1962 and 1963, Moroz (1964a, b) measured the infrared spectra of Venus using the 50-in. reflector of the South Station of the Astronomical Institute in Crimea. Moroz also reported the detection of carbon monoxide (at 2.35 μ) in his spectra: "This depression is equivalent to approximately 4 cm atm in the laboratory spectra. Con-sequently, the CO abundance in a vertical column of the atmosphere of Venus above the cloud layer must be about 1.5 cm." Moroz (1964a, b) remarked that "The intensity of Venus in the region 3.5 μ was found still lower than in Sinton's (1961) previous ob-servations. The minima at 3.18 and 3.57 μ seem to be real... A constituent, which produces this absorption is still unidentified (it is not H_2O), but is of great importance to the physics of Venus. As it absorbes the solar radiation, it absorbs also an intrinsic radiation of the planet. It must create great greenhouse effect, heating the planetary surface to a high temperature..." The absorption features in the neighborhood of 3.5 μ may be due to the clouds. Because of the low intensity of the Venus spectrum in

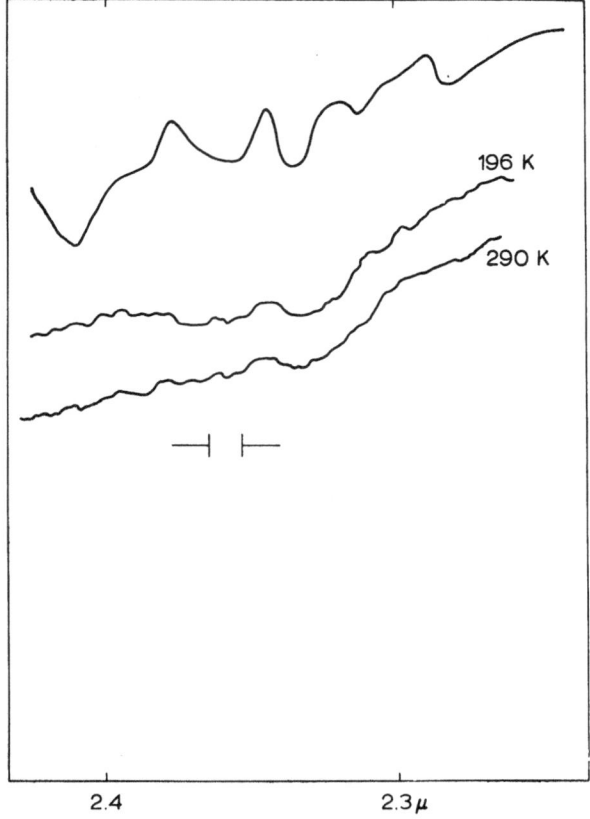

Fig. 13. Upper curve, the mean of 6 ratio spectra of Venus/Sun. The two lower curves are laboratory spectra of 79 cm atm of carbon monoxide. The spectral resolution is indicated on the figure. (From Sinton, 1963, *J. Quant. Spectrosc. Radiat. Transfer* **3**, 551).

this region, it is difficult to make a positive identification of the substance; the proposed H_2SO_4 cloud particles are in qualitative agreement with Moroz's observations.

One of the most exciting observations reported in 1964 was made using a 30-cm aperture telescope carried by balloon to 26.5 km in the Earth's atmosphere. Bottema *et al.* (1964a) measured an absorption in the H_2O band at 1.13 μ of $(10.5\pm0.5)\%$, "the same as that produced by 9.8×10^{-3} gm cm^{-2} of water vapor at atmospheric pressure." They estimated that there was "about 0.7×10^{-3} gm cm^{-2} of (telluric) water above the altitude of the balloon ... The effective slant path through the atmosphere of Vuens is calculated to be 3.82 times the vertical path." This means that if the telluric absorption is ignored, there is about 50 μ of H_2O at 1 atm in a vertical column on Venus. Because the water absorption follows an approximately square root dependence on the pressure, Bottema *et al.* estimated that there would be 222 μ of H_2O above clouds if they were at a pressure level of 90 mb or 52 μ above clouds if they were at the 600 mb pressure level. "The respective mixing ratios would be 2.5×10^{-4} and 0.87×10^{-5}. A choice between these values, or in this range, must await more knowledge about the actual pressures."

Using the same balloon telescope technique, Bottema *et al.* (1964b) "obtained a reflection spectrum of the (Venus) clouds themselves ... The result was compared with reflection spectra of liquid water, ice, silica sand, liquid formaldehyde, oil, solid CO_2,

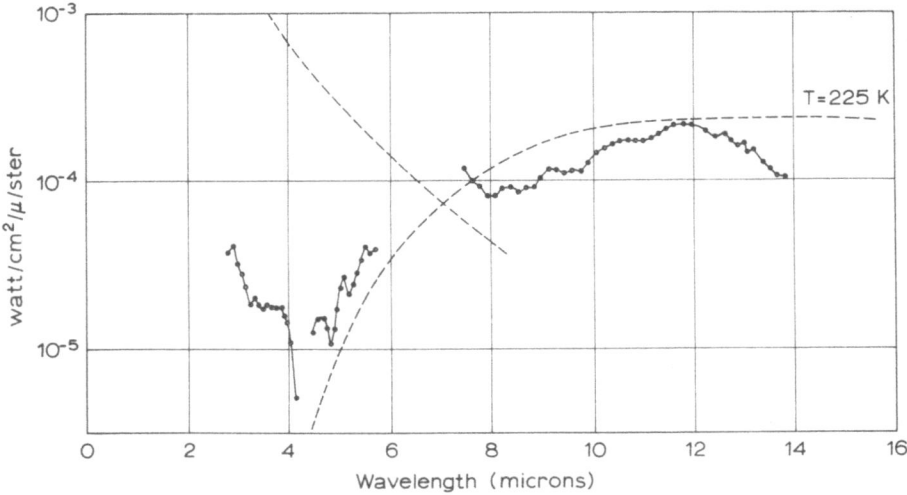

Fig. 14.   Spectrum of Venus between 2.8 and 14 μ. The Planck curve for a blackbody at 225 K is shown and the calculated intensity of reflected sunlight for an albedo of 1 is indicated on the figure. (From Gillett *et al.*, 1968, *J. Atmosphere Sci.* **25**, 594.)

frozen CO_2 cloud, frost, and clouds of water droplets and ice crystals produced at various temperatures in our laboratory." They reported that they had "found agreement with the spectra of ice crystal clouds at temperatures (Sinton and Strong, 1960) comparable to those of the clouds on Venus." Indeed, the two spectra were in qualitative agreement, but it was far from a perfect match.

The amount of water vapor Bottema *et al.* had measured was in good agreement with Dollfus' measurement and their observations made by balloon presumably had the advantage of being taken above most of the telluric water vapor so that only minor corrections needed to be made for the telluric absorption. Öpik (1965) remarked, "The great value of the Strong-Bottema balloon results cannot be enough emphasized. They seem to have proven that water vapor, though in very small quantity, is present above cloud level ..." On the other hand, Öpik disagreed with their identification of the cloud composition as being ice: "While water ice clouds have a high reflectivity in the blue-violet, Venus has a conspicuously low one; its spectral reflectivity curve in the optical region from violet to red, much better determined in absolute value than the relative figures found by Bottema-Strong in the infrared, disagrees completely with the reflectivity of water clouds, solid or liquid." As we mentioned in the section on Cloud Composition, the debate on aqueous clouds on Venus is not over yet!

2.8. MORE SCATTERING IN THE ATMOSPHERE OF VENUS

Although Spinrad's (1962a–e) measurements on the CO_2 band at 7820 Å had apparently shown that the scattering model (Chamberlain and Kuiper, 1956) yielded unrealistic temperatures for the atmosphere of Venus, Chamberlain (1965) attempted to reconcile the apparently conflicting infrared absorption data in terms of a scattering model. Chamberlain used essentially the same data as Kaplan (1961) had used with the reflecting layer model. "With the interpretation offered here, one must frankly admit that the CO_2 abundance and, to a lesser extent, the total gas pressue at the clouds are not nearly so well known as we have come to believe. But by way of compensation, the various conceptual difficulties that exist with the cloud reflection model are now readily disposed of." This time Chamberlain suggested the use of a somewhat different solution to the equation of transfer (Chandrasekhar, 1950, p. 327) than had been used by Chamberlain and Kuiper (1956). Now it referred to the case where the Planck function increased linearly with optical depth, i.e.

$$B_\omega = B^0 + B^1(1 - \varpi)t$$

which has a solution, for the emergent intensity, of the form

$$I(0, \mu) = (1 - \tilde\omega)^{1/2} H(\mu)[B^0 + B^1(1 - \tilde\omega)\mu + B^1(1 - \tilde\omega)^{1/2}\alpha_1/2], \tag{12a}$$

where $\alpha_1 = \int_0^1 H(\mu)\mu\,d\mu$.

The intensity in the continuum is given by

$$I_c(0, \mu) = B^0 + B^1\mu. \tag{12b}$$

While Chamberlain (1965) suggested that these relations could be used to find the temperature at the cloud tops, he did not use them himself.[*]

---

[*] However, one can use this formulation to derive a simple result which illustrates the effect of scattering on an absorption line profile: The residual intensity of the emergent flux in the line to the outward flux from the background continuum, i.e.

$$R \equiv \frac{F(0)}{F_c(0)} = \frac{\int_0^1 I(0, \mu)\mu\,d\mu}{\int_0^1 I_c(0, \mu)\mu\,d\mu}.$$

From (12a) and (12b) we find

$$R = \frac{(1 - \tilde\omega)^{1/2}[\alpha_1(B^0 + B^1\tilde\omega(1 - \tilde\omega)^{1/2}\alpha_1/2 + \alpha_2 B^1(1 - \tilde\omega)]}{B^0/2 + B^1/3}.$$

Chamberlain (1965) used the expression found by Chamberlain and Kuiper (1956) for the line profile:

$$S(J) = 3^{1/2}(\mu + \mu_0)(1 - \varpi)^{1/2}$$

and, after some drastic assumptions about the behavior of collision-broadened lines, arrived at the conclusion that the equivalent width is given by

$$W(J) \simeq 10(\mu + \mu_0)\,\gamma(J)\,\langle 1 - \omega_J \rangle^{1/2}. \tag{13}$$

For the limiting case of pure absorption, $\tilde{\omega}_J \to 0$, this relation is clearly wrong. Since the equivalent width had been 'shown' to depend on the square root of the absorption coefficient, and hence on the square root of the line intensity, Chamberlain suggested that the following relation should hold:

$$W(\text{Venus}) = \text{const}\,[W(\text{lab})]^{1/2}. \tag{14a}$$

The laboratory equivalent widths were assumed to follow a linear absorption law. Chamberlain used Kuiper's (1962) laboratory measurements, which covered the spectral range 1.20–2.16 μ. These spectra had been obtained for a 80 m path of CO_2 at a pressure of 4 atm (or 0.32 km atm of CO_2). The pressure was high enough that all of the bands were unsaturated, with the exception of those near 2 μ, and their equivalent width was directly proportional to the amount of CO_2. Chamberlain found that the relation

$$W(\text{Venus}) \simeq 4.5\,[W(\text{lab})]^{1/2}, \tag{14b}$$

fits the unsaturated bands in Kuiper's laboratory spectra. While Chamberlain interpreted this result as a confirmation of his suggestion that scattering caused the Venus equivalent widths to follow a square root absorption law, it could also be interpreted as showing that the Venus spectra were saturated and followed a square-root absorption law because of a low pressure for line formation. The latter interpretation yields contradictory results, however: For a Venus air mass of 10, (14b) implies a vertical CO_2 abundance of 4 km atm formed at an effective pressure of 0.4 bar; some of the weaker bands in the Venus spectrum would be unsaturated under these conditions.

Chamberlain assumed that the level of line formation for the 8689 Å CO_2 band occurred at an optical depth of $\tau = 10$ and he estimated a total CO_2 abundance of

---

For an isothermal atmosphere, $B_1 = 0$,

$$R = 2(1 - \tilde{\omega})^{1/2}\,\alpha_1.$$

The first moment of the H function varies roughly as

$$\alpha_1 \sim 0.5(1 + \tilde{\omega}^2),$$

where this approximate relation is accurate within 15 % for $0 < \tilde{\omega} < 1$. Hence the residual intensity can be crudely represented by

$$R(\tilde{\omega}) = (1 - \tilde{\omega})^{1/2}(1 + \tilde{\omega}^2).$$

As the scattering increases ($\tilde{\omega} \to 1$), the residual intensity decreases and the line becomes shallower.

5.6×10^3 cm atm. "If one took literally the assumption that the gas and particles are homogeneously mixed, then above the cloud tops ($\tau = 1$) the abundance would be only 5.6×10^2 cm atm." This was considerably smaller than the abundance reported by Spinrad, but Chamberlain cautioned, "It is to be emphasized that no reliability whatever is to be placed on these numerical values. Several arbitrary assumptions and choices of parameters are involved in the calculation, and one could justifiably choose other parameters that would give widely different values. My point here is merely that, if the CO_2 bands are formed through multiple scatterings within the clouds, the CO_2 abundance becomes very uncertain and could easily be far smaller than the values generally quoted."

The high rotational temperature found by Chamberlain and Kuiper (1956) "suggests that the 8689 Å absorption band of CO_2 is normally formed at a mean temperature some 35 or 40 K warmer than the cloud tops. This would, in a dry adiabatic temperature, occur at 4 or 5 km below the (cloud) tops at $\tau \sim 10$ in the near infrared. The fact that many scatterings are required for light to penetrate into and out of this depth means that the first scale height below the cloud top is *far* more effective in line formation than the atmosphere above the clouds."

Chamberlain also found that "Spinrad's measurements of line width, when interpreted by the theory of this paper, yield pressures of almost exactly one half the pressures (Spinrad) derived ... We at least can reaffirm Spinrad's conclusion that carbon dioxide is 'only a minor constituent of (Venus') atmosphere'."

The water vapor observations of Bottema *et al.* (1964a) presented a problem. Chamberlain decided that the relations he had used for weak CO_2 lines were not applicable to strong H_2O lines. "Hence, it may be legitimate to consider the H_2O lines as formed mainly by simple reflection above the cloud deck." The upper limit to the H_2O abundance fixed by Spinrad (1962a–e) and the high CO_2 rotational temperature reported by Chamberlain and Kuiper (1956) would be incompatible with the formation of H_2O lines at the same place in the atmosphere as the CO_2 lines. This implies that the water is not uniformly mixed in the atmosphere of Venus; it certainly isn't in the Earth's atmosphere. However, in the latter case, the same absorption laws apply to H_2O to CO_2. On Venus, things appeared to be different.

2.9. The Search for Minor Constituents Continues

Spinrad and Richardson (1965) attempted to detect oxygen in the spectrum of Venus. They noted "that the pioneering attempt of St. John and Nicholson (1922) led to an extremely severe limit of 1 m atm O_2 on Venus." Once again the oxygen lines failed to appear. "The lack of displaced Cytherian O_2 lines near the tail of the telluric A band leads to a limit of 57 cm atm O_2, less than the terrestrial abundance by a factor of 2800. This result is in disagreement with the tentative detection of Venus oxygen by Prokofiev and Petrova (1962)." Spinrad and Richardson had estimated that a Venus O_2 line in the A band would be detectable if it had an equivalent width of 8 mÅ.

Moroz (1964a, b) continued his observations of the infrared spectrum of Venus in the wavelength ranges 1.2–2.5 μ and 2.8–3.8 μ. His measurements of the absorption by

CO_2 hot bands indicated a vibrational temperature $T_v = 250 \pm 20$ K. Moroz noted that this temperature was somewhat higher than the radiometric temperature (235 K) but lower than the mean rotational temperature (~ 300 K) reported by Spinrad. Moroz's result is in excellent agreement with the best current measurements of the rotational temperature (250 K).

Moroz noticed that the absorption of strong bands in the Venus spectrum appeared to be *linearly* related to their absorption in laboratory spectra. "This means that multiple scattering does not play an important part in the formation of these bands, or in other words, the absorption occurs above the cloud layer." Moroz estimated the CO_2 abundance in a vertical column to be $u = 25$ gm cm^{-2} at a pressure of $p = 0.3$ atm, assuming a carbon dioxide mixing ratio of 0.05 based on Spinrad's (1962a–e) results. (1 gm cm^{-2} = 51 m atm. For a carbon dioxide mixing ratio of 1.0, Moroz's data for Venus implied an abundance of 0.56 km atm at a pressure of 64 mb.)

Spinrad (1966) measured a 'hot' band of CO_2 (at 8736 Å) in the spectrum of Venus and reported a vibrational temperature of $T = 400$ K. He concluded that "The weak hot band lines originate deep in the Venus cloud layer, far below the 'cloud tops' at 230 K – the usually quoted infrared temperature." It is possible that some error was made in measuring the equivalent widths of the lines in this band, since later observations (Connes *et al.*, 1967) have indicated much colder temperatures (~ 250 K) for 'hot' bands.

Ground-based observers continued to search for water on Venus, encouraged by the apparently successful detection of it by the balloon-borne telescope. Belton and Hunten (1966) made photoelectric scans of the water lines at 8189.27 and 8193.00 Å. They noted that "the Venus lines are extremely weak and the spectra contain a considerable amount of noise." but were conviced that the elusive water lines on Venus were indeed visible in their spectra. "A preliminary estimate of the absorption in the red wing of the 8189 Å line on the May (1966) spectrum gives an equivalent width of 20 mÅ. The profile of the observed feature suggests an effective pressure of 5 atm." (The broad line profile could have been due to poor spectral resolution or to a blend with a solar line). Belton and Hunten reported, "we find that the estimated equivalent width corresponds to 317 μ of precipitable water in the total path. This result sets an extreme upper limit [sic] of 125 μ to the amount of precipitable water above the clouds."

Spinrad and Shawl (1966) confirmed the presence of water on Venus: "The equivalent width of the Venus $\lambda 8189$ line is approximately 15 mÅ at the disk center... The Venus component is rather broad and definitely seems stronger near the center of the planet's disk... The center to limb change suspected for this Venus water line differs from the Venus CO_2 line situation; the $\lambda 8689$ band CO_2 lines are quite uniform across the planet..." Spinrad and Shawl reported 250 microns of precipitable water in the line of sight or 60 μ in a vertical path.

Owen (1967) pointed out that "both Belton and Hunten (1966) and Spinrad and Shawl (1966) used the water vapor line at 8189.272 Å as their principal evidence for the presence of water vapor on Venus." Owen's spectra of Venus showed "that a weak absorption indeed appears at the correct position in the spectrum of Venus to correspond to a water vapor line formed in the atmosphere of that planet at an unshifted wavelength of 8180.272 Å ... However, there is no line in the Venus spectrum that corresponds to the telluric (H_2O) feature at 8176.975 Å, although the latter has an intensity virtually identical with that of λ8189.272... In fact, no other Doppler shifted companions to the telluric water vapor lines are evident on this spectrum. It seems difficult to escape the conclusion that the feature appearing in the spectrum of Venus near 8189 Å is simply a solar line that is normally hidden behind the much stronger telluric water vapor absorption (line)." Owen then proceeded to set an upper limit on the water vapor in the atmosphere of Venus, based on the absence of a Doppler shifted companion to the 8176.965 Å line. He set 4 mÅ as an upper limit on the equivalent width, and obtained an upper limit of 64 μ precipitable water in the total path, or 16μ in a vertical column. Hunten *et al.* (1967) replied that no Doppler shifted component of the 8189 Å line was visible in the spectrum of Mercury and hence the feature they had found in the spectrum of Venus could not be a solar line.

2.10. THE DISCOVERY OF HCl AND HF

A surprising discovery about the composition of the atmosphere of Venus was made by W. S. Benedict. P. and J. Connes had obtained in 1966 the first really high resolution spectra of Venus at wavelengths longer than the photographic infrared. The individual rotational lines of the carbon dioxide bands were well resolved and many new CO_2 bands were discovered in the spectrum of Venus (Gray, 1966; Connes *et al*, 1967). In addition to the readily identifiable CO_2 bands, there were some unidentified lines near 5750 cm^{-1} (1.74 μ) which Benedict recognized as belonging to the HCl molecule. Further scrutiny of the spectrum revealed the presence of HF lines; neither HBr nor HI lines appeared. Connes *et al.* (1967) reported, "The HCl lines are consistent with 2 mm Amagat of that gas in the (total) optical path, at temperatures near 240 K and pressures near 0.1 atm, in a spectral region where the effective CO_2 path is 1.7×10^6 times as large. The less extensive HF data indicate about 0.02 mm Amagat of that gas." Connes *et al.* failed to detect any other hydrogen containing gases. "Among those which should have appeared on the spectra if present to more than a part per million are CH_4, CH_3Cl, CH_3F, C_2H_2 and HCN." They did not detect any water vapor on Venus, but the spectra had been taken at a time when the terrestrial abundance of water vapor was large. There was 1 to 2 cm of precipitable (telluric) H_2O in the absorption path, which meant that the minimum amount of water which could be detected on Venus was 20 μ. Hence Connes *et al.* (1967) reported an upper limit to the amount of water in the total path on Venus as 20 μ. This upper limit was considerably smaller than the amount of water that Bottema *et al.* (1964a, b), Belton and Hunten (1966), and Spinrad and Shawl (1966) had reported detecting in the atmosphere of Venus. Connes *et al.* remark that "The evidence concerning H_2O on Venus is not strengthened by our results." They suggested that the low partial pressures of HCl and HF could be explained if these acids were "in weak solutions in clouds of ice water..."

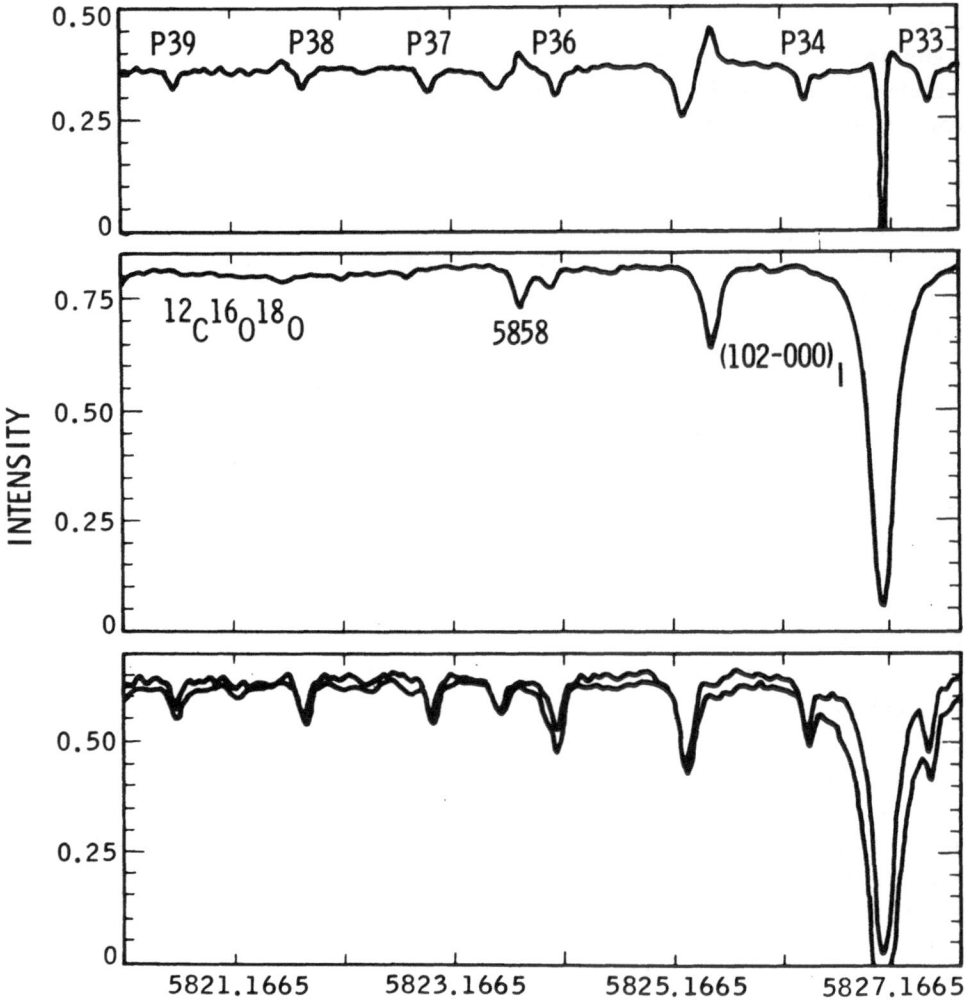

Fig. 15.   A portion of the spectrum of Venus as measured by Connes *et al.* (1969). The upper part of the figure shows the ratio spectrum, Venus/Sun, of a few lines in the carbon dioxide band at 5858 cm^{-1}. The middle part of the figure shows the solar spectrum; the lower part shows two Venus spectra obtained for two different telluric air masses. (From Young, 1972, *Icarus* **17**, 632.).

2.11. AIRCRAFT OBSERVATIONS OF VENUS

The next attempt to detect water on Venus was made by Kuiper and Forbes (1967). They measured the spectrum of Venus in the infrared from 1–2.5 μ with a spectral resolution of 20 cm^{-1}. The data were obtained during two flights of a NASA jet aircraft flying in the lower stratosphere. "The principal result is that in the observable part of the Venus atmosphere ($T \leqslant 320\,K$) water vapor is essentially absent. It followed that the comparatively large amounts of water vapor derived spectroscopically from balloons and at ground-based observatories were spurious; and that atmospheric

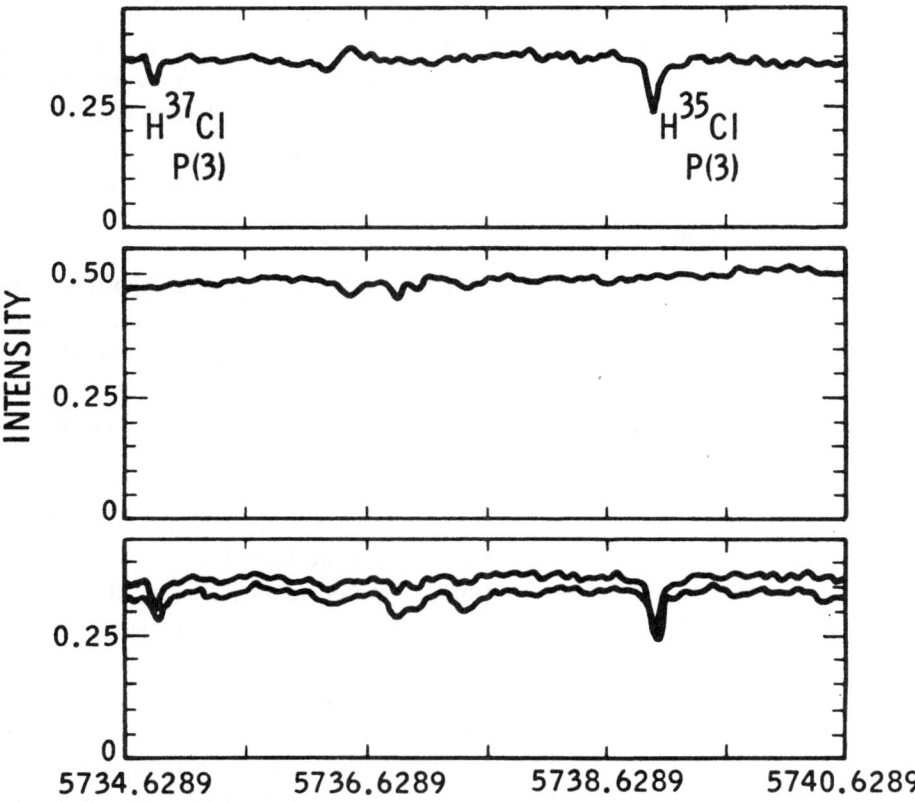

Fig. 16. A portion of the spectrum of Venus as measured by Connes *et al.* (1969). The upper part of the figure shows the ratio spectrum, Venus/Sun, of a few lines of hydrogen chloride. The middle part of the figure shows the solar spectrum; the lower part shows two spectra of Venus obtained for two different telluric air masses. (From Young, 1972, *Icarus* 17, 632.)

models of the planet, based on these earlier observations and invoking a large greenhouse effect by water vapor and water clouds, could not be valid." The result of Kuiper and Forbes "on the near absence of water vapor (amount $<2\,\mu$ in a 2-way transmission)" was in marked contrast with the comparatively large amounts reported by some ground-based observers, and was more in accord with the earlier observations when no detectable water could be found in the spectrum of Venus.

2.12. CARBON DIOXIDE, A MAJOR CONSTITUENT OF VENUS' ATMOSPHERE

One problem about the composition of the atmosphere of Venus was resolved by the spacecraft Venera 4: The atmosphere was found to be composed of $(90\pm10)\%\ CO_2$ with less than $7\%\ N_2$ (Vinogradov *et al.*, 1968). Carbon dioxide was no longer a 'minor' constituent of the atmosphere. Once the mixing ratio became known, it was possible to obtain an upper limit for the pressure of CO_2 line formation simply from the mass of the gas corresponding to the observed absorption features. On this basis, Spinrad's (1962a–e) pressure estimates were a factor of 10–20 too large. Gray (1967) estimated the pressure for line formation to be 40-65 mb and found that "the simple reflecting layer model would yield an amount of (CO_2) that is three times larger than the amount actually above this level." That is, the absorption path traversed by the reflected sunlight would be made three times longer (due to scattering) than the geometric path (without scattering). The influence of scattering did not change the CO_2 abundance in as drastic a way as Chamberlain (1965) had suggested it might.

2.13. WATER ON VENUS

On the other hand, the Venera 4 measurements posed a problem with regard to water. The space probe observations indicated more than 0.7 mg l^{-1} of H_2O (at an altitude where the CO_2 density is 1.25 gl^{-1}) and less than 8 mg l^{-1} of H_2O (where the CO_2 density is 3 g l^{-1}). This meant the H_2O volume mixing ratio could be as high as 1.4×10^{-3}, or orders of magnitude larger than the spectroscopic observations had indicated. Water was detected by Venera 4 at the 700 mb level in the atmosphere of Venus. The carbon dioxide abundance above this level is 4 km atm$_{stp}$. If water were uniformly mixed in the atmosphere of Venus (which seems unlikely), there would be 560 cm atm$_{stp}$, or 4500 μ of precipitable water above the 700 mb level. (1 μ of precipitable water corresponds to 0.1245 cm atm$_{stp}$ of water vapor.) Hence 450 μ of H_2O would be expected to be found above the 'cloudtop' at the 70 mb level. This large abundance could not have avoided being detected spectroscopically. The water on Venus appeared to be hiding below the clouds. Such behavior is not typical for aqueous clouds and is not easily explained.

Pollack and Wood (1968) compared the observed microwave brightness temperatures of Venus at a wavelength of 1.38 cm with theoretical calculations for a water line at that location. They found "an upper limit of 0.8% for the water vapor mixing ratio. This limit is consistent with the amount of water vapor detected by Venera 4, the existence of aqueous ice clouds, and a greenhouse effect caused by water vapor and carbon dioxide." More recent observations of Venus at 1.38 cm have been reported by Janssen et al. (1973). Their results showed "no evidence of water vapor in the lower atmosphere of Venus. The upper limit of 2×10^{-3} for the mixing ratio of water vapor is substantially less than the amounts derived from the Venera space probes (0.5×10^{-2} and 2.5×10^{-2}). This amount of water vapor cannot produce dense clouds, and it is doubtful that it may contribute significantly to a greenhouse effect."

Belton et al. (1968) obtained spectra of Venus for a few H_2O lines in the 8200 Å band and for two CO_2 bands near 1 μ. They chose to fit their observations with synthetic spectra computed for a semi-infinite, homogeneous, isotropically scattering atmosphere. Belton et al. noted that "our model is certainly oversimplified. Scattering in the Venus atmosphere is not isotropic, and the finite optical thickness of the cloud may well be an essential parameter (Sagan and Pollack, 1967). Homogeneity is also a doubtful assumption, which may have led to some disagreement with the observations. We regard our analysis (and the observations) as preliminary in nature, giving a point of departure for future work." Belton et al. reported "We have been able to deduce the following properties of the model which presumably describe the physical state of the atmosphere at a level somewhere near to the visible cloud tops: A temperature of 270 K, an upper limit of 0.2 atm for the total pressure... the clouds on Venus are quite tenuous, the visibility in the cloud being of the order of 4 km. The model also implies that the CO_2 may be the major constituent in the atmosphere." They noted, "It is likely that the H_2O line (at 8189 Å) is formed at a level as low or even lower than the CO_2 lines... Neglecting the fact that the CO_2 and H_2O observations were made at widely separated times, we find (a mixing ratio) $N(H_2O)/N(CO_2) = 10^{-4}$... If the H_2O absorption is actually at a lower level, the ratio will be an upper limit." This mixing ratio was much less than would be expected for water or ice clouds at 270 K; it is compatible with aqueous clouds at 213 K.

The model employed by Belton et al. indicated there were 0.4 km atm of CO_2 in a scattering mean free path $\lambda \geqslant 1$ km. The visual range is 4λ, so presumably their spectra would correspond to 1.6 km atm of CO_2 in an absorption cell.

Belton et al. suggested that one advantage of using synthetic spectra was that the continuum level is unambiguously defined. "Nowhere in the band does the intensity rise all the way to the continuum... It should be noted that this depression occurs in weak bands as well as the relatively strong ones observed here." In bands where the lines are poorly resolved, the instrumental profile can result in spectra that appear to have the 'continuum' depressed only if the 'continuum' is chosen to be at the point where the line absorption is minimum. If the continuum is picked as the level of zero line absorption on either side of the band, this problem is non-existent.

2.14. MEASUREMENT OF CARBON MONOXIDE AND TRACE CONSTITUENTS

Connes et al. (1968) reported, "The lines of the first overtone of CO are clearly resolved and appear prominently in high resolution interferometric spectra of Venus. Analysis of the lines of the principal isotopes shows that they are formed at cloud level, with an effective temperature of 240 K, and a total effective CO path of 13 cm Amagat. Comparison with CO_2 lines yields a CO/CO_2 ratio of 45 parts per million." They remarked that "As with the HCl/CO_2 ratio obtained previously (Connes et al., 1967), this value is independent of the details of the radiation transfer process in the Venus clouds, as long as the CO and CO_2 are uniformly mixed." Connes et al. had obtained an effective pressure of 60 mb but noted that "the effective pressure defined by our analysis procedure is not synonymous with a mean pressure of line formation. It is, in fact, a lower limit."

Belton (1968) questioned the results of Connes et al. (1967, 1968) since they "have interpreted their observations of HCl and HF with the Ladenburg and Reiche curve of growth which applies to the reflecting layer model." In deriving the mixing ratios reported for the minor constituents, Connes et al. had, in fact, compared lines of CO_2 with lines of the minor constituents having the same equivalent width: "It is possible, however, to establish several matches between lines in the 002 band of $^{12}C^{16}O^{18}O$ at 4640 cm^{-1} and the CO band, which are of nearly equal strength for Venus. The most favorable one is the RO line at 4640.21 cm^{-1} whose $W = 0.096$ cm^{-1} is 4 percent less than the corresponding line in CO... We obtain $N_{CO}/N_{CO_2} = 4.4 \times 10^{-5}$. Other comparisons give similar ratios." Belton (1968) thought "it very difficult to obtain precise measurements of the equivalent widths of the individual lines. In fact, some investigators (Belton *et al.*, 1968) have considered the measurement of the equivalent widths in the CO_2 bands to be so imprecise that they abandoned the usual method of interpreting absorption lines, i.e., the curve of growth..." Belton then proceeded to "develop a theory of the curve of growth from pressure broadened lines of arbitrary strength in a semi-infinite, homogeneous, non-conservative, isotropically scattering atmosphere." This approximate theory resulted in a curve of growth remarkably similar to the Ladenburg-Reiche (no scattering) curve, expect the curve predicted smaller equivalent widths for the scattering case when the total path length L is replaced by $\eta_0(\varkappa+\sigma)^{-1}$. Here η_0 is an effective air mass and $(\varkappa+\sigma)^{-1}$ is a mean free path. Belton (1968) had required his theoretical formulation of the curve of growth to approach the Ladenburg-Reiche curve for the case of no scattering, as it should, so the resemblance between the curves was not surprising. He then proceeded to re-analyze the data of Connes *et al.* For HCl, Belton found "the pressures derived here for the scattering model are all *larger* by a factor of 2 or 3, than the effective pressures found by Connes *et al.* ... one expects to derive pressures which are about 2 times *lower* than would be the case for the reflecting layer model."

It should be noted that Belton's 'curve-of-growth' technique required that the particle single scattering albedo be known before the pressure for line formation could be determined. He assumed that the continuum albedo was in the range 0.90 to 0.9999; the pressures ranged from 0.14 atm to 1.2 atm. Belton remarked, "As the continuum albedo at 1.7 μ (where the HCl absorbs) has not been measured, it is difficult to chose which of the pressures... is correct for Venus." However Belton was able to obtain "essentially the same" mixing ratios for the minor constituents as those reported by Connes *et al.* (1967, 1968) by suitable choices of the continuum albedo. Belton (1968) decided that "the applications of the curves of growth developed here leads to a surprisingly consistent picture of the physical conditions in the clouds of Venus. The exception is that this model, while giving the correct sense to the variation of line absorption with phase, does not agree quantitatively with the available observations." It should be recalled that it was the variation of equivalent widths of the lines with phase, which led Chamberlain and Kuiper (1956) to reject the idea that the cloud tops acted as a reflecting layer and to suggest that the observations could be explained if an appropriate scattering model was used. Thus the idealized model used by Belton (1968) obviously required some modification in order to get quantitative agreement with the observations.

2.15. ROTATIONAL TEMPERATURE OF CO_2 BANDS

The next observations of Venus were made by Gray and Schorn (1968) of the three CO_2 bands near 1 μ. These measurements indicated the rotational temperature was $200\,\mathrm{K} < T < 250\,\mathrm{K}$, or significantly lower than the temperatures reported by Chamberlain and Kuiper (1956) or by Belton *et al.* (1968). On the other hand, the temperatures reported by Gray and Schorn were in fairly good agreement with temperatures measured further out in the infrared (Sinton, 1962a, b; Moroz, 1964b; Connes *et al.*, 1967, 1969). These first temperatures obtained by Gray and Schorn assumed that the equivalent widths of the lines followed a square-root absorption law and might be referred to as temperatures for a 'radiative transfer' model.

Gray (1969) then described a general curve of growth technique for finding rotational temperatures which is essentially model independent. The basic idea is to compare two lines in the same CO_2 band, which have identical line strengths, half-widths

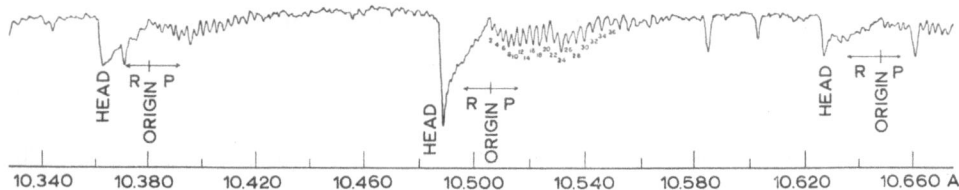

Fig. 17.   Spectrum of Venus showing the three carbon dioxide bands near 1 μ. (From Kuiper, 1962, *Comm. Lunar Planet. Lab.*, No. 15.)

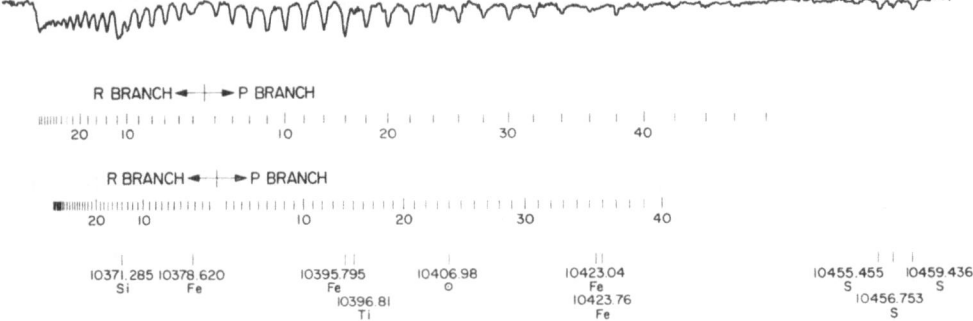

Fig. 18.   Spectrum of Venus showing the carbon dioxide band at 10362 Å, at somewhat higher resolution than is shown in Figure 17. (From Schorn *et al.*, 1970, *Icarus* **12**, 391.)

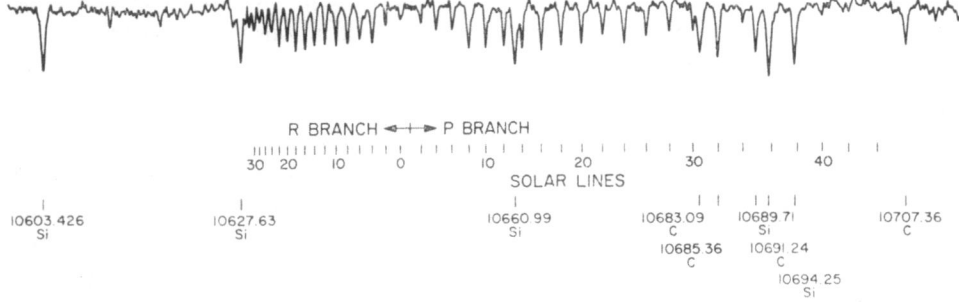

Fig. 19.   Spectrum of Venus showing the carbon dioxide band at 10627 Å, at somewhat higher resolution than is shown in Figure 17. (From Schorn *et al.*, 1971, *Icarus* **14**, 21.).

and equivalent widths. The two lines will have come from different lower state rotational energy levels, $E_{rot}(J)$, with the line intensity given by

$$S(J) = \frac{S(v) F(J)}{Q_{rot}(T)} \exp - \left(\frac{E_{rot}(J)}{kT} \right). \tag{15}$$

Here $S(v)$ is the strength of the band

$$S(v) = \sum_J S(J), \tag{16}$$

and $F(J)$ is related to the f-number of the rotational transition. For the 1μ CO_2 bands:

$$F(J) = J \text{ for the } P \text{ branch}, \tag{17a}$$

and

$$F(J) = J + 1 \text{ for the } R \text{ branch}. \tag{17b}$$

The rotational energy levels are given by

$$E_{rot}(J) = hcBJ(J + 1). \tag{18}$$

Thus when two lines within the same band have the same value of $S(J)$, $S(J_1) = S(J_2)$, the rotational temperature is given by

$$T = \beta [J_1(J_1 + 1) - J_2(J_2 + 1)] [\ln \{F(J_1)/F(J_2)\}]^{-1}, \tag{19}$$

where $\beta = hcB/k = 0.5614$ K for ground state CO_2 bands.

If both of the lines have the same halfwidth, and are formed at the same pressure level in the atmosphere, then they will have identical equivalent widths. In practice, the variation of the line halfwidth $\gamma(J)$ with the rotational quantum number J is easily taken into account (Young, 1970a). Gray (1969) assumed that the equivalent width could be related to the line intensity by

$$W \sim S^b,$$

where b is the local value of the slope of the curve of growth. The major assumption involved in this procedure is that of local thermodynamic equilibrium, and subsequent measurements (Connes et al.) have given no indications to the contrary.

In order to obtain a better value for the rotational temperature than might be found by only comparing two lines in a band, Gray (1969) showed how a least-squares fit could be made to the equivalent widths of all of the lines in the band. This was an iterative procedure, involving both T and b, but it was rapidly convergent. Gray applied the curve of growth analysis to the measurements made by Spinrad (1962a–e) of the ten old Mt. Wilson plates for the 7820 Å band of CO_2. The results indicated that most of the old data followed a nearly square root absorption law and the temperatures were generally lower than those reported by Spinrad. The average temperature found by Gray was $T_{rot} = 293 \pm 20$ K while Spinrad found $T_{rot} = 338 \pm 23$ K. Here the quoted errors refer to one external standard deviation since the old data indicated a pronounced variation in the temperature with time. Subsequent observations (Young, 1972) have not shown any significant variation in the temperature and the average temperature (based on 197 measurements) is 250 ± 2 K.

2.16. THE CURVE OF GROWTH

It is surprising how long it took for a well-known astrophysical technique (van der Held, 1931; Menzel, 1936; Unsöld and Struve, 1949) to 'trickle down' from stellar spectroscopy to the interpretation of planetary spectra. The curve of growth represents the increase in the equivalent width of a line as a function of the number of absorbing molecules, for a fixed line intensity, with the line half-width as a parameter. For a gravitational atmosphere, the effect of scattering is as follows: As a few scatterers are introduced into the atmosphere, the apparent absorption path is increased and the equivalent widths of the lines are increased. The curve appears to be shifted vertically from its position with no scattering. When sufficient scattering particles are added, that optical depth unity is reached at a level in the atmosphere where the pressure is less than the surface pressure, the apparent absorption path will reach a point where it is shorter than in the non-scattering case, and the curve of growth will be shifted below the curve for no scattering. Adding more scatterers to the atmosphere, or increasing the scattering optical depth, will shift the curve of growth even further below the curve for no scattering. In a real atmosphere, there are two reasons for the decrease in equivalent width with the increase of τ_s (for $\tau_s > 2$): One is the decrease in the amount of gas traversed by the reflected light. The second is the decrease in the effective pressure for line formation; this causes the lines to be more saturated and hence makes their equivalent widths smaller. The scattering optical depth has to be known before the amount of gas can be uniquely determined. To date, the scattering model atmospheres which have been used to interpret Venus spectra have used various assumed values for the albedo. This involves assuming both the scattering optical depth and some fixed mixing ratio for the number of scattering particles relative to the number of absorbing molecules. The results obtained from these models clearly depend on how realistic these assumptions are.

2.17. MORE OBSERVATIONS OF VENUS

Cruikshank and Kuiper (1967) looked for SO_2 near 3000 Å, but none could be detected in the atmosphere of Venus. They concluded "that the upper limit of the abundance of this gas in the complete transmission path through the upper Venus atmosphere is 0.05 mm atm," and estimated that the mixing ratio was less than 5×10^{-8}. Cruikshank (1967) also looked for the sulphur compounds COS and H_2S in the spectrum of Venus without success. He estimated that there was less than 0.5 cm atm of COS and less than 1 m atm of H_2S in the total absorption path, and the upper limit to their mixing ratios was 10^{-8} and 2×10^{-4}, respectively.

Owen (1968a, b) searched for carbon suboxide, water vapor, the absorption features reported by Kozyrev, and the night sky emission spectra of Venus. He failed to detect any of the spectral features he had hoped to detect.

Hanel *et al.* (1968) measured the emission spectrum of Venus between 8–13 μ with an interferometer. They observed "the broad absorption-like feature centered at 890 cm^{-1} (11.2 μ)... This observation confirms the existence of this phenomenon. The feature may be caused by a residual ray effect in the material composing the clouds." Hanel *et al.* measured a brightness temperature of "about 250 K." Gillett *et al.* (1968) had obtained spectra of Venus from 2.8–14 μ, with a resolution of $\lambda/\Delta\lambda = = 50$, and reported a brightness temperature of about 225 K for the 8–13 μ region. Presumably the difference in the brightness temperatures reported by various observers is due to their calibration procedures, rather than to Venus herself.

Belton and Hunten (1968) searched for oxygen on Venus. They scanned a 9 Å region in the A band and no features due to O_2 on Venus could be seen. They estimated "that lines of 3 mÅ equivalent width should be readily detected..." Belton and Hunten (1968) found an upper limit of 12.5 cm atm in the total absorption path or less than 3 cm atm$_{stp}$ in a vertical column. Comparison of the upper limit found for O_2 by Belton and Hunten (from $W \leqslant 3$ mÅ) with that found by Spinrad and Richardson (from $W \leqslant 8$ mÅ) reveals that the line intensity used by the former is too large by a factor of 2 (Margolis *et al.*, 1971). The upper limit obtained by Belton and Hunten should be revised upward; there is less than 24 cm atm$_{stp}$ in the total absorption path. Belton and Hunten (1968) found an O_2/CO_2 mixing ratio based on the amount of CO_2 reported by Connes *et al.* (1967) of 3.3 km atm in the total path. They applied an air-mass correction to O_2 but not to CO_2 and reported an O_2/CO_2 mixing ratio of 'less than 2×10^{-5}' (Belton and Hunten, 1969). This upper limit on the mixing ratio must be increased to 7×10^{-5}, if the CO_2 abundance were the same on the dates of the two observations; however, this assumption has no justification. Belton and Hunten (1969) reported a mixing ratio of "less than 8×10^{-5} for the scattering model," the two methods of data reduction gave essentially the same result. Interestingly enough, this upper limit on the mixing ratio is *identical* to that reported by Spinrad and Richardson (1965) which had been obtained from the ratio of the O_2 partial pressure (upper limit) to the "total pressure at the base of the line formation," and the latter had been over-estimated.

Schorn *et al.* (1969a) carried out an extensive series of spectroscopic observations of Venus during 1967, in a search for water vapor. From April to June, their results were negative, giving an upper limit of $< 32 \mu$ in April and May and $< 16 \mu$ in June, for the amount of precipitable H_2O in a vertical path. In November and December, they detected water on Venus, 30–40 μ. "Our conclusion is that the observable amount of water vapor 'above the clouds' of Venus varies significantly... Comparison of H_2O variations with variations in CO_2 bands, UV cloud activity and, possibly, millimeter wavelength observations should be most informative... The questions of why and how these variations occur should have a definite bearing on the composition of Venus' clouds and the structure of its atmosphere... Extensive and homogeneous series of observations are required to make progress in this area, as with most problems in planetary astronomy... 'One shot' observations, no matter how well conceived and executed will not be of much help." Since the question of whether the clouds are aqueous or not had been settled to everyone's satisfaction, more 'one shot' observations would be published.

Schorn *et al.* (1969b) noted, "Compared to the theoretical work done on the subject, relatively few observations have been made of the 7820 Å carbon dioxide band in the spectrum of Venus." They observed this band during 1967 and found no significant variation in the rotational temperature with the phase angle i of Venus for $50° < i < 90°$. Schorn *et al.* reported an average rotational temperature of $T_{rot} = 242 \pm 2$ K (formal standard deviation) from 18 observations. They commented that

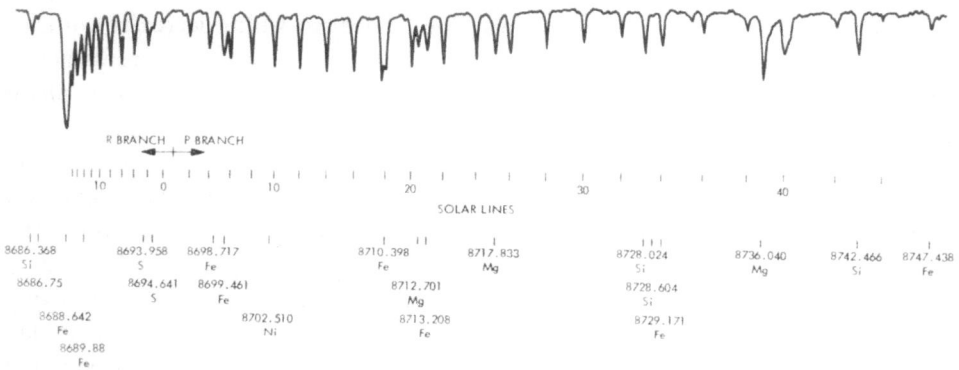

Fig. 20. Spectrum of Venus showing the carbon dioxide band at 8689 Å. (From Young *et al.*, 1969, *Icarus* **11**, 390.).

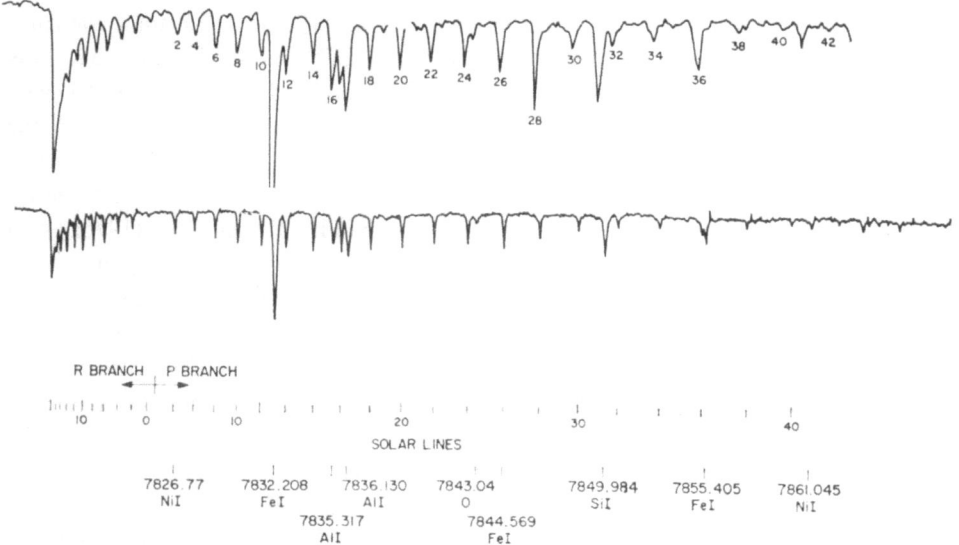

Fig. 21. Spectra of Venus showing the carbon dioxide band at 7820 Å. The upper spectrum was taken by Dunham and the lower spectrum is at somewhat higher resolution. (From Schorn *et al.*, 1969b, *Icarus* **10**, 241.)

"the amount (of CO_2) appeared to vary significantly with time." At a particular phase angle, the abundance varied by almost a factor of two.

Gray et al. (1969) measured rotational temperatures for the CO_2 band at 7883 Å in the spectrum of Venus. From 12 observations, they found an average "rotational temperature of 244 ± 10 K (standard deviation and no significant temporal variation of temperature .. the amount of carbon dioxide in the absorption path appears to vary with time."

Young et al. (1969) found an average rotational temperature for Venus of "$T_{rot}=$ $=238\pm1$ K (standard deviation) based on 23 plates of the 8689 Å CO_2 band... The variation of the equivalent width of the 8689 Å band with Venus phase is seen to agree generally with the observations of Kuiper; the equivalent width decreases with increasing phase angles.... As Kuiper (1952) noted in discussing his measurements, there is considerable fluctuation in the equivalent width at a particular phase angle."

Kuiper et al. (1969) reported that spectra of Venus, which had been taken during flight of the NASA jet aircraft, definitely showed the presence of water vapor on Venus. There was 5 μ of precipitable H_2O in the total absorption path. "Since the corresponding figure for CO_2 is about 4 km atm," the mixing ratio is 1.5×10^{-6}. "The observed amounts of trace constituents, such as H_2O, will, for constant mixing ratios, vary with the amount of CO_2 observed, known to be variable from day to day as well as systematically with phase." Thus, even if H_2O were uniformly mixed in the atmosphere of Venus, the observed amount of water would appear to vary. Kuiper et al. (1969) also identified "numerous vibrational bands of CO_2," and bands belonging to CO, HCl, and HF in their spectra of Venus.

Young (1969) used the Venus spectra of Connes et al. (1969) to determine the rotational temperature of an isotopic CO_2 band at 2.21 μ. The result, $T=245\pm3$ K, was in good agreement with the temperature of $T=240$ K reported by Connes et al. (1967, 1968). Schorn et al. (1970) reported a rotational temperature of $T=237\pm12$ K for Venus from 15 spectra of the CO_2 band at 1.0362 μ. Belton et al. (1968) had observed this band but did not attempt to find a temperature for Venus since "The entire spectrum was not of such high quality as the 1.05 μ band." Young et al. (1970a) also studied the CO_2 band at 1.0488 μ in the spectrum of Venus. From 31 plates, they found an average rotational temperature of $T=237\pm3$ K, which was considerably lower than the temperature of $T=270\pm25$ K, which Belton et al. (1968) had found by matching one spectrum of Venus. The temperatures measured by Young et al. gave no indication of varying with time. Young et al. (1970b) found the CO_2 bands at 1.2030 μ and 1.2177 μ gave similar temperatures: $T=236\pm5$ K based on 10 spectra. Young (1970a) again used the Venus spectra of Connes et al. (1969) to find the rotational temperature of a CO_2 band at 1.71 μ. This band indicated a temperature of $T=242\pm2$ K in agreement with the results found from 0.72 μ to 2.21 μ. All the previous estimates for rotational temperatures, by all investigators, had assumed the variation of the line widths with rotational quantum number could be neglected. Young (1970b) found that including this effect, the CO_2 band at 1.71 μ indicated a temperature of $T = 249 \pm 3$ K. "The results indicate that the effect (of variation in the line widths) on the temperature is slight, increasing it by 3%."

Schorn *et al.* (1971) reported that 17 spectra of Venus for the 1.0627 μ band of CO_2 indicated a rotational temperature of 241 ± 3 K (or 250 ± 3 K when the correction for variable line widths is applied). The temperature did not show any variation for Venus phase angles $26° \leqslant i \leqslant 164°$, while the CO_2 abundance "appeared to vary significantly with the phase of Venus and also with the time of observation."

Young *et al.* (1971) reported on observations of Venus during 1968 and 1969. Twenty-two spectra of the CO_2 bands at 7820 Å and 7883 Å gave an average rotational temperature of 251 ± 2 K and 257 ± 4 K, respectively. These temperatures were found using a variable rotational line half width. They noted that the equivalent widths of these CO_2 bands varied significantly, "but there was little correlation between that variation and the phase angle of Venus" for $10° < i < 126°$. Young *et al.* noted that "theoretical calculations (Chamberlain, 1970) based on observations of the 8689 Å band, predict a much greater variation with phase than we observe. Because of the great care with which we measured the equivalent widths and evaluated the systematic errors involved, we find the difference between theory and observations to be remarkable."

Beer *et al.* (1971) measured the continuum absorption by Venus in the 3–4 μ region. Their spectra had a resolution of $\lambda / \Delta\lambda = 5000$ and "in order to improve the signal-to-noise ratio, an average was made of the three spectra (obtained on one day) together with one obtained a few days earlier. "The resulting spectrum of Venus (2400–3100 cm^{-1}) showed strong absorption in the region 2400–2600 cm^{-1}, minimum absorption between 2880 cm^{-1} and 2980 cm^{-1} and the suggestion of a weak absorption feature at 3050 cm^{-1}. Beer *et al.* (1971) suggested that these features might be due to bicarbonates. They found "that $NaHCO_3$ and $KHCO_3$ show some of the requisite (spectral) characteristics but that the frequency match is inadequate."

Lewis (1972) commented, "Recently Beer *et al.* (1971) have published a high resolution spectrum of Venus in the 3.1–4.2 μ region, which clearly shows a strong cloud absorption band centered at 3.9 μ and a possible weak band centered at $\lambda < 3.2$ μ. They match the Venus reflection spectrum to laboratory transmission spectra of $NaHCO_3$ and $KHCO_3$, which have strong, broad absorption bands centered at 4.03 and 3.83 μ, respectively, and weak absorption bands at 3.45 and 3.40 μ, respectively. It is striking that the *highest* reflectivity found in the Venus spectrum is from 3.35 to 3.48 μ, corresponding precisely to the locations of known *absorption* bands in the alkali bicarbonates." Hence, the suggestion of bicarbonate clouds was ruled out. Lewis went on to note that the spectrum of HCl dissolved in dioxane ($C_4H_{10}O_2$) gave a satisfactory fit to the spectrum of Venus obtained by Beer *et al.*

Young and Young (1973) cautioned, "In any serious attempt to determine spectroscopically the composition of the Venus clouds, the entire spectrum of Venus should be matched, not just a single feature ... To illustrate precisely how weak the spectroscopic evidence is, we have chosen to 'fit' the Venus spectrum at 11.2 μ." Young and Young showed that the Venus absorption band at 11.2 μ could be fit by six substances, several of which had an absorption at 3.9 μ. Of the substances they considered, sulphuric acid had a spectrum which most closely resembled the spectrum of Venus.

Regas *et al.* (1972) observed portions of the 1.0488 μ CO_2 band over several nights. "The region P22-P32 was scanned on October 20, 21, 22 and the region P34–P42 on October 23 and 24," 1967. The "spectra were combined and smoothed..." To interpret their observations, Regas *et al.* computed synthetic spectra. They used the spacecraft results for their model atmosphere: "A tropopause temperature of 238 K, a tropopause pressure of 0.258 atm, a surface temperature of 770 K, a surface pressure of 97 atm, and a composition by volume of 95% CO_2 and 5% N_2. In addition, in our models, we had an isothermal region above the tropopause and an adiabatic region below the tropopause with a lapse rate of 8.83 K km^{-1}." For their scattering model calculations, Regas *et al.* (1972) remarked, "Our clouds have tops and bottoms.... the cloud top is that point in the atmosphere above which there is a negligibly small optical depth, say $\tau < 0.01$." For a dust cloud model with the cloud bottom at the surface of the planet, Regas *et al.* found there was an entire set of pairs of scattering coefficients and cloud top pressures that gave a satisfactory fit to their observations. Similar results were obtained for a condensation cloud model. They reported, "We found that the cloud top pressure was significantly lower than the tropopause pressure for all of our acceptable models." The maximum cloud top pressure, according to Regas *et al.* was 0.157 atm, "... a quantity independent of the scattering model..." They chose "the model of the Venus clouds that would have the weakest possible water line (at 8189 Å) and yet still give a good match to our CO_2 observations. Using a cloud layer model with a cloud top pressure of 0.37 atm, a cloud top temperature of 238 K, and no water vapor above the clouds, we found the 0.8189 μ H_2O line has an equivalent width of 100 mÅ, about 7 times larger than (the largest value reported for) the observed line.... it is unlikely the clouds are pure water ice."

The fact that many observers have reported day-to-day variations in the equivalent width of CO_2 bands in the spectrum of Venus strongly suggests that observations from several days should not be averaged as Regas *et al.* have done. Results obtained from averaged spectra can give us some idea about the conditions at the cloud tops, but it is not clear that this really represents average conditions.

Moroz (1971) reported variations in the CO_2 abundance over different areas of Venus, based on his spectroscopic observations of the 1.58 and 1.61 μ CO_2 bands. Based on 4 days' observations, Moroz concluded "the Venusian cloud layer is lower over the polar than over the equatorial region of the planet... the pressure at the top of the clouds is nearly 3 times larger in polar regions than in equatorial regions... the difference in height of the top of the cloud level between high and low latitudes is $\Delta z \simeq 7$ km." Hunt and Schorn (1971) argued that "the variations over the disk of the planet of the equivalent widths must be associated with the dynamical properties of the Venus atmosphere... but its precise meaning in terms of the variation of the structure of Venusian cloud layers remains unresolved at present." The observations of the CO_2 band at 8689 Å made by Young *et al.* (1969) do not support the suggestion that the cloud layer is consistently lower over the polar regions than over the equatorial region.

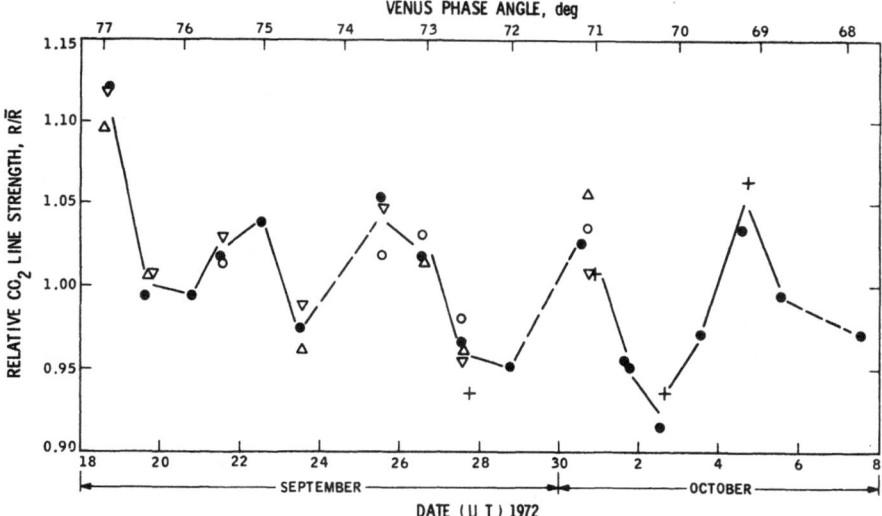

Fig. 22. Measurements of the day-to-day variation in the strength of the carbon dioxide band a 8689 Å, relative to the strength of the neighboring solar lines in the spectrum of Venus. The different symbols refer to different locations on Venus: equator (open circles), northern latitude (triangle, point up) southern latitude (triangle, point down), standard meridian (solid circles and crosses, the latter taken at slightly higher resolution). (From Young *et al.*, 1973, *Astrophys. J.* **181**, L5.).

Fink *et al.* (1972) analyzed infrared spectra of Venus produced by a Fourier spectrometer flown aboard the NASA jet aircraft. They obtained "an improved determination of water vapor for the Venus atmosphere of $1.6 \pm 0.4\,\mu$ of precipitable water in the total path." The CO_2 abundance in the total path was 3.6 km atm which "leads to a volume mixing ratio of 0.6×10^{-6} for water vapor, assuming the two (gases) to be uniformly mixed." Using a homogeneous, isotropic, semi-infinite scattering model with $\tilde{\omega}_c = 0.999$ gave "a mixing ratio of 1.0×10^{-6}, in close accord with the above number." Fink *et al.* remarked, "It is quite obvious that our result is one-to-two orders of magnitude below any other determination except that of Kuiper whose amount was obtained with the same airborne instrument as the present result. We have no physical explanation for this discrepancy except the comment that every 'determination' of the water vapor in the Venus atmosphere appears to have been the experimental limit of detectivity for the particular method. Our results, of course, refer to an average over the illuminated portion of the desk .." Fink *et al.* suggested that "an extremely hydroscopic material" is needed as a cloud constituent if their results are to be reconciled with "a high total water content (as measured by the Russian Venera atmospheric entry probes)." They made calculations for H_2O–HCl droplets and found their water vapor abundance "will not allow a lower cloud deck to be composed of H_2O–HCl droplets." Fink *et al.* remarked, "Although HCl failed in lowering the vapor pressure sufficiently, one can think of more powerful 'drying agents' such as sulfuric acid for example. A rough calculation showed that an 80% by weight solution of sulfuric acid can dry the upper atmosphere sufficiently to give agreement with our measured abundance and yet keep the total amount in the lower atmosphere high enough to yield the Venera spacecraft water measurements. Such an idea must, however, most certainly be rejected, due to other chemical complications this model would introduce." They concluded, "Since neither a cold trap nor a hygroscopic material is presently plausible in the Venus atmosphere, our mixing ratio for water of about 1 ppm must extend through the whole Venus atmosphere... Venus is severely depleted in water."

Owen and Sagan (1972) reported upper limits for the abundances of many minor constituents in the atmosphere of Venus based on ultraviolet spectra obtained from the orbiting astronomical observatory. They noted that, "In all cases, we have adopted a simple reflecting layer formalism. We feel that greater sophistication is unjustified and unnecessary at present (Regas and Sagan, 1970a, b; Sagan and Regas, 1970)." These are included in Table III along with upper limits which have been set for various substances which have not been detected in infrared spectra of Venus.

Young *et al.* (1973) reported that observations of the CO_2 band at 8689 Å showed "the apparent strength of CO_2 absorptions in the spectrum of Venus varies by 20%, in a period of 4 days. The variations are synchronous over the disk, and thus represent a fundamental dynamical mode of the atmosphere... To produce the observed changes

TABLE III

Composition of the atmosphere of Venus

Molecules observed in the atmosphere of Venus above the clouds

Gas	Mixing ratio	Total amount (cm atm$_{stp}$)	Pressure (mb)	Temperature (K)	Reference
CO	$4.5 \pm 1.0 \times 10^{-5}$	13	60	240	Connes *et al.* (1968)
	$5.1 \pm 0.1 \times 10^{-5}$	20	36	249 ± 3	Young (1972)
CO$_2$	0.97 ± 0.04	...	...	...	Vinogradov *et al.* (1970a, b)
	...	$3.3 \pm 0.3 \times 10^5$	100	240 ± 10	Connes *et al.* (1967)
	...	$2.8 \pm 0.4 \times 10^5$	60	240	Connes *et al.* (1968)
	...	$4.1 \pm 0.1 \times 10^5$	36	249 ± 3	Young (1972)
	...	$0.9 - 5.5 \times 10^5$	100	245	Fink *et al.* (1972)
HCl	6×10^{-7}	$1.9 \pm 0.4 \times 10^{-1}$	80	270 ± 30	Connes *et al.* (1967)
	$4.2 \pm 0.7 \times 10^{-7}$	$1.7 \pm 0.3 \times 10^{-1}$	160	240 ± 16	Young (1972)
HF	5×10^{-9}	$2.5 \pm 0.5 \times 10^{-3}$	100	240	Connes *et al.* (1967)
	7×10^{-9}	$2.9 \pm 0.2 \times 10^{-3}$	67	249 ± 3	Young (1972)
H$_2$O	1.1×10^{-2}	...	600	295	Vinogradov *et al.* (1968)
	10^{-6}	...	10000	500	Vinogradov *et al.* (1970a, b)
	10^{-4}	35	...	...	Dollfus (1963a, b)
	2.5×10^{-4}	100	90	...	Bottema *et al.* (1964a, b)
	8.7×10^{-6}	25	600	...	
	10^{-4}	39	...	...	Belton and Hunten (1966)
	10^{-4}	30	...	...	Spinrad and Shawl (1966)
	2×10^{-6}	0.6	...	...	Kuiper (1969)
H$_2$O	2×10^{-5}	5	...	...	Schorn *et al.* (1969a, b)
	0.6×10^{-6}	0.2	100	240	Fink *et al.* (1972)

Table III (Continued)

Molecules not observed in the atmosphere of Venus above the clouds

Gas	Maximum Mixing ratio	Maximum amount (cm atm$_{stp}$)	Reference
CH_2O	10^{-6}	10^{-1}	Owen and Sagan (1972)
CH_3 and aldehydes	10^{-6}	10^{-1}	Owen and Sagan (1972)
CH_4	10^{-6}	3×10^{-1}	Connes *et al.* (1967)
CH_3Cl	10^{-6}	3×10^{-1}	Connes *et al.* (1967)
CH_3F	10^{-6}	3×10^{-1}	Connes *et al.* (1967)
CH_3COCH_3 and ketones	10^{-6}	10^{-1}	Owen and Sagan (1972)
C_2H_2	10^{-6}	3×10^{-1}	Connes *et al.* (1967)
C_2H_4	10^{-5}	2	Kuiper (1952)
C_2H_6	10^{-5}	4	Kuiper (1952)
C_3O_2	10^{-7}	10^{-2}	Owen and Sagan (1972)
HCl	10^{-6}	10^{-1}	Owen and Sagan (1972)
HCN	10^{-6}	3×10^{-1}	Connes *et al.* (1967)
H_2O	6×10^{-6}	2	Connes *et al.* (1967)
	2×10^{-5}	8	Owen (1967)
H_2S	10^{-7}	10^{-2}	Owen and Sagan (1972)
N_2	2×10^{-2}	...	Vinogradov *et al.* (1970a, b)
NH_3	3×10^{-8}	10^{-2}	Kuiper (1962)
NO	10^{-6}	10^{-1}	Owen and Sagan (1972)
NO_2	10^{-8}	10^{-3}	Owen and Sagan (1972)
N_2O	6×10^{-4}	200	Kuiper (1952)
N_2O_4	4×10^{-8}	4×10^{-3}	Owen and Sagan (1972)
O_2	7×10^{-5}	24	Belton and Hunten (1968)
O_3	3×10^{-9}	3×10^{-4}	Owen and Sagan (1972)
SO_2	5×10^{-8}	5×10^{-3}	Cruikshank and Kuiper (1967)
	10^{-8}	10^{-3}	Owen and Sagan (1972)

in line strength, the cloud tops must be moving up and down by 0.2 scale height, or over 1 kilometer, all over the disk at once." Their observations were made when the phase angle of Venus was $68° < i < 72°$.

2.18. INTERPRETATION OF SPECTRA OF VENUS

Sagan and Pollack (1969) used an empirical fit to laboratory measurements of CO_2 bands to intepret published spectra obtained by Kuiper (1962). They found effective pressures for line formation to be between 48 and 110 mb based on Kuiper's spectra for the CO_2 bands near 1.2 μ and 1.6 μ. Sagan and Pollack (1969) suggested that "a multiple scattering model is not the only one capable of explaining the phase dependence of the absorption line strength. Either a single cloud layer with individual clouds having a large range of altitudes or a model with two cloud layers might account for the phase observations."

Young (1970c) used the data of Schorn *et al.* (1969a, b, 1970) and Young *et al.* (1969, 1970a, b) to study the change in the effective pressure for line formation in the atmosphere of Venus as a function i, the Venus phase angle. "The data are analyzed by two techniques: a reflecting layer model for lines with a Voigt line profile and a scattering model for lines with a Lorentz line profile." The effective pressure for line formation for a strong line (P 16) in the 1.0488 μ CO_2 band decreased from 95 mb near superior conjunction ($i = 26°$) to 19 mb near inferior conjunction ($i = 164°$) according to the reflecting layer model. Use of Belton's (1968) scattering model indicated a decrease from 173 mb ($i = 26°$) to 89 mb ($i = 164°$). The trend is the same for the two models, except the scattering model predicts that the lines are formed at higher pressures than is predicted by the reflecting layer model.

Chamberlain (1970) developed "an approximate analytic theory for the formation of spectral absorption lines in a hazy atmosphere with isotropic scattering and a homogeneous mixture of scattering and absorbing matter." He once again used (9a) and van de Hulst's approximation (8) for the H functions. Chamberlain chose to parameterize the absorption lines by two quantities analogous to the parameter x used for pure absorption:

$$q_c \equiv \frac{Su}{\pi\gamma\tau_c}$$

and

$$q_t \equiv \frac{Su}{\pi\gamma(\tau_s + \tau_c)},$$

where τ_s is the scattering optical depth and τ_c is the continuum absorption optical depth (in Chamberlain's notation, $q = q_c$ and $w = q_t$).

Chamberlain (1970) found that the equivalent of a weak line, for $q_c \ll 1$ and $q_t \ll 1$, varied as

$$W = \frac{Su}{(\tau_s + \tau_c)}[1 + f_1(\tau_s, \tau_c, \mu, \mu_0)], \tag{20}$$

where the function $f_1(\tau_s, \tau_c, \mu, \mu_0)$ depends on μ, μ_0 for $\tau_s = 0$. For $\tau_s \gg \tau_c$, so $q_c \gg 1$ and $q_t \ll 1$, Chamberlain found the equivalent width varied as

$$W = \frac{Su}{(\tau_s + \tau_c)}[1 + f_2(\mu, \mu_0)] + \left[\frac{Su\gamma}{\pi(\tau_s + \tau_c)}\right]^{1/2} f_3(\tau_s, \tau_c, \mu, \mu_0). \tag{21}$$

For strong lines, with $q_c \gg q_t \gg 1$, Chamberlain found the equivalent width varied as

$$W = \left[\frac{Su\gamma}{\pi(\tau_s + \tau_c)}\right]^{1/2} f_4(\tau_s, \tau_c, \mu, \mu_0). \tag{22}$$

The functions f_1, f_2, f_3 and f_4 can be found in the paper by Chamberlain (1970) and will not be given here. Chamberlain and Smith (1970) applied "this 'two-parameter theory' to the CO_2 absorptions on Venus to see whether a consistent picture can be derived from all the available data without introducing additional, *ad hoc* assumptions." Chamberlain and Smith reanalyzed the spectrum obtained by Belton *et al.* (1968) for the 1.05 μ CO_2 band. Chamberlain and Smith remarked that the analysis of this spectrum by Belton *et al.* was not unique. They found "for $1 - \tilde{\omega} \ll 1$ that lines that have

$$\left[\frac{Su\gamma}{\pi (\tau_s + \tau_c)} \right]^{1/2} \ln \left(\frac{\tau_c}{\tau_s + \tau_c} \right) = \text{const.}$$

possess nearly identical equivalent widths."

Chamberlain and Smith (1970) concluded that it is "not possible to distinguish among any of the models (which ranged from $\tilde{\omega}_c = 0.95$, $p = 0.422$ atm up to $\tilde{\omega}_c = 0.995$, $p = 0.050$ atm) with the 1.05 μ profile alone." They obtained a qualitative fit to the observed phase variations of the CO_2 band at 8689 Å (Chamberlain and Kuiper, 1956) for $\tilde{\omega}_c = 0.997$, and remarked, "It is clear that the phase variation has little dependence on line strength in this band and that a rather good fit with the data is possible."

In order to fit the phase dependence of the equivalent width in the 1.6 μ bands measured by Moroz (1967), Chamberlain and Smith found that the continuum albedo must be very close to unity. For $\tilde{\omega}_c = 1.000$, the equivalent-width varies as $(\mu + \mu_0)$, in qualitative agreement with Moroz's observations. There is considerable scatter in the measured equivalent widths, but the phase variation computed for $\tilde{\omega}_c = 1.000$ is in much better agreement with Moroz's observations than the phase variation computed for $\tilde{\omega}_c = 0.999$. In order to confirm this large value for the continuum albedo at 1.6 μ, Chamberlain and Smith computed synthetic spectra to match the CO_2 band at 1.71 μ observed by Connes *et al.* (1969). They remarked, "It appears again that we must have $\tilde{\omega}_c \gtrsim 0.999$ to obtain a satisfactory fit for the 1.71 μ band." Belton (1968) had found the same band required $\tilde{\omega}_c = 0.991$, but Chamberlain and Smith (1970) showed that this value of $\tilde{\omega}_c$ predicted less absorption for the strong lines of the band than was observed. Chamberlain and Smith concluded, "For a pure-scattering model atmosphere, the gaseous abundance 'above the cloud tops' is a concept without meaning, since the clouds have no well defined tops... The equivalent abundance of CO_2 cannot be ascertained to an accuracy better than a factor of 2." They also concluded that their scattering model could fit all the data, "provided the scattering albedo is close to unity around 1.7 μ."

Tables of the Bond albedo for isotropically scattering semi-infinite atmospheres are given by Young (1970b) and Chamberlain and Smith (1970); the latter tables are the most extensive.

Chamberlain and Smith (1972) investigated "the hypothesis that the CO_2 absorption lines in Venus's spectrum could be formed between an upper thin cloud and a lower thick one." They found, "The phase effect depends on the amount of incident radiation that is able to penetrate the upper cloud and then emerge in the direction of the observer. It depends not only on the geometry itself but also on the optical thickness of the upper cloud, which acts as a diffusing screen... In view of the day-to-day scatter in W (the equivalent width of a CO_2 band), it would be difficult observationally to distinguish between the two (phase) curves" for a homogeneous model and a two-cloud model. They concluded "that a satisfactory two-cloud model for the Venus atmosphere may be derived from existing data. However, this model is not unique, but is highly dependent on built-in assumptions regarding this type of model. Thus, the two-cloud model fits the observational data about as well as a single cloud model. More sophisticated and accurate spectroscopic data would undoubtedly narrow the acceptable ranges of parameters, but the question of uniqueness will be difficult to establish for any specific case."

Hunt (1972a) suggested that "models based upon homogeneous isotropically scattering atmospheres cannot be used to reproduce observed spectroscopic features of phase effect and the shape of spectral lines for weak *and* strong bands;" an inhomogeneous (gravitational) model of a planetary atmosphere is required. Hunt used a model where the cloud did not extend either to the surface of the planet or to the top of the atmosphere. Based on this model, he arrived at the following conclusions: "For fixed continuum properties, the phase effect increases with increasing line strength. Increasing the mean free path of the cloud particles (which is equivalent to decreasing the particle concentration) increases the phase effect ... The introduction of continuum absorption reduces the phase effect ..."

Hunt (1972b) reported, "We have shown that only observations of the phase variations of the CO_2 bands in the Venus spectrum can provide the information for a unique identification of the structure of the cloud layers. It is proved that Venus cannot have a single dense cloud layer, but must have two scattering layers: a thin aerosol layer situated in the lower stratosphere, overlying a dense cloud deck." Hunt noted that "The effect of an optically thin aerosol on the level of line formation is most noticeable for large phase angles, $i > 110°$, say. The absorption lines are then formed higher in the atmosphere than they would be if only one cloud were present." However, for small phase angles, if the phase curve shows "a positive gradient then this is evidence of a distinct aerosol layer ... If there is a negligible gradient, then the aerosol is continuous down to the main cloud top."

Regas *et al.* (1973) showed that the same kind of phase variation "may be due to the strong backward lobe in the Venus cloud phase function and that the two cloud layers are not necessarily required." Hunt (1973) disagreed.

3. Summary

One may well ask, what is the value of infrared spectroscopy for studying a cloud-covered planet like Venus? Progress in learning about the physical and chemical nature of her atmosphere has not been rapid, using this approach. And, as is true any time complicated phenomena are studied, the infrared spectra have sometimes completely led people astray.

The most sought-after substance in the atmosphere of Venus, H_2O, is either undetectable, variable, or present in such amounts that the clouds of Venus can/

cannot be aqueous. All of these views have been put forth on the basis of Venus spectra. Similarly, spectra of Venus have been used to support various other hypotheses about the cloud composition.

Spectroscopy is a useful tool in the search for minor atmospheric constituents. Besides the controversial observations of water vapor, few minor constituents have been found on Venus: CO, HCl and HF. The major atmospheric constituent, CO_2, discovered spectroscopically in 1932, was shown 'to be a minor constituent' in the 1960's. Only the Venera spacecraft measurements served to dispell that erroneous notion. Various theoretical models have been proposed to intepret the observations but there has been no agreement on the 'right' one. Both the reflecting layer model (Kaplan, 1961; Sagan and Pollack, 1969) and a scattering model (Chamberlain and Smith, 1970) have been shown to 'fit all the observational data' provided certain assumptions are made about the nature of the clouds. Fortunately, both approaches give approximately the same results for the abundance of minor constituents relative to carbon dioxide.

The reasons for the slow progress in understanding Venus are threefold: observational, theoretical, and psychological. On the observational side, it has turned out that spectroscopic data which have an excellent signal-to-noise ratio but low spectral resolution cannot be used to distinguish uniquely between different theoretical models of the Venus atmosphere. Extremely high resolution spectra are required. Unfortunately, some observers have led the theoreticians astray by overestimating the quality of their data. On the other hand, some theoreticians have dismissed perfectly good observations under the assumption that the data were 'noisy' because conditions on Venus appeared to vary on a short time scale, a situation that could not occur in their models. There has been a profusion of crude, oversimplified models which have 'explained' discrepancies between theory and observation as due to effects not included in the theory. Thus there has been a tendency to claim 'agreement' with the observations prematurely. Finally, not only have wrong interpretations of the data been widely accepted at various times, but some correct interpretations have been rejected for long periods of time. What interpretation is 'acceptable' has been colored by prejudices (Venus is/isn't like the Earth, the curve of growth does/doesn't apply to a scattering atmosphere, etc.) so that major questions appear to have been decided more on emotional than on rational grounds.

In the future, it would help if observers would strive to put realistic error estimates on their data; if theoreticians would state clearly the kind and accuracy of observations needed to test their models; and if everyone would adopt a more critical attitude instead of jumping to conclusions.

Acknowledgements

This article was prepared as one phase of research carried out at the Jet Propulsion Laboratory, California Institute of Technology, under contract No. NAS 7-100 sponsored by the National Aeronautics and Space Administration.

Appendix A: The Air-Mass Function

(A. T. Young)

The air-mass for a plane-parallel atmosphere, neglecting refraction, is well known to be proportional to the secant of the zenith angle. But real planetary atmospheres are curved, not flat; and the rays of light are also curved by atmospheric refraction.

Usually, the atmospheric curvature exceeds the ray curvature, especially if we restrict our attention to situations where the atmospheric density (and hence the refraction) is small. Even at sea level on Earth, the curvature of a horizontal ray is only one-sixth that of the atmosphere (Newcomb, 1906), so that neglect of refraction is not a serious error. For the atmosphere of Mars, and the visible part of Venus' atmosphere, the refraction is much less, and the atmospheric curvature somewhat greater; thus the straight-ray approximation is even better for these planets.

Even if we neglect refraction, we must specify the law of density variation with height, for the ray passes more obliquely through the lower than through the upper layers of the atmosphere. If the density variation is $\varrho(h) = \varrho(0) \cdot x(h)$, the air mass is

$$F(z) = \frac{1}{H} \int_0^\infty x(h)\, \mathrm{d}s, \tag{A1}$$

where the height of the homogeneous atmosphere is

$$H = \int_0^\infty x(h)\, \mathrm{d}h, \tag{A2}$$

and the element of path length along the ray, $\mathrm{d}s$, is related to h, z, and the planetary radius R by the law of cosines:

$$(R+h)^2 = R^2 + s^2 + 2Rs \cos z. \tag{A3}$$

We note that $x(h)$ is the density, normalized to unity at $h = 0$; and

$$\mathrm{d}s = \frac{(R+h)\,\mathrm{d}h}{(R^2 \cos^2 z + 2Rh + h^2)^{1/2}}. \tag{A4}$$

Lambert (1760) expanded ds in a power series to obtain

$$F(z) = A \sec z - \tfrac{1}{2} B \sec z \tan^2 z + \cdots, \tag{A5}$$

where the coefficients $A, B, \ldots$ are integrals of functions containing the unknown density law $x(b)$, but not z. Lambert proposed to obtain these coefficients from observations of atmospheric extinction. On the other hand, Bouguer (1760) assumed an exponential atmosphere, i.e.,

$$x(b) = \exp(-h/h_0), \tag{A6}$$

for which $H=h_0$ of course. From this assumed law, Bouguer obtained explicit values of Lambert's coefficients by performing the necessary integrations. However, as Bemporad (1901) pointed out, Bouguer made an unfortunate change of variable in performing the integrations, and thus obtained incorrect results (in error by a factor of 2 even in the second term); the correct series is

$$F(z) = \sec z - \frac{h_0}{R} \sec z \tan^2 z + 3\left(\frac{h_0}{R}\right)^2 \sec^3 z \tan^2 z - \cdots. \tag{A7}$$

In the meanwhile, however, Laplace (1805) had shown the connection between the air-mass problem and the atmospheric-refraction problem, and all subsequent air-mass calculations included the effects of ray curvature. These numerous investigations, which are reviewed by Schoenberg (1929), culminated in the work of Bemporad (1904), which has been enshrined in all subsequent handbooks up to the present as *the* air-mass function.

Bemporad's mean air-mass tables have been so widely reprinted that their limitations are often overlooked. They are based on a mean atmospheric model that is not strictly correct, although the consequent error in $F(z)$ is less than half a percent, according to Abbot et al. (1922). Because the refraction depends strongly on pressure and the effective height of the atmosphere depends strongly on temperature, Bemporad gives corrections that depend on these quantities; unfortunately, they are usually ignored. The calculations assume the absorbing species are uniformly mixed, which is seriously in error for ozone, water vapor, and aerosols; the corrections can be comparable to the difference between the mean $F(z)$ and $\sec z$ (Abbot et al., 1922; Young, 1973). Finally, one must remember that the argument of Bemporad's tables is the apparent (refracted) zenith angle, not the true value.

As Bemporad (1904) points out, expansions like Equation (A7) are unsuitable near the horizon, because each term becomes infinite at $z=90°$. Bemporad uses various auxiliary variables to achieve convergence. However, for grazing incidence it is convenient to develop the air-mass problem somewhat differently, as has been done by Fabry (1929) for a ray that passes completely through an exponential atmosphere, reaching at its lowest point a pressure p:

$$F(p) = \sqrt{2\pi R/h_0} \; p/p_1, \tag{A8}$$

where p_1 is a reference level above which the zenithal air-mass is taken as unity.* For an observer at the level p, the horizontal air-mass is just half the value of Equation (A8). As Fabry was concerned only with the upper atmospheres of planets, he neglected ray curvature. More general formulae, including refraction, are given by Link (1969).

---

* Fabry actually obtains h_0 times (A8) for the equivalent path length at pressure p_1.

The classical problem of the exponential atmosphere without refraction was belatedly solved once again by Chapman (1931), whose air-mass function is

$$f(R/h_0, z) = F(z) \tag{A9}$$

in our notation. Chapman deduces Fabry's (1929) formula for the horizontal air-mass, Bemporad's (1901) corrected version of Bouguer's (1760) formula, and various other series expansions that are somewhat less convenient for numerical evaluation (as they are expressed in terms of higher transcendental functions) than the conventional formulae.

It has unfortunately become the custom to attach Chapman's name to air-mass functions in the English aeronomical literature; the Bouguer-Bemporad function is called 'the ordinary Chapman function', and the air-mass for a nonexponential (but still nonrefracting) atmosphere is called 'a generalized Chapman function' by Green and Martin (1966). They treat an atmosphere whose density profile is a sum of exponential distributions with different scale heights, as well as some other analytical density distributions.

However, as Link (1969) points out, there are many practical problems which require the use of actual, not analytical, densities. The exponential model has long been criticized for neglecting not only the variation of temperature with height, but also that of the local gravitational acceleration. The physical crudeness of such models is sometimes concealed by the deceptive accuracy of published tables, which often extend to 3 or 4 decimal places.

The exponential, nonrefracting atmosphere is again treated by Fesenkov (1955), who expresses the Bouguer-Bemporad series in terms of $\sec z$ alone. As Fesenkov's interest is in in the twilight sky brightness, he also gives the air-masses above various heights [i.e., the integral in Equation (A1) for nonzero lower limits]. The twilight problem for Venus, looking in from space instead of up from the planet, is treated in detail by Schilling and Moore (1967) and by Link (1969, pp 216–225).

Finally, a most practical method of evaluating the Bouguer air-mass is to neglect h in the numerator and h^2 in the denominator of Equation (A4), which allows the integration to be done in closed form, giving

$$F(z) = (\sec z) \left[\pi^{1/2} X \exp(X^2) \operatorname{erfc}(X) \right], \tag{A10}$$

where

$$X = (R/2 h_0)^{1/2} \cos z, \tag{A11}$$

and

$$\operatorname{erfc}(X) = 1 - \operatorname{erf}(X)$$

$$= 2\pi^{-1/2} \int\limits_{X}^{\infty} \exp(-t^2)\, dt. \tag{A12}$$

This approximation is derived at length by Green and Martin (1966), who call this approximation 'the essential part of the Chapman function,' and more simply by

Young (1969), who gives the necessary series expansions for large and small X. The series expansion of the correction factor in brackets in Equation (A10) for large X gives the principal terms in Fesenkov's (1955) series, but with an error of order (h_0/R) in each coefficient; it is equivalent to replacing $\tan^2 z$ by $\sec^2 z$ in the accurate Bouguer-Bemporad series (A7). Green and Martin (1966) compare the approximate and exact values, for three values of R/h_0 (which they call x; they also use a parameter β such that $\beta^2 = 2X^2$): the relative error is, at most, of order h_0/R, as would be expected from the above discussion. For $z < 90°$, the approximate values of F are too small. However, Equation (A10) gives Fabry's value (A8) for twice the horizontal air-mass. Furthermore, the effect of refraction can be approximately included by adopting a larger effective value of R, whose reciprocal is the *difference* in curvatures of the planet and a horizontal ray (Young, 1973).

To illustrate the effects of atmospheric curvature, we consider observations of Venus at large phase angles, i. The *minimum* curvature correction occurs halfway between limb and terminator, where $z_c = i/2$. Above the clouds we have $R \approx 6100$ km, $h_0 \approx 5$ km, so Equation (A8) gives about 88 for twice the horizontal air-mass. (This is the air-mass at the cusp, at all phases.) The curvature correction is one percent at $\sec z_c = 3.5$ or $z_c = 73°3$, i $146°6$; the average correction over the disk will be somewhat larger, of course. The correction exceeds ten percent at $\sec z_c = 11$, $i = 170°6$. The maximum phase angle at which Venus has been observed spectroscopically is near 175°; here $F(z_c) \approx 17$ but $\sec z_c = 23$, a difference of about one third. Thus, contrary to Hunten's (1971) suggestion that $\sec z$ is an adequate approximation, the curvature effects are appreciable.

Finally, let us consider the depth to which a tangential ray can penetrate the atmosphere of Venus. According to Link (1969, p. 214) the maximum refraction observed in transits of Venus is about a minute of arc, which corresponds to a maximum pressure of about 5 millibars of CO_2. This is an order of magnitude *lower* than the cloud-top pressure deduced spectroscopically. Equation (A8) then tells us that we are seeing the Sun through some 8.8 times as much gas as lies above the cloud tops, which is not much more than we see at moderate phase angles; we cannot expect to find minor constituents by looking tangentially through Venus' atmosphere.

Presumably the tangential ray is being cut off by aerosols above the main cloud deck. Schilling and Moore (1967) find from twilight phenomena that appreciable aerosol extends at least 15 km (3 scale heights) above the clouds. If the aerosol were distributed with the same scale height as the gas, we would have $e^{-3} \approx 1/20$ as much at this height as at the cloud tops. This agrees reasonable well with the pressure deduced from the refraction, about 1/10 of the cloud-top value. Since we are seeing nearly a hundred times as much of this material at grazing incidence, the normal-incidence optical depth of the aerosol would be about 0.01 – that is, the 'high scattering layer' observed at inferior conjunction is practically invisible under ordinary circumstances.

From the above examples, it can be seen that observations near inferior conjunction can, by the relation (A1) between $\rho(h)$ and $F(z)$, provide information on the vertical structure of Venus' atmosphere.

Appendix B: Preliminary Spectroscopic Data for Carbon Dioxide Bands Observed in High Resolution Spectra of Venus

(P. Connes, J. Connes, W. S. Benedict, and L. D. Gray)

The spectrum of Venus was used to find spectroscopic constants for comparatively weak bands of CO_2. These bands are difficult to measure in the laboratory, whereas the atmosphere of Venus provides an absorption path more than a km long at an effective pressure on the order of a tenth of an atmosphere. We have tabulated preliminary values of spectroscopic data obtained from the spectrum of Venus measured by Connes *et al.* (1969). Additional data for spectral regions not measured by Connes *et al.* can be found in the tabulations of McClatchey *et al.* (1973), Courtoy (1959), and Young *et al.* (1970c).

The data we have tabulated allow one to compute the intensity and position of any carbon dioxide line which appears in the spectrum of Venus, for the spectral regions covered by the atlas of Connes *et al.* We have used the following expression for the rotational energy levels of a non-rigid rotator (the energy levels of a rigid rotator are given by Equation (18)):

$$E_{rot}(J) = hc(B_v J(J+1) - D_v J^2 (J+1)^2 + H_v J^3 (J+1)^3 + \cdots.$$

For most vibrational levels the constant H_v was taken to be equal to zero. The rotational line positions are given by

$$\omega(m) = \omega_0 + am + bm^2 + cm^3 + dm^4 + \cdots,$$

where

$$a = (B' + B''),$$
$$b = (B' - B'') - (D' - D''),$$
$$c = -2(D' + D''),$$

and

$$d = -(D' - D'');$$

double primes refer to the lower vibrational state and primes to the upper vibrational state. The index $m = J'' + 1$ for the R branch and $m = -J''$ for the P branch. The rotational constants for the ground state of the $^{12}C^{16}O_2$ isotope are accurately known: $B(00^00) = 0.390218$ cm^{-1} and $D(00^00) = 13.3 \times 10^{-8}$ cm^{-1}. The vibrational energy levels are identified by the vibrational quantum numbers $v_1 v_2^1 v_3$.

Many isotopic bands of carbon dioxide can be seen in the spectrum of Venus, and we will use the following notation for the isotopes: $626 = {}^{12}C^{16}O_2$, $636 = {}^{13}C^{16}O_2$, $628 = {}^{12}C^{16}O^{18}O$, $627 = {}^{12}C^{16}O^{17}O$, $638 = {}^{13}C^{16}O^{18}O$, etc.

The rotational line intensities, $S(J)$ or $S(m)$, are found from Equation (15). The factor $F(J)$, or $F(m)$, is given by Herzberg (1950): For parallel bands, $l' = l''$, we have

$$F(m) = (m^2 - l^2)/m$$

for the P and R branches ($J' = J'' - 1$ and $J' = J'' + 1$, respectively);

$$F(J) = l^2 (2J'' + 1)/(J''(J'' + 1))$$

TABLE B-I

Rotational constants for different lower vibrational states

(a) Nonsymmetrical molecules of CO_2

Vibrational state	example	Rotational constant	Rotational quantum number
Σ	00^00	B^c	$J = 0, 1, 2, 3, 4, \ldots$
Π	01^10	B^c	$J = 1, 2, 3, 4, 5, \ldots$
		B^d	$J = 1, 2, 3, 4, 5, \ldots$
Δ	02^20	B^c	$J = 2, 3, 4, 5, 6, \ldots$
		B^d	$J = 2, 3, 4, 5, 6, \ldots$

(b) Symmetrical molecules of CO_2

Σ_g	00^00	B^c	$J = 0, 2, 4, 6, 8, \ldots$
Σ_u	00^91	B^c	$J = 1, 3, 5, 7, 9, \ldots$
Π_u	01^10	B^c	$J = 1, 3, 5, 7, 9, \ldots$
		B^d	$J = 2, 4, 6, 8, 10, \ldots$
Δ_g	02^20	B^c	$J = 2, 4, 6, 8, 10, \ldots$
		B^d	$J = 3, 5, 7, 9, 11, \ldots$

TABLE B-II

Absorption intensity units

Units for (path × concentration)	Units for intensity, S (frequency/length × concentration)
1.0 cm $\mathrm{atm}_{300\,K} = 0.91$ cm atm_{stp}	1.0 cm^{-1}/cm atm_{stp} $= 0.91$ cm^{-1}/cm $\mathrm{atm}_{300\,K}$
1.0 cm $\mathrm{atm}_{296\,K} = 0.922$ cm atm_{stp}	1.0 cm^{-1}/cm atm_{stp} $= 0.922$ cm^{-1}/cm $\mathrm{atm}_{296\,K}$
1.0 km atm_{stp} $= 10^5$ cm atm_{stp}	1.0 cm^{-1}/cm atm_{stp} $= 10^5$ cm^{-1}/km atm_{stp}
1.0 mole/cm^2 $= 22.415$ cm atm_{stp}	1.0 cm^{-1}/cm atm_{stp} $= 22,415$ cm/mole
1.0 mole/cm^2 $= 0.24303$ km $\mathrm{atm}_{296\,K}$	1.0 cm^{-1}/km $\mathrm{atm}_{296\,K} = 0.24303$ cm/mole
1.0 cm atm_{stp} $= 2.69 \times 10^{19}$ molecules cm^{-2}	1.0 cm/molecule $= 2.69 \times 10^{19}$ cm^{-1}/cm atm_{stp}
1.0 km $\mathrm{atm}_{296\,K} = 2.48 \times 10^{24}$ molecules cm^{-2}	1.0 cm/molecule $= 2.48 \times 10^{24}$ cm^{-1}/km $\mathrm{atm}_{296\,K}$
	1 'intensity unit' $= 10^7$ cm/mole
	1 'intensity unit' $= 446$ cm^{-1}/cm atm_{stp}
	1.0 cm^{-1}/cm atm_{stp} $= 30$ GHz/cm atm_{stp}

Relationships between integrated intensity and transition probabilities at 273.16 K

Electric dipole matrix element, $\langle M \rangle$ (Debye or 10^{-18}esu)	$S = 11.1908 \, \omega \, \langle M \rangle^2 \, (273/T) \, (N_1/N)$
f-number (dimensionless)	$S = 2.3789 \times 10^7 \, f \, (273/T)(N_1/N)$
Einstein absorption coefficient, B (sg^{-1})	$S = 1.7801 \times 10^7 \, \omega \, B(273/T)(N_1/N)$
Einstein emission coefficient, A (s^{-1})	$S = 3.5670 \times 10^7 \, \omega^{-2} \, A \, (273/T)(N_1/N)$

Here S has the units cm^{-1}/cm atm_T, ω is the wavenumber of the band in cm^{-1} and (N_1/N) is the fraction of the molecules in the lower energy state.

TABLE B-III

Carbon dioxide bands in the Connes et al. (1969) infrared spectra of Venus

Band origin ω_0 cm^{-1} (in vacuum)	Rotational Constants				Isotope	Transition	Branch	Intensity S_v cm^{-1}/km atm$_{296\,K}$	Interaction term, ζ_r
	$a \times 10^5$ (cm^{-1})	$b \times 10^5$	$c \times 10^8$	$d \times 10^8$					
3980.60	77621.3	−711.7	−53.6	0.6	626	$01^12\text{-}02^20$	Rc	0.05	+0.1
3980.60	77681.3	−651.7	−55.8	−0.9	626	$01^12\text{-}02^20$	Rd	0.05	+0.1
3980.60	−651.7	−651.7	−1.4	−0.7	626	$01^12\text{-}02^20$	Qc	0.05	+0.1
3980.60	−711.7	−711.7	0.8	0.4	626	$01^12\text{-}02^20$	Qd	0.05	+0.1
3987.61	73633.0	1.0	−40.4	2.0	628	$(30^00\text{-}00^00)_{II}$	Rc	4.0	0
4005.94	−719.0	−719.0	0.4	0.2	626	$00^02\text{-}01^10$	Qd	2.0	+0.12
4005.94	77469.0	−658.0	−52.8	0.2	626	$00^02\text{-}01^10$	Rc	2.0	+0.12
4023.48	75698.0	−30.0	−48.0	0	627	$(30^00\text{-}00^00)_{III}$	Rc	0.2	0
4030.32	77503.2	−593.6	−57.4	2.5	626	$(01^12\text{-}10^00)_{III}$	Rc	0.05	+0.1
4030.32	−533.6	−533.6	2.4	1.2	626	$(01^12\text{-}10^00)_{III}$	Qc	0.05	+0.1
4167.91	73508.0	125.0	−36.0	8.0	628	$(30^00\text{-}00^00)_{I}$	Rc	0.03	0
4416.150	78157.5	113.9	−61.0	−4.3	626	$(31^10\text{-}00^00)_{IV}$	Rc	0.1	+0.15
4416.150	286.0	286.0	−6.0	−3.0	626	$(31^10\text{-}00^00)_{IV}$	Qc	0.1	+0.15
4508.749	73078.0	−557.8	−44.4	0	638	$00^02\text{-}00^00$	Rc	0.4	0
4524.88	75155.0	−577.0	−48.2	0	637	$00^02\text{-}00^00$	Rc	0.04	0
4529.870	77985.5	−58.1	−63.4	−5.3	626	$(40^00\text{-}01^10)_{IV}$	Rc	0.01	+0.2
4529.870	−119.1	−119.1	−10.6	−5.3	626	$(40^00\text{-}01^10)_{IV}$	Qd	0.01	+0.2
4578.09	78214.0	85.5	−52.8	0.2	626	$(32^20\text{-}01^10)_{III}$	Rc	0.01	+0.2
4578.09	78275.0	24.5	−56.8	−1.8	626	$(32^20\text{-}01^10)_{III}$	Rd	0.01	+0.2
4578.09	85.5	85.5	−3.6	−1.8	626	$(32^20\text{-}01^10)_{III}$	Qc	0.01	+0.2,
4578.09	24.5	24.5	0.4	0.2	626	$(32^20\text{-}01^10)_{III}$	Qd	0.01	+0.2
4591.118	78014.3	−29.3	−56.4	−2.0	626	$(31^10\text{-}00^20)_{III}$	Rc	0.25	+0.18
4591.118	112.0	112.0	−3.0	−1.5	626	$(31^10\text{-}00^00)_{III}$	Qc	0.25	+0.18
4611.31	77836.0	−186.5	−59.4	−4.5	626	$(31^11\text{-}11^10)_{IV}{}^{I}$	Rc	0.05	0
4611.31	78162.0	−113.0	−62.2	−4.9	626	$(31^11\text{-}11^10)_{IV}{}^{I}$	Rd	0.05	0
4614.779	73132.0	−574.0	−43.8	0.5	628	$01^12\text{-}01^10$	Rc	1.4	0
4614.779	73240.8	−577.2	−43.8	0.5	628	$01^12\text{-}01^10$	Rd	1.4	0
4630.37	75217.0	−591.0	−48.0	0	627	$01^12\text{-}01^10$	Rc	0.13	0
4630.37	75330.0	−594.0	−48.0	0	627	$01^12\text{-}01^10$	Rd	0.13	0
4639.502	73053.6	−579.2	−44.4	0	628	$00^02\text{-}00^00$	Rc	35.0	0
4655.205	75131.5	−596.7	−48.2	0	627	$00^02\text{-}00^00$	Rc	3.4	0
4673.68	78154.0	−169.0	−56.0	0	636	$(22^21\text{-}02^20)_{III}$	Rc	0.2	0
4673.68	78154.0	−169.0	−60.0	−3.0	636	$(22^21\text{-}02^20)_{III}$	Rd	0.2	0

ν						Assignment			
4683.12	77048.0	−99.0	−52.0	0	636	$(31^10{-}00^00)_{II}$	Rc	0.005	+0.2
4685.77	73057.7	−126.5	−78.2	−8.5	636	$(30^01{-}10^00)_{IV}^{II}$	Rc	0.5	0
4687.798	77909.4	−132.4	−64.4	−7.6	626	$(30^01{-}10^00)_{IV}^{I}$	Rc	0.90	0
4692.18	73468.0	−168.0	−52.4	−4.0	638	$(20^01{-}00^00)_{III}$	Rc	0.7	0
4708.52	77924.4	−198.4	−56.6	−1.7	636	$(21^11{-}01^10)_{III}$	Rc	4.0	0
4708.52	78135.6	−112.8	−60.2	−3.5	636	$(21^11{-}01^10)_{III}$	Rd	4.0	0
4718.35	75567.0	−165.0	−54.0	−3.0	637	$(20^01{-}00^00)_{III}$	Rc	0.1	0
4721.92	69100.0	−282.0	−41.6	0.0	828	$(20^01{-}00^00)_{III}$	Rc	0.25	0
4732.90	78275.0	−201.0	−63.0	−4.0	626	$(23^31{-}03^30)_{III}$	Rcd	1.8	0
4743.70	73436.5	−269.5	−50.8	−3.0	628	$(21^11{-}01^10)_{III}$	Rc	4.5	0
4743.70	73596.0	−222.0	−46.8	−1.0	628	$(21^11{-}01^10)_{III}$	Rd	4.5	0
4748.05	77907.0	−140.0	−61.8	−4.7	636	$(20^01{-}00^00)_{III}$	Rc	72.0	+0.30
4753.41	77993.0	−51.0	−48.2	1.6	626	$(31^10{-}00^00)_{II}$	Rc	0.3	+0.30
4753.41	88.0	88.0	0.0	0	626	$(31^10{-}00^00)_{II}$	Qc	0.3	0
4757.705	77928.0	−218.4	−62.2	−3.1	626	$(31^11{-}11^10)_{IV}^{II}$	Rc	4.8	0
4757.705	78192.0	−143.6	−64.6	−3.7	626	$(31^11{-}11^10)_{IV}^{II}$	Rd	4.8	0
4768.54	78109.0	−224.0	−52.8	1.0	626	$(22^21{-}02^20)_{III}$	Rc	35.0	0
4768.54	78109.0	−224.0	−60.0	−3.0	626	$(22^21{-}02^20)_{III}$	Rd	35.0	0
4784.66	77242.0	−194.0	−60.4	−4.0	626	$(20^02{-}00^01)_{III}$	Rc	0.4	0
4786.688	77734.9	−347.3	−54.0	−1.8	626	$(31^11{-}11^10)_{III}^{I}$	Rc	1.6	0
4786.688	77970.6	−304.0	−55.4	−1.5	626	$(31^11{-}11^10)_{III}^{I}$	Rd	1.6	0
4790.573	77936.9	−159.9	−71.0	−4.3	626	$(30^01{-}10^00)_{IV}^{II}$	Rc	42.0	0
4791.26	73388.0	−243.0	−52.2	−3.9	628	$(20^01{-}00^00)_{II}$	Rc	126.0	0
4807.692	78067.6	−183.0	−60.2	−3.5	626	$(21^11{-}01^10)_{III}$	Rd	900.0	0
4807.692	77879.0	−250.0	−53.0	0	626	$(21^11{-}01^10)_{III}$	Rc	900.0	0
4808.186	78073.0	−56.0	−46.0	3.6	626	$(40^00{-}01^10)_{II}$	Rc	~0	0
4808.186	−117.0	−117.0	7.0	3.6	626	$(40^00{-}01^10)_{II}$	Rc	~0	0
4814.57	73299.0	−337.0	−44.4	0	638	$(20^01{-}00^00)_{II}$	Qd	3.6	0
4821.50	75505.0	−223.0	−56.0	−4.0	627	$(20^01{-}00^00)_{III}$	Rc	20.0	0
4839.731	77692.0	−349.8	−57.6	−4.2	626	$(30^01{-}10^00)_{III}^{I}$	Rc	37.0	0
4853.62	77841.0	−203.0	−61.2	−4.4	626	$(20^01{-}00^00)_{III}$	Rc	21700.0	0
4871.46	77767.0	−356.0	−49.8	1.7	636	$(21^11{-}01^10)_{II}$	Rc	32.0	0
4871.46	77945.0	−303.0	−52.6	0.3	636	$(21^11{-}01^10)_{II}$	Rc	32.0	0
4887.39	77708.0	−339.0	−55.2	−1.4	636	$(20^01{-}00^00)_{II}$	Rd	800.0	0
4887.97	77918.0	−126.0	−52.4	0	626	$(12^21{-}00^00)_{II}$	Rc	2.2×10^{-5}	⋯
4896.185	73379.0	−327.0	−45.0	0	628	$(21^11{-}01^10)_{II}$	Rc	12.0	0
4896.185	73530.0	−288.0	−45.0	0	628	$(21^11{-}01^10)_{II}$	Rd	12.0	0
4904.85	73301.0	−332.0	−44.4	0	628	$(20^01{-}00^30)_{III}$	Rc	300.0	0
4924.98	73366.0	−270.0	−38.4	3.0	638	$(20^01{-}00^00)_{I}$	Rc	1.2	0
4828.91	75464.0	−344.0	−48.0	0	627	$(21^11{-}01^10)_{II}$	Rc	2.0	0

Table B-III (Continued)

Band origin ω_0 cm⁻¹ (in vacuum)	Rotational Constants $a\times10^5$ (cm⁻¹)	$b\times10^5$	$c\times10^8$	$d\times10^8$	Isotope	Transition	Branch	Intensity S_v cm⁻¹/km atm₂₉₆ₖ	Interaction term, ζ_v
4828.91	75626.0	−298.0	−48.0	0	627	$(21^11\text{-}01^10)_{II}$	*Rd*	2.0	0
4931.083	77767.0	−369.2	−56.8	−0.4	626	$(31^11\text{-}11^10)_{III}^{II}$	*Rc*	13.0	0
4931.083	78001.3	−334.7	−57.6	−0.3	626	$(31^11\text{-}11^10)_{III}^{II}$	*Rd*	13.0	0
4939.33	75377.0	−351.0	−48.0	0	627	$(20^11\text{-}00^10)_{II}$	*Rc*	62.0	0
4942.506	77719.5	−377.3	−64.2	−0.9	626	$(30^11\text{-}10^10)_{III}^{II}$	*Rc*	380.0	0
4946.807	77721.0	−361.4	−53.8	−1.7	626	$(31^11\text{-}11^10)_{II}^{I}$	*Rc*	8.0	0
4946.807	77947.0	−326.8	−47.4	2.3	626	$(31^11\text{-}11^10)_{II}^{I}$	*Rd*	8.0	0
4953.363	78013.0	−320.0	−54.6	0.1	626	$(22^21\text{-}02^20)_{II}$	*Rc*	140.0	0
4953.363	78013.0	−320.0	−51.8	1.2	626	$(22^21\text{-}02^20)_{II}$	*Rd*	140.0	0
4959.668	77666.3	−375.5	−44.0	2.8	626	$(30^11\text{-}10^10)_{II}^{I}$	*Rc*	225.0	0
4965.36	77766.1	−362.5	−51.4	0.9	626	$(21^11\text{-}01^10)_{II}$	*Rc*	3570.0	0
4965.36	77938.1	−312.5	−52.4	0.4	626	$(21^11\text{-}01^10)_{II}$	*Rd*	3570.0	0
4977.83	77674.5	−369.1	−52.0	0.2	626	$(20^11\text{-}00^10)_{II}$	*Rc*	94000.0	0
4991.35	77694.0	−353.5	−44.6	3.9	636	$(20^11\text{-}00^10)_{I}$	*Rc*	570.0	0
5013.783	77746.7	−376.1	−48.6	2.3	636	$(21^11\text{-}01^10)_{I}$	*Rc*	23.0	0
5013.783	77918.6	−329.8	−46.6	3.3	636	$(21^11\text{-}01^10)_{I}$	*Rd*	23.0	0
5028.78	77836.0	−350.0	−56.0	−0.2	636	$(22^21\text{-}02^20)_{I}$	*Rc*	0.4	0
5028.78	77836.0	−350.0	−49.4	2.3	636	$(22^21\text{-}02^20)_{I}$	*Rd*	0.4	0
5042.58	73429.0	−204.0	−39.4	2.5	628	$(20^11\text{-}00^10)_{I}$	*Rd*	61.0	0
5061.776	77873.0	−171.0	−51.4	0.5	626	$(12^21\text{-}00^20)_{I}$	*Rc*	2.5×10^{-5}	⋯
5062.443	77693.8	−403.0	−50.2	6.1	626	$(30^11\text{-}10^10)_{III}^{II}$	*Rc*	64.0	0
5064.68	73442.0	−264.0	−45.0	0	628	$(21^11\text{-}01^10)_{I}$	*Rc*	3.5	0
5064.68	73619.0	−199.0	−39.0	3.0	628	$(21^11\text{-}01^10)_{I}$	*Rd*	3.5	0
5068.93	75496.0	−232.0	−44.0	2.0	627	$(20^11\text{-}00^10)_{I}$	*Rc*	17.0	0
5099.66	77770.6	−273.0	−44.2	4.1	626	$(20^11\text{-}00^10)_{I}$	*Rc*	30200.0	0
5114.892	77819.7	−222.0	−39.6	4.8	626	$(30^11\text{-}10^10)_{I}^{I}$	*Rc*	83.0	0
5123.20	77806.0	−323.0	−48.6	2.3	626	$(21^11\text{-}01^10)_{I}$	*Rc*	1430.0	0
5123.20	77987.0	−264.0	−46.6	3.3	626	$(21^11\text{-}01^10)_{I}$	*Rd*	1430.0	0
5139.40	78032.0	−301.0	−57.6	−1.4	626	$(22^21\text{-}02^20)_{I}$	*Rc*	55.0	0
5139.40	78032.0	−301.0	−50.8	1.6	626	$(22^21\text{-}02^20)_{I}$	*Rd*	55.0	0
5168.60	77493.0	−554.0	−52.4	0	636	$01^12\text{-}00^00$	*Rc*	0.5	0
5168.60	−493.0	−493.0	0	0	636	$01^12\text{-}00^00$	*Qc*	0.5	0
5217.667	77847.0	−249.6	−46.2	8.1	626	$(30^11\text{-}10^10)_{II}^{II}$	*Rc*	6.3	0
5584.391	77119.0	−922.0	−50.8	−0.8	626	$(00^03\text{-}10^00)^{I}$	*Rc*	1.9	0

5670.08	77243.0	−923.0	−55.0	0.5	626	$(01^13-11^10)_{II}$	Rc	0.02	o
5670.08	77374.0	−962.0	−55.0	1.0	626	$(01^13-11^10)_{II}$	Rd	0.02	o
5687.166	77147.4	−949.4	−57.4	2.5	626.	$(00^03-10^00)_{II}$	Rc	2.0	o
5830.792	78025.7	−17.9	−61.4	−4.5	626	$(41^00-00^00)_{IV}$	Rc	0.20	o
5858.022	73065.4	−567.4	−49.3	−2.47	628	$(10^22-02^20)_{II}$	Rc	1.04	o
5885.336	75154.0	−574.0	−48.2	0	627	$(10^22-00^00)_{II}$	Rc	0.04	o
5904.47	77959.0	−164.0	−59.	−3.0	636	$(31^11-01^10)_{IV}$	Rc	0.05	o
5904.47	78214.0	−34.0	−63.0	−5.0	636	$(31^11-01^10)_{IV}$	Rd	0.05	o
5933.8	...	−223.0	...	...	628	$(30^01-00^00)_{IV}$	Rc	0.08	o
5951.59	77989.0	−58.0	−73.8	−10.7	636	$(30^01-00^00)_{IV}$	Rc	0.50	o
5955.8	73000.0	−280.0	...	...	628	$(11^22-01^10)_{I}$	Rc	0.08	o
5959.954	73076.9	−555.9	−40.1	2.15	628	$(10^22-01^10)_{I}$	Rc	0.95	o
5972.54	78159.0	−174.0	−47.4	3.6	626	$(32^21-02^20)_{IV}$	Rc	0.7	o
5972.54	78159.0	−174.0	−59.2	−2.6	626	$(32^21-02^20)_{IV}$	Rd	0.7	o
5987.0	...	...	...	...	627	$(10^22-00^00)_{I}$	...	0.02	o
5993.581	73450.1	−182.7	−182.7	−6.7	628	$(30^01-00^00)_{IV}$	Rc	1.0	o
5998.569	78003.4	−93.4	−77.0	−7.3	626	$(40^01-10^00)_{IV,I}$	Rc	1.3	o
6000.52	77940.0	−104.0	−52.4	0.0	626	$(41^00-00^00)_{III}$	Rc	0.2	+
6020.795	77919.0	−209.6	−60.8	−3.8	626	$(31^11-01^10)_{IV}$	Rc	12.6	o
6020.795	78139.7	−100.9	−62.6	−4.7	626	$(31^11-01^10)_{IV}$	Rd	12.6	o
6027.6	...	...	...	...	638	$(30^00-00^00)_{III}$	...	0.1	o
6033.478	75569.0	−159.0	−60.0	−6.0	627	$(30^00-00^00)_{IV}$	Rc	0.12	o
6072.34	77752.0	−289.0	−62.8	−6.8	626	$(40^01-10^00)_{IV,I}$	Rc	0.2	o
6075.98	77911.0	−133.0	−66.0	−6.8	626	$(30^00-00^00)_{III}$	Rc	128.0	o
6088.21	77800.0	−323.0	−59.6	−3.0	636	$(31^11-01^10)_{III}$	Rc	0.6	o
6088.21	78015.0	−233.0	−58.2	−2.5	636	$(31^11-01^10)_{IV}$	Rd	0.6	o
6103.67	77967.0	−79.0	−51.6	0.4	626	$(22^21-00^00)_{III}$	Rc	0.2	o
6119.61	77783.0	−264.0	−61.6	−4.6	636	$(30^01-00^00)_{III}$	Rc	8.1	o
6127.782	73265.9	−366.9	−48.4	−2.0	628	$(30^01-00^00)_{III}$	Rc	6.8	o
6170.09	78028.0	−305.0	−51.6	1.6	626	$(32^21-02^20)_{III}$	Rc	3.6	o
6170.09	78028.0	−305.0	−53.2	0.4	626	$(32^21-02^20)_{III}$	Rd	3.6	o
6175.12	77780.0	−317.0	−69.4	−3.5	626	$(40^01-10^00)_{IV,II}$	Rc	5.0	o
6175.950	75364.0	−364.0	−52.0	−2.0	627	$(30^01-00^00)_{III}$	Rc	0.9	o
6179.01	77986.0	−58.0	−52.4	0	626	$(41^10-00^00)_{II}$	Rc	0.2	o
6196.174	77758.2	−370.4	−55.4	−1.1	626	$(31^11-01^10)_{III}$	Rc	75.0	+
6196.174	77958.6	−292.0	−55.8	−1.3	626	$(31^11-01^10)_{III}$	Rd	75.0	o
6205.60	77577.0	−465.0	−54.8	−2.8	626	$(40^01-10^00)_{III,I}$	Rc	3.0	o
6227.91	77689.0	−355.0	−57.4	−2.5	626	$(30^01-00^00)_{II}$	Rc	1200.0	o
6241.96	77609.0	−438.0	−47.6	2.4	636	$(30^01-00^00)_{II}$	Rc	13.0	o
6243.57	77702.0	−422.0	−53.0	0	636	$(31^11-01^10)_{II}$	Rc	1.1	o

Table B-III (*Continued*)

Band origin ω_0 cm⁻¹ (in vacuum)	Rotational Constants $a\times10^5$ (cm⁻¹)	$b\times10^5$	$c\times10^8$	$d\times10^8$	Isotope	Transition	Branch	Intensity S_v cm⁻¹/km atm$_{296\,\mathrm{K}}$	Interaction term, ζ_v
6243.57	77895.0	−353.0	−50.0	1.5	636	$(31^11{-}01^10)_{\mathrm{II}}$	Rd	1.1	0
6254.592	73344.0	−259.0	−41.4	1.5	628	$(30^01{-}00^00)_{\mathrm{II}}$	Rc	4.0	0
6288.492	77869.0	−175.0	−53.4	−0.5	626	$(22^21{-}00^00)_{\mathrm{II}}$	Rc	0.5	0
6298.07	75404.0	−324.0	−46.2	1.0	627	$(30^01{-}00^00)_{\mathrm{II}}$	Rc	0.8	0
6308.28	77604.0	−493.0	−61.4	0.5	626	$(40^01{-}10^00)_{\mathrm{III,II}}$	Rc	5.3	0
6318.17	77696.0	−450.0	−54.2	0.9	626	$(41^11{-}11^10)_{\mathrm{III,II}}$	Rc	0.5	0
6318.17	77949.0	−387.0	−54.8	1.2	626	$(41^11{-}11^10)_{\mathrm{III,II}}$	Rd	0.5	0
6346.27	77724.0	−318.0	−50.4	−0.6	626	$(40^01{-}10^00)_{\mathrm{I}}$	Rc	3.0	0
6347.855	77667.4	−376.4	−45.2	3.6	626	$(30^01{-}00^00)_{\mathrm{I}}$	Rc	1150.0	0
6356.293	77740.9	−387.7	−51.2	1.0	626	$(31^11{-}01^10)_{\mathrm{II}}$	Rc	90.0	0
6356.293	77936.8	−313.8	−48.2	2.5	626	$(31^11{-}01^10)_{\mathrm{II}}$	Rd	90.0	0
6359.32	77814.0	−348.0	−50.6	2.1	626	$(32^21{-}02^20)_{\mathrm{II}}$	Rc	2.5	0
6359.32	77814.0	−348.0	−49.2	2.4	626	$(32^21{-}02^20)_{\mathrm{II}}$	Rd	2.5	0
6363.62	77727.0	−320.0	−44.2	4.1	636	$(30^01{-}00^00)_{\mathrm{I}}$	Rc	3.5	0
6388.085	78050.8	7.2	−44.4	4.0	626	$(41^10{-}00^00)_{\mathrm{I}}$	Rc	6.0	+0.005
6397.545	77749.0	−374.0	−51.0	1.0	636	$(31^11{-}01^10)_{\mathrm{I}}$	Rc	0.2	0
6397.545	77948.0	−300.0	−45.0	4.0	636	$(31^11{-}01^10)_{\mathrm{I}}$	Rd	0.2	0
6429.172	73478.0	−154.8	−34.2	5.1	628	$(30^01{-}00^00)_{\mathrm{I}}$	Rc	0.3	0
6449.05	77751.0	−346.0	−56.2	3.1	626	$(40^01{-}10^00)_{\mathrm{III,II}}$	Rc	0.01	0
6463.48	75542.0	−186.0	−32.0	8.0	627	$(30^01{-}00^00)_{\mathrm{I}}$	Rc	0.1	0
6466.44	77593.0	−535.5	−61.2	−4.0	626	$(20^02{-}01^10)_{\mathrm{II,III}}$	Rc	0.4	+0.03
6466.44	−595.5	−595.5	−8.0	−4.0	626	$(20^02{-}01^10)_{\mathrm{II,III}}$	Qd	0.4	+0.03
6474.53	77888.0	−156.0	−56.2	−2.0	626	$(22^21{-}00^00)_{\mathrm{I}}$	Rc	0.4	0
6498.67	77666.0	−462.0	...	2.0	626	$(12^22{-}01^10)_{\mathrm{II}}$	Rc	0.6	+0.02
6498.67	77727.0	−523.0	...	0	626	$(12^22{-}01^10)_{\mathrm{II}}$	Rd	0.6	+0.02
6498.67	−462.0	−462.0	...	0	626	$(12^22{-}01^10)_{\mathrm{II}}$	Qc	0.6	+0.02
6498.67	−523.0	−523.0	4.0	2.0	626	$(12^22{-}01^10)_{\mathrm{II}}$	Qd	0.6	+0.02
6503.079	77820.6	−223.0	−41.2	5.6	626	$(30^01{-}00^00)_{\mathrm{I}}$	Rc	135.0	0
6532.65	77877.0	−165.0	−36.2	6.5	626	$(40^01{-}10^00)_{\mathrm{I}}$	Rc	0.1	0
6536.45	77825.0	−307.0	−47.2	3.0	626	$(31^11{-}01^10)_{\mathrm{I}}$	Rc	13.0	0
6536.45	78039.0	−212.0	−43.2	5.0	626	$(31^11{-}01^10)_{\mathrm{I}}$	Rd	13.0	0
6537.958	...	−542.4	−51.6	0.40	626	$(11^12{-}00^00)_{\mathrm{II}}$	Rc	6.0	+0.005
6537.958	−449.6	−449.6	−5.02	−2.51	626	$(11^12{-}00^00)_{\mathrm{II}}$	Qc	6.0	+0.005
6562.444	78058.0	−275.0	...	...	626	$(32^21{-}02^20)$	Rc	0.7	0

6562.444	78056.0	−277.0	…	…	626	(32^21-02^20)	*Rd*	0.7	0
6635.43	77904.0	−192.5	−42.8	9.8	626	$(40^01-10^00)_{II}$	*Rc*	0.05	0
6670.77	77613.0	−515.0	−52.8	0.2	626	$(12^22-01^10)_{I}$	*Rc*	0.7	+0.03
6670.77	77674.0	−576.0	−52.8	0.7	626	$(12^22-01^10)_{I}$	*Rd*	0.7	+0.03
6670.77	−515.0	−515.0	1.4	0.7	626	$(12^22-01^10)_{I}$	*Qc*	0.7	+0.03
6670.77	−576.0	−576.0	0.4	0.2	626	$(12^22-01^10)_{I}$	*Qd*	0.7	+0.03
6679.703	77452.8	−590.8	−50.1	1.15	626	$(11^22-00^00)_{I}$	*Rc*	8.0	+0.01
6679.709	…	−509.1	…	2.16	626	$(11^22-00^00)_{I}$	*Qc*	8.0	+0.01
6710.32	77505.0	−624.0	−51.2	1.0	626	$(20^02-01^10)_{I}$	*Rc*	0.2	0
6710.32	−685.0	−685.0	2.0	1.0	626	$(20^02-01^10)_{I}$	*Qd*	0.2	0
6728.36	72800.0	−837.0	−44.4	0	638	00^03-00^00	*Rc*	0.2	0
6745.12	77243.3	−879.5	−53.2	0	636	01^13-01^10	*Rc*	3.5	0
6745.12	77363.9	−884.5	−53.2	0	636	01^13-01^10	*Rd*	3.5	0
6780.215	77160.0	−887.0	−53.4	−0.5	636	00^03-00^00	*Rc*	44.0	0
6804.404	77177.0	−864.8	−55.7	−3.2	626	$(10^03-10^00)_{II}$	*Rc*	…	0
6867.281	77167.6	−904.8	−48.8	0.8	626	$(11^13-11^10)_{I}$	*Rc*	0.6	0
6867.281	77340.7	−933.9	−49.4	1.5	626	$(11^13-11^10)_{I}$	*Rd*	0.6	0
6870.795	77259.9	−886.3	−55.3	+0.3	626	$(11^13-11^10)_{II}$	*Rc*	0.3	0
6870.795	77442.4	−893.6	−58.2	−0.5	626	$(11^13-11^10)_{II}$	*Rd*	0.3	0
6897.751	77426.5	−905.5	−56.2	−0.9	626	02^33-02^20	*Rcd*	11.0	0
6897.80	76505.0	−920.0	−53.2	0	626	00^04-00^01	*Rc*	0.18	0
6905.770	77101.4	−940.4	−46.7	1.25	626	$(10^03-10^00)_{I}$	*Rc*	5.0	0
6907.144	77204.5	−892.3	−62.3	0.06	626	$(10^03-10^00)_{II}$	*Rc*	7.0	0
6922.21	72764.3	−868.5	−44.4	0	628	00^03-00^00	*Rc*	15.0	0
6935.15	77214.6	−914.0	−53.6	−0.2	626	01^13-01^10	*Rc*	310.0	0
6935.15	77331.6	−919.0	−53.2	0	626	01^13-01^10	*Rd*	310.0	0
6945.0	…	…	−48.2	0	627	00^03-00^00	*Rc*	2.0	0
6972.578	77120.8	−922.8	−52.4	0	626	00^03-00^00	*Rc*	4000.0	0
7460.53	77754.0	−290.0	−64.4	−6.0	626	$(40^01-00^00)_{IV}$	*Rc*	12.0	0
7481.51	77567.0	−480.0	−32.0	10.0	636	$(40^01-00^00)_{III}$	*Rc*	0.3	0
7583.265	77688.0	−441.0	−53.2	0	626	$(41^11-01^10)_{III}$	*Rc*	2.5	0
7583.265	77907.0	−344.0	−53.2	0	626	$(41^11-01^10)_{III}$	*Rd*	2.5	0
7593.69	77578.0	−466.0	−56.4	−2.0	636	$(40^01-00^00)_{III}$	*Rc*	34.0	0
7600.13	77609.0	−438.0	−38.4	7.0	626	$(40^01-00^00)_{II}$	*Rc*	0.2	0
7616.62	77975.0	−69.0	−43.2	4.6	626	$(51^10-00^00)_{II}$	*Rc*	0.05	+
7734.46	77725.0	−319.0	−51.2	0	626	$(40^01-00^00)_{II}$	*Rc*	9.0	0
7743.70	77551.0	−493.0	−50.0	1.0	626	$(21^12-00^00)_{III}$	*Rc*	0.6	+0.03
7743.70	−373.0	−373.0	−4.0	−2.0	626	$(21^12-00^00)_{III}$	*Qc*	0.6	+0.03
7757.626	77789.0	−339.5	−40.0	6.6	626	$(41^11-01^10)_{II}$	*Rc*	0.8	0
7757.626	77978.0	−272.5	−45.4	3.9	626	$(41^11-01^10)_{II}$	*Rd*	0.8	0

Table B-III (Continued)

Band origin ω_0 cm⁻¹ (in vacuum)	Rotational Constants				Isotope	Transition	Branch	Intensity S_v cm⁻¹/km atm$_{296\,K}$	Interaction term, ζ_v
	$a \times 10^5$ (cm⁻¹)	$b \times 10^5$	$c \times 10^8$	$d \times 10^8$					
7897.573	...	−132.0	...	...	626	$(32^21{-}00^00)_I$	Qc	0.1	0
7901.479	77418.0	−626.0	...	...	626	$(21^12{-}00^00)_{II}$	Rc	0.7	+0.015
7901.479	−514.0	−514.0	...	...	626	$(21^12{-}00^00)_{II}$	Qc	0.7	+0.015
7920.84	77878.0	−166.0	−37.8	7.3	626	$(40^11{-}00^00)_I$	Rc	0.6	0
7929.92	77302.0	−821.0	−55.0	−1.0	636	$(11^13{-}01^10)_{II}$	Rc	0.05	0
7929.92	77460.0	−798.0	−57.0	−2.0	636	$(11^13{-}01^10)_{II}$	Rd	0.05	0
7961.29	77892.0	−236.0	−46.0	3.0	626	$(41^11{-}01^10)_I$	Rc	0.06	0
7961.29	78104.0	−146.0	−44.0	4.0	626	$(41^11{-}01^10)_I$	Rd	0.06	0
7981.29	77251.0	−796.0	−56.2	−1.9	636	$(10^03{-}00^00)_{II}$	Rc	0.8	0
8000.80	77257.0	−785.0	−58.8	−4.8	626	$(20^03{-}10^00)_{III,I}$	Rc	0.006	0
8056.024	77463.0	−581.0	−48.0	2.2	626	$(21^12{-}00^00)_I$	Rc	0.6	+0.01
8056.024	−469.3	−469.3	5.0	2.5	626	$(21^12{-}00^00)_I$	Qc	0.6	+0.01
8070.91	77190.0	−933.0	−52.0	0.5	636	$(11^13{-}01^10)_I$	Rc	0.15	0
8070.91	77324.0	−924.0	−53.0	0.0	636	$(11^13{-}01^10)_I$	Rd	0.15	0
8084.06	77472.0	−861.0	−50.8	2.0	626	$(12^23{-}02^20)_{II}$	Rc	0.3	0
8084.06	...	...	−53.2	0.4	626	$(12^23{-}02^20)_{II}$	Rd	0.3	0
8089.04	77097.0	−950.0	−51.0	0.7	636	$(10^03{-}00^00)_I$	Rc	1.9	0
8103.58	77284.0	−812.0	−65.4	−1.5	626	$(20^03{-}10^00)_{III,II}$	Rc	0.4	0
8120.104	72785.0	−848.0	−47.4	−1.5	628	$(10^03{-}00^00)_{II}$	Rc	0.52	0
8128.78	77069.0	−973.0	−52.8	−1.8	626	$(20^03{-}10^00)_{II,I}$	Rc	0.2	0
8135.886	77251.1	−877.5	−53.9	−0.36	626	$(11^13{-}01^10)_{II}$	Rc	10.0	0
8135.886	77399.7	−850.9	−57.0	−1.89	626	$(11^13{-}01^10)_{II}$	Rd	10.0	0
8154.47	74870.0	−858.0	−44.0	2.0	627	$(10^03{-}00^00)_{II}$	Rc	0.1	0
8192.556	77187.0	−865.7	−57.3	−2.44	626	$(10^03{-}00^00)_{II}$	Rc	114.0	0
8220.363	72782.0	−850.8	−41.4	1.5	628	$(10^03{-}00^00)_I$	Rc	0.52	0
8231.56	77096.0	−1000.0	−59.4	1.5	626	$(20^03{-}10^00)_{II,II}$	Rc	0.3	0
8243.16	77148.0	−894.0	−47.0	1.0	626	$(20^03{-}10^00)_{I,I}$	Rc	0.2	0
8254.80	77412.0	−921.0	−54.8	0	626	$(12^23{-}02^20)_I$	Rc	0.5	0
8254.80	...	...	−52.0	1.0	626	$(12^23{-}02^20)_I$	Rd	0.5	0
8255.39	74835.0	−893.0	−44.0	2.0	627	$(10^03{-}00^00)_I$	Rc	0.1	0
8276.767	77190.7	−937.9	−50.2	1.5	626	$(11^13{-}01^10)_I$	Rc	17.0	0
8276.767	77328.7	−921.9	−49.8	1.7	626	$(11^13{-}01^10)_I$	Rd	17.0	0
8293.955	77101.0	−942.2	−47.4	2.5	626	$(10^03{-}00^00)_I$	Rc	165.0	0

for the Q branch ($J' = J''$). The isotopes 626 and 636 are symmetrical linear molecules; isotopes like 638 and 627 are non-symmetrical linear molecules. The symmetry, of lack thereof, has the following practical result that some rotational levels are forbidden for the symmetrical molecules and they are allowed for the unsymmetrical molecules. The allowed rotational levels and hence the vibration-rotation transitions which are allowed, by the selection rules, depend on the vibrational state of the molecule. It is discussed in detail by Herzberg (1945) and we will not repeat that material here. We simply mention it for the benefit of the reader who may not be familiar with molecular spectra. The bands of the symmetrical molecules, whose lower state is the vibrational ground state, only have lines for even values of the rotational quantum number, e.g. $R(4)$, $R(6)$, $R(8)$; alternate lines are missing from the spectrum. For the non-symmetrical isotopes, all the lines will be present in the spectrum, e.g. $R(4)$, $R(5)$, $R(6)$. Various tables are available for the carbon dioxide molecule which permit the intensity of a rotational line to be found directly from the band intensity. For example, Gray (1965) has tabulated relative line intensities, $S^0(J)/S^0(v)$, for the 626 molecule at 300 K for all the allowed transitions from the ground state and the first excited state (01^10). These are applicable to laboratory data measured at room temperature. Gray (1967) has also tabulated relative line intensities for the 626 molecule for the temperature range 160 K to 280 K at intervals of 20 K, for ground state (00^00) transitions. Young (1970) gives relative line intensities for the 636, 628, 627, and 638 molecules at 200 K, 250 K and 300 K, for ground state transitions.

For approximate intensity calculations, the rotational partition function, $Q_{rot}(T)$, (which appears in Equation (15) for the line intensity) can be assumed to be given by the expression for a rigid rotator:

$$Q_{rot}(T) = (kT/hcB'')(1/s),$$

where s is the symmetry number of the molecule ($s = 2$ for 626 and 636; $s = 1$ for 627, 628 and 638). Gray and Young (1969) have tabulated internal partition functions for the isotopes of CO_2 from 180 K to 300 K at intervals of 10 K; the 626 isotope has the internal partition functions tabulated from 180 K to 1230 K at intervals of 10 K.

The vibrational energy states are called Σ, Π, Δ, Φ, ... states for values of the vibrational angular momentum $l = 0, 1, 2, 3, \ldots$. All vibrational states are doubly degenerate for $l \neq 0$, since there are two equivalent directions for the vibrational angular momentum vector l. This causes each rotational level to be split into two components; they are separated by an energy difference

$$\delta E(J) = hc(qJ(J + 1) - \mu J^2(J + 1)^2),$$

where $q = B^d - B^c$ and $\mu = D^d - D^c$. For many rotational levels the splitting may be so slight that $B^d = B^c$ for all practical purposes. The vibrational energy states of symmetrical molecules (626, etc.) have an additional label which refers to the behavior of the eigenfunctions with respect to the operation of inversion. A subscript g indicates that the eigenfunctions are symmetric and a subscript u indicates that they are antisymmetric. These subscripts tell us whether B^d or B^c is the appropriate constant to use for a particular rotational line. For the unsymmetrical isotopes (e.g. 628) Table B-Ia summarizes the rules for using B^d and B^c. Table B-Ib gives the rules for symmetrical isotopes (e.g. 626): not all possible states are listed, but all of the lower vibrational states giving rise to bands in the spectrum of Venus are included.

We have attempted to limit the theoretical discussion to a minimum and still provide some explanation of the quantities that are tabulated. The last quantity which we shall mention is the constant ζ_v. A vibration-rotation interaction occurs for some bands and it causes the rotational intensity distribution to be modified by a factor $(1 + \zeta_v m)^2$. We denote the unperturbed line intensity by $S^0(m)$ and the perturbed line intensity by $S_v(m)$ where $S_v(m) = S^0(m)[1 + m\zeta_v]^2$.

In Equation (16) we defined the band intensity $S(v)$ as the sum of the individual line intensities. For bands with no Coriolis interaction, i.e. $\zeta_v = 0$, this definition still applies:

$$S^0(v) = \sum_\omega S^0(m).$$

However for bands with a vibration-rotation interaction, it is convenient to use this same definition of band intensity even though it is not the actual band intensity, $S_v(v) = \sum_m S_v(m)$. We have tabulated $S^0(v)$ in units of cm^{-1} per km atm of carbon dioxide at a pressure of 1 atm and a temperature of 296 K (room temperature).

We give some conversion factors for various units which have been used to report band intensities in Table B-II. These may be used to convert our measurements to other systems of units.

DISCUSSION

Traub: Doppler shift measurements in the 8700 Å CO_2 band have been made on a number of occasions by N. P. Carleton and myself. The result is that on most days for which we have good data, the equatorial wind speed appears to be near 100 m s⁻¹, retrograde, with errors of the order of 10 m s⁻¹. The same speed has been measured by us for a Fraunhofer line, indicating that both the visible clouds, and the CO_2 gas participate in this rotation.

References

Abbot, C. G., Fowle, F. E., and Aldrich, L. B.: 1922, *Ann. Astrophys. Obs. Smithsonian Inst.* **4**, 334–344.
Adams, W. S. and Dunham, T.: 1932, *Publ. Astron. Soc. Pacific* **44**, 243.
Adel, A.: 1937, *Astrophys. J.* **86**, 337.
Adel, A. and Dennison, D. M.: 1933, *Phys. Rev.* **43**, 716; *ibid.* **44**, 99.
Adel, A. and Slipher, V. M.: 1934, *Phys. Rev.* **46**, 240.
Adel, A., Slipher, V. M., and Barker, E. F.: 1935, *Phys. Rev.* **47**, 580.
Angstrom, K.: 1890, *Ann. Phys.* **39**, 267.
Arrhenius, S.: 1908, *Worlds in the Making*, Harper, New York, p. 58.
Arrhenius, S.: 1918, *The Destinies of the Stars*, Putnam, New York, p. 250.
Avduevsky, V. S., Marov, M. Ya., and Rozhdestvensky, M. K.: 1968, *J. Atmospheric Sci.* **25**, 537.
Avduevsky, V. S., Marov, M. Ya., and Rozhdestvensky, M. K.: 1970, *J. Atmospheric Sci.* **27**, 561.
Barnard, E. E.: 1897, *Astrophys. J.* **5**, 299.
Barrett, A. H.: 1961, *Astrophys. J.* **133**, 281.

Barrett, A. H. and Staelin, D. H.: 1964, *Space Sci. Rev.* **3**, 109.
Belopolsky, A.: 1900, *Astron. Nachr.* **152**, 263.
Beer, R., Norton, R. H., and Martonchik, J. V.: 1971, *Astrophys. J.* **168**, L121.
Belton, M. J. S.: 1968, *J. Atmospheric Sci.* **25**, 596.
Belton, M. J. S. and Hunten, D. M.: 1966, *Astrophys. J.* **146**, 307.
Belton, M. J. S. and Hunten, D. M.: 1968, *Astrophys. J.* **153**, 970.
Belton, M. J. S. and Hunten, D. M.: 1969, *Astrophys. J.* **156**, 797.
Belton, M. J. S., Hunten, D. M., and Goody, R. M.: 1968, in J. C. Brandt and M. B. McElroy (eds.) *The Atmospheres of Venus and Mars*, Gordon and Breach, New York, p. 69.
Bemporad, A.: 1901, *Mem. Soc. Spett. Ital.* **30**, 217.
Bemporad, A.: 1904, *Mitt. Grossh. Sternwarte Heidelberg* **4**, 1.
Bottema, M., Plummer, W., and Strong, J.: 1964a, *Astrophys. J.* **139** 1021.
Bottema, M., Plummer, W., and Strong, J.: 1965a, *Ann. Astrophys.* **28**, 225.
Bottema, M., Plummer, W., Strong, J., and Zander, R.: 1964b, *Astrophys. J.* **140**, 1640.
Bottema, M., Plummer, W., Strong, J., and Zander, R.: 1965b, *J. Geophys. Res.* **70**, 4401.
Bouguer, P.: 1760, *Traité d'optique sur la gradiation de la lumière*, ouvrage posthume, publié par l'abbé de Lacaille, Paris.
Boyer, C.: 1965, *Astronomie* **79**, 323.
Boyer, C. and Camichel, H.: 1961, *Ann. Astrophys.* **24**, 531.
Boyer, C. and Camichel, H.: 1967, *Compt. Rend. Acad. Sci.* **264**, 990.
Boyer, C. and Guerin, P.: 1966, *Compt. Rend. Acad. Sci.* **263**, 253.
Boyer, C. and Guérin, P.: 1969, *Icarus* **11**, 338.
Boyer, C. and Guerin, P.: 1971, *Compt. Rend. Acad. Sci.* **273**, 154.
Caldwell, J.: 1972, *Icarus* **17**, 608.
Carpenter, R. L.: 1970, *Astron. J.* **75**, 61.
Chamberlain, J. W.: 1965, *Astrophys. J.* **141**, 1184.
Chamberlain, J. W.: 1970, *Astrophys. J.* **159**, 137.
Chamberlain, J. W. and Kuiper, G. P.: 1956, *Astrophys. J.* **124**, 399.
Chamberlain, J. W. and Smith, G. R.: 1970, *Astrophys. J.* **160**, 755.
Chamberlain, J. W. and Smith, G. R.: 1972, *Astrophys. J.* **173**, 469.
Chandrasekhar, S.: 1950, *Radiative Transfer*, Oxford University Press
Chapman, S.: 1931, *Proc. Phys. Soc.* **43**, 483.
Chase, S. C., Kaplan, L. D., and Neugebauer, G.: 1963, *J. Geophys. Res.* **68**, 6157.
Claydon, A. W.: 1909, *Monthly Notices Roy. Astron. Soc.* **69**, 195.
Claydon, A. W.: 1918, *Monthly Notices Roy. Astron. Soc.* **79**, 507.
Coblentz, W. W.: 1905, *Investigations of Infrared Spectra*, Carnegie Inst., Washington, D. C.
Coblentz, W. W. and Lampland, C. O.: 1924, *Publ. Astron. Soc. Pacific* **36**, 272.
Coblentz, W. W. and Lampland, C. O.: 1925a, *J. Franklin Inst.* **199**, 785.
Coblentz, W. W. and Lampland, C. O.: 1925b, *J. Franklin Inst.* **200**, 103.
Coblentz, W. W.and Lampland, C. O.: 1927, *Nat. Bur. Stand. Sci. Papers* **22**, 237.
Coffeen, D. L.: 1968, Thesis, Univ. Arizona.
Coffeen, D. L.: 1969, *Astron. J.* **74**, 446.
Coffeen, D. L. and Gehrels, T.,: 1969, *Astron. J.* **74**, 433.
Connes, P., Connes, J., Benedict, W. S., and Kaplan, L. D.: 1967, *Astrophys. J.* **147**, 1230.
Connes, P., Connes, J., Kaplan, L. D., and Benedict, W. S.: 1968, *Astrophys. J.* **152**, 731.
Connes, J., Connes, P., and Maillard, J. P.: 1969, *Atlas des spectres dans le proche infrarouge de Venus, Mars, Jupiter et Saturne*, CNRS, Paris.
Courtoy, C. P.: 1959, *Ann. Soc. Sci. Bruxelles* **73**, 5.
Cruikshank, D. P.: 1967, *Comm. Lunar Planet. Lab.*, No. 98.
Cruikshank, D. P. and Kuiper, G. P.: 1967, *Comm. Lunar Planet. Lab.*, No. 97.
Danielson, R. E. and Solomon, P. M.: 1966, *Astron. J.* **71**, 382.
Danilov, A. D. and Yatsenko, S. P.: 1963, *Geomagnetizm i Aeronomiya* **3**, 585.
Danilov, A. D.: 1964, *Kosmich. Issled., Akad. Nauk SSSR* **2**, 188.
Davidson, G. T. and Anderson, A. D.: 1967, *Science* **156**, 1729.
Dayhoff, M. O., Eck, R. V., Lippincott, E. R., and Sagan, C.: 1967, *Science* **155**, 556.
Deirmendjian, D.: 1964, *Icarus* **3**, 109.
Dickel, J. R.: 1967, *Icarus* **6**, 417.

Dollfus, A.: 1953, *Astronomie* **67**, 61.

Dollfus, A.: 1955a, *Astronomie* **69**, 413.

Dollfus, A.: 1955b, Thése, Univ. Paris, *Suppl. Ann. Astrophys*, 1955.

Dollfus, A.: 1958, *Compt. Rend. Acad. Sci.* **246**, 2345.

Dollfus, A.: 1963a, *Compt. Rend. Acad. Sci.* **256**, 1920.

Dollfus, A.: 1963b, *Compt. Rend. Acad. Sci.* **256**, 3250.

Dollfus, A.: 1966, *Proc. Caltech-JPL Lunar and Planetary Conference* (JPL TM 33–266).

Dollfus, A.: 1968, in J. Brandt and M. B. McElroy (eds.), *Atmospheres of Venus and Mars*, Gordon and Breach, New York, p. 147.

Dollfus, A. and Coffeen, D. L.: 1970, *Astron. Astrophys.* **8**, 251.

Drake, F. D.: 1967, *Astrophys J.* **149**, 459.

Dunham, T., Jr.: 1949 in G. P. Kuiper (ed.), *The Atmospheres of the Earth and Planets*, 1st ed. University of Chicago Press, p. 286.

Eshleman, V.: 1968, *Science* **162**, 661.

Espinola, R. P. and Blau, H. H.: 1965, *J. Geophys. Res.* **70**, 6263.

Evershed, J.: 1918, *Observatory* **41**, 371.

Evershed, J.: 1919a, *Observatory* **42**, 51.

Evershed, J.: 1919b, *Monthly Notices Roy. Astron. Soc.* **79**, 257.

Evershed, J.: 1919c, *Monthly Notices Roy. Astron. Soc.* **80**, 7.

Fabry, Ch.: 1929, *J. Obs.* **12**, 1.

Fesenkov, V. G.: 1955, *Astron. Zh.* **32**, 265.

Fink, U., Larson, H. P., Kuiper, G. P., and Poppin, R. F.: 1972, *Icarus* **17**, 617.

Flammarion, C.: 1894, *Observatory* **17**, 354.

Gehrels, T. and Samuelson, R. E.: 1961, *Astrophys. J.* **134**, 1022.

Gerasimovič,: 1937, *Pulkova Obs. Bull.*, No. 127.

Gillett, F. C., Low, F. J., and Stein, W. A.: 1968, *J. Atmospheric Sci.* **25**, 594.

Gray, L. D.: 1965, *J. Quant. Spectrosc. Radiat. Transfer* **5**, 795.

Gray, L. D.: 1966, *J. Opt. Soc. Am.* **56**, 1455.

Gray, L. D.: 1967, *Icarus* **8**, 513.

Gray, L. D.: 1969, *Icarus* **10**, 90.

Gray, L. D. and Schorn, R. A.: 1968, *Icarus* **8**, 409.

Gray, L. D., Schorn, R. A., and Barker, E.: 1969, *Appl. Opt.* **8**, 2087.

Gray, L. D. and Young, A. T.: 1969, *J. Quant. Spectrosc. Radiat. Transfer* **9**, 569.

Green, A. E. S. and Martin, J. D.: in A. E. S. Green (ed.), 1966, *The Middle Ultraviolet: Its Science and Technology*, John Wiley and Sons, New York, Chap. 7.

Hall, R. W. and Branson, N. J. B. A.: 1971, *Monthly Notices Roy. Astron. Soc.* **151**, 185.

Hanel, R., Forman, M., Stamback, G., and Meilleur, T.: 1968, *J. Atmospheric Sci.* **25**, 586.

Hansen, J. E. and Arking, A.: 1971, *Science* **171**, 669.

Hansen, J. E. and Cheyney, H.: 1968, *J. Geophys. Res.* **73**, 6136.

Hartech, P., Reeves, R. R., Jr., and Thompson, B. A:. 1963, NASA TN D-1984.

Hastings, C. S.: 1883, *Sidereal Messenger* **1**, 273.

Held, E. F. M. van der: 1931, *Z. Physik* **70**, 508.

Herschel, W.: 1793, *Phil. Trans.* **100**, 201.

Herzberg, G.: 1945, *Molecular Spectra and Molecular Structure, II. Infrared and Raman Spectra of Polyatomic Molecules*, Van Nostrand, New York.

Herzberg, G.: 1950, *Molecular Spectra and Molecular Structure, I. Spectra of Diatomic Molecules*, Van Nostrand, Princeton, (2nd Ed.), p. 169.

Herzberg, G.: 1951, *J. Roy Astron. Soc. Can.* **45**, 100.

Herzberg, G.: 1966, *Molecular Spectra and Molecular Structure, III. Electronic Structure of Polyatomic Molecules*, Van Nostrand, Princeton, p. 507.

Heyden, F. S., Kiess, C. C., and Kiess, H. K.: 1959, *Science* **130**, 1195.

Ho, W., Kaufman, I. A., and Thaddeus, P.: 1966, *J. Geophys. Res.* **71**, 5091.

Howard, J. N., Burch, D. L., and Williams, D.: 1951, AFCRC-TR-55-213, Cambridge, Massachusetts.

Hoyle, F.: 1955, *Frontiers of Astronomy*, Harper, New York, p. 68.

Huggins, W.: 1864, *Phil. Trans.* **154**, 422.

Hulst, H. C. van de : 1952, in G. P. Kuiper (ed.), *The Atmospheres of the Earth and Planets*, (2nd ed.) Univ. of Chicago Press, Chap. 3.

Hunt, G. E.: 1972a, *J. Quant. Spectrosc. Radiat. Transfer* **12**, 387.
Hunt, G. E.: 1972b, *J. Quant. Spectrosc. Radiat. Transfer* **12**, 405.
Hunt, G. E.: 1973, *J. Quant. Spectrosc. Radiat. Transfer* **13**, 465.
Hunt, G. E. and Schorn, R. A. J.: 1971, *Nature Phys. Sci.* **233**, 39.
Hunten, D. M.: 1971, *Space Sci. Rev.* **12**, 539.
Hunten, D. M., Belton, M. J. S., and Spinrad, H.: 1967, *Astrophys. J.* **150**, L125.
Hunten, D. M. and Goody, R. M.: 1969, *Science* **165**, 1317.
Janssen, M. A., Hills, R. E., Thornton, D. D., and Welch, W. J.: 1973, *Science* **179**, 994.
Jenkins, E. B., Morton, D. C., and Sweigart, A. V.: 1969, *Astrophys. J.* **157**, 913.
Jones, D. E.: 1961, *Planetary Space Sci.* **5**, 166.
Kaplan, L. D.: 1961, *Planetary Space Sci.* **8**, 23.
Kaplan, L. D.: 1962, *Mem. Soc. Roy. Sci. Liége* **7**, 323.
Kaplan, L. D.: 1963, in *Encyclopaedic Dictionary of Physics, Supplementary*, Vol. I, Pergamon, Oxford.
King, A. S.: 1922, *Astrophys. J.* **55**, 411.
Kliore, A., Levi, G. S., Cain, D. L., Fjeldbo, G., and Rasool, S. I.: 1967, *Science* **158**, 1683.
Kliore, A. and Cain, D. L.: 1968, *J. Atmospheric Sci.* **25**, 549.
Kourganoff, V.: 1952, *Basic Methods in Transfer Problems*, Oxford University Press.
Kozyrev, N. A.: 1954a, *Publ. Crimean Astrophys. Obs.* **12**, 169.
Kozyrev, N. A.: 1954b, *Publ. Crimean Astrophys. Obs.* **12**, 177.
Kozyrev, N. A.: 1966, *Publ. Main Astron. Obs. Pulkova* **24**, 76.
Kuiper, G. P.: 1947, *Astrophys. J.* **106**, 251.
Kuiper, G. P.: 1949, *Astrophys. J.* **109**, 540.
Kuiper, G. P.: 1952, in G. P. Kuiper (ed.), *The Atmospheres of the Earth and Planets*, Univ. of Chicago Press, p. 371.
Kuiper, G. P.: 1954, *Astrophys. J.* **120**, 603.
Kuiper, G. P.: 1957, in M. Zelihoff (ed.), *The Threshold of Space*, Pergamon Press, New York, p. 85.
Kuiper, G. P.: 1962, *Comm. Lunar Planet. Lab.*, No. 15.
Kuiper, G. P.: 1967, *Newsweek*, 12 June, p. 65.
Kuiper, G. P.: 1969, *Comm. Lunar Planet. Lab.*, No. 101.
Kuiper, G. P. and Forbes, F. F.: 1967, *Comm. Lunar Planet. Lab.*, No. 95.
Kuiper, G. P., Forbes, F. F., Steinmetz, D. L., and Mitchell, R. I.: 1969, *Comm. Lunar Planet. Lab.*, No. 100.
Kuiper, G. P., Sill, G. T., Cruikshank, D. P., and Fink, U.: 1967, *Comm. Lunar Planet. Lab.*, No. 99,
Kuzmin, A. D.: 1964, in M. Florkin and A. Dollfus (eds.), 'Life Sciences and Space Research', North. Holland, Amsterdam, Vol. II, p. 211.
Kuzmin, A. D.: 1967, *Radiophysics 1965–1966*, Acad. Sci. USSR, Moscow.
Ladenburg, R. and Reiche, F.: 1913, *Ann. Phys.* **42**, 18.
Lambert, J. H.: 1760, *Photometria sine de Mensura et Gradibus Luminis, Colorum et Umbrae, Augustae Vindelicorum*.
Langley, S. P. and Abbott, C. G.: 1900, *Ann. Astrophys. Obs. Smithsonian Inst.* **1**.
Laplace, P. S.: 1805, *Mechanique Celeste* **4**, Book 10, Chap. 3.
Lemansky, M. S.: 1871, *Monatsberichte der Königlichen Akademie der Wissenschaften zu Berlin*.
Lewis, J. S.: 1968a, *Icarus* **8**, 434.
Lewis, J. S.: 1968b, *Astrophys. J.* **152**, L79.
Lewis, J. S.: 1969, *Icarus* **11**, 367.
Lewis, J. S.: 1971a, *Nature* **230**, 295.
Lewis, J. S.: 1971b, *Am. Scientist* **59**, 557.
Lewis, J. S.: 1972, *Astrophys. J.* **171**, L75.
Link, F.: 1969, *Eclipse Phenomena in Astronomy*, Springer Verlag, New York.
Lowell, P.: 1905, *Lowell Obs. Bull.* No. 17.
Lyman, T.: 1866, *Am. J. Sci.* (2) **43**, 129.
Lyman, T.: 1874, *Am. J. Sci.* (3) **9**, 47.
Lyot, B.: 1929, *Ann. Obs. Paris (Meudon)* **8**, 1.
Mädler, I. H.: 1849, *Astron. Nachr.* **29**, 107.
Margolis, H.: 1964, *Bull. Atomic Scientists*, April, p. 38.
Margolis, J. S., Schorn, R. A. J., and Young, L. D. G.: 1971, *Icarus* **15**, 197.
Marov, M. Ya.: 1972, *Icarus* **16**, 415.

Mayer, C. H., McCullough, T. P., and Sloanaker, R. M.: 1958, *Astrophys. J.* **127**, 1.

McClatchey, R. A.: 1967, *Astrophys. J.* **148**, L93.

McClatchey, R. A., Benedict, W. S., Clough, S. A., Burch, D. E., Calfee, R. F., Fox, K., Rothman, L.S., and Garing, J. S.: 1973, *Atmospheric Absorption Line Parameters Compilation*, AFCRL-TR-73-0096.

Menzel, D. H.: 1923, *Astrophys. J.* **58**, 65.

Menzel, D. H.: 1936, *Astrophys. J.* **84**, 462.

Menzel, D. H., Coblentz, W. W., and Lampland, C. O.: 1926, *Astrophys. J.* **63**, 177.

Menzel, D. H. and Whipple, F. L.: 1955, *Publ. Astron. Soc. Pacific* **67**, 161.

Mintz, Y.: 1961, *Planetary Space Sci.* **5**, 141.

Möller, J.: 1900, *Astron. Nachr.* **152**, 76.

Moore, P.: 1959, *The Planet Venus*, Mcmillan, New York.

Moroz, V. I.: 1963, *Astron. Zh.* **40**, 144.

Moroz, V. I.: 1964a, *Astron. Zh.* **41**, 411.

Moroz, V. I.: 1964b, *Mem. Soc. Roy. Sci. Liège* **9**, 406.

Moroz, V. I.: 1967, *Astron. Zh.* **44**, 816.

Moroz, V. I.: 1971, *Nature Phys. Sci.* **231**, 36.

Mueller, R. F.: 1964, *Icarus* **3**, 285.

Murray, B. C., Wildey, R. L., and Westphal, J. A.: 1963a, *Science* **140**, 391.

Murray, B. C., Wildey, R. L., and Westphal, J. A.: 1963b, *J. Geophys. Res.* **68**, 4813.

Newcomb, S.: 1906, *A Compendium of Sperical Astronomy*, Macmillan, New York, p. 199.

Newkirk, G.: 1959, *Planetary Space Sci.* **1**, 32.

Ohring, G.: 1966, *Icarus* **5**, 329.

Ohring, G.: 1969, *Icarus* **11**, 171.

O'Leary, B.: 1966, *Astrophys. J.* **146**, 754.

O'Leary, B.: 1970, *Icarus* **13**, 292.

Öpik, E. J.: 1955, *Irish Astron. J.* **3**, 192.

Öpik, E. J.: 1956, *Irish Astron. J.* **4**, 37.

Öpik, E. J.: 1965, *Irish Astron. J.* **7**, 130.

Owen, T.: 1962, *Comm. Lunar Planet. Lab.* **1**, 29–31.

Owen, T.: 1967, *Astrophys. J.* **150**, L121.

Owen, T.: 1968a, *Science* **161**, 915.

Owen, T.: 1968b, *J. Atmospheric Sci.* **25**, 583.

Owen, T. and Sagan, C.: 1972, *Icarus* **16**, 557.

Paschen, F.: 1894, *Ann. Physik* **53**, 324.

Payne-Gaposchkin, C.: 1950, *The Reporter*, March 14, p. 37.

Petit, E. and Nicholson, S. B.: 1923, *Publ. Astron. Soc. Pacific* **35**, 194.

Petit, E. and Nicholson, S. B.: 1924, *Publ. Astron. Soc. Pacific* **36**, 227, 269.

Petit, E. and Nicholson, S. B.: 1927, *Publ. Am. Astron. Soc.* **5**, 184.

Petit, E. and Nicholson, S. B.: 1930, *Astrophys. J.* **71**, 102.

Petit, E. and Nicholson, S. B.: 1936, *Astrophys. J.* **83**, 84.

Petit, E. and Nicholson, S. B.: 1955, *Publ. Astron. Soc. Pacific* **67**, 203.

Plummer, W. T.: 1969a, *Science* **163**, 1191.

Plummer, W. T.: 1969b, *J. Geophys. Res.* **74**, 3331.

Plummer, W. T.: 1970, *Icarus* **12**, 233.

Plummer, W. T. and Carson, R. K.: 1970, *Astrophys. J.* **159**, 159.

Plummer, W. T. and Strong, J.: 1965, *Astronaut. Acta* **11**, 375.

Plummer, W. T. and Strong, J.: 1966, *Astrophys. J.* **144**, 422.

Plummer, W. T. and Strong, J.: 1967, *Astrophys. J.* **149**, 463.

Pollack, J. B.: 1969, *Icarus* **10**, 314.

Pollack, J. B. and Morrison, D.: 1970, *Icarus* **12**, 376.

Pollack, and J. B. Sagan, C.: 1968, *J. Geophys. Res.* **73**, 5943.

Pollack, J. B., and Wood, A. T., Jr.,: 1968, *Science* **161**, 1125.

Priester, W., Roemer, M., and Schmidt-Kaler, T.: 1962, *Nature* **196**, 465.

Prokofiev, V. K. and Petrova, N. N.: 1962, *Mem. Soc. Roy. Sci. Liège* **7**, 311.

Rasool, S. I.: 1970, *Radio Sci.* **5**, 367.

Rea, D. G.: 1972, *Rev. Geophys. Space Sci.* **10**, 369.

Rea, D. G. and O'Leary, B. T.: 1968, *J. Geophys. Res.* **73**, 664.

Regas, J. L., Boese, R. W., Giver, L. P., and Miller, J. H.: 1973, *J. Quant. Spectrosc. Radiat. Transfer* **13**, 461.

Regas, J. L., Giver, L. P., Boese, R. W., and Miller, J. H.: 1972, *Astrophys. J.* **173**, 711.

Regas, J. L. and Sagan, C.: 1970a, *Comments Astrophys. Space Phys.* **2**, 116.

Regas, J. L. and Sagan, C.: 1970b, *Comments Astrophys. Space Phys.* **2**, 138.

Richardson, E. H.: 1960, *Astron. J.* **65**, 56.

Richardson, R. S.: 1955, *Publ. Astron. Soc. Pacific* **67**, 304.

Richardson, R. S.: 1958, *Publ. Astron. Soc. Pacific* **70**, 251.

Robbins, R. C.: 1965, *Planetary Space Sci.* **12**, 1143.

Ross, F. E.: 1928, *Astrophys. J.* **68**, 57.

Ross, F. E.: 1931, *Publ. Am. Astron. Soc.* **6**, 30.

Rubens, H. and Aschkinass, E.: 1898, *Ann. Physik* **64**, 584.

Russell, H. N.: 1899, *Astrophys. J.* **9**, 297.

Russell, H. N.: 1916, *Astrophys. J.* **43**, 190.

Sagan, C.: 1960, Caltech/JPL Tech. Rept, No. 32–34.

Sagan, C.: 1961, *Science* **133**, 849.

Sagan, C.: 1967a, *Astrophys. J.* **149**, 731.

Sagan, C.: 1967b, *Nature* **216**, 1198.

Sagan, C.: 1968, *Astrophys. J.* **152**, 1119.

Sagan, C.: 1971, in C. Sagan, T. C. Owen, and H. J. Smith (eds.), 'Planetary Atmospheres', *IAU Symp.* **40**, 116.

Sagan, C. and Kellogg, W. W.: 1963, *Ann. Rev. Astron. Astrophys.* **1**, 235.

Sagan, C. and Pollack, J. B.: 1967, *J. Geophys. Res.* **72**, 469.

Sagan, C. and Pollack, J. B.: 1969, *Icarus* **10**, 247.

Sagan, C. and Regas, J.: 1970, *Comments Astrophys. Space Phys.* **2**, 161.

Scheiner, J.: 1894, *A Treatbise on Astronomical Spectroscopy*, Ginn and Co., Boston.

Schiaparelli, G. V.: 1890, *Publ. Astron. Soc. Pacific* **2**, 246.

Schilling, G. F. and Moore, R. C.: 1967, Memorandum RM-5386-PR, 'The Twilight Atmosphere of Venus', Rand Corporation, July 1967.

Schoenberg, E.: 1929, in *Handbuch der Astrophysik*, Band II, 'Grundlagen der Astrophysik', Julius Springer, Berlin, pp. 171–190.

Schorn, R. A. and Young, L. G.: 1971, *Icarus* **15**, 103.

Schorn, R. A., Barker, E. S., Gray, L. D., and Moore, R. C.: 1969a, *Icarus* **10**, 98.

Schorn, R. A., Gray, L. D., and Barker, E. S.: 1969b, *Icarus* **10**, 241.

Schorn, R. A., Young, L. G., and Barker, E. S.: 1970, *Icarus* **12**, 391.

Schorn, R. A., Young, L. D. G., and Barker, E. S.: 1971, *Icarus* **14**, 21.

Schröter, J.: 1792, *Phil. Trans.* **100**, 309.

Scott, A. H. and Reese, E. J.: 1972, *Icarus* **17**, 589.

Sill, G. T.: 1973, *Bull. Am. Astron. Soc.*, **5**, 299.

Sinton, W. M.: 1953, Dissertation, Johns Hopkins University.

Sinton, W. M.: 1961, *Proc. of 11th General Assembly IAU*, Berkeley, 246.

Sinton, W. M.: 1962a, *Mem. Soc. Roy. Sci. Liège* **7**, 300.

Sinton, W. M.: 1962b, *Appl. Opt.* **1**, 105.

Sinton, W. M.: 1963, *J. Quant. Spectrosc. Radiat. Transfer* **3**, 551.

Sinton, W. M.: 1964a, AFCRL-64-926, Lowell Observatory, Flagstaff.

Sinton, W. M.: 1964b, *Appl. Opt.* **3**, 175.

Sinton, W. M. and Strong, J.: 1960, *Astrophys. J.* **131**, 470.

Slipher, E. C., and Edson, J. B.: 1939 *Publ. Am. Astron. Soc.* **9**, 229.

Slipher, V. M.: 1903a, *Astron. Nachr.*, p. 35.

Slipher, V. M.: 1903b, *Lowell Obs. Bull.*, No. 3.

Slipher, V. M.: 1908, *Astrophys. J.* **28**, 397.

Slipher, V. M.: 1921, *Lowell Obs. Publ.*, No. 84.

Smith, B.: 1967, *Science* **158**, 115.

Spinrad, H.: 1962a, *Publ. Astron. Soc. Pacific* **74**, 156.

Spinrad, H.: 1962b, *Publ. Astron. Soc. Pacific* **74**, 187.

Spinrad, H.: 1962c, *Icarus* **1**, 226.

Spinrad, H.: 1962d, *Mem. Soc. Roy. Sci. Liège* 7, 322.

Spinrad, H.: 1962e, *Astrophys. J.* **135**, 651.

Spinrad, H.: 1966, *Astrophys. J.* **145**, 943.

Spinrad, H. and Richardson, E. H.: 1965, *Astrophys. J.* **141**, 282.

Spinrad, H. and Shawl, S. J.: 1966, *Astrophys. J.* **146**, 328.

St. John, C. E. and Nicholson, S. B.: 1920, *Publ. Astron. Soc. Pacific* **32**, 332.

St. John, C. E. and Nicholson, S. B.: 1921, *Astrophys. J.* **53**, 380.

St. John, C. E. and Nicholson, S. B.: 1922, *Publ. Am. Astron. Soc.* **4**, 326.

Strong, J.: 1965, *Sci. Am.* **212**, 28.

Struve, O.: 1948, *Astron. J.* **53**, 161.

Struve, O.: 1954, *Sky Telesc.* **13**, 118.

Tacchini and Ricco: 1882, *Mem. Spettr. Italiani*.

Thompson, H. W. and Healy, N.: 1936, *Proc. Roy. Soc. London* **A157**, 331.

Tolbert, C. and Straiton, A. W.: 1962, *J. Geophys. Res.* **67**, 1741.

Unsöld, A. and Struve, O.: 1949, *Astrophys. J.* **110**, 455.

Vakhnin, V. M. and Lobedinskii, A. J.: 1966, *Zemlya i Vselennaya*, No. 1, p. 79.

Vaucouleurs, G. de and Menzel, D. H.: 1960, *Nature* **188**, 28.

Velikovsky, I.: 1950, *Worlds in Collision*, Doubleday, New York, p. 369.

Veverka, J.: 1971, *Icarus* **14**, 282.

Vinogradov, A. P., Surkov, Yu. A. and Florensky, C. P.: 1968 *J. Atmospheric Sci.* **25**, 535.

Vinogradov, A. P., Surkov, A., and Andreichikov, B. M.: 1970a, *Dokl. Akad. Nauk SSSR* **190**, 552.

Vinogradov, A. P., Surkov, Yu. A., Andreichikov, B. M., Kalinkina, O. M., and Grechisheheva, I. M.: 1970b, *Cosmic Res.* **8**, 533.

Vogel, H.: 1874, *Untersuchungen über die Spectra der Planeten*, s. 10.

Walker, R. G. and Sagan, C.: 1966, *Icarus* **5**, 105.

Warner, B.: 1960, *Monthly Notices Roy. Astron. Soc.* **121**, 271.

Wildt, R.: 1940a, *Astrophys. J.* **91**, 266.

Wildt, R.: 1940b, *Astrophys. J.* **92**, 247.

Wildt, R.: 1942, *Astrophys. J.* **96**, 312.

Wood, A. T., Wattson, R. B., and Pollack, J. B.: 1968, *Science* **162**, 114.

Wright, W. H.: 1927, *Publ. Astron. Soc. Pacific* **39**, 220.

Young, A. T.: 1973, *Icarus* **18**, 564.

Young, A. T. and Gray, L. D.: 1968, *Icarus* **9**, 74.

Young, C. A.: 1885, *Am. J. Sci. Arts.* **35**, 328.

Young, L. G.: 1969, *Icarus* **11**, 66.

Young, L. D. G.: 1970a, *J. Quant. Spectrosc. Radiat. Transfer* **10**, 99.

Young, L. D. G.: 1970b, *Icarus* **13**, 270.

Young, L. D. G.: 1970c, *Icarus* **13**, 449.

Young, L. G.: 1972, *Icarus* **17**, 632.

Young, L. G. and Young, A. T.: 1972, *Astrophys. J.* **176**, 533.

Young, L. D. G. and Young, A. T.: 1973, *Astrophys. J.* **179**, L 43.

Young, L. D. G., Schorn, R. A., Barker, E. S., and McFarlane, M.: 1969, *Icarus* **11**, 390.

Young, L. D. G., Schorn, R. A. J., and Barker, E. S.: 1970a, *Icarus* **13**, 58.

Young, L. D. G., Schorn, R. A. J., and Smith, H. J.: 1970b, *Icarus* **13**, 74.

Young, L. D. G., Young, A. T., and Schorn, R. A.: 1970c, *J. Quant. Spectrosc. Radiat. Transfer* **10**, 1291.

Young, L. D., Schorn, R. A. J., Barker, E. S., and Woszczyk A.: 1971, *Acta Astron.* **21**, 329.

Young, L. G., Young, A. T., Young, J. W., and Bergstralh, J. T.: 1973, *Astrophys. J.* **181**, L5.

RADIO INTERFEROMETRY OF VENUS AT SHORT
WAVELENGTHS

MICHAEL A. JANSSEN

*Jet Propulsion Laboratory, California Institute of Technology,
Pasadena, Calif., U.S.A.*

Abstract. From short wavelength ($\lambda < 4$ cm) interferometry of Venus interpreted on the basis of the physical structure and bulk composition of its atmosphere as given by spacecraft measurements, evidence for the presence of a source of opacity in addition to CO_2 is found. This opacity presumably is due to unknown minor constituents of the atmosphere of Venus.

1. Introduction

The source of the microwave opacity of the atmosphere of Venus is strictly unknown. Pressure-induced absorption by the bulk constituent CO_2 can explain many features of the emission observed at radio wavelengths. It is likely, however, that additional microwave absorption is present beneath the visible cloud layer. Trace amounts of strongly polar molecules such as H_2O and NH_3 can contribute measurably to the opacity. In concentrations sufficient to affect the microwave emission such molecules can have important implications for the thermal and chemical properties of the atmosphere. The question of the source of the microwave opacity is therefore an interesting one to pursue.

The identification of this source cannot be made uniquely since no molecular resonances are observed. Rather, the radio observations must be compared with predictions made by an emission model based on independent evidence. The composition and physical structure of the atmosphere are known from the Mariner and Venera spacecraft experiments (Fjeldbo *et al.*, 1971; Avduevsky *et al.*, 1970, 1971). The microwave properties of the surface may be deduced from the radar cross-section measurements (Muhleman, 1969). Using the absorption coefficient of CO_2 found from laboratory measurements (Ho *et al.*, 1966) one may specify a detailed model for the microwave emission. The presence of additional absorption may be inferred from differences between the predictions of this model and the observed emission. Pollack and Morrison (1970) have examined the radar and disk temperature measurements on this basis and find a mixing ratio of water vapor in the range 0.3% to 1.0% to be compatible with the existing data. A more recent series of measurements in the neighborhood of the H_2O $\lambda 1.35$ cm resonance shows no evidence for water vapor to an upper limit of 0.2% for the mixing ratio, and generally limits the absorption which may be present in the uppermost region of the atmosphere beneath the visible cloud deck (Janssen *et al.*, 1973; Janssen, 1973). From the frequency dependence of the radar cross-section Muhleman finds the total opacity of the atmosphere to be consistent with the law $\tau = (14.6 \pm 0.6)/\lambda^2$, where λ is the wavelength in centimeters. Sinclair *et al.* (1972) interpret their interferometric measurements at $\lambda 11$ cm to be consistent

Woszczyk and Iwaniszewska (eds.), Exploration of the Planetary System, 161–169. All Rights Reserved

with a larger opacity $\tau = 22.7/\lambda^2$. For microwave absorption by CO_2 alone, the model calculation described below predicts $\tau = 11/\lambda^2$. Berge *et al.* (1972) find no simple model for the surface and atmosphere which is fully consistent with their interferometric measurements in the wavelength range 3–21 cm. A larger opacity than given by CO_2 alone is suggested, however.

We consider here radio interferometric observations at wavelengths less than λ 4 cm The atmospheric emission dominates the surface contribution at these wavelengths, and observations in this region of the spectrum are particularly appropriate for the study of the atmospheric absorption. Here the frequency dependence of the emission contains information both on the total atmospheric opacity and the distribution of absorption with altitude. Further, uncertainties about the surface contribution to the emission are of less significance. Interferometric measurements may be made with great precision. Although the interpretation of interferometric data is less straightforward than for disk temperature or radar cross-section measurements it is demonstrated here that it is possible to discriminate more clearly among plausible cases for the absorption. The few existing measurements examined here give evidence for the presence of a source of opacity in addition to CO_2, with some indication about its distribution with altitude. The rapid development of the technique of interferometry at millimeter wavelengths will permit more conclusive measurements to be made in the future.

2. Discussion

The response of an ideal two-element interferometer to an extended source of emission is called the 'visibility function'. The argument of this function is the baseline, in wavelengths, as projected onto the sky in the direction of the source. This describes the 'spatial frequency' of the finely spaced interferometer fringes on the plane of the sky, and for present purposes it is convenient to normalize this spatial frequency to the variable planet radius as

$$\beta = \frac{\text{ephemeris semi-diameter of Venus}}{\text{angular fringe spacing}}.$$

The visibility function is related to the brightness distribution of the source through a Fourier transform. In principle, if data is obtained for a sufficiently extensive sampling of spatial frequencies (baseline separations) a detailed map of the source in spatial coordinates may be obtained. In practice this is difficult to achieve. Further, it is largely unnecessary since many important questions may be answered by a very restricted sample of the data.

We consider here two particular measurements; that of the first zero of the unpolarized visibility function β_1, and of the polarized visibility $V_\perp - V_\parallel$ in the neighborhood of this zero. The quantity β_1 is proportional to the baseline separation at which the source is just resolved, and depends upon the overall size of the source. In the case of Venus the boundaries of the emission are well known, while the distribution of brightness across the disk is not. In effect β_1 gives a measure of the overall limb darkening, which in turn depends upon the temperature lapse rate of the atmosphere, the distribution of absorption with altitude, and the surface dielectric constant. We define the polarized visibility $V_\perp - V_\parallel$ as the difference between the visibilities with the linear polarizations at the two antennas perpendicular and parallel respectively to the projected baseline. Since the surface contribution is polarized while the atmospheric contribution is not, the polarized visibility can be interpreted to give the total opacity of the atmosphere. Both quantities can be precisely measured. The determination of the first zero β_1 is effectively a null measurement, and is limited in practice by the signal-to-noise ratio. The polarized visibility in the neighborhood of β_1 is uncontaminated by instrumental polarization errors.

The existing measurements of the first visibility zero are shown in Figure 1. The error bars depict one standard deviation. The $\lambda 1.35$ cm point combines results from the 1970 and 1972 conjunctions reported by Janssen et al. (1973) and Janssen (1973). The point at $\lambda 3.1$ cm was communicated by Berge (1973), and is based on observations described in Berge and Greisen (1969). The $\lambda 3.7$ cm point was determined from data taken by E. T. Olsen during the 1972 inferior conjunction with the National

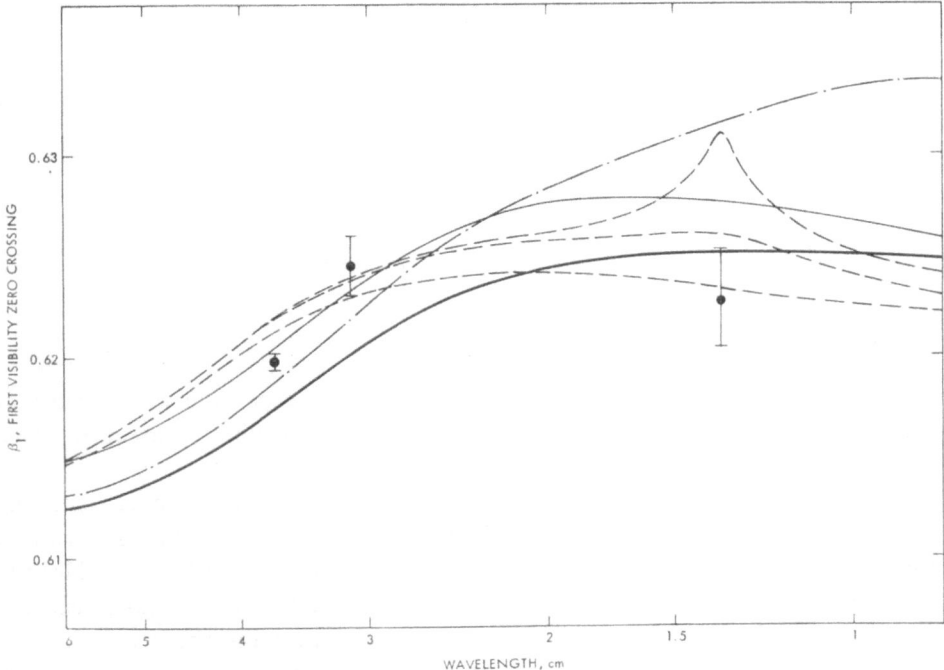

Fig. 1. Limb darkening of Venus in the wavelength range 0.8–6 cm as measured by the first zero crossing of the visibility function. The heavy solid curve is calculated for a model in which the only source of microwave opacity is atmospheric CO_2. The remaining curves were calculated for models to include the effect of additional absorption by water (dashed curves), a constant absorber between 35 km and 50 km altitude (light solid curve), and an absorber at 60 km (dot-dash curve) according to details in the text. The points are from existing observations shortward of $\lambda 4$ cm.

Radio Astronomy Observatory three-element interferometer (Olsen, 1973). Two days of data were obtained, which included two first zero crossings each day with the baseline projections approximately east–west and north–south respectively. The determination of β_1 was made by averaging all data in the interval $0.59 < \beta < 0.63$, with the resulting small formal error. These observations are presently being examined in more detail by Olsen.

A computer program was written to calculate the visibility function and disk temperature of Venus for several different cases for the microwave absorption. In the computation a thermal, non-scattering transfer of radiation is assumed. Geometric effects such as the finite thickness of the atmosphere and refraction of the emergent rays are taken into account. The surface is assumed to be smooth and horizontally uniform with a dielectric constant $\varepsilon = 4.8$. The atmospheric profile is based on measurements obtained by the Venera 4–7 atmospheric probes (Avduevsky et al., 1970, 1971). The program makes use of the pressures and temperatures listed in Table I, inter-

TABLE I

Basic atmospheric model

Altitude (km)	Pressure (atm)	Temperature (K)
76	0.008	195
64	0.1	243
57	0.33	275
51	0.86	312
45.9	1.7	355
37.3	4.8	432
28.5	11.5	515
19.6	24.2	588
9	53.1	676
0	96.0	747

polating between points in the table using an isentropic expansion with a constant temperature lapse rate. The bulk composition of the atmosphere is assumed to be 95% CO_2 and 5% N_2.

The heavy solid curve of Figure 1 was calculated for this basic model, assuming CO_2 to be the only source of microwave opacity. The $\lambda 1.35$ cm point is consistent with this curve; however, the measurements of β_1 at $\lambda 3.1$ and $\lambda 3.7$ are significantly greater than the basic model predicts. The remaining curves correspond to calculations for particular cases, described below, in which other absorbers are added to this model. The intent of these cases is primarily to examine the behavior of the visibility for plausible absorption models, rather than to search for a particular explanation for the present data.

A uniformly bright disk of the radius of Venus would yield a horizontally straight line at $\beta_1 \approx 0.61$ in Figure 1. The model curves show differing amounts of limb darkening with a strong wavelength dependence. Note that the solid curve of the basic model shows a generally wavelength-independent behavior shortward of $\lambda 2$ cm. In this region the emission originates in the atmosphere and the limb darkening is controlled by the approximately constant lapse rate. At longer wavelengths there is a tendency towards limb brightening, i.e., β_1 tends to smaller values, as the surface begins to become visible near the sub-Earth point. The surface, at the same temperature as the adjacent atmosphere, appears somewhat cooler due to its reduced emissivity.

The data indicate an increase in limb darkening over the basic model at the longer wavelengths considered here. The addition of opacity to the model generally increases the limb darkening longward of approximately $\lambda 2$ cm, while the exact wavelength dependence is sensitive to the way in which the additional opacity is distributed with altitude. The polarization data impose a constraint on the total opacity. We first consider the simple case in which the absorption of the basic model is uniformly increased by a factor f. The resulting emission is in every way identical after the wavelength transformation $\lambda \rightarrow f^{1/2} \lambda$, which follows from the λ^{-2} dependence of the CO_2 absorption coefficient. By transposing the heavy solid curve of Figure 1 to fit the data, we find $f = 1.5$.

The lighter curves of Figure 1 illustrate the effect produced by several other plausible models for the additional absorption. The three dashed curves show the effect of adding water vapor to the atmosphere. In the uppermost curve water vapor is added at a constant mixing ratio of 0.5% by volume up to the cloud top level at 60 km altitude. The non-resonant absorption increases the limb darkening for $\lambda > 2$ cm, and would provide a best fit to the data for a mixing ratio of 0.3%. The $\lambda 1.35$ cm point is in disagreement with the limb darkening produced by the H_2O resonance, however, and yields an upper limit for this case of 0.2% for the mixing ratio. If water vapor is removed from the upper layers of the atmosphere where the $\lambda 1.35$ cm line is produced, a more consistent fit to the data is possible. The lower two dashed curves were generated from model calculations in which the 0.5% mixing ratio was arbitrarily set to zero above 50 km and 40 km, respectively. Such a drying out of the upper atmosphere is not consistent with simple water vapor saturation, but is possible with chemical reactions. The affinity of H_2SO_4 for water can produce such an effect, for example (Young, 1973).

The addition of absorption at very high altitudes produces a generally poor fit to the data. The dot-dash curve of Figure 1 shows a case in which a thin layer of optical depth $\tau = 0.2/\lambda^2$ is placed at 60 km altitude. The λ^{-2} dependence would approximate that for small liquid water droplets. A large increase in limb darkening is produced at short wavelengths, while little effect is apparent beyond $\lambda 3$ cm. A source of opacity at intermediate altitudes can provide a more acceptable fit. The light solid curve was calculated for the case of a constant absorption $\alpha = 0.012/\lambda$ km^{-1} between 35 km and 50 km altitude. This case is consistent with the anomalous attenuation found from the Mariner V occultation experiments (Fjeldbo et al., 1970) if the attenuation is produced by an absorption varying with wavelength to the inverse first power. The atmosphere below 35 km altitude was not measured in the occultation experiments, and the lower altitude cut-off in absorption was arbitrarily assumed for this case.

Figure 2 shows the polarized visibility $V_\perp - V_\parallel$ measured at $\lambda 3.7$ and $\lambda 3.1$ cm along with calculations for the models considered above. The polarized visibility is sensitive only to the total opacity. The polarized visibility measured at $\lambda 3.7$ cm is about two-thirds of that predicted by the basic model (heavy solid curve), implying an actual increase in opacity over the basic model by a factor of 1.2. This may be compared with the factor $f = 1.5$ found above. While absorption at high altitudes has a strong effect on the limb darkening, there is little effect on the total opacity. The case of absorption at 60 km altitude coincides with the heavy solid curve of Figure 2. A slight

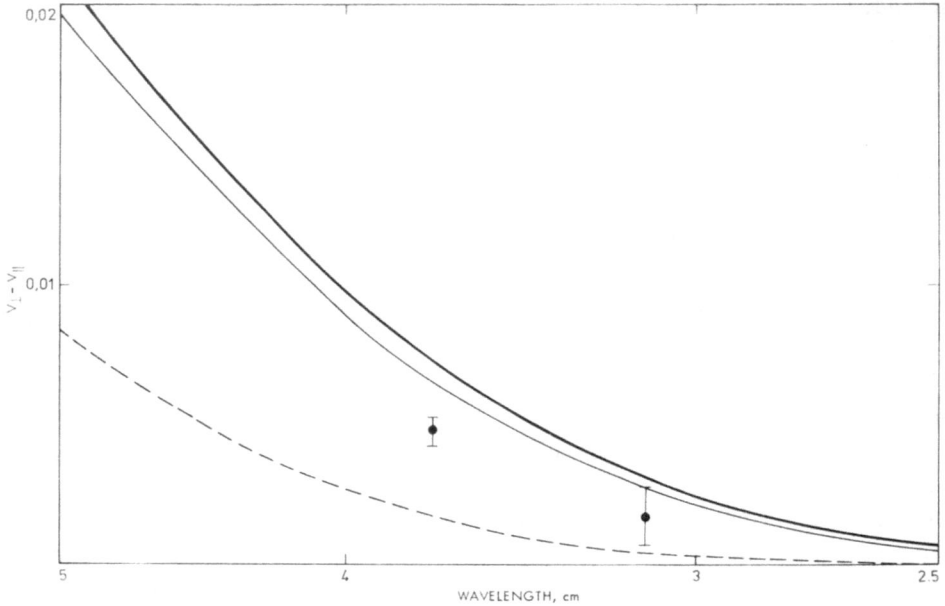

Fig. 2.   The polarized visibility of Venus in the neighborhood of $\beta = 0.61$. The curves correspond to the cases considered in Figure 1.

decrease in total opacity is seen for the case of an absorber at intermediate altitudes, as shown by the light solid curve. The models with water vapor, which coincide to give the single dashed curve of Figure 2, are seen to produce far too much opacity. The $\lambda 3.7$ cm measurement calls for a mixing ratio of only 0.15%.

These inferences depend on the credibility of the basic model used in the calculations. Several features of this model may be altered to provide an explanation of the interferometric data without invoking an additional source of opacity. For example, a computation was made in which the surface dielectric constant was reduced from 4.8 to 2.5. Although in disagreement with the radar data, this provided an acceptable fit to both the limb darkening and polarization data. Also, if the CO_2 absorption coefficient is incorrect the error would have to be as large as 20% or 50% from the above determinations. We may further consider systematic errors in the spacecraft measurements. Increasing the CO_2 content from 95% to 100% gives a 10% increase in opacity. The CO_2 absorption coefficient is proportional to P^2/T^5; the limb darkening data would require a reduction in temperature by nearly 50 K throughout the atmosphere, or an increase of pressures by a factor of 1.2, to increase the absorption coefficient of CO_2 to the required value. We note that where they overlap, temperatures from the S-band occultation measurement are about 20 K higher than the corresponding Venera measurements. Finally, upon lowering the mean surface of the model by 4 km, the required increase in opacity is provided by the extra layer of atmosphere. The surface temperature and pressure would be raised by 30 K and 25 atm, respectively, as a result. None of these choices appear consistent with present knowledge.

Disk temperature spectra calculated for the above models are shown in Figure 3. Except for the case in which water vapor extends to 60 km altitude, there is a relative lack of distinguishing structure among the cases considered. Also, not apparent in the figure is the increased sensitivity to model parameters. If atmospheric composition and temperature are varied within the uncertainty permitted by the spacecraft measurements, a range in disk temperature of more than 5% is possible at a given wavelength. This is sufficient to obliterate differences between all but the most extreme

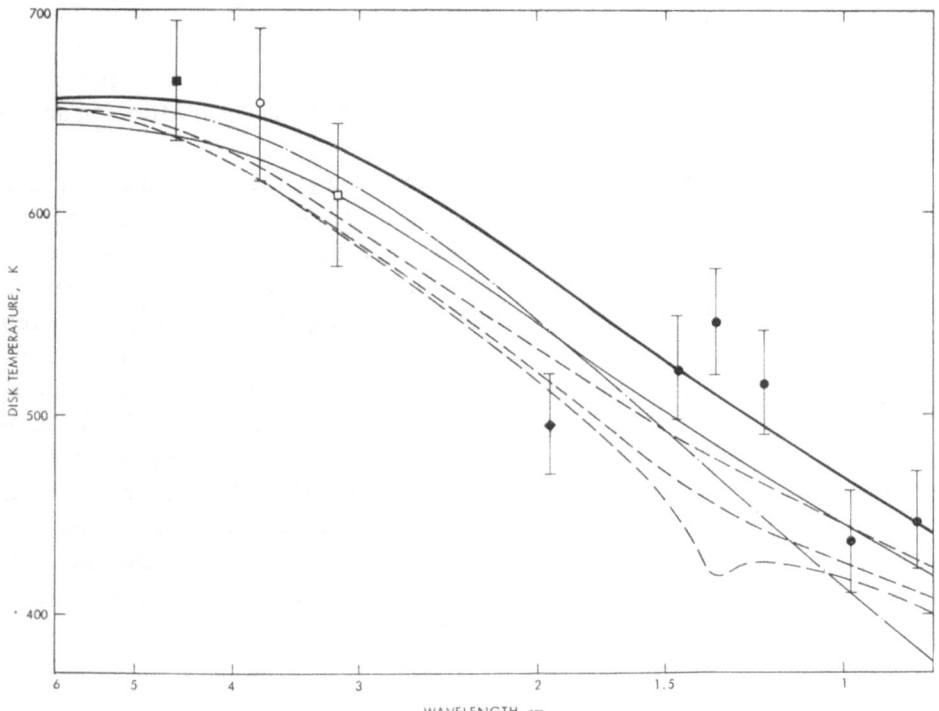

Fig. 3. The disk temperature spectrum of Venus at short wavelengths. The curves correspond to the cases considered in Figure 1. The data points are from Warnock and Dickel, 1972, ■; Klein, 1973, ○; Berge *et al.*, 197□; Pollack and Morrison, 1970, ◆; Janssen *et al.*, 1973, ●.

cases of Figure 3. The data points represent a sampling of recent measurements in this wavelength range. The error bars indicate the expected standard deviations of the measurements and primarily reflect the uncertainties of the absolute flux calibrations. These uncertainties amount to at least 5%. Again, all but the more extreme cases are in acceptable agreement with the measurements. In contrast to the potential of milli- meter wavelength interferometry, the flux calibration at these wavelengths is not likely to be greatly improved in the near future.

3. Conclusions

On the basis of present knowledge of the atmosphere and surface of Venus, the inter- ferometric measurements at wavelengths less than $\lambda 4$ cm reveal the presence of a source of microwave opacity in addition to CO_2. If CO_2 were indeed the only absorber the measurements would imply significant errors in either the spacecraft measurements of the atmospheric structure, the interpretation of the radar measurements, or the labo- ratory measurements of the CO_2 absorption coefficient. The limb darkening, as measured from the first zero of the visibility function at $\lambda 3.1$ and $\lambda 3.7$ cm, indicates an amount of additional absorption equivalent to increasing the CO_2 absorption by a factor of 1.5 in the basic model given by spacecraft, radar, and laboratory measure- ments. The polarized visibility at $\lambda 3.7$ cm is consistent with an increase in the total opacity by a factor of 1.2 over the case for CO_2 absorption alone. These amounts are consistent with Muhleman's interpretation of the radar data giving an increase in opacity by a factor of 1.35, but less than that found by Pollack and Morrison and by Sinclair *et al.*

The apparent discrepancy between the opacity found from the limb darkening measurements and from the polarization are easily resolved if we consider possible variations in the absorption with altitude. Absorption at or near the cloud top level is ruled out by the limb darkening at $\lambda 1.35$ cm. Strong absorption near the surface would mostly affect the total opacity, thereby increasing the discrepancy. A concen- tration of absorption at intermediate levels relative to CO_2 is preferred, although it is difficult to draw definite conclusions because of present data and model uncertainties. Particular cases such as water vapor at concentrations in the range 0.15–0.3% or an absorber at intermediate altitudes provide acceptable agreement with the observations. Water vapor is, of course, not specifically preferred over other non-resonant molecular absorbers. If the absorber at intermediate levels is to be identified with the source of attenuation of the S-band occultation signal, however, a dependence of the absorption approximately on the inverse first power of the wavelength is required. This would imply a particulate, rather than a molecular absorber.

Acknowledgements

I wish to thank Dr E. T. Olsen of the Jet Propulsion Laboratory for making available his interferometric data on Venus and several useful discussions about their interpretation. I would also like to thank D. O. Muhleman and G. L. Berge of the California Institute of Technology, for useful discussions and the use of their unpublished data. The present research was supported through a National Research Council Resident Research Associateship.

References

Avduevsky, V. S., Marov, M. Ya., and Rozhdestvensky, M. K.: 1960, *J. Atmospheric Sci.* **27**, 561.
Avduevsky, V. S., Marov, M. Ya., Rozhdestvensky, M. K., Borodin, N. F., and Kerzhanovich, V. V.: 1971, *J. Atmospheric. Sci.* **28**, 263.
Berge, G. L. and Greisen, E. W.: 1969, *Astrophys. J.* **156**, 1125.
Berge, G. L., Muhleman, D. O., and Orton, G. S.: 1972, *Icarus* **17**, 675.
Berge, G. L.: 1973, private communication.
Fjeldbo, G., Kliore, A. J., and Eshleman, V. R.: 1971, *Astron. J.* **76**, 123.
Ho, W., Kaufman, I. A., and Thaddeus, P.: 1966, *J. Geophys. Res.* **71**, 5091.
Janssen, M. A.: 1973, *Bull. Am. Astron. Soc.* **5**, 302.
Janssen, M. A., Hills, R. E., Thornton, D. D., and Welch, W. J.: 1973, *Science* **179**, 994.
Klein, M. J.: 1973, private communication.
Muhleman, D. O.: 1969, *Astron. J.* **74**, 57.
Olsen, E. T.: 1973, private communication.
Pollack, J. B. and Morrison, D.: 1970, *Icarus* **12**, 376.
Sinclair, A. C. E., Basart, J. P., Buhl, D., and Gale, W. A.: 1972, *Astrophys. J.* **175**, 555.
Warnock, W. W. and Dickel, J. R.: 1972, *Icarus* **17**, 684.
Young, A. T.: 1973, *Icarus* **18**, 564.

PROFONDEUR DE PENETRATION ET FORMATION DES RAIES DANS UNE ATMOSPHERE DIFFUSANTE

Y. FOUQUART

*Laboratoire d'Optique Atmosphérique, Université de Sciences
et Techniques de Lille, France*

Résumé. L'étude systématique de la fonction de distribution du chemin optique des photons diffusés par une atmosphère diffusante finie permet de connaître la profondeur de pénétration du rayonnement dans un nuage; on peut alors obtenir une information précise sur l'épaisseur optique de la couche nuageuse dans laquelle a été formée une raie spectrale. La relation qui existe entre la fréquence de la raie observée et l'altitude de la région de l'atmosphère correspondante est ainsi précisément exprimée. La méthode qui est présentée ici, ainsi que quelques résultats préliminaires de Vénus, Jupiter ou Saturne permettent d'améliorer la connaissance de la structure interne des nuages.

L'application de cette méthode à l'interprétation des raies formées sur Vénus a permis de préciser que l'épaisseur optique des nuages, dans le cas du modèle de Hansen serait de l'ordre de 40 et leur sommet situé vers 60 km d'altitude.

Abstract. The systematic study of the distribution of the photon optical path in a finite scattering atmosphere gives the penetration depth of the radiation in a cloud; one can get then an accurate knowledge about the layer thickness in which the spectral line has been formed. The relation between the observed line frequency and the corresponding altitude is thus precisely expressed. The method is presented here with some preliminary results, it could be applied to the Venus, Jupiter and Saturne atmospheres to improve the knowledge of their clouds internal structure.

Using this method in case of Venus spectral lines it has been possible to deduce that for the Hansen's model, the optical depth of Venus clouds should be of the order of 40 and the cloud top closed to 70 km height.

1. Introduction

Ces dernières années de nombreux travaux ont été consacrés à l'étude de la formation des raies spectrales en milieu diffusant, citons ceux de Chamberlain (1965), Belton (1968), Sagan et Regas (1970), et Fouquart et Lenoble (1973). Ces travaux ont permis de rendre compte de l'effet de phase c'est-à-dire de l'augmentation de la largeur équivalente lorsque diminue l'angle de phase et de justifier l'existence des régimes limites. Cependant l'une des questions essentielles qui restent posées est celle du niveau de formation de la raie observée; en effet, dans le cas d'une atmosphère claire ce niveau peut être précisément défini mais dans le cas d'une atmosphère nuageuse l'impossibilité de déterminer le chemin suivi par les photons a conduit à considérer la pression et la température obtenues comme celles qui règnent au sommet des nuages, si ceux-ci sont très denses c'est certainement une bonne approximation mais si le libre parcours moyen des photons est grand c'est très certainement faux. La méthode que nous proposons ici permet de déterminer la pénétration du rayonnement observé, on peut ainsi définir le niveau de formation de la raie et quelques résultats intéressants concernant la couche nuageuse peuvent être déduits.

Woszczyk and Iwaniszewska (eds.), Exploration of the Planetary System, 171–178. All Rights Reserved
Copyright © 1974 by the IAU

2. La distribution du chemin optique

Dans ce but nous utilisons la distribution du chemin optique des photons diffusés. Considérons une couche plane parallèle d'épaisseur optique τ, σ est le coefficient de diffusion, k_c le coefficient d'absorption du continu, k_v le coefficient d'absorption du gaz à la fréquence v.

L'albédo de diffusion du continu est défini par la relation (1)

$$\omega_c = \frac{\sigma}{\sigma + k_c} \tag{1}$$

et l'albédo de diffusion à la fréquence v par la relation (2)

$$\omega_v = \frac{\sigma}{\sigma + k_c + k_v}. \tag{2}$$

Dans ces conditions, et pour une direction d'incidence (μ_0, φ_0) et d'émergence (μ, φ), la distribution $p(\lambda)$ des photons ayant parcouru dans le continu un chemin optique λ est défini par la relation (3)

$$I(\omega_v) = I(\omega_c) \int_0^\infty p(\lambda)\, e^{-r\lambda}\, d\lambda, \tag{3}$$

où r est défini par la relation (4)

$$r = \frac{k_v}{\sigma + k_c}. \tag{4}$$

La fonction $p(\lambda)$ est donc la transformée de Laplace inverse de la fonction $I(r)/I(\omega_c)$, cette dernière fonction est interpolée au moyen des approximants de Pade (relation (6)).

$$\frac{I(r)}{I(\omega_c)} = \frac{\sum\limits_{i=1}^{N-1} b_i r^i}{\sum\limits_{j=1}^{N} c_j r^j}. \tag{6}$$

On obtient alors $p(\lambda)$ sous la forme d'une série d'exponentielles (relation (7)).

$$p(\lambda) = \sum_{m=1}^{N} A_m e^{\gamma_m \lambda} \tag{7}$$

les γ_m étant les racines du dénominateur et les A_m les coefficients de la décomposition en éléments simples. La figure 1 représente les fonctions de distribution correspondant à diverses épaisseurs optiques pour les directions $\mu = \mu_0 = 1$. Il est clair que si un photon a pénétré jusqu'à une profondeur optique τ il a parcouru un chemin optique au moins égal à deux fois cette profondeur, dans ces conditions si $p_\tau(\lambda)$ est normalisée suivant la relation (8)

$$\int_0^\infty p_\tau(\lambda)\, d\lambda = \frac{I(\omega_c, \tau)}{I(\omega_c, \infty)} \tag{8}$$

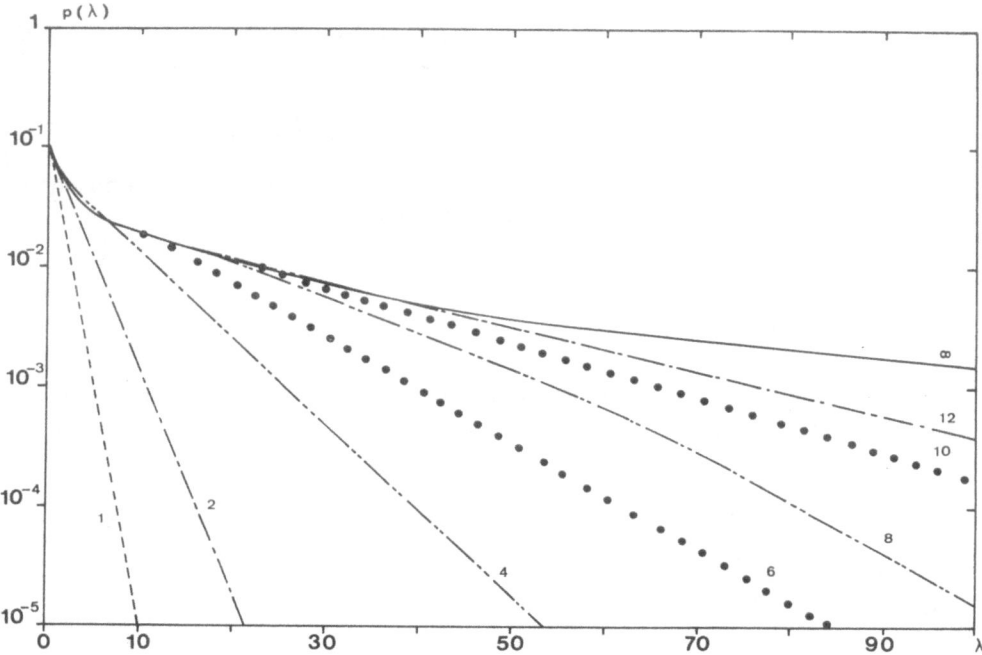

Fig. 1. Fonctions de distribution $p(\lambda)$ des photons correspondant à diverses épaisseurs optiques pour les directions $\mu = \mu_0 = 1$.

on aura :

$$p_\tau(\lambda) \equiv p_\infty(\lambda) \quad \text{si} \quad \lambda \leqslant 2\tau. \tag{9}$$

La connaissance des distributions correspondant à diverses épaisseurs optiques permet d'obtenir la probabilité de sortie d'un photon ayant parcouru au total un chemin optique λ et pénétré jusqu'à une profondeur optique donnée. A titre d'exemple le tableau I présente, pour un chemin optique égal à 40, le pourcentage de

TABLEAU I

Pourcentage de photons provenant de couches
d'épaisseurs optiques τ

$\tau \leqslant$	4	6	8	10	12	∞
%	2	20	57	76	94	100

photons provenant de couches de diverses épaisseurs optiques. On notera que dans ce cas 60% des photons ont pénétré jusqu'à une profondeur optique maximum comprise entre 6 et 10. On peut donc définir le chemin optique moyen correspondant à une pénétration jusqu'à une profondeur optique comprise entre deux épaisseurs optiques consécutives τ_1 et τ_2 (relation 10).

$$\langle \lambda_{12} \rangle = \frac{I(\omega_c, \infty)}{I(\omega_c, \tau_2) - I(\omega_c, \tau_1)} \int\limits_0^\infty \lambda \left(p_{\tau_2}(\lambda) - p_{\tau_1}(\lambda) \right) d\lambda . \tag{10}$$

3. Etude des raies formées en atmosphère diffusante

Pour une raie formée au cours de la réflexion du rayonnement dans une atmosphère diffusante, la largeur équivalente s'exprime par la relation (11)

$$W = \int\limits_0^\infty p(\lambda)\, W(\lambda b)\, d\lambda , \tag{11}$$

où $W(\lambda b)$ est la largeur équivalente d'une raie formée par absorption le long d'un trajet géométrique $L = \lambda/(\sigma + k_\tau)$ et b est défini par (12)

$$b = \frac{SM\omega_c}{2\pi\alpha_0 p} \tag{12}$$

S – est l'intensité de la raie,

α_0 – sa demi largeur,

p – la pression,

M – la 'quantité spécifique' c'est-à-dire le nombre de cm-atm de gaz absorbant contenu dans une colonne de longueur $1/\sigma$.

Compte tenu de l'expression de la distribution $p(\lambda)$, pour une raie de Lorentz, W s'exprime sous la forme (13):

$$W = 2\pi\alpha_0 p \sum_{m=1}^N \frac{A_m}{(-\gamma_m)^{3/2}} (2b - \gamma_m)^{-1/2} \tag{13}$$

en régime faible (b petit) on obtient l'expression (14)

$$W \simeq SM\omega_c \sum_{m=1}^N \frac{A_m}{(-\gamma_m)^2} = SM\omega_c \langle \lambda \rangle \tag{14}$$

et en régime fort l'expression (15)

$$W \simeq (\pi\alpha_0 p M S \omega_c)^{1/2} \sum_{m=1}^N \frac{A_m}{(-\gamma_m)^{3/2}} = 2 (\alpha_0 p M S \omega_c \bar{\lambda})^{1/2} , \tag{15}$$

où $\langle \lambda \rangle$ est le chemin optique moyen défini par (16)

$$\langle \lambda \rangle = \int\limits_0^\infty \lambda p(\lambda)\, d\lambda \tag{16}$$

et $\bar{\lambda}$ est le chemin optique efficace défini par (17)

$$\bar{\lambda}^{1/2} = \tfrac{1}{2} \int\limits_{0}^{\infty} \lambda^{1/2} p(\lambda)\,\mathrm{d}\lambda. \tag{17}$$

4. Application au cas de Vénus

Cette méthode a été appliquée au cas des raies formées sur Vénus aux environs de 8000 Å. On a donc considéré un nuage de particules sphériques d'indice $n = 1,46$ correspondant aux résultats de Hansen et Arking (1971), de granulométrie de type C_1 de Deirmendjann (1969), de rayon critique 0,8 μ avec un albedo continu de 0,999 en bon accord avec les valeurs d'albédo sphérique trouvées par Irvine (1968). Les calculs ont été effectués par la méthode des harmoniques sphériques avec la véritable fonction de phase déduite de la théorie de Mie.

On a tout d'abord calculé, pour un milieu d'épaisseur optique infinie les chemins optiques moyens et efficaces correspondant aux divers angles de phase et présentés au tableau II.

Les maximum situés vers 45° correspondent à la pointe arrière de la fonction de phase qui est assez forte et donne une diffusion primaire plus importante au voisinage de 0° diminuant ainsi les chemins optiques moyens correspondant.

Compte tenu des expressions des largeurs équivalentes des raies faibles et des raies fortes, il ressort clairement de ce tableau que:

(a) les raies faibles correspondant à des chemins optiques plus grands correspondent aussi à des pénétrations plus grandes que les raies fortes.

TABLEAU II

Chemins optiques moyens et efficaces correspondant aux divers angles de phase, calculé pour: $n = 1,46$, $\tau = \infty$, $r_c = 0,8\ \mu$, $\lambda = 0,8\ \mu$, $\omega_c = 0,999$

α	$\langle \lambda \rangle$	$\bar{\lambda}$
168	1,97	1,24
145	13,35	4,89
120	37,15	15,59
106	51,65	23,85
91	66,49	33,11
74	77,57	41,54
52	85,55	47.58
45	89,53	48,25
36	87,78	48,04
25	81,00	44,36
0	81,74	41,77

(b) l'effet de phase, pour une atmosphère diffusante homogène, produit d'abord une augmentation de la largeur équivalente jusque vers 45° puis une diminution.

(c) cet effet de phase est plus important pour les raies faibles que pour les raies fortes.

On a ensuite calculé les chemins optiques moyens correspondant à une pénétration comprise entre deux épaisseurs consécutives et les résultats sont présentés au tableau

TABLEAU III

Chemins optiques moyens $\langle \lambda_{12} \rangle$ pour deux épaisseurs consécutives et différents angles de phase

τ_1	τ_2	Angles de phase						
		168	145	120	91	52	36	0
0	0,25	1,32	0,94	0,70	0,70	0,60	0,27	0,26
0,25	0,5	6,18	2,96	2,09	1,87	1,49	1,17	0,86
0,5	1	8,81	5,49	4,06	3,55	2,88	2,40	1,90
1	2	11,8	10,0	7,9	6,8	5,8	5,3	4,9
2	4	19,3	18,0	15,6	13,8	11,7	11,6	10,8
4	6			28,6	23,6	22,8	22,2	21,6
6	8					35,9	35,4	31,7
					42,5			
8	10			47		56,9	52,6	51,9
10	12				108	69	66	67

III pour différents angles de phase. De ces résultats on peut déduire le niveau moyen de formation des raies faibles et des raies fortes par comparaison des tableaux II et III. Puisqu'il s'agit là d'une information statistique moyenne on pourra considérer que la pression efficace de formation des raies est la pression qui règne à leur niveau moyen de formation.

Bien que ce tableau ne soit pas très détaillé, on peut déjà remarquer que la profondeur optique de formation des raies est sensiblement identique entre 0 et 52°, de l'ordre de 8 pour les raies fortes et de l'ordre de 12–13 pour les raies faibles.

Si l'on admet que les raies observées par Gray-Young (1970) aux environs de 8000 Å sont des raies fortes, on peut déduire la pression et le produit M_p de la 'quantité spécifique' par la pression à partir de l'expression des largeurs équivalentes de raies fortes, on obtient alors $M_p \simeq 0,03$ km atm^2 et p de l'ordre de 200 mb, résultats déjà obtenus[*] de 1 μ. Dans ces conditions il est possible de déterminer l'altitude du sommet des nuages et le libre parcours moyen des photons. Nous avons, pour cela, utilisé le modèle d'atmosphère de Marov (1972) et les résultats obtenus en supposant les raies formées à diverses altitudes sont présentés au tableau IV. Les mesures de flux effectuées par la sonde Venera VIII ont montré que la limite inférieure des nuages était située vers 35 km et qu'il y subsistait environ 1% du flux solaire incident, ce qui correspondrait d'après Herman *et al.* (1974) à une épaisseur optique de l'ordre de la centaine dans le

---

[*] Par L. Gray et Belton mais aux environs.

TABLEAU IV

Altitude du sommet des nuages d'après le modèle de Marov

Z (km)	T (K)	p (atm)	n (m^{-3})	n/N_0 (atm)	$\sigma + k_c$ (km^{-1})	Altitude du sommet (km)
56	285	0,485	$1,2 \times 10^{25}$	0,448	7,244	57,1
58	271	0,340	$9,1 \times 10^{24}$	0,350	3,960	60
60	260	0,250	6,8	0,254	2,116	63,8
62	250	0,178	5,0	0,187	1,107	69,2
64	240	0,124	3,6	0,134	0,556	78,5
66	237	0,085	2,5	0,094	0,265	96,2
68	234	0,059	1,7	0,0635	0,125	132
70	231	0,040	1,2	0,0448	0,062	203

visible. Dans ces conditions, le meilleur accord semble être obtenu en fixant le niveau de formation des raies vers 60 à 62 km. Compte tenu des températures de rotation observées, voisines de 240 km et de la pression obtenue, nous avons fixé le niveau de formation à 62 km, ce qui donne un libre parcours moyen de 1,1 km et fixe le sommet des nuages vers 70 km puisque la profondeur optique de formation est d'environ 8. Dans ces conditions, la pression efficace de formation est de 178 mb et l'épaisseur optique des nuages de l'ordre de 40. Cette dernière valeur est sans doute trop faible mais la valeur du produit Mp est inversement proportionnelle au chemin optique

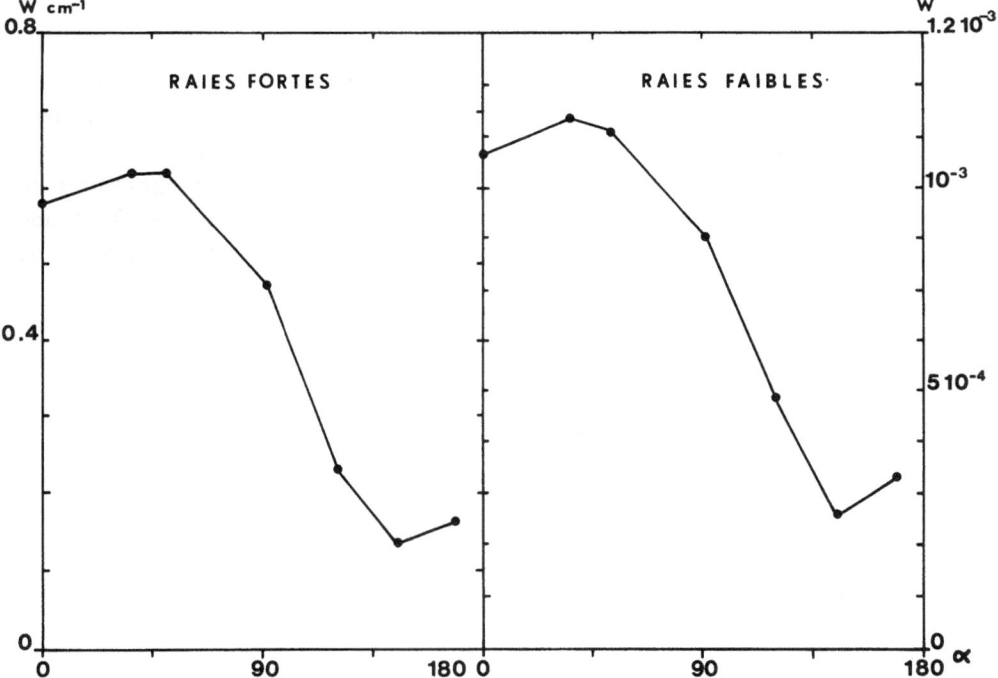

Fig. 2. Courbes de phase de la largeur équivalente tracées pour une raie d'intensité 0,67 km^{-1} (cm-atm)$^{-1}$ et une raie faible d'intensité 1000 fois plus petite.

efficace qui augmente lorsque l'albédo continu augmente; en choisissant pour ω_c la valeur 0,9999 le chemin optique efficace serait de 94 et le produit Mp serait voisin de 0,015, le libre parcours moyen serait alors de 4 à 500 m et l'épaisseur optique de nuage de l'ordre de 80. La profondeur optique de formation n'a pas été calculée, faute de temps, mais sa variation ne serait pas considérable et compte tenu du fait que le libre parcours moyen serait divisé par 2, l'altitude du sommet des nuages ne varierait pas sensiblement. Cependant cette dernière valeur de l'albédo continu est sans doute un peu trop grande si l'on s'en réfère aux valeur d'albédo sphérique, il serait sans doute préférable de choisir une valeur voisine de 0,9995, mais étant donné la précision des diverses mesures effectuées, l'ordre de grandeur des résultats obtenus ici semble satisfaisant.

Les courbes de phase de la largeur équivalente ont été tracées dans le cas sélectionné (niveau de formation 62 km, albédo continu 0,999) et sont présentées (figure 2) pour une raie d'intensité 0,67 km^{-1} (cm-atm)$^{-1}$ et une raie faible d'intensité 1000 fois plus petite. La pression efficace de formation dans le nuage est donc 178 mb pour la raie forte et 270 mb pour la raie faible, la pression efficace au dessus du nuage, en atmosphère claire étant de 20 mb. On remarquera que l'allure des courbes respecte les observations de Gray-Young *et al.* (1971) sans qu'il soit nécessaire de faire intervenir un deuxième nuage ainsi qu'avait cru devoir le conclure Hunt (1972); l'effet de remontée de la courbe de phase étant essentiellement dû à la pointe arrière du diagramme de diffusion.

5. Conclusion

En conclusion, l'utilisation de la distribution du chemin optique des photons diffusés permet de fixer avec précision la profondeur moyenne de pénétration du rayonnement dans une couche diffusante, le niveau de formation des raies fortes ou des raies faibles peut alors être déterminé et relié à la pression efficace de formation. L'application de cette méthode au cas de Vénus a permis de déterminer approximativement l'altitude du sommet des nuages et le libre parcours moyen des photons et de rendre compte de l'effet de phase.

Bibliographie

Belton, M.: 1968, *J. Atmospheric Sci.* **25**, 596.
Chamberlain, J. W.: 1965, *Astrophys. J.* **141**, 1184.
Deirmendjann, D.: 1969, *The Electromagnetic Scattering on Spherical Polydropersions*, Elsevier, New York.
Fouquart, Y. et Lenoble, J.: 1973, *J. Quant. Spectrosc. Radiat. Transfer* **13**, 447.
Gray-Young, L. D.: 1970, *Icarus* **13**, 449.
Gray-Young, L. D., Schorn, R. A., Barker, E. S., et Woszczyk, A.: 1971, *Acta Astron.* **21**, 329.
Hansen, J. E. et Arking, A.: 1971, *Science* **171**, 669.
Herman, M., Devaux, C., et Deuze, J. L.: 1974, sous presse.
Hunt, G. E.: 1972, *J. Quant. Spectrosc. Radiat. Transfer* **12**, 405.
Irvine, W. M.: 1968, *J. Atmospheric Sci.* **25**, 610.
Marov, M. Ya.: 1972, *Icarus* **16**, 415.
Sagan, C. et Regas, J. L.: 1970, *Comments Astrophys. Space Phys.* **2**, 116; 138; 161.

ETUDE THEORIQUE DE LA REPARTITION DE LUMINANCE SUR LE DISQUE DE VENUS

C. BIGOURD, J. L. DEUZE, C. DEVAUX, M. HERMAN, et J. LENOBLE

Laboratoire d'Optique Atmosphérique, Université des Sciences et Techniques de Lille, France

Abstract. The analysis of the sunlight scattered by Venus gives some insight upon its clouds. The measurements of polarized light are probably more sensitive to the nature of their constituents, and some recent studies seem to be able to give a satisfactory interpretation of this part of the scattered light. But the polarized light concerns the upper part of the clouds, and it is interesting to compare these results with intensity measurements. The phase curves, for the integrated light, leave some indeterminations, so we have studied if the intensity distribution on the Venus disk could give more accurate informations. This work, based on some plates kindly communicated by A. Dollfus, and analysed at Meudon Observatory, is more a preliminary investigation of the sensitivity of the method than an interpretation of the partial results presented. A simple model, of homogeneous plane parallel cloud, has been used, and the influence of various parameters has been tested (single scattering albedo, refractive index of particles and size distribution, optical depth of the cloud).

Nous avons étudié la distribution relative, sur Vénus, de la lumière solaire diffusée. Ce travail a été entrepris en collaboration avec le Professeur Dollfus, qui nous a communiqué quelques réseaux expérimentaux des courbes d'égale luminance, ou isophotes, déduites des clichés de Vénus. On ne poussera toutefois pas la comparaison entre ces réseaux et les calculs, les clichés devant encore être déconvolués des effets de diffraction et de la turbulence atmosphérique, avant leur publication définitive. On considérera donc plutôt ce qui suit comme une étude préalable de ce que peut apporter ce type d'analyse. L'idée est que la luminance, beaucoup moins sensible que la polarisation aux propriétés du milieu, permet par contre un sondage nettement plus pénétrant du nuage.

Nous supposerons d'abord l'épaisseur optique infinie, et partirons du modèle déduit par Arking et Hansen (1971), des mesures de polarisation; soit un nuage de particules sphériques, d'indice 1,46 environ, et de granulométrie de la forme

$$n(r) = n_0 r^{p_1} \exp(p_1 r/p_2), \qquad p_1 = 6, p_2 = r_c = 0,8 \text{ à } 1 \mu$$

dont on peut déduire la fonction de phase, sous la forme

$$p(\theta) = \sum \beta_l P_l(\cos\theta).$$

Il reste à préciser l'albédo pour une diffusion, des particules, ω_0^∞. Les mesures d'Irvine (1968) donnent un albédo sphérique A^* de 0,9 environ, pour la longueur d'onde 0,58 μ correspondant aux clichés étudiés. On en déduit facilement ω_0^∞ à l'aide de la formule approchée de Sobolev, ou de celle établie par Wang (1972), dans l'approximation du noyau exponentiel, soit

$$\omega_0^\infty = 1 - \frac{(3 - \beta_1)(1 - A^*)^2}{4(1 + A^*)^2} = 0,9994. \tag{1}$$

On peut alors évaluer la luminance diffuse (on suppose la géométrie plane applicable). Le calcul est fait suivant la méthode des Harmoniques Sphériques en quelques centaines de points sur le disque (de 200 à 500 points suivant l'angle de phase considéré). L'exploitation reste raisonnable, l'essentiel des calculs ne dépendant pas du point considéré, ou ne dépendant que de la direction d'incidence. On a ainsi pu étudier l'influence des paramètres du milieu sur les réseaux d'isophotes.

La figure 1 montre à titre d'exemple les résultats, pour une absorption variable,

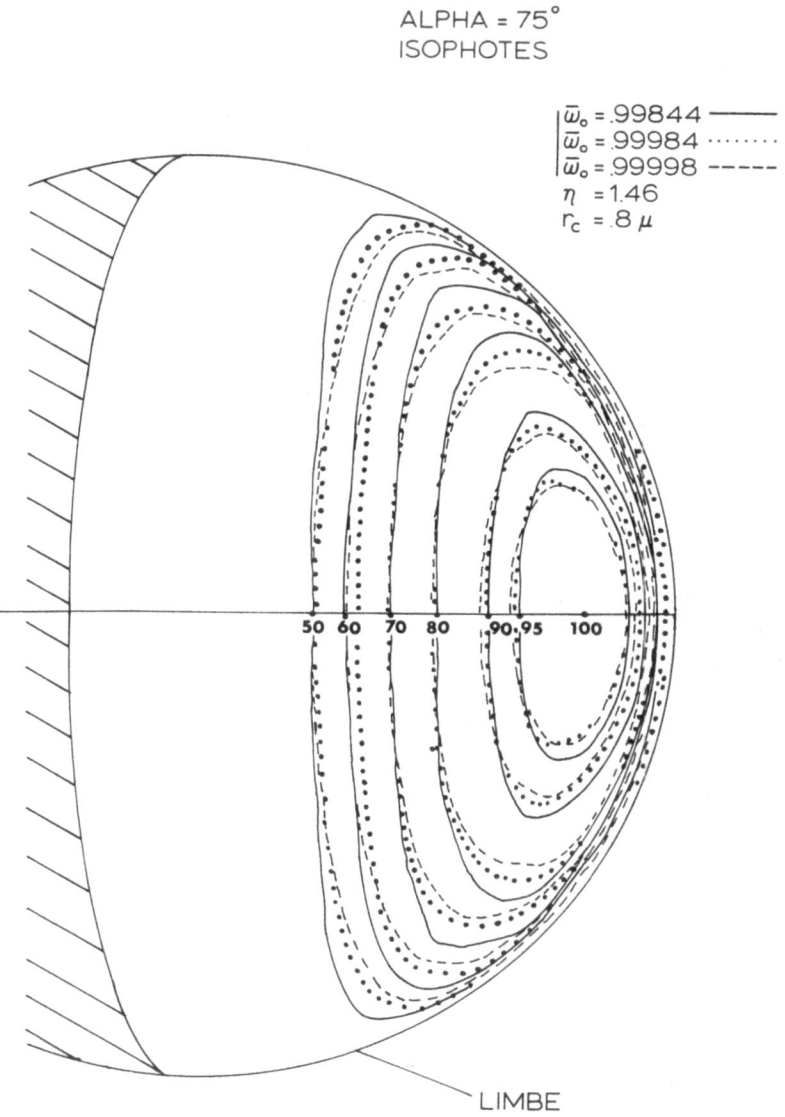

Fig. 1.   Influence des paramètres du milieu sur les réseaux d'isophotes pour l'angle de phase de 75°.

dans le cas d'un angle de phase ($\alpha = 75°$); les valeurs extrêmes de ω_0^∞ donneraient des albédos sphériques de 0,64 et 0,94, qui encadrent nettement la valeur probable (les maxima des réseaux sont toujours ramenés arbitrairement à 100). Les luminances relatives varient très peu. Pour une variation de r_c de 0,8 à 0,3 μ les modifications sont du même ordre, bien qu'on se déplace vers les petites particules, ce qui déforme au maximum la fonction de diffusion. Enfin, on a comparé au modèle d'indice 1,46 un milieu d'indice 1,20. Les écarts s'accentuent quand α décroît (avec l'influence croissante de la rétrodiffusion, par laquelle se différencient surtout les deux milieux); mais ils restent encore faibles, et finalement, aucun des 3 paramètres ne modifie radicalement le phénomène. On met donc bien en évidence la très faible influence des propriétés optiques du nuage sur la luminance relative.

Tenons compte maintenant de l'épaisseur τ_1 du nuage, et donc, simultanément, de la nature du sol; on supposera simplement qu'il suit la loi de Lambert, avec un coefficient de réflexion ϱ. Pour fixer les idées, nous avons utilisé les résultats de Venera 8, et tenté de déterminer, suivant la valeur supposée de ϱ, les ordres de grandeur de τ_1 et ω_0 compatibles avec les valeurs mesurées de A^* et du flux solaire transmis au sol, au point d'impact de Venera 8 (Marov *et al.*, 1973). Si on se donne ϱ et l'indicatrice des particules, la solution du problème est unique; en remontant le fond, on ne maintiendra en effet un flux réfléchi constant qu'en diminuant l'absorption propre des particules, qu'on devra simultanément augmenter, au contraire, pour conserver un même flux transmis; d'où 2 courbes $\tau_1(\omega_0)$ qui se couperont si les mesures et l'hypothèse sur ϱ sont cohérentes. On peut obtenir une solution analytique approchée du problème. On a pour cela développé, dans l'approximation du noyau exponentiel, les calculs de l'albédo sphérique A^*, du flux plan transmis au sol $\phi^-(\tau_1)$ (et du flux remontant $\phi^+(0)$), pour la géométrie considérée (Bigourd *et al.*, 1973). La validité des formules obtenues a été testée. Si on suppose $\omega_0 \sim 1$, $\tau_1 \gg 1$ et $(1-\varrho) > 0,5$ environ, ce qui doit être raisonnable, on peut déduire de ces expressions de A^* et $R = \phi^-(\tau_1)/\mu_0 f$ (f est le flux solaire)

$$\text{th}(C\tau_1) = \frac{1 - [(1-\varrho)\sqrt{(3-\beta_1)}]/[(1+\varrho)\sqrt{4(1-\omega_0^\infty)}]}{\sqrt{\frac{(1-\omega_0)}{(1-\omega_0^\infty)}} - [(1-\varrho)\sqrt{(3-\beta_1)}]/[(1+\varrho)\sqrt{4(1-\omega_0)}]} \qquad (2)$$

$$\text{sh}(C\tau_1) = [(3\mu_0 + 2)\sqrt{(1-\omega_0)}]/[R(1-\varrho)\sqrt{(3-\beta_1)}], \qquad (3)$$

où

$$C = \sqrt{(3-\beta_1)(1-\omega_0)} \ll 1.$$

et l'on déduit facilement de ces équations que, pour la solution cherchée, on a, en fonction de R et de A^*, avec la relation (1)

$$\omega_0 = \omega_0^\infty + \frac{(3-\beta_1)(1-\varrho)^2 R^2}{(3\mu_0 + 2)^2} \qquad (4)$$

et donc τ_1 par (2) ou (3). Nous nous sommes contentés d'un calcul à la seule longueur d'onde 0,63 μ, correspondant au maximum de sensibilité de la cellule de Venera 8. Avec $\mu_0 \sim 0,1$, $R \sim 1\%$ et $A^* \sim 0,94$ (soit $\omega_0^\infty \sim 0,9998$) on obtient

$$\varrho = 0 \qquad\qquad 0,5$$
$$\tau_1 = 50 \qquad\qquad 200$$
$$\omega_0 = 0,999\,816 \qquad 0,999\,804$$

Ces valeurs sont à considérer avec réserve, compte tenu des approximations et des incertitudes de mesure, mais avec de tels ordres de grandeur, il serait inutile de calculer

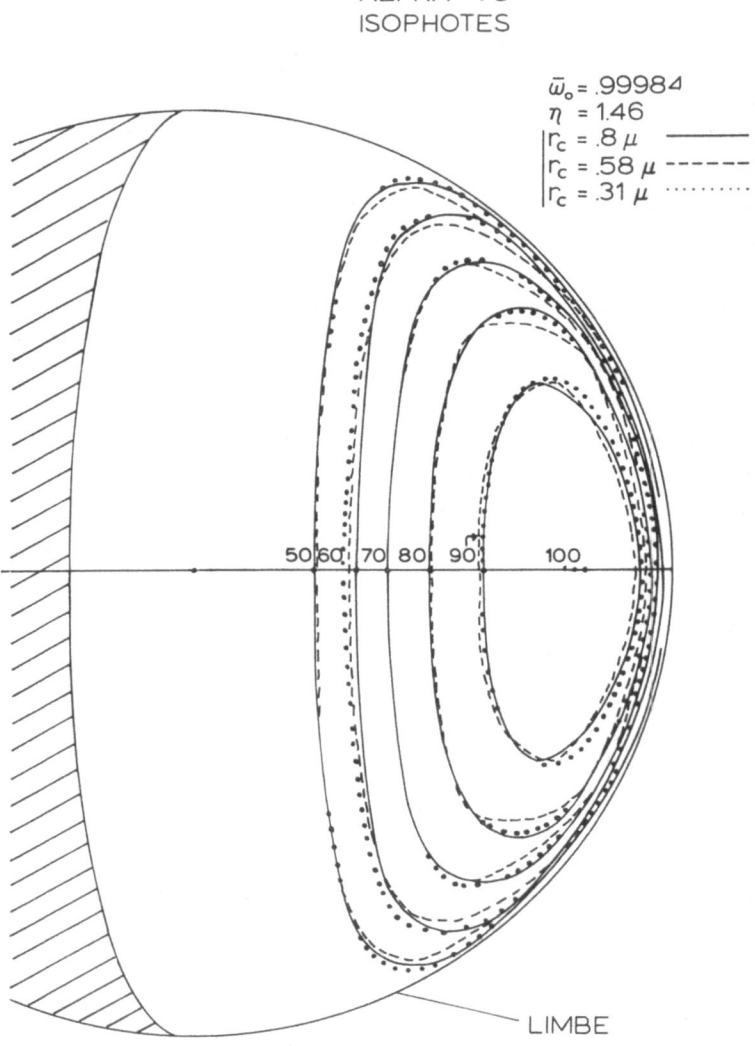

Fig. 2.   Comparaison des réseaux d'isophotes obtenus pour divers diamètres.

les nouveaux réseaux d'isophotes. Dans l'approximation précédente, en effet, le flux réfléchi en un point du disque s'écrit

$$\frac{\phi^+_{(0)}}{\mu_0 f} = 1 - \frac{3C\,(3\mu_0 + 2)}{2\,(3 - \beta_1)\,\mathrm{th}\,(C\tau_1)}.\qquad(5)$$

La variation relative de ce flux, lorsqu'on passe d'une couche infinie à une épaisseur

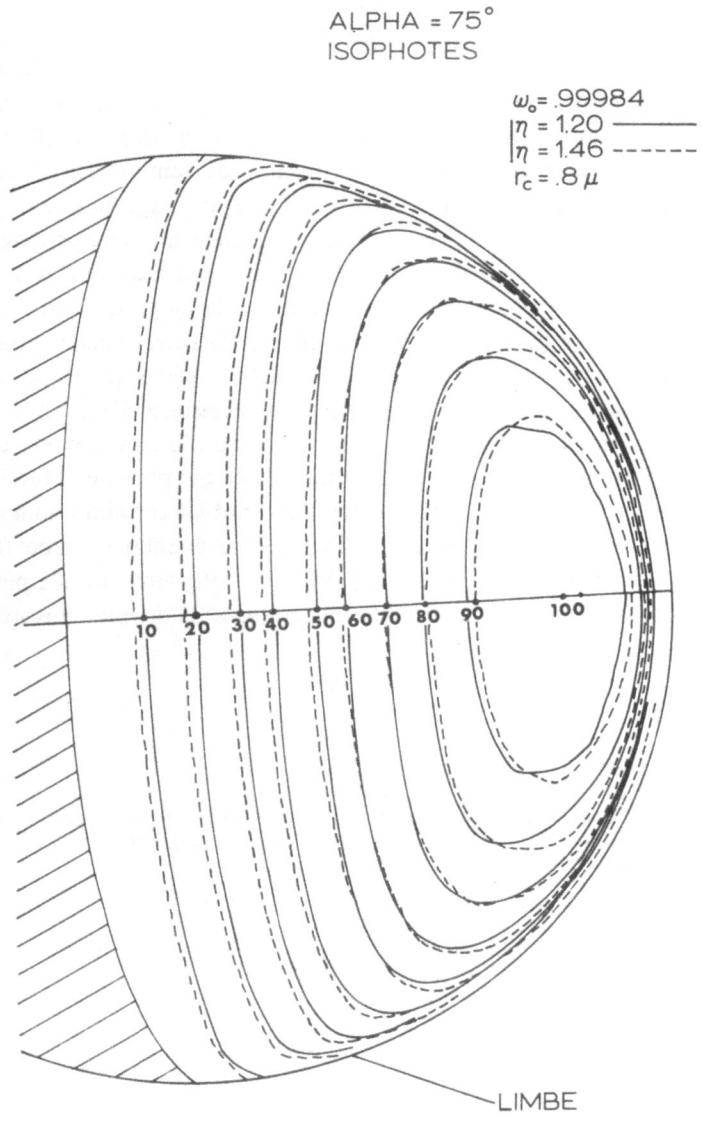

Fig. 3.   Comparaison entre les réseaux d'isophotes théoriques pour $n = 1,20$ et $1,46$.

de l'ordre de 100, ne dépasse alors pas 0,5% au point subsolaire où elle est maximum. La luminance varierait dans le même ordre de grandeur, et les réseaux seraient indiscernables.

En conclusion, si nous supposons que les nuages conservent sur toute la planète, les caractéristiques déduites de la polarisation, et une épaisseur supérieure ou égale à celle qu'on peut envisager au point sondé par Venera 8, la distribution de luminance relative est calculable avec précision, et une comparaison précise devrait être possible.

Si l'on s'interroge sur la possibilité de détecter la présence éventuelle, sous la brume sondée en polarisation, d'un nuage de nature différente mais d'absorption propre voisine, la conclusion semble négative. La figure 3 montre l'ordre de grandeur des écarts maxima observés pour un nuage d'indice 1,20 au lieu de 1,46, dans le cas où $\alpha = 22°$. Ces écarts seraient encore atténués par la superposition d'une couche, même fine, d'indice 1,46. De plus on doit être à la limite de la précision des mesures. (Il s'agirait de mesures diurnes, et les faibles niveaux de luminance, ou se localisent justement les plus gros écarts, seraient perturbés par le voisinage du soleil).

Il n'est pourtant pas exclu que ces mesures présentent une intéressante extension. Les comparaisons entre réseau théorique et expérimental montrent des divergences qui paraissent très fortes, et on peut se demander si la correction des effets de turbulence et de diffraction pourra suffire à les effacer. D'autres clichés présentent par ailleurs de nettes dissymétries équatoriales, et parfois même des isophotes fermées sur un seul hémisphère. Il peut donc se révéler nécessaire d'élargir le modèle en introduisant une inhomogénéité horizontale. Pour qu'elle soit capable d'influencer la répartition de luminance relative, compte tenu de ce qui précède, il faudrait supposer que l'épaisseur du nuage supérieur varie, devenant en certaines zônes, plutôt de l'ordre de 10 que de 100. Par l'analyse de réseaux expérimentaux, on devrait pouvoir déterminer ces variations; et des mesures à plusieurs longueurs d'onde permettraient peut-être de préciser si c'est le sol même, ou un autre nuage plus absorbant, qu'on détecterait ainsi.

Bibliographie

Bigourd, C., Devaux, C., et Herman, M.: 1973, rapport interne.
Hansen, J. E. et Arking, A.: 1971, *Science* **171**, 669.
Irvine, W. M.: 1968, *J. Atmospheric Sci.* **25**, 610.
Marov, M. Ya., Avduevsky, V. S., Borodin, N. F., Kerzhanovich, V. V., Lysov, V. P., Moshkin, B. Ye., Rozhdestvenskiy, M. K., Ryabov, O. L., et Ekonomov, A. P.: 1973, *Proc. NASA Technical Translation* **F14**, 909.
Wang, L.: 1972, *Astrophys. J.* **174**, 671.

NUMERICAL MODELLING OF THE REFLECTION SPECTRUM
OF VENUS IN THE VISUAL AND NEAR INFRARED RANGES

E. M. FEIGELSON

Institute of Optics of the Atmosphere, Tomsk, U.S.S.R.

N. L. LUKASHEVICH

Institute of Applied Mathematics, Moscow, U.S.S.R.

and

G. M. KREKOV and G. A. TYTOV

Institute of Optics of the Atmosphere, Tomsk, U.S.S.R.

Abstract. The influence of a dense molecular atmosphere under a cloud layer on spectral albedo of a planet is investigated. The lengths of paths of protons in the clouds are calculated. Experimental spectral transmission functions were used to calculate gas absorption. The results are compared with experimental data.

Our attempts of numerical modelling of the reflection spectrum have, among numerous other attempts, the following particular features:

(a) the influence of the molecular underlayer is investigated,

(b) the absorption in bands of CO_2 gas is taken into account rather completely, together with multiple scattering and absorption by cloud particles.

1. The Spectral Albedo in the Visual Range

Almost the whole optical thickness due to molecular scattering is in the layer $z \leqslant 50$ km. If the lower boundary of the cloud is not below 50 km, it is reasonable to consider a two-layer system: the cloud, and the pure molecular atmosphere below it.

The problem was: given the known optical properties of the molecular layer (the Rayleigh optical thickness, $\tau_{R,\lambda}$; the scattering function, $\gamma_R(\varphi)$; and the single scattering albedo, $\omega = 1$), and for given models of the cloud layer (the size distribution of cloud particles and their refractive index), to find the spectral optical thicknesses of the cloud that reproduce the experimental albedo spectrum.

Two models were considered:

1.1. A cloud layer similar to terrestrial clouds: $r_{mod} = 10$ μm, having Deirmendjian's scattering function with mean cosine of the scattering angle 0.83. In this case $\tau_\lambda = $ = const, and with $\tau_\lambda = 60$ the calculated albedo corresponds to the measured one, as Figure 1 shows. The decrease of the albedo A_λ with wavelength above 0.6 μm is due to to include absorption. The cross on the curve at $\lambda_0 = 0.55$ μm denotes the value of A_{λ_0} with $\omega = 0.998$.

1.2. The second model has $r_{mod} = 1.1$ μm; $n = 1.44$, as the polarization measurements suggest. In this case the optical thickness of the cloud and the scattering function

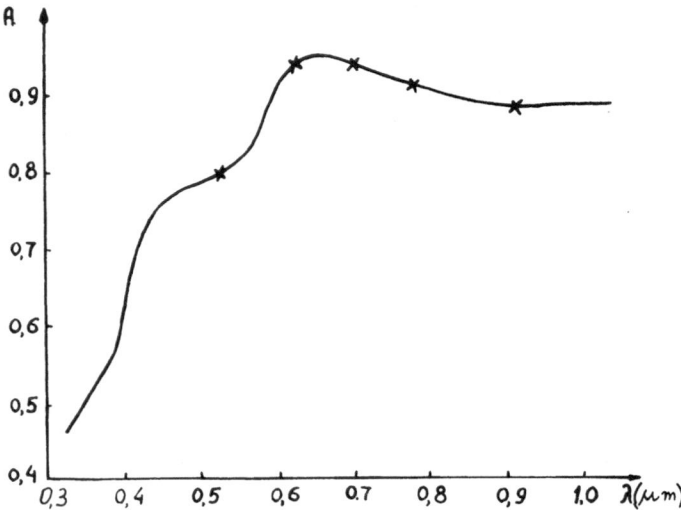

Fig. 1.   The measured (full line) and the calculated (crosses) albedo of Venus for $\tau = 60$.

TABLE 1

Wavelength dependence of cloud parameters

λ (μm)	$\tau_{R,\lambda}$	τ_1	$N \times 10^{-8}$ (part cm^{-2})
0.69	7.9	13.0 ± 1.5	1.5
0.86	3.3	8.5 ± 0.5	1.2
1.14	1.1	13.5 ± 0.5	1.0

change significantly with the wavelength. The scattering functions are less elongated than for larger particles. Therefore, the optical thicknesses obtained are smaller, as Table I shows. The last colum in the table gives the number of cloud particles per unit cross section in the whole cloud column,

$$N = \frac{\tau}{K},$$

where K is the scattering cross section.

The spectral dependence of N is weak, which confirms our approach.

2. The Infra-Red Range

The difficulty in modelling the infra-red reflection spectrum comes from the line structure of the absorption spectra of gases. The concepts of the absorption coefficient and the radiative transfer equation retain their meanings only for monochromatic radiation. Therefore, calculations of the reflection spectrum of Venus include, as a rule, multiple scattering and absorption by cloud particles, but only indirect evaluations of the gas absorption are presented. Such evaluations are insufficient to examine the possibility of water clouds.

We used the approach to study the paths of photons in a scattering medium of van de Hulst, Irvine, Romanova, and the Monte-Carlo technique. The reflected flux at the upper boundary of the atmosphere is equal to:

$$F_{\Delta\lambda} = \int\limits_0^\infty J_{\Delta\lambda}(l)\, P_{\Delta\lambda}\left[M_{\Delta\lambda} + \varrho_{cl}l\right] \exp\left[-\alpha_{\Delta\lambda}l\right]\, dl.$$

Here $J_{\Delta\lambda}$ is the distribution of paths l of photons for pure scattering of the reflected light. This function was calculated by the Monte-Carlo method. $P_{\Delta\lambda}(M)$ is the transmission function for CO_2. Experimental data for $\Delta\nu = 25$ cm^{-1} were used. $M_{\Delta\lambda}$ is the gas mass along the slant path above the cloud:

$$M_{\Delta\lambda} = (\sec\zeta + \sec\bar\zeta) \int\limits_{z_2}^\infty \varrho(z)\left(\frac{p(z)}{p_0}\right)^{n_{\Delta\lambda}} dz,$$

where z_2 is the upper boundary of the cloud; p, the pressure, $p_0 = 1$ atm; ζ, the zenith distance of the Sun; $\bar\zeta$, the mean zenith angle of propagation of the reflected light above the cloud; ϱ_{cl} is the mean density of CO_2 within the cloud, or in the upper part of the cloud; $\alpha_{\Delta\lambda}$, the spectral absorption coefficient of cloud particles.

An ice cloud model was considered. In Figures 2 and 3 models of ice clouds ($n = 1.33$; $r = 1.1\ \mu$m) are given, without (Figure 2) and with (Figure 3) absorption by CO_2 molecules. In both cases $M = 0$, i.e. the absorption above the cloud is neglected.

The spectral geometric albedo is considered relative to its value at $\lambda = 1.85\ \mu$m. The

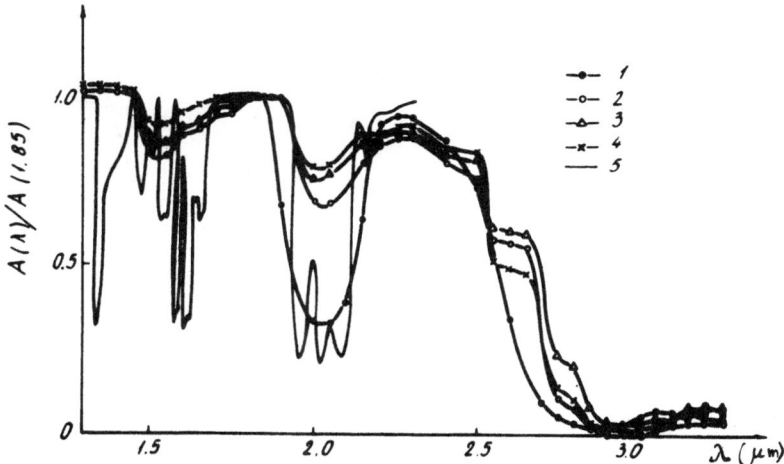

Fig. 2.   The spectral albedo without CO_2 absorption. Curves 1 and 5 — the averaged data of Kuiper and Bottema; 2 — calculations for $\zeta = 45°$, $\tau = 20$; 3 — ditto but $\zeta = 70°$; 4 — ditto but $\zeta = 45°$, $\tau = 10$.

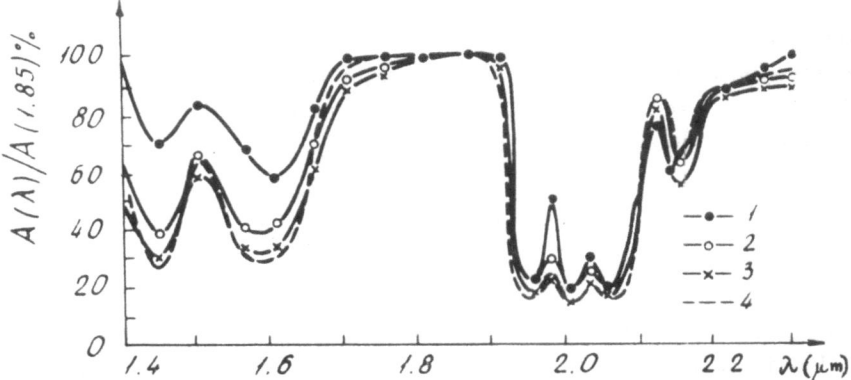

Fig. 3.   The spectral albedo with CO_2 absorption. Curves: 1 – experimental data as in Figure 2;
2 – calculations for $\zeta = 70°$; $\tau = 20$; 3 – ditto but $\zeta = 45°$; 4 – ditto but $\zeta = 10°$.

thick line is the averaged experimental data of Kuiper (at $\lambda \leqslant 1.8 \ \mu$m) and Bottema
(at $\lambda > 1.8 \ \mu$m). The figures show that the ice absorption is small compared to CO_2
absorption. So the suggestion of an ice cloud may not contradict the reflection
spectrum.

The optical depth of the cloud was varied in the range 10–20 with little influence on
the reflection. It is still less influenced by the medium under the cloud. The figures
show the effect of the zenith distance of the Sun – the bands become deeper as the Sun
rises. In Figure 4 the calculations are compared with the corresponding experimental
data and the fine spectral structure is shown. Figure 5 gives the results of calculations

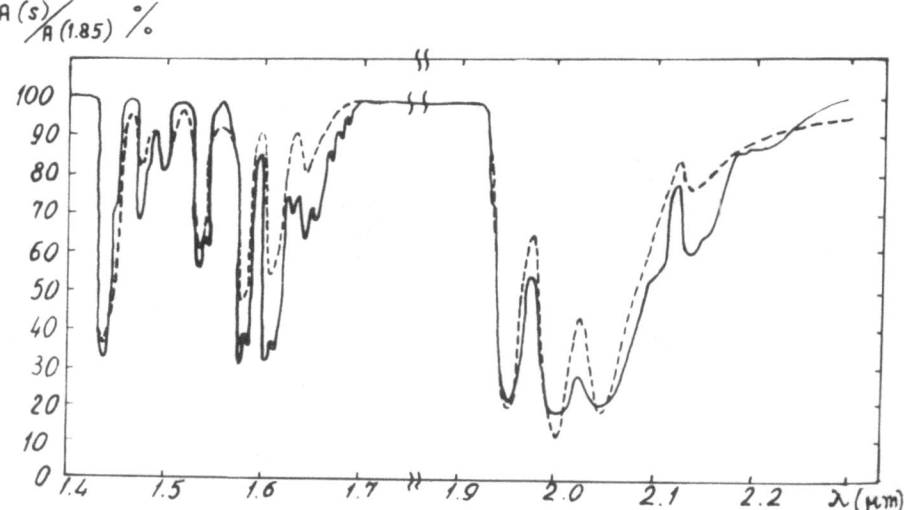

Fig. 4.   The reflection spectrum with fine spectral structure. Full line – measurements; dashed
line – calculations.

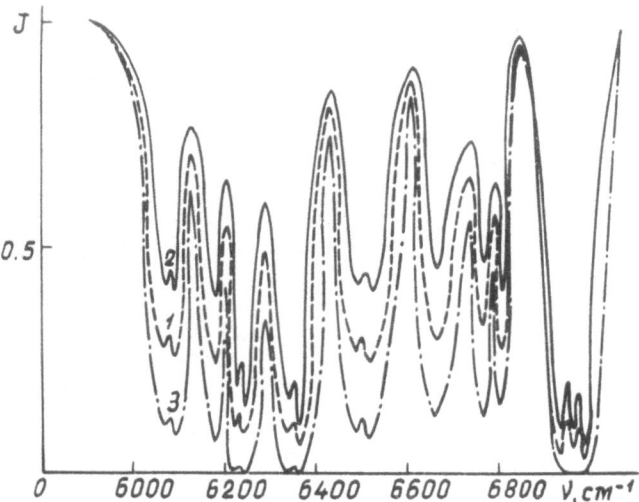

Fig. 5.   The calculated reflection spectrum for $\alpha = 67°$. Curves 1 and 2: two cloud layers with upper boundaries at $z_1 = 70$ km and $z_2 = 72$, curve 1: $\tau^* = 1$, curve 2: $\tau^* = 2$, curve 3: only the lower cloud.

for two-layer cloud models. The absorption of CO_2 above and between the clouds is also taken into account. The calculations were made for a phase angle α of 67°.

3. Conclusions

The following conclusions were drawn from the consideration of the two cloud layers:

The addition of a thin upper cloud leads to a decrease of the band equivalent widths, compared with the case of one lower cloud. The intensity of the reflected radiation is very sensitive to the small optical thickness τ^* of the upper cloud. The sensitivity is especially significant for strong bands.

The explanation of the phenomena described is that the number of photons reflected from the upper cloud increases with τ^*, and for these photons the abundance of CO_2 along the path is small.

In dealing with Venus we have to remember that there may be several cloud layers, and the situation is much more complicated than is usually considered in the spectro-scopic evaluations of the water vapour abundance.

The optical thickness of the lower cloud is 30, its upper boundary at $z_1 = 70$ km. The lower cloud is in the layer 71–72 km, its optical thickness is 1 or 3.

INTERPRETATION OF THE ILLUMINATION MEASUREMENTS BY THE AUTOMATIC INTERPLANETARY STATION 'VENERA 8'

E. M. FEIGELSON

Institute of Atmospheric Physics, Moscow, U.S.S.R.

N. L. LUKASHEVICH

Institute of Applied Mathematics, Moscow, U.S.S.R.

and

M. Ya. MAROV

Academy of Sciences of USSR, Moscow, U.S.S.R.

Abstract. Measurements of the downward light flux during the descent from an altitude of 50 km to the ground have been made. Atmospheric layers with different optical thicknesses have been discovered. The optical properties of Venus atmosphere for different cloud models have been studied.

The scientific program of 'Venera 8' included measurements of the downward light flux – the illumination $F_\downarrow(z)$ during descent from an altitude $z = 50$ km to the ground, $z_z = 0$. The sensor was a cadmium-sulfide photoresistor; its spectral range was 0.4–0.8 μm, and the effective wavelength $\lambda_{eff} = 0.63$ μm. The duration of the measurements was 56 min, the mean zenith distance of the Sun $\zeta = 84.5° \pm 2.5°$.

The main direct result of the illumination measurements is that the thick Venus atmosphere transmits light. The measurements have given:

$$100 \frac{F_\downarrow(0)}{\mathscr{S}_{0,\lambda_{eff}} \cos \zeta} = 1\%^{+1}_{-0.5}.$$

$\mathscr{S}_{0,\lambda}$ is the spectral solar constant.

A second direct result of the measurements is the discovery of layers with different optical thicknesses: the light extinction per km is equal to 0.15 in the layer from 0 to 35 km, 0.20 from 35 to 50 km, and 0.35 between 50 and 70 km (if the extinction at $z > 70$ km is neglected). Thus the optical thickness increases with decreasing density.

The authors have attempted to study the optical properties of the Venus atmosphere by selecting an atmospheric model with a calculated flux $F_\downarrow(z)$ near the measured one. Such a model is obviously not unique, because of the large number of parameters influencing the flux.

The scattering properties of the pure molecular Venus atmosphere are known (Galin *et al.*, 1971). Therefore, it seemed reasonable to calculate $F_\downarrow(z)$ for a pure molecular scattering atmosphere with albedo of the underlying surface $A_0 = 0$, and then for an increasing succession of the scattering σ and absorption α coefficients, the albedo and the elongation of the scattering function $\gamma(\varphi)$. The parameters used were (instead of the ones mentioned): the optical thickness τ, the single scattering albedo,

$\omega = \sigma/(\alpha + \sigma)$ and the mean cosine of the scattering angle

$$x_1 = \tfrac{3}{2} \int\limits_{-1}^{1} \gamma(\mu)\, \mu\, d\mu, \qquad \mu = \cos\varphi$$

or the elongation

$$\Gamma = \tfrac{1}{2} \int\limits_{0}^{1} \left\{ \int \gamma(\mu, \mu')\, d\mu' - \int\limits_{-1}^{0} \gamma(\mu, \mu')\, d\mu' \right\} d\mu;$$

μ and μ' here are the cosines of the vertical angles. The elongation is the difference between the parts of the light scattered into the upper and lower hemispheres. The parameter x_1 (or $l = 4/(3 - x_1)$) is used in asymptotic methods of the theory of transfer for large optical depths (Sobolev, 1972); and the parameter Γ, in two-stream approximations. In the present work both methods are used.

The main part of the calculations was made in the Schwarzschild approximation, generalized for the case of anisotropic scattering:

$$\frac{1}{2} \frac{d(F_\uparrow - F_\downarrow)}{d\tau} = -(1 - \omega)(F_\uparrow + F_\downarrow), \tag{1}$$

$$-\frac{1}{2} \frac{d(F_\uparrow + F_\downarrow)}{d\tau} = q(F_\uparrow - F_\downarrow). \tag{2}$$

$F_\uparrow$ and $F_\downarrow$ are here the upward and downward fluxes;

$$q = 1 - \omega\Gamma.$$

These equations were chosen because they give exact solutions for Rayleigh scattering and an error not bigger than 15% for strongly elongated scattering functions and thick layers.

Figure 1 represents examples of numerical solutions of the exact equations of transfer, of the asymptotic formula, and of the Schwarzschild equations for a homogeneous model: $\omega = \text{const}$ with different values of

$$l > \tfrac{4}{3}; \qquad q \leqslant 1; \qquad \omega \leqslant 1$$

and

$$0 \leqslant A_0 < 1$$

where A_0 is the albedo of the underlying surface.

It may be seen that the albedo does not influence the flux appreciably. The three sets of solutions (exact, asymptotic and Schwarzschild) give the same result in the Rayleigh case. No choice of ω and l or Γ brings the calculations close to the measurements. Therefore, the calculations were made for two-layer models. In each layer the solutions of Equations (1) and (2) were tabulated for values of the parameters ω_i, q_i and τ_i (the optical thickness of the ith layer). On the boundary between the layers, at

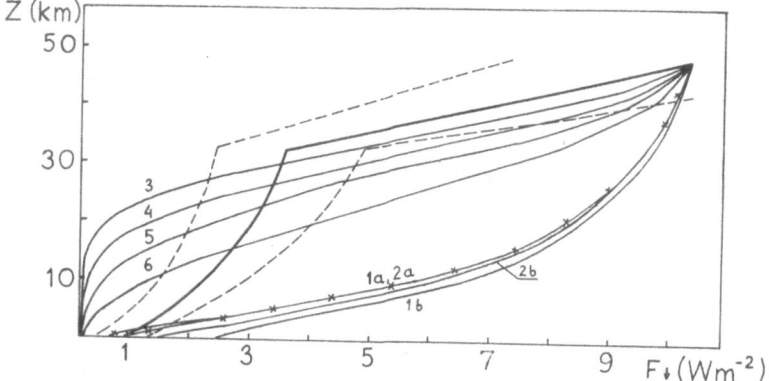

Fig. 1.   Variation of the illumination with altitude. Thick line – experimental data; dashed line – error limits. Crosses – the numerical solution of the exact radiative transfer equations for $\omega = 1$, $A_0 = 0$ and $\gamma_R(\varphi)$. Curve 1a – calculations using the asymptotic formula with $\omega = 1$, $A_0 = 0$, $l = \frac{4}{3}$. Curve 1b – ditto with $l = 4$. Curve 2a – solution of Schwarzschild Equations (1) and (2) with $\omega = 1$, $A_0 = 0$, $\Gamma = 0$. Curve 2b – Schwarzschild solution with $\omega = 1$ and $A_0 = 0.85$. Curves 3–6 – ditto with $\omega = 0.90$; 0.95; 0.975; 0.999 respectively.

$z_0 = 32$ km, the conditions of continuity of the fluxes were used. At the upper boundary of the upper layer ($i = 2$), the flux $F_\downarrow(z)$ was taken equal to the measured value at the upper point, $z = 49$ km. At the lower boundary of the lower layer $A_0 = 0$ was assumed. The final condition was

$$\tau^{(i)}(z) = c^{(i)} p^{(i)}(z),$$

where $p^{(i)}$ is the pressure and $c^{(i)}$ is a contant.

Figure 2 and Figure 3 give the results of calculations for two models. For the first

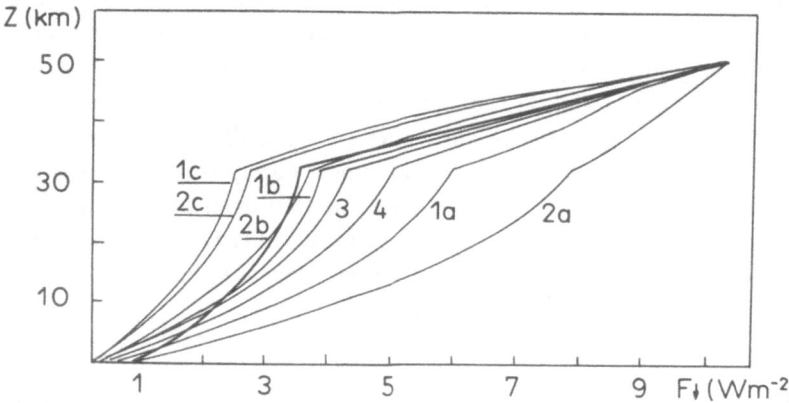

Fig. 2.   Illumination profiles for two-layer model I. Thick line – experimental data. In the lower layer in each case $\omega^{(1)} = 1$; $\Gamma^{(1)} = 0$; $\tau^{(1)} = \tau^{(1)}_{R,\lambda\,eff} = 9$. In the upper layer in all cases $\tau^{(2)} = 3$ (except curve 3, with $\tau^{(2)} = 5$). Curves 1a, 1b, 1c, and 3 – $\Gamma^{(2)} = 0.73$. Curves 2a, 2b, 2c – $\Gamma^{(2)} = 0$. Curve 4 – $\Gamma^{(2)} = 0.5$ Curves 1a, 2b, 3, 4 – $\omega^{(2)} = 0.95$. Curves 1b, 2c – $\omega^{(2)} = 90$. Curves 1c – $\omega^{(2)} = 0.80$.

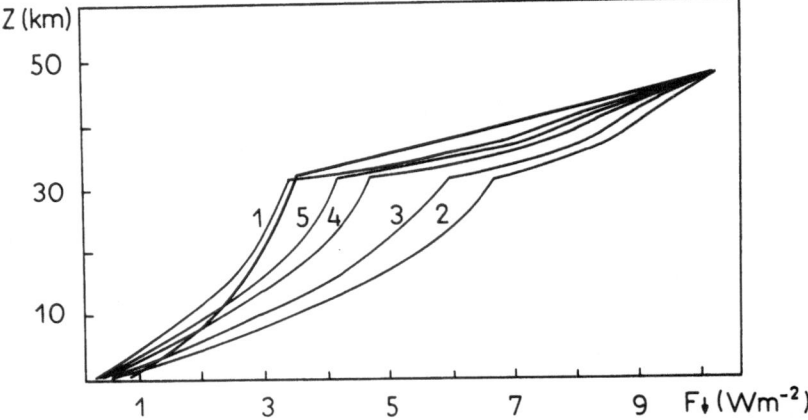

Fig. 3.   Illumination profiles for model II. In all cases $\omega^{(2)} = 1$. Curves 1, 2 – $\tau^{(2)} = 20$. Curve 3 – $\tau^{(2)} = 25$. Curve 4 – $\tau^{(2)} = 40$. Curve 5 – $\tau^{(2)} = 50$. Curve 1 – $\Gamma^{(2)} = 0$. Curves 2–5 – $\Gamma^{(2)} = 0.73$.

model the upper layer (32 km $\leqslant z \leqslant 49$ km) is absorbing ($\omega_1 \approx 0.95$) and optically rather thin, close to the Rayleigh case:

$$\tau^{(2)} = 3, \quad \text{with} \quad \tau^{(2)}_{R, \lambda_{eff}} \approx 1.$$

The second model gives

$$\tau^{(2)} \approx 20 - 50; \quad \omega \approx 1,$$

In both cases the lower layer is purely molecular scattering. Figures 2 and 3 allow us to conclude:

(a) The lower part of the Venus atmosphere ($z \leqslant 32$ km) is not significantly hazy, which corresponds to the small wind velocities here (Marov *et al.* 1973). There is no significant absorption. The correspondence between the calculations and the measurements for the very lowest layers may be improved by selecting $A_0 > 0$ (see Figure 1).

(b) Two alternative models are equally probable for the upper layer ($i = 2$; $32 \leqslant z \leqslant 49$ km). The first (model I) is an absorbing and optically thin medium: the second (model II) represents an optically thick and almost purely scattering layer. As $\tau_1^{(2)} \approx 3\tau^{(2)}_{R, \lambda_{eff}}$, the scattering function must differ somewhat from the Rayleigh case; i.e. Γ must be taken > 0. Figure 2 shows that this supposition is consistent with $\omega < 0.95$. So the radiative transfer conditions for the first model appear to be:

$$\tau \approx 3 - 5; \quad 0.9 \leqslant \omega \leqslant 0.95.$$

The second model is naturally associated with strongly elongated scattering functions. The calculations for the case ($\omega = 1$; $\Gamma = 0.73$; $\tau = 50$) corresponds satisfactorily to the mesurements (see Figure 3).

The first model suggests the penetration of a small number of cloud particles below a cloud layer with a lower boundary higher than 50 km. The significant absorption in this case may be connected with an assumed absorption (see for instance Galin *et al.*, 1971) in the cloud layer for $\lambda \leqslant 0.6 \ \mu$m. Galin obtained $\omega = 0.998$ for $\lambda = 0.55$. As $\sigma_{cl} \gg \sigma^{(2)}$, we get $\omega_{cl} > \omega^{(2)}$ (the index 'cl' refers to the cloud layer). The idea of a cloud extending downward to the level ≈ 32 km is natural for model II. We note that the boiling point of water occurs at 35 km.

An independent check of the above evaluations is given by the equation for the illumination at depth in the scattering medium, where

$$F_\downarrow (z) \sim \exp \left[-k\tau (z) \right]$$

and

$$k = 2 \left(\frac{1 - \omega}{l\omega} \right)^{1/2} .$$

For

$$k = \frac{1}{\tau^{(2)}} \ln \frac{F_\downarrow (49)}{F_\downarrow (32)}, \qquad l \geqslant \tfrac{4}{3},$$

and $\tau^{(2)}$ corresponding to models I and II, we get:

$$\omega_1 \leqslant 0.96 \quad \text{and} \quad \omega_2 \approx 0.999 .$$

in agreement with the evaluations by the selection method.

The mean scattering coefficient $\sigma^{(2)} = \tau^{(2)}/H$, with $H = 17$ km, is approximately 0.2 km^{-1} for model I and 2.5 km^{-1} for model II. The corresponding absorption coefficients $\alpha^{(2)} = (\sigma/\omega)(1 - \omega)$ are 0.01 km^{-1} for model I with $\omega = 0.95$, and 0.005 km^{-1} for model II with $\omega = 0.999$. We may regard both values of $\alpha^{(2)}$ as the same because of the low accuracy of the evaluations.

Let us consider one more evaluation of the absorption coefficient. The visual albedo of the planet is 80%. The transmission is only a few per cent. Therefore, the absorption is 15–20%. The optical thickness of the combined cloud (1) and under cloud layers (2) is of the order of 10–50 for model I, and 30–100 for model II. The mean path length of photons $l = nH$, where $n \approx 3$–6 for $\tau = 10$–100, and $H = 20$–30 km is the thickness of the system. The absorption $A \approx 0.2 = \exp(-\alpha l)$, so $\alpha = 0.01$–0.005 km^{-1}, in agreement with the previous results. The number of particles per unit volume of layer (2) may be obtained from the equation

$$N = \frac{\sigma}{\pi r^2 K} .$$

Supposing that the layer consists of large particles $r \geqslant 10 \ \mu$m with scattering cross section $K = 2$, we find:

$$N_1 \lesssim 0.3 \text{ part cm}^{-3}, \quad \text{or} \quad N_2 \lesssim 5 \text{ part cm}^{-3} .$$

In conclusion we remark that both models may agree with the known reflection spectrum of the planet. A cloud layer with an optical thickness of tens and $0.99 \leqslant \omega \leqslant 1$ agrees with the measured value of $F_{\downarrow}(z)$ at $z = 49$ km.

References

Galin, V. Ya., Lukashevich, N. L., and Feigelson, E. M.: 1971, *Doklady Akad. Nauk SSSR* **197**, No. 2.
Marov, M. Ya., Avduevsky, V. S., Borodin, N. F., Ekonomov, A. P., Kerzanovich, V. V., Lysov, V.P., Moshkin, B. Ye., Rozhdestvensky, M. K., and Ryabov, O. L.: 1973, *Icarus* **20**, 407.
Sobolev, V. V.: 1972, *Light Scattering in Planetary Atmospheres*, Moscow (in Russian).

NATURE OF THE VENUS CLOUDS AS DERIVED FROM THEIR POLARIZATION

JAMES E. HANSEN

Goddard Institute for Space Studies, New York, N.Y., U.S.A.

and

J. W. HOVENIER

Vrije Universiteit, Amsterdam, The Netherlands

Abstract. The linear polarization of sunlight reflected by Venus is analyzed by comparing observations with extensive multiple scattering computations. The analysis establishes that Venus is veiled by a cloud or haze layer of particles which have a narrow size distribution with a mean radius $\sim 1\ \mu$. The refractive index of the particles is 1.44 ± 0.015 at $\lambda = 0.55\ \mu$ with a small normal dispersion, the refractive index decreasing from the ultraviolet toward the infrared. The particles exist at a high level in the atmosphere, with the optical thickness unity occurring where the pressure is about 50 mb.

The particle properties deduced from the polarization eliminate all but one of the cloud compositions which have been proposed for Venus. A concentrated solution of sulfuric acid (H_2SO_4–H_2O) provides good agreement with the polarization data.

Polarization observations at present provide the most sensitive means for determination of the cloud particle characteristics on Venus. In the analysis of the polarization it is essential that the effects of multiple scattering be accurately accounted for, as has been demonstrated by Hansen and Hovenier (1971). Thus in order to help extract the available information on the cloud particles we have made a series of multiple scattering computations and compared them with observations of Venus made by Lyot (1929), Kuiper (1957), Marin (1965), Dollfus (1966), Coffeen and Gehrels (1969), Dollfus and Coffeen (1970), Forbes (1971), and Veverka (1971). Our computations are much more extensive than those of Hansen and Arking (1971); we have included many observations not employed by Hansen and Arking and we have analyzed observations at all available wavelengths. In addition we have related the derived cloud particle refractive index and its dispersion to the proposed cloud compositions for Venus. A paper giving a detailed documentation of our method of analysis is in preparation and will be submitted for publication in the *Journal of the Atmospheric Sciences*.

Our computations for the degree of linear polarization involve three steps. First the well-known Mie theory is used to compute the single scattering properties of a size distribution of spherical particles. We have found that in practice it is usually possible to describe the effect of the size distribution by two parameters which describe the mean particle size and the width of the size distribution. The second part of the computations is for multiple scattering by a homogeneous plane-parallel atmosphere. This we accomplish with the doubling method which was developed for the case of polarized light by Hansen (1971) and Hovenier (1971). Finally we integrate the multiple scattering results over the planetary disk, assuming a spherical but locally-plane-

Woszczyk and Iwaniszewska (eds.), Exploration of the Planetary System, 197–200. All Rights Reserved
Copyright © 1974 by the IAU

parallel atmosphere. The basic method of integration is the same as that described by Horak (1950). For the limited number of observations referring to a small area on the planet we of course used the computations without integration over the disk.

Several prominent features in the theoretical polarization depend on the spherical shape of the particles. These include the 'rainbow', the 'glory' and a feature of 'anomalous diffraction'. The existence of all these features in the observed polarization of Venus, in close agreement with the theory, is sufficient to prove that the particles in the visible clouds of Venus are in fact spherical. In addition we find that the effective radius for the particle size distribution, defined as

$$r_{\text{eff}} = \frac{\int\limits_0^\infty r\pi r^2 n(r)\, dr}{\int\limits_0^\infty \pi r^2 n(r)\, dr},$$

is $1.05\pm0.1\ \mu$, and that the effective variance, defined as

$$v_{\text{eff}} = \frac{\int\limits_0^\infty (r - r_{\text{eff}})^2 \pi r^2 n(r)\, dr}{r_{\text{eff}}^2 \int\limits_0^\infty \pi r^2 n(r)\, dr},$$

is 0.07 ± 0.02. $n(r)\, dr$ is the number of particles with radius between r and $r+dr$. The spherical particle shape and the narrow size distribution strongly suggest that the cloud particles on Venus are liquid.

The particle refractive index is perhaps the most important information which we extract from the polarization. For this we obtain $n_r(\lambda=0.55\ \mu)=1.44\pm0.015$, where n_r decreases slightly with increasing wavelength, from $n_r(\lambda=0.365\ \mu)=1.46\pm0.015$ to $n_r(\lambda=0.99\ \mu)=1.43\pm0.015$. In order to compare these values with laboratory values of n_r for proposed cloud constituents we must be able to interpolate (or extrapolate) the laboratory values in wavelength and temperature. We handle this problem by using two formulae of classical physics (cf. Born and Wolf, 1959): the Lorentz-Lorenz equation, which yields the dependence of the refractive index on temperature, and a dispersion formula for classical oscillators, which yields the dependence of the refractive index on wavelength. We are able to demonstrate that this procedure is satisfactory for substances which do not have a strong absorption band in the visible part of the spectrum.

Of all the cloud compositions which have been proposed in the literature for Venus, we find that only a concentrated solution of sulfuric acid is in agreement with the polarization data. Other evidence in support of sulfuric acid on Venus has been presented by Young (1973) and Sill (1972). In particular we find that the refractive indices of hydrochloric acid solutions and carbon suboxide are definitely not compatible with the values of n_r for the Venus cloud particles. These materials had previously been considered as candidates for the cloud composition, in large part because their refractive indices were in the general range of values indicated by the polarization.

Finally, the polarization also indicates that the cloud particles occur high in the atmosphere of Venus, with the optical depth unity being at a pressure ~ 50 mb. This conclusion, unlike those for the particle shape, size distribution and refractive index, is not independent of the homogeneous model atmosphere which we employ. However we estimate that the uncertainty in the value 50 mb is at most ~ 25 mb.

It is notable that there are some marked similarities between the particles in the atmosphere of Venus and the aerosol layer on the Earth at an altitude ~ 20 km (the Junge layer), as we have previously pointed out (cf. Hunten, 1971). The Junge layer is at the pressure level ~ 50 mb and the particle size distribution is apparently very narrow (Friend, 1966; Mossop, 1965), corresponding to $v_{eff} = 0.06\text{--}0.08$. On the other hand the particles in the Junge layer are smaller ($r_{eff} = 0.3\text{--}0.4 \, \mu$; Friend, 1966) and have a smaller optical thickness ($\sim 2 \times 10^{-2}$; Elterman *et al.*, 1973) than the particles on Venus. Nevertheless the above similarities, together with the fact that the Junge layer contains a large fraction (perhaps more than 50%; Rosen, 1971, Lazrus *et al.*, 1971) of sulfuric acid, offer additional support for the conclusion that the particles in the visible clouds of Venus are composed of strong sulfuric acid solutions.

Acknowledgements

J.E.H. was supported during the course of part of this work by NASA Grant 33-008-012 through Columbia University. J.W.H. was supported by the Netherlands Organization for the Advancement of Pure Research (Z.W.O.) during a one year stay at the Institute for Space Studies in 1970–71 when this work was originated.

References

Born, M. and Wolf, E.: 1959, *Principles of Optics*, Pergamon, London, p. 86.
Coffeen, D. L. and Gehrels, T.: 1969, *Astron. J.* **74**, 433.
Dollfus, A.: 1966, Contribution au Colloque Caltech-JPL sur la lune et les planètes: Venus. JPL Tech. Memo. No. 33-226, 187.
Dollfus, A. and Coffeen, D. L.: 1970, *Astron. Astrophys.* **8**, 251.
Elterman, L., Toolin, R. B., and Essex, J. D.: 1973, *Appl. Opt.* **12**, 330.
Forbes, F. F.: 1971, *Astrophys. J.* **165**, L21.
Friend, J. P.: 1966, *Tellus* **18**, 465.
Hansen, J. E.: 1971, *J. Atmospheric Sci.* **28**, 120.
Hansen, J. E. and Arking, A.: 1971, *Science* **171**, 669.
Hansen, J. E. and Hovenier, J. W.: 1971, *J. Quant. Spectrosc. Radiat. Transfer* **11**, 809.
Horak, H. G.: 1950, *Astrophys. J.* **112**, 445.
Hovenier, J. W.: 1971, *Astron. Astrophys.* **13**, 7.
Hunten, D. M.: 1971, *Space Sci. Rev.* **12**, 539.
Kuiper, G. P.: 1957, in M. Zelikoff (ed.), *Threshold of Space*, Pergamon, New York, p. 78.
Lazrus, A. L., Gandrud, B., and Cadle, R. D.: 1971, *J. Geophys. Res.* **76**, 8083.
Lyot, B.: 1929, *Ann. Obs. Paris (Meudon)* **8**, [available in English as NASA TT F-187, 1964].
Marin, M.: 1965, *Rev. Optique* **44**, 115.
Mossop, S. C.: 1965, *Geochim. Cosmochim. Acta* **29**, 201.
Rosen, J. M.: 1971, *J. Appl. Meteorol.* **10**, 1044.
Sill, G. T.: 1972, *Comm. Lunar Planet. Lab.*, No. 171, 191–198.
Veverka, J.: 1971, *Icarus* **14**, 282.
Young, A. T.: 1973, *Icarus* **18**, 564.

DISCUSSION

Icke: The straight-line relationships which you found are determined by the slope and intercept of the lines. Which physical properties of the molecules determine these two parameters?

Hovenier: These parameters are essentially determined by the mean resonance frequency and by the mean oscillator strength of the molecules.

Van de Hulst: There seems to exist a very fundamental relation: 'much money ≪ good ideas'!

WATER VAPOR IN VENUS DETERMINED BY AIRBORNE OBSERVATIONS OF THE 8200 Å BAND

THEODORE R. GULL* and C. R. O'DELL**

Yerkes Observatory, Williams Bay, Wis., U.S.A.

and

R. A. R. PARKER‡

Johnson Space Center, Houston, Tex., U.S.A.

Abstract. The region of the 8200 Å Band of H_2O was studied in spectra of Venus obtained with an echelle grating spectrograph operated at an altitude of 14.6 km in the NASA Learjet research aircraft. Taking advantage of low foreground absorption, observing at a time of velocity quadrature, differential spectroscopy with respect to lunar spectra, and spectrum averaging, we establish a value of H_2O of 3 ± 20 μ for the total path over the entire disk. This value differs from earlier studies of the integrated disk but supports the low values recently derived from infrared bands and by very high spectral resolution groundbased studies.

These results establish that the average value over the entire disk is quite low but allows that locally larger values may exist as reported by Barker. A more complete description of this work appeared in *Icarus* **21** (1974), 213.

DISCUSSION

Barker: Comment; The McDonald data have high spatial resolution on the disk of Venus. Some positions show large amounts of H_2O whereas others show non-detectable amounts similar to those of O'Dell, and Traub and Carleton, who looked basically at the integrated disk values. The average or integrated disk values must be lower than the largest amounts.

Baum: Comment; I hope that the spectroscopic observations of Venus reported at this symposium will be repeated during other time intervals. The features in ultraviolet images are found to change in strength considerably, particularly from one year to another. It would be helpful to know whether spectra reveal similar (perhaps correlated) variations.

---

* Now at the Kitt Peak National Observatory
** Now on leave at the George C. Marshall Space Flight Center
‡ On leave from the University of Wisconsin

Woszczyk and Iwaniszewska (eds.), Exploration of the Planetary System, 201. *All Rights Reserved*
Copyright © 1974 by the IAU

GROUND-BASED OBSERVATIONS OF MARS AND VENUS WATER VAPOR DURING 1972 AND 1973

E. S. BARKER

University of Texas, McDonald Observatory, Fort Davis, Tex., U.S.A.

Abstract. The Venus water vapor line at 8197.71 Å has been monitored at several positions on the disk of Venus and at phase angles between 22° and 91°. Variations in the abundance have been found with both position and time. The total two-way transmission has varied from less than 5 to 77 μ of water vapor. Comparisons will be made between the water vapor abundances, presence of UV features, and the CO_2 abundances determined from near simultaneous observations of CO_2 bands at the same positions on the disk of Venus.

The amount of Martian atmospheric water vapor has been monitored during the past two years at McDonald Observatory using the échelle coudé scanner of the 272 cm reflector. Two periods of the Martian year have been monitored. The first period was during and after the great 1971 dust storm ($L_s = 280°$ to 20° or summer in the southern hemisphere). The results obtained will be compared to the Mariner 9 IRIS and Mars 3 observations made during the same period.

During the second period ($L_s = 124°$ to 266°) observations were made to follow the seasonal latitudinal and diurnal changes in the water vapor abundance in the Martian atmosphere. The water vapor abundance declines from a maximum of 20–35 μm at $L_s = 125°$ to the 5–15 μm level at $L_s = 180°$. Then it remained relatively constant until $L_s = 250°$ when the increase to 20–25 μm occurred in the southern latitudes. Studies of the latitudinal and diurnal water vapor distributions indicate the location of maximum and minimum abundances for this season are positively correlated with surface temperature variations.

1. Introduction

As part of the continuing planetary observational program at McDonald Observatory, the several lines in the 8200 Å water vapor band are monitored as frequently as possible to study variations in the observed abundance of Venus water vapor 'above the clouds', and to follow the seasonal and spatial variations in the water vapor abundance in Martian atmosphere. Both 2 Å mm^{-1} spectrographic plates and photoelectric scanner observations have been taken since completion of the 107-in. (272 cm) telescope in the spring of 1969. Other papers presented at this IAU meeting cover the historical evolution of the search for water vapor in planetary atmospheres (Schorn and Barker, 1973; Traub and Carleton, 1973). This paper will describe only the McDonald results obtained during 1972 and up to August 15, 1973.

2. Instrumentation and Reductions

Photoelectric spectrum scans have been made using the rapid scanner at the coudé focus of the 107-in. (272 cm) telescope which has an image scale of 2.3″ mm^{-1}, ideally suited for spatial resolution on planetary disks. We have used essentially the same system for the Mars and Venus observations; a 79 groove per mm échelle grating in double pass which gives a very high dispersion of 8 mm Å^{-1} at 8200 Å. The only difference is the choice of slit widths used which define the resolution obtained. With 200 μm slits, a final resolution of about 30 mÅ or 275000 at 8200 Å (FWHM)

Woszczyk and Iwaniszewska (eds.), Exploration of the Planetary System, 203–222. *All Rights Reserved* is attainable; similarly, 400 μm slits give about 60 mÅ. The entire system is described in more detail by Tull (1972) and Wells (1972).

RCA GaAs (31034 and 31034 A) photomultipliers have made possible these high resolution and photometrically accurate scans. Their high quantum efficiency (10 to 20% at 8200 Å coupled with the low dark count of 2 counts/s at dry ice temperatures) makes it possible to obtain high quality photoelectric scans of a single line within 10 min for Venus and 40 min for Mars. Scans can be made about 50 times faster now than when the échelle scanner was put into operation two years ago. Different slit widths or effective resolutions have been used to compensate for or take advantage of the quantum efficiency of the photomultiplier available at the time. In general, all the Venus scans have a resolution (FWHM) of 30 mÅ and the Mars scans have 60 mÅ. The Mars scans obtained since the beginning of June, 1973, are an exception and are at the higher resolution. This was necessitated because the Doppler shift decreased to less than 0.25 Å and terrestrial humidity increased during the summer months at McDonald Observatory.

Most strong water lines in the 8200 Å band are contaminated to some degree by weak Fraunhofer lines primarily due to CN transitions. All usable water lines (i.e., free from major blends and having laboratory line strengths) have been observed on Mercury, as a source of variable Doppler shift, to look for 'hidden' Fraunhofer lines. Lines at 8141, 8193, and 8256 Å are so seriously blended that they are not usable for future studies. Weak lines in the wings of 8176, 8189, and 8197 Å have been measured and can be used if the solar component of the Doppler shift does not place them at the Doppler shifted wavelengths of Mars and Venus water lines. The work presented in this paper uses primarily the line at 8176.9 Å for Mars and the line at 8197.7 Å for Venus.

An image rotator allowed the slit of the spectrograph to be oriented with respect to the intensity equator on Venus or Mars and the Martian north–south direction. On a few occasions a decker was used to shorten the slit to cover only part of the planet, but usually the slit length was just slightly larger than the planet's diameter to minimize the amount of sky background. The counting ratio between the planet-plus-sky and sky was monitored several times during each scan. With this ratio the appropriate amount of sky background was subtracted from each planet scan by using a sky or solar scan of the same wavelength region obtained on the same day.

Each photoelectric scan is processed through several steps to remove effects of sky background, dark count level, and vignetting in the échelle system. The final plot of the spectrum consists of data points representing the summation of the forward and reverse scans and a smooth curve through these data points which has been drawn by the computer as a result of a Fast Fourier Transform smoothing and interpolating operation on the data. All equivalent widths are measured with respect to such a smoothed curve. The local continuum curve has been transferred from a solar scan of similar water content taken on the same day. Attempts have been made to produce a synthetic solar spectrum scan to calculate a ratio spectrum with the planetary scan leaving only the planetary component of the water line. Both methods have the same major source of error (namely the placement of the continuum level), and we feel that we are adding an additional source of error by trying to produce a solar scan having identical water content and instrumental resolution.

The digital form of the photoelectric data makes computer processing of data almost routine and processing presently is being carried out within a few days after the observation using an IBM 1800 computer at McDonald Observatory. Lack of manpower is the only reason for not always producing reduced spectra immediately after an observation, although we do so when the occasion warrants. Compared to the former time lag of weeks and the necessary travel to a microdensitometer to reduce regular photographic plates, our present system is far superior and, most important, we can modify our observing program on a day to day basis to best utilize available observing time.

3. Venus Observations

The 107-in. échelle-coudé scanner has proved to be well suited for spatial studies of Venus water vapor, mostly because of its speed, a prime requirement for spatial studies requiring good seeing over a long period of time. With a resolution of 30 mÅ

TABLE I

Physical data for Venus H_2O observations

Date	$\Delta\lambda$ (Å)	i (°)	Diameter (arcsec)	CO_2	UV
8/24/72	+ 0.369	91.0	24.4	+	+
9/23/72	+ 0.350	74.3	18.4	+	+
9/24/72	+ 0.350	73.8	18.2	+	+
9/26/72	+ 0.347	73.3	17.9	+	+
9/29/72	+ 0.344	71.4	17.5	+	+
11/25/72	+ 0.262	47.2	12.7		
11/30/72	+ 0.253	45.2	12.5		
12/01/72	+ 0.251	42.9	12.4		
12/12/72	+ 0.231	40.4	11.9		
12/13/72	+ 0.229	40.0	11.8		
12/14/72	+ 0.227	39.6	11.8		
12/16/72	+ 0.225	38.9	11.7		
6/05/73	− 0.139	21.2	10.2		
6/14/73	− 0.164	24.8	10.4		?
6/16/73	− 0.169	25.6	10.4		?
6/17/73	− 0.171	26.0	− 10.4		?
7/07/73	− 0.221	34.1	11.0	+	?
7/08/73	− 0.223	34.5	11.0		?
8/03/73	− 0.276	45.1	11.3		?
8/04/73	− 0.278	45.5	11.4		?
8/10/73	− 0.288	47.9	11.7	+	?
8/12/73	− 0.291	48.7	11.8	+	?
8/14/73	− 0.294	49.5	12.7	+	?
8/15/73	− 0.296	49.9	12.8	+	?

(FWHM) which is better than the resolution of a 2 Å mm^{-1} spectrographic plate, we can obtain a water line scan in 15 min, compared to a plate exposure of one hour. This means we can observe at many locations on the disk of Venus during a few hours of good seeing.

The physical data for the Venus scanner observations of the 8197.7 Å and 8176.9 Å lines are summarized in Table I. Table I contains the physical data pertinent to Venus observations and indicates that phase angle coverage is fairly good between 20° and 90°. It is difficult to observe near phase angles of 0° and 180° because of the small Doppler shifts involved. Only those observations reported by Traub and Carleton (1973, 1974) have been made at phase angles greater than 120°. Schorn et al. (1969) presented observations over phase angles of 52° to 92°. The scanner resolution and terrestrial water vapor limit the useful Doppler shift range to greater than ±0.15 Å which corresponds to approximate phase angle limits of 20° < i < 150°.

Total abundances ($\eta\omega$) for a two-way transmission through the Venus atmosphere were calculated on assumptions of an effective pressure of 100 mb and a temperature of 250 K. The H_2O line strengths (Farmer, 1971) at 250 K were used to calculate total abundances from the observed equivalent widths. The internal error involved in a scan is around ±5 μm of precipitable H_2O.

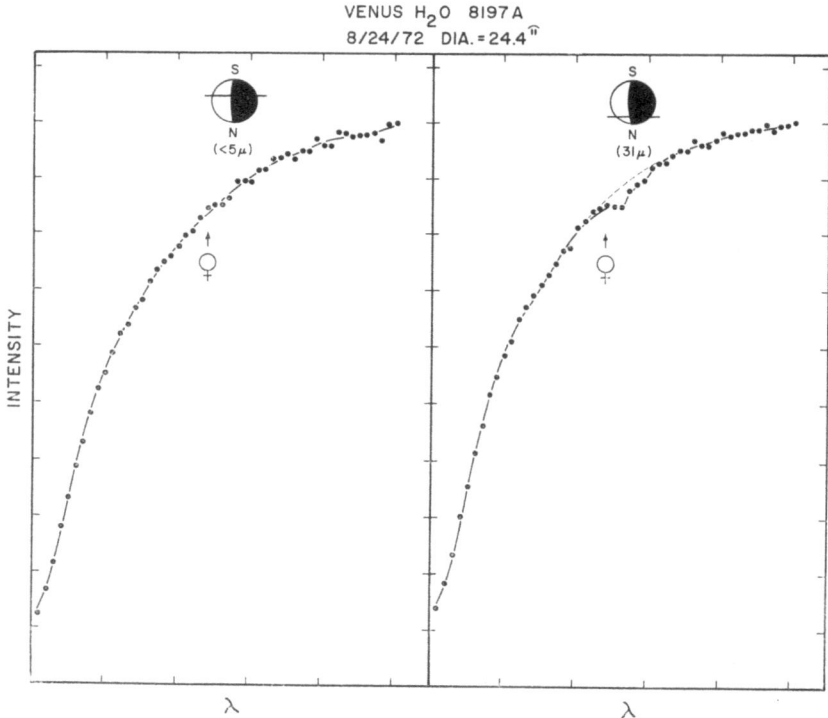

Fig. 1.   The position of 8197 Å water vapor line on Venus is indicated by the ♀ symbol. The slit position is shown in the small insert.

Putting the amounts of <5 to 77 μm of water vapor in perspective with previous measurements, they fall between the total disk values of Fink *et al.* (1972) who found 1.6 μm for the water bands at 1.4, 1.9 and 2.7 μ, and those found by Schorn *et al.* (1969) and Owen (1968) of <64 to 120 μm using several lines in the 8200 Å band.

Variability of amount of water vapor with phase angle and position on the disk is illustrated in Figures 1 and 2. Figure 1 shows the red wing of the 8197 Å water line for two different spectrograph slit positions, S1/4 and N3/4. The slit in most cases was placed parallel to the intensity equator and the N–S notation is for convenience only to indicate a position N or S of the intensity equator. It does not refer to the actual N–S direction on Venus. An upper limit of 5 μm at the S1/4 position contrasts with the easily detectable amount of 30 μm at the N3/4 position. Scans shown in Figure 2 again graphically show that we can find different amounts of water vapor at different locations on the disk. The 56 μm in the equator scan is the largest amount detected during the August, 1972, through August, 1973, period and would have barely been detectable on 2 Å mm^{-1} spectrographic plates. Comparable amounts have been seen on many spectrographic plates taken at McDonald since 1967, but the spectrographic plates lacked the resolution and photometric accuracy needed to detect smaller amounts.

A summary of spatial distribution data is presented in Figure 3 with various slit

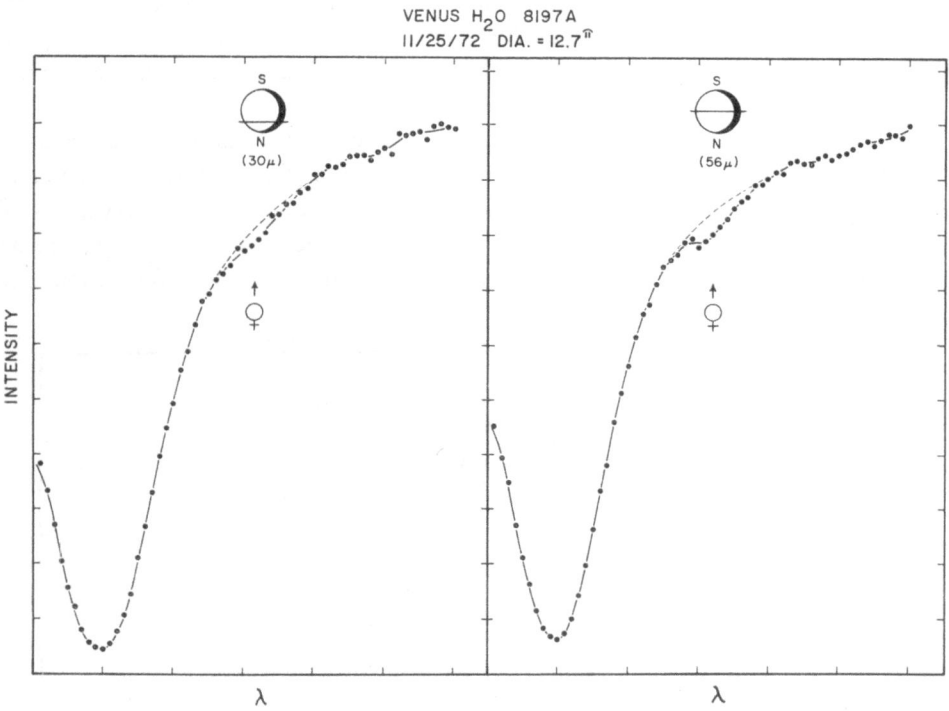

Fig. 2.   The position of 8197 Å water vapor line on Venus is indicated by the ♀ symbol. The slit position is shown in the small insert.

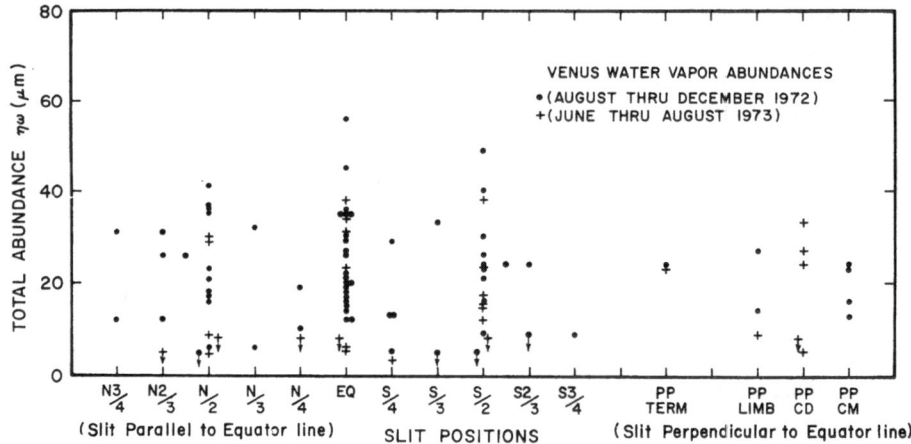

Fig. 3. Spacial distribution of water vapor determinations over the disk of Venus. Slit Positions: Parallel to intensity equator at various fractions of disk diameter toward poles; N3/4, N2/3, N/2, N/3, N/4, EQ, S/4, S/3, S/2, S2/3, S3/4. Perpendicular to intensity equator or pole to pole at various fractions of disk diameter; T designates terminator, CM – central meridian, CD – center of disk, L – limb; PP T/3, PP T/2, PP CM, PP C/3, PP CD, PP L/2, PP L/3.

positions explained in the caption. There does not seem to be any preferential location for high or low abundances. The detection limit (indicated by vertical arrows in Figure 3) was around 5 to 8 μm, a value frequently reached at various phase angles and points on the disk.

Variation of the equivalent width of a line as a function of phase angle can be used as a parameter in models to explain the vertical structure of cloud layering on Venus (Hunt, 1972a, b; Margolis and Hunt, 1973). In Figure 4 the 1972–73 observations of total water vapor abundance on Venus are plotted as a function of phase angle and/or time. Several points at one phase angle indicate abundance variations for various slit

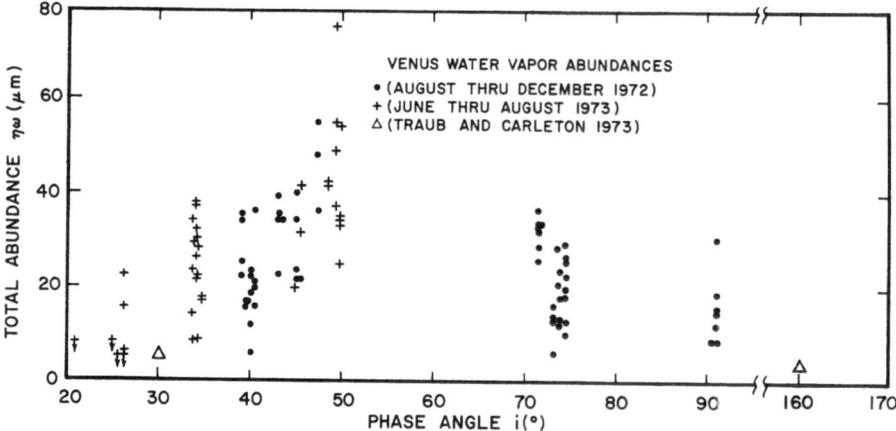

Fig. 4. Venus water vapor determinations on various dates or at different phase angles, i.

positions on that day. A typical internal error bar is ± 5 μm; these have been left off the figure for clarity. Some scans exhibited no Venus water line at all and are indicated by vertical arrows representing an upper limit to the abundance for that scan.

No compelling correlation between phase angle and water vapor abundance is present in Figure 4; in fact, the large scatter at any one phase angle could suggest the conclusion that we are observing a random event. However, there is a possibility of preferential appearance of the larger abundances at intermediate phase angles 35°–90°. Observations of lower abundances at phase angles greater than 140° and less than 30° are consistent with the two observations of Traub and Carleton (1973, 1974) in which they report no line greater than 0.1 mÅ or a total abundance of 1 μm at $i = 1.56°$ and a detectable line of 0.5 mÅ or a total abundance of 5 μm at $i = 30°$. As indicated earlier, the Doppler shifts are small in these phase angle ranges ($i < 30°$, $i > 150°$) and may lead to large uncertainties in the measured equivalent widths of non-terrestrial water lines which are masked by strong telluric components only 0.1 Å away. Only observations with very high resolution (> 250000) and dry terrestrial conditions can be used when

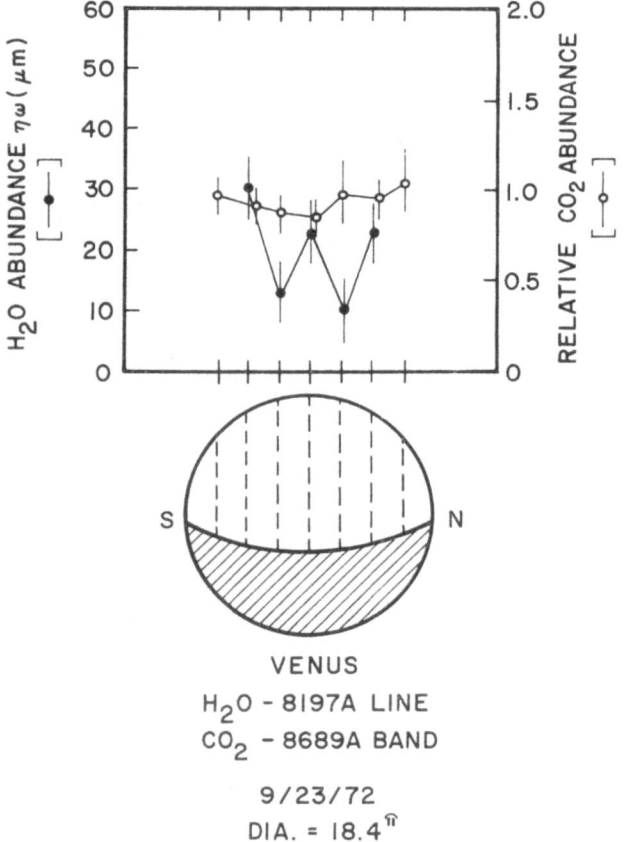

Fig. 5.   Venus water vapor and CO_2 abundances for the same slit positions on September 23, 1972.

the Doppler shifts are so small. Attempts will be made at McDonald in January of 1974 during our dry season to observe at phase angles greater than 150°. Since we will have a very large crescent (>40″), we hope to see any variations in the spatial distribution along the crescent similar to the changes of a factor of two or three noted in abundance during the 1972 inferior conjunction (Barker and Schorn, 1973).

As seen from Table I, it will be possible to intercompare the CO_2 and H_2O abundances and the ultraviolet (UV) features for several days during September, 1972. As a sample, Figure 5 shows the slit positions for H_2O and CO_2 abundance determinations made within hours on the same day. Note that CO_2 is relatively constant on this day whereas H_2O abundances vary by a factor of 3.

On this day the ultraviolet 'Y' feature was present on the disk, but offset to the south making any comparison with UV features difficult for slits placed parallel to the intensity equator. But high and low water vapor abundance values fall over both light and dark areas.

Note in Figure 6 that on this day the variations in CO_2 and H_2O abundances may

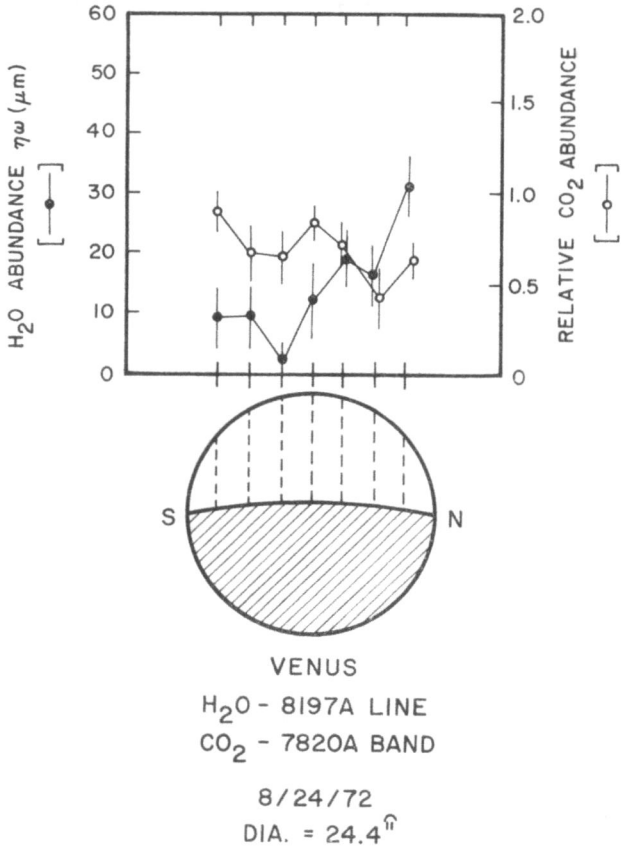

Fig. 6.   Venus water vapor and CO_2 abundances for the same slit positions on August 24, 1972.

be positively correlated. However, the systematic trends were of decreasing CO_2 and of increasing H_2O from south to north across the disk.

Young *et al.* (1973) have noted a four-day periodicity in the relative CO_2 absorption strength in their observations of the 8689 Å CO_2 band during September and early October of 1972. Similar observations were obtained at McDonald Observatory with the 107-in. (272 cm) and 82-in. (208 cm) coudé spectrographs (Schorn and Barker, 1973). The 107-in. data on the 7820 and 8700 Å CO_2 bands show a similar periodicity. Unfortunately, the coverage of the 8197 Å water vapor abundance during the same period of time was not as complete as the CO_2 data, being only four days out of 13. Detailed intercomparison between days and slit positions may reveal a slight periodicity, but the CO_2 photographic spectral data are not yet fully reduced.

The Venus water vapor observations during 1972–73 show quite conclusively a day-to-day abundance variation and also a significant abundance variation with position on the disk of Venus on any single day. Consequently, the level or altitude of line formation must change, consistent with vertical and/or horizontal movement of the altitude of 'tops' of the lower cloud deck.

4. Mars Observations

The search for Martian water vapor has long been pursued, with first spectrographic detection being obtained by Spinrad *et al.* (1963). Subsequently, further ground-based studies were presented by Schorn *et al.* (1967), Owen (1967, 1969), Schorn *et al.* (1969) Tull (1970), Barker *et al.* (1970), Barker (1971), Tull and Barker (1972), and Larson (1973). We also have *in situ* measurements made by Mariner 9 and Mars 2 and 3 reported by Hanel *et al.* (1972), Kunde (1973), Moroz and Ksanfomaliti (1972), and Moroz and Petrov (1973). Despite all these observations, there still remain parts of the Martian year which have not been covered observationally, because a period of 15 years is required to fully observe from the Earth all seasons on Mars. Also we have no *a priori* reason to believe that the weather on Mars is the same every year. The great dust storm of 1971–1972 is an obvious example which affected the water vapor content detected in the Martian atmosphere both during and after the storm.

This paper presents two results: (1) additional data obtained during the 1971–72 apparition after dissipation of the great dust storm; and (2) data obtained in the 1972–74 apparition through August, 1973, which cover a seasonal period ($150° < L_s < 230°$) which had not previously been observed. The later results bear on interesting questions: (a) Does the water vapor disappear between the northern hemisphere summer-fall season and the reappearance of larger abundance values (20–50 μm) during the southern hemisphere late summer season? (b) If the water does not disappear, how is the transition made between hemispheres and where is the water vapor located spatially with respect to aerographic latitude and diurnal distribution of the water vapor.

4.1. WATER VAPOR OBSERVATIONS DURING AND AFTER THE GREAT DUST STORM OF 1971

Tull and Barker (1972) presented 1971–72 data in preliminary form. This section is

TABLE II

Mars H_2O abundances for the 1971–72 apparition

L_s	Amount (μm)	Location	Reference
232°	20	Total Disk	Larson (1973)
293°	10–25	SSolar Rev 20	Hanel et al. (1972)
297°	5–20	SPC Rev 30	Hanel et al. (1972)
314°	10–20	SSolar Rev 92	Hanel et al. (1972)
321°	5–20	SPC Rev 116	Hanel et al. (1972)
336°	5–15	SSolar Rev 174	Hanel et al. (1972)
270–360°	8	NPC	Kunde (1973)
60°	20–30	Northern hemisphere	Kunde (1973)
352°	1–3	S–N track ± 50° lat.	Moroz (1972)
358°	20	Equatorial	Moroz (1972)
286°	16.5	PPCM	Tull (1972)
285°	12.0	PPCM	Tull (1972)
286°	10.5	PPCM	Tull (1972)
288°	13.5	PPCM	Tull (1972)
299°	13.0	PPCM	Tull (1972)
314°	18.25	PPCM	Tull (1972)
316°	7.5	EQ	Tull (1972)
324°	10.75	PPCM	Tull (1972)
327°	7.0	PPCM	Tull (1972)
332°	6.0	PPCM	Tull (1972)
348°	13.0	PPCM	Tull (1972)
350°	13.5	PPCM	Tull (1972)
352°	14.0	PPCM	Tull (1972)
353°	15.0	PPCM	Tull (1972)
353°	16.5	PPCM	Tull (1972)
11°	9.3	PPCM	This paper
21°	5.0	PPCM	This paper

devoted to combining their data with a few additional ground-based observations made by Larson (1973) and with spacecraft determinations made by Mars 2 and 3 and by Mariner 9.

A summary of 1971–72 observations appears in Table II and Figure 7 in which cross-hatched areas refer to previous determinations and other points are identified in the caption. An obvious conclusion is that after the 1971 dust storm the reappearance of atmospheric water vapor during the southern summer season ($270° < L_s < 360°$) was different from the usual reappearance pattern. Two primary reasons have been advanced by Hanel et al. (1972) and others for the much lower abundance (5–20 μm vs 20–50 μm) during the great dust storm summer: (1) a larger amount of water may have been trapped in the northern cap (since the north polar hood extended further south than usual); and/or (2) the release of water vapor in the southern hemisphere may

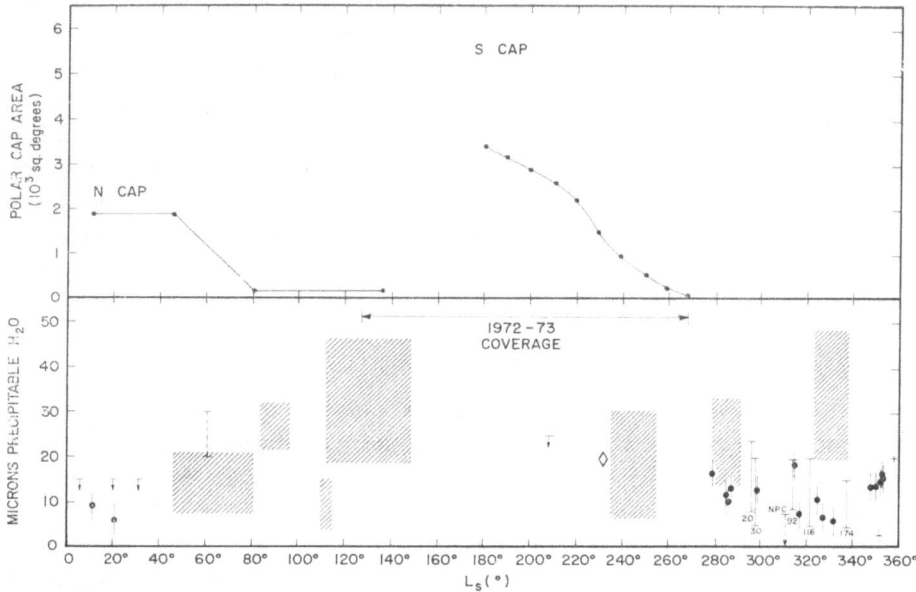

Fig. 7.   Mars H₂O abundances for the 1971–72 apparition. The cross-hatched areas refer to pre-1971 determinations. ◇ Larson (1973); | Mariner 9 data from Hanel *et al.* (1972) with the revolution number attached to the bar; | Mariner 9 data from Kunde (1973); + Mars 3 data from Moroz and Ksanfomaliti (1972); ● ground-based data from Tull and Barker (1972). Surface area of polar caps calculated from Baum and Martin (1973).

have been curtailed by the dust storm which occurred at a season preceding the time when water vapor would be expected to increase in the southern hemisphere (Barker *et al.*, 1970; Barker, 1971). This curtailment could be caused, for example, by adsorption on the dust particles or simply by a thick dust covering on moisture-laden soil.

Agreement among the three sets of data, ground-based (Tull and Barker, 1972), Mariner 9 (Hanel, 1972; Kunde, 1973), and Mars 3 (Moroz and Ksanfomaliti, 1972), is quite good and indicates that the higher spatial resolution spacecraft measurements were representative of the entire planetary atmosphere during the southern summer season in 1971–72.

The airborne observation at $L_s = 232°$ obtained near opposition by Larson (1973) was made before the onset of the major dust storm and before the period of time when the Doppler shift was large enough for ground-based observations. The value of 20 μm agrees with previous determination at this L_s by Barker *et al.* (1970) indicating that water vapor was beginning its normal reappearance in the southern spring season. The major dust storm commenced at an L_s of 263° and the atmospheric water vapor content observed above the dust during the storm and then subsequent to it was less than 15 to 25 μm throughout the remainder of the southern summer, instead of increasing to a maximum near $L_s = 330°$ as it did in 1969.

The single Mariner 9 data point of 20 to 30 μm at an $L_s = 60°$ (Kunde, 1973) during the northern spring suggests that water vapor content had returned to normal by that time and no longer showed the effects of the dust storm.

Surface area of the polar caps has been calculated from data presented by Baum and Martin (1973) and Baum (1973) on the behavior of boundaries of polar caps since 1905. The standstill period for the north cap occurs between an L_s of 10° and 50°. Then the area decreases to the size of the permanent cap at $L_s = 80°$. Due to the eccentric boundary of the south cap, the standstill period is not as obvious and there appears to be a second standstill period between $L_s = 180°$ and 220°. Then the south cap retreats to its minimum size. Note that maximum water vapor abundance does not occur until the surface area of the north and south cap reaches a minimum value.

4.2. 1972–73 MARTIAN WATER VAPOR MEASUREMENTS

Starting in late November, 1972, when Mars was only 4.5″ diam, we have carried out a very extensive patrol of water vapor abundance variations on Mars. About 250 photoelectric scans or water vapor determinations have been made through August 15, 1973 providing an L_s coverage of $118° < L_s < 263°$ which has been indicated in Figure 7. Most important is the filling in of a completely unobserved range of L_s between 150° and 230° with spectra taken pole-to-pole along the Martian central meridian. By $L_s = 139°$ we were able to begin a study of latitudinal distribution of water vapor as the subsolar point moved southward into the southern hemisphere. Latitudinal spectra were taken with the slit placed parallel to the Martian equator at various fractions of the polar disk diameter from the equator such as N/2, S2/3. Diurnal observations of water vapor content were begun at an $L_s = 170°$. Diurnal data consist of spectral scans taken with the entrance slit parallel to the terminator and placed respectively near the termination, at the center of the disk, and near the limb. Terminator and limb scans were taken with the center of the slit 1.5″ in form of the visible terminator and limb and were made only when atmospheric conditions were stable with seeing about 1″. To reduce the effects of differential atmospheric dispersion, all Mars scans were guided through a Schott RG-5 filter so that the effective guiding wavelength was about 6500 Å.

Equivalent widths for the Martian water vapor lines were measured by methods described earlier. Assuming a temperature of 225 K, Voigt profiles and a value of $a = 0.05$ corresponding to a surface pressure of 6 mb; these equivalent widths were converted into total abundances of precipitable water by using water line strengths at 225 K given by Farmer (1971) and the tables of Jansson and Korb (1968). Using methods presented by Woodman and Barker (1973), airmass value for each observation was calculated from Martian physical data and estimates of atmospheric seeing during each observation. This method and its computer program were slightly modified to work with photoelectric data, and proved very informative as to different values of the airmass near the terminator (~ 5.0) and limbs (~ 2.8) of Mars due primarily to large phase angles ($\sim 45°$) for Mars.

A typical days' set of observations is shown in Figure 8, along with a diagram indicating spectrograph slit placement on the disk of Mars. Measured and calculated

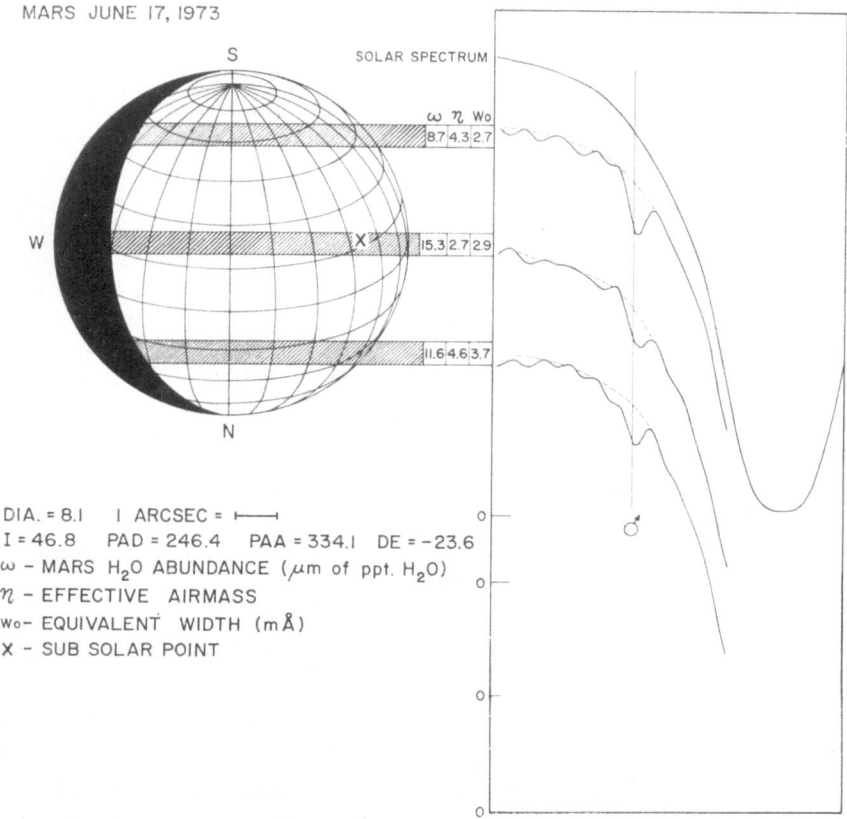

MARS JUNE 17, 1973

DIA. = 8.1   1 ARCSEC = ⊢——⊣
I = 46.8   PAD = 246.4   PAA = 334.1   DE = −23.6
ω – MARS H₂O ABUNDANCE (μm of ppt. H₂O)
𝜂 – EFFECTIVE AIRMASS
Wo– EQUIVALENT WIDTH (mÅ)
X – SUB SOLAR POINT

Fig. 8. Spectral scans of the 8176 Å line and slit positions for Mars on June 17, 1973.

values of the airmass, abundance and equivalent width are given for each slit position. Note that the smaller airmass nearer the center of the disk, unlike the airmass at 1.5″ from the polar limbs, more than compensates for the larger equivalent widths on the polar scans, resulting in a vertical equatorial abundance which is 50% greater than vertical polar abundances.

We can divide the 1972–73 data into three sets: PPCM (pole-to-pole on the central meridian), EQ (parallel to Mars equator) at various latitudes, and TERM (parallel to the terminator line) at positions near the limb and terminator and at the center of the disk. The various slit positions used are explained in the captions for Figures 9, 11, and 15.

Figure 9 is similar to Figure 7 but contains only 1972–73 data over the L_s range $126° < L_s < 270°$. The water vapor abundance seen along an entire central meridian quickly declines from a maximum of 25–40 μm near the start of the observations to a nearly constant level of 10–20 μm at an $L_s = 180°$. Then it continues to decline more slowly to a level of 6–10 μm up to $L_s = 250°$. The two relatively high PPCM observations after $L_s = 250°$ probably reflect real increases in planet-wide abundance asso-

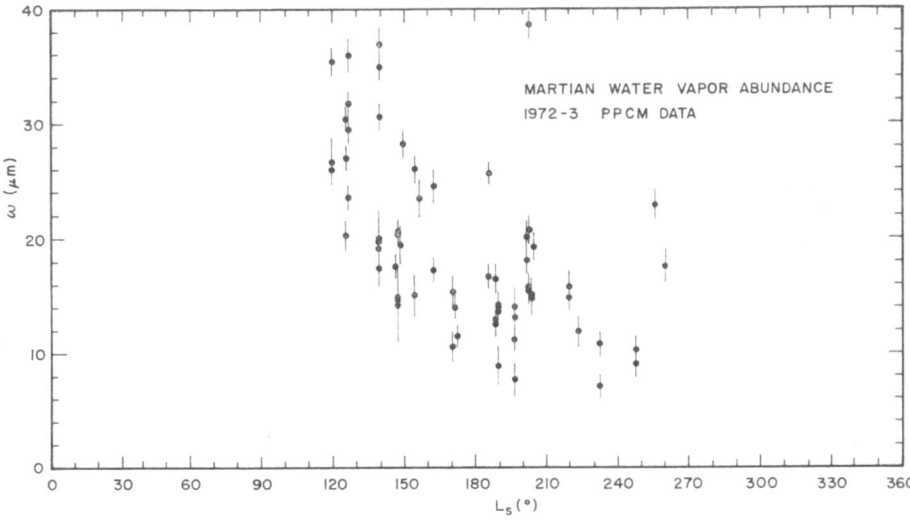

Fig. 9.   1972–73 Mars H₂O abundances for slits placed pole-to-pole on the central meridian as a
function of Martian season, L_s.

ciated with the increase in water vapor seen at higher southern latitudes at the L_s values
greater than 250° (see next section and Figure 11).

The conclusion can be drawn from Figure 9 that the atmospheric H_2O average
over the entire planet decreased to a very low level of 8–10 μm around $L_s = 170°$ to
250°. So low a level could not have been measured before the use of the échelle coudé
scanner, previous upper limits or detection limits based on high-dispersion spectrographic plates being in the range of 20 μm. The scatter in points is indicative of external errors or effects, since the typical internal errors are only of the order of one to two
microns. In particular, the scatter may be representative of local weather or topography
on Mars. To check the effect of topography and gravity variations, the data shown in
Figure 9 are presented as a function of the central meridian longitude (L_{CM}) in Figure
10. No pronounced correlation can be noted with respect to L_{CM}, but a future
analysis of the data will try to remove the effects of topography and gravity anomalies
from the seasonal data.

Tull (1970) found, at $L_s = 132°$ and 148°, a north-south latitude distribution of
water vapor abundance decreasing from a maximum in the north to a minimum in the
south. The northern cap had just finished its regression and had reached its minimum
area. The 1972–73 data which can be analyzed for latitudinal distribution were taken
later in the Martian season than Tull's study. When plotted vs L_s in Figure 11, the
equatorial data or latitudinal data show the same general decrease in abundance as
the pole-to-pole or central meridian data until L_s reached 250°; when plotted vs L_{CM}
we again find little indication of water vapor variation with longitude. But when the
equatorial data are plotted as a function of Martian latitude, we see definite trends in
the data for different seasons or values of L_s. Equatorial data can be broken into three

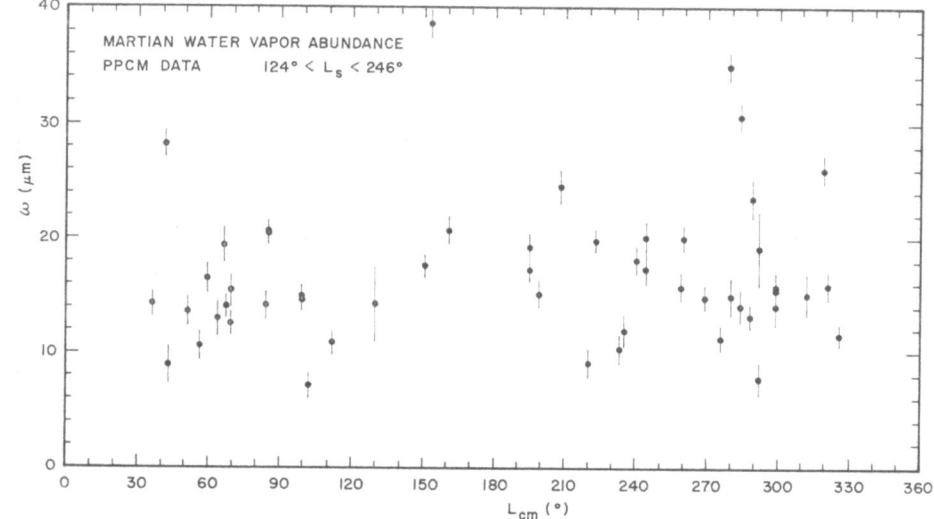

Fig. 10.   1972–73 Mars H₂O abundances for slits placed pole-to-pole on the central meridian as a function of central meridian longitude, L_{CM}.

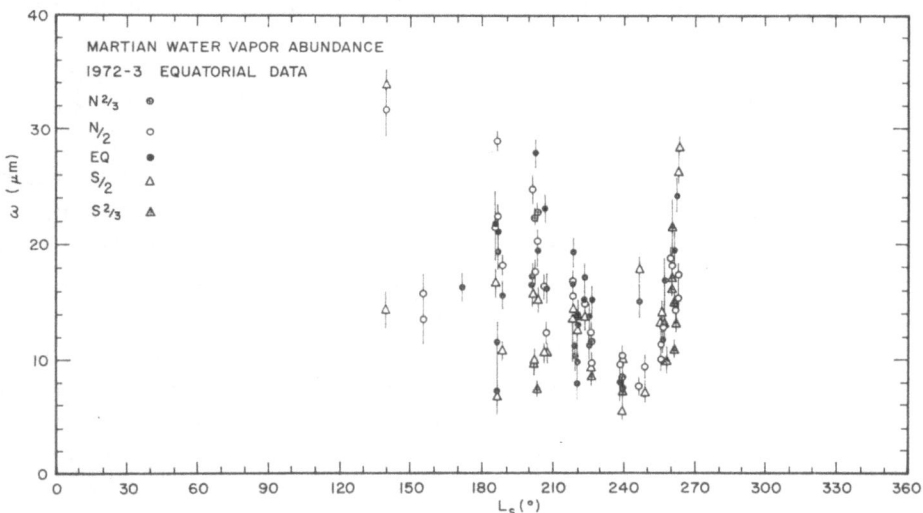

Fig. 11.   1972–73 latitudinal abundances of Mars water vapor as function of Martian season, L_s. Slit orientations are parallel to the Martian equator at various fractions of the disk diameter toward the poles; N2/3, N/2, EQ, S/2, and S2/3.

ranges of L_s, $185° < L_s < 208°$ and $217° < L_s < 249°$, which are plotted in Figure 12. The first period shows the same kind of N–S distribution found by Tull (1970) at earlier seasonal dates, $L_s = 132°$ and $148°$. The second period has lower abundances with the decline in water occurring mostly in the northern and subsolar latitudes; the amount near the south polar cap (the surface area of which continues to shrink until $L_s = 270°$)

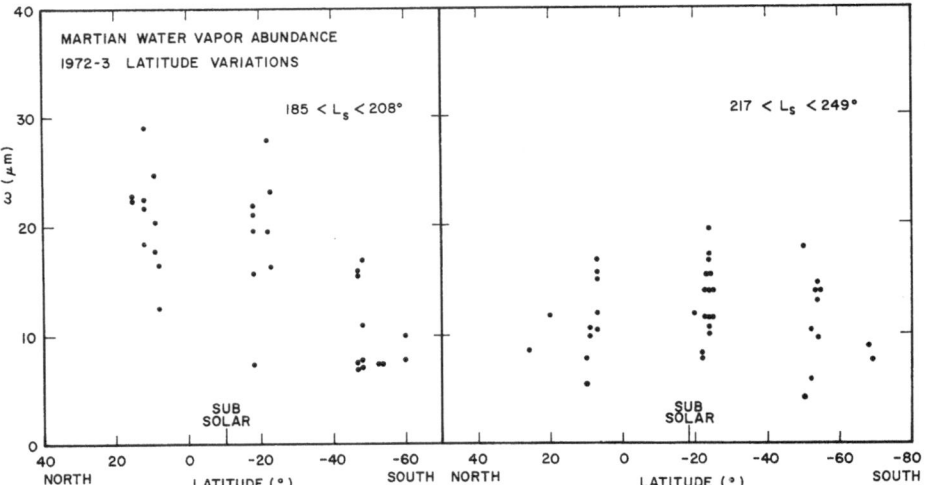

Fig. 12.   1972–73 latitudinal abundances as function of aerographic latitude for $185° < L_s < 208°$ and $217° < L_s < 249°$. Subsolar indicates the mean latitude or declination of the Sun as seen from Mars for this range of L_s.

remaining the same. The abundance over the subsolar latitude may be slightly higher, but a large scatter is also present at this latitude.

It is interesting to note that abundance at the edge of the south polar cap (at a latitude between $-60°$ and $-69°$) does not show any change between $L_s = 202°$ and $L_s = 239°$, over one seasonal month on Mars, indicating that the polar cap between $-60°$ and $-69°$ must be almost entirely solid CO_2 and not a solid CO_2-H_2O mixture.

The next range of L_s, $256° < L_s < 266°$, Figure 13, shows that around summer solstice, as the south polar cap is receding to its remnant size near $-80°$ latitude, the water vapor abundance increases markedly at middle southern latitudes, but those taken on the same dates at the edge of the cap show only a small increase.

Attempting to find out how water vapor behaves on a diurnal basis in the Martian atmosphere, we have obtained several sets of scans of three localities which represent the diurnal temperature variation on the Martian disk. By placing the spectrograph slit parallel to the terminator line near the terminator and limb (1.5″ onto the illuminated disk) and at the center of the disk as in Figure 14, we sampled areas each having roughly its own local Martian time and hence diurnal temperature phase. (Ideally, one should break the slit length down into smaller increments, but the signal level was not high enough to do so; consequently there is a temperature gradient along the slit length and possibly a seasonal effect with respect to latitude as discussed in the previous section.)

Due to the large phase angle of about 45° during the observing period, the subsolar point was under the limb scan most of the time. Slit positions refer to the following times during a day; Limb – midday, CD – mid-afternoon, and Terminator – late evening. Results of these sets of scans are presented in Figure 15 for three ranges of L_s in order to minimize any seasonal effect. Results are similar for the three periods

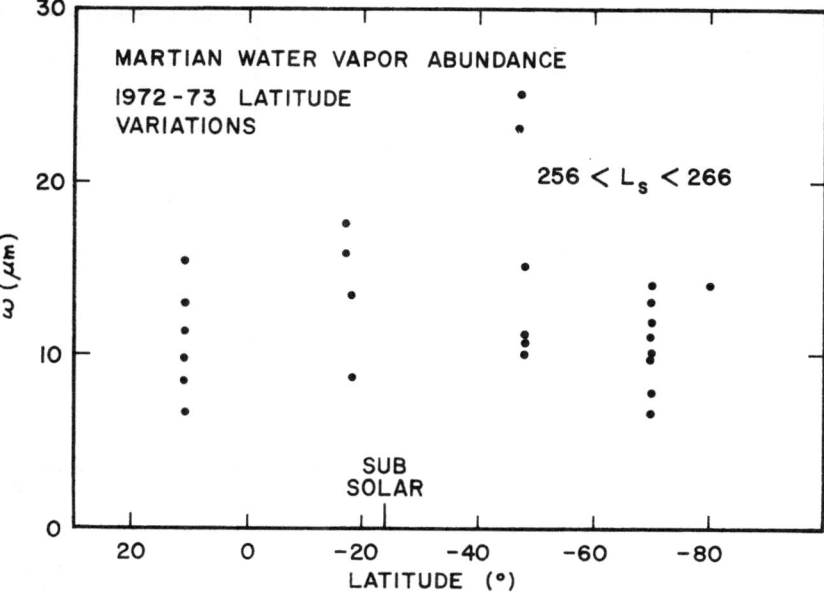

Fig. 13.   1972–73 latitudinal abundances as function of aerographic latitude for $246° < L_s < 266°$. Subsolar indicates the mean latitude or declination of the Sun as seen from Mars for this range of L_s.

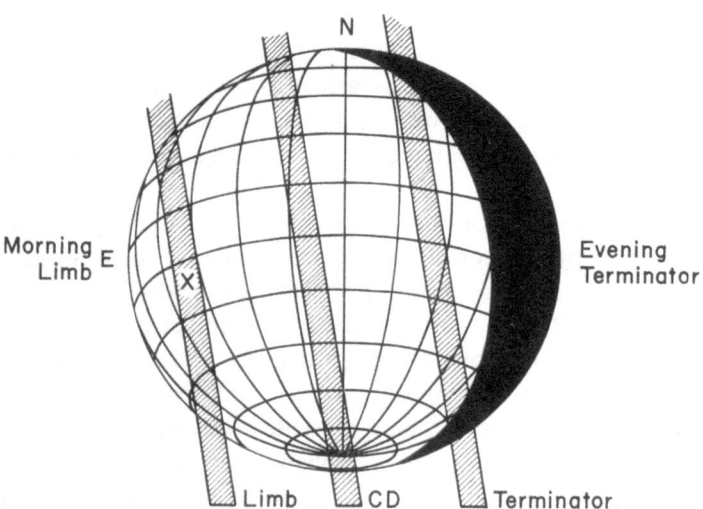

MARS   JULY 7, 1973

DIA. = 10.1   I ARCSEC = ⊢—⊣
I = 47.0   PAD = 247.1   PAA = 328.1   DE = –22.0
X – SUB SOLAR POINT

Fig. 14.   Schematic representation of slit positions for diurnal data.

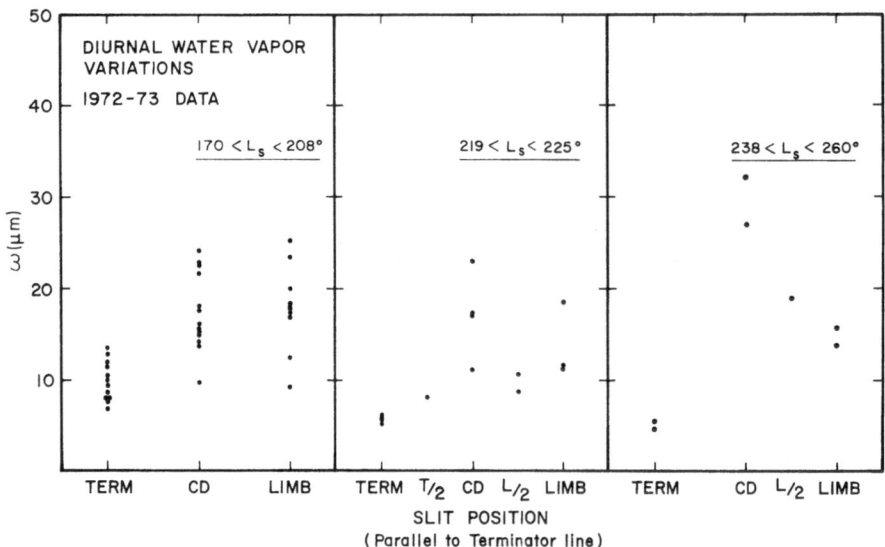

Fig. 15. 1972–73 diurnal water vapor abundances for $170° < L_s < 208°$, $219° < L_s < 225°$, $238° < L_s < 260°$. Slit orientations are perpendicular to the intensity equator at various fractions of the illuminated disk; TERM – terminator, T/2 – half way to the terminator, CD – center of the disk, L/2 – half way to limb, LIMB.

of L_s. The calculated airmass factor plays an important role in determination of water vapor abundance for these slit orientations. The limb and center of the disk values are similar because of their symmetry around the subsolar point and the sub-earth point, but the terminator airmass values can be quite large with 5.0 being a typical value. As the seeing improves, the airmass value increases because of lack of smearing which would bring more regions of lower airmass than higher airmass under the slit.

One definite conclusion can be drawn: the amount of water vapor near the evening terminator is smaller by a factor of two or more; water condenses out of the atmosphere as daytime temperature drops. The center-of-the-disk and limb abundances are not significantly different, probably due to the similarity between mid-day and early afternoon (time of maximum heating) temperatures. These diurnal variations need to be observed under the opposite orbital conditions of an evening limb and morning terminator. But from the data presented, we can plausibly infer that the amount of atmospheric water vapor is probably very small at sunrise, reaches a mid-day maximum, and is condensing out rapidly by late afternoon. This statement must be taken with the caution that these results refer only to the low water vapor abundance period of $170° < L_s < 260°$ and not necessarily to times of the Martian year when much larger amounts are present in the atmosphere, $L_s \sim 130°$ and $\sim 330°$.

Although water abundance seems to vary with many spatial parameters, an investigation was made into the assumption of an effective temperature of 225 K for water vapor. Four lines of differing strengths were used and scans were made in the EQ slit position to minimize any topographical, local and diurnal variations. Abundances were calculated by using the line strengths given by Farmer (1971) for 200, 225 and 250 K. Using lines at 8176, 8181, 8234, and 8282 Å, observations were obtained at $L_s = 225°$. Another set was obtained at $L_s = 123°$ using the lines at 8176, 8189, and 8197 Å. The sigma for the rms fit of determined abundances indicated that at L_s of 123° the best temperature was 250 K; and similarly at $L_s = 220°$, $T = 200°$; and $L_s = 225°$, $T = 225$ K. The fit was not significantly better for any temperature; this indicates the method is not very sensitive when only a few lines are used and the abundance is low due to seasonal effects.

5. Conclusions

The new Mars water data, obtained during the first half of the 1973 apparition and presented here, represent more than three times the number of previous determinations of water vapor from all observers, and the internal error of each point is only about a third of that characteristic of the older photographic data because of sharply increased spectral and spatial resolution plus photon-counting photometry. Also, observations during a very important and previously unobserved seasonal period between $150° < L_s < 230°$ showed that the water vapor decreased to a small but observable 5–15 μm level by an L_s of 180°. Studies of the latitudinal and diurnal water vapor distributions indicate the location of maximum and minimum abundances for this season are positively correlated with surface temperature variations.

Acknowledgements

The author wishes to express his thanks to Mr Michael Perry who assisted in obtaining the majority of the observations and Mrs Amber Woodman for computer processing the digital data into final spectrum plots. Discussions with Drs Robert Tull, Ronald Schorn, Harlan Smith, C. B. Farmer, and Mr Jerry Woodman in various areas were quite useful. This work was supported by NASA NGR 44-012-152.

References

Barker, E. S.: 1971, *Bull. Am. Astron. Soc.* **3**, 277 (abstract).
Barker, E. S.: 1973, *Bull. Am. Astron. Soc.* **5**, 300 (abstract).
Barker, E. S. and Schorn, R. A.: 1973, *Bull. Am. Astron. Soc.* **5**, 301 (abstract).
Barker, E. S., Schorn, R. A., Woszczyk, A., Tull, R. G., and Little, S. J.: 1970, *Science* **170**, 1308.
Baum, W. A.: 1973, personal communication.
Baum, W. A. and Martin, L. J.: 1973, *Bull. Am. Astron. Soc.* **5**, 296 (abstract).
Farmer, C. B.: 1971, *Icarus* **15**, 190.
Fink, U., Larson, H. P., Kuiper, G. P., and Poppen, R. R.: 1972, *Icarus* **17**, 617.
Hanel, R., Conrath, B., Hovis, W., Kunde, V., Lowman, W., McGuire, E., Pearl, J., Pirraglia, J., Prabhakara, C., and Schlachtman, B.: 1972, *Icarus* **17**, 423.
Hunt, G. E.: 1972a, *J. Quant. Spectrosc. Radiat. Transfer* **12**, 405.
Hunt, G. E.: 1972b, *Bull. Am. Astron. Soc.* **4**, 360 (abstract).
Jansson, P. A. and Korb, C. L.: 1968, *J. Quant. Spectrosc. Radiat. Transfer* **8**, 1399.
Kunde, V. G.: 1973, *Bull. Am. Astron. Soc.* **5**, 297 (abstract).

Larson, L. P., Fink, U., and Michel, G.: 1973, *Bull. Am. Astron. Soc.* **5**, 297 (abstract).

Margolis, J. S. and Hunt, G. E.: 1972, *Bull. Am. Astron. Soc.* **4**, 359 (abstract).

Moroz, V. I. and Ksanfomaliti: 1972, *Icarus* **17**, 408.

Moroz, M. Ya. and Petrov, G. 1.: 1973, *Icarus* **19**, 163.

Owen, T.: 1969, *Astrophys. J.* **150**, 121.

Owen, T. and Mason, H. P.: 1969, *Science* **165**, 893.

Rank, K. H., Fink, U., Foltz, J. V., and Wiggins, T. A.: 1964, *Astrophys. J.* **140**, 366.

Schorn, R. A. and Barker, E. S.: 1973, *Bull. Am. Astron. Soc.* **5**, 300 (abstract).

Schorn, R. A., Barker E. S., Gray, L. D., and Moore, R. C.: 1969, *Icarus* **10**, 98.

Schorn, R. A., Farmer, C. B., and Little, S. J.: 1969, *Icarus* **11**, 283.

Schorn, R. A., Spinrad, H., Moore, R. C., Smith, H. J., and Giver, L. P.: 1967, *Astrophys. J.* **147**, 743.

Spinrad, H., Munch, G., and Kaplan, L. D.: 1963, *Astrophys. J.* **137**, 1319.

Traub, W. B. and Carleton, N. P.: 1973, *Bull. Am. Astron. Soc.* **5**, 299 (abstract).

Traub, W. B. and Carleton, N. P.: 1974, this volume, p. 223.

Tull, R. G.: 1970, *Icarus* **13**, 43.

Tull, R. G.: 1972, in S. Iaustsen and A. Reiz (eds.), *Proc. ESO/CERN Conference on AuxiliaryIn-strumentation*, Geneva, p. 259.

Tull, R. G. and Barker, E. S.: 1972, *Bull. Am. Astron. Soc.* **4**, 372 (abstract).

Wells, D.: 1972, *Publ. Astron. Soc. Pacific* **84**, 203.

Woodman, J. H. and Barker, E. S.: 1973, *Icarus*, **19**, 327.

Young, L. D. G., Young, A. T., Young, J. W., and Bergstralh, J. T.: 1973, *Astrophys. J.* **181**, L5.

OBSERVATIONS OF O_2, H_2O AND HD IN PLANETARY ATMOSPHERES

W. A. TRAUB and N. P. CARLETON

Smithsonian Astrophysical Observatory and Harvard University,
Cambridge, Mass., U.S.A.

Abstract. We have searched for molecular lines in planetary atmospheres using the PEPSIOS spectrometer with an instrumental width about equal to that of the expected absorption line, and in the case of Jupiter, with the additional feature of Doppler compensation over the planetary disc.

Following our earlier detection of O_2 (7635 Å) on Mars, where the mixing ratio is $O_2/CO_2 \simeq 1.3 \times 10^{-3}$, we attempted to observe the same lines on Venus. The observed spectra are shown in Figure 1, where the lowermost tracing is the sum of scans taken during one day. The strongly saturated absorption features are due to two lines of

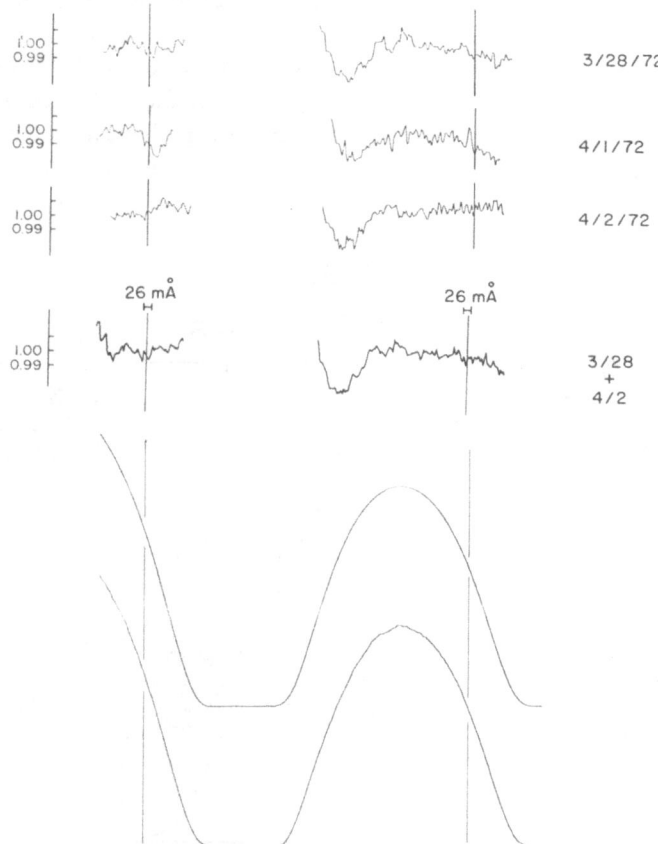

Fig. 1.   Observed spectra of Venus in the region of the 7635 Å line of O_2.

Woszczyk and Iwaniszewska (eds.), Exploration of the Planetary System, 223–228. All Rights Reserved

terrestrial O_2 in the Fraunhofer A band. Also shown is a theoretical profile for these lines, which when divided into the data yields the spectra as they appear outside the Earth's atmosphere, except in the core of the lines where there is zero flux. Our three best summed scans are shown normalized at the top of the figure. The vertical line indicates the Doppler shifted position of the Venusian O_2 lines. From the sum of the two best spectra, we obtain an upper limit of 0.5% on the depth of a possible feature, which together with an expected line width of 26 mÅ, yields an equivalent width of less than 0.13 mÅ.

Using our radiative transfer computer program, which follows the method of calculation described by Grant and Hunt (1969), we have computed a corresponding abundance of O_2. In this calculation, we used a cloud distribution with altitude which was developed to interpret our observations of CO_2 on Venus; it consists of a dense lower cloud at 300 mbar, along with a thin upper cloud at about 10 mbar, with optical depth 1.8. A forward peaked (Sobolev) scattering function was used. The computed equivalent width (EW) as a function of phase angle is shown in Figure 2 (upper curve). Scaling this curve down to the position of the O_2 observation, we see that the mixing ratio $O_2/CO_2 < 1 \times 10^{-6}$.

If this O_2 were solely due to dissociation of CO_2 into O and CO, with $O + O \rightarrow O_2$, we would then expect to have $CO/CO_2 < 2 \times 10^{-6}$, but in fact the value 45×10^{-6} has been observed. Clearly the O_2 is being tied up elsewhere, perhaps in water vapor.

We have also attempted to measure H_2O on Venus with these same techniques, using

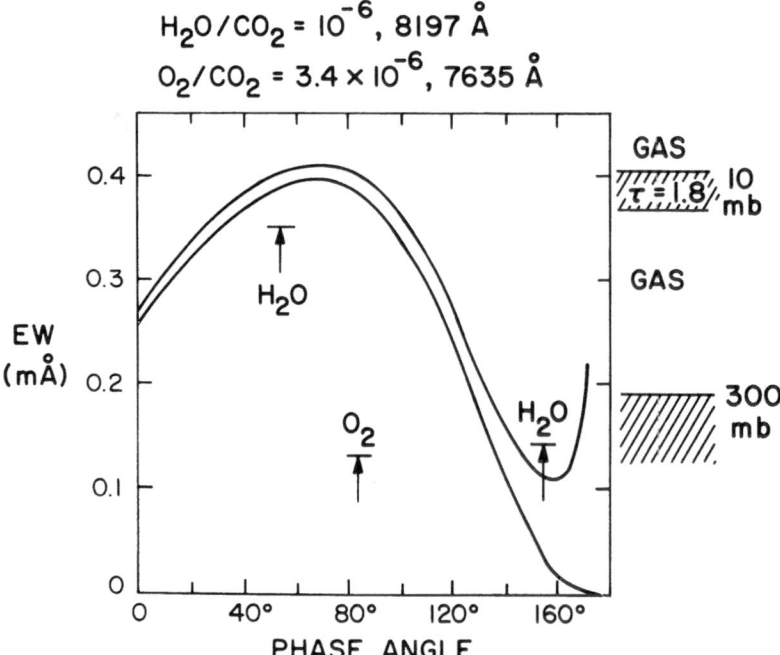

Fig. 2.   Computed equivalent width of O_2 vs phase angle on Venus.

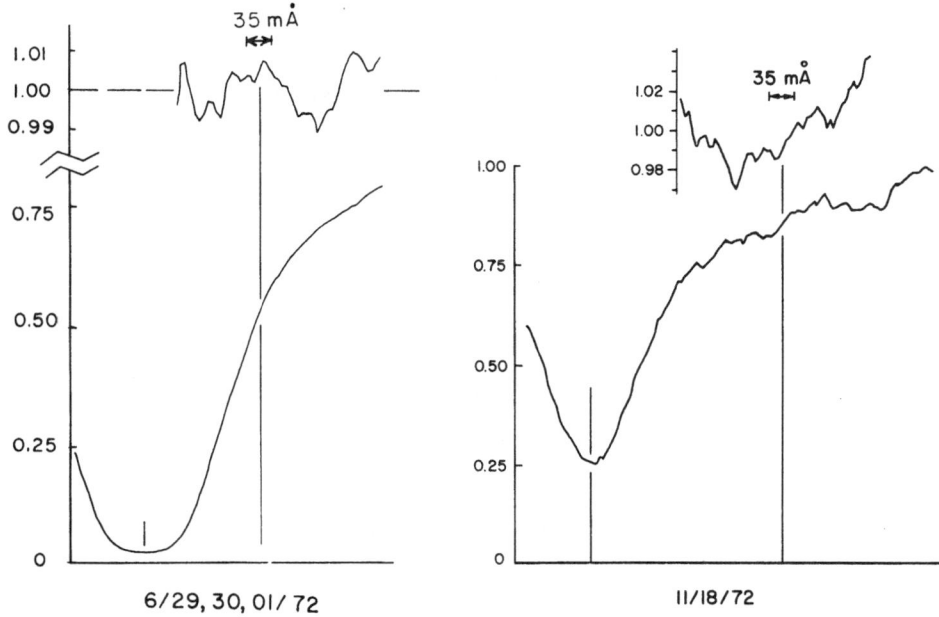

Fig. 3. Observed spectra of Venus near the 8197 Å H₂O line, for two different phase angles. The Doppler-shifted position and full-width at half-intensity of the expected Venusian feature are indicated.

the 8197 Å line. In Figure 3 we show the observed spectrum at two different phase angles, and also the curves which result from dividing out the telluric water vapor line profile. Again, we see no evidence for a line as deep as 0.4 and 1.0% for these respective curves. With an estimated width of 35 mÅ, the resulting EW values are 0.14 and 0.35 mÅ respectively. Referring to Figure 2 again, we see that this corresponds to a mixing ratio $H_2O/CO_2 < 1 \times 10^{-6}$. Note that the upper curve refers to a condition of uniform mixing of H_2O with CO_2, which is not too likely, since the water will be frozen out at lower levels and will not extend above the upper cloud. The lower curve applies if H_2O is confined to the region between cloud layers. Since this H_2O line has been successfully detected at various times by Barker (to be reported at this symposium), it is clear that this is a variable entity, perhaps depending upon the extent of the upper cloud. Possibly there is a connection with the four-day variation in observed CO_2 equivalent width, reported by Young *et al.* (1973).

We have also searched for water vapor on Jupiter; the observed spectrum is shown in Figure 4. For this experiment we operated the Fabry-Perot etalons in an off-axis mode, in order to precisely cancel the rotational smearing from Jupiter. Also, since we were observing near opposition, it was necessary to utilize only the limb spectrum in order to produce a significant shift from the terrestrial water vapor line. A concomitant property of the off-axis PEPSIOS is that it also yields broadened, non-symmetrically distorted spectra of stationary sources, and in particular, the telluric H_2O line. Since we have not yet incorporated this asymmetry into our calculations, the divided spectrum will not be quite accurate. This is probably why the upper curve oscillates

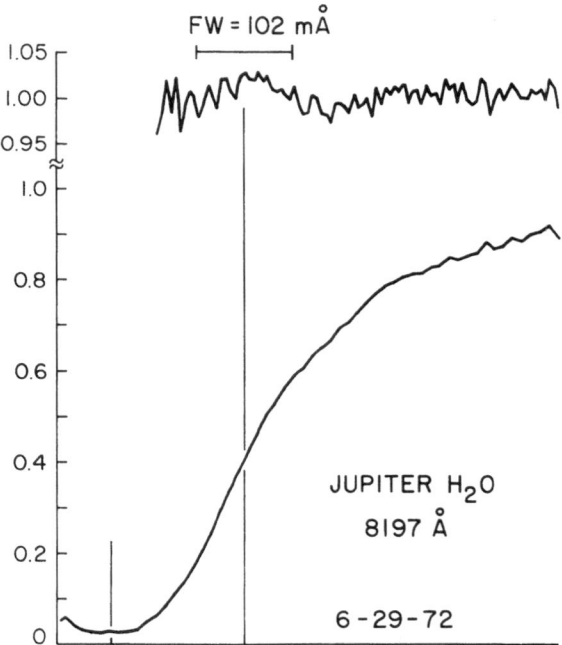

Fig. 4.   Observed 8197 Å line of H₂O on Jupiter.

near the line core. We expect that this line will be severely pressure broadened to about 102 mÅ, since the absorption will almost surely take place near the lower, warmer, cloud deck, where the gas pressure is of the order of 1 atm. For this width, we estimate an upper limit on the line depth to be 2%, so that EW < 2 mÅ. This is a factor of 5 smaller than the previously reported value due to Fink and Belton (1969). We intend to extend these observations by using the center of the disc at a time away from opposition, and by using a wider instrument function, in order to match the pressure-broadened line. Recently, Keay *et al.* (1973) have shown that 'hot spots' on the disc are correlated with blue and brown features; it is possible that H_2O will be more easily detected in these regions than in the colder surroundings.

Using the same Doppler compensation technique, we have recently measured the HD (7467 Å) 4–0 $P(1)$ absorption line on Jupiter (Trauger *et al.*, 1973). In Figure 5 we see this line, along with a nearby Jovian line which is probably due to CH_4. The identification of this line as HD follows from both the wavelength coincidence (within 7 mÅ of the laboratory line) as well as its width, which is comparable to the pressure-narrowed H_2 lines. We also searched for the 4–0 $R(0)$ and R(1) lines, but found these regions to have a number of stronger, interfering lines which completely obscured the HD. For comparison, we also measured the H_2 4–0 $S(1)$ line during the same period, finding an EW = 8.1 ± 0.2 mÅ. Note that both of these observations are for the full, integrated disc.

To derive the number ratio HD/H₂, we assume that the atmosphere is homogeneous and has an effective rotational temperature of 150 K. We also assume that mixing

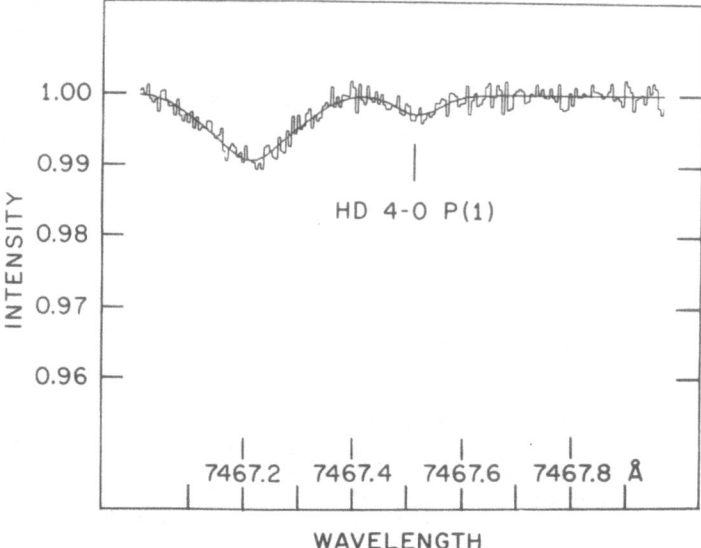

Fig. 5.   Spectrum of Jupiter in the region of the 7467 Å line of HD.

occurs down to a level where the temperature is sufficiently great to establish a 3:1 ortho-to-para-H_2 ratio. For this case then, calculated absorption coefficients are available for the H_2 line; in addition, we apply a saturation correction of 1.10. For HD, we can combine recent theoretical work on the 1–0 through 4–0 bands (Ford and Browne, 1973), with laboratory observations of the 4–0 $R(0)$ line (McKellar, 1973). We then have $D/H = (HD/H_2)/2$, neglecting the isotope fractionation effects of minor species such as CH_3D. The result is $D/H = (2.1 \pm 0.4) \times 10^{-5}$, where the uncertainty is one standard deviation and includes both observational and laboratory error estimates.

Recently Beer and Taylor (1973) analyzed the Jovian CH_3D spectrum and found $D/H = (5.1 \pm 2.2) \times 10^{-5}$, using a model dependent calculation of radiative transfer and deuterium fractionation. Since our result is nearly model independent, this suggests that the CH_3D observations can be used instead to determine that the deuterium enrichment of methane is between 1.3 and 5.6 times the stochiometric ratio. For such strong fractionation to have occurred, the effective temperature must lie in the range of about 240 to 520 K. However, in this range, the thermal equilibration time is about 10^7 yr or more. Therefore, there must exist a catalytic agent in the main cloud deck of Jupiter which drastically increases the rate of fractionation of deuterium.

Identifying the Jovian D/H ratio with that of the primitive solar nebula, we can compare our direct determination with independent estimates of this quantity. From analysis of meteoritic gases, Black (1973) has estimated a protosolar $D/H = 1.5$ $(+1.5, -0.7) \times 10^{-5}$, and from consideration of helium isotopes in the Sun and solar wind, Geiss and Reeves (1972) have suggested $D/H = 2.5 (\pm 0.5) \times 10^{-5}$. Both values are in good agreement with our determination.

If the observed D/H ratio can be taken to be directly representative of the production in a Big Bang (Wagoner, 1973), then the present density of the Universe is about $(4.3 \pm 0.3) \times 10^{-31}$ g cm^{-3}. If, on the other hand, there is significant D production in supernovae shock waves (Colgate, 1973), then the present density of the Universe can be greater than indicated, and the possibility is raised that the density may be high enough to allow the Universe to be gravitationally closed.

References

Beer, R. and Taylor, F. W.: 1973, *Astrophys. J.* **179**, 309.
Black, D. C.: 1973, *Icarus* **19**, 154.
Colgate, S. A.: 1973, *Astrophys. J.* **181**, L53.
Fink, U. and Belton, M. J. S.: 1969, *J. Atmospheric Sci.* **26**, 952.
Ford, A. L. and Browne, J. C.: 1973, preprint.
Geiss, J. and Reeves, H.: 1972, *Astron. Astrophys.* **18**, 126.
Grant, I. P. and Hunt, G. E.: 1969, *Proc. Roy. Soc.* **A313**, 183.
Keay, C. S. L., Low. F. J., Rieke, G. H., and Minton, R. B.: 1973 *Astrophys. J.* **183**, 1063.
McKellar, A. R. W.: 1973, *Astrophys. J.* **185**, L53.
Trauger, J. T., Roesler, F. L., Carleton, N. P., and Traub, W. A.: 1973, *Astrophys. J.* **184**, L137.
Wagoner, R. V.: 1973, *Astrophys. J.* **179**, 343.
Young, L. G., Young, A. T., Young, J. W., and Bergstralh, J. T.: 1973, *Astrophys. J.* **181**, L5.

DISCUSSION

Beer: Do you have direct laboratory measurements of the 4–0 *P*(1) line of HD to compare to the astronomical data?

Traub: We used the very recent laboratory work on HD by McKellar, who has measured lines in the 1–0 through 4–0 bands. Also there exist calculations of these lines by Ford and Browne (1973). However the calculations are quite sensitive to cancellation effects, and it is therefore not surprising that it is found that the observations and calculations are different by an approximately constant factor, for all bands. Therefore, since in fact the 4–0 *P*(1) line was not observed in the laboratory, we have scaled the calculated 4–0 *P*(1) intensity by the empirical constant just mentioned. Details of this appear in *Astrophys. J.* **184**, L137, 1973.

Trafton: What is the basis for your estimate that the Jovian H_2O line has a half width of 0.1 Å at 1 atm pressure? Did you use room temperature results?

Traub: I took as a basis the results for H_2O published by Farmer. Room temperature results have been used.

Fox: Recent calculations by Tejwani of broadening of CH_3D lines by H_2, as a function of temperature, indicate significant differences from the values used by Beer *et al.* in deriving a Jovian D/H ratio from CH_3D/CH_4.

Beer: This would not affect our D/H ratio significantly because our most serious source of error lies in the model atmospheres employed, not in the data themselves.

THE SPECTRUM OF MARS IN THE REGION 1800–3200 cm⁻¹

REINHARD BEER

Space Sciences Division, Jet Propulsion Laboratory, Pasadena, Calif., U.S.A.

Abstract. During the 1971 opposition of Mars, new infrared spectra covering the region 1800–3200 cm^{-1} (3.1–5.6 μ) were taken at a resolution of 0.095 cm^{-1} using a Connes'-type Fourier spectrometer on the 2.7 m telescope, McDonald Observatory. Spectra were obtained near 6° and 33° phase and were calibrated against the Sun, standard stars and an internal black body.

No new trace constituents have, as yet, been found in the spectra, but several previously unobserved combination and isotopic bands of CO_2 are visible. It has also been found possible to fit fairly well defined kinetic temperatures and Bond albedos to the two sets of data. The kinetic temperatures have been determined by a new technique. It is found that the albedo at 33° phase, which was determined a few days after the onset of the great dust storm of 1971, was significantly higher than for the clear atmosphere. The explanation for this phenomenon must await detailed radiative transfer calculations for a dust-laden atmosphere.

1. Introduction

The possibility of finding new trace constituents, particularly of the lighter molecules, has provided much of the impetus to achieve better spectral resolution in the infrared spectra of the planet Mars. All such searches have proven to be negative, even from spacecraft (Beer *et al.*, 1971a; Horn *et al.*, 1972; Conrath *et al.*, 1973). Nevertheless, it was thought worthwhile to pursue the topic somewhat further than before (Beer *et al.*, 1971a), particularly in view of the fact that our earlier spectra (which were only of moderate quality) had suggested that there might be residual, unidentified, lines in the spectrum. Furthermore, our earlier spectra were not flux-calibrated and, as a result, our efforts to determine an albedo at these long wavelengths were subject to gross errors. We, therefore, hoped to produce spectra at better resolution, over a wider wavelength range and flux-calibrated. In this we were successful.

2. Observations

All the observations were made during August and September–October of 1971 using a Connes'-Type Fourier spectrometer at the coudé focus of the 2.7 m telescope, McDonald Observatory, University of Texas. The instrumentation has been described elsewhere (Beer *et al.*, 1971b). Most of the data were taken at 0.095 cm^{-1} resolution but it was found, 7 weeks later, that the power level from Mars had fallen appreciably and, as a consequence, the signal-to-noise ratio was much poorer. Later, of course, we discovered that the great dust storm of 1971 that interfered with the early phases of the Mariner 9 mission was the cause of the surprisingly low temperatures observed.

The spectra are all in the integrated light from the planet and the details of the observations are given in Table I.

Woszczyk and Iwaniszewska (eds.), Exploration of the Planetary System, 229–239. All Rights Reserved
Copyright © 1974 by the IAU

TABLE I

Conditions of the observations

Run number	Julian Date of Z.P.D. 2 441 000.000 +	Phase angle (deg)	Longitude of central meridian (deg)	Resolution (cm^{-1})	Signal Noise	Air mass at ZPD
304	168.763	6.8	182	0.165	83	1.76
305	169.792	6.3	183	0.095	73	1.84
311	171.729	5.4	144	0.515	130	1.95
315	220.670	32.6	56	0.095	26	1.61
316	221.607	33.0	10	0.129	26	1.71
319	222.611	33.4	2	0.095	20	1.70
322	223.580	33.8	341	0.115	20	1.85
325	227.583	35.3	305	0.095	17	1.73

3. The Spectrum of Mars

Figure 1 is a portion of an average of runs 304 and 305 (taken on successive days) together with a solar spectrum taken somewhat later, with the same spectrometer but not with the same telescope.* Unfortunately, it transpired after the observations in the fall of 1971, that all our solar spectra had digitization errors and only one was at all useful: that presented in the figure. It, too, had errors but could be truncated to produce a satisfactory spectrum at $0.12 \ cm^{-1}$. However the difference both in the date of observation and in the resolution has made the correction for atmospheric transmission quite difficult and full correction must await implementation of the AFCRL atmospheric absorption line catalog (McClatchey *et al.*, 1973).

It is intended to publish the entire series of spectra in the form of an atlas, together with the relevant calibrations in order that other workers may have access to these data. Consequently, no upper limits for possible trace constituents have been determined.

Clearly visible in the spectra are the 'hot' $10°1–02°0$ band of normal CO_2, centered on $2429.36 \ cm^{-1}$, the $00°0–04°0$ band of $^{12}C^{16}O^{18}O$ centered at $2500.74 \ cm^{-1}$ and the $00°0–12°0$ band of $^{12}C^{16}O^{18}O$ centered on $2614.24 \ cm^{-1}$, a portion of which is shown in Figure 1. The 'hot' band is of particular interest because some strengths have been measured for this band (Plyler *et al.*, 1962) and, consequently, it will be possible to deduce a kinetic temperature from the relative population of the lower state. The band is exceedingly weak. For example, the strength of the $P16$ line is listed by Plyler *et al.* as $5.5 \times 10^{-5} \ cm^{-1}$ cm A and yet this line, having an equivalent width of about $0.005 \ cm^{-1}$, is quite clearly visible in the spectrum, indicating the great power of high resolution spectroscopy for the measurement of weak absorptions.

4. The Flux Calibrations

For flux calibration purposes, we deliberately operate at truncated resolution ($2 \ cm^{-1}$)

* All our solar spectra are obtained with an auxiliary 15 cm telescope which can feed sunlight to our interferometer.

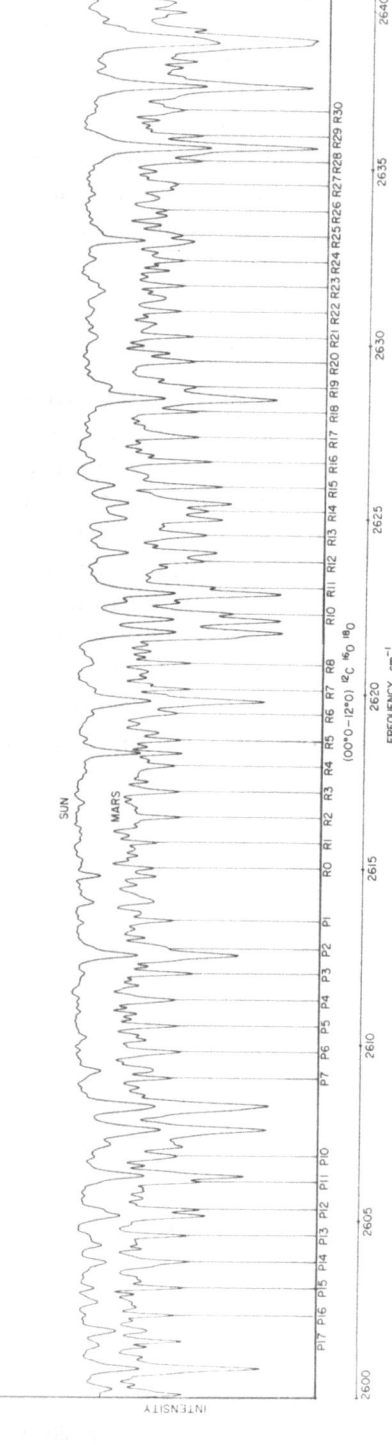

Fig. 1. A portion of the spectrum of Mars and of the Sun near the center of the $(00^\circ0–12^\circ0)$ ^{12}C ^{16}O ^{18}O band. The upper curve is of the Sun at a resolution of 0.125 cm⁻¹. The lower curve is of Mars at a resolution of 0.095 cm⁻¹.

in order to avoid difficulties with variable airmass during a given observation. The instrumental response is determined by observation of an internal black body and the external transmittance by observation of the Sun and standard stars (Betelgeuse, in this instance). The solar flux distribution is obtained from the data of Labs and Neckel (1968) and the L and M magnitudes for Betelgeuse from various sources (Johnson, 1966, 1967; Gillett *et al.*, 1968; Low and Krishna Swamy, 1970). There is significant disagreement between the authors as to the absolute calibration for Betelgeuse. However, the most complete study is that of Gillett *et al.* (1968), who also give an energy distribution over the entire 3–6 μ region. Their values are:

$$L(2940 \text{ cm}^{-1}) = 1.655 \times 10^{-22} \text{ W m}^{-2} \text{ Hz}^{-1}$$
$$M(2000 \text{ cm}^{-1}) = 8.25 \times 10^{-23} \text{ W m}^{-2} \text{ Hz}^{-1}$$

and these are the values employed in the present study. Observations of Betelgeuse extending over many months give absolutely consistent results, so that we are satisfied as to the stability of the system. Much less satisfactory is the correction for atmospheric transmission. It is, operationally, very difficult ever to obtain enough data under identical conditions in a program of high-resolution spectroscopy to insure that the transparency has not altered. That is, a typical high-resolution run requires 2–4 h for completion, but, for flux-calibration purposes, we abstract only the first 10–20 min worth of data. Consequently, at best, there is likely to be a 2–4 h delay before the acquisition of a calibration spectrum. In this interval, there can have been drastic changes in the atmospheric conditions. We believe that the final solution must await implementation of an extensive atmospheric transmission program using the, recently-

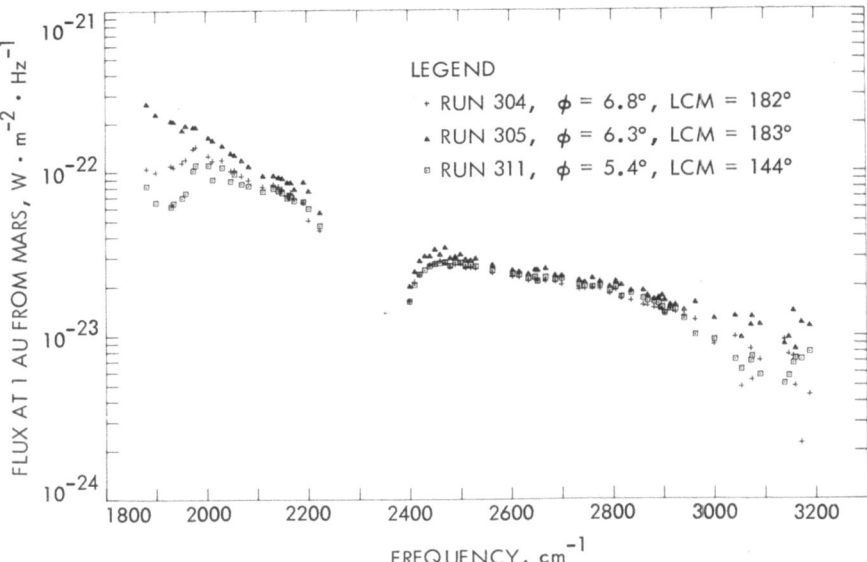

Fig. 2.   The absolute, integrated-disk, spectrum of Mars near opposition (mean phase angle = 6.2°).

available, AFCRL line parameter tape (McClatchey *et al.*, 1973). However, there is no doubt that production of such transmissions at sufficient resolution will be exceedingly expensive in terms of computing costs.

In order to reduce the problem, somewhat, we chose to sample the spectra only in regions where the transmission of the atmosphere is high (in 'micro-windows') and the errors in applying the correction are not too great. Figures 2 and 3 show the result of this for the two series of spectra. As may be seen, the scatter is fairly small except at the end-points, where the 6.3 and 2.7 μ telluric water vapor bands become dominant.

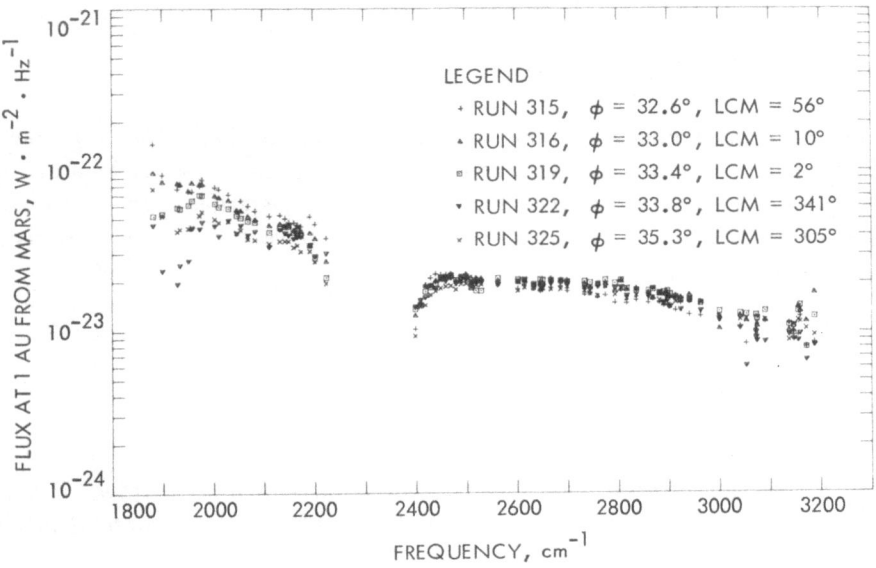

Fig. 3. The absolute, integrated-disk, spectrum of Mars about 1 week after the onset of the great dust-storm of 1971 (mean phase angle = 33.6°).

The gap in the middle is due to the telluric 4.2 micron CO_2 bands. However, the 'turn-down' at the edges of the gap is real and a consequence of the, even stronger, Martian CO_2 bands. That is, in these wings, the observed flux is probably due to atmospheric emission, rather than flux from the surface.

5. The 3–6 μ Albedo of Mars

It is easily verified that, for any reasonable temperatures, there is almost no point in the 3–6 μ region wherein either the thermal emission from the surface or the solar reflection become negligible. It was therefore necessary to construct a model for the combination of reflection and emission.

It may be shown that

$$L_1(\nu) = \Omega_1 \left[(1 - A) B_M(\nu, T) + A \bar{B}_s(\nu) F(\phi) \Omega_s \right] \text{Wm}^{-2} \text{ Hz}^{-1},$$

where

$L_1(v)$ = the flux at 1 AU from Mars
Ω_1 = the solid angle subtended by Mars at 1 AU
A = the Bond albedo
$B_M(v, T)$ = the Planck function at temperature T [W m^{-2} ster^{-1} Hz^{-1}]
$\bar{B}_s(v)$ = the mean brightness of the Sun [W m^{-2} ster^{-1} Hz^{-1}]
Ω_s = the solid angle subtended by the Sun at Mars
$F(\phi)$ = a function of the phase angle ϕ and for a Lambert surface (the model assumed here)

$$F(\phi) = \frac{2}{3\pi} \left[(\pi - \phi)\cos\phi + \sin\phi\right].$$

Under normal circumstances, one is faced with an insoluble problem: that of finding both T and A from a single observation. Consequently, the temperature normally reported is a *brightness temperature* which can only be related to the true, kinetic, temperature if the albedo is otherwise determinable. However, it can be seen that, in the above expression, if

$$B_M(v, T) = \bar{B}_s(v) F(\phi) \Omega_s$$

then the terms in A cancel and

$$L_1(v') = \Omega_1 B_M(v', T)$$

independent of any unknown parameters. *That is, the kinetic temperature T may be found explicitly.* A similar effect occurs even if the surface is non-Lambertian (Beer, 1973).

Fortunately, this cross-over condition occurs within our spectral region. The principal sources of error are as follows:

(a) If $F(\phi)$ does not have the simple Lambertian form assumed here, the 'cross-over' point between the emitted and reflected energy will move. Michaux and Newburn (1972) have made a critical study of all the available empirical phase data and conclude that the 'opposition effect' on Mars is certainly small (less than 0.1 mag.) and may be zero. In our region, at some ten times longer wavelengths, we might reasonably expect such effects to be even smaller. Even if the effect is the same, it will affect only the 6° phase data by 10%, an amount much smaller than the scatter in the data at the cross-over region, and the 33° data by a negligible amount.

(b) Systematic errors in either our flux calibration or in the solar flux tables of Labs and Neckels (1968). The Labs and Neckels' data is based upon a model fit to all known flux measurements from the ultra-violet to the sub-millimeter region and is entirely self-consistent. The error is unlikely to exceed 1%. Our calibrations are, of course, subject to errors of 10–20% but again, the scatter of the data is much greater than this.

(c) The principal source of error is in the telluric transmission correction and the errors are as great as a factor of 2 in the relevant region.

Notwithstanding the large scatter, we have succeeded in making acceptable fits only over quite narrow temperature ranges:

$$6° \text{ phase}: \bar{T} = 270 \pm 5\,\text{K}$$
$$33° \text{ phase}: \bar{T} = 253 \pm 7\,\text{K}.$$

$\bar{T}$ is a global mean temperature which will be, of course, strongly weighted towards the hottest point on the disk. The values are entirely consistent with other workers, both from Earth and from spacecraft (Kieffer *et al.*, 1973; Conrath *et al.*, 1973; Michaux and Newburn, 1972).

Using these temperatures, we proceeded to fit a large series of albedo models, graphically, to the data in Figures 2 and 3. For each temperature $\bar{T}$ within the range permitted above, an upper and lower bound and a central value was calculated as a function of frequency. Some representative results are presented in Figures 4 and 5, superposed upon the data of Figures 2 and 3. The fit is quite surprisingly good.

However, by the same token that the flux from Mars at the low-frequency end is independent of albedo, it is also true that the albedo, here, is virtually indeterminate. We can do little more than bound the values at frequencies not far removed from the cross-over and assume that it changes little or not at all at the cross-over.

We present the results in Figures 6 and 7 for two mean effective temperatures, 270 K and 253 K. If it should transpire that the 'correct' values differ from these, the correction for the mean Bond albedo may be determined from Figure 8, which contains plots of $-dA/dT$ for the two cases. The corrections are substantially linear for a few degrees on either side of the mean. The error bounds move with the mean and do

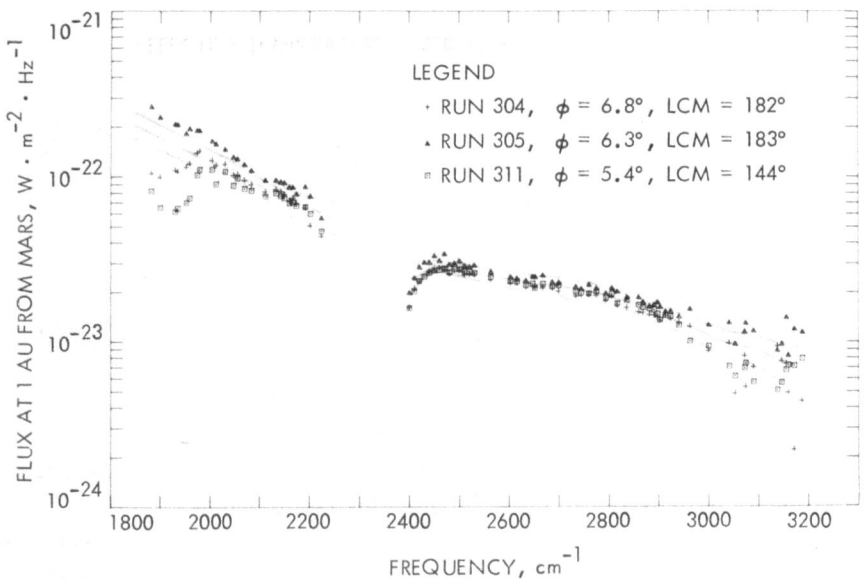

Fig. 4.   The same data as Figure 2 with some representative temperature-albedo models superposed.

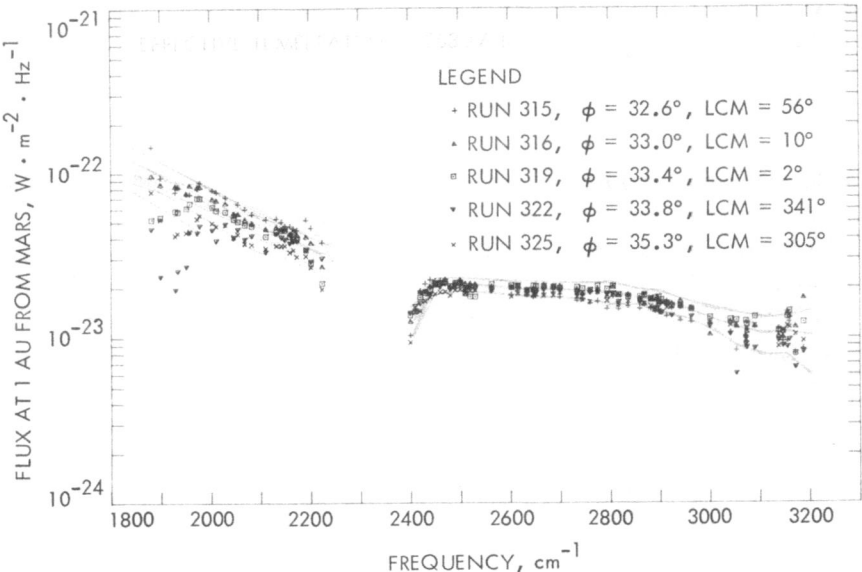

Fig. 5.   The same data as Figure 3 with some representative temperature-albedo models superposed.

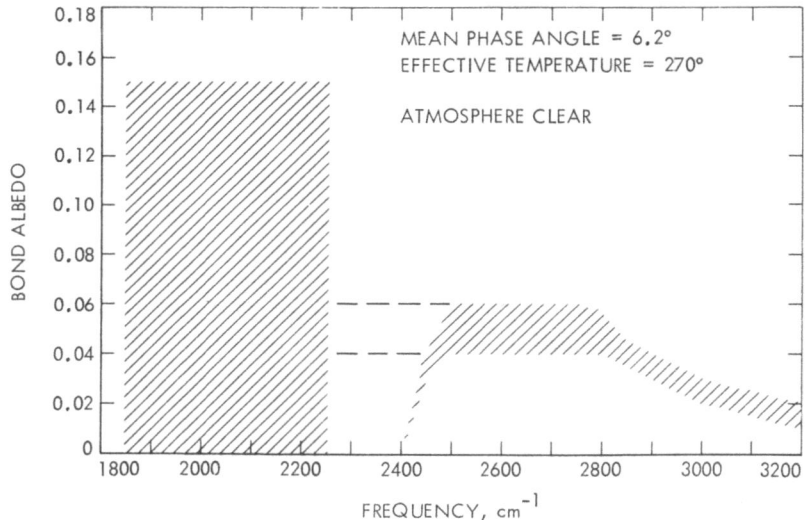

Fig. 6.   The albedo model for a clear Martian atmosphere.

not change in relative magnitude, with the exception that, in the low frequency region where $A = 0$ cannot be excluded, the lower bound remains zero.

Taking Figures 6 and 7 at their face value, it is notable that, in the 3–4 μ region (2500–3200 cm^{-1}), the albedo during the dust storm was significantly higher than during the 'clear atmosphere' period. At no possible temperature does the 6° albedo

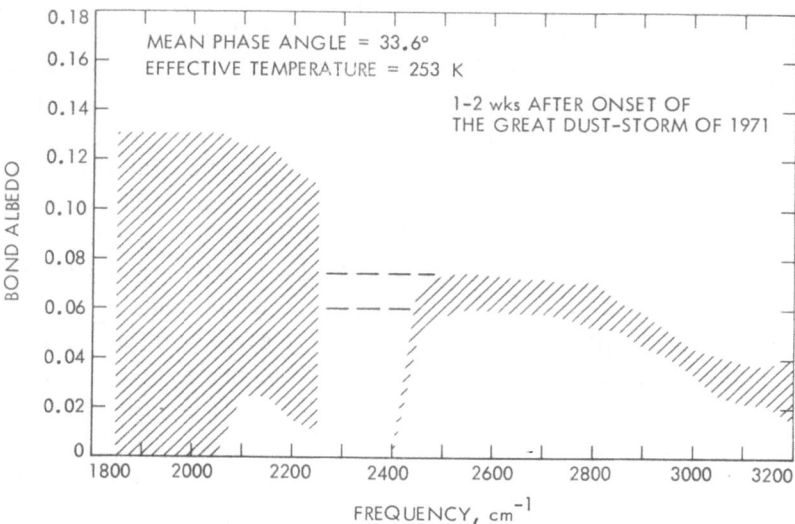

Fig. 7. The albedo model for a dusty Martian atmosphere.

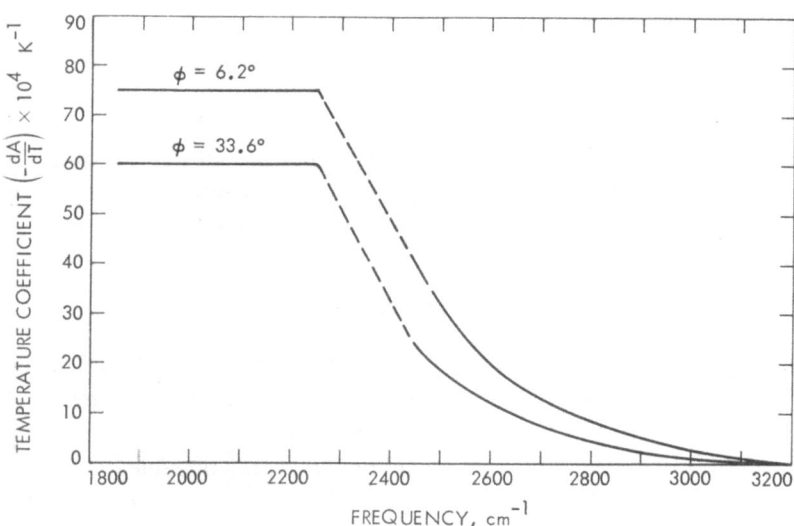

Fig. 8. Curves for the correction of the albedo models in Figures 6 and 7 for other global mean temperatures. The horizontal branches are approximate only.

curve fit the 33° data so that the difference is probably real. It is also worthy of note that the visual albedo appeared to be higher during the dust storm, based on evidence of the shorter exposure times required by the Mariner 9 cameras during this period as compared to later in the mission (Briggs, 1973). The turn-down to 2400 cm⁻¹ is probably spurious in both cases and a consequence of the Martian 4.2 μ CO_2 band. However, the decay from 2800 to 3200 cm⁻¹ is real and has been noted by several observers, previously (Sinton, 1967; Beer *et al.*, 1971; Houck *et al.*, 1973). It is commonly held that this is due to water of hydration in some form bound in the Martian surface material. It is interesting to note that the character of the band remains substantially the same in the two series of spectra. If the dust cloud was optically thick at these wavelengths (as it certainly was at visible wavelengths), we may draw the conclusion that the 'water band' is intimately associated with the surface material and is not due to either free water at the surface or some other volatile species. That is, the absorbing agent is, almost certainly, molecularly bound to the surface material.

The absolute values for the Bond albedos are also of interest: they are extremely low. In no circumstance wherein the albedo is reasonably bounded does it exceed about 7%. That is, the surface material is absorbing at least 93% of the incident energy, a remarkably high value. The values in the 5μ (2000 cm^{-1}) region are much more poorly known. Even here, however, values in excess of 15% are excluded and a value of zero is possible, howbeit unlikely.

The explanation for the rise in albedo during the dust storm must depend upon a detailed analysis of the radiative transfer under conditions of particulate matter being suspended in the atmosphere. This, in turn, would demand detailed knowledge of the optical constants, size and distribution of the material, most of which information is conjectural in advance of Martian lander missions.

6. Conclusion

We have presented evidence stemming from the highest resolution spectra yet obtained in the 3–6 μ region that demonstrates the power of high resolution in observing weak absorptions and, for the first time, have succeeded in determining albedos and kinetic temperatures from such data.

Acknowledgements

Our thanks are due to Dr Harlan J. Smith of the University of Texas for observing time on the 2.7 m telescope, McDonald Observatory, and to T. G. Barnes, D. L. Lambert, J. V. Martonchik, and R. H. Norton for unstinting aid in the observations and analyses. Without their aid, this paper would not have been possible.

This paper presents the results of one phase of research in the joint Jet Propulsion Laboratory-University of Texas Infrared Astronomy Program, supported, in part, by National Science Foundation Grant GP-32322X with the University of Texas at Austin and, in part, by National Aeronautics and Space Administration Contract NAS7-100 with the Jet Propulsion Laboratory, California Institute of Technology.

References

Beer, R.: 1973, in preparation.
Beer, R., Norton, R. H., and Martonchik, J. V.: 1971, *Icarus* **15**, 1.
Beer, R., Norton, R. H., and Seaman, C. H.: 1971, *Rev. Sci. Instr.* **42**, 1393.
Briggs, G. A.: 1973, private communication.

Conrath, B., Curran, R., Hanel, R., Kunde, V., Maguire, W., Pearl, J., Pirraglia, J., Welker, J., and Burke, T.: 1973, *J. Geophys. Res. (Surfaces)* **78**, 4267.

Gillett, F. C., Low, F. J., and Stein, W. A.: 1968, *Astrophys. J.* **154**, 677.

Horn, D., McAfee, J. M., Winer, A. M., Herr, K. C., and Pimentel, G. C.: 1972, *Icarus* **16**, 543.

Houck, J. R., Pollack, J. B., Sagan, C. Schaak, D., and Decker, J. A.: 1973, *Icarus* **18**, 470.

Johnson, H. L.: 1966, *Ann. Rev. Astron. Astrophys.* **4**, 193.

Johnson, H. L.: 1967, *Astrophys. J.* **149**, 345.

Kieffer, H. H., Chase, S. C., Miner, E., Münch, G., and Neugebauer, G.: 1973, *J. Geophys. Res. (Surfaces)* **78**, 4291.

Labs, D. and Neckel, H.: 1968, *Z. Astrophys.* **69**, 1.

Low, F. J. and Krishna Swamy, K. S.: 1970, *Nature* **227**, 1333.

McClatchey, R., Benedict, W. S., Clough, S. A., Burch, D. E., Calfee, R. S., Fox, K., Rothman, L. S., and Garing, J. S.: 1973, 'AFCRL Atmospheric Absorption Line Parameters Compilation', Report AFCRL-TR-73-0096, Air Force Cambridge Research Laboratories, Bedford, Massachusetts.

Michaux, C. M. and Newburn, R. L.: 1972, 'Mars Scientific Model', Jet Propulsion Laboratory Document 606-1, Pasadena, California.

Plyler, E., Tidwell, E. D., and Benedict, W. S.: 1962, *J. Opt. Soc. Am.* **52**, 1017.

Sinton, W. M.: 1967, *Icarus* **6**, 222.

RESULTS OF CURRENT MARS STUDIES AT THE IAU
PLANETARY RESEARCH CENTER

WILLIAM A. BAUM

Planetary Research Center, Flagstaff, Ariz., U.S.A.

Abstract. The purpose of this paper is to give an updated report on several recent findings concerning Mars obtained from studies now in progress at the Planetary Research Center of Lowell Observatory. In this work, extensive use has been made of ground-based images obtained almost hourly by seven observatories cooperating under the International Planetary Patrol Program (Baum, 1973).

Martin has recently completed a detailed hour-by-hour mapping of the first twenty days of the 1971 global dust storm, and Figure 1 shows an interesting behavior that we were not so much aware of at the time of Capen and Martin's (1971, 1972a) original reports a little over a year ago. In Figure 1 you see, for illustration, the outline of storm-brightened areas at two-hour intervals on the 11th day of the storm. Each small map represents approximately the area visible from Earth at that hour, with the morning terminator at the left-hand edge and the afternoon limb at the right-hand edge. The solid line identifies the central meridian, while the dashed line represents where it is noon on the planet.

If the dust storm were associated mainly with particular regions on Mars where the dust has been stirred up, evolving gradually in the course of weeks, its shape on the map would appear to change only a little in the course of one day. Figure 1 clearly shows that it does not behave that way. Nor could the apparent changes of shape and position be easily explained as some kind of optical scattering effect. The degree of activity of the dust storm depends both on the region and on the time of day.

We find, in fact, that the dust storm seems to be locally regenerated in this manner about midday each day during its developing stages. These maps for two-hour intervals can be superimposed upon one another to produce a single map describing the day's progress of the storm and this can be done for each day of the storm. Figure 2 shows ten such maps representing the first ten days of the storm. The contours show outlines of the active areas at two-hour intervals, and the numbers labelled on them indicate the time of day at 0° longitude. In other words, the numbers are approximately a Martian equivalent of Greenwich Mean Time. A very similar set of maps has been made for days 11 through 20. After the 20th day, the accumulated general haze had become so dense that the daily pattern of regional dust generation became too difficult to map. This kind of hour-to-hour history of the storm would have been totally impossible without a complete network of observatories participating full time in the Patrol Program.

Figure 3 sums up the regions that were active during the developing stages of the storm. The main core and the secondary core include the great bowls of Hellas and Argyre, and extend westward from them. Of the five regions identified as recurring bright spots, the one at the right is centered on the great depression just east of Syrtis

Woszczyk and Iwaniszewska (eds.), Exploration of the Planetary System, 241–251. All Rights Reserved
Copyright © 1974 by the IAU

Major, the one near the center of the map also lies in a region of low elevation, and the small one at about 70° longitude in the vicinity of Melas Lacus straddles the enormous equatorial canyon discovered by Mariner 9. However, the remaining two recurring bright spots do not seem to be associated with unusual topography.

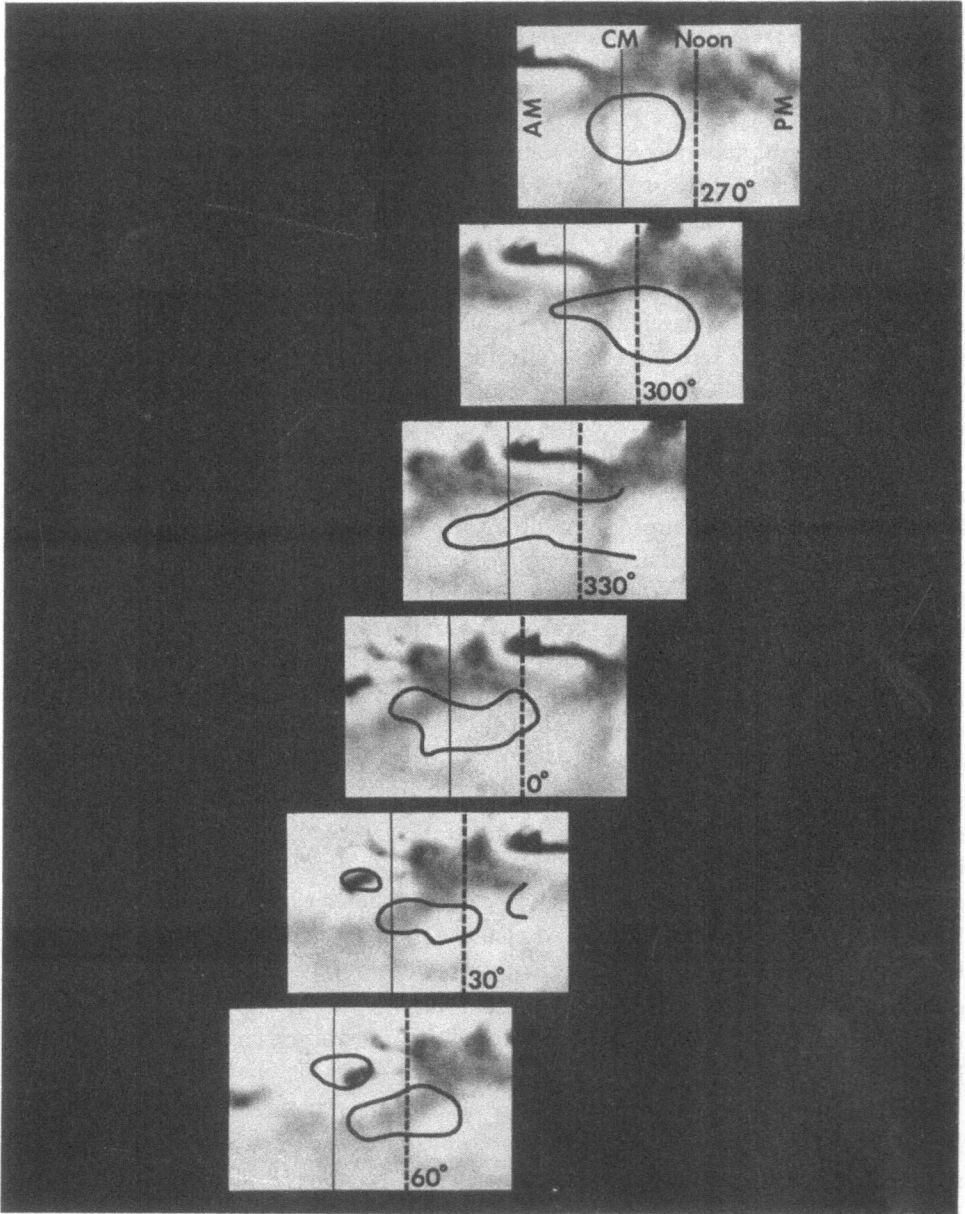

Fig. 1. Outline of Mars storm brightened areas at two-hour intervals on the 11th day of the storm.

Neither does the initial cloud, but as Figure 4 by Capen and Martin (1972b) shows, the three best recorded dust storms of the Martian perihelion season have all started at about the same place. In addition to the remarkable coincidence of position of these three initial clouds, they share another unusual trait, namely, the suddenness and

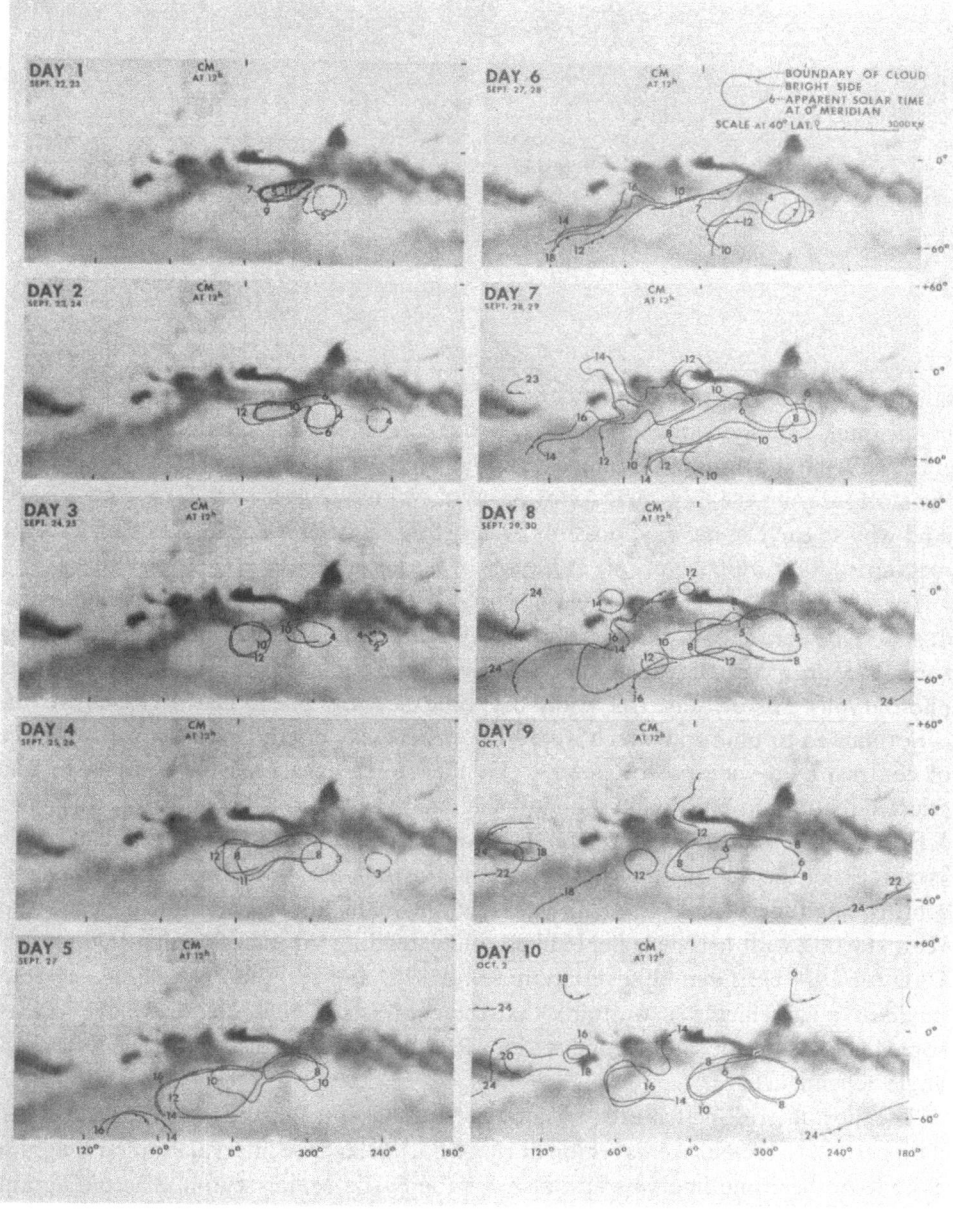

Fig. 2.   Mars – daily progress of the storm for the first ten days.

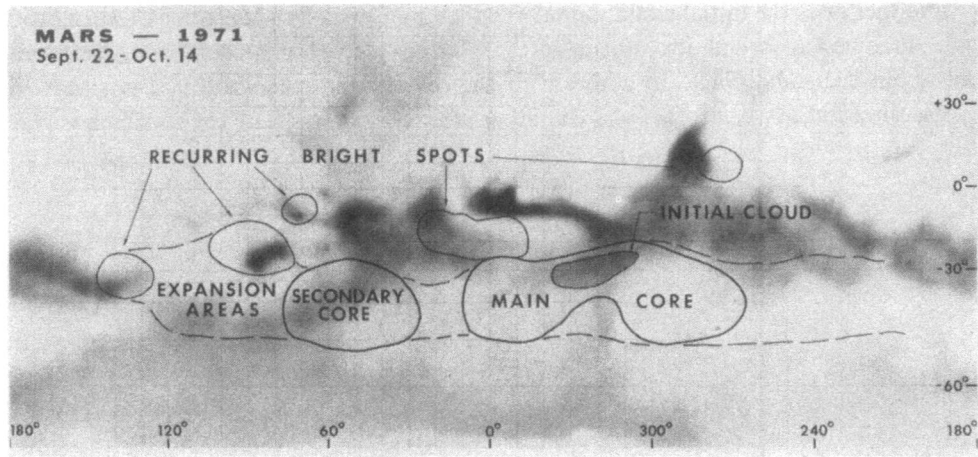

Fig. 3.   Mars active regions during the developing stages of the storm.

intensity of their onset. In September 1971, for example, there was no visible evidence of any unusual activity on September 21st, but early the following morning as Noachis emerged from the morning terminator, it was brighter and whiter than any features of the storm during the days that followed. What would trigger such a vigorous onset? And why so early in the day, possibly even before dawn? If condensates play a role in making it rather white, why only at the site of the initial cloud?

There is an entirely different kind of observation that may be evidence for persistent dust activity throughout the Martian year. Figure 5 shows data obtained by Thompson (1973) from the statistical analysis of regional contrast variations on our Patrol photographs. Historically, this is the old 'blue clearing' phenomenon, which we find is not limited to blue and which we do not believe has anything to do with dilution of contrast by a widespread overcast. Thompson has measured the contrast of four selected features in blue light on about 6000 image sets from 1969 and 1971. In Figure 5 the ordinate represents the brightness of neighboring light areas divided by the brightness of the dark feature indicated, and the data are plotted against the planetocentric longitude of the Sun, L_s. This diagram spans about one-third of a Martian year, starting with Martian 'September' and extending through Martian 'December'. Data for 1971 considerably overlap those for 1969 and seem to show some degree of qualitative agreement, so we think that the contrasts of these features have at least some dependence on the Martian season. Residual differences may be due to a phase angle dependence.

The most interesting feature of Thompson's analysis is illustrated in Figure 6. These data pertain to the Nilokeras region during a particular time interval. Similar diagrams exist for other time intervals and also for the Syrtis Major region. These diagrams confirm our earlier finding that there is a systematic trend of regional contrast with time of the Martian day and that the afternoon is not symmetric with the morning.

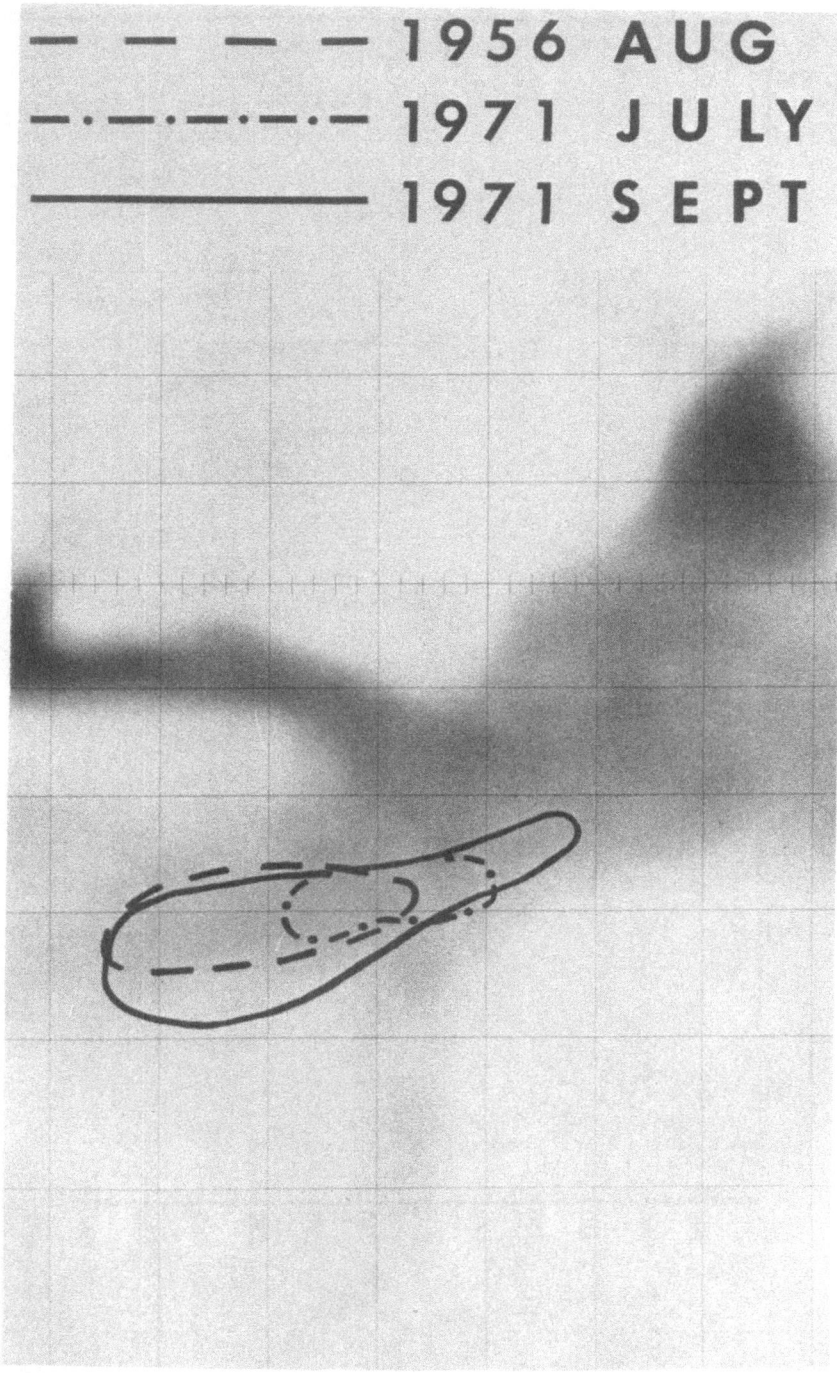

Fig. 4.   Mars area where three best recorded dust storms have started.

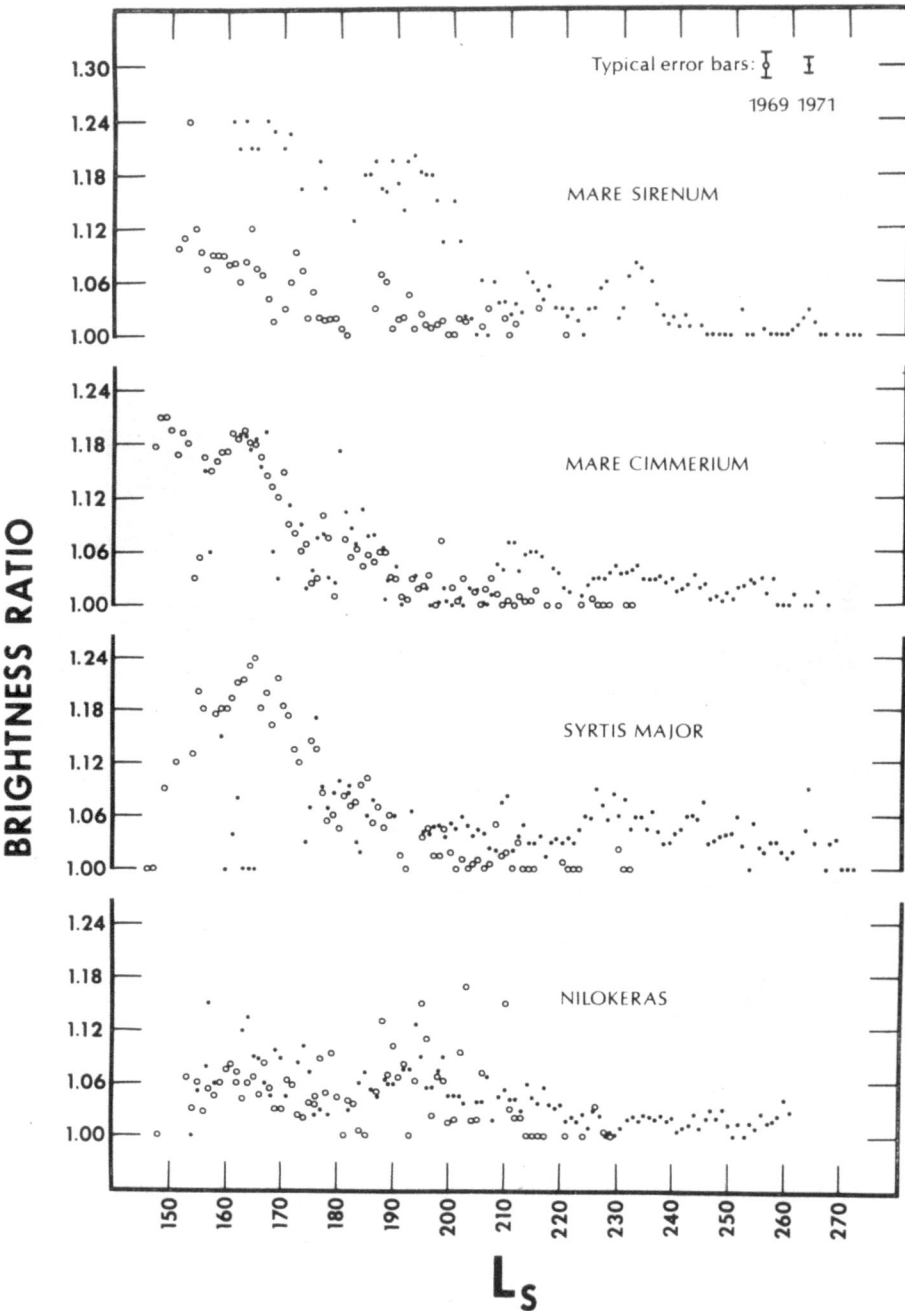

Fig. 5.   Contrast of four selected dark features versus planetocentric longitude of Sun.

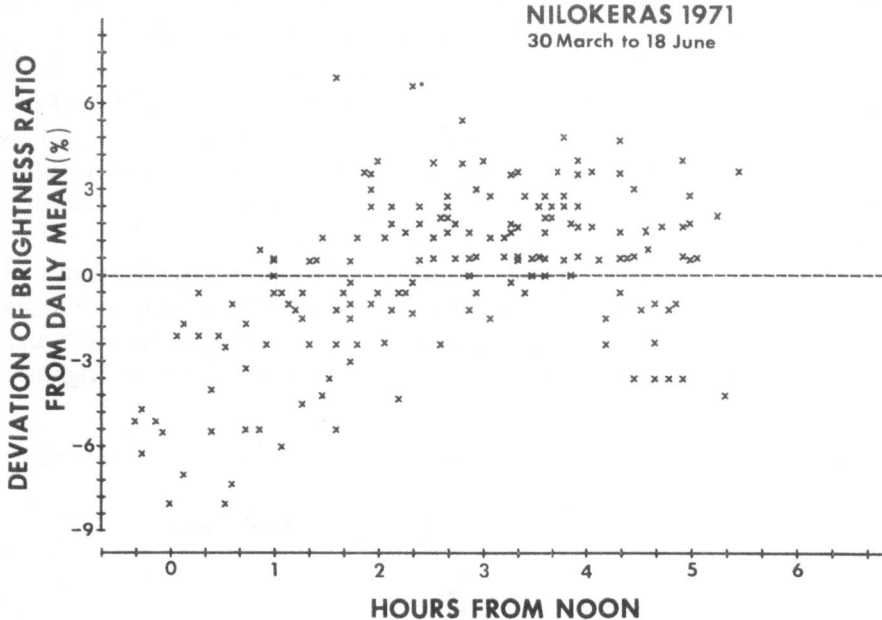

Fig. 6. Systematic trend of regional contrast with time of the Martian day, shown for Nilokeras region.

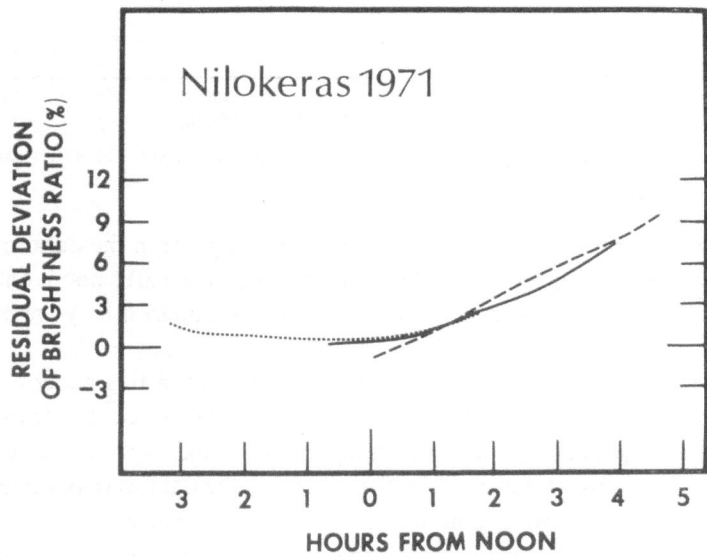

Fig. 7. Systematic afternoon climb of albedo contrast ascribed to the atmosphere, after a surface contribution has been subtracted.

If we assume that the contribution of the surface can be represented by a symmetric Minnaert function having slightly different coefficients for light and dark areas, we are then left with the non-symmetric residual contribution shown in Figure 7. We think it is reasonable to assume that this contribution is due to the atmosphere. The three curves in Figure 7 represent three different time intervals – early, middle, and late – during the 1971 Mars apparition. The agreement of the three curves shows that the apparent difference between morning and afternoon has nothing to do with any phase angle effect. It is really due to something going on on Mars.

The afternoon upturn illustrated here is an order of magnitude greater than statistical uncertainties in the data. It amounts to a daily rise of 8% between noon and 4 p.m. local Martian time, and it is occurring when Mars is supposedly 'normal' and no actual dust storm is in progress. We interpret this as mainly a rise in the brightness of

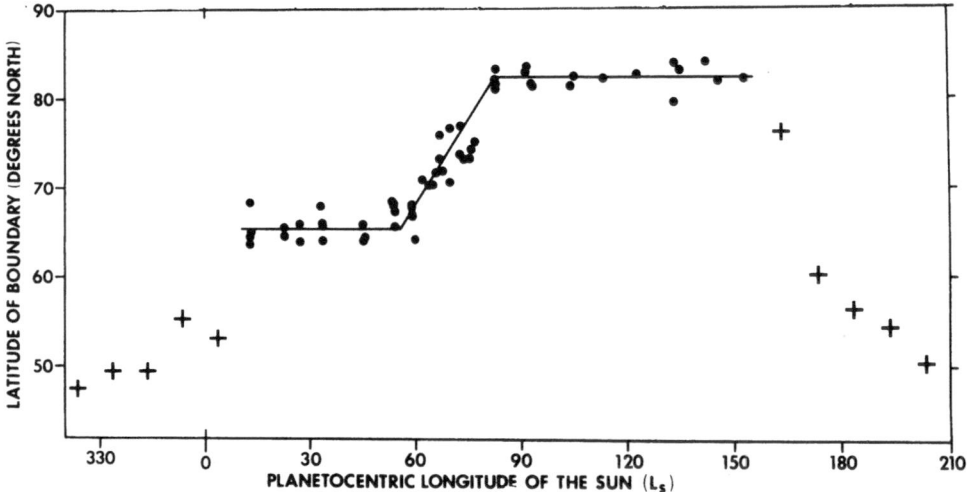

Fig. 8.   Mean latitude of the north polar cap boundary versus Martian season.

the light areas, with less change (perhaps none) taking place in the dark areas. If this phenomenon is due to a ground haze that is regenerated each afternoon, one can show that it may limit the horizontal visibility range of Mars landers if they are put down in light areas.

The third finding (Baum and Martin, 1973) to mention in this paper concerns the Martian polar caps and is illustrated in Figure 8. We have plotted the mean latitude of the north polar cap boundary against the Martian season, expressed in terms of L_s. This diagram spans two-thirds of a Martian year, but the data were obtained from the Lowell plate collection covering more than 60 years (Fischbacher et al., 1969), plus more recent data from Patrol films; and the solid dots represent means for different years. Since the scatter of the dots is not significantly greater than the estimated errors, the regression of the surface cap evidently runs on a very predictable schedule.

Now, what does this diagram say? Until a few weeks after the vernal equinox (until a bit past $L_s = 0°$), the boundary is vague, rapidly variable, and visible mainly on blue or ultraviolet images. It commonly reaches down to latitudes as low as 40° north. We identify this as the north polar *hood*, and we use the crosses here to represent approximate long-term means. As the hood clears away soon after the vernal equinox, a new and different looking boundary becomes visible much further north. It is sharp, it is seen in all colors, and its mean latitude is 65 to 66 deg north. We identify it as the *surface cap*, and we suspect that it never extends much below 65°, because it is not initially receding. In fact, it does not start receding rapidly until 80 days later. The boundary again becomes almost stationary when it reaches the edge of the permanent cap at 82 to 83 deg north. About 150 days after that, a new polar hood starts to obscure the cap.

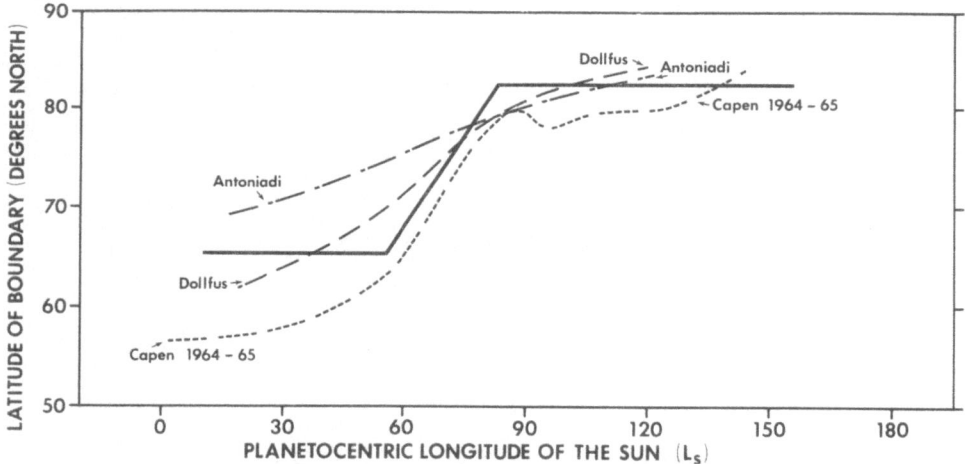

Fig. 9.   Comparison of data of Figure 8 with other observations.

In Figure 9, our result is again represented schematically by the bent solid line. For comparison, the three dashed lines show north cap regression curves published previously by others. These dashed curves do not give any clear indication of the approximate stillstand that we find near 65° latitude, and the overall disagreement amongst the curves is large. We believe that the differences arise from the method of measurement. We superimpose images optically onto an orthographic grid and read the latitude of the boundary of the cap near the central meridian of the image. Others have instead measured the apparent angular width of the cap near the polar limb, where the hood would be harder to distinguish from a surface cap and where any atmospheric haziness would bias the readings.

The situation in the southern hemisphere shown in Figure 10 is much the same, except that the cap is more eccentric with respect to the pole. When the south surface

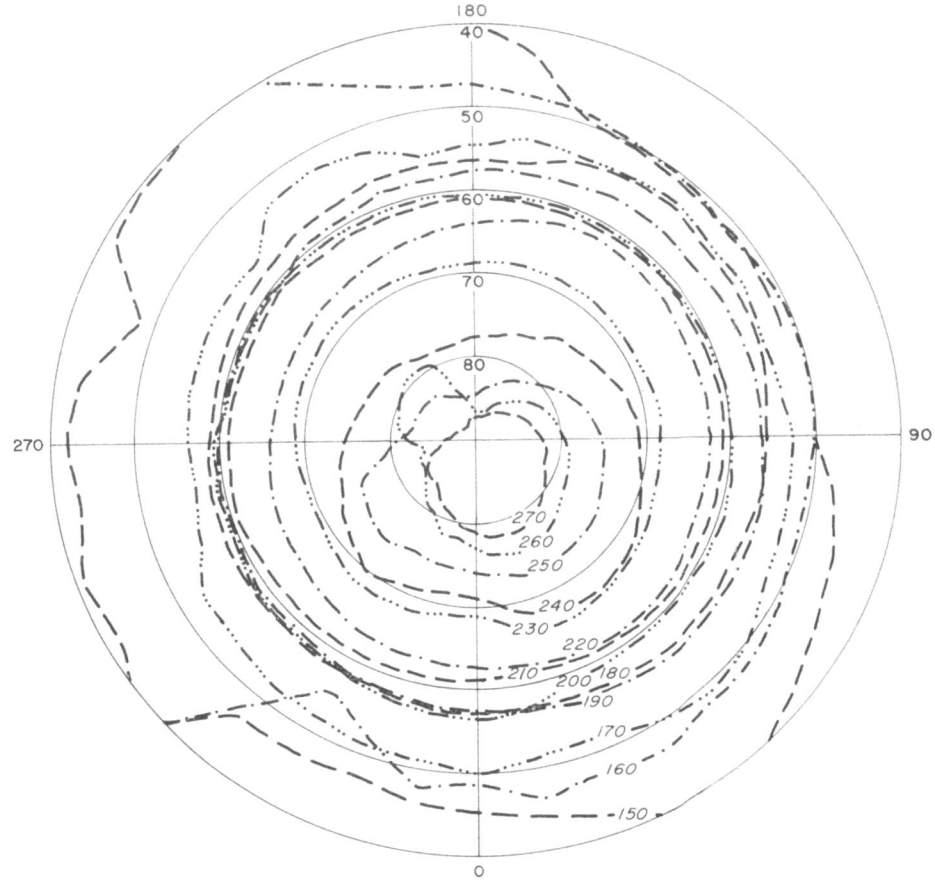

Fig. 10.   Regression of the south polar cap as a function of L_s (Martian season).

cap first becomes visible, its boundary runs from about 56°S latitude on one side to 60°S on the other. Like the north cap, it appears to have been at a stillstand, but it starts receding sooner. It reaches the eccentric permanent cap when L_s is about 280°. There is no evidence to support the widely held belief that the south cap dissipates completely.

References

Baum, W. A.: 1973, *Planetary Space Sci.* **21**, 1511.
Baum, W. A. and Martin, L. J.: 1973, *Bull. Am. Astron. Soc.* **5**, 296.
Capen, C. F. and Martin, L. J.: 1971, *Lowell Obs. Bull.*, No. 157.
Capen, C. F. and Martin, L. J.: 1972a, *Sky Telesc.* **43**, 276.
Capen, C. F. and Martin, L. J.: 1972b, *Bull. Am. Astron. Soc.* **4**, 374.
Fischbacher, G. E., Martin, L. J., and Baum, W. A.: 1969, 'Mars Polar Cap Boundaries', Report under JPL Contract No. 951547, May 1969.
Thompson, D. T.: 1973, *Icarus* **18**, 164.

DISCUSSION

Barrow: Is there anything to be learned about the motions of the dust storm features from time-lapse photography?

Baum: In a sense, our International Planetary Patrol material *is* time-lapse photography, with a network of observatories trying to cover Mars every hour, every day. We do not find the storm-brightened areas to move in a continuous manner, like a weather 'front'. Instead, they arise and fade in widely separated places (in addition to changing shape) as a function of the local time of the Martian day.

Vsehsvyatsky: Can the coincidence of the places of the beginning of the 1956 and 1971 dust storms be taken as evidence for the eruptive (volcanic) nature of the storms?

Baum: The place of beginning of the 1956 and 1971 storms does not appear to be the kind of terrain where one should expect volcanic events.

Icke: Has the possibility been considered that the dust might be the cause of the storms, as well as a consequence, since the dust heats up the atmosphere? In other words, can the dust cause a thermal instability which shows up as a storm?

Baum: Yes. I can well imagine a 'feedback' mechanism, but there are others better qualified than I to develop a theoretical model.

Editor's Note: Such a theory has been developed by Golitsyn.

UBV PHOTOMETRY OF MARS

A. T. YOUNG

*Jet Propulsion Laboratory, California Institute of Technology, Pasadena, Calif., U.S.A.**

Abstract. A critical analysis of selected high-quality photometric observations of Mars indicates that: (1) The phase function is concave upward out to at least 40° phase. No *sudden* brightening occurs at opposition, but the curvature increases at small phase. (2) Large systematic differences (0.1–0.2 mag.) exist between different observers' data. However, the small random scatter attributable to Mars (0.01–0.02 mag.) in the better series suggests that these differences represent systematic errors in data reduction, not variations in the planet's brightness. (3) The disentangling of seasonal, diurnal, and phase effects leaves considerable ambiguity; more observations are needed, over a long time, with a stable instrumental system. However, even the present data are sufficient to expose substantial errors in published phase curves of Mars (and consequently, in interpretations based on them).

1. Introduction

Ground-based photometry of Mars is primarily useful for determining the phase function of the planet. It can also be used to study large-scale atmospheric and surface changes, such as 'blue clearings' and seasonal variations.

Accurate photoelectric observations of Mars were first made by Guthrick and Prager (1914), who discovered the rotational effect, with maximum brightness near longitude 100°. Unfortunately it is not possible to reduce their data to any modern photometric system. The first UBV observations were made by Johnson and Gardiner (1955), who misinterpreted their data owing to an error in computing central meridians. In re-analyzing their data, I suggested 'a real brightening of Mars by one- or two-tenths of a magnitude at phase angles less than about 12°' (Young, 1957). This suggestion was repeated by de Vaucouleurs (1960), who made additional observations in 1958; however, his data (which cover only 3 nights) were too limited to establish the effect with certainty. I therefore made additional observations, especially at small phase angles, during the 1960–61 opposition. Even from these data I was unable to reach a definite conclusion, so I resolved to postpone a full discussion until the completion of the extensive Harvard-NASA program of planetary photometry (Young and Irvine, 1967). These data are now available (Irvine *et al.*, 1968a, b) and, although a definitive solution still is not possible, the problem is clear enough to permit a preliminary discussion.

It must be admitted at once that the UBV system is poorly suited to Martian photometry. The V band is situated in the steepest part of the Martian reflectance spectrum, so that slight errors in reducing 'V' data to a common system produce considerable systematic errors in the results. The V band is also poorly placed to display the rotational effect (Irvine *et al.*, 1968), which is stronger at longer wavelengths. Furthermore, Mars is so red ($B-V \approx +1.4$) that red-leak corrections (Shao and Young, 1965) are important in the ultraviolet; if corrections are merely estimated from the colors, systematic errors due to the peculiar spectral energy distribution of the planet can be

---

* Present address: Dept. of Physics, Texas A. and M. University, College Station, Tex. 77843, U.S.A.

Woszczyk and Iwaniszewska (eds.), Exploration of the Planetary System, 253–285. All Rights Reserved
Copyright © 1974 by the IAU

expected. On the other hand, the UBV data are the only ones with enough time span to allow one to look for (Martian) seasonal effects, as several different oppositions have been observed.

2. Photometric Central Meridian

Previous investigators have analyzed the rotational light curve in terms of the areographic longitude of the center of the disc, which is tabulated for 0^h UT of each day in the Astronomical Ephemeris. However, the phase of Mars near quadrature is so large that the center of the disc is not suitable as a reference meridian for photometry, since an appreciable portion of the terminator side of the disc is in shadow. We need to define a *photometric* central meridian.

It is easy to show that the effective photometric center of the disc lies halfway between the subsolar point and the center of the disc. For, consider a photometric longitude l measured from this 'mirror' meridian, which is perpendicular to the photometric equator (the great circle through the subsolar point and the center of the disc); at phase angle α, the photometric longitude runs from $(90° - \alpha/2)$ at the limb to $-(90° - \alpha/2)$ at the terminator. We want to find the mean longitude, weighting each element of area dA on the planet by its apparent projected area and surface brightness. If i and ε are the angles of incidence and emission, measured from the local normal, the projected area is $\cos \varepsilon \, dA$.

Now consider two similar elements of area $dA = dA'$ related by the reciprocity principle (Minnaert, 1962), so that $i' = \varepsilon$ and $\varepsilon' = i$; they are symmetrically placed about $l = 0$, so that $l' = -l$. But the reciprocity principle requires that $B \cos \varepsilon = B' \cos \varepsilon'$ if their surface brightnesses are B and B'. Hence the primed and unprimed areas make equal but opposite contributions to the integral of (l B $\cos \varepsilon \, dA$) over the whole visible surface, and the photometric mean longitude $\bar{l}$ is zero. Note that the argument is based only on the reversibility of the light rays and does not require any further assumptions about the photometric function of the surface. [†]

We can readily correct the geometric central meridian ω to the photometric central meridian ω^*, using quantities tabulated in the Astronomical Ephemeris:

$$\omega^* = \omega - \frac{(A_S - A_E)}{2}, \tag{1}$$

where A_S and A_E are the areocentric right ascensions of the Sun and Earth, respectively. Since the correction term can be as large as $\pm 20°$, its neglect could produce relative misplacements of points by as much as $40°$ in longitude.

3. Analytic Representation

We can expect three types of regular, intrinsic variation in the brightness of Mars: the phase effect, the rotation effect, and possibility a seasonal effect.

[†] Stated another way, the reciprocity principle requires dA to send the observer the same amounts of light when it is at l and l'.

Previously, a linear dependence of magnitude on phase angle seemed adequate, although several authors (Young, 1957; de Vaucouleurs, 1960; O'Leary and Rea, 1968) have suggested a sudden brightening on the order of 0.1 mag. at small phase angles. To represent this brightening, we must introduce additional powers of the phase angle. After some unsatisfactory experiments with a cubic, I decided on terms in α^2 and $1/\alpha$; the latter seems to represent the brightening around opposition quite well over the range of the data. Of course, we cannot interpret the constant term as the absolute magnitude at zero phase, or the linear coefficient as the slope of the phase curve in this case.

We can represent the rotational effect analytically as a Fourier series in ω^*. To find the number of terms required for adequate representation, consider a perfectly black planet with a small white spot on one side. As the projected size of the spot varies as $\cos\omega$, the light curve is a half-wave rectified sinusoid if there is no limb darkening, or a half-wave of $\cos^2\omega$ if the spot is a perfect (Lambert) diffuser; Mars is intermediate, between these two cases (Young and Collins, 1971). For the undarkened (lunar) case, the series contains the terms

$$\cos\omega + \frac{4}{3\pi}\cos 2\omega - \frac{4}{15\pi}\cos 4\omega \ldots ,$$

where I have normalized the series to the amplitude of the fundamental term. The corresponding series for the Lambert (fully darkened) surface is

$$\cos\omega - \tfrac{3}{16}\pi\cos 2\omega + \tfrac{1}{5}\cos 3\omega - \tfrac{1}{35}\cos 5\omega \ldots .$$

As the visual rotation effect on Mars has an amplitude on the order of a tenth of a magnitude, and we can neglect terms on the order of a hundredth, we see that a fully-darkened Mars would just require the inclusion of the 3ω term. However, the limb darkening is actually less, so we can probably neglect this term as well and use terms in ω and 2ω only.

Two kinds of seasonal effects may be anticipated: a semiannual variation, with a maximum near each solstice, and an annual effect related to the asymmetry of the two hemispheres and the eccentricity of the planet's orbit. Hence we expect to include Fourier series terms in L_S, the areocentric longitude of the Sun measured along the planet's orbit from its vernal equinox. As Mars traverses about 30–40° of L_S from quadrature to opposition, we may expect these seasonal effects to produce slightly different apparent phase coefficients and opposition brightnesses at different oppositions.

We therefore adopt, as equation of condition,

$$\begin{aligned}
m_1 = \frac{\mu_{-1}}{\alpha} &+ m_1(0) + \mu_1\alpha + \mu_2\alpha^2 + a_1\cos\omega^* + b_1\sin\omega^* + \\
&+ a_2\cos 2\omega^* + b_2\sin 2\omega^* + c_1\cos L_S + \\
&+ d_1\sin L_S + c_2\cos 2L_S + d_2\sin 2L_S ,
\end{aligned} \tag{2}$$

where m_1 is an observed magnitude reduced to unit distances from both Sun and Earth, μ_i is the coefficient of the ith power phase term, and the other literal coefficients describe the rotational and seasonal effects. We have thus to determine the 12 parameters $(m_1(0), \mu_i, a_1, ..., d_2)$ for each color.

4. UBV Observations

There are five series of UBV observations of Mars available for analysis: 21 observations made at Flagstaff in 1954 (Johnson and Gardiner, 1955); 18 at Flagstaff in 1958 (de Vaucouleurs, 1960); 27 at Agassiz Station in 1960–61, reported here for the first time; 76 at Le Houga in 1963–65 (Irvine *et al.*, 1968a); and 124 at Boyden in 1963–65 (Irvine *et al.*, 1968b). Our first concern is to place all the data on the same system, as it has become increasingly clear that systematic errors approaching a tenth of a magnitude can occur in published UBV data (Lawrence and Reddish, 1965; Fernie and Watt, 1967).

Johnson and Gardiner (1955) compared Mars to the stars σ Sgr = HR 7121 and λ Sgr = HR 6913. Both these stars and Mars in 1954 were visible from Flagstaff only at large zenith distances, so that extinction errors are important. Johnson and Gardiner estimated probable errors of 0.04, 0.02, and 0.04 in V, $B-V$, and $U-B$ respectively, in tying the stars to UBV standards, and errors of 0.01 mag. in comparing Mars to the stars. We can now use more accurate UBV data for the two stars to estimate systematic corrections to these Mars data.

Table I gives Johnson and Gardiner's values, together with more extensive measurements made at the Royal Cape Observatory (Cousins and Stoy, 1963) and at Catalina and Tonantzintla (Johnson *et al.*, 1966; the mean of only the four new observations is given, not the value in their Table 2 which includes the 1954 values.) The modern

TABLE I

UBV data for 1954 comparison stars

Star	V	$B-V$	$U-B$	Source
σ Sgr = HR 7121	2.10	-0.20	-0.72	Cousins and Stoy (1963)
	2.07	-0.21	-0.72	Johnson *et al.* (1966)
	2.09	-0.20	-0.72	Adopted values
	2.00	-0.22	-0.76	Johnson and Gardiner (1955)
	$+0.09$	$+0.02$	$+0.04$	Correction to 1954 values
λ SGR = HR 6913	2.845	1.04	$+0.91$	Cousins and Stoy (1963)
	2.83	1.04	$+0.91$	Johnson *et al.* (1966)
	2.84	1.04	$+0.91$	Adopted values
	2.80	1.04	$+0.89$	Johnson and Gardiner (1955)
	$+0.04$	0.00	$+0.02$	Correction to 1954 values

TABLE II

1954 Johnson and Gardiner data

Date	UT	V	$B - V$	$U - B$	α	ω	ω^*	L_S	V_1
1954 April	23.48	− 0.530	1.380	0.590	35.12	200.7	218.4	154.0	− 0.850
1954 April	27.46	− 0.710	1.430	0.600	34.18	156.2	171.8	156.1	− 0.930
1954 May	4.46	− 0.980	1.460	0.620	32.18	90.9	105.5	159.8	− 1.020
1954 May	6.44	− 1.030	1.440	0.600	31.52	65.3	79.8	160.8	− 1.020
1954 May	11.43	− 1.130	1.390	0.560	29.69	15.5	29.1	163.5	− 0.990
1954 May	31.38	− 1.730	1.390	0.540	19.35	175.1	184.2	174.4	− 1.100
1954 June	2.37	− 1.830	1.390	0.540	18.04	153.5	162.0	175.5	− 1.160
1954 June	4.37	− 1.910	1.420	0.550	16.67	135.5	143.4	176.6	− 1.190
1954 June	8.35	− 2.060	1.460	0.610	13.82	92.6	99.2	178.9	− 1.270
1954 June	10.35	− 2.060	1.400	0.620	12.32	74.7	80.7	180.0	− 1.230
1954 June	19.31	− 2.380	1.320	0.660	5.50	340.6	343.3	185.1	− 1.420
1954 June	20.39	− 2.410	1.300	0.650	4.75	359.8	2.1	185.7	− 1.430
1954 June	21.37	− 2.460	1.330	0.680	4.14	343.9	345.8	186.3	− 1.470
1954 June	29.27	− 2.410	1.300	0.610	5.15	238.0	236.8	190.8	− 1.370
1954 June	30.32	− 2.390	1.300	0.600	5.92	246.7	245.1	191.4	− 1.350
1954 July	1.34	− 2.360	1.340	0.620	6.72	244.9	242.9	192.0	− 1.310
1954 July	2.31	− 2.370	1.360	0.620	7.49	225.5	223.1	192.6	− 1.320
1954 July	5.35	− 2.380	1.370	0.570	9.98	212.9	209.3	194.4	− 1.320
1954 July	6.33	− 2.360	1.370	0.590	10.79	197.0	193.1	195.0	− 1.310
1954 July	7.32	− 2.380	1.380	0.600	11.60	184.7	180.4	195.6	− 1.330
1954 July	15.23	− 2.240	1.350	0.630	17.96	81.8	74.5	200.4	− 1.230

TABLE III

1958 de Vaucouleurs data

Date	UT	V	$B - V$	$U - B$	α	ω	ω^*	L_S	V_1
1958 Oct	23.188	− 2.034	1.308	0.579	21.32	94.8	104.9	315.5	− 1.381
1958 Oct	23.193	− 2.027	1.326	0.613	21.31	96.6	106.7	315.5	− 1.374
1958 Oct	23.325	− 1.994	1.287	0.562	21.21	142.9	153.6	315.6	− 1.340
1958 Oct	23.381	− 1.993	1.275	0.542	21.17	162.6	172.7	315.6	− 1.338
1958 Oct	23.497	− 2.074	1.260	0.644	21.08	203.3	213.3	315.7	− 1.419
1958 Oct	23.503	− 2.062	1.237	0.626	21.08	205.4	215.4	315.7	− 1.406
1958 Oct	23.528	− 2.070	1.255	0.641	21.06	214.2	224.2	315.7	− 1.414
1958 Oct	23.533	− 2.037	1.237	0.616	21.05	215.9	225.9	315.7	− 1.381
1958 Nov	24.219	− 2.147	1.318	0.710	6.83	183.3	180.1	333.4	− 1.557
1958 Nov	24.260	− 2.119	1.300	0.700	6.87	197.7	194.5	333.4	− 1.529
1958 Nov	24.377	− 2.050	1.260	0.679	6.97	238.8	235.6	333.5	− 1.462
1958 Nov	24.469	− 2.018	1.271	0.748	7.05	271.1	267.9	333.5	− 1.431
1958 Nov	24.474	− 2.010	1.310	0.670	7.05	272.8	269.6	333.5	− 1.423
1958 Nov	28.115	− 2.098	1.323	0.679	10.14	111.4	106.7	335.5	− 1.569
1958 Nov	28.122	− 2.106	1.334	0.697	10.14	113.9	109.2	335.5	−- 1.577
1958 Nov	28.219	− 2.081	1.316	0.686	10.22	147.9	143.2	335.6	− 1.553
1958 Nov	28.353	− 2.122	1.305	0.650	10.33	195.0	190.2	335.6	− 1.597
1958 Nov	28.413	− 2.082	1.293	0.642	10.38	216.0	211.2	325.7	− 1.558

data are quite accordant, but differ appreciably from the 1954 values. Since the Cape values contain at least twice as many observations as the North American ones, and since these stars pass nearly through the zenith at the Cape but must be observed at about 2 air masses by the Arizona observers, we must give practically all the weight to the Cape data. The corrections are much larger for the blue star and increase steadily toward shorter wavelengths, which strongly suggests systematic extinction errors in the 1954 values. Rather than extrapolate the correction to the color of Mars $(B-V\sim 1.4)$ I simply adopt the corrections for the redder star $(B-V=1.04)$. The corrected values, reduced to unit distance from Sun and Earth, are given in Table II along with the necessary data from the physical ephemeris of Mars.

TABLE IV

Extinction stars and errors for each observing run at Agassiz Station

Dates	Star	V	$B-V$	$U-B$	n	Std
Nov. 18–	ι Ari	+ 1.982	+ 1.150	+ 1.129	3	*
Dec. 28, 1960	β Tau	1.639	− 0.137	− 0.478	5	*
	ο Per	3.806	+ 0.060	− 0.765	6	*
	ι CMi	0.351	+ 0.411	− 0.013	7	*
	β Cnc	3.533	+ 1.487	+ 1.776	5	*
	β Gem	1.150	+ 0.984	+ 0.837	11	*
	ν And	4.531	− 0.146	− 0.590	1	*
	10 Lac	4.898	− 0.199	− 1.037	1	*
	ι Gem	1.569	+ 0.048	− 0.031	9	
	ι Tau	0.886	+ 1.538	+ 1.934	8	
	R	0.019	0.020	0.016		
	S	0.014	0.020	0.018		
Mar. 4–	β Tau	+ 1.645	− 0.138	− 0.482	4	*
Mar. 21, 1961	α CMi	0.352	+ 0.419	+ 0.014	6	*
	ι Gem	1.580	+ 0.019	− 0.020	7	
	β Gem	1.141	+ 0.993	+ 0.861	7	*
	β Cnc	3.507	+ 1.485	+ 1.820	5	*
	109 Vir	3.716	+ 0.004	− 0.036	1	*
	ζ Aql	3.002	+ 0.002	+ 0.001	4	*
	ι Leo	1.360	− 0.112	− 0.381	3	*
	μ Cnc	5.273	+ 0.636	+ 0.240	3	
	47 UMa	5.028	+ 0.611	+ 0.127	3	
	α Cr B	2.245	− 0.014	− 0.053	3	
	ε Cr B	4.158	+ 1.223	+ 1.348	4	
	β Ser A	3.669	+ 0.069	+ 0.080	2	
	γ Ser	3.868	+ 0.471	− 0.033	2	*
	BD + 4° 4048	9.140	+ 1.486	+ 1.086	2	*
	BD + 13° 3832	8.971	+ 1.215	+ 1.081	1	
	BD + 13° 3827	8.878	+ 1.497	+ 1.429	2	
	χ¹ Ori	4.420	+ 0.541	+ 0.126	2	
	BD + 13° 3816	9.143	+ 0.191	− 0.278	1	
	BD + 13° 3826	7.257	+ 0.257	+ 0.248	1	
	R	0.006	0.007	0.017		
	S	0.015	0.017	0.039		

The 1958 observations of de Vaucouleurs (1960) were made on 3 nights with a total of 42 standard star observations. The planet was north of the equator, so extinction errors should not be serious. I accept these data at face value; they are listed in Table III.

My 1960–61 observations were made with the 24-in. Clark reflector at the George R. Agassiz Station of Harvard College Observatory. The reductions were made by the same program used to reduce the later Harvard-NASA observations (Young and Irvine, 1967; Irvine et al., 1968a, b). The observations were made before I realized the importance of the red leak in the U filter; the raw U data have been corrected by the method of Shao and Young (1965), using constants for the same equipment deduced from observations in 1964–65. This is rather unsatisfactory, but the best that can be done. The average red-leak correction for Mars was about 0.04 mag. The standard stars and standard errors are given for each observing period in Table IV. R is the rms residual in the extinction solution for each filter, and hence represents the internal error of a single observation at one air mass. S is the rms residual in fitting the instrumental system, reduced to extra-atmospheric magnitudes, to the standard stars indicated by asterisks. The R and S values in the blue and ultraviolet columns refer to

TABLE V

1960–61 ATY data

Date	UT	V	$B-V$	$U-B$	α	ω	ω^*	L_S	V_1
1960 Nov	18.178	−0.801	1.418	0.549	30.70	352.2	6.4	353.4	−1.052
1960 Nov	18.257	−0.762	1.368	0.517	30.70	20.0	34.1	353.4	−1.011
1960 Nov	18.321	−0.770	1.402	0.499	30.60	42.4	56.5	353.5	−1.018
1960 Nov	18.408	−0.826	1.444	0.533	30.60	72.9	87.0	353.5	−1.073
1960 Nov	31.248	−1.043	1.322	0.539	23.40	259.0	269.9	0.0	−1.113
1960 Dec	19.112	−1.431	1.298	0.544	10.00	51.4	56.3	8.8	−1.353
1960 Dec	19.210	−1.494	1.322	0.524	10.00	85.8	90.7	8.8	−1.415
1960 Dec	19.262	−1.532	1.352	0.513	9.90	104.1	108.9	8.9	−1.453
1960 Dec	19.314	−1.530	1.335	0.521	9.90	122.4	127.2	8.9	−1.451
1960 Dec	19.385	−1.505	1.312	0.535	9.80	147.3	152.1	8.9	−1.426
1960 Dec	19.418	−1.463	1.290	0.549	9.80	158.9	163.7	8.9	−1.384
1960 Dec	23.031	−1.452	1.332	0.519	6.80	347.8	351.2	10.7	−1.367
1960 Dec	23.085	−1.468	1.306	0.499	6.80	6.8	10.2	10.7	−1.383
1960 Dec	23.256	−1.477	1.275	0.474	6.60	66.9	70.2	10.8	−1.392
1960 Dec	23.335	−1.558	1.328	0.511	6.60	94.6	97.9	10.8	−1.473
1960 Dec	23.401	−1.559	1.306	0.437	6.50	117.8	121.0	10.9	−1.474
1960 Dec	27.988	−1.517	1.276	0.523	3.10	288.9	290.3	13.1	−1.443
1960 Dec	28.022	−1.535	1.292	0.563	3.10	300.9	302.3	13.1	−1.462
1960 Dec	28.090	−1.556	1.301	0.540	3.10	324.8	326.2	13.1	−1.483
1960 Dec	28.144	−1.567	1.292	0.546	3.00	343.7	345.1	13.1	−1.494
1961 Mar	4.160	0.075	1.460	0.557	34.40	113.0	97.2	43.6	−1.040
1961 Mar	18.062	0.527	1.432	0.458	36.20	307.1	290.1	49.7	−0.866
1961 Mar	18.110	0.523	1.407	0.581	36.20	323.9	306.9	49.7	−0.871
1961 Mar	18.172	0.501	1.442	0.615	36.20	345.6	328 6	49.8	−0.894
1961 Mar	21.084	0.541	1.475	0.504	36.50	286.5	269.3	51.1	−0.910
1961 Mar	21.156	0.584	1.407	0.588	36.50	311.7	294.5	51.1	−0.868
1961 Mar	21.195	0.553	1.450	0.571	36.50	325.4	308.2	51.1	−0.900

A. T. YOUNG

TABLE VI
Le Houga data

Date	UT	V	B − V	U − B	α	ω	ω*	Ls	V₁
1964 Aug	31.190	−	1.350	0.570	28.52	4.6	17.5	358.0	− 0.990
1964 Sept	9.201	−	1.400	0.570	29.71	281.2	294.8	2.0	− 0.950
1964 Sept	10.189	−	1.370	0.540	29.83	267.3	280.9	2.5	− 0.950
1964 Sept	23.182	−	1.450	0.590	31.47	138.8	153.4	8.9	− 1.020
1964 Dec	8.189	−	1.470	0.640	36.70	128.3	147.4	44.1	− 0.920
1964 Dec	8.261	−	1.450	0.600	36.70	153.5	172.6	44.1	− 0.950
1964 Dec	9.140	−	1.460	0.630	36.68	101.6	120.7	44.5	− 0.960
1965 Jan	11.072	−	1.430	0.610	32.60	126.4	143.8	59.0	− 1.000
1965 Jan	11.163	−	1.400	0.590	32.58	158.3	175.7	59.1	− 0.980
1965 Feb	3.020	−	1.340	0.530	24.29	257.3	270.4	69.1	− 1.040
1965 Feb	3.111	−	1.380	0.560	24.24	289.2	302.3	69.1	− 1.030
1965 Feb	4.003	−	1.350	0.570	23.79	242.3	255.1	69.5	− 1.010
1965 Feb	4.120	−	1.350	0.540	23.73	283.3	296.1	69.5	− 1.060
1965 Feb	4.217	−	1.370	0.590	23.68	317.4	330.2	69.6	− 1.100
1965 Feb	5.004	−	1.350	0.530	23.28	233.6	246.1	69.9	− 1.030
1965 Feb	5.110	−	1.340	0.550	23.22	270.8	283.3	70.0	− 1.050
1965 Feb	10.011	−	1.420	0.620	20.49	191.3	202.3	72.1	− 1.210
1965 Feb	10.106	−	1.380	0.580	20.43	224.6	235.6	72.2	− 1.080
1965 Feb	24.947	−	1.290	0.540	10.44	36.3	41.9	78.6	− 1.330
1965 March	10.023	−	1.280	0.570	2.42	309.8	309.6	84.4	− 1.500
1965 March	25.853	−	1.360	0.540	13.34	110.6	103.5	91.3	− 1.330
1965 March	25.964	−	1.390	0.590	13.42	149.6	142.5	91.4	− 1.320
1965 March	27.826	−	1.320	0.500	14.78	83.6	75.7	92.2	− 1.270
1965 March	27.932	··	1.400	0.550	14.86	120.8	112.9	92.3	− 1.320
1965 March	28.046	−	1.380	0.590	14.94	160.8	152.8	92.3	− 1.310
1965 March	28.115	−	1.380	0.580	14.99	185.1	177.1	92.3	− 1.290
1965 March	28.846	−	1.290	0.490	15.52	81.8	73.5	92.7	− 1.310
1965 March	28.959	−	1.380	0.560	15.60	121.5	113.2	92.7	− 1.300
1965 March	29.046	−	1.400	0.590	15.66	152.0	143.6	92.8	− 1.290
1965 March	29.125	−	1.420	0.610	15.72	179.8	171.4	92.8	− 1.240
1965 March	29.817	−	1.280	0.490	16.20	62.8	54.1	93.1	− 1.250
1965 March	29.923	−	1.340	0.530	16.28	100.0	91.3	93.1	− 1.300
1965 March	30.029	−	1.400	0.580	16.35	137.2	128.5	93.2	− 1.320
1965 March	30.104	−	1.410	0.560	16.40	163.6	154.8	93.2	− 1.280
1965 March	30.937	−	1.340	0.510	16.99	96.1	87.0	93.6	− 1.260
1965 March	31.042	−	1.390	0.570	17.06	133.0	123.9	93.6	− 1.270
1965 April	3.835	−	1.330	0.510	19.59	24.8	14.3	95.3	− 1.140
1965 May	8.853	−	1.300	0.480	34.93	71.7	52.7	111.1	− 0.910
1965 May	10.866	−	1.310	0.490	35.42	57.6	38.3	112.0	− 0.900
1965 May	10.978	−	1.320	0.480	35.44	96.8	77.5	112.1	− 0.970
1965 May	11.863	−	1.330	0.490	35.65	47.1	27.7	112.5	− 0.900
1965 May	12.003	−	1.350	0.480	35.68	96.2	76.8	112.6	− 0.990
1965 May	12.867	−	1.320	0.500	35.87	39.2	19.7	112.9	− 0.860
1965 May	13.866	−	1.360	0.510	36.09	29.4	9.8	113.4	− 0.900
1965 May	19.872	−	1.360	0.520	37.23	335.1	314.9	116.2	− 0.870
1965 May	21.883	−	1.340	0.510	37.55	320.0	299.6	117.1	− 0.840
1965 May	23.902	−	1.300	0.470	37.86	307.7	287.1	118.1	− 0.870
1965 May	23.976	−	1.390	0.500	37.87	333.7	313.1	118.1	− 0.830
1965 June	17.897	−	1.280	0.480	39.80	67.1	45.8	129.9	− 0.840
1965 July	9.878	−	1.460	0.560	39.61	207.7	187.0	140.6	− 0.850
1965 July	10.882	−	1.470	0.590	39.57	199.4	178.7	141.1	− 0.870

TABLE VII

Boyden data

Date	UT	V	B−V	U−B	α	ω	ω*	L_S	V_1
1963 May	28.734	−	1.310	0.510	37.46	39.4	19.2	100.2	−0.830
1963 May	28.815	−	1.280	0.490	37.46	67.8	47.6	100.2	−0.950
1963 May	29.784	−	1.340	0.540	37.42	47.3	27.1	100.6	−0.900
1963 June	2.694	−	1.330	0.500	37.25	337.2	317.0	102.4	−0.880
1963 June	2.796	−	1.370	0.610	37.24	13.0	352.8	102.4	−0.830
1963 June	3.737	−	1.370	0.500	37.19	342.7	322.5	102.9	−0.890
1963 June	8.760	−	1.260	0.460	36.91	302.4	282.3	105.1	−0.850
1963 June	9.740	−	1.200	0.440	36.86	285.8	265.8	105.6	−0.820
1963 June	14.740	−	1.340	0.550	36.51	237.4	217.5	107.8	−0.830
1963 June	15.732	−	1.350	0.510	36.44	224.9	205.0	108.3	−0.900
1963 June	23.791	−	1.320	0.590	35.79	167.9	148.4	112.0	−0.910
1963 June	23.803	−	1.270	0.610	35.79	172.1	152.6	112.0	−0.860
1963 June	24.707	−	1.360	0.510	35.71	128.8	109.3	112.4	−1.000
1963 June	24.801	−	1.290	0.640	35.70	161.7	142.2	112.4	−0.900
1965 April	12.713	−	1.250	0.560	24.84	261.4	248.0	99.3	−1.050
1965 April	12.805	−	1.240	0.420	24.89	293.7	280.3	99.3	−1.060
1965 April	12.918	−	1.330	0.480	24.95	333.4	319.9	99.4	−1.040
1965 April	27.707	−	1.380	0.520	31.53	122.6	105.5	106.0	−1.070
1965 April	27.801	−	1.410	0.560	31.57	155.5	138.4	106.1	−1.080

B and U magnitudes, not colors, as the reduction is all done in terms of magnitudes. The systematic error in a run should be on the order of S divided by the square root of the number of standard stars; this is generally less than 0.01 mag. The random error of one observation should be on the order of R times the air mass, or typically 0.02 or 0.03 mag. Table V gives the observations and parameters as in Tables II and III.

The Harvard-NASA photometry has been described by Irvine et al. (1968a, b). These data are so numerous that we can afford to reject observations affected by clouds or with R or S greater than 0.03 for B or V, or 0.05 for U. The remaining data are listed in Tables VI and VII. The B and U magnitudes have been converted to $B−V$ and $U−B$ colors for consistency with the other data. Note that only the magnitude at unit distance (V_1) is given, not V.

We may now hope that all the data are on the same photometric system. However, we must fear that substantial systematic errors remain, particularly in U and B. There is first of all the problem of correcting so red an object for atmospheric extinction. The ultraviolet extinction corrections applied to all the observations after 1959 may differ systematically from those used by Johnson, but this difference should not exceed about 0.03 mag. A more serious problem arises from the fact that Mars does not resemble a normal star or a black body in the $U−B$, $B−V$ diagram (Figure 1); it corresponds roughly to a late B star seen through about 1.5 mag. of interstellar reddening ($\sim$4.5 mag. of visual absorption). Relative to main-sequence stars of similar $B−V$ color, Mars appears to have an 'ultraviolet excess' of about 0.6 mag. Schmidt-Kaler (1961) has shown that the normal transformation of instrumental systems to UBV, which is largely based on unreddened stars, often gives incorrect values for

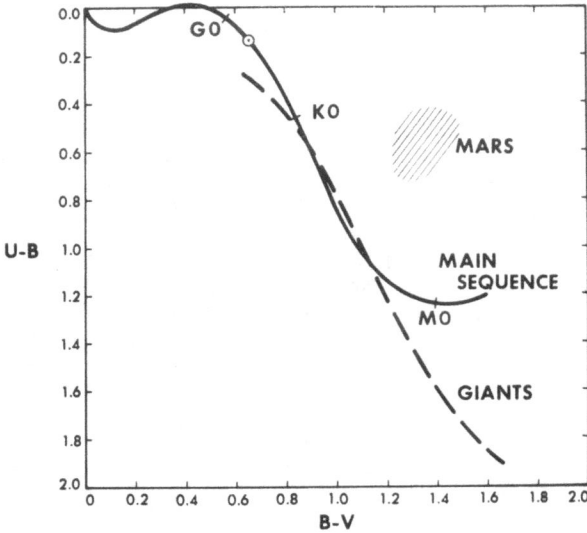

Fig. 1.   Location of Mars data in the $(U-B)$, $(B-V)$ plane.

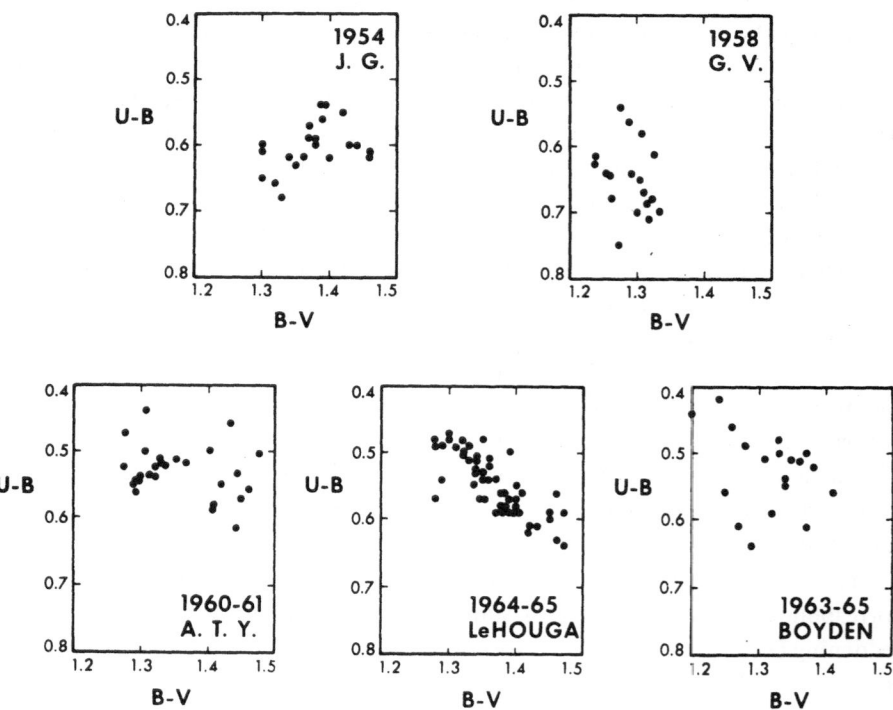

Fig. 2.   Distribution of different observers' Mars data in the color-color diagram.

heavily-reddened objects. Systematic errors in $(B-V)$ exceeding a tenth of a magnitude per magnitude of reddening can occur, so we may expect systematic errors in $(B-V)$ of 0.15 mag. for Mars. The transformation difficulties in the ultraviolet are even more severe. Basically, the UBV 'system' is not defined for objects like Mars.

In fact, Figure 2 suggests that systematic errors of the expected size do exist. We must therefore include a systematic correction term for each set of observations in fitting the data to Equation (2). This greatly weakens the determination of seasonal effects, but appears necessary because of the impossibility of placing Mars uniquely on a UBV-type system.

5. Analysis of the Data

The geometric data in Tables II, III, V, VI, and VII are approximately on the modern system defined in U.S. Naval Observatory Circular 98 (Meiller, 1964), which has been adopted by the Astronomical Ephemeris beginning with 1968. Because the new tabulation for past oppositions only goes to phase angles of about 25°, the old Ephemeris values for each opposition have been systematically corrected by the values given in Table VIII.

TABLE VIII

Corrections to be added to AE values
for reduction to new physical
ephemeris of Mars

Year	$\Delta\omega$	ΔL_S
1954	$-0°6$	$+3°8$
1958	-1.1	$+3.8$
1960–61	$+0.2$	$+3.8$
1963	0.0	$+3.8$
1965	-0.4	$+3.8$

To see whether the data are well enough distributed to separate the phase, rotational, and seasonal effects, I have plotted their distribution in ω^* and L_S (see Figure 3). Open symbols mark points with $i \leqslant 12°$. Although there are gaps in L_S, the distribution appears fairly satisfactory.

I therefore fit the data by least squares to Equation (2), determining the four additional zero-point adjustments required to reduce all series to the Le Houga system. (I adopt the Le Houga data as standard not only because they are most numerous, but also because they show the best-defined color-color relation in Figure 2.) The least-squares fits were performed in double precision, using the Los Alamos least-squares routine LEAST.

Initially, all observations were given equal weight, but some data sets are clearly better than others, so the standard error of one observation was computed from the initial residuals for each set and used to determine an average weight. The solution was repeated, weighing each set accordingly.

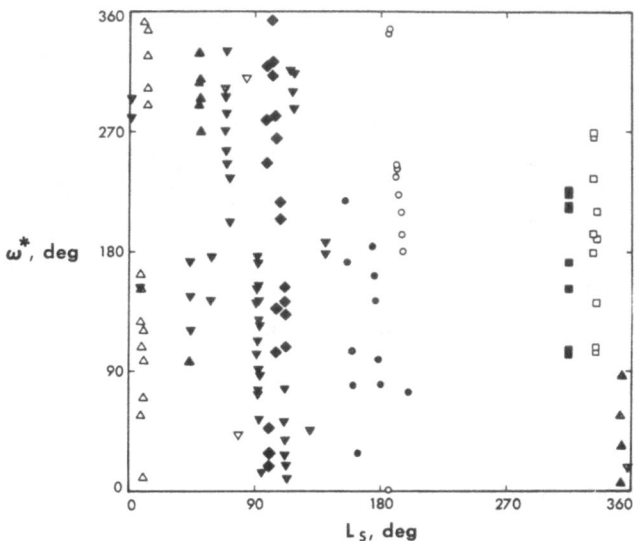

Fig. 3.   Distribution of Mars data with L_S and ω^*. Observations at phase angles less than 12° are indicated by open circles. Note that small-phase points occur at all central meridians, and that data are well-distributed in ω^* at each L_S covered.

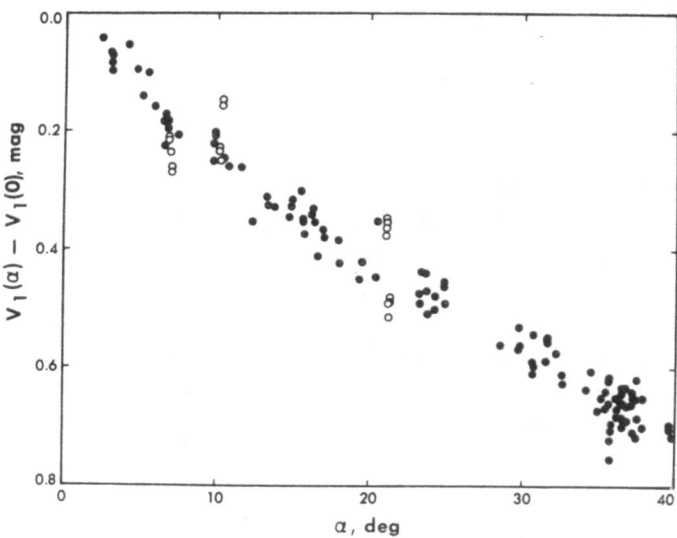

Fig. 4. V magnitudes of Mars, approximately corrected for rotation and seasonal effects, as a function of phase. The 1958 data are indicated by open circles; note their large scatter and systematic displacement from night to night.

The weights for most sets differ by a factor of 2 or less, but the 1958 data have an unusually low weight. Inspection of these residuals in the phase and L_S diagrams (Figures 4 and 5) shows large systematic shifts from night to night, as well as large scatter within each night. Because of these large systematic effects, I have tried solutions both with and without the 1958 data. Unfortunately, the 1958 data have a large influence on the seasonal effect in spite of their low weight. I have therefore tried solutions both with and without the seasonal effects. Thus, there are four solutions altogether: (I) all data, full solution; (II) all data, no seasonal effect; (III) no 1958 data, no seasonal effect; and (IV) full solution without 1958 data.

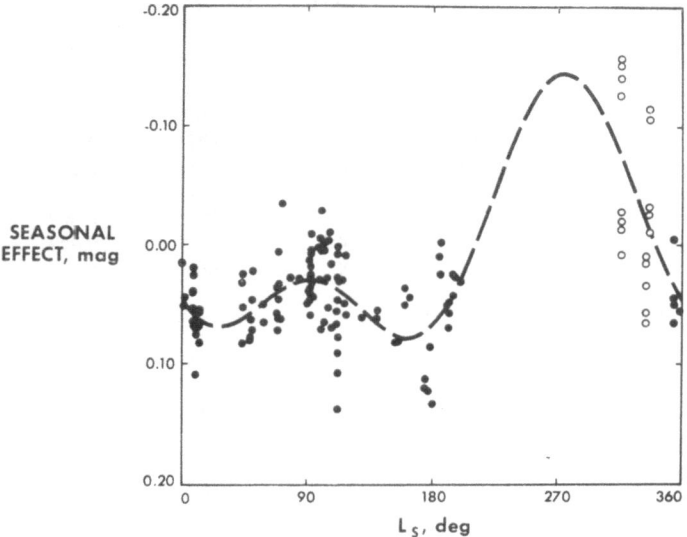

Fig. 5. V magnitudes of Mars, approximately corrected for rotation and phase effects, as a function of Martian season. The 1958 data are open circles. The least-squares seasonal effect is represented by a dashed line. The influence of the 1958 data on the determination of the seasonal effect is obvious.

Tables IX through XII give the results of the (weighted) solutions I through IV, respectively. Each solution has been done separately for the V, B, and U magnitudes and the $(B-V)$ and $(U-B)$ colors. The coefficients (and their corresponding terms) are listed in order from Equation (2). The zero-point term $Z_{(set)}$ is such that

$$m_{(set)} = m_{(Le\ Houga)} + Z_{(set)},\tag{3}$$

i.e., it appears on the right-hand side of Equation (2) as an additive constant. The standard errors of all coefficients and zero points are also given. The tabulated weights, *used* in each solution, are inversely proportional to the error variances (mean-square residuals) found from the corresponding unweighted solution; but the tabulated rms residuals are those from the weighted solution. Thus the weights are not quite proportional to the inverse squares of the rms residuals given below them in

TABLE IX

Solution I: full solution, all data

Quantity	V	B	U	$B-V$	$U-B$
Phase coefficients (and terms):					
μ_{-1} $(1/\alpha)$	-0.235	-0.218	-0.237	-0.087	$+0.012$
	± 0.105	± 0.166	± 0.248	± 0.130	± 0.126
$m_1(0)$ (1)	-1.508	-0.286	$+0.285$	$+1.254$	$+0.553$
	± 0.042	± 0.061	± 0.087	± 0.046	± 0.049
μ_1 (α)	$+0.0171$	$+0.0256$	$+0.0211$	$+0.0047$	-0.0023
	± 0.0031	± 0.0045	± 0.0065	± 0.0034	± 0.0036
μ_2 (α^2)	-0.000040	-0.000164	-0.000070	-0.000056	$+0.000046$
	± 0.000062	± 0.000089	± 0.000128	± 0.000067	± 0.000070
Rotational coefficients (and terms):					
a_1 $(\cos\omega^*)$	$+0.022$	-0.001	-0.010	-0.018	-0.015
	± 0.004	± 0.006	± 0.009	± 0.005	± 0.005
b_1 $(\sin\omega^*)$	-0.051	-0.037	-0.032	$+0.013$	$+0.002$
	± 0.004	± 0.006	± 0.009	± 0.004	± 0.004
a_2 $(\cos 2\,\omega^*)$	0.000	$+0.011$	$+0.018$	$+0.004$	$+0.016$
	± 0.004	± 0.006	± 0.008	± 0.004	± 0.004
b_2 $(\sin 2\,\omega^*)$	$+0.017$	-0.008	-0.020	-0.022	-0.013
	± 0.004	± 0.006	± 0.009	± 0.004	± 0.004
Seasonal coefficients (and terms)					
c_1 $(\cos L_S)$	-0.010	-0.033	-0.014	-0.015	$+0.011$
	± 0.021	± 0.030	± 0.042	± 0.022	± 0.023
d_1 $(\sin L_S)$	$+0.085$	$+0.128$	$+0.163$	$+0.078$	$+0.038$
	± 0.026	± 0.036	± 0.048	± 0.020	± 0.027
c_2 $(\cos 2 L_S)$	$+0.056$	$+0.090$	$+0.111$	$+0.050$	$+0.025$
	± 0.018	± 0.026	± 0.036	± 0.017	± 0.019
d_2 $(\sin 2 L_S)$	$+0.006$	$+0.041$	$+0.077$	$+0.035$	$+0.035$
	± 0.011	± 0.015	± 0.022	± 0.011	± 0.012
Zero points for:					
Flagstaff, 1954	$+0.001$	$+0.007$	$+0.075$	$+0.015$	$+0.054$
	± 0.037	± 0.051	± 0.071	± 0.037	± 0.040
Flagstaff, 1958	-0.117	-0.091	$+0.053$	$+0.051$	$+0.144$
	± 0.034	± 0.049	± 0.069	± 0.032	± 0.036
Agassiz Station, 1960–61	-0.039	-0.027	-0.088	-0.002	-0.053
	± 0.010	± 0.017	± 0.025	± 0.013	± 0.013
Le Houga, 1964–65	(0.000)	(0.000)	(0.000)	(0.000)	(0.000)
Boyden, 1963–65	$+0.014$	-0.040	-0.040	-0.051	$+0.001$
	± 0.012	± 0.014	± 0.024	± 0.013	± 0.013
Weights used (and rms residuals):					
Flagstaff, 1954	1.143	1.338	1.871	1.832	1.024
	(0.034)	(0.042)	(0.053)	(0.024)	(0.044)
Flagstaff, 1958	0.360	0.441	0.596	1.821	0.788
	(0.066)	(0.084)	(0.099)	(0.022)	(0.047)
Agassiz Station, 1960–61	2.213	1.149	1.020	0.883	0.861
	(0.023)	(0.046)	(0.071)	(0.043)	(0.045)
Le Houga, 1964–65	2.578	1.854	1.710	1.363	2.757
	(0.021)	(0.033)	(0.050)	(0.029)	(0.019)
Boyden, 1963–65	0.745	1.067	0.691	0.535	0.589
	(0.044)	(0.047)	(0.086)	(0.051)	(0.050)

TABLE X

Solution II: all data, no seasonal effect

Quantity		V	B	U	B − V	U − B
Phase coefficients (and terms):						
μ_{-1}	$(1/\alpha)$	− 0.282	− 0.249	− 0.221	− 0.124	+ 0.084
		± 0.097	± 0.159	± 0.250	± 0.125	± 0.138
$m_1(0)$	(1)	− 1.436	− 0.206	+ 0.360	+ 1.308	+ 0.543
		± 0.033	± 0.053	± 0.079	± 0.038	± 0.045
μ_1	(α)	+ 0.0135	+ 0.0221	+ 0.0200	+ 0.0029	0.0000
		± 0.0024	± 0.0037	± 0.0056	± 0.0027	± 0.0031
μ_2	(α^2)	+ 0.000041	− 0.000080	− 0.000037	− 0.000021	− 0.000001
		± 0.000044	± 0.000069	± 0.000105	± 0.000050	± 0.000058
Rotational coefficients (and terms):						
a_1	$(\cos\omega^*)$	+ 0.020	− 0.006	− 0.016	− 0.023	− 0.018
		± 0.004	± 0.007	± 0.010	± 0.005	± 0.006
b_1	$(\sin\omega^*)$	− 0.051	− 0.042	− 0.042	+ 0.008	− 0.005
		± 0.004	± 0.006	± 0.009	± 0.005	± 0.005
a_2	$(\cos 2\omega^*)$	0.000	+ 0.012	+ 0.016	+ 0.004	+ 0.014
		± 0.004	± 0.006	± 0.009	± 0.004	± 0.005
b_2	$(\sin 2\omega^*)$	+ 0.015	− 0.013	− 0.031	− 0.028	− 0.017
		± 0.004	± 0.006	± 0.010	± 0.005	± 0.005
Zero points for						
Flagstaff, 1954		+ 0.020	+ 0.066	+ 0.119	+ 0.039	+ 0.049
		± 0.010	± 0.014	± 0.018	± 0.009	± 0.011
Flagstaff, 1958		− 0.196	− 0.244	− 0.147	− 0.055	+ 0.090
		± 0.016	± 0.022	± 0.031	± 0.012	± 0.014
Agassiz Station, 1960–61		− 0.028	− 0.001	− 0.024	+ 0.017	− 0.018
		± 0.007	± 0.012	± 0.019	± 0.010	± 0.011
Le Houga, 1964–65		(0.000)	(0.000)	(0.000)	(0.000)	(0.000)
Boyden, 1963–65		+ 0.003	− 0.068	− 0.090	− 0.070	− 0.021
		± 0.011	± 0.013	± 0.024	± 0.013	± 0.013
Weights used (and rms residuals):						
Flagstaff, 1954		0.867	1.161	1.987	1.968	1.059
		(0.041)	(0.049)	(0.057)	(0.027)	(0.047)
Flagstaff, 1958		0.374	0.398	0.483	0.961	0.704
		(0.069)	(0.104)	(0.135)	(0.039)	(0.057)
Agassiz Station, 1960–61		2.339	1.415	1.291	0.934	1.031
		(0.023)	(0.043)	(0.064)	(0.042)	(0.042)
Le Houga, 1964–65		2.276	1.623	1.341	1.405	1.658
		(0.023)	(0.037)	(0.066)	(0.030)	(0.029)
Boyden, 1963–65		0.804	1.210	0.845	0.550	0.703
		(0.043)	(0.047)	(0.088)	(0.054)	(0.051)

TABLE XI

Solution III: all data but 1958; no seasonal effect

Quantity		V	B	U	$B - V$	$U - B$
Phase coefficients (and terms):						
μ_{-1}	$(1/\alpha)$	-0.218	-0.083	$+0.180$	$+0.091$	$+0.310$
		± 0.099	± 0.153	± 0.236	± 0.126	± 0.132
$m_1(0)$	(1)	-1.469	-0.280	$+0.177$	$+1.214$	$+0.431$
		± 0.035	± 0.053	± 0.080	± 0.042	± 0.046
μ_1	(α)	$+0.0161$	$+0.0278$	$+0.0339$	$+0.0100$	$+0.0084$
		± 0.0024	± 0.0037	± 0.0057	± 0.0030	± 0.0033
μ_2	(α^2)	-0.000006	-0.000185	-0.000294	-0.000149	-0.000154
		± 0.000045	± 0.000068	± 0.000106	± 0.000055	± 0.000060
Rotational coefficients (and terms):						
a_1	$(\cos\omega^*)$	$+0.018$	-0.010	-0.027	-0.027	-0.023
		± 0.004	± 0.006	± 0.010	± 0.005	± 0.006
b_1	$(\sin\omega^*)$	-0.052	-0.042	-0.043	$+0.008$	-0.001
		± 0.004	± 0.006	± 0.009	± 0.005	± 0.005
a_2	$(\cos 2\omega^*)$	$+0.002$	$+0.017$	$+0.033$	$+0.010$	$+0.021$
		± 0.004	± 0.006	± 0.009	± 0.005	± 0.005
b_2	$(\sin 2\omega^*)$	$+0.017$	-0.010	-0.026	-0.026	-0.018
		± 0.004	± 0.006	± 0.010	± 0.025	± 0.005
Zero points for:						
Flagstaff, 1954		$+0.022$	$+0.068$	$+0.127$	$+0.044$	$+0.058$
		± 0.010	± 0.013	± 0.020	± 0.009	± 0.013
Agassiz Station, 1960–61		-0.023	$+0.008$	-0.002	$+0.027$	-0.005
		± 0.007	± 0.011	± 0.017	± 0.009	± 0.010
Le Houga, 1964–65		(0.000)	(0.000)	(0.000)	(0.000)	(0.000)
Boyden, 1963–65		$+0.003$	-0.067	-0.089	-0.069	-0.020
		± 0.010	± 0.013	± 0.022	± 0.013	± 0.012
Weights used (and rms residuals):						
Flagstaff, 1954		0.711	0.766	0.861	1.314	0.637
		(0.039)	(0.050)	(0.077)	(0.033)	(0.060)
Agassiz Station, 1960–61		1.340	1.025	1.249	1.052	1.120
		(0.025)	(0.044)	(0.061)	(0.039)	(0.038)
Le Houga, 1964–65		1.839	1.422	1.318	1.518	1.851
		(0.022)	(0.035)	(0.057)	(0.027)	(0.025)
Boyden, 1963–65		0.558	0.885	0.720	0.528	0.709
		(0.043)	(0.046)	(0.082)	(0.054)	(0.048)

TABLE XII

Solution IV: all data but 1958; full solution

Quantity		V	B	U	$B-V$	$U-B$
Phase coefficients (and terms):						
μ_1	$(1/\alpha)$	-0.208	-0.064	$+0.199$	$+0.147$	$+0.201$
		±0.107	±0.162	±0.232	±0.133	±0.124
$m_1(0)$	(1)	-1.522	-0.319	$+0.172$	$+1.198$	$+0.509$
		±0.041	±0.060	±0.089	±0.048	±0.049
μ_1	(α)	$+0.0184$	$+0.0296$	$+0.0342$	$+0.0111$	$+0.0040$
		±0.0031	±0.0046	±0.0067	±0.0037	±0.0036
μ_2	(α^2)	-0.000066	-0.000226	-0.000287	-0.000159	-0.000061
		±0.000062	±0.000089	±0.000133	±0.000073	±0.000071
Rotational coefficients (and terms):						
a_1	$(\cos\omega^*)$	$+0.021$	-0.004	-0.018	-0.023	-0.019
		±0.004	±0.007	±0.010	±0.005	±0.005
b_1	$(\sin\omega^*)$	-0.052	-0.036	-0.029	$+0.015$	$+0.006$
		±0.004	±0.006	±0.009	±0.005	±0.004
a_2	$(\cos2\omega^*)$	$+0.001$	$+0.016$	$+0.035$	$+0.011$	$+0.021$
		±0.004	±0.005	±0.008	±0.005	±0.004
b_2	$(\sin2\omega^*)$	$+0.019$	-0.006	-0.020	-0.024	-0.016
		±0.004	±0.006	±0.009	±0.005	±0.004
Seasonal coefficients (and terms):						
c_1	$(\cos L_S)$	-0.015	-0.033	-0.028	-0.016	-0.004
		±0.021	±0.030	±0.045	±0.024	±0.023
d_1	$(\sin L_S)$	$+0.088$	$+0.069$	$+0.016$	-0.009	-0.043
		±0.028	±0.043	±0.061	±0.032	±0.037
c_2	$(\cos2L_S)$	$+0.059$	$+0.055$	$+0.031$	$+0.001$	-0.013
		±0.019	±0.028	±0.041	±0.023	±0.022
d_2	$(\sin2L_S)$	$+0.008$	$+0.040$	$+0.079$	$+0.032$	$+0.040$
		±0.010	±0.016	±0.023	±0.012	±0.012
Zero points for:						
Flagstaff, 1954		-0.007	$+0.012$	$+0.068$	$+0.022$	$+0.034$
		±0.036	±0.052	±0.078	±0.041	±0.041
Agassiz Station, 1960–61		-0.036	-0.012	-0.043	$+0.022$	-0.032
		±0.010	±0.016	±0.023	±0.013	±0.012
Le Houga, 1964–65		(0.000)	(0.000)	(0.000)	(0.000)	(0.000)
Boyden, 1963–65		$+0.015$	-0.047	-0.057	-0.061	-0.006
		±0.012	±0.014	±0.023	±0.014	±0.013
Weights used (and rms residuals):						
Flagstaff, 1954		0.911	0.826	0.830	1.278	0.680
		(0.033)	(0.046)	(0.071)	(0.031)	(0.051)
Agassiz Station, 1960–61		1.479	1.105	1.416	1.318	1.044
		(0.023)	(0.040)	(0.052)	(0.032)	(0.036)
Le Houga, 1964–65		1.805	1.476	1.506	1.447	2.638
		(0.021)	(0.033)	(0.048)	(0.028)	(0.018)
Boyden, 1963–65		0.503	0.842	0.658	0.544	0.605
		(0.044)	(0.045)	(0.078)	(0.050)	(0.046)

parentheses. The zero point for Le Houga is defined as zero, as it is the reference system.

Although I have shown in Section 3 that the terms used should represent the rotational effect quite accurately, there is no reason to believe that the terms used to represent the phase and seasonal variations are an accurate representation of these effects; they are merely analytically convenient interpolation formulae. Each effect can be truly displayed by plotting the sum of the computed least-squares value and the residual for each observational datum, however; this sum is simply the observed value *minus* the analytical representation of all other effects. Thus, we need only

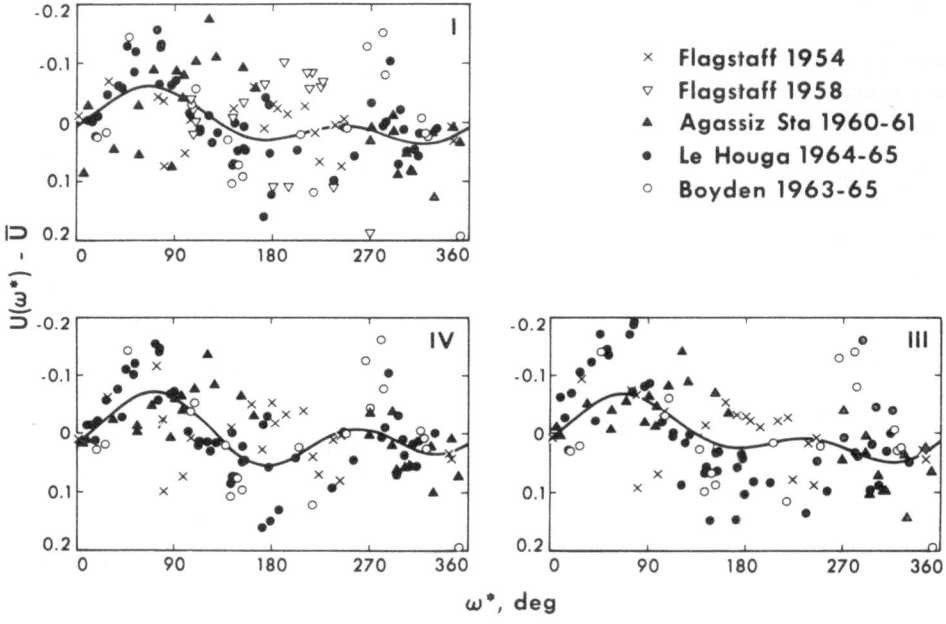

Fig. 6. Ultraviolet rotational effect, from Solutions I–IV (see Tables IX–XII). The formats of Figures 6–14 are similar: the upper two plots (Solutions I and II) contain the 1958 data, which the lower two (Solutions III and IV) exclude; and the left two solutions (I and IV) include the seasonal terms, which are omitted from the right two (Solutions II and III). In the present case, the scatter for Solution II was too large to fit in the space available, so it has been omitted.

require that the interpolation formula represents the bulk of each effect, so that the remainder does not introduce appreciable scatter or systematic error into the graphs for the other effects.

The graphs of the rotational, phase, and seasonal effects in U, B, and V (constructed as described above) are given in Figures 6 through 14. The least-squares representation for each effect is drawn in for comparison. As we should expect, the least-squares formula fits the rotational effect very well, but the reader may choose to draw other lines through the points that give the phase and seasonal effects.

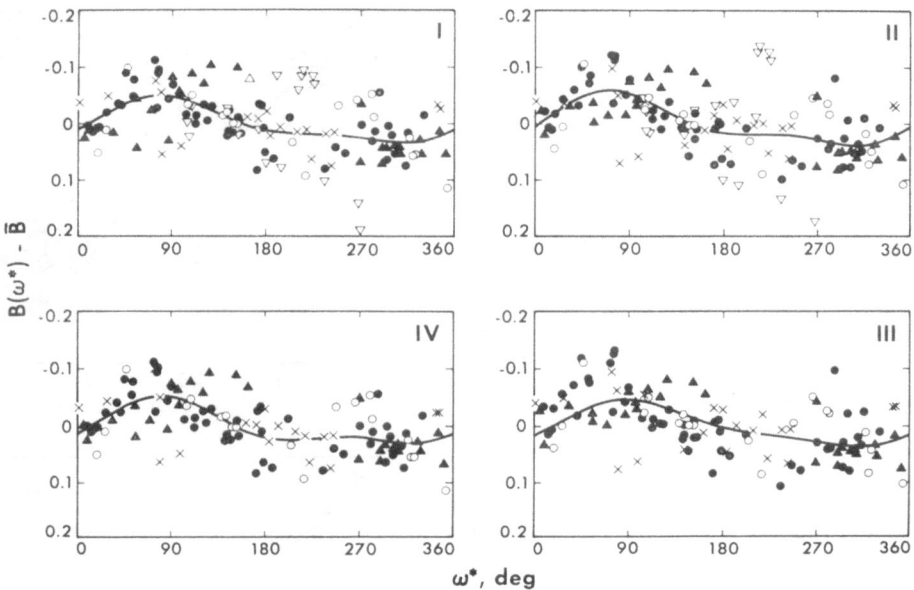

Fig. 7.   Blue rotational effect. (See legend and caption of Figure 6 for explanation.)

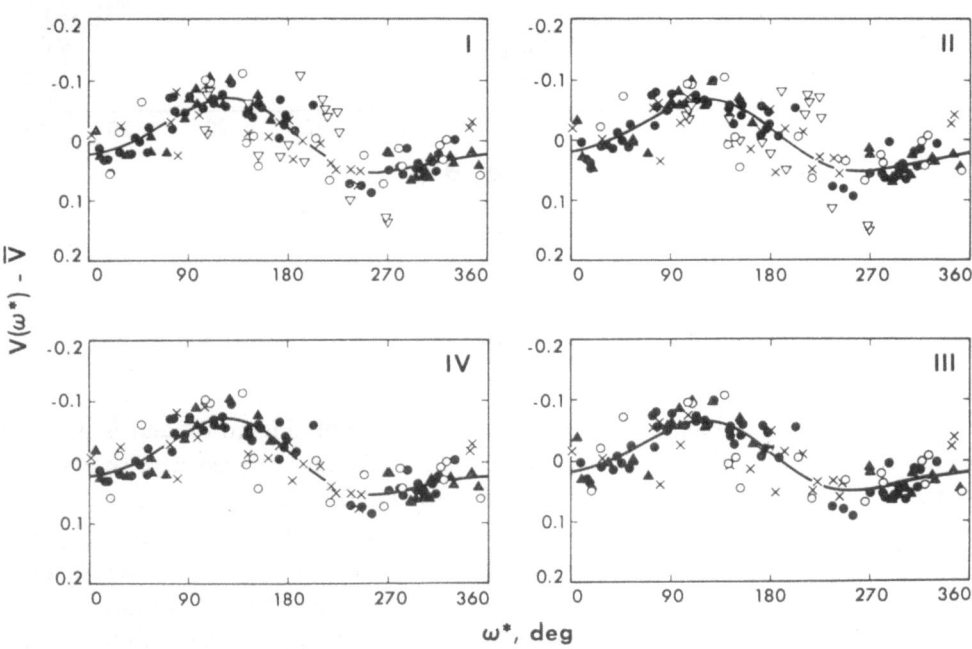

Fig. 8.   Visual rotational effect. (See Figure 6.)

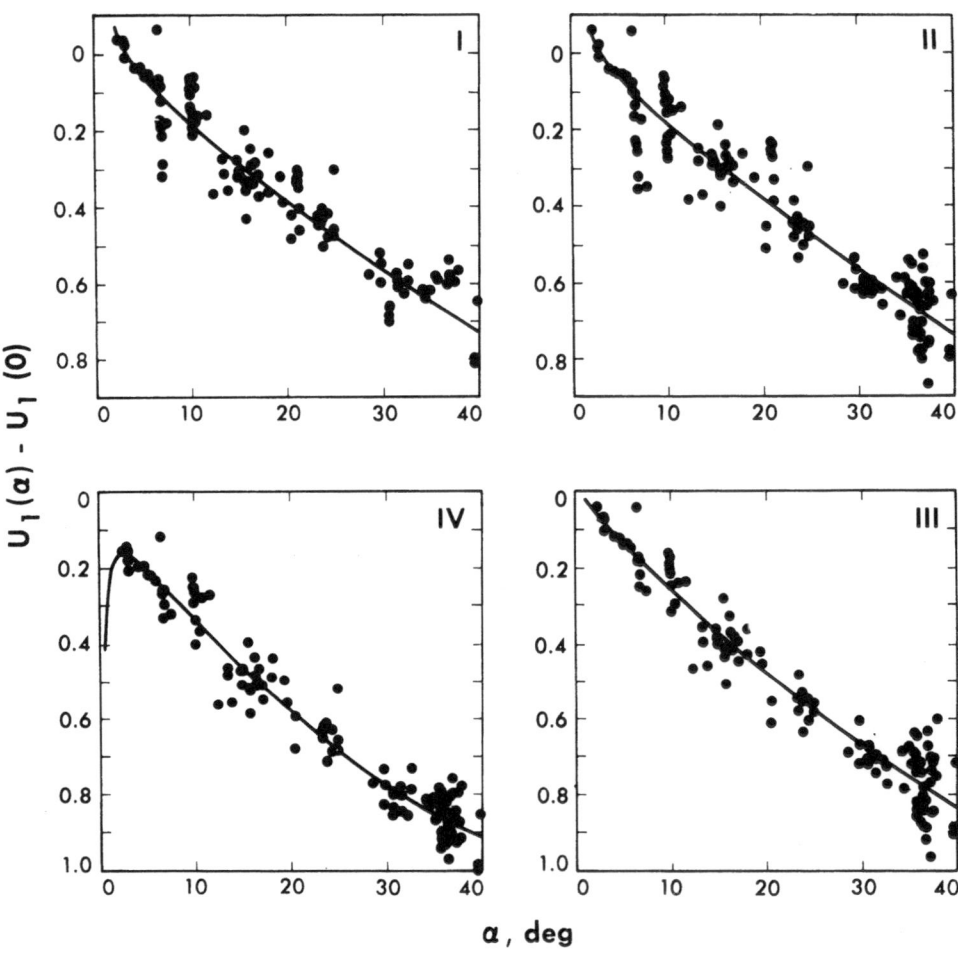

Fig. 9.   Ultraviolet phase effect. (See Figure 6.)

6. Discussion

6.1. THE ROTATIONAL EFFECT

The rotation curve is very well determined in visual light (Figure 8), and is not in doubt by as much as 0.01 mag. The large volume of data analyzed here shows an appreciable rotational effect even in blue (Figure 7) and ultraviolet (Figure 6) light; the progressive changes in the curves with changing wavelength confirm the reality of these results. The roughly constant longitude of the maximum suggests that the surface features retain appreciable contrast even in the ultraviolet; the shift toward lower longitudes at shorter wavelengths may be due to the increasing influence of the North Polar Cap, which is displaced toward longitude 30°.

6.2. THE SEASONAL EFFECT

The reality of the seasonal effect may be judged by comparing Solutions II and III (no seasonal effect) with Solutions I and IV, respectively. The effect is clearly present in the ultraviolet (Figure 12), and probably present in the blue data (Figure 13), whether the 1958 observations are included or not. Furthermore, the similar shape in all wavelengths, and the steady increase in amplitude toward the ultraviolet, strongly suggest a real effect.

However, the sign of the effect is puzzling if it represents real seasonal changes on Mars. We would expect the planet to appear brightest when the polar caps appear largest, as the caps are the brightest part of the disc, especially in the ultraviolet. Unfortunately, the observed effect is brightest after the Martian solstices, when the caps have disappeared, and faintest after the equinoxes, when the visible cap is large. Two possible explanations are that (a) we are observing seasonal changes of 10 to 20%

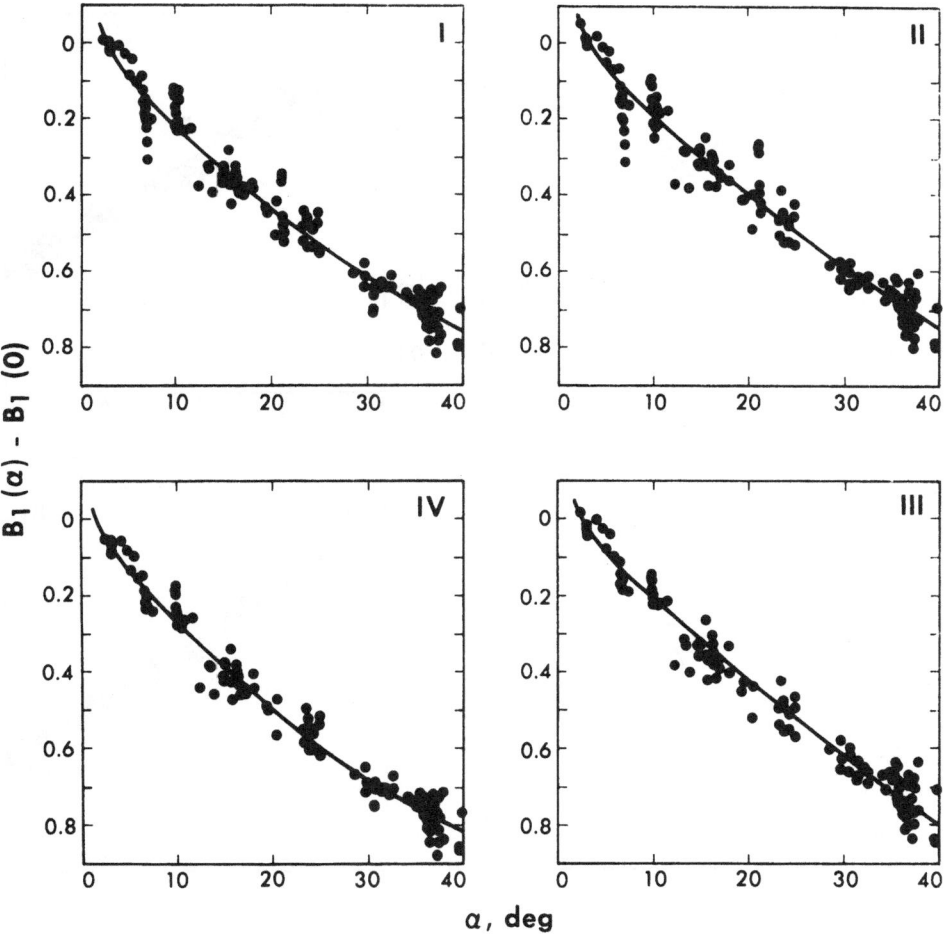

Fig. 10. Blue phase effects. (See Figure 6.)

in the brightness of the bright areas, which contribute most of the observed light in any case, or that (b) the light contributed by Martian aerosols increases after the solstices, due either to an increase in atmospheric pressure (if the caps are CO_2) or to an increase in wind speed at the peak of the annual temperature cycle. The 1971 dust storm, which peaked near $L_S = 280°$, supports this interpretation.

Alternatively, one may imagine that the effect, though real, is in the data but not on Mars – e.g., some systematic, time-dependent error in the Le Houga photometer, due perhaps to secular or (terrestrial) seasonal effects. However, inspection of the zero-point terms in the tables shows no regular, progressive change with wavelength, so the regular progression with wavelength argues against such an explanation.

The maximum at $L_S = 300°$ (near perihelion) is alarming. However, it appears whether the 1958 data are included or not (compare Solution I with Solution IV, or II with III). Furthermore, the inclusion of the seasonal effect reduces not only the

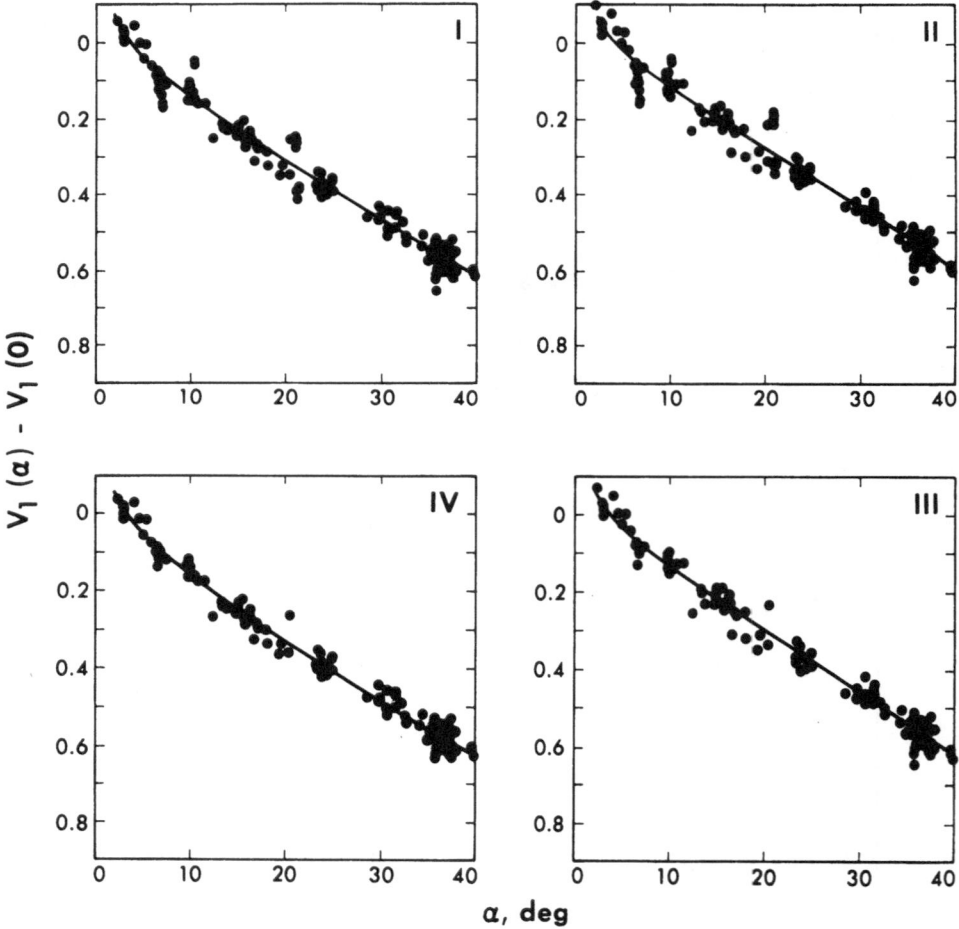

Fig. 11.   Visual phase effect. (See Figure 6.)

scatter of the 1958 data, but also the large zero-point terms. This is indirect evidence that the maximum, though uncertain in amplitude, is probably real. The actual shape and size of the maximum cannot be determined until a homogeneous set of good observations covering the gap in L_S (200°–350°) is available. However, the shape of the seasonal effect in this gap is not needed for the correction of the existing data, so it should not greatly influence the determination of the phase and rotational effects in the present work.

6.3. Phase effect

All the curves show a sharp brightening at small phase angles, except for Solutions III and IV in the ultraviolet. The shapes of the phase curves are sensitive to the inclusion of the seasonal effect: they bend up more at large phase angles (30°–40°) in the solutions that include the seasonal effect (I and IV) than in those that do not (II and III).

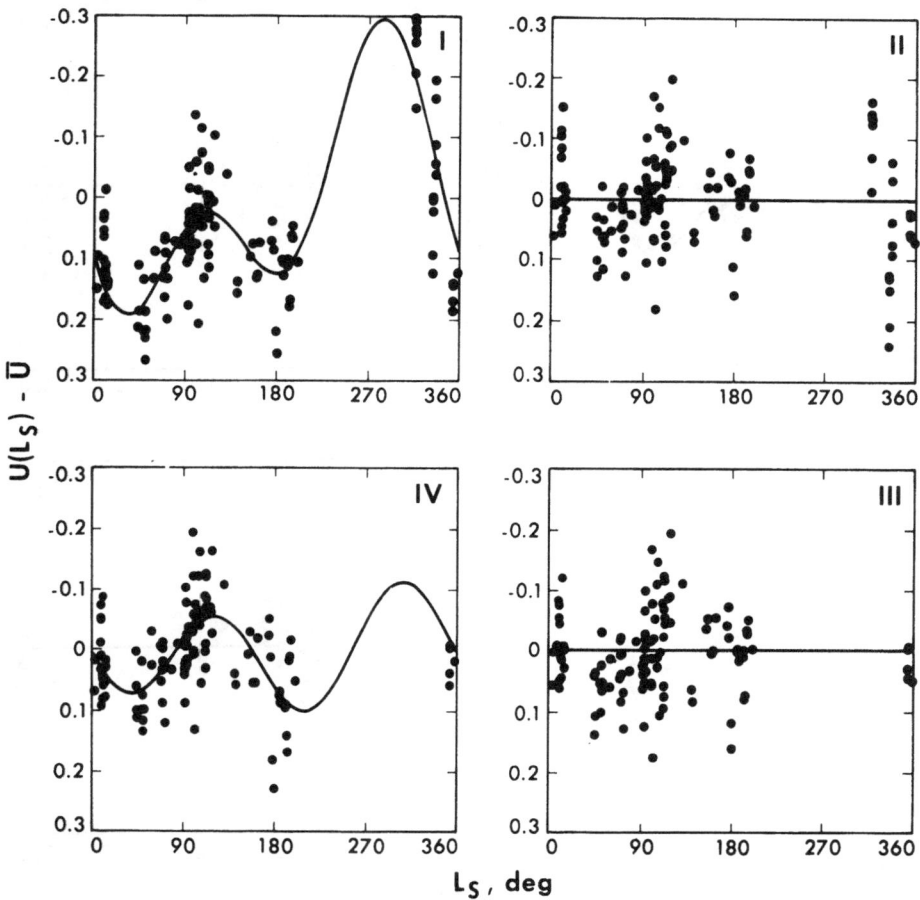

Fig. 12.   Ultraviolet seasonal effect. (See Figure 6.)

I have argued above that the seasonal effects are probably real (whether Martian or terrestrial), so I favor the more strongly-curved Solutions I and IV.

Because of the uncertainty in the seasonal effect, and the mutual dependence of the phase and seasonal effects, the uncertainty in the phase effects remains fairly large. However, the uncertainty in the phase effect is not as large as would appear from the standard errors of the coefficients, because the terms used are not linearly independent; thus a small change of one coefficient would be largely compensated by changes in the others. Only in the rotational effect, where the terms are linearly independent and the observations well distributed, do the tabulated errors really indicate the uncertainty of the curve. A realistic estimate of the uncertainty near the middle ($\alpha = 20°$) of the phase curve would be 0.02 mag, in V, 0.03 mag. in B, and 0.05 mag. in U; at the ends ($\alpha \approx 0°$ or $40°$) the uncertainty is about twice as big. Thus the slopes near $\alpha = 20°$ are still uncertain by about 0.001 mag deg^{-1} in V, and more in B and U.

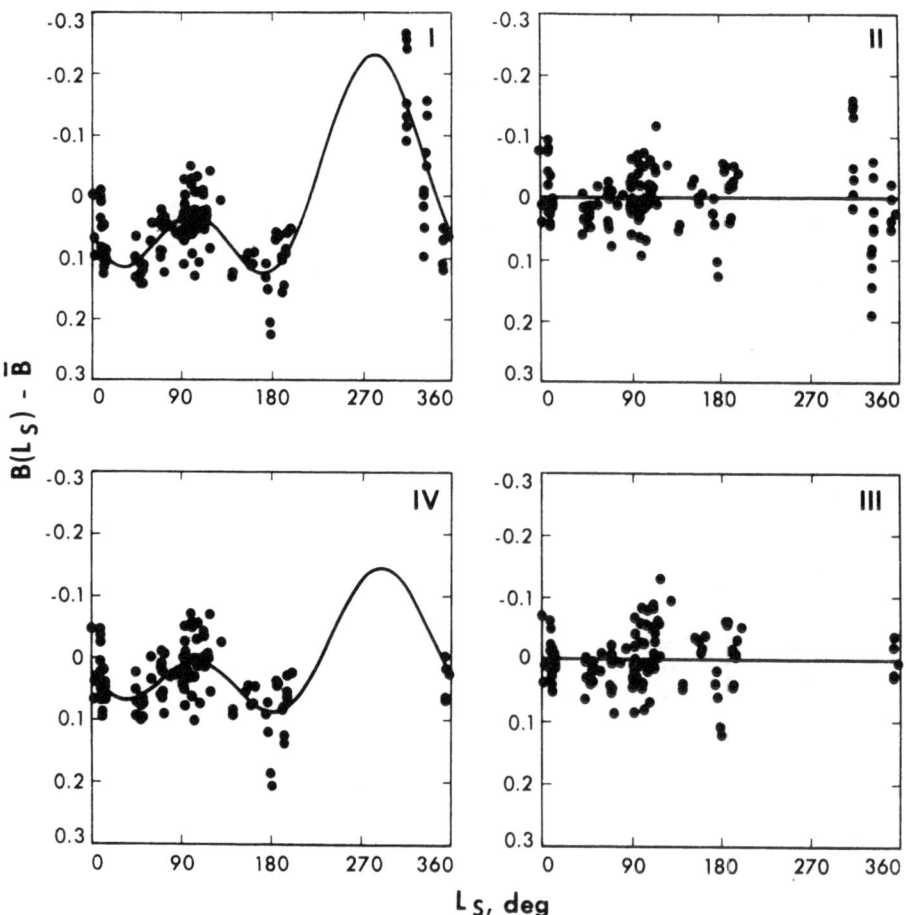

Fig. 13.   Blue seasonal effect. (See Figure 6.)

6.4. OBSERVATIONAL ERRORS AND MARTIAN VARIABILITY

The striking differences between the rms residuals of different series of observations raise a number of questions. The first is whether the residual scatter represents observational errors or real fluctuations in the brightness of Mars.

The figures show that the scatter is not localized to a particular range of central meridian or phase angle. Aside from the 1958 data, the scatter does not appear concentrated to any range of L_S; since the scatter is quite small for the data at slightly larger L_S, a Martian seasonal effect seems unlikely. Furthermore, there is considerable overlap in L_S between series with markedly different scatter. Thus, one's initial impression is that the scatter is not Martian and must therefore be terrestrial.

We can go further and ask how much of the scatter *could* be due to Mars, by removing the estimated errors of observation. This is a fairly straight forward process for

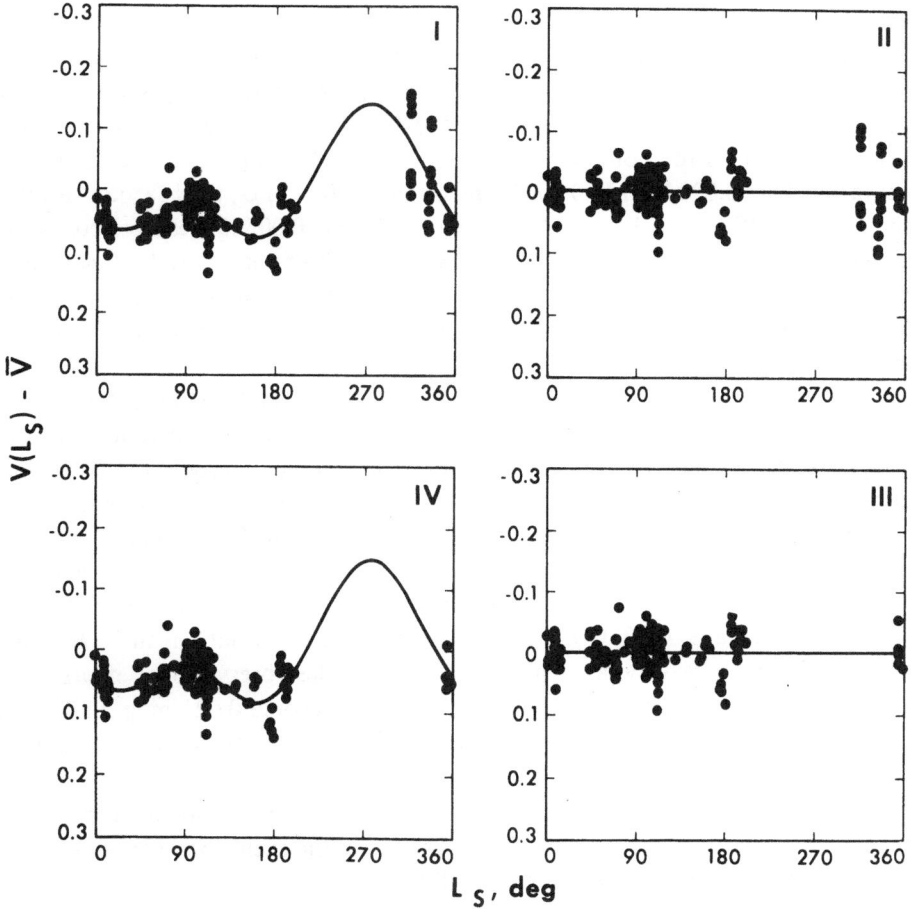

Fig. 14.   Visual seasonal effect. (See Figure 6.)

the later data, which have all been reduced with the same computer program. For these series we can estimate errors in two ways: from the internal scatter R of the extinction solution, and from the external scatter S of the fit to UBV standard stars. The latter method may give a more realistic assessment of night-to-night errors, but also includes the errors in the UBV standards, which are comparable to the Martian residuals. It is also complicated by the averaging of several observations for each star whereas the extinction residuals, like the residuals in the tables, refer to single observations. I therefore choose to use only the extinction residuals; this should lead to an upper limit for the intrinsic variability of Mars, because the random errors of transformation to the UBV system are not considered.

In writing the extinction program (Young and Irvine, 1967), I assumed that the observational error is proportional to air mass; the errors given are reduced to unit air mass. Although a steeper law is probably more accurate (Young, 1969), the errors at the mean air masses used can be correctly reproduced by using the relation assumed in the program. Thus, the rms error of observations is the product of the zenith error with the mean air mass of Mars.

My 1960 observations have a mean air mass of 1.452, so the expected rms errors are 0.028 in V, 0.029 in B, and 0.023 in U (see Table IV). The corresponding values for the 1961 observations (mean $\sec Z = 1.542$) are 0.009, 0.011, and 0.026. The mean-square observational errors for both sets combined (weighted according to the number of observations) are 0.00050 in V, 0.00059 in B, and 0.00064 in U; the Martian variances from Table IX (Solution I) are 0.00053, 0.00212, and 0.00503, respectively; so the possible Martian contributions are 0.00003 in V, 0.00153 in B, and 0.00439 in U, corresponding to rms Martian variations of $\sigma_V = 0.005$, $\sigma_B = 0.039$, and $\sigma_U = 0.066$. (The value for σ_U is probably much too large, owing to the very approximate red-leak corrections.)

Similarly, one can estimate values from the Boyden and Le Houga data. As the exact air masses are not readily available, I have assumed the mean value (1.5) from the Agassiz Station data; this is probably an underestimate – especially for Boyden, because of the northern declination of Mars during these oppositions. For Le Houga, the expected rms observational errors in V, B, and U are 0.022, 0.026, and 0.032 respectively, which allow corresponding Martian variations of 0.000, 0.020, and 0.038 mag. For Boyden, the observational errors reduced to 1.5 air masses are 0.029, 0.031, and 0.032 which leave 0.033, 0.036, and 0.080 for Martian variations in V, B, and U. However, if a more realistic air mass of 2.0 is used, the observational errors rise to 0.019, 0.041, and 0.042, and the Martian variations become 0.021 mag. in V, 0.023 mag. in B, and 0.075 mag. in U; these figures agree better with the Agassiz Station and Le Houga results.

The low rms variation of the Le Houga data is particularly convincing, because they are the most numerous. Considering all three sets, and the exclusion of transformation errors, it seems safe to say that Mars actually varies randomly by no more than 0.01 in V, 0.02–0.03 in B, and 0.04 or so in U.

However, if we interpret the tabulated rms residuals as primarily due to observational errors, we have the remarkable conclusion that the best observations were made near sea level and the worst were made at the high-altitude observatories, which are generally regarded as having much better photometric conditions. In fact, both the Agassiz Station (except for U) and the Le Houga data compare favorably with best of the high-altitude data, which is all the more surprising because the former represent all-sky photometry, whereas the latter were made differentially.

The large errors in the Flagstaff magnitudes may be due in part to the large photomultiplier temperature variations that occur in Johnson's cold boxes if a heat-transfer liquid is not used (Young, 1963, 1966). Such liquids were used in the Le Houga and Boyden observations; the Agassiz Station data were taken at ambient temperature, a procedure which Stock has found to give very good results at Cerro Tololo. (The low quality of the Boyden data may be due to the difficulty of keeping trained observers at Boyden during the Harvard/NASA program.)

The Flagstaff *colors* are much better than the magnitudes, and generally better than the later colors, possibly because the Flagstaff B and U data were reduced as colors rather than magnitudes. Again, the low scatter of the Le Houga colors is surprising, as these data were reduced as magnitudes and differenced to obtain the colors.

The correlation of the residuals in adjacent bands can be estimated by comparing the rms residuals in the two magnitudes with the rms residual for the corresponding color. For example, the B and V residuals for 1958 are highly correlated, as the $(B-V)$ residuals are small; but the B and V residuals for Boyden are almost uncorrelated, as the $(B-V)$ residuals are larger than those in either magnitude alone. Such differences in correlation from one series to another are additional evidence that the residuals are largely observational error, because any Martian variations should be similarly correlated in all series.

7. Comparison with O'Leary

O'Leary (1967a, b) has also made UBV observations of Mars, at small phase angles near the 1967 opposition. His data can in principle be used to strengthen the determination of the phase effect at small phase angles, and to choose the most satisfactory of the Solutions I–IV discussed in the previous sections. Unfortunately his observations were all made differentially with respect to Spica, a close binary variable star with a most complex light curve; Shobbrock *et al.* (1969) have shown that one component is a β Canis Majoris star and the system is strongly affected by ellipticity effects. Thus Spica has variations, reflected in O'Leary's Mars data, both on a scale of a few hours and a few days, with a total variation of a tenth of a magnitude. These cannot be removed in detail because O'Leary does not give the individual observations. Finally, O'Leary's data were taken with non-standard filters and photomultiplier so that the transformation difficulties are certain to be more severe than those already discussed; indeed, O'Leary (1967a) says that his Kitt Peak data differ systematically from simultaneous Cerro Tololo observations by two tenths of a magnitude. At first sight it appears hopeless to salvage anything of value from these observations.

However, because O'Leary observed for several hours on each night, one can hope that the β-Cephei-like variations of Spica have been partially averaged out in his tabulated nightly means. These can then be corrected for the known orbital variations of Spica, because O'Leary indicates the phase at which he observed.

As a preliminary check on his corrections for the Martian rotational effect, I have compared his mean rotational light curve (Figure 12 of his thesis) with the V light curves derived in Solutions I–IV (see Figure 15). The agreement is reasonably good; in view of the complications involved, no conclusion can be drawn from O'Leary's slightly smaller amplitude, which could be due to a Martian phase effect, a systematic transformation error (wrong effective wavelength), residual Spica variations, or other circumstances.

From O'Leary's (1967a) phase data and the visual light curve of Spica (Shobbrock *et al.*, 1969), I have determined approximate systematic corrections to O'Leary's nightly means. These are +0.018 mag. for April 14, 18, and 22; +0.07 for April 15 and 23; +0.010 for April 16, 20, and 24; and zero for April 13, 17, and 21. I have rejected the data for two cloudy nights (April 10 and 12). The corrections should probably be larger in B and U, but Shobbrock *et al.* do not give information for these colors. I have not corrected O'Leary's data for the seasonal effect, because they cover only 5° of L_S very close to the maximum at $L_S \approx 120°$; even in U, the variation over this range is less than a hundredth of a magnitude.

The corrected nightly means are plotted in Figures 16–18. Only Solutions I and IV are shown for comparison, because these include the seasonal effect. In V (Figure 18), the two agree equally well with O'Leary's data; in B (Figure 17) Solution IV agrees a little better, but the difference is small. In U (Figure 16), Solution I is clearly better than IV, which bends the wrong way at small phase angles; Solution I gives the larger

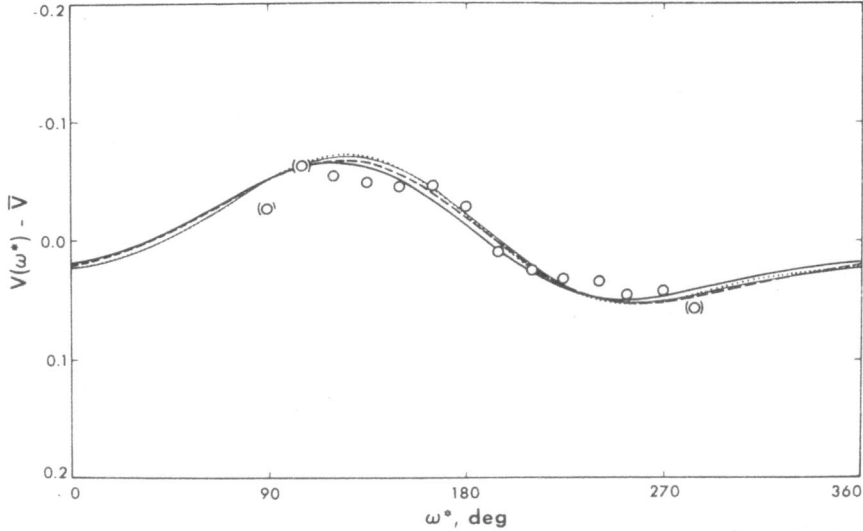

Fig. 15. Comparison of O'Leary's V light curve of Mars (circles) with the least-squares curves from Solutions I–IV. (Figure 8.)

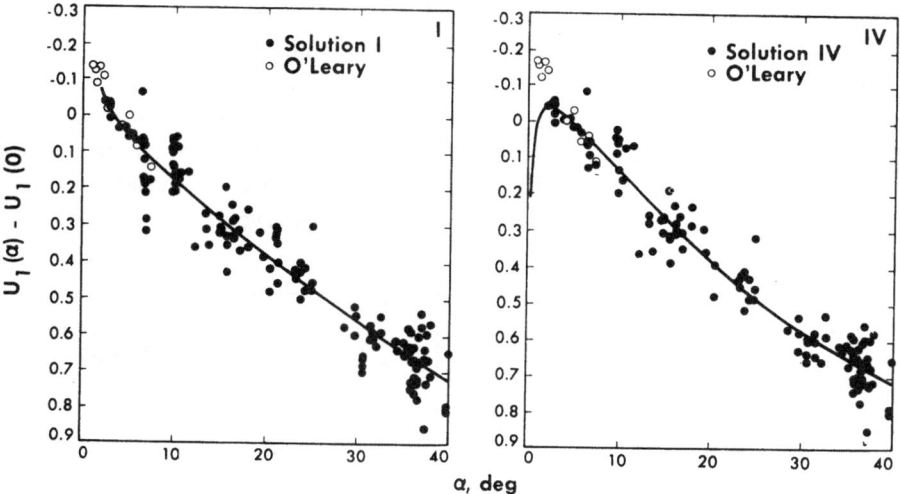

Fig. 16.   Comparison of O'Leary's U phase data (open circles) with those for Solutions I and IV (curves and filled circles). Solution I fits the (corrected) O'Leary data quite well.

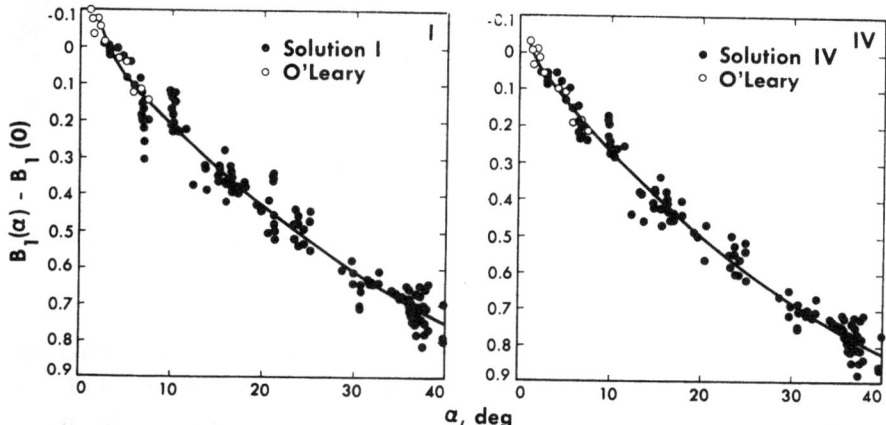

Fig. 17.   Comparison of blue phase curves (see Figure 16). Here Solution IV fits O'Leary's data slightly better than Solution I, but the difference is small.

seasonal effect, so O'Leary's data give indirect support for the larger seasonal effect found by including the 1958 data (Solution I). I conclude that Solution I is probably closest to the truth.

The agreement of the extrapolation of this solution to small phase angles with O'Leary's data is quite unexpected. Thus the analytic interpolation formulae represent the data well even for phase angles as small as 2°. Because the least-squares solution is based on observations down to $\alpha = 3°$, the overlap with O'Leary's data (which extend to $\alpha = 7.5°$) is good, and there is no ambiguity in fitting the two together; this represents a substantial improvement over the fitting done by O'Leary on the basis of a *linear* extrapolation from larger phase angles.

The smooth, continuous curvature of the phase curves, especially in B and U, disagrees with O'Leary's assertion of a *sudden* brightening at small (5–10°) phase angles. Indeed, if we regard the coefficient of the $1/\alpha$ term as representing the 'opposition effect', we see that it still amounts to a hundredth of a magnitude at $\alpha \approx 23°$. This also ignores the continued upward curvature of the phase relation at larger α, indicated by the negative α^2 terms. Thus, the name 'opposition effect' is misleading, and we should instead talk about the general *shape* of the phase curve.

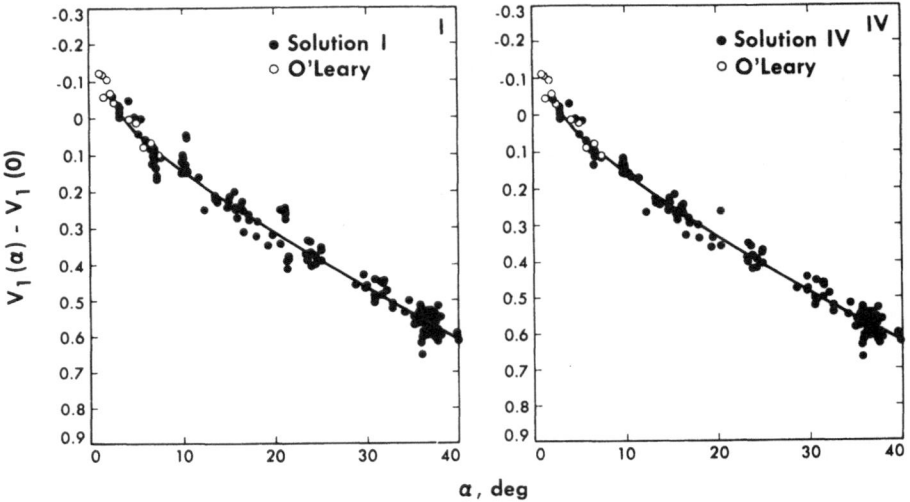

Fig. 18.   Comparison of visual phase curves (see Figure 16). Solutions I and IV are similar, and both fit O'Leary's data well.

Furthermore, the present analysis does not support O'Leary's conclusion that the 'opposition effect' (interpreted as the $1/\alpha$ term) increases steadily from V to U. Indeed, this term is practically the same in all three colors. Differences would show most plainly in the $(B-V)$ and $(U-B)$ colors (Figure 19); although the *solutions* indicate a slightly greater effect in B than in V or U, the *data points* (including O'Leary's) do not indicate any opposition effect in the colors at all. In fact, although the $(B-V)$ color shows the well-known reddening with phase, the $(U-B)$ color is practically constant over the entire range of phase.

The phase curves for the colors are difficult to understand. Atmospheric scattering should be more important, relative to the surface, at short wavelengths and large phase angles. Thus one would expect Mars to become *bluer* at large phase – especially in $(U-B)$. Instead it becomes redder, but only in $(B-V)$. However, as the present data show that the marked wavelength dependence for the 'opposition effect' claimed by O'Leary does not exist, the various attempts to *explain* this fictitious phenomenon (O'Leary and Rea, 1968; Egan and Foreman, 1971; Mead, 1971) must be rejected. Furthermore, the phase independence of $(U-B)$ color (Figure 19b), together with the

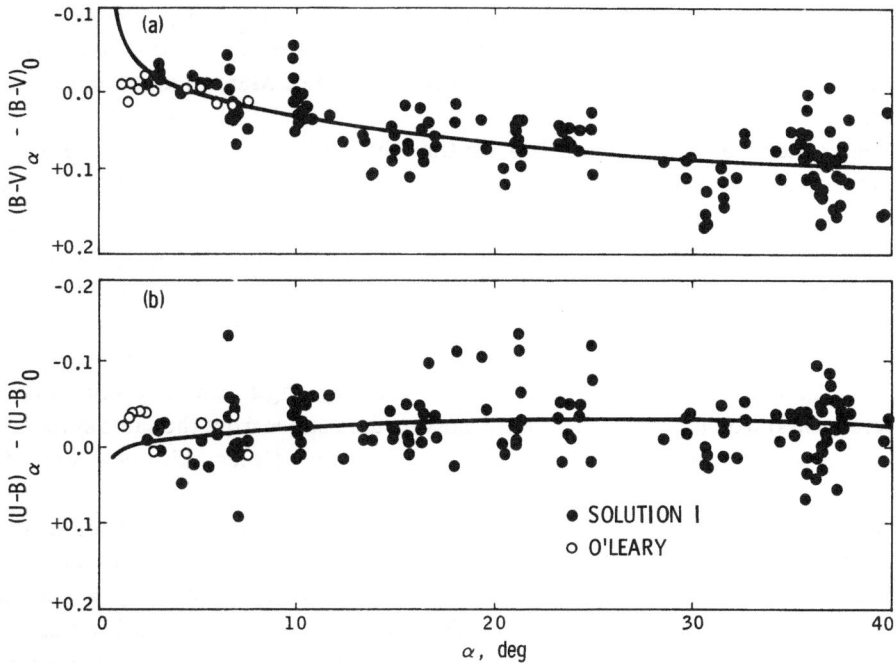

Fig. 19. (a) $(B-V)$ phase curve from Solution I, compared with O'Leary's data (open circles).
(b) $(U-B)$ phase curve from Solution I, compared with O'Leary's data (open circles).

steep phase curves (Figure 16–18), suggests that, contrary to the claims of Ingersoll (1971), only a negligible fraction of the ultraviolet light received from Mars at large phase angles is due to Rayleigh scattering.

It is clear that a *realistic* interpretation of the Martian phase curves, allowing for the errors in the data, is still badly needed.

8. Conclusions

Let us adopt Solution I (Table IX) as the best. We see that the phase curves (Figure 16–18) are fairly well determined from 40° to about 2° phase angle. However, the extrapolation to zero phase is very uncertain. On the other hand, the mean rotation curves are very well determined; they change markedly with wavelength, as the $B-V$ and $U-B$ solutions show.

The seasonal effects and the absolute magnitudes and colors of the planet are very poorly determined. Even if we attribute the large scatter and zero-point terms of the 1958 data entirely to seasonal dust-storm effects, we are left with a wide range of zero-points: 0.05 in V, 0.05 in B, 0.16 in U, 0.07 in $(B-V)$, and 0.11 in $(U-B)$. These represent the uncertainties due to systematic differences between different UBV photometers.

These uncertainties are inherent in the UBV 'system', which is poorly defined for objects like Mars. Thus the fundamental ambiguities of this photometric system prevent us from ever giving the brightness (or albedo) of Mars to an accuracy better than 0.05 mag.

These systematic errors are much larger than the random fluctuations of the planet's brightness, at least outside of the dusty season near $L_S \approx 270°$. Thus, although additional long, homogeneous series of UBV data would improve our knowledge of seasonal (and, to some extent, phase) effects, the albedo problem requires an entirely new approach. To determine the albedo of Mars accurately, we would need a long series of observations in a well-defined photometric system; the means of setting up such a system are described elsewhere (Young, 1973).

The large size of the seasonal effect shows that a very lengthy observational program (several years) would be required. As all existing data suffer from the same systematic-error problem, it may not be practical to combine *any* of them with new, error-free data.

Finally, one must consider the consequences of such systematic errors in other planetary albedoes. If one allows for the additional uncertainties in the magnitudes and colors of the Sun, one must conclude that any planetary albedo may be uncertain by at least ten percent, apart from the effect of errors in size. For dark objects like Mars, Moon, and Mercury, the albedo error has a small effect on the planetary heat budget, as a fractional albedo error of $(\Delta A/A) = f$ is only an error of $[fA/(1-A)]$ in absorbed energy. Thus if $A = 0.2$, an error f of 10% is only a 2.5% error in absorbed radiation. But for a bright planet like Venus, with $A \approx 0.8$, $f = 0.1$ corresponds to a 40% error in the calculated heat budget. Better data are certainly needed.

Acknowledgements

This paper presents the results of one phase of research carried out at the Jet Propulsion Laboratory, California Institute of Technology, under Contract No. NAS 7-100, sponsored by the National Aeronautics and Space Administration.

References

Cousins, A. W. J. and Stoy, R. H.: 1963, *Roy. Obs. Bull.*, No. 64.
De Vaucouleurs, G.: 1960, *Planetary Space Sci.* **2**, 26.
Egan, W. G. and Foreman, K. M.: 1971, in C. Sagan, T. C. Owen, and H. J. Smith (eds.), 'Planetary Atmospheres', *IAU Symp.* **40**, 156.
Fernie, J. D. and Watt, V.: 1967, *Astrophys. J.* **150**, L113.
Guthnick, P. and Prager, R.: 1914, *Veröff. Kgl. Sternw. Berlin-Babelsberg* **1**, 1.
Ingersoll, A. P.: 1971, in C. Sagan, T. C. Owen, and H. J. Smith (eds.), 'Planetary Atmospheres', *IAU Symp.* **40**, 170.
Irvine, W. M., Simon, T., Menzel, D. H., Charon, J., Lecomte, G., Griboval, P., and Young, A. T.: 1968a, *Astron. J.* **73**, 251.
Irvine, W. M., Simon, T., Menzel, D. H., Pikoos, C., and Young, A. T.: 1968b, *Astron. J.* **73**, 807.
Johnson, H. L. and Gardiner, A. J.: 1955, *Publ. Astron. Soc. Pacific* **67**, 74.

Johnson, H. L., Mitchell, R. I., Iriarte, B., and Wisniewski, W. Z.: 1966, *Comm. Lunar Planet. Lab.* **4**, No. 63.

Lawrence, L. C. and Reddish, V. C.: 1965, *Publ. Roy. Obs. Edinburgh* **3**, No. 9, 280.

Mead, J. M.: 1971, in C. Sagan, T. C. Owen, and H. J. Smith (eds.), 'Planetary Atmospheres', *IAU Symp.* **40**, 166.

Meiller, V.: 1964, *U.S. Naval Obs. Circ.*, No. 98, 1964.

O'Leary, B. T.: 1967a, Technical Report on NASA Grant NsG 101-61, Space Sciences Laboratory Series 8, Issue 103 (Ph.D. thesis, Astronomy Department, University of Calif., Berkeley).

O'Leary, B. T.: 1967b, *Astrophys. J. Letters* **149**, L147.

O'Leary, B. T. and Rea, D. G.: 1968, *Icarus* **9**, 405.

Schmidt-Kaler, T.: 1961, *Observatory* **81**, 246.

Shao, C. Y., and Young, A. T.: 1965, *Astron. J.* **70**, 726.

Shobbrock, R. R., Herbison-Evans, D., Johnston, I. D., and Lomb, N. R.: 1969, *Monthly Notices Roy. Astron. Soc.* **145**, 131.

Young, A. T.: 1957, *Publ. Astron. Soc. Pacific* **69**, 568.

Young, A. T.: 1963, *Appl. Opt.* **2**, 51.

Young, A. T.: 1966, *Observatory* **86**, 71.

Young, A. T. and Irvine, W. M.: 1967, *Astron. J.* **72**, 945.

Young, A. T.: 1969, *Appl. Opt.* **8**, 869.

Young, A. T. and Collins, S. A.: 1971, *J. Geophys. Res.* **76**, 432.

Young, A. T.: 1973, in *Methods of Experimental Physics*, Academic Press, New York, Vol. 12A, Chapter 3.

PHOTOMETRIC DATA FROM SOME PHOTOGRAPHS OF MARS OBTAINED WITH THE AUTOMATIC INTERPLANETARY STATION 'MARS 3'

V. V. BOTVINOVA, O. I. BUGAENKO, I. K. KOVAL, M. K. NARAJEVA,

and

A. S. SELIVANOV

Main Astronomical Observatory, Kiev, U.S.S.R.

Abstract. The results of detailed photometric treatment of Mars photographs obtained with the Automatic Interplanetary Station 'Mars 3' in three wavelengths are given. Photometric maps of the Martian surface have been constructed; a thin layer observed near the limb has been investigated.

As is well known (*Pravda*, 1971), the scientific equipment on the Soviet automatic interplanetary stations 'Mars 2' and 'Mars 3' included phototelevision devices for obtaining both a limited number of whole-disk planetary photographs in the visible region of the spectrum, and photographs of separate regions of the Martian surface. According to the program the photographic investigations were mainly executed by 'Mars 3', whose orbit allowed surveys of the planet from both small and relatively great distances (distance in periares 1500 km, distance in apoares 200000 km). Photography of the Martian surface has been also made at phases not observed from the Earth. Each of the stations had two phototelevision devices (FTU) equipped with objectives having focal distances $F = 52$ mm (FTU I) and $F = 350$ mm (FTU II). In front of the wideangle objective FTU I was a wheel with three colour filters (Figure 1). The picture taking has been carried out cyclically at the rate of 140 or 35 s and with 12 frames in a cycle.

In the television instruments the pictures were recorded on film (width 25.4 mm, frame size 24 × 24 mm), with subsequent chemical processing. The image obtained was read out by an optical-mechanical television device. The number of lines in a frame can be set equal to 64, 250, or 1000.

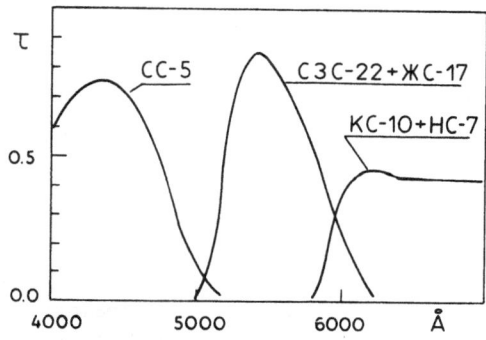

Fig. 1. Graphs of filter transmissions used in FTU II of 'Mars 3'.

Woszczyk and Iwaniszewska (eds.), Exploration of the Planetary System, 287–292. All Rights Reserved

A series of Mars pictures taken by 'Mars 3' on February 28, 1972, at a distance of 18 000 km, has been selected for detailed photometric study. These photographs were of interest because of the phase angle ($\alpha \approx 103°$), at which the Martian surface has not been observed before. Besides, the photographs showed that one of the Martian cusps had a weakly visible thin haze next to the limb, that rapidly decreased in brightness off the limb. The phenomenon was similar to those observed with 'Mariner 6, 7' (Leovy *et al.*, 1971) and 'Mariner 9' (Masursky *et al.*, 1972).

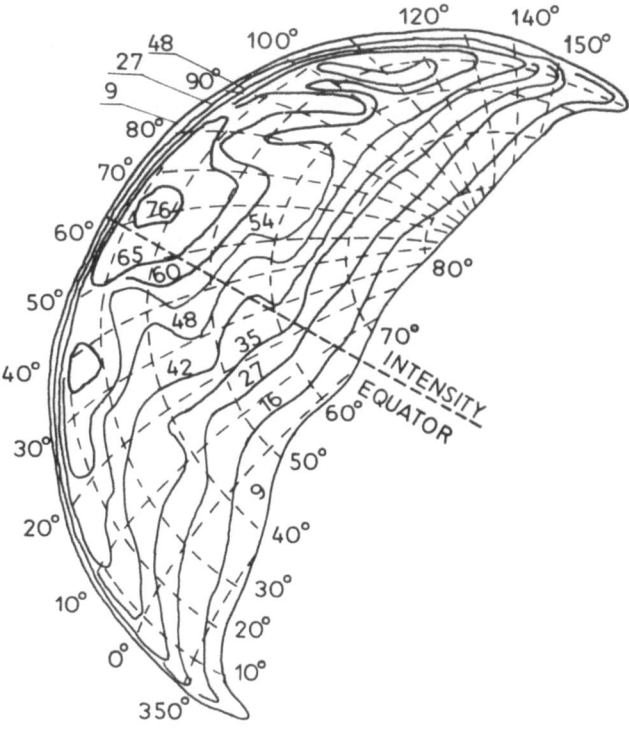

Fig. 2.   Photometric map of Mars obtained on February 28, 1972 (red filter).

On the ground of some considerations the photometric analysis of the Martian surface has been carried out not on the pictures themselves but using the photo-television lines recorded on the band of the loop oscillograph.

From photometric treatment of the selected pictures, we got the photometric maps of Mars for three spectral regions, presented (Botvinova *et al.*, 1974) as isophotes (relative intensities in Figure 2). From these maps, the authors estimated the contrast between Mare Acidalium and a continental region placed symmetrically with respect to the intensity equator. In red light the contrast appeared to be 0.25 (lower limit) which agrees with ground-based measurements (Barabashov and Koval, 1959; de Vaucouleurs, 1967). In blue light the contrast reaches 0.20 and has the opposite sign because of the cloudy formation extending over Mare Acidalium.

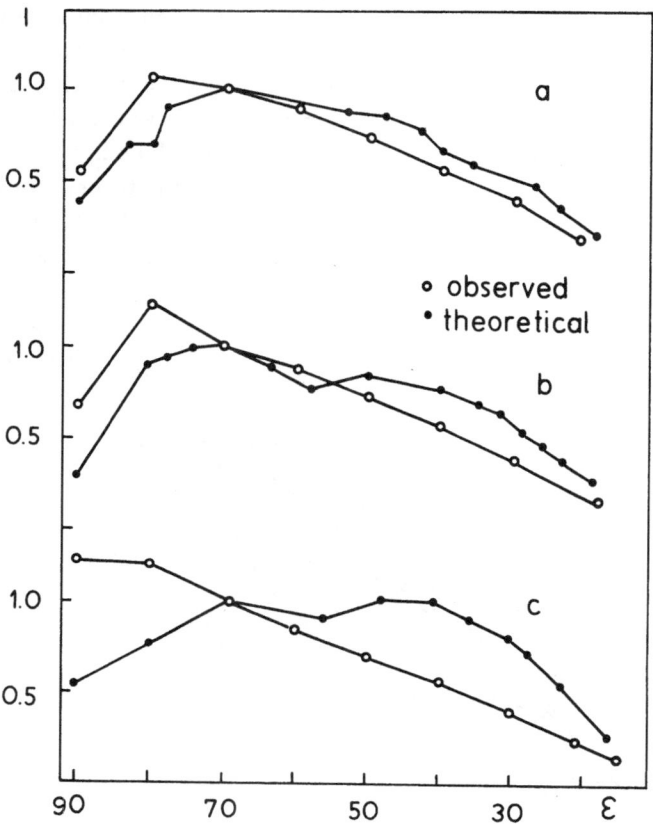

Fig. 3.   Brightness distribution along the Mars intensity equator obtained on February 28, 1972.
a – red filter, b – green filter, c – blue filter.

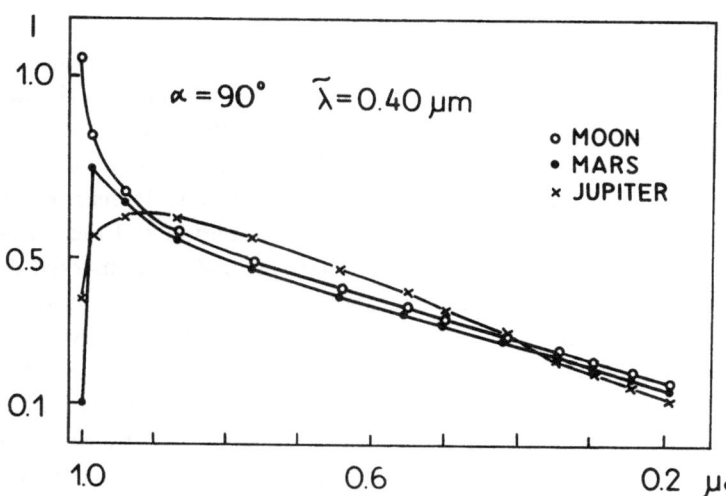

Fig. 4.   Calculated distribution of brightness along the intensity equator of Moon,
Mars and Jupiter ($\mu_0 = \cos\varepsilon$).

For the same pictures the graphs of brightness variation from limb to terminator were plotted. These were compared with theoretical intensity curves calculated by Morozhenko and Yanovitsky (1971) on the assumption that the Hapke model is true for the Martian surface (Figure 3). The comparison showed satisfactory agreement between the theoretical and observed curves in red and blue light (within the limits of measuring errors and the approximate character of the theory). In blue light the calculated curve differs from the observed one mainly due to the cloudy formation in the centre of the picture. The brightness distribution along the Martian intensity

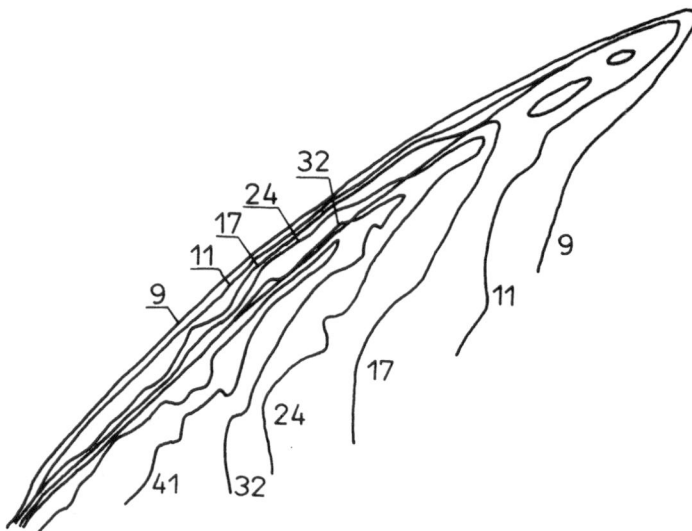

Fig. 5.   Photometric map of a cusp with cloud formation in the limb zone (red filter).

equator when $\alpha \approx 103°$ is not very informative. However, its comparison with a theoretical model which is in a good agreement with observations obtained with 'Mariner 6, 7' (Hord, 1972) gives an additional criterion for estimating the accuracy of the photographic experiment obtained with 'Mars 3' (Figure 4).

Study of the cloudy formation near a cusp showed that its upper limit is at 30–40 km and its extension along the limb reaches 20 areographic degrees. From detailed photometry of a cusp the isophotes of the cloudy formation and the surface border Figure 5 have been constructed. The authors have estimated the marked decrease of the cloud brightness relative to the planetary limb when passing from red to blue.

To study the distribution of cloud intensity with height, six photometric sections were made in each of the three filters along the radius of the image. The measurements were reduced to absolute units by tying a photometric section made along the planetary intensity equator to the theoretical ones, calculated by Morozhenko and Yanovitsky for the usual case of high transparency of the Martian atmosphere. Analysing the height profiles of the Martian atmosphere, one can conclude that the scattering particles forming the cloud at different heights have the same size spectrum, if this cloud is optically thin and its particles are purely scattering (Figure 6). The intensity relations of a cloud in blue and green light to that in the red (Figure 7) showed that the cloud formation, throughout its height, had evidently selective reflection, i.e., its brightness in the red is three or four times more than in the blue. This means that the cloud under investigation consists of large particles throughout its height (according to the above model).

Qualitative characteristics of particles in the cloud formations under investigation (mean radius, the number in vertical column of atmosphere) can be estimated only from an investigation of the spectral dependence of their optical thickness. But this is the subject of another study.

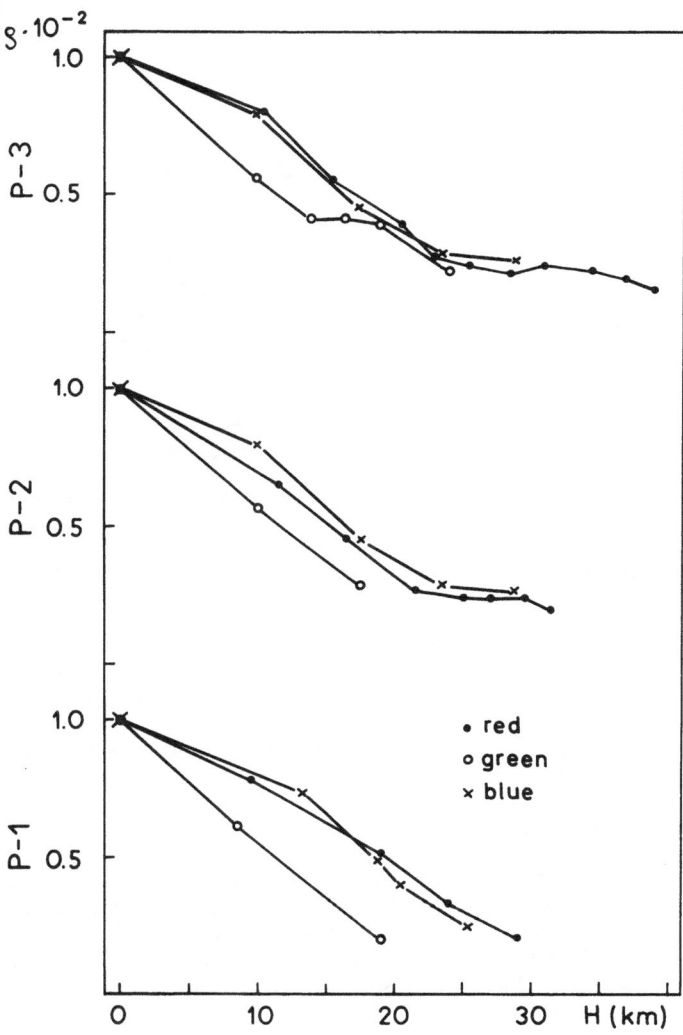

Fig. 6.   Height-brightness profiles of different regions of cloud formations in a cusp limb zone (graphs photometrically connected near the base of the cloud).

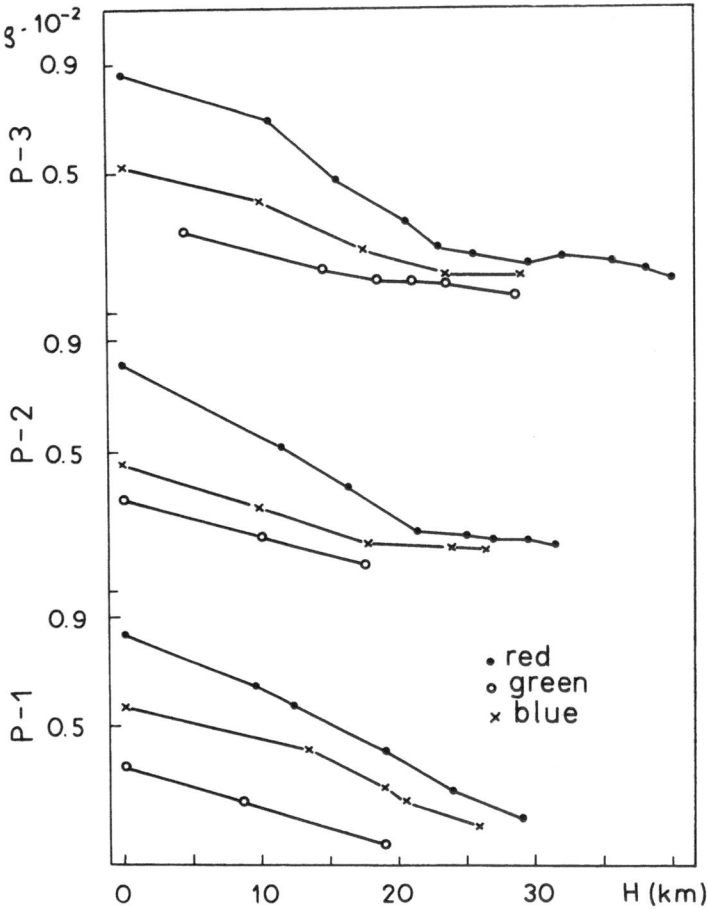

Fig. 7. Height-brightness profiles of different regions of the cloud formations in the cusp limb zone (in absolute units).

References

Barabashov, N. P. and Koval, I. K.: 1959, *Photographic Photometry of Mars with Colour Filters in 1956*, Izd. Kharkiv. Univ. (in Russian).

Botvinova, V. V., Bugaenko, O. I., Koval, I. K., Narajeva, M. K., and Selivanov, A. S.: 1974, *Photometric Data from Photographs Obtained with 'Mars 3' on February 28, 1972* (in Russian), in press

De Vaucouleurs, G.: 1967, *Icarus* **7**, 310.

Hord, C. W. 1972, *Icarus* **16**, 253.

Leovy, C. B., Smith, B. A., Young, A. T., and Leighton, R. B.: 1971, *J. Geophys. Res.* **76**, 297.

Masursky, H., Batson, R. M., McCauley, J. F., Soderblom, L. A., Wildey, R. L., Carr, M. H., Milton, D. J., Wilhelms, D. E., Smith, B. A., Kirby, T. B., Robinson, J. C., Zeovy, C. B., Briggs, G. A., Duxbury, T. C., Acton, C. H., Jr., Murray, B. C., Cutts, J. A., Sharp, R. P., Smith, S., Leighton, R. B., Sagan, C., Veverka, J., Norald, M., Zederberg, J., Zevinthal, E., Pollack, J. B., Moore, J. T., Jr., Hartmann, W. K., Shipley, E. N., de Vaucouleurs, G., and Davies, M. E.: 1972, *Science* **175**, 294.

Morozhenko, A. V. and Yanovitsky, E. G.: 1971, *Astron. Zh.* **48**, 798.

Pravda: 1971, No. 353, 2 (in Russian)

RESULTS FROM THE INFRARED SPECTROSCOPY
EXPERIMENT ON MARINER 9

J. PEARL, B. CONRATH, R. CURRAN, R. HANEL, V. KUNDE, and J. PIRRAGLIA

Goddard Space Flight Center, Greenbelt, Md., U.S.A.

Abstract. Over 20 000 thermal emission spectra of Mars have been obtained, providing extensive diurnal, seasonal and spatial coverage of the planet. Each spectrum covers the spectral range from 200 to 2000 cm^{-1} with an apodized spectral resolution of 2.4 cm^{-1}; the noise equivalent radiance of the instrument is 0.5×10^{-7} W cm^{-2} sr^{-1} (cm^{-1})$^{-1}$.

Data obtained during the dust storm ($290° < L_s < 320°$) contain broad spectral features due to entrained dust which are centered near 480 and 1090 cm^{-1}; from these features the SiO$_2$ content of the dust material is estimated to be $60 \pm 10\%$. Radiative transfer calculations for model dust clouds suggest particle radii of a few micrometers. By using a Stokes-Cunningham-settling and turbulent-atmospheric-mixing model, the observed secular cooling of the atmosphere can be explained in terms of dust particles having radii of $\sim 1\ \mu$m and an atmospheric eddy diffusion coefficient somewhat in excess of 10^6 cm^2 s^{-1} during the settling phase of the dust storm. Atmospheric temperature profiles during this period showed large diurnal excursions (± 15K) and strongly subadiabatic lapse rates (~ 1.5 K km^{-1}). Wind fields derived from these data show a strong tidal behavior, with wind speeds above the surface boundary layer of roughly 40 m s^{-1} for midlatitudes; the accompanying diurnal surface pressure variations were about $\pm 5\%$. Large scale temperature wave structures have been discovered in the atmosphere with ~ 5K amplitudes; these are found to be caused by the interaction of the diurnal heating wave with the strong wavenumber 2 component of the planetary topography. Atmospheric water vapor concentrations of ~ 10 precipitable micrometers were observed at southern and mid-northern latitudes, with no detectable amounts over the north polar region.

Following the clearing of the dust ($320° < L_s < 350°$) atmospheric temperatures in the stratosphere decreased ~ 30K while the lapse rate increased to about 3 K km^{-1}. Between latitudes of 25°N and 65°S, surface pressure mapping indicates that the highest pressures (~ 9 mb) occur in Hellas and in Isidis Regio. The lowest pressure observed (~ 1.5 mb) was near the top of South Spot ($-9°$, 120°); pressures could not be determined for the other shield volcanoes. Clouds in the north polar hood have been conclusively identified as being composed of water ice. They are found at pressures exceeding 2.0 mb and are generally in the temperature range of 160–180 K.

Later less extensive coverage ($45° < L_s < 80°$) indicated further cooling of the upper atmosphere to 155 K, but a stabilization of the lapse rate at 3 K km^{-1}. Water vapor vanished over the south polar region, remained nearly constant at $\sim 10\ \mu$m in the equatorial region, increased to as much as 50 μm at mid-northern latitudes, and increased to $\sim 30\ \mu$m over the north polar region. Clouds observed in the Tharsis area

Woszczyk and Iwaniszewska (eds.), Exploration of the Planetary System, 293–294. All Rights Reserved

have also been conclusively identified as composed of water ice particles a few micrometers in size.

DISCUSSION

Beer: Would you explain the basis upon which the 60% silicate content of the dust was found?

Pearl: The features in the standard deviation of the spectra which I have shown give us a measure of the dust transmittance as a function of wavelength. From the general positions of these features we conclude that they arise from silicate minerals rather than other minerals, such as calcium carbonate, which has a much different spectral behaviour. From a more detailed comparison of the widths of the features, and the positions of these maxima and minima with laboratory data, we conclude that the dust is of largely intermediate silica content.

Icke: Does the direction of your derived wind velocity field coincide with the direction of the black-and-white streaks on the surface?

Pearl: Yes.

Low: Do your spectra give information about surface composition?

Pearl: In principle, yes. However, the persistence of the atmospheric dust has hampered the interpretation, because the spectral features of the dust occur in the same region as the spectral features of the surface.

Low: Does this mean that the method is invalidated as a tool for remote sensing of surface composition on Mars?

Pearl: The question is unresolved. By making nearly simultaneous observations of a given area at different viewing angles, the effects of the dust can be eliminated in the analysis. Unfortunately, we have very few such measurements, and their reduction is incomplete.

Golitsyn: What is the mean radius of the particles, and what amount of dust is in a unit column of the atmosphere?

Pearl: The mean radius is 1–2 μm, and the mean dust content is about 200 μg cm^{-2}.

Kliore: Was there any evidence of dust in spectra taken during the latter part of the mission?

Pearl: In some there appears to be little dust, in others a considerable amount. In general, there was much less than in the early part of the mission.

Gehrels: Did you observe Phobos and Deimos?

Pearl: Phobos was observed, but as it filled only a small portion of the field of view, the signal-to-noise ratio is low; our reductions of the Phobos data are still in progress.

RADIO OCCULTATION EXPLORATION OF MARS

A. J. KLIORE

Jet Propulsion Laboratory, California Institute of Technology, Pasadena, Calif., U.S.A.

Abstract. The radio occultation technique, consisting of the observation of changes in the phase, frequency, and amplitude of a radio signal from a spacecraft as it passes through the atmosphere of a planet before and after occultation, was first applied to measure the atmosphere of Mars with the Mariner IV spacecraft in 1965. The interpretation of these changes in terms of refraction of the radio beam by the neutral atmosphere and ionosphere of the planet provided the first direct and quantitative measurement of its vertical structure and established the surface atmospheric pressure of Mars as lying between 5 and 9 mb. The presence of a daytime ionosphere with a peak electron density of about 10^5 el cm^{-3} was also measured. The Mariner VI and VII spacecraft flew by Mars in 1969 and provided an additional four measurements of the atmosphere and surface radius of the planet. They confirmed the surface pressure values measured by Mariner IV and provided data for a crude estimate of the shape of the planet.

By far the greatest volume of radio occultation information on the atmosphere and surface of Mars was returned by the Mariner IX orbiter which was placed in orbit about Mars in November of 1971. During three occultation episodes in November–December 1971, May–June 1972, and September–October 1972, the Mariner IX mission provided 260 successful radio occultation measurements.

The early measurements, made at the time of the Martian dust storm of 1971, showed greatly reduced temperature gradients in the daytime troposphere, indicating the heating effect of the dust. The temperature gradients that were measured later in the mission, when the atmosphere was apparently free of dust, were still much lower than expected under conditions of radiative-convective balance, indicating that dynamics may play a large part in determining the temperature structure of the Martian troposphere. Temperatures taken at night near the winter poles were consistent with the condensation of carbon dioxide.

The surface atmospheric pressure was observed to vary widely with topography ranging from about 1 mb at the summit of the Middle Spot volcano (Pavonis Mons) to over 10 mb in the North circumpolar region. In the South equatorial region the highest surface pressure of about 9 mb was measured at the bottom of the Hellas basin.

The radius of the planet was measured with accuracies ranging from about 0.25 to about 2.1 km over latitudes ranging from 86° to − 80°. These measurements have shown that Mars has pronounced equatorial and north–south asymmetries, which make it difficult to represent its shape by a simple triaxial figure.

The daytime ionosphere measurements indicated that the main ionization peak was similar in behavior to a terrestrial F_1 layer and is probably produced by photoionization of carbon dioxide by solar extreme ultraviolet. Comparison of the heights of the maximum between the early data taken in November–December, 1971, and the Extended Mission of May–June 1972, showed that the lower atmospheric temperatures decreased by about 25 %, which is consistent with clearing of the atmosphere.

The experience gained from Mars radio occultation experiments suggests that the quality of data can be significantly improved by such features of the spacecraft radio system as a stable oscillator, dual frequency downlink capability, and a steerable high-gain antenna.

1. Introduction

The technique of spacecraft radio occultation, although less than a decade old, has already produced a large body of results on the atmosphere and topography of Mars. This technique, which is well known and has been described previously, is based on the observation of very small changes in the phase and amplitude of a radio signal from a spacecraft introduced by the effects of refraction by the neutral and electrically charged portions of a planetary atmosphere during the times immediately prior

Woszczyk and Iwaniszewska (eds.), Exploration of the Planetary System, 295–316. All Rights Reserved
Copyright © 1974 by the IAU

to and immediately after the occultation of the spacecraft by the planetary body. Such experiments were made possible not only by the advent of interplanetary space-craft, which brought radio apparatus to the vicinity of planets, but also by the development of very precise spacecraft tracking instrumentation and techniques by the NASA/JPL Deep Space Net. These, in conjunction with advances in celestial mechanics and digital computer technology and techniques, made it possible to unravel the very small effects of refraction in the ionosphere and atmosphere of Mars from the received Doppler frequency data, thus enabling the recovery of the results presented herein.

The refractive effects of an atmosphere upon the propagation of radio waves has been known and understood for some time (Bean and Thayer, 1963). However, the application of this knowledge to spacecraft radio occultation experiments was not proposed until approximately a decade ago (Fjeldbo, 1964; Kliore et al., 1964). The first radio occultation measurement of the atmosphere of Mars was performed with Mariner IV in 1965 (Kliore et al., 1965a). Further measurements were performed with Mariners VI and VII in 1969 (Kliore et al., 1969). These first two opportunities produced six individual measurements. In contrast, the Mariner IX orbiter collected 260 successful individual measurements during 1971 and 1972, which provided the ma-

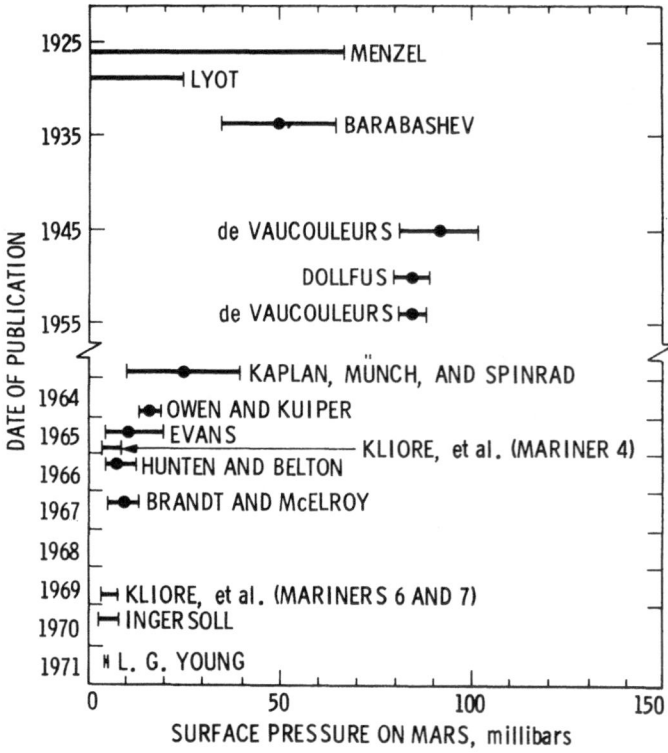

Fig. 1.   Historical review of Martian surface pressure estimates.

jority of the important results regarding the atmosphere, ionosphere, and topography of Mars that are discussed in this paper (Kliore *et al.*, 1970a). The Soviet Mars 2 spacecraft was also used to perform a radio occultation experiment (Kolosov *et al.*, 1972).

The scientific importance of the Mars radio occultation measurements is evident from Figure 1 which displays the range of surface pressure estimates on Mars from a variety of sources. Prior to about 1964, various Earth-based photometric, polarimetric and spectroscopic measurements had indicated a value for the surface pressure of Mars of approximately 85 mb. After 1963, improved spectroscopic techniques were used to drastically reduce the surface pressure estimates, which were confirmed by the results of the Mariner IV Radio Occultation experiment. Since then, a number of re-evaluations of the spectroscopic data as well as radio occultation experiments have indicated a mean surface pressure on Mars of about 5–6 mb, with a wide variation due to topographical height variations of the Martian surface.

2. The Atmosphere

The phase delay data produced by refraction in a planetary atmosphere, together with the precisely known ephemeris of the spacecraft, can be inverted to produce a vertical profile of refractivity as a function of distance from the center of the planet (Fjeldbo and Eshleman, 1968; Kliore, 1972). Such inversion procedures generally assume spherical symmetry in the planetary atmosphere. Using well known relationships from magnetoionic theory, the refractivity in the ionized upper atmosphere can be directly converted to electron density. It should be pointed out that no direct information on the ion species is provided.

In the neutral lower atmosphere a composition must be known or assumed in order to derive other atmospheric parameters from the refractivity. In the case of Mars, this problem is simplified considerably by the fact that no major constituent in addition to carbon dioxide has been discovered by spectroscopic spacecraft observations (Barth *et al.*, 1972). Consequently, all computations of atmospheric parameters are made under the assumption of a carbon dioxide atmosphere. Once the composition has been established, the refractivity can be converted to mass density, and the hydrostatic equation can be integrated vertically to obtain the pressure profile. The temperature profile is then obtained by applying the perfect gas law. It should be noted that an assumption of an upper level initial temperature must be made.

The first two radio occultation measurements of Mars were performed with Mariner IV in July of 1965 (Fjeldbo and Eshleman, 1968; Kliore *et al.*, 1965b, 1968). The entry, or immersion, measurement was made at the latitude of $-50.5°$ in the Electris-Mare Chronium region, and indicated a surface pressure of about 4.5–5.0 mb. It should be pointed out that most of the entry measurements are performed with the spacecraft in the so-called coherent two-way mode, in which the spacecraft transponder coherently transmits back a signal which it receives from the ground station, where the reference frequency is supplied by a very stable frequency standard. Thus

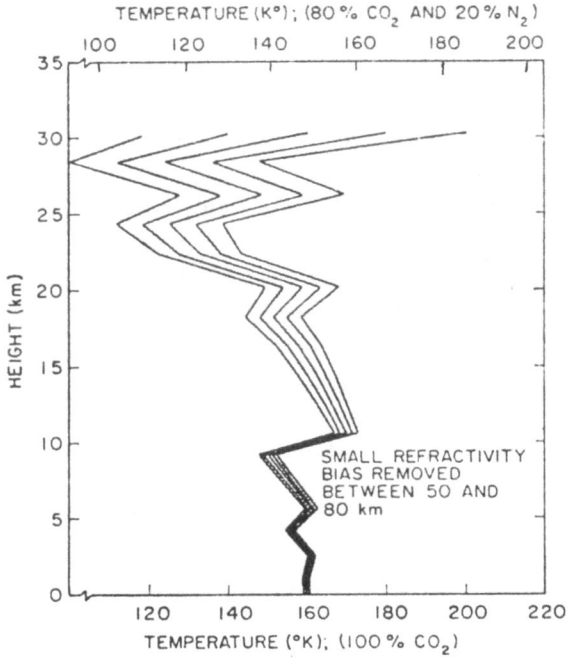

Fig. 2.   Mariner IV entry temperature profile for the lower atmosphere measured during Mariner IV
entry (from Fjeldbo and Eshleman, 1968).

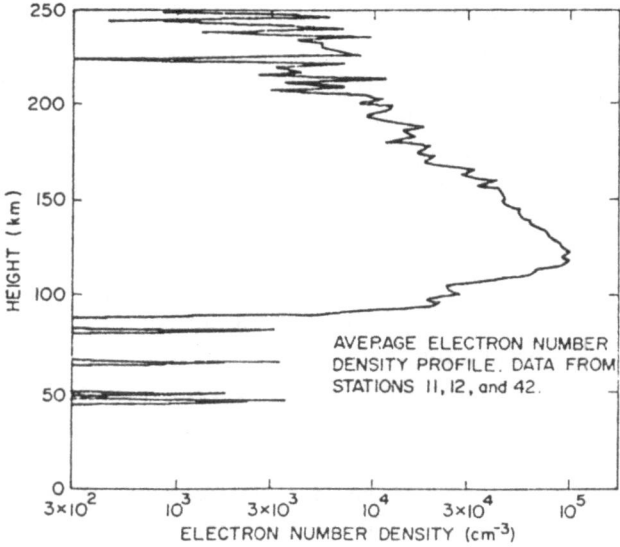

Fig. 3.   Ionospheric electron density profile for the Martian upper atmosphere measured during
Mariner IV entry (from Fjeldbo and Eshleman, 1968).

the entry data display frequency stability of at least one part in 10^{11}, which is sufficient for the analysis of temperature profiles in the atmosphere. In contrast, during emersion, or the exit phase of occultation, the spacecraft transmitter is referenced to its own crystal oscillator, the short-time stability of which is about an order of magnitude lower, and hence the exit data are not very reliable. In the case of Mariner IV exit, the two-way coherent mode was re-established after the radio beam had traversed about the first 10 km of the atmosphere, and hence two separate kinds of data had to be arbitrarily conjoined during analysis. Nevertheless, it was possible to establish that the surface pressure at the location of the exit measurement at 60° latitude in Mare Acidalium was between 8 and 9 mb. Figure 2 shows a temperature profile derived from the Mariner IV entry data. The five different curves correspond to different choices of the initial temperature and, the temperature profile is remarkably isothermal for the first 30 km. The entry measurement, which was performed approximately at 13:00 h Mars local time in late winter at a solar zenith angle of 67° also produced a profile of the electron density in the ionosphere, which is shown in Figure 3. A main electron peak of about 10^5 el cm^{-3} was observed at an altitude of about 125 km with a subsidiary layer of density 3×10^4 el cm^{-3} at an altitude of about 95 km.

The Mariner IV results established a rather low surface pressure for Mars, which had been predicted from precise Earth-based spectroscopic measurements, and a large disparity between the entrance and exit surface pressures was attributed to local elevation differences. The topside electron scale height of about 22 km was far lower than expected and indicated a plasma temperature of about 280 K if CO_2^+ was the principal ion.

The Mariner VI and VII spacecraft flew by Mars in July and August of 1969 and produced four more occultation measurements (Kliore *et al.*, 1969, 1970b, 1971; Rasool *et al.*, 1970; Fjeldbo *et al.*, 1970). The Mariner VI entry measurement occurred at a latitude of about 4° in the area Meridiani Sinus where a pressure of about 5 mb was measured. The exit occurred at night at about 80° latitude in the area of Boreosyrtis, where the surface pressure was about 6.9 mb. Mariner VII produced a more nearly diametrical occultation, measuring a surface pressure of about 4.2 mb at −68.2° near Hellespontus and about 7.3 mb at 38.1° in the Arcadia-Amazonis area. The temperature profiles derived from the Mariner VI and VII measurements are plotted against pressure in Figure 4. The smooth solid line on the left of the figure is the carbon dioxide saturation curve. The near surface temperature about 250 K indicated at the point of Mariner VI entry was found to be in good agreement with that predicted theoretically on the basis of a surface temperature of about 275 K observed by the infrared radiometer experiment of Mariner VI. In the case of the Mariner VI exit measurement, occurring in the north polar region at night, the CO_2 condensation temperature was reached at an altitude of about 15 km. It was also observed that the average temperature gradients for both of the daytime measurements were of the order of −2.9 to −3.0 K km^{-1}.

Both daytime occultations produced measurements of the structure of the ionized

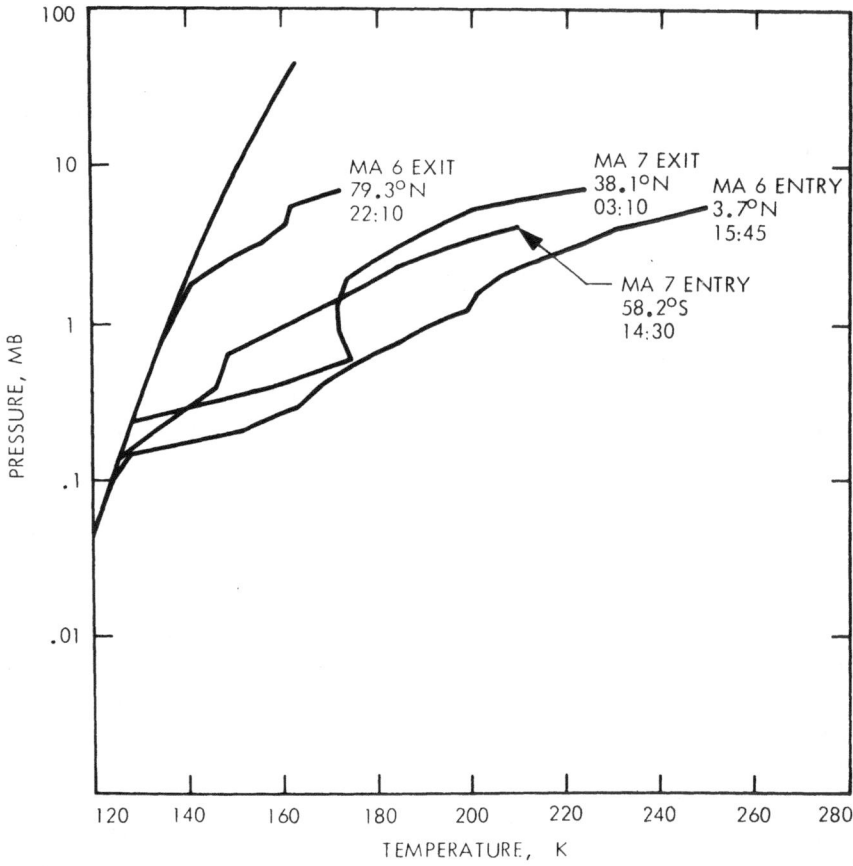

Fig. 4.   Mariner VI and VII temperature profiles plotted vs. pressure. The smooth solid line is the
carbon saturation curve (from Kliore *et al.*, 1971).

upper atmosphere. The main ionization layer was observed at 135 km altitude with
a peak density of approximately 1.7×10^5 el cm^{-3}. A minor layer was observed about
25 km below the main peak. The main ionization peak was interpreted as a F_1 layer
consisting primarily of CO_2^+ ions, and the topside plasma temperature was deduced
to be 400–500 K, with a neutral density of approximately 10^{10} m cm^{-3} at an altitude
of about 135 km.

By far the largest and most important body of radio occultation data on Mars
has been provided by the Mariner IX mission in 1971 and 1972 (Kliore *et al.*, 1972a, b,
1973; Cain *et al.*, 1972, 1973). There were three separate episodes of Mariner IX
radio occultations during which data were taken. The first, referred to as the Standard
Mission, began with the arrival of Mariner IX at Mars and its injection into orbit on
November 14, 1971 and ended on December 23, 1971. The second, called Extended
Mission I, began on May 7, 1972 and continued until June 25, 1972. The third, called
Extended Mission II, began on September 27, 1972 and continued until October 26,

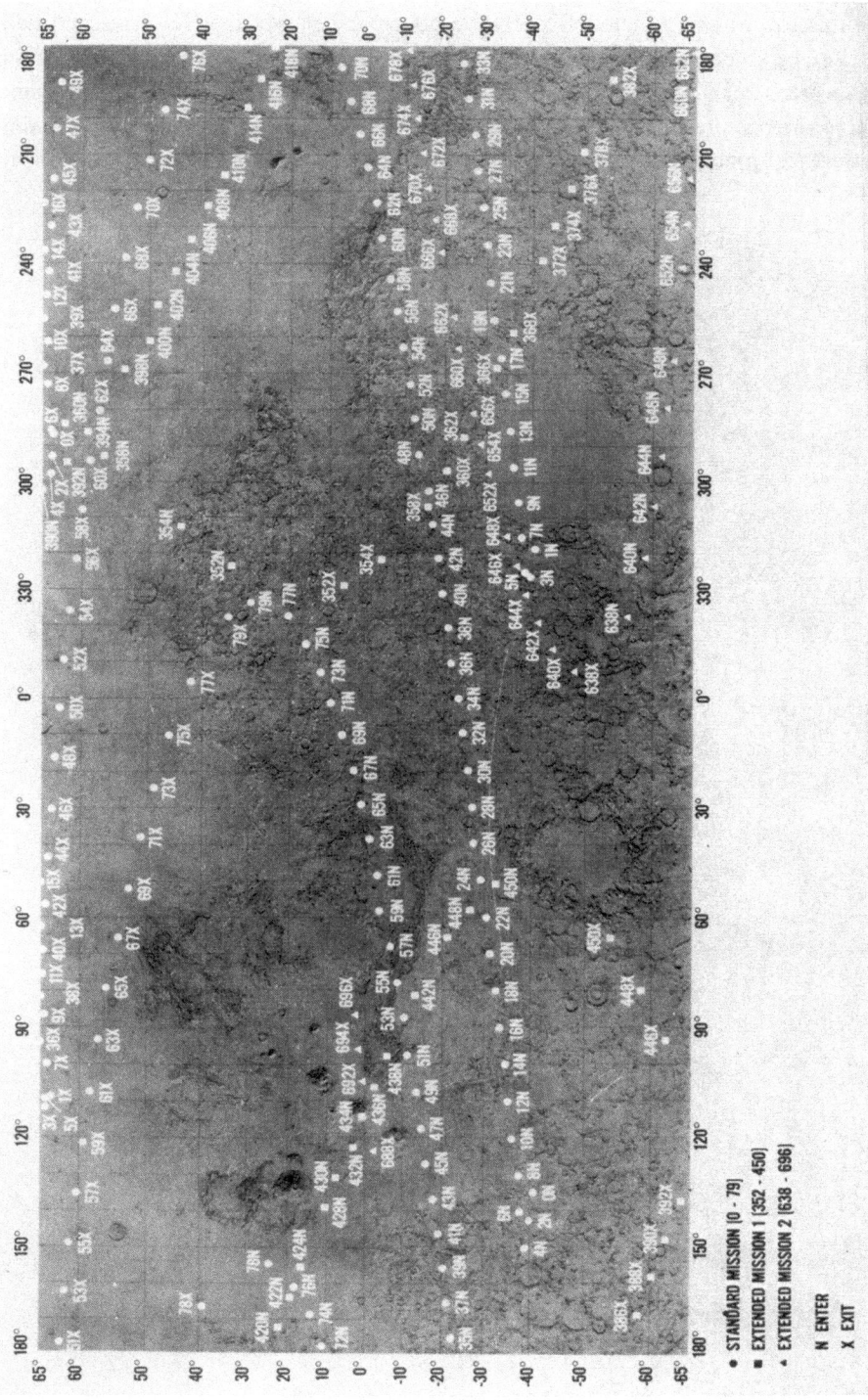

Fig. 5. Locations of all Mariner IX occultation points between the latitudes of ± 65 degrees. The latitudes were computed under the assumption of a spherical planet of radius 3387 km (from Kliore *et al.*, 1973).

at which time the Mariner IX spacecraft exhausted its attitude control gas supply and the mission was terminated. The twelve hour orbit of Mariner IX produced two occultation pairs per day. During the Standard Mission, the spacecraft high gain antenna was pointed in the direction of the Earth, and the signal level was sufficient to observe the occultation from the Deep Space Net stations in Australia and Spain, which were equipped with 26-m diameter antennas, as well as the DSN station at

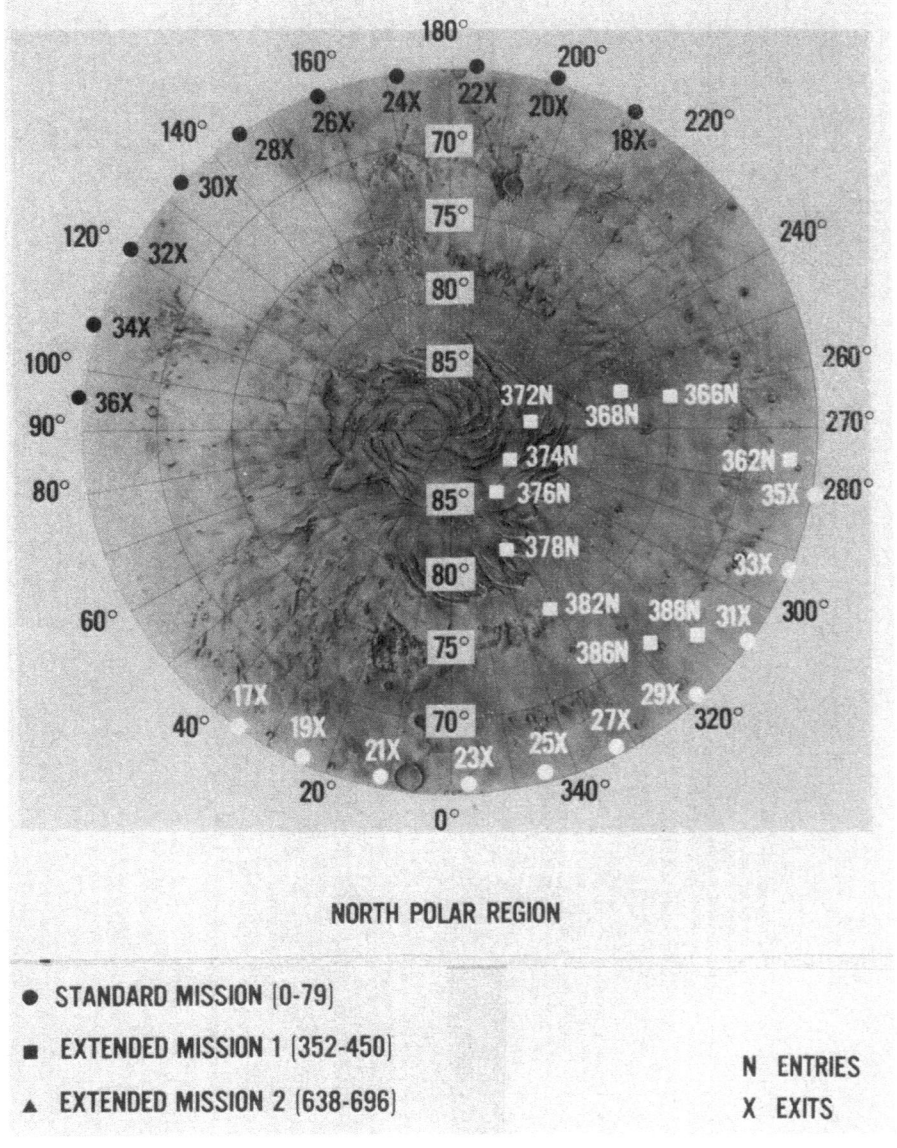

NORTH POLAR REGION

● STANDARD MISSION (0-79)

■ EXTENDED MISSION 1 (352-450)

▲ EXTENDED MISSION 2 (638-696)

N ENTRIES

X EXITS

Fig. 6.   Locations of Mariner IX occultation points in the north polar area (from Kliore *et al.*, 1973).

Goldstone, California which has a 64-m diameter antenna. During both of the extended missions, however, the high gain antenna was not oriented toward the Earth, and hence occultation observations could only be taken during those orbits which were visible to the Goldstone tracking station. The locations of Mariner IX occultation measurements on the surface of Mars are shown in Figures 5, 6 and 7. Figure 5 shows all occultation points lying within the latitudes of $\pm 65°$, and Figures 6 and 7

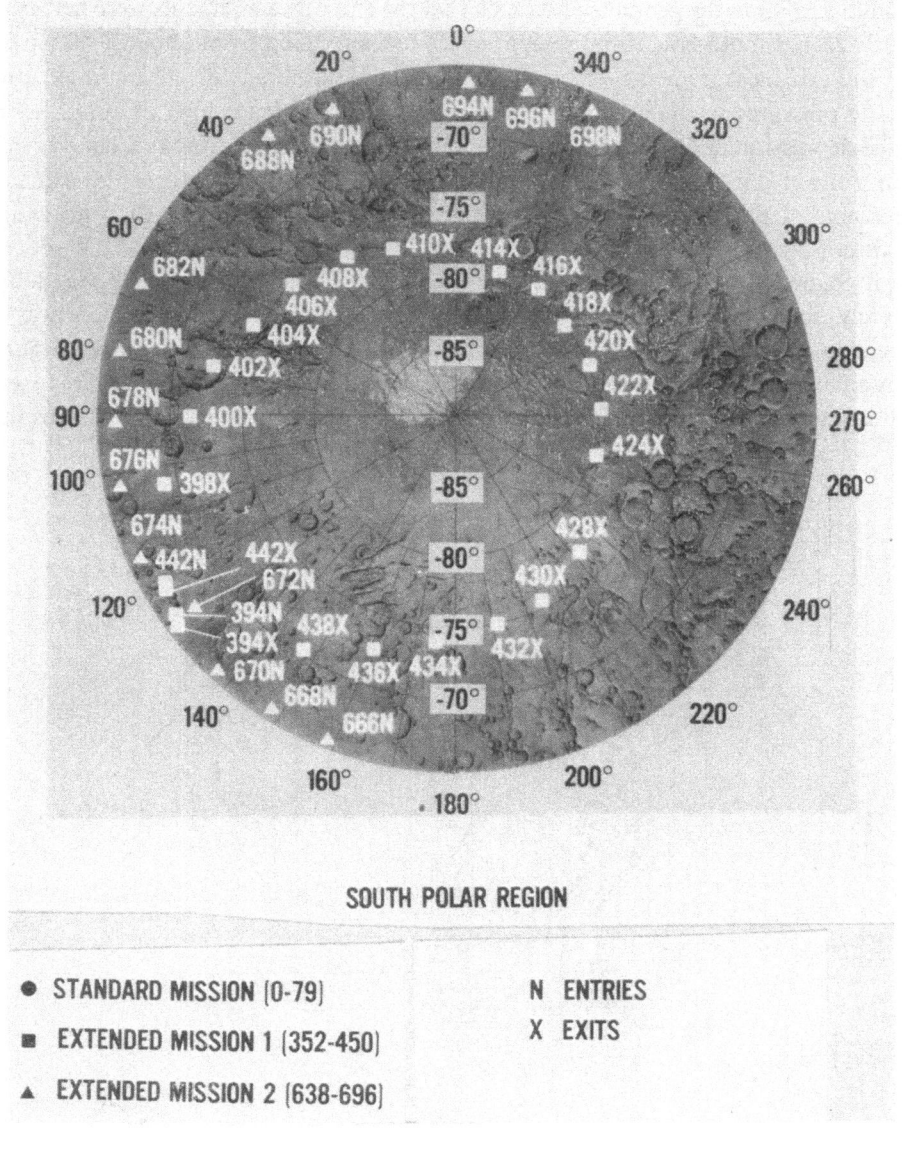

SOUTH POLAR REGION

- STANDARD MISSION (0-79)   N   ENTRIES
- EXTENDED MISSION 1 (352-450)   X   EXITS
▲ EXTENDED MISSION 2 (638-696)

Fig. 7.   Locations of Mariner IX occultation points in the south polar area (from Kliore *et al.*, 1973).

describe the locations of occultation points in the north and south polar areas. The locations are labeled with the appropriate orbital revolution numbers, and are identified as being entry or exit measurements with an *N* or an *X*, respectively. The circular marks refer to the Standard Mission (Rev 0–79). The square marks identify measurement points during the Extended Mission I (Rev 352–450) and the triangular marks show the locations of Extended Mission II measurements (Rev 639–696). The relief maps which form the backgrounds of Figures 5, 6 and 7 have been produced by the U.S. Geological Survey from Mariner IX television photographs. During the Standard Mission the geometry was such that the entry measurements were performed mostly in the south equatorial region, at latitudes ranging from about −40° to 20°, and the exit measurements occurred mostly in a band about 65°. The surface atmospheric pressures in the near equatorial region ranged from a high of about 8.9 mb in the depression of Hellas to a low of about 2.8 mb in the Claritas and Tharsis areas, with a mean pressure of about 4.95 mb. The pressures derived from the exit measurements at 65° latitude were much higher, ranging from 7.2 to 10.3 mb, with a mean of 8.9 mb.

All Standard Mission measurements were taken when Mars was covered by a severely obscuring global dust storm. The effect of dust in the atmosphere was apparent in the temperature profiles obtained during the Standard Mission. Figure 8 shows daytime temperature profiles for Revs 0–9. The first three profiles are essentially isothermal, suggesting that large amounts of dust entrained in the atmosphere

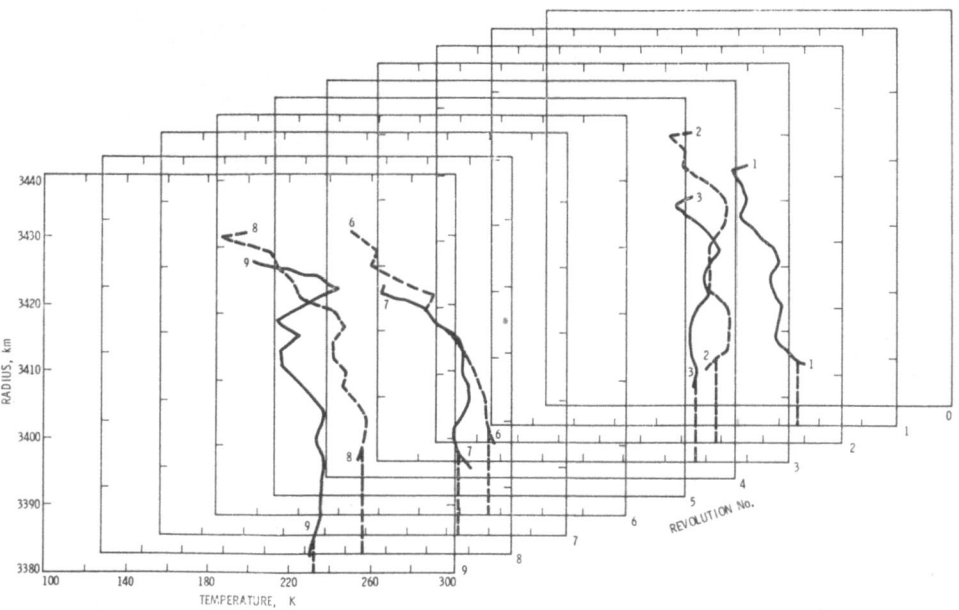

Fig. 8.   Temperature profiles derived from Mariner IX entry data for revolutions 0–9 (from Kliore *et al.*, 1972b).

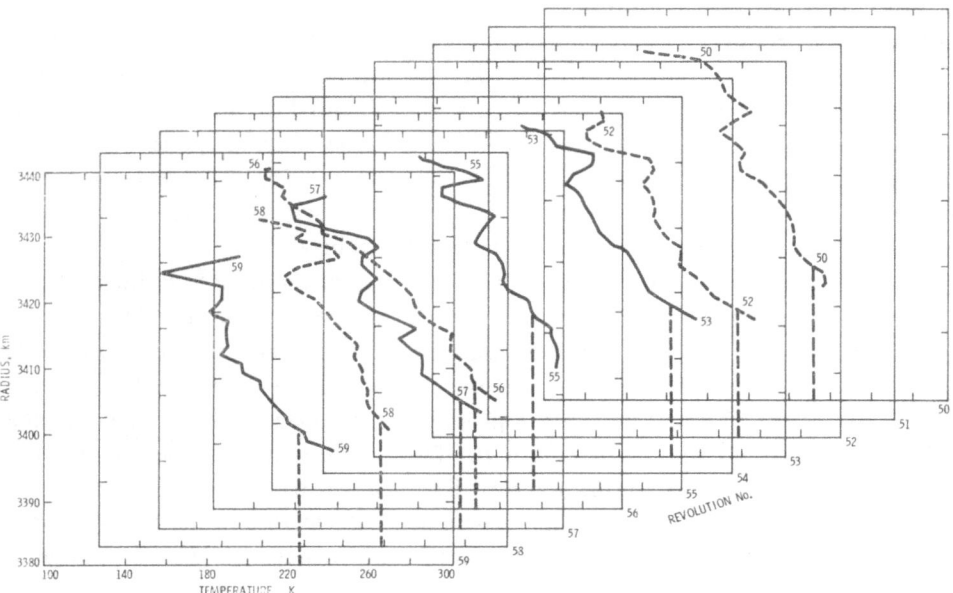

Fig. 9.   Temperature profiles from Mariner IX entry occultation data for revolutions 50–59 (from Kliore *et al.*, 1972b).

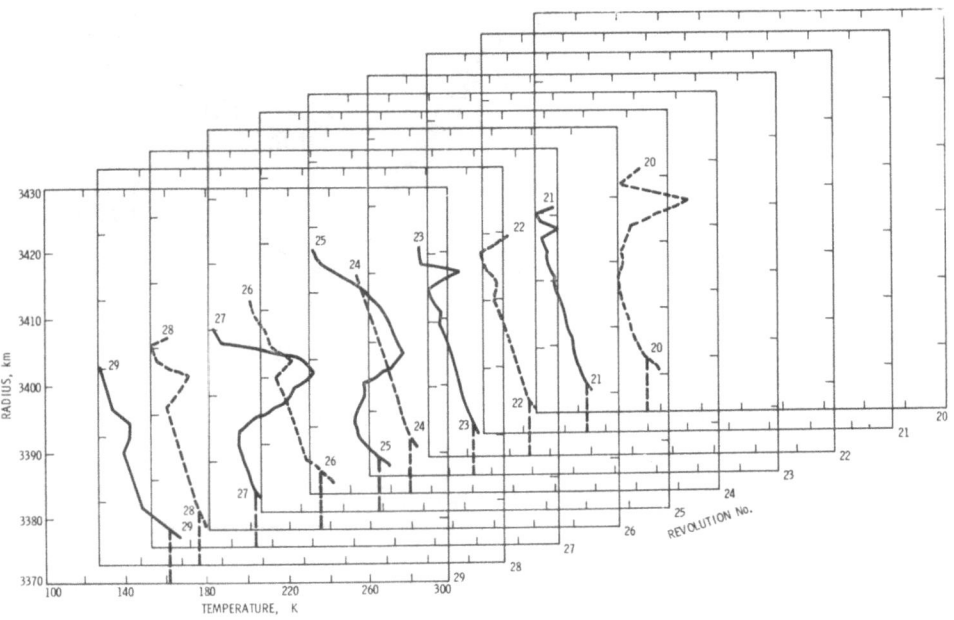

Fig. 10.   Temperature profiles derived from exit occultation measurements of Mariner IX for revolutions 20–29 (from Kliore *et al.*, 1972b).

to levels o at least 30 km are absorbing solar radiation, thus heating the upper portions of the troposphere and reducing the temperature of the lower layers. Figure 9 portrays temperature profiles measured in daytime during revolutions 50–59, twenty-five days later. Although Mars was still visually obscured at this time, as evident from television photographs, the atmosphere nevertheless had undergone some degree of clearing and was no longer isothermal in character. The average temperature gradient, however, was still only about $-2.5\,K\,km^{-1}$, which at that time was attributed to the presence of residual dust in the atmosphere.

Standard Mission exit measurements were obtained at around 65° latitude during northern hemisphere mid-winter on Mars, and the temperature profiles derived from these measurements show very low temperatures. Surface temperatures of 150–160 K are indicated, with temperatures reaching the carbon dioxide saturation temperature at altitudes below 10 km. Temperature profiles for exit measurements taken during revolutions 20–29 are shown in Figure 10. The erratic nature of the temperature profiles for revolutions 25 and 27 are most likely caused by instability of the space-craft oscillator. Measurements of the electron density structure in the daytime iono-sphere were obtained for all entry measurements which were taken over a range of

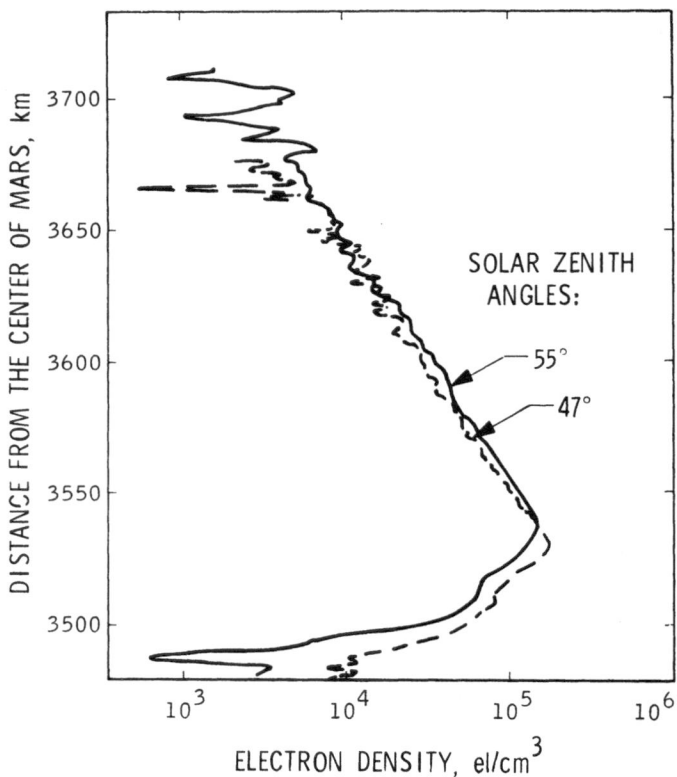

Fig. 11.   A comparison of electron density profiles measured by Mariner IX at solar zenith angles of 55 and 47 deg (from Kliore *et al.*, 1973).

a solar zenith angles of 56 to 47 deg. Figure 11 shows two typical electron density profiles, one obtained during revolution 12 at a solar zenith angle of approximately 55°, and the other during revolution 67 at a solar zenith angle of about 47°. As the solar zenith angle gradually decreased in the course of the Standard Mission, the density of ionization increased and the altitude of the maximum became lower.

The second episode of Earth occultations, called Extended Mission I, took place in May and June of 1972 (Kliore *et al.*, 1973; Cain *et al.*, 1973). In contrast to the measurements made during the Standard Mission, these Extended Mission I measurements were remarkable because the latitude coverage ranged from about +86 to −80°, thus providing the first occultation measurements of the north and south polar regions of Mars.

The nine entry measurements obtained at latitudes above 65° indicate surface atmospheric pressures ranging from 4.4 to 7.4 mb, with the average lying about 5.7 mb. The temperature profiles obtained in this region, all taken at low solar elevation angles, did not show appreciable differences in temperature gradients, which were about −2 to −2.5 K km^{-1}. The near surface temperatures, measured directly over the north polar cap, showed temperatures ranging from about 178 K to about 191 K, all substantially above the freezing point of carbon dioxide. Since the surface temperature there was at this time measured to be about 150 K (Kieffer, 1972) by the Infrared Radiometer instrument on Mariner IX, it is strongly suggested that a sharp temperature discontinuity existed in the lower 1 km or so of the atmosphere, which could not be resolved by the radio beam. Some representative temperature profiles from the mid-latitude entry measurements during Extended Mission I are shown in the Figure 12. It is seen that the temperature gradients are quite similar and have an average value of about −2.3 K km^{-1}. The mean temperature gradients

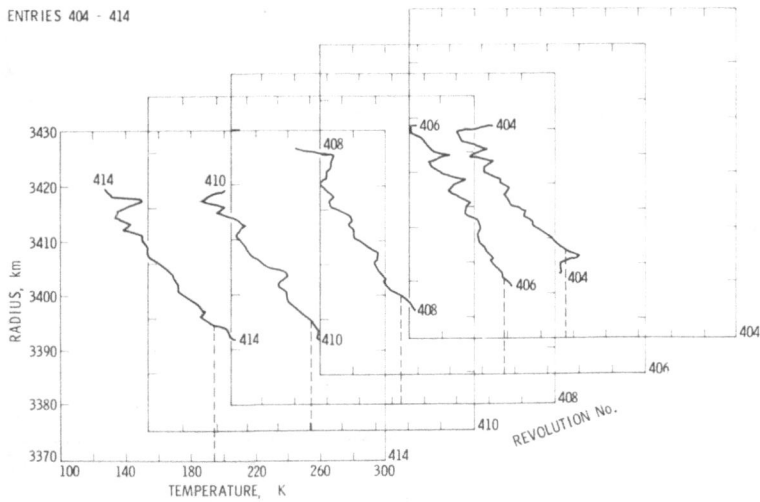

Fig. 12. Temperature profiles from Mariner IX occultation entry measurements during revolutions 404–414 (from Kliore *et al.*, 1973).

in the troposphere for all entry measurements during Extended Mission I range from about 0 to $-3.8\,\mathrm{K\,km^{-1}}$, with no apparent correlation with latitude, Mars local time, or solar elevation angle. These gradients are only slightly steeper than the ones observed during the Standard Mission, when the atmosphere still held a considerable amount of dust, and are far below the theoretical adiabatic gradient of about $-5\,\mathrm{K}$ $\mathrm{km^{-1}}$ (Gierasch and Goody, 1968). Thus it is evident that of all the temperature profiles measured in the Martian atmosphere with Mariners IV, VI, VII and IX, there is not a single case of temperature profile with an adiabatic lapse rate. In fact, they are significantly sub-adiabatic, and the average gradient of about $-2.3\,\mathrm{K\,km^{-1}}$ is in very good agreement with gradients deduced for a radiative-dynamical model of the lower atmosphere of Mars described by Stone (1972).

The data taken in the southern polar area consist of one-way exit data, the quality of which was adversely affected by the instabilities of the spacecraft auxiliary oscillator. Most of the deduced surface pressures clustered around 4–5 mb, with temperatures consistent with expectation for nighttime in the winter season at the South Pole.

In the upper atmosphere, data on the structure of the ionosphere was obtained for solar zenith angles greater than 72° during Extended Mission I, in contrast to the 47–57° range covered during the Standard Mission. The observed changes in the

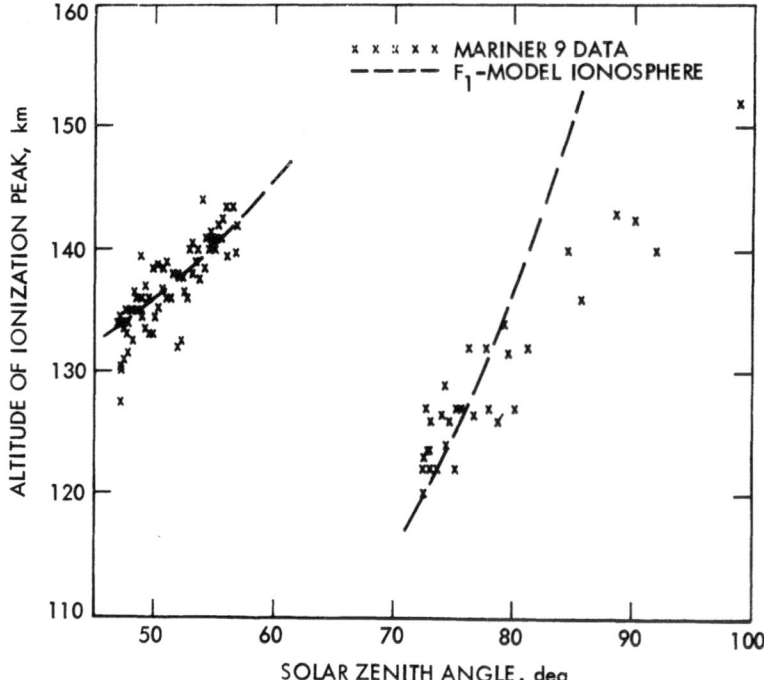

Fig. 13. Variation in the height of the ionosphere peak between the Standard Mission and the Extended Mission I (from Kliore et al., 1973).

altitude of the ionization peak as a function of the solar zenith angle are shown in Figure 13. The discontinuity in the altitude between 57° and 72° is most likely caused by a cooling of the atmosphere between the two sets of measurements. A reduction of the order of 25–30% in the average atmospheric temperature below the ionization peak would be required to explain the entire altitude change, which would be consistent with a clearing of the atmosphere. The model calculation for a F_1 type layer, shown by dashed curve in Figure 13, assumes atmospheric cooling to be the cause of the discontinuity.

The last Mariner IX occultation episode, Extended Mission II, took place in September and October of 1972. In complete contrast to the Standard Mission, the Extended Mission II entry measurements were in a band about −65° latitude, with the exit measurements falling into the mid-latitude regions. Unfortunately, the stability of the spacecraft oscillator had further deteriorated since Extended Mission I, and most of the exit measurements provided unreliable temperature profiles. Some parameters describing the Extended Mission II entry measurements are given in Figure 14. The latitudes are seen to range from about −56 to −67° while the solar elevation angle changes from about +2 to −2° and Mars local time ranges from about 09:30 for the early measurements to about 13:00 at the end. Some temperature profiles derived from Extended Mission II entry measurements are shown in Figure 15. The temperatures are all seen to be quite low, with near-surface temperatures ranging from about 140 to 150 K, consistent with frozen carbon dioxide on the south polar cap at a time closely approaching southern winter solstice. The data for revolution 648 was taken in the one-way mode and displays the effects of oscillator instability.

The surface pressure readings for Extended Mission II entry measurements are

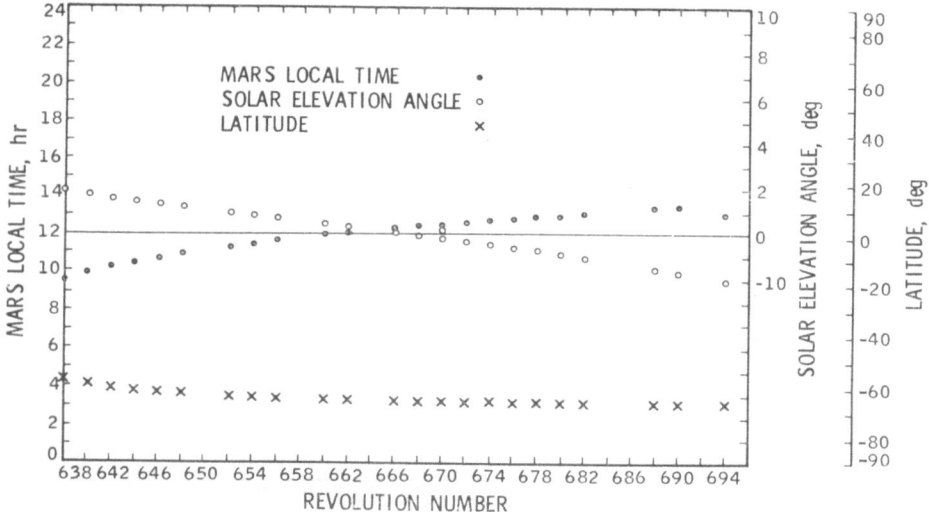

Fig. 14. Latitudes, Mars Local Time, and Solar Illumination angles for entry measurements during Extended Mission II.

shown in Figure 16. The pressures all seem to lie mostly between 4 and 5 mb, in very good agreement with the Extended Mission I exit measurements.

Radio occultation experiments were also performed by the Soviet Mars 2 and Mars 3 spacecraft in 1971 and 1972 (Kolosov *et al.*, 1972). Several measurements of the atmospheric pressure and electron density profiles have been described, generally in agreement with the Mariner IX results.

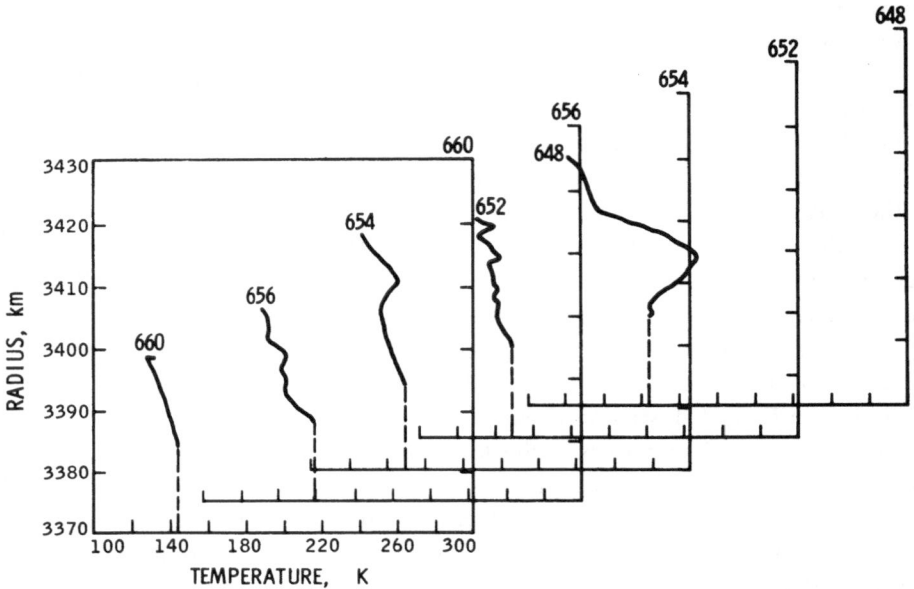

Fig. 15.   Temperature profiles from entry occultation measurements during revolutions 648–660.

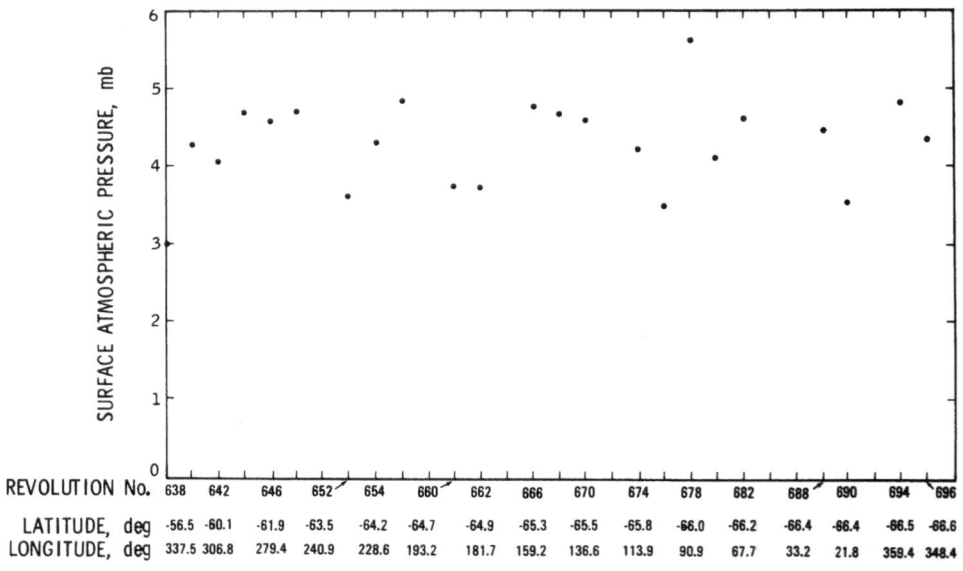

Fig. 16.   Surface atmospheric pressures from Extended Mission II entry occultation measurements.

3. Topography

Topographical information is obtained from radio occultation data through the precise measurement of the radius of closest approach of the refracted radio beam. This is made possible by the precise timing of the occultation event, accurate knowledge of the ephemeris of the spacecraft near the time of occultation, and the inversion of the phase changes due to refraction to obtain the refractivity profile. The timing of the occultation is established by analyzing a spectrum of the signal at intervals sufficient to provide an adequate signal-to-noise ratio in the spectrum, and by performing an interpolation to obtain the point in time when the signal is approximately 6 dB below its normal pre-occultation value. This value corresponds closely to the value expected in the Fresnel pattern produced by a knife edge when the source is directly behind the edge. During the Standard Mission the signal-to-noise ratio was sufficient to analyze the signal spectrum at intervals of 0.068 s. During the Extended Missions, the signal-to-noise ratio had deteriorated to the point where 0.68 s sample intervals were required, and hence the uncertainty was greater by about an order magnitude. The ephemeris of the spacecraft at the time of occultation was obtained from precise determination of the orbit. It was estimated that the uncertainty in the time of periapsis passage was less than 0.1 s during the Standard Mission and may have been as large as 0.6 s during the Extended Missions. Hence, under the assumption that the uncertainties in occultation timing and of the trajectory time are independent, one obtains a timing uncertainty for the Standard Mission of about 0.1 s and for the Extended Missions about 0.9 s. The timing uncertainty maps into radius error through the magnitude of the radial velocity of the radio beam with respect to the surface of Mars. During the Standard Mission, the radial velocity was less than 2.5 km s^{-1}, and hence, the radius uncertainty due to timing was less than 0.25 km. During the first Extended Mission, the radial velocity was about 1 km s^{-1}, thus yielding a radius error of about 0.9 km. During the second Extended Mission in September and October of 1972, the timing uncertainty was still 0.9 s, but the velocity of the beam had increased about 2.3 km s^{-1}, resulting in a total radius error of about 2.1 km.

The first two Mars radius measurements with Mariner IV showed radii different by about 5 km, at roughly corresponding north and south latitudes, which immediately indicated the presence of large topographic elevation differences on Mars. Mariner VI and Mariner VII provided four more radius measurements, including measurements near the equator and the North Pole. These six points were then used to produce an oblate spheroid approximation to the physical shape of Mars, having an equatorial radius of 3394.5 km and a polar radius of 3378.2 km, for a flattening of 0.0048 (Kliore, 1971).

The large number of radius measurements taken with Mariner IX has provided much significant information on Martian global topography. The Standard Mission provided radius as well as pressure altitude measurements in the south equatorial regions and a narrow band of latitudes along 65° latitude. Altitudes in the south equatorial region ranged from about −4.4 km in Hellas to a high of 9.6 km in Claritas, with a net excursion of 14 km and a mean altitude of 2.7 km referred to the 6.1 mb pressure level. The measurements in the vicinity of 65° latitude showed uniformly negative altitudes with a mean of −2.6 km. This disparity in surface pressure and elevation between the equatorial and circumpolar regions at first was thought to indicate that the physical shape of Mars was more oblate than the shape of a gravitational equipotential surface, or a geoid, leading to higher atmospheric pressures at high latitudes than at the equator. This implication was supported by a comparison of a triaxial ellipsoids which were fit to the first raw radius data from the Standard Mission, and then to radii of the 6.1 mb pressure level, which were computed from the measured radii as well as the deduced surface pressures and temperature profiles (Cain *et al.*, 1972). The flattening coefficients obtained from an average equatorial radius were 0.0074 in the case of the physical surface ellipsoid, and 0.0051 for the 6.1 mb isobaric surface, which was in fairly good agreement with the dynamical flattening of 0.005238, obtained from the motion of the natural satellite of Mars (Sinclair, 1972).

Occultation measurements were taken near the North Pole during Extended Mission I, and it was found that the radii at these locations averaged to about 3377.8 km, indicating that the radius of Mars only changes by about 2 km from a latitude of 65° to the North Pole. These radii were also consistent with exit measurement of Mariner 6 (Kliore *et al.*, 1969, 1971). Thus it is obvious that the region around 65° latitude is substantially and systematically lower than other areas on Mars. In addition, all locations of occultation measurements in the northern hemisphere in Aetheria, Phlegra, and Amazonis exhibited altitudes from 1 to 3 km below the 6.1 mb isobaric level. However, when the measurements crossed into the Tharsis area (Figure 6) the elevations rose and the surface pressures dropped proportionately. Measurement points 432 and 436, lying on the Tharsis plateau, indicated pressure altitudes of about 6.7 km and measurement 438 taken on the 'Chandelier' region of the Coprates canyon indicates a pressure altitude of about 10 km and a surface pressure of 2 mb.

By coincidence, the location of measurement 434 entry fell very close to the top of the volcanic feature known as Middle Spot (Pavonis Mons), which was one of the four prominent features first discovered in Mariner IX televison pictures during the Martian dust storm (Masursky *et al.*, 1972). Although the location of the occultation tangency point did not fall within the caldera of the volcano, the geometry was such that the line of sight practically bisected the entire volcanic shield, thus making it virtually certain that the beam was actually intercepted by the highest feature along the track, which is likely to have been the summit area. The radius that was measured here was 3417.1 km which is about 13.6 to 13.8 km above adjacent occultation measurements. On the basis of pressure altitudes, the height of Middle Spot was 12.5 km, and the pressure at the top was about 1 mb.

The entire southern hemisphere of Mars was found to be 3 to 4 km higher than corresponding latitudes in the northern hemisphere and the South polar area was

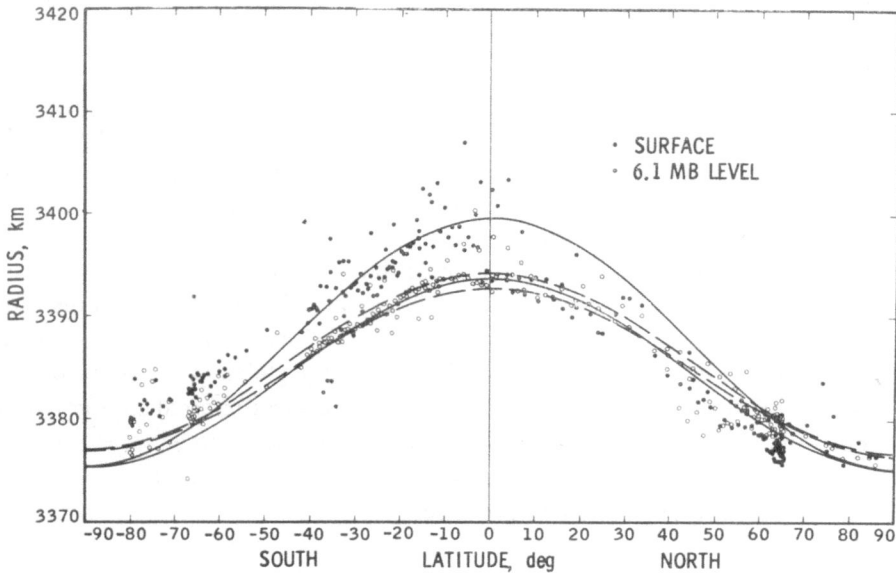

Fig. 17. Occultation radii measured by Mariner IX. The closed circles indicate radii of the solid surface and the open circles are derived radii of the 6.1 mb pressure level. The solid and dashed lines define the radii of fitted tri-axial ellipsoids approximating the solid surface and a 6.1 mb isobaric surface, respectively.

found to be higher than the North Pole by about 3.4 km. This asymmetry is obvious from an examination of Figure 17 which shows the measured radii of the solid surface, (solid dots) and the 6.1 mb level, (open dots) plotted vs the latitude of measurement. The solid lines show the maximum and minimum radii of an ellipsoid fitted to the measured solid radius data. The parameters of this ellipsoid are: A major equatorial axis) = 3400.12 km; B (minor equatorial axis) = 3394.19 km; and C (polar axis) = = 3375.45. The longitude of the major axis is 99.57°W. The broken lines represent radii of a triaxial ellipsoid fitted to the 6.1 mb level radii. For this ellipsoid $A = 3396.67$ km; $B = 3395.23$ km, $C = 3377.22$, and $\theta = 108.1°$W (Cain *et al.*, 1973). It is immediately obvious that radii measured in the southern hemisphere are obviously and systematically displaced about the upper limit of the triaxial ellipsoid, with the exception of four points in Hellas, which lie below the ellipsoid boundaries. At the same time, all northern hemisphere measurements, with the exception of a few points near the North Pole, lie within or below the ellipsoid boundaries. In particular, the large concentration of measurements at 65° and $-65°$ latitudes clearly illustrate the asymmetry between the two hemispheres of Mars. In contrast, the 6.1 mb radii cluster fairly closely around the contours of their ellipsoid, with the exception of a few measurements in the southern hemisphere which were undoubtedly affected by the poor quality of the one-way data. It should be pointed out that the flattening corresponding to the solid surface ellipsoid is 0.0064 compared to a flattening for the 6.1 mb ellipsoid of 0.0055.

It is clear from the previous discussion that a triaxial ellipsoid is not a very good approximation for the physical surface of Mars. For that reason, models using a spherical harmonic representation were used to portray the shape of Mars (Cain *et al.*, 1973). For a second order spherical harmonic approximation, the best fit is displaced southward from the mass center by 2.85 km and there is an equatorial displacement of about 1.7 km in the direction of about 100° W. The resulting mean equatorial and polar radii are 3400.6 km and 3379.6 km, giving a flattening of 0.00607.

A spherical harmonic fit was also performed with the 6.1 mb isobaric radii. The resulting surface should correspond to a geoid, but only under quiescent conditions. Care must be taken in the interpretation of the isobaric data derived from the occultation experiment, because of the unknown effects of dynamical processes in the Martian atmosphere as well as seasonal effects upon the global atmospheric pressure of Mars. When an attempt was made to compare pressures referred to a gravitationally determined geoid of Mars in the circumpolar and the south equatorial regions, somewhat higher pressures than expected were noted in the circumpolar region, possibly indicating the effect of dynamical processes.

4. Conclusions and Recommendations

From the radio occultation results produced by Mariner IV, VI and VII and especially Mariner IX, several conclusions can be drawn regarding the nature and structure of the atmosphere and surface of Mars:

(1) The structure of the atmospheric temperature profiles in the daytime are quite sensitive to the presence of dust in the atmosphere, and thus could be used to gauge the relative clarity of the Martian lower atmosphere. For example, the nearly isothermal temperature profile obtained from the Mariner IV daytime measurement may indicate the presence of substantial dust in the atmosphere, which could partially explain the low contrast of the Mariner IV television pictures.

(2) Temperature gradients obtained from occultation measurements by Mariners VI, VII and IX while the atmosphere was apparently clear of dust, do not approach the adiabatic gradient predicted by theory for a clear Martian atmosphere in radiative-convective balance. It may be suggested that the vertical temperature structure in the daytime atmosphere of Mars is strongly modified by dynamical processes.

(3) The surface atmospheric pressure on Mars changes very drastically from one location on Mars to another in correspondence with changes in the topography. For example, pressure from about 10.3 mb in the North circumpolar region down to about 1 mb near the top of the Middle Spot volcano have been measured. In the near equatorial regions, the relative elevation difference between the Tharsis plateau and the Hellas basin amounts to some 20 km.

(4) The main peak of the daytime ionosphere behaves like a F_1 type layer produced by photoionization of carbon dioxide by extreme solar ultraviolet, with a subsidiary peak, possibly of the E-type, produced by solar X-rays.

(5) The physical body of Mars displays very pronounced equatorial and north-

south asymmetries, amounting to about 5 km and 3.5 km, respectively, thus rendering a simple triaxial figure representation impractical, and calling for more sophisticated spherical harmonic modelling.

(6) An isobaric surface derived from occultation data does not precisely agree with a geoid figure derived from spacecraft tracking data, indicating the presence of possible atmospheric dynamical effects upon the surface pressure.

The experience of the Mariner IV, VI, VII and IX radio occultation experiments have suggested several possible improvements in spacecraft radio system design that would greatly improve the quality of radio occultation data. Among these are the following:

(a) A stable spacecraft oscillator with a short-term stability of about 1 part in 10^{11} would almost double the amount of radio occultation data for any given mission by improving the quality of the one-way exit data to the level of the two-way entry data.

(b) The presence of a second coherent downlink frequency would enable the unambiguous separation of charged particles from a neutral atmosphere. Such a system will be flown on the Viking orbiters in 1975–76.

(c) A steerable spacecraft high gain antenna and an expanded attitude control system would extend the lifetime of the spacecraft useful for radio occultation measurements and would provide data on possible seasonal changes in atmospheric surface pressures.

Acknowledgements

The author gratefully acknowledges the help of colleagues and associates who have contributed to the success of the Mariner Radio Occultation Experiments. These include: T. W. Hamilton, D. L. Cain, G. Fjeldbo, B. L. Seidel, G. S. Levy, S. I. Rasool, J. F. Jordan, M. J. Sykes, P. M. Woiceshyn, and various Mariner project personnel at JPL as well as NASA, together with the personnel of the Deep Space Net. The majority of the work described in this paper was performed at the Jet Propulsion Laboratory, under NASA Contract NAS 7-100 and at Stanford University under NASA grant NGR 05-020-065.

References

Barth, C. A., Stewart, A. I., Hord, C. W., and Lane, A. L.: 1972, *Icarus* **17**, 457.
Bean, B. R. and Thayer, G. D.: 1963, *J. Res. NBS-D; Radio Propagation* **07D**, No. 3, 273.
Cain, D. L., Kliore, A. J., Seidel, B. L., and Sykes, M. J.: 1972, *Icarus* **17**, 517.
Cain, D. L., Kliore, A. J., Seidel, B. L., Sykes, M. J., and Woiceshyn, P. M.: 1973, *J. Geophys. Res.* **78**, 4352.
Fjeldbo, G.: 1964, Report SU-SEL-64-025, Stanford Electronics Laboratories, Stanford, California.
Fjeldbo, G. and Eshleman, V. R.: 1968, *Planetary Space Sci.* **16**, 1035.
Fjeldbo, G., Kliore, A., and Seidel, B.: 1970, *Radio Sci.* **5**, 381.
Gierasch, P. and Goody, R.: 1968, *Planetary Space Sci.* **16**, 615.
Kieffer, H. H.: 1972, private communication.
Kliore, A. J., Cain, D. L., and Hamilton, T. W.: 1964, JPL-TR-32-674, Jet Propulsion Laboratory, Pasadena, Calif.
Kliore, A. J. Cain, D. L. Levy, G. S., Eshleman, V. R., Drake, F. D., and Fjeldbo, G.: 1965a, *Astron. Aeron.* **7**, 72.
Kliore, A. J., Cain, D. L., Levy, G. S., Eshleman, V. R., Fjeldbo, G., and Drake, F. D.: 1965b, *Science* **149**, 1243.

Kliore, A. J., Cain, D. L., Levy, G. S.: 1968, in *Space Research VII, Moon and Planets*, North Holland Publishing Company, Amsterdam, p. 220.

Kliore, A. J., Fjeldbo, G., Seidel, B. L., and Rasool, S. I.: 1969, *Science* **166**, 1393.

Kliore, A. J., Cain, D. L., Seidel, B. L., and Fjeldbo, G.: 1970a, *Icarus* **12**, 82.

Kliore, A. J., Fjeldbo, G., and Seidel, B. L.: 1970b, *Radio Sci.* **5**, 373.

Kliore, A. J.: 1971, *Bull. Am. Astron. Soc.* **3**, 498 (abstract).

Kliore, A. J., Fjeldbo, G., and Seidel, B. L.: 1971, *Space Research* **10**, 165.

Kliore, A. J.: 1972, in L. Colin (ed.), *Proc. Workshop on the Mathematics of Profile Inversion*, NASA TM-X-62, 150, 3-2, 3-16.

Kliore, A. J., Cain, D. L., Fjeldbo, G., Seidel, B. L., and Rasool, S. I.: 1972a, *Science* **175**, 313.

Kliore, A. J., Cain, D. L., Fjeldbo, G., Seidel, B. L., Sykes, M. J., and Rasool, S. I.: 1972b, *Icarus* **17**, 484.

Kliore, A. J., Fjeldbo, G., Seidel, B. L., Sykes, M. J., and Woiceshyn, P. M.: 1973, *J. Geophys. Res.* **78**, 4331.

Kolosov, M. A., Yakovlev, O. I., Kruglov, Yu. M., Trusov, B. P., Yefimov, A. I., and Kerzhanovich, V. V.: 1972, *Radiotechnika i Elektronika* **12**, 2483 (in Russian).

Masursky, H., *et al.*: 1972, *Science* **174**, 1321.

Rasool, S. I., Hogan, J. S., Stewart, R. W., and Russell, L. H.: 1970, *J. Atmospheric Sci.* **27**, 841.

Sinclair, A. T.: 1972, *Monthly Notices Roy. Astron. Soc.* **155**, 249.

Stone, P. H.: 1972, *J. Atmospheric Sci.* **29**, 405.

DISCUSSION

Pearl: I have two comments. First, such strong temperature inversions as you suggest over the north polar cap have terrestrial analogs: such inversions are observed in the antarctic. Secondly, computations based on Mariner 9 IRIS temperature profile data indicate that 15–20% of the incident solar radiation is absorbed by the dusty atmosphere during the dust storm.

MARS: LOCAL STRUCTURE OF DUST STORMS

G. I. BARENBLATT

Institute of Mechanics of Moscow University, Moscow, U.S.S.R.

and

G. S. GOLITSYN

Institute of Atmospheric Physics, Academy of Sciences of the U.S.S.R., Moscow, U.S.S.R.

Abstract. A hydronamical theory describing the motion of the dusty flow including the effects of thermal stratification is presented. It is shown that the dust, while decreasing the intensity of turbulence in the flow, can accelerate the wind. Applied to Mars, the theory reveals that these effects may be important there in understanding the nature and the long duration of Martian dust storms.

Woszczyk and Iwaniszewska (eds.), Exploration of the Planetary System, 317. All Rights Reserved

PART III

OUTER PLANETS AND THEIR SATELLITES

THE HYDROGEN TO HELIUM MIXING RATIO
IN THE GIANT PLANETS

DANIEL GAUTIER

Groupe Planètes, Observatoire de Paris-Meudon, France

Abstract. The theoretical bases for deducing the solar abundance of helium from the determination of the H₂/He mixing ratio in the giant planets are examined. The present state of the determination of the helium abundance is given. Experimental methods used to evaluate the H₂/He ratio in the giant planets are critically reviewed. Methods in prospect are also briefly exposed.

1. The Importance of the Knowledge of the H_2/He Mixing Ratio in the Giant Planets

From the low density of Jupiter and Saturn, it has been thought for a long time that these planets are mainly composed of hydrogen and helium. Moreover, high gravitational fields and low temperatures reduce the escape of molecules and atoms. Thus, the composition of these planets should be similar to the composition of the primitive nebula.

Hence many people believe that the cosmic abundance of hydrogen and helium will be simply and accurately given by determining the atmospheric composition of the giant planets.

I feel we should be more cautious. In fact, this assumption implies that the whole of each planet has a constant composition everywhere. Clearly, the problem can only be solved by a correct theory of the interiors of these planets. Such theories have to be checked by measurements of the boundary conditions including, among others, measurements of H_2/He in the atmospheres.

Sophisticated models of the interiors of Jupiter and Saturn have thus been calculated as a function of the H_2/He mixing ratio (for a review, see Hubbard and Smoluchowski, 1973).

Roughly, there are two theses proposed in the recent models.

In the model of Smoluchowski (1967, 1971), the ratio of H_2/He is not constant as a function of the radius of the planet and, accordingly, the atmosphere of the planet could be enriched in helium relative to the solar composition.

On the contrary, the model of Hubbard (1968, 1969) implies that Jupiter and Saturn should be chemically homogeneous and, as a consequence, the measurements of H_2/He in the atmosphere would give the relative abundance everywhere in the planet.

It is clear that the final choice between the interior models of Jupiter and Saturn cannot be made before an accurate determination of the H_2/He ratio in their atmospheres is obtained.

Because new determinations of the radii of Uranus and Neptune have diminished the mean density of these planets, previous models of the interiors of these planets are obsolete. But their composition should be similar to that of Jupiter and Saturn, with possible addition of heavier elements in their cores (according to Prinn (1973), solar

Woszczyk and Iwaniszewska (eds.), Exploration of the Planetary System, 321–327. All Rights Reserved
Copyright © 1974 by the IAU

composition objects of the size of Uranus and Neptune would have maximum mean densities of 0.2 and 0.18 g cm^{-3} while the present measured values are 1.31 and 1.66 g cm^{-3}).

In any case, measurements of the H_2/He ratio in all the giant planets would confirm or invalidate the hypotheses that a large fractionation of hydrogen and helium could have occurred at the beginning of the evolution of the solar system.

2. The Present State of the Evaluation of the Solar Helium Abundance

Because of our lack of information on the exact composition of the giant planets it is usually assumed that the H_2/He mixing ratio in these planets is close to the solar or cosmic abundance. In fact, the information contained in this assertion is less than is generally suspected.

For cosmic abundance the situation looks rather confusing (Danziger, 1970); the helium to atomic hydrogen ratios (per volume) quoted are between 0.06 and 0.16, with a mean value of 0.1 or 0.11 for the majority of normal stars and for the interstellar medium. Adopting these values for the giant planets will lead to a helium percentage of between 11% and 27%.

For the Sun, the question is very uncertain since the different methods of determining the helium abundance refer to different regions of the solar atmosphere (except for measurements of neutrino flux which unfortunately cannot be correctly interpreted because of the lack of satisfactory theoretical solar models).

From a recent and careful analysis Hirsberg (1973) concludes that the most accurate measurement comes from helium line intensity measurements (5 to 8% of helium by number); for the above mentioned reasons, this author estimates that this value is uncertain to a factor of 2 or 3. As a consequence, the helium percentage in the giant planets as deduced from this solar composition can be anywhere between 10% and 26%, with some preference for the range 10%–20%.

3. Present Information from Experimental Data

What can be deduced from the experimental data? We used at the present time two indirect methods to obtain some upper (or lower) limit of helium percentage in the giant planets.

3.1. ABSORPTION SPECTROSCOPY

The first method is the spectroscopic method; it is based on the measurement of the line-width of CH_4 which is broadened by hydrogen and helium. From an independent measurement of the hydrogen abundance, the partial pressure of helium is then estimated. This method is very crude because both the measurements of CH_4 line widths and of hydrogen abundance are very inaccurate, because scattering is neglected and because the depths of line formation are assumed to be the same for both CH_4 and H_2, which is extremely unlikely.

Owen and Mason (1969) deduced from such an analysis of 6200 Å CH_4 band that H_2/He should be larger than 4.5.

However, because of the uncertainties of the method, Hunten and Munch (1973) consider that it is safer to set an upper limit of about 1 for H_2/He.

To illustrate the uncertainties only due to inaccurate measurements of line-width and hydrogen abundance, we did an analysis of the H_2/He determination from the measurements of the line-widths of the 3 v_3 band of CH_4 located at 1.1 μ, recently made by Maillard *et al.* (1973) for Jupiter and De Bergh *et al.* (1973) for Saturn, with a Connes interferometer at Saint Michel de Provence Observatory.

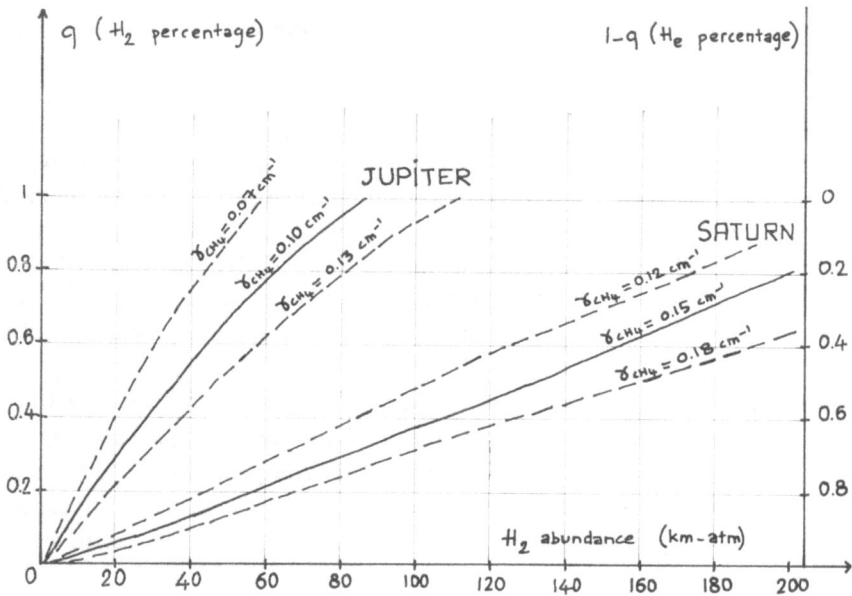

Fig. 1.   Percentage, q, of hydrogen in Jupiter and Saturn vs abundance of hydrogen, for the mean value and extreme limits of the measured line-width.

On Figure 1, is plotted the percentage, q, of hydrogen in Jupiter and Saturn as a function of the abundance of hydrogen, for the mean value and the extreme limits of the measured line-widths γ, in cm^{-1}; $1 - q$ is the percentage of helium. In his analysis of 6190 Å CH_4 band, Owen took a Jovian hydrogen abundance of 85 km-atm. From a more recent analysis of the earlier observations, Margolis and Hunt (1973) deduced an abundance of 65 ± 10 km-atm, a value in agreement with recent measurements of Trafton (1972).

For Saturn, ancient estimations of H_2 abundance were recently considerably reduced by new measurements. Trafton (1973) mentions the range 85–150 km-atm while Encrenaz and Owen (1973) announce 76 ± 20 km-atm.

All these results come from observations of quadrupole lines of H_2. On the other hand, De Bergh *et al.* (1974) give upper limits of 48 km-atm for Jupiter and 92 km-atm for Saturn, based on an evaluation of the hydrogen absorption, between 1.04 and 1.33 μ, due to the first overtone (2–0) pressure induced band centered at 1.2 μ.

The result for Jupiter has to be compared to the abundance of 45 km-atm deduced by Danielson (1966) from the fundamental (1–0) induced band of H_2 at 2.4 μ, observed at low resolution with Stratoscope II.

Clearly, quadrupole lines and pressure-induced bands of H_2 refer to different levels of formation.

In any case, even if we adopt only the results of quadrupole line measurements, from the examination of the diagram of Figure 1, we are faced with the following alternatives:

– either, because the above mentioned approximations are too crude, this method of inferring the H_2/He gives meaningless results;

– or a larger percentage of helium than the upper limit of the solar abundance cannot be excluded in the atmopheres of Jupiter and Saturn.

Hunt (1972) made a sophisticated analysis, including scattering, of the hydrogen quadrupole lines and of the $3\nu_3$ CH_4 band on Jupiter. He verified that the equivalent widths of the $3\nu_3$ band could be retrieved from a two-layer atmospheric model, with a mixing ratio H_2/He equal to 6 and a mixing ratio of CH_4/H_2 of $(7\pm1)\times10^{-4}$. However, it is not clear if the fitting of the $3\nu_3$ equivalent widths is sensitive to the H_2/He ratio, and what the exact influence of the atmospheric model and of each scattering parameter is.

For instance, Bergstralh (1973) in an analysis with scattering of the same $3\,\nu_3$ CH_4 band, mentions that five of the six combinations of continuum single scattering albedo ω_{oc} and total optical thickness τ_c of the upper level layer yield satisfactory fits to the observed equivalent widths with proper choices of other scattering parameters. Our lack of information on the properties of the Jovian clouds is so large that, in my opinion, the question remains open.

3.2. OCCULTATION EXPERIMENTS

Contrarily to a rather common belief, it is very difficult to simply infer the H_2/He ratio from measurements of star occultation by a giant planet. In fact, to deduce the mean molecular weight (and thus the H_2/He ratio) from the scale height we need to know the temperature profile, a quantity which is not known with any precision. The best thing we can do then is to infer a set of temperature profiles as a function of the H_2/He ratio. Very important results on the upper atmosphere of Jupiter were obtained by different teams from the 1971 β Scorpii occultation measurements. Both the results of Texas Group (Hubbard *et al.*, 1972) and the Meudon Group (Vapillon *et al.*, 1973) infirm the previous results of Baum and Code (1953) deduced from the 1952 σ Arietis occultation.

From the set of temperature profiles deduced by Vapillon (1974), it can be seen that the inferred thermal profiles become hotter and hotter with increasing values of helium percentage.

Admitting that a maximal value of 400 K can hardly be exceeded, we can take as upper limit of the helium percentage the value 30%. Hubbard *et al.* (1972) consider that helium should be less than 25%.

An interesting experiment based on a method suggested by Brinckman (1971) was made by the Harvard-Cornell team during the same occultation. The method exploits the fact that the light curve is interrupted by a large number of spikes which are presumably due to inhomogeneities in the planetary atmosphere.

By measuring the difference of the two spike arrival times at two different wavelengths, Brinckman estimated that one could infer the ratio of the refractivity of the atmospheres at these two wavelengths and deduce from that the H_2/He mixing ratio. However, Wasserman and Veverka (1973) pointed out that the complete occultation curve should be also measured, which in fact reduces the accuracy of the method. At rhis time, results of Harvard-Cornell teams have not yet been published, but from a teport of Veverka *et al.* (1973) the inferred percentage of helium would be between 5% and 42%.

High temperatures of the upper atmosphere of Neptune inferred by Kovalevsky and Link (1969) from the 1968 occultation measurement of BD $- 17°4388$ by Neptune seem to indicate that the atmosphere of this planet is rich in hydrogen.

4. Methods in Prospect

Two other methods of remote determination of the helium abundance will be tried out in the next years from spatial missions of possible airborne experiments.

The first one consists of measuring the emissivity of the helium resonance line at 584 Å. Carlson and Judge (1971) showed that this dayglow depends on the H_2/He ratio in the lower atmosphere and the number density of the homopause. This latter quantity can be determined from $L\alpha$ measurements. Pioneer 10 and 11 carry ultraviolet photometers for both the $L\alpha$ and the helium resonance line. Simultaneous radio occultation measurements would bring information on thermal profiles. However from the analysis of Carlson and Judge (1971), it can be seen that the determination of helium will be very inaccurate for an abundance less than 50%.

Infrared experiments look more promising. They consist of measuring the infrared thermal spectrum of the studied planet at several wavelengths properly chosen in the spectral range where H_2 and He are only responsible for the absorption. It is the pressure induced spectrum due to collisions, between 18 and 50 μ for Jupiter and for wavelengths superior to 9 μ for the other giant planets. The H_2/He mixing ratio is then inferred from spectral measurements by an iterative method (Gautier and Grossman, 1972; Encrenaz and Gautier, 1973). Such measurements can be tried from an aircraft flying at the altitude of the tropopause where good atmospheric windows exist in the far infrared, or from a spacecraft in a fly-by or an Orbiter Mission.

Several experiments using Michelson interferometers are planned next year to measure the infrared spectrum of Jupiter from the NASA airborne infrared laboratory (C-141 aircraft). From a numerical analysis, taking into account the expected signal to noise ratios, it is hoped that the H_2/He ratio can be obtained from these experiments (Encrenaz and Gautier, 1973) but with a possible systematic error due to the lack of spatial resolution. An infrared radiometer with two broad band channels (channel S covering the range 13–26 μ and channel L covering 29–60 μ) is carried by Pioneer 10 and also by Pioneer 11. These radiometers should measure 'limb darkening' of Jupiter in the two channels. Hunten and Munch (1973) did a detailed analysis of the theory of this experiment and their conclusion is rather pessimistic. They are probably right mainly because it will not be possible to properly evaluate the influence of the emissivity due to the pure rotational bands of NH_3 in the L channel.

A much more sophisticated experiment is planned for the 1977 Mariner Jupiter/ Saturn Mission which will carry a radiometer or an interferometer.

A detailed analysis of the possibilities of remote sounding of the atmospheres of the four giant planets by infrared technics has been made by Taylor (1972). From this work, it appears that the H_2/He ratio could be inferred from far infrared measurements at four wavelengths for any Jovian planet with an accuracy of $\pm 1\%$ if $H_2/He \geqslant 10$; but this accuracy decreases with decreasing H_2/He ratios. Errors of calibration may have been underestimated in this work.

If these experiments do not succeed the final answer may be given by *in situ* measurements which are not planned before 1980's years.

5. Conclusion

At the present time there is no experimental evidence that the H_2/He mixing ratio in the giant planets should correspond to the solar composition, but it is probably the best choice to assume it. The solar helium abundance is itself very imprecise and the corresponding percentage of helium in the giant planets could be between 10% and 26%.

It can be reasonably expected that the H_2/He ratio will be accurately determined during the next ten years for Jupiter and Saturn. And the possibility cannot be excluded that in the future the 'cosmic abundance' will be deduced from the composition of these planets.

References

Baum, W. A. and Code, A. D.: 1953, *Astron. J.* **58**, 108.
Bergstralh, J.: 1973, *Icarus* **19**, 390.
Brinckman, R. T.: 1971, *Nature* **230**, 515.
Carlson, R. W. and Judge, D. L.: 1971, *Planetary Space Sci.* **19**, 327.
Danielson, R. E.: 1966, *Astrophys. J.* **143**, 949.
Danziger, I. J.: 1970, *Ann. Rev. Astron. Astrophys.* **8**, 161.
De Bergh, C., Combes, M., Encrenaz, Th., Lecacheux, J., Vion, M., and Maillard, J. P.: 1974, this volume, p. 357.
De Bergh, C., Combes, M., Lecacheux, J., and Maillard, J. P.: 1973, *Astron. Astrophys.* **28**, 457.
Encrenaz, Th. and Gautier, D.: 1973, *Astron. Astrophys.* **26**, 143.
Encrenaz, Th. and Owen, T.: 1973, *Astron. Astrophys.* **28**, 119.
Gautier, D. and Grossman, K.: 1972, *J. Atmospheric Sci.* **29**, 788.
Hirsberg, J.: 1973, *Rev. Geophys. Space Phys.* **11**, 115.

Hubbard, W. B.: 1968, *Astrophys. J.* **152**, 745.
Hubbard, W. B.: 1969, *Astrophys. J.* **155**, 333.
Hubbard, W. B. and Smoluchowski, R.: 1973, *Space Sci. Rev.* **14**, 599.
Hubbard, W. B., Nather, R. E., Evans, D. S., Tull, R. G., Wells, D. C., Van Atters, G. W., Warner, B., and Vanden Bout, P.: 1972, *Astron. J.* **77**, 41.
Hunt, G. E.: 1973, *Monthly Notices Roy. Astron. Soc.* **161**, 347.
Hunten, D. M. and Munch, G.: 1973, *Space Sci. Rev.* **14**, 433.
Kovalevsky, J. and Link F.: 1969, *Astron. Astrophys.* **2**, 398.
Maillard, J. P., Combes, M., Encrenaz, Th., and Lecacheux, J.: 1973, *Astron. Astrophys.* **25**, 219.
Margolis, J. S. and Hunt, G. E.: 1973, *Icarus*, **18**, 593.
Owen, T. and Mason, H. P.: 1969, *J. Atmospheric Sci.* **26**, 870.
Prinn, R. G.: 1973, *Planetary Space Sci.* **21**, 1501.
Smoluchowski, R.: 1967, *Nature* **215**, 691.
Smoluchowski, R.: 1971, *Astrophys. J.* **147**, 765.
Taylor, F. W.: 1972, *J. Atmospheric Sci.* **29**, 950.
Trafton, L. M.: 1972, *Bull. Am. Astron. Soc.* **4**, 367.
Trafton, L. M.: 1973, *Astrophys. J.* **182**, 615.
Vapillon, L.: 1974, this volume, p. 391.
Vapillon, L., Combes, M., and Lecacheux, J.: 1973, *Astron. Astrophys.* **29**, 135.
Wasserman, L. and Veverka, J.: 1973, *Icarus* **18**, 599.
Veverka, J., Wasserman, L., Elliot, J., Sagan, C., and Liller, W.: 1973, Cornell Univ. Report CRS-R-556.

DISCUSSION

Irvine: What do you consider to be the best current estimates of the density of Uranus and Neptune?

Gautier: From the measurements of the radius of Uranus made by Stratoscope, the mean density of this planet would be about 1.31 g cm^{-3}. From the measurement of the radius of Neptune inferred from occultation measurements by Kovalevsky and Link (1969) its density would be 1.49 g cm^{-3}.

Beer: Is there, in fact, any real, direct, evidence for there being any helium whatsoever in the outer planets?

Gautier: No, there is no direct evidence, but it would be extremely difficult to explain how helium of mass 4 would have escaped out of the atmosphere and not hydrogen of mass 2.

Fox: Are spectroscopic data on collision-induced absorption by mixtures of H_2 and He (as functions of temperature and ortho-para ratio, for example), and on pressure-broadening of CH_4 lines by H_2 and He in the 1.1 μ vibration-rotation band, sufficiently accurate for a precise determination of the H_2-to-He ratio?

Gautier: I took the values of the coefficients of pressure broadening in Varanasi *et al.* (*Astrophys. J.* **179**, 977, 1973). Even if the values are not very accurate, the causes of error come mainly from the assumptions I have mentioned above: using a reflecting-layer model, neglecting scattering, assuming the depths of line formation are the same for CH_4 and H_2.

Traub: In the near-infrared, at least, the greatest difficulty with measuring line widths on Jupiter is the contribution of image motion due to atmospheric 'seeing' effects. For example, in our H_2 quadrupole line observations on Jupiter, the seeing width is larger than the combined instrumental and intrinsic widths, even after the overall rotational broadening has been removed by proper spectrometer alignment.

ON THE EQUATION OF STATE OF HYDROGEN AND ITS USE IN
MODELS OF MAJOR PLANETS

R. S. HAWKE*

University of California Lawrence Livermore Lab.**, Calif., U.S.A.

Abstract. Since Jupiter and Saturn are considered to be composed primarily of hydrogen, its pressure-density equation of state is needed for computational models of their interiors. Until recently, experimental data were limited to 20 kbar statically and 40 kbar dynamically. Since the majority of a major planet is at a pressure in excess of this, there were only theoretical calculations available for modeling.

Wigner-Seitz type calculations have been shown to be accurate at determining the equations of state of the alkali metals. Hence, it has been assumed that the equation of state of metallic hydrogen can be calculated in the same way with fair confidence. However, the molecular hydrogen equation of state has been much more elusive. The many attempts at modeling the interatomic forces have led to rather scattered pressure density relationships.

The planetary model situation is further complicated by the expectation that the transition from the molecular to the metallic phase will be in conjunction with a relatively large density change.

Recently, data from new experiments have become available; in one case up to 8 Mbar. The data are not in disagreement with many calculations on hydrogen, but the resolution is not yet adequate to determine accurately and confidently, the pressure and the densities of the molecular to metallic phase transition. The accuracy of these parameters in turn affect the models of the planetary interiors, such as the radii of the metallic sphere and high density core.

This paper will discuss the details of these relations and the possible affects of the speculative properties of metallic hydrogen.

1. Introduction

First I would like to discuss the theoretical and experimental status of the Equation of State (EOS) of molecular and metallic hydrogen. Second, I would like to discuss their application to the models of Jupiter and Saturn. And finally, I would like to mention some of the proposals of the properties of metallic hydrogen as they apply to the major planets or at least Jupiter and Saturn.

For nearly 40 years it has been expected that molecular hydrogen will transform to the metallic phase at a pressure of the order of a Megabar (Mbar) (Wigner and Hunington, 1935). Until recently, pressures of that magnitude were not attainable in the laboratory in a form that could allow measurements of the EOS and properties of soft and low density materials. This was especially true of hydrogen. Direct experiments on hydrogen at very high pressures, have only recently been feasable. The existence of metallic hydrogen in the interiors of the major planets is well established (Ramsey, 1950; Kronig *et al.*, 1946; Abrikosov, 1954; Demarcus, 1958; Peebles, 1964; Smoluchowski, 1967; Hubbard, 1968). In fact, in the case of Jupiter, nearly all of the mass is in the metallic phase. As far as models for major planets go, the largest uncertainties about hydrogen have been the EOS of its molecular phase. Figure 1 shows the pressure in Mbars plotted as a function of density in g cm^{-3}, at 0 K, for many theoretical cal-

* On leave at the Max-Planck Institut für Festkörperforschung, Stuttgart, West Germany.
** Work performed under the auspices of the U.S. Atomic Energy Commission.

Woszczyk and Iwaniszewska (eds.), Exploration of the Planetary System, 329–335. All Rights Reserved
Copyright © 1974 by the IAU

culations. The right most curves above about a megabar, are the calculations for the
EOS of the metallic phase, while the left most curves at lower pressure are those of the
molecular phase. The metallic calculations vary less than 10% in density. The molecular calculations vary by as much as 50%. The transition from the molecular to the
metallic phase is expected to be accompanied by a large change in density (20 to 50%).
The range in pressure at which the transition is expected (P_t), varies from a little less
than a megabar to many megabars. The effects of these uncertainties will be considered as they influence the planetary models. There then remains two further interesting
questions about hydrogen. First is the question of the solvability of helium in the
various phases and throughout the pressure range of the planetary interiors, say 0 to

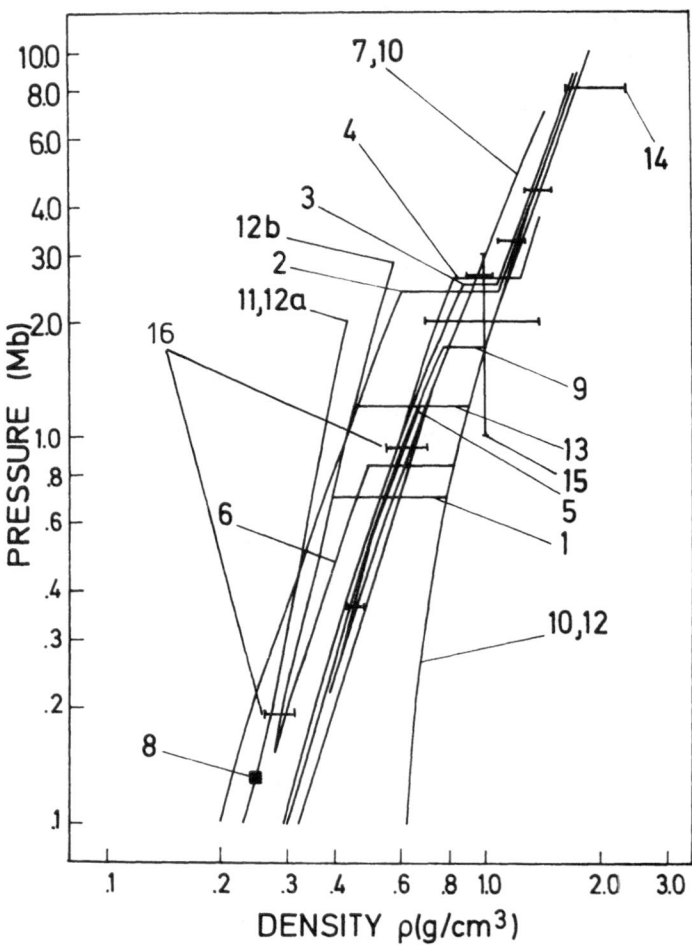

Fig. 1.   The pressure as a function of density at 0K for many theoretical calculations. *Curves:*
1 – Kronig *et al.* (1946), 2 – Abrikosov (1954), 3 – DeMarcus (1958), 4 – Schneider (1969), 5 – Ross
(1970), 6 – Neece *et al.* (1971), 7 – Liberman (1972), 8 – Dick (1972), 9 – Hoover *et al.* (1972), 10
– Trubitsyn (1966), 11 – Etters *et al.* (1971), 12 – Pollack (1972), 13 – Østgaard (1972), 14 – Grigoryev
et al. (1972), 15 – Hawke *et al.* (1972), 16 – Van Thiel *et al.* (1972).

40 Mbar. This subject just began to be investigated in the laboratory (Street, 1973) and has upset the simplifying assumptions that have aided modelling (Smoluchowski, 1973). This question is already being considered and will not be included here. The other remaining question concerns the possible effects of the unusual properties that metallic hydrogen might have. In particular, what could be the effects of the hoped for high temperature superconducting transition temperature (Ashcroft, 1968; Schneider, 1969) and the even more hoped for metastability of the metallic phase (Schneider, 1969; Brovman *et al.*, 1971).

2. The Equation of State of Hydrogen

Metallic hydrogen is expected to be the simplest of all metals. To calculate its EOS, people have used increasing refinements of the Wigner-Sietz method (Wigner and Sietz, 1933). This method leads to the energy, per atom, as a function of atomic volume by assuming three contributions, (1) the ground state energy of an electron in a periodic field caused by the hydrogen nuclei and other electrons, (2) the Fermi-energy, which represents the average energy of motion per electron above the ground state due to the energy band being half filled and, (3) a correction for electron exchange and correlation of their positions. The volume derivative then gives the pressure-volume EOS. Wigner and Hunington (1935) were the first to use this method and agreement has been fairly good since then as indicated in Figure 1. One of the refined calculations (Neece *et al.*, 1971) applied to the alkali metals, as a check, was found to be quite successful. Schneider has made extensive calculations of the possible crystal structures and obtained a slightly denser EOS.

In contrast to the relative simplicity and agreement of the calculations for metallic hydrogen, the case for molecular hydrogen is less precise. This is partly because the iterations have not yet been completely described from first principles by reason of the extreme complexity; hence, more approximate calculations are used. Part of the complication is due to the large change in the intermolecular spacing that occurs as the pressure increases from zero to a few megabars. At room pressure the spacing between the molecules is large compared to the atomic spacing within the molecule, but these distances become comparable in the high pressure region. Generally the attractive and repulsive forces are based on two types of approaches, Lennard-Jones and Buckingham. The Lennard-Jones form takes the repulsive potential to be proportional to the inverse twelfth power of the spacing, while the Buckingham (exponential-six) assumes the repulsive potential to be proportional to the inverse exponent. Both methods take the attractive potential as proportional to the inverse sixth power. The differences in the results are due to the choice of different coefficients in the relationships and sometimes added terms. Usually these calculations are made to agree with the experimental data of Stewart (1956), which give the pressure volume relationship up to 20 kbars. Pollack *et al.* (1972) used the Domb-Slater method with both types of repulsion and got results that are stiffer than others in both cases. Recently there have been shock wave experiments of two types, plane and cylindrical, and a magnetic technique used to compress hydrogen. In Figure 1 the shock results of Dick (1972) and van Thiel *et al.* (1972) and the multiple shock results of Grigoryev *et al.* (1972) and the magnetic work of Hawke *et al.* (1972) are shown. Due to the low initial density and great compressibility of hydrogen, single shock experiments are limited to pressures of a couple hundred kilobars. From preliminary analysis of the plane wave shock results, it has been found that good agreement can be made (van Thiel *et al.*, 1972) with the exponential-six model calculations of Ross (1970). It is nice to note that the agreement between Ross (1970), Trubitsyn (1966), and DeMarcus (1958) is good, as many of the planetary models have used the later two. The maximum discrepancy is about 6% in density. The experimental data of Grigoryev *et al.* (1972) are also in agreement. There is some reservation about the support that data gave to the understanding of hydrogen's EOS (Al'tshuler *et al.*, 1973). Also the pressure was apparently calculated and there is no support for the lack of uncertainty in the results. In any case, the data leave the pressure of the transition open over a range of at least a couple of megabars, and a 30% or more uncertainty in the density of the molecular phase at the transition pressure (D_t).* The magnetic compression has not yet been precised enough to reduce the uncertainties. Effort is being made to measure the pressure and volume and also to measure the metallization directly and electrically. The shock experiments also leave an open limit on the upper end of P_t but indicate a lower limit of 1.4 Mbar. (In summary, of the several experiments done in the megabar range, none can pin down P_t nor the molecular transition density.) As of yet, no one has reported static experiments beyond 20 kbar, although laboratories in the U.S. and Russia are striving to that end. It is extremely difficult and likely to be several years before a pressure in excess of a megabar will be statically exerted on liquid or solid hydrogen with the facility to make EOS or conductivity measurements.

3. Effect of Uncertainty in EOS on Planetary Models

The fortunate part of the situation is that the overall dependence of planetary models on the EOS of hydrogen is about the same order of magnitude as the uncertainty. Demarcus (1958) extrapolated two EOS's of molecular hydrogen that differed by 2% at 20 kbar. In the megabar range the simple extrapolation of Stewart's data is shown as curve 3 in Figure 1. The alternate stiffer form of molecular hydrogen is similar to curve 6 with the phase transition at 1.9 Mbar instead of 2.5 Mbar as for the softer form. The density difference at 1 Mbar is nearly 10%. The effect of the change in EOS on the planetary model was not stated as it was concluded that the alternate EOS was the most suitable at the time, and the reality of a non-isothermal model was added. Peebles (1964) continued the calculations and used two equations of state. In one case

---

* The 30% error is the sum of two sources of error: the uncertainty in the density data (approx. 15%), and the variation in the density due to an optimistic uncertainty in the transition pressure of only 1 Mbar (again 15% error in density). A 2 Mbar uncertainty in the transition would lead to a total uncertainty in the molecular transition density of 45%. The errors in the metallic density are comparable.

he used a less dense EOS, by 3% relative to Stewart's data at 20 kbar. That lead to a 25% lower density at 1.3 Mbar and would require a larger heavy core at the center of Jupiter to provide the necessary total observed mass. It was found that if the hydrogen density were everywhere 2% greater than the DeMarcus value, no heavy core at all would be needed, otherwise the core constitutes about 3% of the total mass. This 2% variation in density is less than the uncertainty in the EOS, even in the metallic phase. Peebles also found that, using the normal DeMarcus EOS extrapolated from Stewart and an adiabatic temperature distribution and a cloud pressure of 1 atmosphere at 150 K, no high density core is required. If the pressure of the observed cloud layer at 150 K is raised to 5 atm, then 3% of the mass of the planet is needed as a core. Hence the effect of varying the equation of state of hydrogen by 2% and the pressure of the cloud layer that is observed both lead to comparable effects as far as their effects on the need for a high density core is concerned.

Lowering the density of the molecular hydrogen allows the molecular to metallic phase transition to occur at a lower pressure which leads to a larger radius and mass of the planet in the higher density metallic phase which in turn diminishes the need for a high density core. The calculated mass of the core in Saturn is considerably larger (approx. 20%) and varies in the same way.

Now let us consider the size of the metallic hydrogen sphere and its dependence on the EOS. If the model and density distribution of Peebles is used, a variation in the molecular transition density from 0.8 to 1.0 g cm^{-3} (less than the uncertainty), leads to the radius of the metallic sphere (R_M), to vary from 0.84 to 0.79 and the mass fraction from 0.86 to 0.80, respectively. Hence, a 20% variation in D_t leads to about a 4% variation in R_M and the mass fraction in the metallic phase. In the case of Saturn the same variation in D_M allows R_M to vary from 0.61 to 0.53 (15%) and the mass fraction from 0.59 to 0.49 (10%). Overall, while these uncertainties are not small, they are probably no more than comparable to the uncertainties arising from the possibly erroneous assumption of a uniform helium distribution throughout the planetary interior.

4. The Effects of the Unusual Properties of Metallic Hydrogen

Recently, the speculations on the nature of metallic hydrogen have included its possibly being a high temperature superconductor and/or metastable at pressures below its phase transition. It has been suggested that the possible superconducting nature of metallic hydrogen could lead to the generation of Jupiter's magnetic field (Ashcroft, 1968, Schneider, 1969). Their calculations have concluded that the Debye and critical superconducting transition temperatures could be as high as about 3000 K and 200 K, respectively. Schneider also found that the critical temperature is lower at both higher and lower pressures compared to the metallic transition pressure, if the interiors of the planets were isothermal and equal to the cloud temperature of 150 K, it would seem feasible. However, at a sufficient depth for the metallic phase to exist, the realistic interior temperature is more likely to be 3000 K to 4000 K. Hence a superconducting magnetic generator seems unlikely.

Perhaps the proposed metastability of metallic hydrogen is of interest in terms of the major planets. Because of the possibly high superconducting transition temperature of metallic hydrogen, the question of it possibly being metastable at ambient pressure has been raised. Calculations of metallic hydrogen's crystal structure have been made and indicate the possibility (Schneider, 1969; Brovman *et al.*, 1971). Estimates of its lifetime at zero pressure have been considered as limited by surface evaporation (Chapline, 1972) and by bulk effects (Estrin, 1971) and found to be probably less than a second, unless a protective environment is provided. In one case it is argued that it cannot be metastable above 14 K (Liberman, 1972). However it is likely that the lifetime will increase a little (Chapline, 1972) or greatly (Salpeter, 1973) with pressure. At a 100 kbar its lifetime could be considerable. If that is true, it is possible that metallic hydrogen could be present in droplets or as waves on the metallic surface at heights greater than R_M. If the metallic phase could exist down to a pressure of about 100 kbar, then it could be found in the region between 0.8 and 0.96 of Jupiter's radius and between 0.6 and 0.9 in Saturn. In terms of heights, this amounts to about 10^4 km for Jupiter and twice that for Saturn. This could be important in considerations about convection and instabilities. The energy content of metastable metallic hydrogen would be quite high (about 1 Mbar-cm^3 g^{-1}, at $P=0$), and could be a high energy density transport mechanism.

Another speculation about metallic hydrogen is that it could be a quantum super fluid (Schneider, 1969), at a pressure of a little more than 100 Mbar. Although Jupiter's central pressure was estimated at one time to be of that magnitude (DeMarcus, 1958) more recent estimates lead to pressures of about 30 to 40 Mbar (Peebles, 1964), hence, it is not likely that such a superfluid is in abundance.

5. Summary

In summary, I would like to say that progress is being made in generating EOS data that describe molecular hydrogen and experiments have progressed into the megabar region in the last 2 years and should lead to quite refined data in the next 5. Probably the importance of this information will be much less than the need for accurate helium-hydrogen phase diagrams, but it appears that area of experimentation is about where megabar pressures were in 1954 when, with great difficulty, pressures of 20 kbar were reached.

The possible spectacular properties of metallic hydrogen should be kept in mind but so far it does not appear any problems of planetary models are better understood by their consideration. Probably the gathering of additional information about the planets via satellites, will be of great stimulus to reduce the uncertainties to a unique model; however, even here on Earth, with the *corpus delicti* in hand, there is not yet a single agreed upon model.

References

Abrikosov, A. A.: 1954, *Astron. Zh.* **31**, 112.
Al'tschuler, L. V., Dynin, E. A., and Svidinski, V. A.: 1973, *Zh. Experim. Theor. Fiz.* **17**, 20.

Ashcroft, N. W.: 1968, *Phys. Rev. Letters* **21**, 1748.
Brovman, E. G., Kagan, Yu., and Kholas, A.: 1971, *Zh. Experim. Theor. Fiz.* **61**, 2429.
Chapline, G. F.: 1972, *Phys. Rev.* **B6**, 2067.
DeMarcus, W. C.: 1958, *Astron. J.* **63**, 2.
Dick, R.: 1972, in G. L. Kerley (ed.), *High Pressure Physics and Planetary Interiors*, North Holland Publ. Co., p. 78.
Estrin, E. I.: 1971, *Soviet Phys. JETP Letters* **13**, 510.
Etters, R. D., Raich, J. C., and Chand, P. J.: 1971, *Low Temp. Phys.* **5**, 6711.
Grigoryev, F. V., Kormer, S. B., Mikhailova, O. L., Tolochko, A. P., and Urlin, V. D.: 1972, *JETP Letters* **5**, 286.
Hawke, R. S., Duerre, D. E., Huebel, J. G., Keeler, R. N., and Klapper, H.: 1972, in G. L. Kerley (ed.), *High Pressure Physics and Planetary Interiors*, North Holland Publ. Co., p. 44.
Hoover, W. G., Keeler, R. N., Van Thiel, M., Hord, B. L., Hawke, R. S., Olness, R. J., Ross, M., Rogers, F. J., and Bender, C. F.: 1973, *Adv. Cryogenic Engin.* **18**, 447.
Hubbard, W. B.: 1968, *Astrophys. J.* **152**, 745.
Kronig, R., DeBoer, J., and Korringa, J.: 1946, *Physica*, **12**, 245.
Liberman, D. A.: 1972, Los Alamos Sci. Lab., Report LA 4727-MS.
Rogers, F. J., and Neece, G. A., Hoover, W. G.: 1971, *J. Compt. Phys.* **7**, 621.
Østgaard, E.: 1972, *Z. Phys.* **252**, 95.
Peebles, P. J. E.: 1964, *Astrophys. J.* **140**, 328.
Pollack, E. L., Bruce, T. A., Chester, G. V., and Krumhansl, J. A.: 1972, *Phys. Rev.* **5**, 4180.
Ramsey, W. H.: 1950, *Monthly Notices Roy. Astron. Soc.* **110**, 325.
Ross, M.: 1970, Lawrence Livermore Report, UCRL 50911.
Salpeter, E. E.: 1972, *Phys. Rev. Letters* **28**, 560.
Salpeter, E. E.: 1973, *Astrophys. J.* **181**, L83.
Schneider, T.: 1969, *Helv. Phys. Acta* **42**, 957.
Smoluchowski, R.: 1967, *Nature* **215**, 691.
Smoluchowski, R.: 1973, *Astrophys J.* **185**, L95.
Stewart, J. W.: 1956, *J. Phys. Chem. Solids* **1**, 145.
Street, W. P.: 1973, *Astrophys. J.* **186**, 1107.
Trubitsyn, V. P.: 1966, *Soviet Phys. Solid State* **7**, 2708.
Van Thiel, M., Ross, M., Hord, B. L., Mitchel, A. C., Gust, W. A., D'Addario, M. J., Keeler, R. N., and Boutnell, K.: 1972, *Phys. Rev. Letters* **31**, 979.
Wigner, E. P. and Sietz, F.: 1933, *Phys. Rev.* **43**, 804.
Wigner, E. P. and Hunington, H. B.: 1935, *J. Chem. Phys.* **3**, 764.

DISCUSSION

Hide: Are there any plans to extend the work you have described to hydrogen-helium mixtures?

Graboske: The great difficulty and expense of these high pressure experiments requires that they be carried out on standard materials having a wide range of applications. To my knowledge, there are no current plans to study H–He mixtures at such high pressures. However, other materials pertinent to planetary atmospheres, such as methane and ammonia, are being studied at Livermore. Once pure hydrogen and pure helium have been experimentally investigated, I feel it would be important to study H–He mixtures, and high pressure researches in all countries should be encouraged to consider them. I might mention here the recent extensive work by Street at West Point, soon to appear in the *Astrophysical Journal*, on static studies of the complex phase structure of H–He mixtures at temperatures and pressures appropriate to the giant planet atmospheres.

THE INFLUENCE OF THE SURFACE BOUNDARY LAYER
ON EVOLUTIONARY MODELS OF JUPITER*

HAROLD C. GRABOSKE, JR.

Lawrence Livermore Laboratory, Livermore, Calif., U.S.A.

Abstract. A recent theoretical study of the structure and evolution of Jupiter (Graboske *et al.*, 1974b) is based on a three-stage model of Jovian evolution. The central phase, gravitational contraction of an adiabatic, homogeneous convective fluid system, begins early in solar system evolution and lasts for times of the order of 2×10^9 yr. Good agreement with observed radius and luminosity is achieved for a model with a solar mixture composition. The surface boundary layer has a dominant influence on the evolutionary timescale. Surface boundary factors which are important are the solar energy input, a function of the solar luminosity and the planetary albedo, and the detailed physics of the superadiabatic zone, which depends on the variation of opacity and ∇_{ad} with depth. The evolutionary study demonstrates that the current planet cannot be an adiabatic homogeneous fluid throughout. The inclusion of a superadiabatic zone is necessary, and the existence of a heterogeneous (gravitationally layered) fluid interior is possible.

1. Introduction

The structure and evolution of the planet Jupiter can be investigated by two different approaches. The original method used by DeMarcus (1958) to Hubbard (1970) involves construction of static planetary models utilizing the observed mass and radius. These static models have been successful in allowing estimation of chemical composition and semi-quantitative determinations of interior densities and temperatures. The second approach, recently carried out by Graboske, Pollack, Grossman and Olness (1974b), hereinafter referred to as GPGO), uses a stellar structure method to calculate an evolving model of the planet, specifying only the mass and chemical composition. This evolutionary method has produced models of the planet which not only agree well with observational characteristics such as mass, radius and luminosity, but give us significant information on the energetics and interior conditions of the Jovian system.

2. Construction of Evolutionary Model Sequences

The evolutionary study deals with the central phase of a three stage process. The first phase, the assembly stage, postulates either a stellar gaseous collapse origin or a cold accretion origin, this protoplanetary phase culminating in a stable quasistatic configuration at the top of the Hayashi track. The central phase is the fluid contraction state, gravitational contraction along the Hayashi track for times exceeding 10^9 yr. A third phase is suggested, the post-fluid contraction stage, where additional energy sources become important, adding to the energy released from stored heat of prior contraction and from current contraction. These additional sources may be energy released by phase changes (Smoluchowski, 1967) or by gravitational separation (Salpeter, 1973;

---

* Work performed under the auspices of the U.S. Atomic Energy Commission.

Woszczyk and Iwaniszewska (eds.), Exploration of the Planetary System, 337–344. All Rights Reserved
Copyright © 1974 by the IAU

Smoluchowski, 1973). The evolutionary study of GPGO is concerned only with the fluid contraction stage.

The basic assumptions are that the planetary model is spherical nonrotating, non-magnetic, and is composed of a solar mixture composition ($X = 0.74$, $Z = 0.02$). The fundamental structure chosen is that originally proposed by Hubbard, an adiabatic, homogeneous (fully mixed) fluid mixture. The calculation is started from an extended gaseous condition ($R = 15.6\ R_J$) a stable, convective configuration high on the Jovian Hayashi track. It is evolved to a time approaching 10^{10} yr, using a Henyey-type star code, which solves the stellar structure equations for a convective system. The surface boundary condition is established from planetary model atmospheres calculated with the method of Pollack and Ohring (1973). These flux corrected atmospheres cover a temperature range from 50 K to 2000 K, including as principal opacity sources hydrogen, ammonia, water vapor and methane. An additional feature is the inclusion of a solar energy component, characterized as $T_\odot$ an effective insolation temperature, which is a function of the solar luminosity and the planetary orbit and albedo. The thermodynamic properties for the solar mixture are taken from Graboske *et al.* (1974a), where fully mixed fluids composed of a solar mixture ($X = 0.74$, $Z = 0.02$) were calculated by two theoretical approaches. The high density (metallic) regime is studied using a version of Thomas-Fermi theory developed for mixtures and strongly coupled coulomb systems. The low density (molecular, atomic and plasma) regime is studied using a free energy minimization model, which incorporates recent high pressure experimental results for the molecular fluid, as well as Monte Carlo results for the dense plasma.

The results of the study are evolutionary models for a Jovian mass object that have two different phases. The early evolution produces a stellar object, behaving in a manner similar to the low mass premain sequence stars ($0.01 \leqslant M/M_\odot \leqslant 0.20$), with relatively high effective temperatures (1600 K), luminosities ($\log L/L_\odot = -2.3$ initially), and internal temperatures ($T_{c\ max} = 40600$ K at 4.1×10^4 yr). This phase is succeeded by a degenerate dwarf cooling phase, during which the model contracts, cools and approaches very close to the observed radius and luminosity of the present Jupiter ($R/R_J = 0.98$, $L/L_J = 1.00$). A number of parametric studies were made, computing evolutionary sequences while varying quantities whose values have a range of uncertainty, but the best agreement is achieved for the standard constitutive physics and the ($X = 0.74$, $Z = 0.02$) solar mix composition. A significant result was found, obtainable only from an evolutionary calculation, concerning the duration of the fluid contraction stage. The age of the model which agrees best with the current R_J, L_J values is only 1.87×10^9 yr. The sensitivity of this relatively short lifetime to various parameters such as chemical composition and $T_\odot$ were investigated, and the results are instructive.

3. Effects of the Surface Boundary Layer

The construction of evolutionary sequences for different constitutive physics resulted in a range of model tracks in the $\log L$, $\log T_e$ plane. The location of the 4.5×10^9 yr models are shown in Figure 1. The solar mix standard model ($T_\odot = 89\,\mathrm{K}$), track passes very close to the observational values for Jupiter ($\oplus$) at 1.87×10^9 yr, but is over 3 times less luminous at 4.5×10^9 yr. The other models are the same epoch models for different values of $T_\odot$ (3 K, 108 K) representing the extremes of the solar energy deposition, for different chemical composition (pure hydrogen), for a modified set of

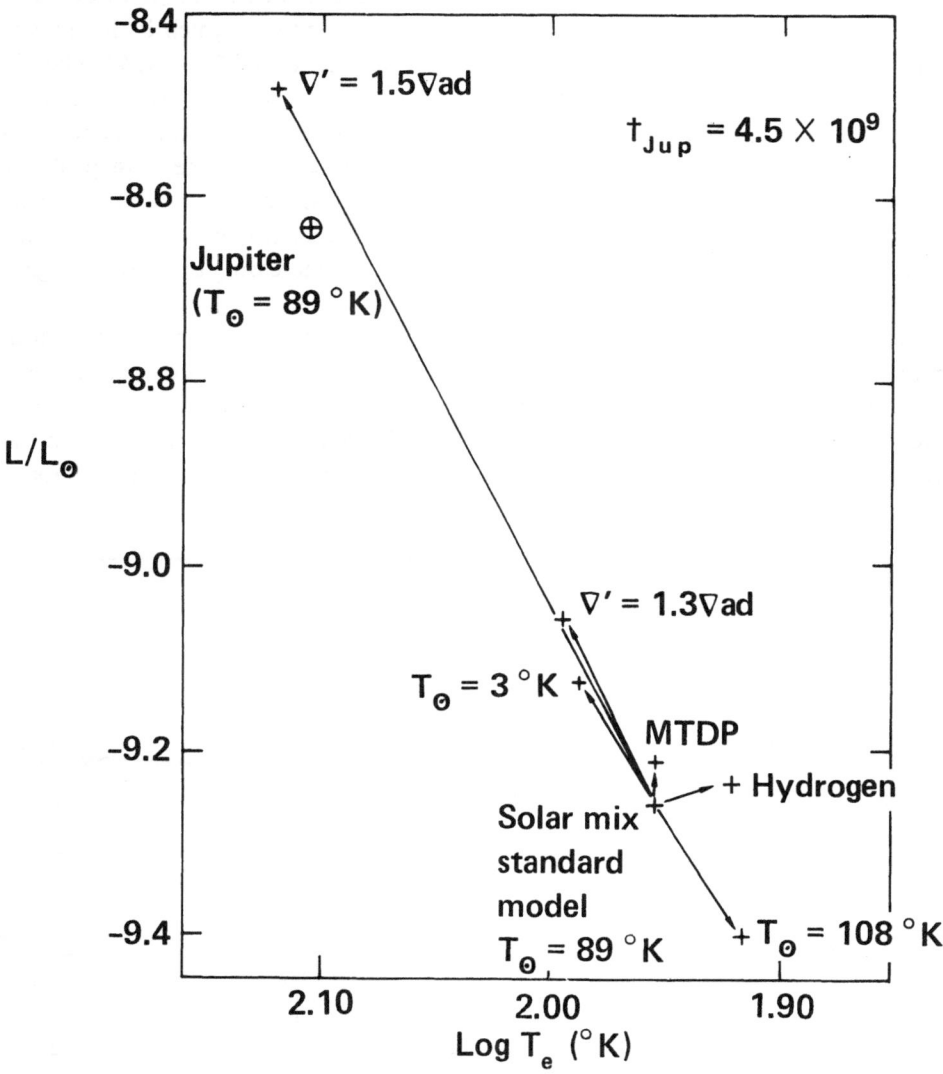

Fig. 1.   Planetary models in the $\log L - \log T_e$ plane at $t = 4.5 \times 10^9$ years. The observed values for Jupiter ($\oplus$) lie above the standard model (solar mix, $T_\odot = 89\,\mathrm{K}$). The positions of the models at 4.5×10^9 yr for different $T_\odot$ values, different chemical composition (pure hydrogen), modified constitutive physics (MTDP), and arbitrary superadiabatic zones (∇') are indicated.

thermodynamic properties (MTDP), for a modified set of model atmospheres (change too small to be discriminated from the standard model) and for an artificial super-adiabatic zone of varying extent (∇'). The primary conclusion is that no change in the model atmospheres, thermodynamics, chemical composition or solar energy compo-nent can extend the lifetime of the fluid contraction stage enough to yield an adiabatic homogeneous fluid model which has R_J, L_J and a lifetime (t_J) comparable to that of Jupiter.

The surface boundary layer effects, $T_\odot$ and ∇', are presented in more detail in Figures 2 and 3. In Figure 2, the effective age of the model at which its luminosity equals that of the present Jupiter ($t_J^{L_J}$) is given as a function of $T_\odot$ and ∇'. The 3 K value of $T_\odot$ represents an absolute minimum, a 0.00095 $M_\odot$ object evolving without a stellar companion, while the 89 K and 108 K values represent the maximum and mini-mum planetary albedo values. A temperature of 102 K would be the most probable value of $T_\odot$. Variations in $t_J^{L_J}$ of approximately 10% are achieved, and hence no reasonable variations in $T_\odot$, due to secular changes in the solar luminosity or changing albedo of the contracting cooling planet are capable of increasing the lifetime signif-icantly.

The other surface boundary layer effect is more difficult to assess, and it is only done in an arbitrary and approximate manner in GPGO. Since an accurate treatment of a stellar model containing a superadiabatic zone requires use of the mixing length theory incorporating both opacity and ∇_{ad} as a function of depth, a simple approximation was used. The superadiabatic gradient ∇' was set equal to a multiple of the adiabatic

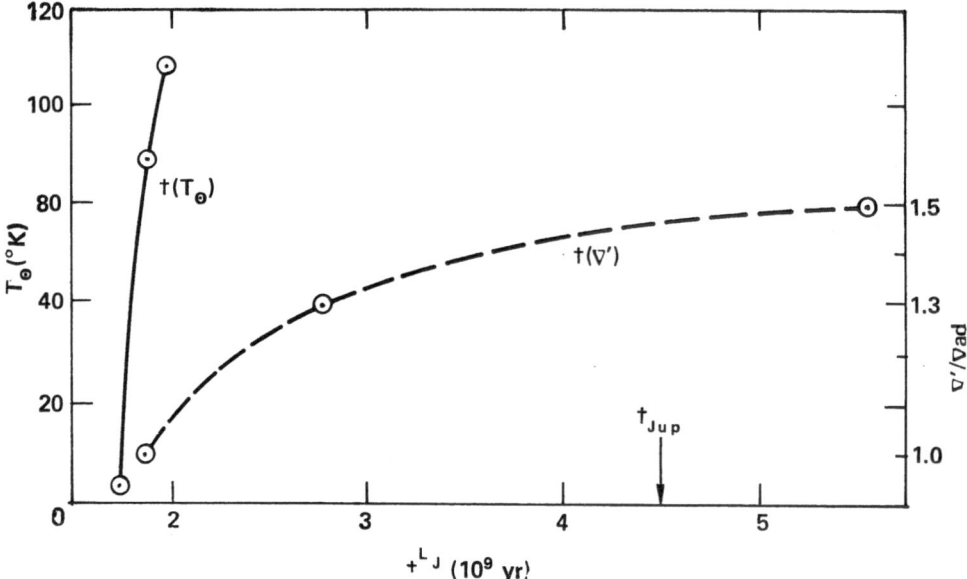

Fig. 2.   The dependence of model age at $L = L_J$ as a function of insolation temperature and of superadiabatic gradient.

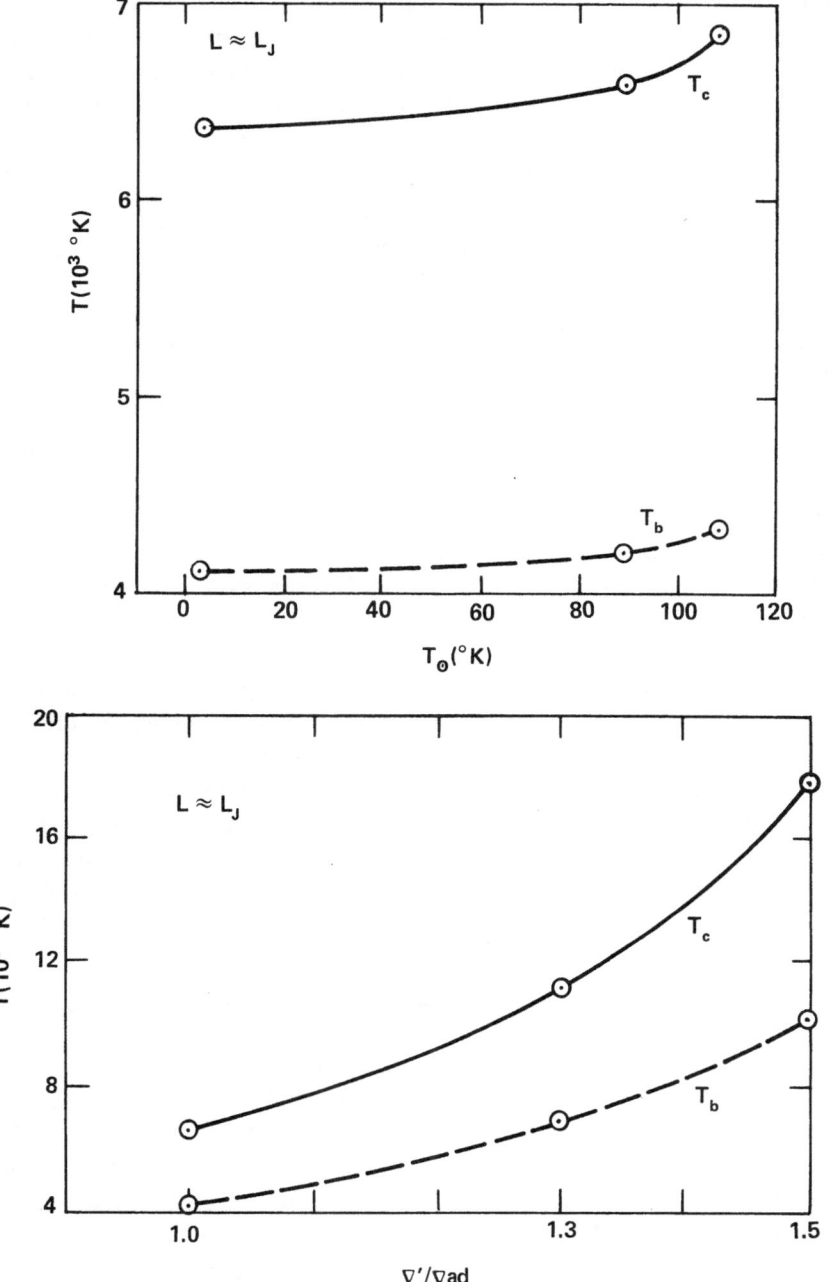

Fig. 3.   The dependence of model central temperature (T_c) and core boundary temperature (T_b, where $M/M_J = 0.97$) are given as a function of $T_\odot$ and ∇', for the track location $L = L_J$.

gradient over a mass range ΔM in the outer envelope of the model. The size (ΔM) and strength (∇'/∇_{ad}) of the artificial superadiabatic zone were chosen by comparison with the surface layers of low mass premain sequence stars (0.01 $M_\odot$–0.10 $M_\odot$), with $\Delta M = 10^{-4} M_{total}$. Three values of ∇'/∇_{ad} were chosen, 1.0 (the standard, adiabatic model), 1.3 and 1.5. The dependence of $t_J^{L_J}$ on ∇' is shown in Figure 2 to be very strong, by far the dominant influence on the planetary structure. As determined from Figures 1 and 2, a value of $\nabla'/\nabla_{ad} \approx 1.45$ would yield an adiabatic homogeneous fluid model with R, L and t all near the Jovian values. The track would lie slightly to the right of the planetary values, indicating that possibly a slightly higher He content is necessary.

In addition to modifying the observational characteristics of the models, the $T_\odot$ and ∇' parameters also modify the internal structure. The modifications in internal temperature are shown in Figure 3, for models with $L \approx L_J$. These models have widely differing ages, since $L = L_J$ at very different times for various values of $T_\odot$ and ∇'. The central temperatures (T_c) for the $T_\odot$ variation range from 6368 K to 6842 K, while the temperatures at the core boundary (T_b), where $M/M_J = 0.97$, range from 4072 K to 4339 K. As was the case for the lifetime variation, the influence of ∇' is far stronger than that of $T_\odot$. The central temperature as a function of ∇' increases from 6600 K (adiabatic) to 16893 K, while the temperature at the core boundary increases from 4204 K to 10018 K. The direction and relation between the lifetime and internal temperature changes are obviously correlated. The stronger the superadiabatic zone, the slower the energy release from the interior, so that the increasing lifetime with increasing ∇' is accompanied by higher thermal content of the core.

The construction of evolutionary model sequences incorporating a realistic super-adiabatic zone, using mixing length theory with appropriate outer envelope opacities and thermodynamic properties for the dissociating, ionizing gases, is currently being carried out. The results of this further step in the evolutionary study should provide a definitive model sequence for the fluid contraction stage.

4. Evolutionary Structure Subsequent to the Fluid Contraction Stage

The inclusion of a realistic superadiabatic zone will give a specific age for the $L = L_J$ model, within the limits allowed by the uncertainties in $T_\odot$, chemical composition and the constitutive physics, all of which are seen to have fairly small effects. If $t_J^{L_J}$ is 4.5×10^9 yr or greater, the superadiabatic, homogeneous fluid model can be used to describe the structure of present Jupiter. The size of the effective ∇' required to yield such lifetimes $(\nabla'/\nabla_{ad} \approx 1.45)$ will produce a dense (4 g cm^{-3}) and hot $(T_c \approx 16000$ K, $T_b \approx 9500$ K) interior, which would correlate well with a model assuming ionized (metallic) fully mixed components. However, if the superadiabatic model sequence has $t_J^{L_J} < 4.5 \times 10^9$ yr, then an additional source of energy must be sought to achieve agreement with the observed planet.

Two such sources have been proposed, by Smoluchowski and Salpeter. Smoluchowski (1967) suggested that as the planetary interior cooled, hydrogen would solidify, the fluid helium would diffuse out of the hydrogen solid under the influence of the gravitational field and a multilayer structure would result. Two additional sources of energy are available in this model: the latent heat of fusion released by the freezing hydrogen plus the gravitational energy released by the inward-diffusing helium. Salpeter (1973) proposed a different mechanism for gravitational layering, pointing out that a neutral helium fluid might be able to diffuse out of a metallic hydrogen fluid against a convective mixing flow. This immiscible fluid model would produce a differentiated model with consequent gravitational energy release. As recently estimated by Smoluchowski (1973), gravitational layering can produce the observed Jovian luminosity from the separation of a small fraction of the Jovian mass (1%).

The accurate evolutionary model of the superadiabatic homogeneous fluid will give quantitative values for the energy produced by current gravitational contraction and that produced by release of heat stored from prior contraction. These two sources contribute approximately equivalent amounts for the time period $t > 10^9$ yr. This will in turn allow an accurate estimation of the energy derived from gravitational differentiation, and provide information on the size of the planetary mass involved in the layering process. A second advantage of obtaining an accurate homogeneous fluid sequence is that, in combination with recent advances in the theory of dense fluids and fluid mixtures, it will allow a determination of possible phase transitions occurring in the planet. The molecular fluid-metallic fluid phase transition has been accepted since DeMarcus' study (1958). The possibility of a metallic solid phase, at least for time-scales of 10^{10} yr or less is almost certainly ruled out by the high temperatures existing in the central zone of the planet. Using a recent study of the dense hydrogen fluid region by Ross (1973), an analysis of the pure hydrogen model sequence (adiabatic, homogeneous) indicated that a thin shell of solid molecular hydrogen would form at $R/R_J \approx 0.6$ and slowly grow outward, at $t = 2.0 \times 10^9$ yr. The inclusion of even a modest superadiabatic zone in the pure hydrogen model would raise the internal temperatures sufficiently so that no solid molecular hydrogen could form anywhere in the interior. The appearance of solid phase in a hydrogen-helium mixture is more difficult to study, but it is unlikely to be stable at significantly higher temperatures than the hydrogen solid. Thus, the possibility of solid phases in a Jovian model with $t \lesssim 4.5 \times 10^9$ yr is improbable, and if gravitational layering is found to be required by planetary energetics, Salpeter's immiscible fluid is the favored mechanism.

References

DeMarcus, W.: 1958, *Astron. J.* **63**, 2.
Graboske, H. C., Jr., Olness, R. J. and Grossman, A. S.: 1974a, *Astrophys. J.*, to be published.
Graboske, H. C., Jr., Pollack, J. B., Grossman, A. S., and Olness, R. J.: 1974b, *Astrophys. J.*, to be published.
Hubbard, W. B.: 1970, *Astrophys. J.* **162**, 687.
Pollack, J. B. and Ohring, G.: 1973, *Icarus* **19**, 34.
Ross, M.: 1973, *J. Chem. Phys.*, to be published.
Salpeter, E. E.: 1973, *Astrophys. J.* **181**, L83.
Smoluchowski, R.: 1967, *Nature* **215**, 691.
Smoluchowski, R.: 1973, *Astrophys. J.* **185**, L95.

DISCUSSION

Trafton: Do experiments involving such transient phenomena really give the *equilibrium* properties of hydrogen?

Graboske: Yes. The time resolution, for example, is some 50 times that required to determine the transient behaviour.

Williams: With a graduate student, R. Donnison, I attempted a similar calculation and similar problems to what you mentioned occurred, namely the radius being too short. The problems you outlined are therefore real I think.

Graboske: As shown by DeMarcus, cold hydrogen models are not capable of reproducing the observed radius of either giant planet, and these more recent studies show that adiabatic models with hot interiors are also too large. In addition to the surface boundary layer effects mentioned, our study of the evolutionary sequences included extensive analyses of the possible effects of changes in the thermodynamic properties, in the model atmospheres and in chemical composition. Changes in lifetime at $L = L_J$ are typically no greater than 20 to 30% for any of these variations, with the exception of the inclusion of the superadiabatic zone.

Hide: The presence of a thin 'crust' of high viscosity would greatly simplify the problem of interpreting the variable rotation of Jupiter (as evinced by observations of the radio-period and atmospheric features). It will be important in your future calcultions to set a crust on the basis of various theoretical models and not simply to consider completely fluid models.

Graboske: This is an important point I would like to respond in three parts. (1) The structure method is not limited to fluid models. The fluid model was adopted for simplicity and because the recent models of Hubbard and Trubitsyn indicated high internal temperatures incomptatible with the presence of solid material. In future studies we intend to allow for the formation of solid zones in the interior if the physical conditions require them.

(2) The liquid-solid high pressure phase diagram of H–He is totally unknown at present, and that of pure hydrogen is only partially complete. Using the hydrogen phase diagram of M. Ross of the Lawrence Livermore Laboratory, I have observed that conditions favouring the appearance of a thin shell of solid molecular hydrogen occur in the pure hydrogen model sequence. This shell would form at 0.6 R_J, at about 2.5×10^9 yr, and slowly grow outward. However, this is an adiabatic fluid model; inclusion of a superadiabatic zone will increase the internal temperatures in the sense shown above for the solar mixture models. This would delay or completely prevent the appearance of a solid layer. Its possible formation in a superadiabatic, H–He model is even less probable.

(3) Even if final evolutionary models yield a fluid planetary structure, the presence of a thin crust in the outermost layers might be permitted. It is possible it could be inserted without significantly perturbing the first model, although its calculation would require special procedures.

OBSERVATIONS OF SPATIAL AND TEMPORAL VARIATIONS OF THE JOVIAN H₂ QUADRUPOLE LINES

N. P. CARLETON and W. A. TRAUB

Smithsonian Astrophysical Observatory and Harvard University, Cambridge, Mass., U.S.A.

Abstract. We have made high resolution scans of the Jovian H_2 4–0 S(1) line at 6367.75 Å and the 3–0 S(1) line at 8150.67 Å, in an attempt to establish improved equivalent widths and to measure any spatial-temporal variations across the disc. Preliminary analysis indicates that temporal changes in the 3–0 S(1) line can be as large as about 40% over a period of about a week; furthermore, this line is systematically weaker at the limb (about 10%) and stronger at the poles (about 20%) than it is at the center of the disc. The 4–0 S(1) line shows very little temporal change, but does have a similar center to limb variation. Current theoretical models of line formation and cloud structure on Jupiter are not able to explain the observed variations although it is clear that some sort of two-layer model is necessary.

We used a PEPSIOS spectrometer, consisting of three Fabry-Perot etalons in series with an interference filter, at the coudé focus of the Smithsonian Observatory's 60-in. telescope. Following Trauger and Roesler (1972) we operated the PEPSIOS in an off-axis mode, so that the large rotational smearing of Jovian spectral lines is exactly compensated.

Examples of spectra taken at various points on the disc and averaged over one night's observing are shown in Figure 1. Each profile has been fitted with a Gaussian line shape, shown by plus signs. Although a fairly good fit has been obtained, there are systematic departures which can be seen in the 3–0 S (1) profiles; in particular, the wings of the line are slightly stronger than allowed by a pure Gaussian profile, and also an asymmetry in the line core can be detected. The observed 3–0 S (1) line profiles have widths (FWHM) of about 85 mÅ, which can be accounted for if we combine the intrinsic line width (about 47 mÅ) and instrumental width (about 27 mÅ) with the effects of image motion due to seeing on the order of FWHM = 3.7″ or less (63 mÅ). However, we are basically concerned here with the equivalent widths, which are plotted in Figures 2 and 3.

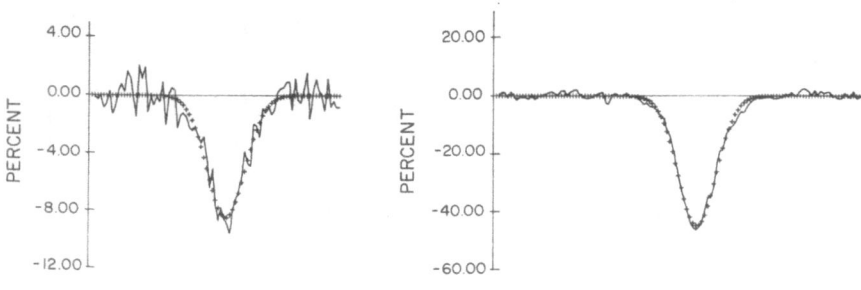

Fig. 1. Examples of the scans of the Jovian H_2 4–0 S(1) and 3–0 S(1) lines obtained at various points of the disc.

Woszczyk and Iwaniszewska (eds.), Exploration of the Planetary System, 345–349. All Rights Reserved

In Figure 2, the longitudinal extent of our various sized circular apertures is shown by crosshatching, and the observed equivalent widths are plotted above in corresponding positions. The observations at various points on the disc for June 27, 1972 were made sequentially and cycle repeated several times, so time dependence is suppressed. First note that the equivalent width (EW) at the center of the disc in the south equatorial zone (SEZ) is slightly less than that at the same meridian but in the north equatorial belt (NEB) and great red spot (GRS). There is a further significant change in intensity (about -10%) toward the limbs, but still within the SEZ. We shall see later that it is difficult to account for this near-constancy in strength on the basis of current theoretical models, which predict either much greater or much

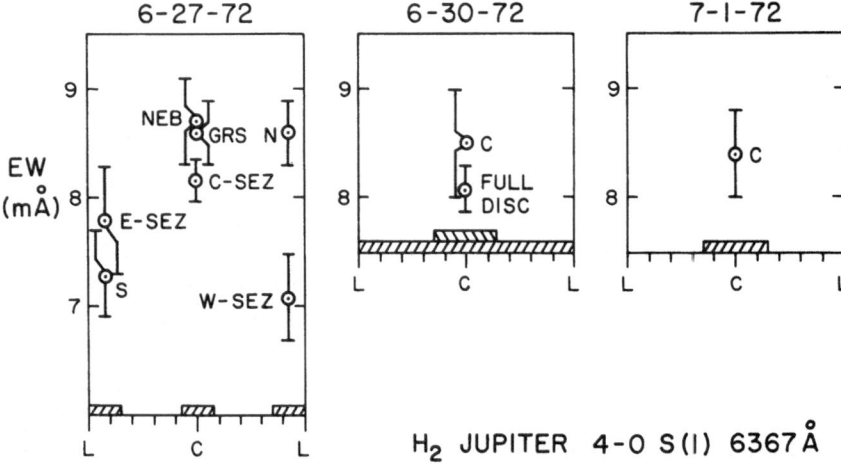

Fig. 2.   Observed equivalent widths of Jovian H_2 4–0 S(1) line at various points of the disc.

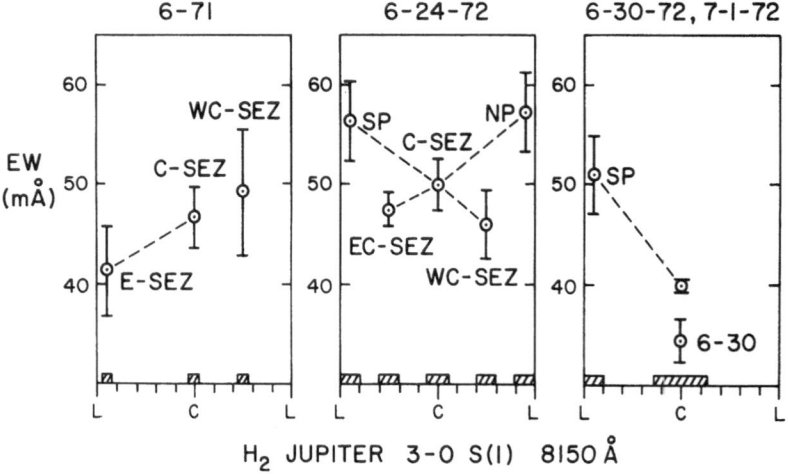

Fig. 3.   Observed equivalent widths of Jovian H_2 3–0 S(1) line at various points of the disc.

weaker lines toward the limbs. Note also that at the north and south poles (N, S) the line strength is not symmetric; we have no ready explanation for this behavior. Measurements at the center of the disc taken 3 and 4 days later are also shown; within our measurement errors the EW appears to have made no significant change in this period.

In Figure 3 we plot the EW's of a number of 4–0 S(1) observations. First, we show some averaged EW data from 1971 which were obtained over several days, and are useful mainly for showing that the EW at the limb is again only about 10% weaker than at the center. The error bar on the mid-radius observation is large, and should not necessarily be taken as evidence for an increase in EW.

The data for June 24, 1972 are of especially good quality, and show first that in the SEZ the greatest EW is on the central meridian. Altogether, the data so far described indicate a monotonic decrease in EW (in the SEZ) from center to limb. An entirely different trend is seen toward the poles, where the 3–0 S(1) EW increases dramatically. Six days later, the EW in the center of the SEZ has dropped by about 30%. The next day it has partially recovered, but to a value still clearly outside the indicated measurement error (one standard deviation). The difference between the first and last of these 3–0 S(1) observations is even more remarkable in that they both represent averages over nearly the same longitude limits, but separated in time by 17 revolutions of the planet. Finally, we note that again on the last set of observations, that the EW over the SP is significantly larger than at the center of the SEZ.

We now wish to compare these observations with the recent theoretical analysis by Hunt (1973) and by Margolis and Hunt (1973), where both the line profile and radiative transfer effects are accurately computed, for the chosen model atmosphere. Since the time variation in EW of the 3–0 S(1) line is clearly so large, it is most profitable to first discuss center to limb behavior. Hunt has calculated that the EW will increase strongly toward the limb in a clear gas atmosphere, but will decrease to nearly zero at the limb for the two-cloud model he has chosen; this is true for several values of anisotropy parameter in the adopted Henyey-Greenstein particle scattering function. This upper cloud effectively blocks sunlight at the limb from penetrating to the underlying atmosphere, and so causes a strong decrease in EW. Clearly, neither model describes the observations, which show nearly constant EW, with only a small (about 10%) drop toward the limb.

We suggest that there exists a cloud model which probably will explain the observations. Although we have not yet made any calculations for the case of H_2 on Jupiter, we have done a number of similar model studies for CO_2 on Venus, and we expect that this experience may carry over to some degree. For this purpose, we note that Hunt's upper cloud is actually a relatively extensive haze, and not a thin layer. It is well known that in a homogeneous infinite haze model, that EW will strongly decrease toward the limb. We believe that that is effectively what is occurring here, since as Hunt shows, the effective level of line formation goes up rapidly as the limb is approached. If, on the other hand, we use instead a physically thin, as well as optically thin, layer for the upper cloud, we achieve the desired 'shielding' effect near the limb, but still allow some light to penetrate to the gas and subsequently escape. Although it is extremely difficult to produce a constant EW from center to limb with this sort of cloud (which appears to be nearly the case with Venus), it is not hard to imagine only a 10% drop. We expect to be able to continue our calculations and produce such a model in the near future.

The polar EW behavior is quite different, appearing to increase for 3–0 S(1) at least. This suggests that the upper cloud is thinner or even absent at the poles. Support for this comes from Gehrels' (1969) polarization studies which indicate that the atmosphere is clearer over the poles than elsewhere.

TABLE I

H_2 Jupiter equivalent widths (mÅ)

	4-0 S(1)	3-0 S(1)
SPINRAD AND TRAFTON (1963)	8.0 ± 2.0	28, 40
BECKMAN (1967)	6.0 ± 2.5	
OWEN AND MASON (1968)	8.5 ± 1.5	49 ± 6
EMERSON, EDDY, DULK (1969)	7.8 ± 1.2	
FINK AND BELTON (1969)	9.0 ± 2.0	71 ± 8
TRAFTON (1972)		53 ± 2
TRAUGER (1972)	8.1 ± 0.2	
CARLETON AND TRAUB (1974)	8.4 ± 0.4 (-10% at limb)	34 to 50 (variable) (-10% at limb) (+20% at pole)

Finally, the problem of the constancy of 4–0 S(1) and variability of 3–0 S(1) must be discussed. This behavior is possibly even more dramatic than we have indicated, according to several observers whose results are compiled in Table I. Since 4–0 S(1) is weak (intrinsic depth ≈ 20%), it is formed deep in the atmosphere at about optical depth unity. The presence or absence of a thin haze at high altitude has no strong effect on the line. However, the 3–0 S(1) line is nearly saturated (about 80% deep), and hence is effectively formed at higher altitudes; introducing a scattering cloud at this higher level could conceivably affect this line noticeably. Clearly more calculations are required to investigate this possibility.

References

Beckman, J. E.: 1967, *Planetary Space Sci.* **15**, 1211.
Emerson, J. P., Eddy, J. A., and Dulk, G. A.: 1969, *Icarus* **11**, 413.
Fink, U. and Belton, M. J. S.: 1969, *J. Atmospheric Sci.* **26**, 952.
Gehrels, T.: 1969, *Icarus* **10**, 410.

Hunt, G. E.: 1973, *Icarus* **18**, 637.
Hunt, G. E. and Margolis, J. S.: 1973, *J. Quant. Spectrosc. Radiat. Transfer* **13**, 417.
Owen, T. and Mason, H. P.: 1968, *Astrophys. J.* **154**, 317.
Spinrad, H. and Trafton, L. M.: 1963, *Icarus* **2**, 19.
Trafton, L.: 1972, *Bull. Am. Astron. Soc.* **4**, 358.
Trauger, J. T.: 1972, Ph.D. Thesis, University of Wisconsin.
Trauger, J. T. and Roesler, F. L.: 1972, *Appl. Opt.* **11**, 1964.
Trauger, J. T., Roesler, F. L., Carleton, N. P., and Traub, W. A.: 1973, *Astrophys. J.* **184**, L137.

DISCUSSION

Fox: What is the laboratory strength of the H_2 4–0 S(1) line you discussed?

Traub: This line has not been measured in the laboratory, but certainly should be measured if possible. We use the theoretical line strength calculated by Dalgarno *et al.*, and Birnbaum and Poll in the September, 1969, *J. Atmospheric Sci.*

TD1A SATELLITE SPECTROSCOPIC OBSERVATIONS
OF JUPITER IN THE ULTRAVIOLET

R. DUYSINX and M. HENRIST

Institut d'Astrophysique, Université de Liège, Belgium

Abstract. Three spectra of Jupiter were recorded by TD1A satellite between 1350 and 2600 Å. From these data, the geometrical albedo of the planet was computed.

From 2600 to 2100 Å, the value of the albedo matches the simple diffusing models of Jupiter's atmosphere (Greenspan and Owen, 1967).

From 2100 to 1800 Å, a broad absorption can be assigned to NH_3 molecule (2168 Å transition).

Another absorption commences below 1800 Å. The absorbing molecule involved will be determined at a later date.

1. Introduction

Planetary ultraviolet spectroscopy, particularly for Jupiter, is based at the present time on a limited number of original measurements (Newburn and Gulkis, 1973).

Up to the present time, primary data have been published by Stecher (1965), Evans (1966), Moos *et al.* (1969), Jenkins (1969), Jenkins *et al.* (1969), Anderson *et al.* (1969), Kondo (1971), and Wallace *et al.* (1972).

The latter observations are the only ones recorded by satellite. The particular interest for such measurements is that almost no correction must be made for terrestrial atmospheric opacity.

For this reason, it has been considered of interest to show herewith three spectra of Jupiter, which were recorded from 1350 to 2600 Å by S2/68 experiment aboard TD1A (European Astronomical Satellite).

2. Experimental

Various scientific experiments are mounted aboard TD1A satellite, among which is the S2/68 telescope. This device was designed cooperatively by the Astrophysical Institute of Liége University, the Royal Observatory of Edinburgh and the Astrophysical Division of Culham Laboratory. The telescope features a 275 mm clear aperture, a three-channel spectrometer and a photometric channel.

Detailed descriptions have been published elsewhere, related to the TD1A satellite (Tilgner, 1971) and to the S2/68 experiment (Gardier *et al.*, 1973; Malaise *et al.*, 1972).

It will be simply recalled here that the spectrometer covers the range 1350–2600 Å with a resolution of 36 Å. The photometric channel records a fixed 440 Å band centered on 2800 Å.

Relative calibration of the whole instrument was made in Liége, by means of a specially built vacuum calibration bench (Jamar *et al.*, 1971). Furthermore, unique absolute calibration techniques were used (Marette, 1973) so that the measurements reported here have an overall accuracy better than 20%.

Woszczyk and Iwaniszewska (eds.), Exploration of the Planetary System, 351–355. All Rights Reserved
Copyright © 1974 by the IAU

3. Results

The three spectra of Jupiter, which are shown in Figure 1, were recorded on March 27th 1972. At that time, the Jovian parameters were the following:

- heliocentric distance   5.203 AU
- geocentric distance   5.283 AU
- phase angle   10°.

For the sake of clarity, the zero level of the spectra was offset in the figure. The first left point of each spectrum represents the zero level.

The three curves are very similar, and this is a proof of the photometric accuracy of the observations.

Albedo calculations were performed as follows:

(a) From the three individual spectra, a best-fit curve has been defined.

(b) Solar data were taken from Broadfoot (1972)'s photoelectric spectrum for $\lambda > 2150$ Å and reduced to the S2/68 spectral resolution of 36 Å. Below 2150 Å, the data of Detwiler *et al.* (1961) have been chosen.

(c) The attenuation factor of the solar flux was computed for the path Sun-Jupiter-Earth. The former planet was treated as a Lambert disk, the diameter of which is supposed to be known. So the geometrical albedo obeys the relation:

$$\text{Albedo} = \frac{\text{measured flux}}{\text{attenuated solar flux}}.$$

Figure 2 shows the obtained albedo curve. Figure 3 shows the results of the calcula-

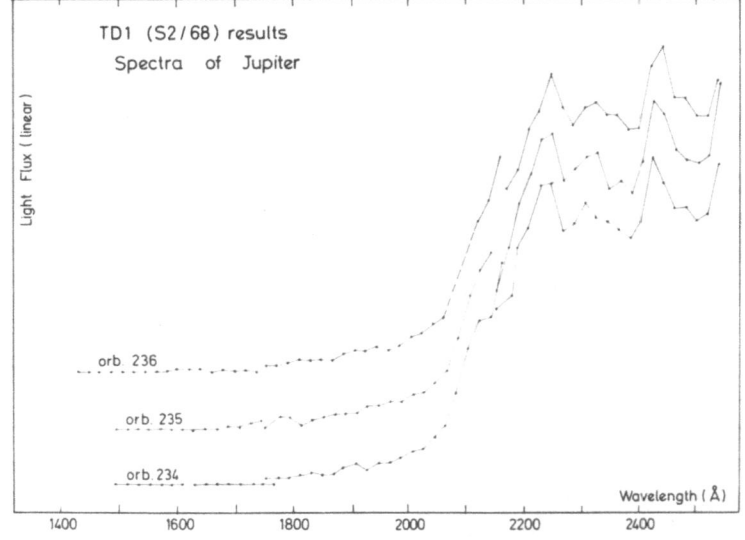

Fig. 1.   Three spectra of Jupiter, recorded on March 27, 1972.

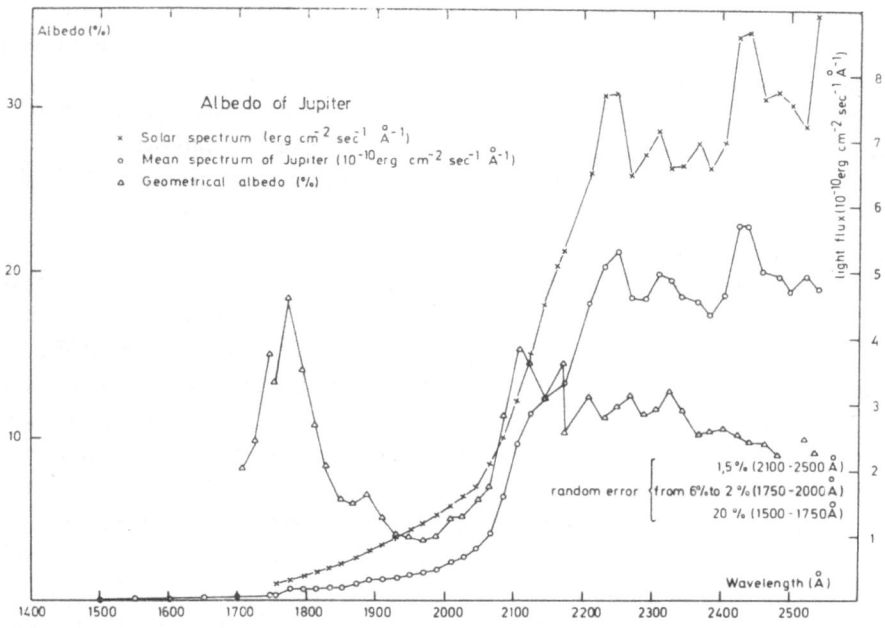

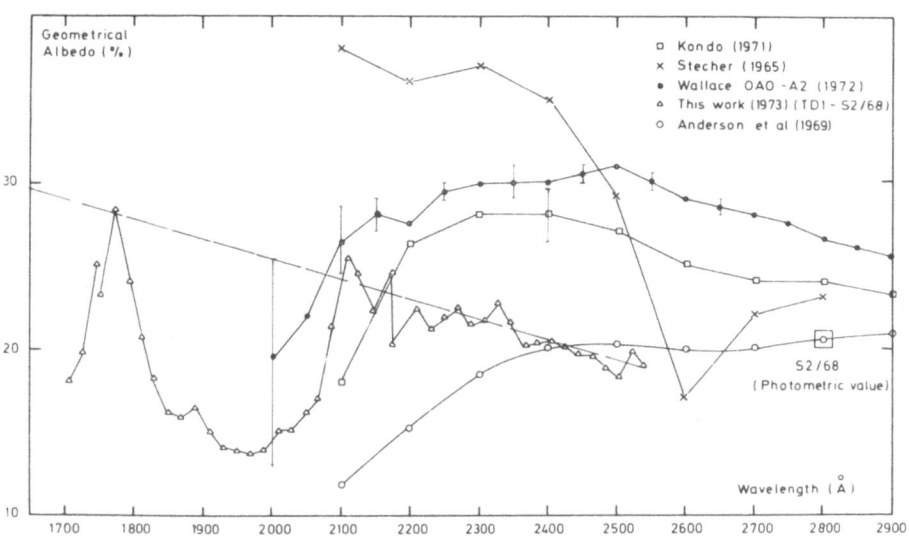

Fig. 3.   Comparison of different albedo calculations. The dashed line represents the best-fit line for
our observations between 2100 and 2550 Å.

tions, together with similar data published elsewhere. It can be noticed that the curves obtained by the two satellites TD1A and OAO-2 are particularly similar in shape, whereas the absolute values from OAO-A2 are definitely greater.

Another remark to be made on our curve is related to some sawteeth which appear between 2600 and 2100 Å. These probably do not have any physical significance, although their amplitude is far greater than our measurement accuracy. Indeed, all local maxima correspond to local minima in the solar spectrum which was used.

Below 2100 Å, the albedo curve is of special interest.

(1) From 1800 to 2100 Å, a deep absorption occurs in the albedo. This absorption can undoubtedly be identified as caused by the ammonia molecule, the presence of which was proved by other means (Kuiper, 1952; Mason, 1969; Owen, 1970). This conclusion was arrived at as follows: The absorption feature ends on a local maximum, around 1770 Å. This maximum falls on a straight line which is a best-fit for the points between 2100 and 2550 Å. That such a line exists seems to confirm the well-known models of Jupiter's upper atmosphere (Greenspan and Owen, 1967), i.e. a diffusing layer located over a layer of reflecting clouds. From this reference level (the dashed line drawn on Figure 3), an absorption profile due to the NH_3 molecule, in the region 1800–2100 Å, has been computed.

In the next step we have compared this coefficient with the inverse cross section of NH_3 for the 2168 Å transition (Watanabe, 1954). Figure 4 shows the result of the comparison.

Since the curves behave similarly, the origin of the absorption must be attributed to the ammonia molecule.

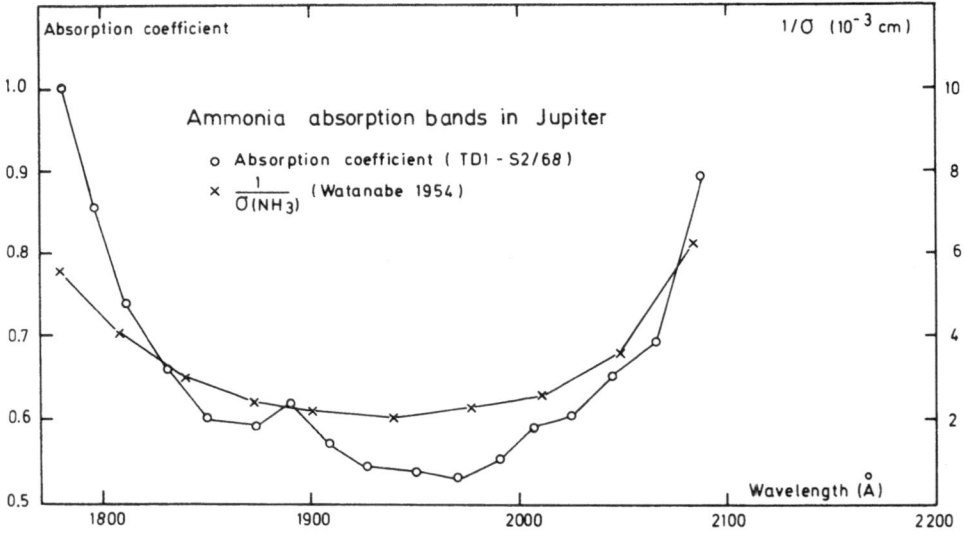

Fig. 4. Comparison of absorption coefficients obtained from NH_3 profile, calculated from the dashed-line reference level (Figure 3) with Watanabe's inverse cross section of NH_3 for the 2168 Å transition.

(2) Figure 3 also shows a further absorption below 1770 Å. However, the measurements ranging from 1700 to 1350 Å have not been reduced yet, because of a lower signal-to-noise ratio. Nevertheless:

– this absorption feature appears to be quite real. Indeed, the detected light flux for this region is much lower than the predicted flux for a simple Rayleigh scattering atmosphere;

– further spectra of Jupiter will be soon available, so that the signal-to-noise ratio will be appreciably increased and the relevant absorbing species will be identified.

Acknowledgements

The collaboration of Messrs C. and J. Jamar is gratefully acknowledged.

References

Anderson, R. C., Pipes, J. G., Broadfoot, A. L., and Wallace, L.: 1969, *J. Atmospheric Sci.* **26**, 874.
Beeckmans, F., Macau, D., and Malaise, D.: 1973, *Bolometric Corrections for B Stars*, COSPAR 1973.
Broadfoot, A. L.: 1972, *Astrophys. J.* **173**, 681.
Detwiler, C. R., Garret, D. L., Purcell, J. D., and Tousey, R.: 1961, *Ann. Geophys.* **17**, 265.
Evans, D. C.: 1966, NASA Goddard Space Flight Center Report Nr X-613-66-172.
Gardier, S., Jamar, C., and Malaise, D.: 1973, *Rev. Universelle des Mines* **1**, 19.
Greenspan, J. A. and Owen, T.: 1967, *Science* **156**, 1489.
Jamar, C., Malaise, D., and Monfils, A.: 1971, *ELDO-CECLES/ESRO-CERS Scient. and Tech. Rev.* **3**, 427.
Jenkins, E. B.: 1969, *Icarus* **10**, 379.
Jenkins, E. B., Morton, D. C., and Sweigart, A.: 1969, *Astrophys. J.* **157**, 913.
Kondo, Y.: 1971, *Icarus* **14**, 269.
Kuiper, G. P. (ed.): 1952, in *Atmospheres of the Earth and Planets*, University of Chicago Press, Chicago, pp. 304–345.
Malaise, D., Macau, J. P., and Jamar, C.: 1972, 'Description, but scientifique, premiers résultats de S2/68 de TD1', Colloque d'Aussois.
Marette, G.: 1973, *Optics Communications*, in press.
Mason, H. P.: 1969, Thesis, University of Illinois.
Moos, H. W., Fastie, W. G., and Bottema, M.: 1969, *Astrophys. J.* **155**, 887.
Newburn, R. L. and Gulkis, S.: 1973, *Space Sci. Rev.* **3**, 179.
Owen, T.: 1970, *Science* **167**, 1675.
Stecher, T. P.: 1965, *Astrophys. J.* **142**, 1186.
Tilgner, B.: 1971, *ELDO-CECLES/ESRO-CERS Scient. and Techn. Rev.* **3**, 567.
Wallace, L., Caldwell, J. J., and Savage, B. D.: 1972, *Astrophys. J.* **172**, 755.
Watanabe, K.: 1954, *J. Chem. Phys.* **22**, 1564.

DISCUSSION

Low: Do you not expect large scale changes in the UV albedo of Jupiter?

Duysinx: The answer will be possible only when we have computed the albedo for the sets of measurements. The set which is presented here is the first one, and we expect three or four of them.

Beer: Is the peak at 1790 Å real or is it an artefact of low signal-to-noise?

Duysinx: The peak at 1790 Å is quite real because the signal to noise ratio is still very good at that wavelength. The measured intensities were not apparent on the slide because of the linear scale used to show the whole spectrum.

NEW INFRARED SPECTRA OF THE JOVIAN PLANETS:
STUDY OF JUPITER AND SATURN IN THE $3v_3$
METHANE BAND BY FOURIER TRANSFORM
SPECTROSCOPY

C. DE BERGH, M. COMBES, TH. ENCRENAZ, J. LECACHEUX, and M. VION

Groupe Planètes, Observatoire de Paris-Meudon, France

and

J. P. MAILLARD

Laboratoire Aimé Cotton, Orsay, France

Abstract. High resolution spectra of Jupiter and Saturn were obtained with a Fourier Transform Michelson interferometer. A comparison of the observed spectra, after elimination of the solar and terrestrial contributions to absorption, with synthetic profiles for the reflecting layer model has permitted new determinations of the Lorentz half-width, the methane abundance, the rotational temperature and the pressure at the level of formation of the methane lines for both Jupiter and Saturn.

We recorded high-resolution spectra of Jupiter and Saturn from 4000 to 12000 cm^{-1}, respectively, in May and December 1972, with a Fourier Transform Interferometer at the 193 cm telescope of the Haute Provence Observatory (France). The resolution obtained is 0.22 cm^{-1} for the Jupiter spectrum and 0.26 cm^{-1} for the Saturn spectrum. Spectra of the Sun and the Moon were also recorded for comparison. For both planets the region of the R-branch of the $3v_3$ methane band 9050–9150 cm^{-1} has been analysed in order to determine the average Lorentz half-width, the abundance of methane, the rotational temperature and the pressure at the level of formation of the methane lines.

The equivalent widths and the average Lorentz half-width (assuming that all lines had the same half-width) were deduced from a comparison of synthetic profiles of the methane J-manifolds with the observed ones ($R(0)$ to $R(7)$), after the elimination of telluric and solar absorptions. The wavenumbers of the J-manifolds used in the computations were determined accurately by recording laboratory spectra of CH_4 at 0.02 cm^{-1} resolution. The rotational temperature and abundance of methane were then obtained by using the method of Margolis and Fox (1969a, b), assuming that the reflecting-layer model was appropriate. The air-mass factor η was taken equal to 2.5 for Jupiter and 2.8 for Saturn. The results obtained are shown in Table I.

The continuum level used to do the analysis was defined by comparing maxima of Jupiter and Saturn spectra with maxima of the solar spectrum in a large spectral range (9000–9500 cm^{-1}), which permitted us to find a continuum gaseous absorption in the $3v_3$ region. Dipole-induced hydrogen absorption (2–0 band) is probably responsible for the observed increase in absorption from 9300 cm^{-1} towards lower wavenumbers. Indeed, it was found that, by using laboratory absorption coefficients of hydrogen (Hunt, 1959; Welsh, 1969) for the temperature range considered, 28 to

Woszczyk and Iwaniszewska (eds.), Exploration of the Planetary System, 357–358 All Rights Reserved
Copyright © 1974 by the IAU

TABLE I

Average Lorentz half-width, rotational temperature, abundance of
methane and effective pressure of the R-branch of the $3v_3$ methane
band for Jupiter and Saturn

Jupiter	Saturn
$\gamma_{\text{average}} = 0.10 \pm 0.03 \text{ cm}^{-1}$	$\gamma_{\text{average}} = 0.15 \pm 0.03 \text{ cm}^{-1}$
$T_{\text{rotational}} = 150 \pm 15 \text{ K}$	$T_{\text{rotational}} = 134 \pm 15 \text{ K}$
$\eta a = 95 \pm 20$ m-atm	$\eta a = 118 \pm 30$ m-atm
$a = 38 \pm 8$ m-atm	$a = 42 \pm 11$ m-atm
$P_{\text{effective}} = 1.0 \pm 0.3$ atm	$P_{\text{effective}} = 1.3 \pm 0.3$ atm

48 km-atm of hydrogen (depending on the ortho-para hydrogen ratio) in the Jupiter
atmosphere would explain the pseudo-continuum absorption which shows up in the
range 9300–9000 cm^{-1}. For Saturn, the upper limit for the hydrogen amount
is 54 to 92 km-atm. The results shown in Table I correspond to such an amount of
hydrogen absorption.

$^{13}CH_4$. In the region of the R-branch of the $3v_3$ band of $^{12}CH_4$ we have detected
absorption lines which cannot be $^{12}CH_4$, telluric or solar lines, and appear precisely
at the wavelengths of the $R(2)$, $R(3)$ and $R(6)$ lines of $^{13}CH_4$ $3v_3$ band of the laboratory spectrum recorded by Pugh, Owen and Rao (Stony Brook).

References

De Bergh, C., Vion, M., Combes, M., Lecacheux, J., and Maillard, J. P.: 1973, *Astron. Astrophys.*
    28, 157.
Bergstralh, J. T.: 1973, *Icarus* **18**, 605.
Connes, J., Delouis, H., Connes, P., Guelachvili, G., Maillard, J. P., and Michel, G.: 1970, *Nouv.
    Rev. d'Opt. App.* **1**, 1,3.
Guelachvili, G. and Maillard, J. P.: 1970, in G. A. Vanosse, A. T. Stair, and D. J. Baker (eds.),
    Aspen Int. Conf. on Fourier Transform Spectroscopy.
Hunt, J. L.: 1959, Ph.D. Thesis, University of Toronto.
Maillard, J. P., Combes, M., Encrenaz, T., and Lecacheux, J.: 1973, *Astron. Astrophys.* **25**, 219.
Margolis, J. S. and Fox, K.: 1968, *J. Chem. Phys.* **49**, 2451.
Margolis, J. S. and Fox, K.: 1969a, *Astrophys. J.* **157**, 935.
Margolis, J. S. and Fox, K.: 1969b, *Astrophys. J.* **158**, 1183.
Trafton, L.: 1973, *Astrophys. J.* **182**, 615.
Varanasi, P., Sarangi, S., and Pugh, L.: 1973, *Astrophys. J.* **179**, 977.
Welsh, H. L.: 1969, *J. Atmospheric Sci.* **26**, 835.

DISCUSSION

Trafton: What is the basis of your estimate that 28–48 km-amagats of H_2 exists above the reflecting
layer?

De Bergh: I assumed several abundances, and chose the one which gave the observed distortion to
the continuum in the $3v_3$ CH_4 region.

Traub: Can you deduce an ortho-to-para hydrogen ratio for Jupiter from your spectra?

De Bergh: The estimates on hydrogen abundances we get depend on the ortho-para hydrogen ratio,
and it is certainly not possible to deduce the ortho-para ratio from our spectra only.

ON THE MICROWAVE SPECTRUM OF
METHANE IN THE ATMOSPHERES OF THE OUTER PLANETS

KENNETH FOX

*Earth Resources and Astrophysics Laboratory, Dept. of Physics and Astronomy,
University of Tennessee, Knoxville, Tenn., U.S.A.*

Abstract. The existence of a small dipole moment in the vibronic ground state of methane leads to transitions in the microwave spectral range. Frequencies and line strengths appropriate to the atmospheres of the outer planets are presented.

1. Introduction

Methane is known to be an important constituent of the atmospheres of Jupiter, Saturn (and its satellite Titan), Uranus, and Neptune (McElroy, 1969; Fox, 1972a; Newburn and Gulkis, 1973). Spectra of CH_4 in the infrared and visible regions have been studied for a long time, and have yielded information on the characteristics of these atmospheres (Fox, 1972a). The optical spectra of CH_4 usually arise from vibration-rotation transitions in the ground vibrational and electronic ('vibronic') state. In these transitions, the change of electric dipole moment 'μ' with vibration permits absorption (or possibly emission) of electromagnetic radiation (Herzberg, 1945).

A tetrahedral ('T_d') molecule like CH_4 was conventionally assumed (Herzberg, 1945) to have no permanent μ in its ground vibronic state, by virtue of the high symmetry in this state. Consequently, CH_4 was not expected to have pure rotational transitions, i.e., transitions in which only its rotational quantum numbers change. Spectra of CH_4 in the far-infrared and microwave regions were thus thought to be forbidden.

Difference vibration-rotation spectra of CH_4 may occur in the far-infrared (see Herzberg, 1945), but are expected to be relatively weak because of the Boltzmann factor.

Collision-induced far-infrared absorption spectra of CH_4 have been discussed extensively by Ozier and Fox (1970); this paper contains references to earlier experimental and theoretical work. Such spectra depend on molecular collisions to permit absorption, and their strengths are not great at low gas pressures. However, collision-induced absorption by mixtures of CH_4 and H_2 has recently been considered as a source of thermal opacity in the atmospheres of Jupiter, Saturn, Uranus, and Neptune (Fox and Ozier, 1971) and Titan (Pollack, 1973).

Pure rotational transitions in an *excited* vibrational state of CH_4 were predicted by Mizushima and Venkateswarlu (1953) and one such transition was recently observed directly by Curl and Oka (1973); the latter paper contains references to earlier experimental and theoretical work. These excited-state transitions are also expected to be relatively weak because of the Boltzmann factor, as in difference spectra.

However, in 1971, it was theoretically predicted that pure rotational transitions in the vibronic ground state of T_d molecules could occur as the result of a small μ arising

Woszczyk and Iwaniszewska (eds.), Exploration of the Planetary System, 359–365. All Rights Reserved

from vibration-rotation interactions (Fox, 1971) or, alternatively, from centrifugal distortion effects (Watson, 1971; Aliev, 1971). This μ was measured (Ozier, 1971), by molecular-beam techniques, to have the magnitude $(5.38 \pm 0.10) \times 10^{-6}$ D for $^{12}CH_4$. Subsequently, a far-infrared absorption spectrum corresponding to pure rotational $\Delta J = +1$ transitions (here J represents the quantum number for the total angular momentum $\mathbf{J}$), arising from μ in the ground vibronic state of $^{12}CH_4$ was also reported (Rosenberg et al., 1972). More recently, a pure rotational $\Delta J = 0$ transition of $^{12}CH_4$, produced by the same mechanism (Fox, 1971; Watson, 1971; Aliev, 1971) was observed by infrared-radio frequency double resonance (Curl et al., 1973).

In 1972, it was noted (Fox, 1972b), that in the absence of an NH_3 atmospheric consituent on Saturn, Uranus, or Neptune, CH_4 may be promoted as a source of microwave opacity there. This remark would apply to Titan also. A microwave spectrum was calculated and tabulated with absolute intensities corresponding to $T = 0\,°C$ by Fox (1972b). Some similar results for $T = 300$ K were presented by Dorney and Watson (1972). As experimental evidence for pure rotational transitions in methane has increased, and because more refined spectroscopic constants needed to calculate spectral line positions with microwave accuracy are now available (Curl, 1973), it seems appropriate and useful to discuss in more detail, with calculations tailored to the atmospheres of the outer planets, microwave spectra of methane in those environments.

2. Theory

The strength of an individual absorption line is given by Fox (1972b)

$$\alpha_{abs} = (8\pi^3/3hc)\,(N/\mathscr{Z})\,(E_f - E_i) < \mu_{JK} >^2 \varepsilon_{JK}\,(e^{-E_i hc/kT} - e^{-E_f hc/kT}),$$

(1)

where h and k are Planck's and Boltzmann's constants, respectively, and c is the speed of light in vacuum; N is the number of molecules per cm^3, and ε_{JK} is the nuclear-spin statistical weight factor for the initial state of the transition. The dipole moment for $\Delta J = 0$ transitions in the ground vibronic state is

$$\langle \mu_{JK} \rangle^2 = (3/2)\,C_{34}^2\,(2J - 2)\,(2J - 1)\,(2J + 1)\,(2J + 3) \times$$
$$\times (2J + 4)\,[f(JK)]^2,$$

(2)

where $f(JK)$ is a linear combination of vector-coupling coefficients; K is the projection of $\mathbf{J}$ (total angular momentum) on the molecule-fixed z-axis. The coefficient C_{34} depends only on molecular parameters, in addition to universal constants.

The partition function $\mathscr{Z}$ may be approximated well for $T \leqslant 300$ K by the rotational partition function $\mathscr{Z}_r$, which in turn may be expressed (Fox, 1970) as

$$\mathscr{Z}_r = (4\pi^{1/2}/3)\,(B_0 hc/kT)^{-3/2}\,e^{B_0 hc/kT},$$

(3)

a form accurate to 1% for $T > 40$ K, for CH_4; B_0 is the ground-state rotational constant. The energies E_i and E_f refer to the initial and final states, respectively. The difference of Boltzmann factors in Equation (1) may be rewritten as

$$e^{-E_i hc/kT} - e^{-E_f hc/kT} \equiv e^{-E_i hc/kT} \left[1 - e^{-(E_f - E_i) hc/kT} \right]. \tag{4}$$

A useful approximation to Equation (4) may be obtained as follows:

$$(E_f - E_i) hc/kT = 4.8 \times 10^{-5} (E_f - E_i)/T, \tag{5}$$

where E_f and E_i are in units of MHz, and T is in K. For $J \leqslant 20$, the range of the tabulation by Fox (1972b), the transition energies $E_f - E_i < 130000$ MHz. Consequently, for $T > 45$ K, a possible effective temperature for Neptune (Newburn and Gulkis, 1973),

$$(E_f - E_i) hc/kT < 0.14. \tag{6}$$

Equation (4) may then be replaced, to an accuracy of at least 7%, by

$$e^{-E_i hc/kT} - e^{-E_f hc/kT} = e^{-E_i hc/kT} (E_f - E_i) hc/kT. \tag{7}$$

As the absorption coefficient α_{abs} has already been tabulated by Fox (1972b) at $T = 0\,°C$, for the present purposes it is essentially necessary only to recalculate its temperature-dependent part $\alpha_{abs}(T)$. From Equations (1), (3), and (7),

$$\alpha_{abs}(T) = \mathscr{Z}_r^{-1} e^{-E_i hc/kT} hc/kT. \tag{8}$$

The initial-state energy is given in the usual form (Herzberg, 1945),

$$E_i = B_0 J(J + 1) - D_s J^2 (J + 1)^2, \tag{9}$$

where D_s is the centrifugal distortion constant appropriate (Hecht, 1960) to T_d molecules. Higher-order contributions (Hecht, 1960; Moret-Bailly, 1961) have been neglected in Equation (9). These include the T_d fine-structure terms which determine the microwave transition energies, but which are characteristically (Moret-Bailly et al., 1965) much less than 1% of Equation (9) for methane.

Then from Equations (3), (8), and (9)

$$\frac{\alpha_{abs}(T)}{\alpha_{abs}(273.15)} = \left(\frac{273.15}{T} \right)^{5/2} \exp \left\{ - \left[B_0 (J + \tfrac{1}{2})^2 - D_s J^2 (J + 1)^2 \right] \times \right.$$
$$\left. \times (hc/kT) \left(1 - \frac{T}{273.15} \right) \right\}. \tag{10}$$

This is the factor by which the absolute intensities in (Fox, 1972b) are to be multiplied to make them correspond to temperatures appropriate for the atmospheres of the outer planets.

Spectral line positions may be readily determined from the ground-state T_d fine-structure terms expressed (Moret-Bailly, 1965), to fourth order, as

$$E^0_{(J, p)} = \left[(2J - 3) \cdots (2J + 5) \right]^{1/2} \left[\varepsilon^0 + \varrho^0 J (J + 1) \right] (-1)^J F^{(4JJ)}_{A_1 pp} +$$
$$+ \left[(2J - 5) \cdots (2J + 7) \right]^{1/2} \xi^0 (-1)^J F^{(6JJ)}_{A_1 pp}, \tag{11}$$

where ε^0, ϱ^0, and ζ^0 are spectroscopic constants for the vibronic ground state. The F functions are vector-coupling coefficients adapted to cubic symmetry. These functions have been tabulated numerically (Moret-Bailly *et al.*, 1965) for $J \leqslant 21$. The subscripts p label the T_d sub-levels in a J multiplet. Each transition frequency may be obtained from an energy difference of the form $E^0_{(J, p)} - E^0_{(J, p')}$.

3. Results

Of the many microwave transitions tabulated previously (Fox, 1972b), only those which fall in the frequency range observed for the outer planets (Newburn and Gulkis, 1973) are considered in the present study. Moreover, only the stronger transitions with $\alpha_{abs} > 10^{-12}$ cm^{-2} amagat^{-1} at the relevant atmospheric temperatures are given here.

TABLE I

Spectral line positions and absolute line intensities for microwave absorption by methane in the atmospheres of the outer planets, for a characteristic temperature of 134 K

Transition[a]	Frequency (MHz)	Absolute intensity (cm^{-2}amagat^{-1})[b]
$6 \rightarrow \leftarrow 6$		
$A2 \rightarrow \leftarrow A1$	760.921	1.51 (−12)
$7 \rightarrow \leftarrow 7$		
$F1(1) \rightarrow \leftarrow F2(2)$	1 562.879	1.62 (−12)
$8 \rightarrow \leftarrow 8$		
$F1(1) \rightarrow \leftarrow F2(2)$	2 774.606	5.07 (−12)
$F1(1) \rightarrow \leftarrow F2(1)$	1 563.809	1.06 (−12)
$9 \rightarrow \leftarrow 9$		
$A2 \rightarrow \leftarrow A1$	1 257.944	2.33 (−12)
$F1(1) \rightarrow \leftarrow F2(2)$	2 546.742	1.52 (−12)
$F2(2) \rightarrow \leftarrow F1(2)$	1 791.730	1.31 (−12)
$10 \rightarrow \leftarrow 10$		
$A2 \rightarrow \leftarrow A1$	5 014.758	4.44 (−12)
$F2(1) \rightarrow \leftarrow F1(1)$	3 946.940	1.68 (−12)
$F2(2) \rightarrow \leftarrow F1(2)$	2 688.583	1.50 (−12)
$11 \rightarrow \leftarrow 11$		
$E(1) \rightarrow \leftarrow E(2)$	4 599.229	3.14 (−12)
$F1(2) \rightarrow \leftarrow F2(3)$	5 169.506	2.77 (−12)
$F1(1) \rightarrow \leftarrow F2(2)$	7 468.747	2.38 (−12)
$12 \rightarrow \leftarrow 12$		
$A1(1) \rightarrow \leftarrow A2$	13 276.228	2.83 (−11)
$F1(1) \rightarrow \leftarrow F2(2)$	10 797.232	3.80 (−12)
$F1(2) \rightarrow \leftarrow F2(3)$	6 940.206	1.62 (−12)
$14 \rightarrow \leftarrow 14$		
$E(1) \rightarrow \leftarrow E(3)$	27 137.337	2.55 (−12)
$E(1) \rightarrow \leftarrow E(2)$	19 281.566	1.30 (−12)
$15 \rightarrow \leftarrow 15$		
$A1(1) \rightarrow \leftarrow A2$	21 287.410	1.17 (−12)

[a] For convenience in tabulation, $F2(1)$ denotes $F_2^{(1)}$, etc. (see Table II of Fox, 1972b).

[b] Powers of 10 are indicated in parentheses, e.g., $1.51 (-12) \equiv 1.51 \times 10^{-12}$.

These restrictions may be relaxed as necessitated by future observations and/or models of the atmospheres of the outer planets and their satellites. Also, the atmospheres may be characterized by many different values of temperature, so that it is difficult to choose a particular one or even a small set; and it would be prohibitive to present the results for all possible temperatures here. Consequently, one probable characteristic temperature $T = 134$ K has been selected. The frequencies given in Table I are likely to include most of the stronger transitions for Jupiter, Saturn, Uranus, Neptune, and Titan. Modifications of the absolute intensities in Table I may readily be made by the application of Equation (10). With $T = 134$ K, and the following values of ground-state rotational constants obtained in the analysis of a high-resolution vibration-rotation infrared spectrum (Barnes *et al.*, 1972): $B_0 = 5.24059$ cm^{-1} and $D_s = 1.0855 \times \times 10^{-4}$ cm^{-1}; Equation (10) becomes

$$\alpha_{abs}(134)/\alpha_{abs}(273.15) = (5.933) \times$$
$$\times \exp\left[-0.0287(J + 1/2)^2 + 5.94 \times 10^{-7}J^2(J + 1)^2\right]. \qquad (12)$$

The squared magnitude of the dipole moment is taken (Ozier, 1971) to be $C_{34}^2 = = (5.38 \times 10^{-6}D)^2$, as in Table II of Fox (1972b).

Microwave transition frequencies had been obtained previously (Fox, 1972b) on the basis of a single (Hecht, 1960) ground-state spectroscopic constant $D_t \equiv -3.05505$ ε^0. A mean value of $D_t = 4.403 \times 10^{-6}$ cm^{-1} was adopted on the basis of three determinations made directly from experimental data (Barnes *et al.*, 1972; Husson and Dang Nhu, 1971; Ozier *et al.*, 1970). More recently, however, a refined value of ε^0, together with accurate values of ϱ^0 and ζ^0, have been deduced (Curl, 1973) from infrared-radio frequency double resonance observations of two $\Delta J = 0$ transitions in the vibronic ground state of $^{12}CH_4$; the molecular-beam measurements of Ozier, (1971) were used in the analysis of Curl (1973). The values employed in Equation (11) of the present work are $\varepsilon^0 = -43\,516.819_1$ Hz, $\varrho^0 = 5.514_1$ Hz, and $\zeta^0 = -0.452_9$ Hz. These are only slightly different from Curl (1973). The spectral line positions in Table I have been determined from energy differences based on Equation (11) with the current values of ε^0, ϱ^0, and ζ^0 shown above. These tabulated frequencies may be expected to be accurate to at least 0.1 MHz$\equiv 10^5$ Hz. The notation for T_d sub-levels in Moret-Bailly (1965) and Moret-Bailly *et al.* (1965) differs from that in Table II (see footnote a) of Fox (1972b). The conventions used in the latter are followed here in Table I.

Acknowledgements

I am grateful to R. F. Curl, Jr. for preprints and useful discussions of his papers. Partial support from the Fowler-Marion Fund for travel to the Extraordinary General Assembly of the International Astronomical Union in Poland (1973) is deeply appreciated.

Note added in proof. Recently some new work has appeared which is related to the present paper. Takami, M., Uehara, K., and Shimoda, K. (1973, *Japan J. Appl. Phys.*

12, 924) observed two pure rotational transitions in the same excited vibrational state of CH_4 utilized by Curl and Oka (1973). An explicit generalization of Mizushima-Venkateswarlu (1953) transitions to excited states of all infrared-active transitions in T_d molecules was given by Fox, K. (1974, *J. Chem. Phys.* **60**, 337). The coincidence of Er:YAG laser emission with methane absorption was considered by Fox, K. (1974, *Appl. Phys. Letters* **24**, 24) from the viewpoint of experiments and calculations related to the polar character of the ground and excited vibrational states. Holt, C. W., Gerry, M. C. L., and Ozier, I. (1973, *Phys. Rev. Letters* **31**, 1033) observed four $\Delta J = 0$ transitions in the ground vibronic state of methane in a microwave absorption spectrum. None of these measured lines corresponds to the transitions in Table I. However, it may be inferred that the predicted frequency for $E(2) \to \leftarrow E(3)$ of $J = 14$ would differ from the measured value by a few MHz, so that the frequencies given in Table I may be expected to be accurate to only several MHz. The sixth-order analogue of Equation (11) has been developed by Michelot, F., Moret-Bailly, J., and Fox, K. (1974, *J. Chem. Phys.*, to be published). Tarrago, G., Dang Nhu, M., and Poussigue, G. (1974, *Compt. Rend. Acad. Sci. Paris* **278B**, 207) have determined values of ε^0, ϱ^0, and ζ^0 which differ somewhat from those used in the present work, and which may lead to transition frequencies varying from those in Table I by several MHz. The parameters B_0 and D_s in Equations (3), (9), and (10) have been re-evaluated by Tarrago, G., Dang Nhu, M., and Poussigue, G. (1974, *J. Mol. Spectrosc.* **49**, 322) from new high-resolution infrared spectra; however, their changes are not expected to have any effect on the absolute intensities in Table I to the stated accuracy.

References

Aliev, M. R.: 1971, *Zh. Eksperim. Teor. Fiz. Pis'ma Red.* **14**, 600.
Barnes, W. L., Susskind, J., Hunt, R. H., and Plyler, E. K.: 1972, *J. Chem. Phys.* **56**, 5160.
Curl, R. F., Jr. and Oka, T.: 1973, *J. Chem. Phys.* **58**, 4908.
Curl, R. F., Jr., Oka, T., and Smith, D. S.: 1973, *J. Mol. Spectrosc.* **46**, 518.
Curl, R. F., Jr.: 1973, *J. Mol. Spectrosc.* **48**, 165.
Dorney, A. J. and Watson, J. K. G.: 1972, *J. Mol. Spectrosc.* **42**, 135.
Fox, K.: 1970, *J. Quant. Spectrosc. Radiat. Transfer* **10**, 1335.
Fox, K.: 1971, *Phys. Rev. Letters* **27**, 233.
Fox, K. and Ozier, I.: 1971, *Astrophys. J.* **166**, L95.
Fox, K.: 1972a, in K. N. Rao and C. W. Mathews (eds.), *Molecular Spectroscopy: Modern Research*, Academic Press, New York, p. 79.
Fox, K.: 1972b, *Phys. Rev.* **A6**, 907.
Hecht, K. T.: 1960, *J. Mol. Spectrosc.* **5**, 355, 390.
Herzberg, G.: 1945, *Infrared and Raman Spectra of Polyatomic Molecules*, Van Nostrand, Princeton, N.J.
Husson, N. and Dang Nhu, M.: 1971, *J. Phys. (Paris)* **32**, 627.
McElroy, M. B.: 1969, *J. Atmospheric Sci.* **26**, 798.
Mizushima, M. and Venkateswarlu, P.: 1953, *J. Chem. Phys.* **21**, 705.
Moret-Bailly, J.: 1961, *Cahiers Phys.* **15**, 237.
Moret-Bailly, J.: 1965, *J. Mol. Spectrosc.* **15**, 344.
Moret-Bailly, J., Gautier, L., and Montagutelli, J.: 1965, *J. Mol. Spectrosc.* **15**, 355.
Newburn, R. L., Jr. and Gulkis, S.: 1973, *Space Sci. Rev.* **3**, 179.
Ozier, I. and Fox, K.: 1970, *J. Chem. Phys.* **52**, 1416.
Ozier, I., Yi, P. N., Khosla, A., and Ramsey, N. F.: 1970, *Phys. Rev. Letters* **24**, 642.

Ozier, I.: 1971, *Phys. Rev. Letters* **27**, 1329.
Pollack, J. B.: 1973, *Icarus* **19**, 43.
Rosenberg, A., Ozier, I., and Kudian, A. K.: 1972, *J. Chem. Phys.* **57**, 568.
Watson, J. K. G.: 1971, *J. Mol. Spectrosc.* **40**, 536.

DISCUSSION

Irvine: Can any of the microwave lines be measured with existing instrumentation?

Fox: Definitely *yes* in the laboratory (see references in my paper). For atmospheres, I don't know yet. The answer will depend on atmospheric conditions such as the abundance of methane and masking by other molecules.

JUPITER'S MICROWAVE SPECTRUM:
IMPLICATIONS FOR THE UPPER ATMOSPHERE

S. GULKIS, M. J. KLEIN, and R. L. POYNTER

Jet Propulsion Laboratory, California Institute of Technology, Pasadena, Calif., U.S.A.

Abstract. It is shown through the use of weighting functions that Jupiter's brightness temperature in the wavelength range 0.8–1.5 cm contains information on the thermal structure and abundance of ammonia in and above the tropopause in Jupiter's atmosphere. We present new data of Jupiter's brightness temperature in this wavelength range, and compare the results with theoretical spectra. The pressure in the Jovian atmosphere is estimated from these data to be 0.48 atm at 130K.

1. Introduction

It has long been recognized (Barrett, 1962) that microwave spectroscopy at short centimeter and millimeter wavelengths is a potentially effective means of studying the upper atmospheres of the major planets. Trace amounts of ammonia distributed in and above the clouds of the major planets are expected to produce line features in the planets thermal spectra near 1.25 cm wavelength due to the inversion splitting of the molecule's rotational energy states. The widths of these lines are dependent upon the pressure at the altitudes where they are formed. Within the ammonia clouds the pressure is thought to be sufficiently high ($\geqslant 1$ atm) that the individual lines are collision broadened into a single absorption feature a few GHz wide. In the atmospheric region above the clouds, where the pressure is a factor of ten or more less than the cloud pressures, narrow ammonia lines may form a fine structure superimposed on the broad absorption feature. Depending upon the thermal structure of the atmosphere, these narrow lines might appear either in emission or in absorption. Unfortunately, attempts to detect microwave spectral lines in the major planet spectra have been largely unsuccessful because of limitations imposed by sensitivity and calibration accuracy.

The most accurate data published to date are the broad band measurements of the thermal spectrum of Jupiter in the frequency range 20.5 GHz to 35.5 GHz by Wrixon *et al.* (1971). These measurements suggest the existence of the broad absorption feature discussed above, but the measurement uncertainties are too large to allow a quantitative interpretation. In this paper we report new observational data for Jupiter in the 20 GHz to 24 GHz band which 1) accurately define the slope of the low frequency ammonia absorption-line wing, and 2) set upper limits to the peak amplitude intensity of narrow (~ 50 MHz) emission or absorption lines relative to the continuum which might form in Jupiter's upper atmosphere. The pressure in the Jovian ammonia cloud is derived from the broad band data.

2. Theoretical Considerations

Previous theoretical investigations (Winter, 1964; Law and Staelin, 1968; Wrixon

Woszczyk and Iwaniszewska (eds.), Exploration of the Planetary System, 367–374. All Rights Reserved

et al., 1971; Gulkis and Poynter, 1972) have shown that a broad absorption feature centered near 1.25 cm should be present in Jupiter's microwave spectrum. We have made a series of model calculations to investigate how the spectral characteristics of such a feature depend upon specific atmospheric parameters. Our analysis shows that the slopes in the wings of the broad absorption depend primarily on the ammonia cloud pressure and only slightly on the ammonia mixing ratio and temperature below this cloud.

The models used in our analysis are in hydrostatic equilibrium throughout, with the troposphere in convective equilibrium, and with an isothermal stratosphere at 112K. We assume that the major atmospheric constituent is H_2 and that trace amounts of ammonia (tropospheric mixing ratio $= 2 \times 10^{-4}$) provide the radio opacity. We

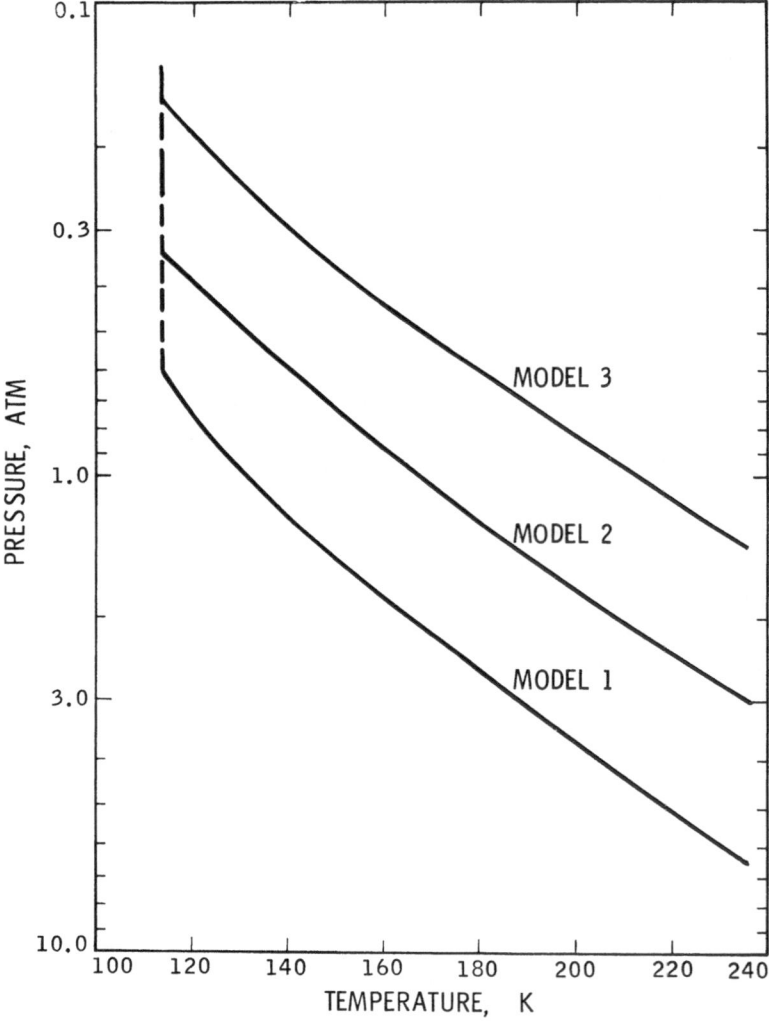

Fig. 1.   Pressure-temperature models for the Jovian atmosphere.

consider the effects of including He as an additional major atmospheric constituent along with H_2 in the next section. The models are basically similar to those used by Gulkis and Poynter (1972) in their analysis of the longer wavelength portion of the microwave spectrum. In Figure 1 we show the basic model for three different cloud pressures typical of those found in the literature.

In order to define the slopes in the wings of the broad absorption line we solved the equation of transfer for a plane parallel atmosphere using the ammonia absorption coefficients computed by one of us (RLP). Scattering due to aerosols was assumed to be negligible for the wavelengths under consideration. We determined the frequency range most sensitive to the cloud pressure by computing the weighting functions for

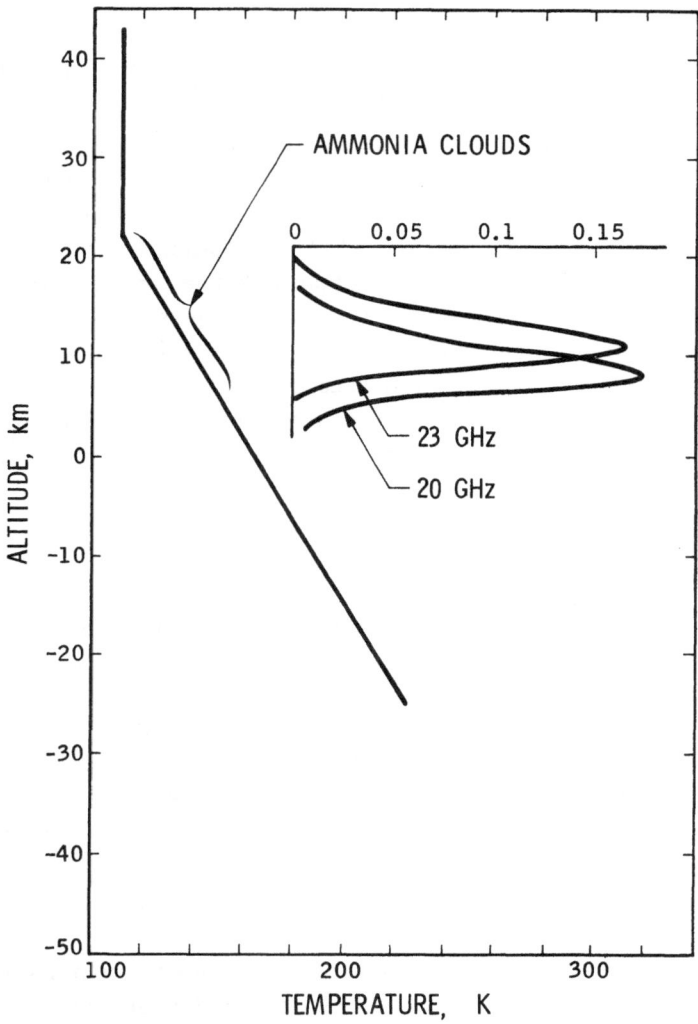

Fig. 2.   Figure gives temperature-altitude profile for a model Jovian atmosphere. Insert shows the temperature weighting functions expressed in units of km^{-1} for two closely spaced frequencies.

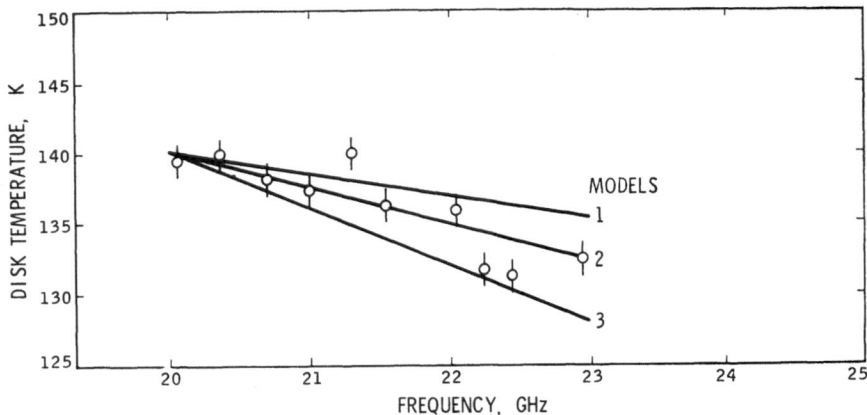

Fig. 3.   Solid lines give theoretical spectral slopes for Models 1, 2, and 3. Experimental data are
shown with one sigma uncertainties indicated.

a representative set of frequencies. The weighting function is a measure of the contribution which the kinetic temperature makes to the brightness temperature from each
altitude increment. Weighting functions for two frequencies which are near the ammonia band center are shown in Figure 2. The weighting functions for these frequencies have an altitude resolution of less than 10 km and are primarily confined to
the cloud forming region. Qualitatively, increasing the pressure in the clouds causes
the weighting functions to converge toward a common altitude and tends to flatten
the microwave spectrum. In contrast, decreasing the pressure causes the weighting
functions to diverge and tends to increase the spectral slopes. The spectral slopes
predicted by Models 1, 2, and 3 are shown in Figure 3. Their relationship with the
observations is discussed in the next section.

3. Broad-Band Observations

In the spring of 1970 and again in 1971 we observed Jupiter to confirm the existence
of the broad ammonia absorption feature and to search for discrete spectral features
at the frequencies of the more intense ammonia lines. We used the 9-m antenna at the
Venus Station of the NASA Deep Space Instrumentation Facility, Goldstone, California. The receiver consisted of two broad-band (20–24 GHz) traveling-wave-tube amplifiers followed by a parallel set of 13 tuned r.f. filters and separate detectors. The
filter bandwidths ranged from 20 MHz to 180 MHz. The relative precision of the
spectral measurements was enhanced with this system because all thirteen frequencies
were simultaneously observed. The radio source Cassiopeia A was observed each
night for calibration purposes.

   The principal results of these observations are: (a) the Jovian brightness temperature
decreases approximately 7 K between 20 GHz and 23 GHz and either remains constant or increases slightly between 23 GHz and 24 GHz; (b) the disk temperature at
23.5 GHz is 132 ± 10 K (total standard error). This temperature is in good agreement

with the absolute brightness temperature measurements of Wrixon *et al.* (1971), and with theoretical calculations based on model atmospheres in which the upper atmosphere is saturated with ammonia.

The results most relevant to the determination of the pressure in the ammonia cloud are the observed temperatures between 20 GHz and 23 GHz. These data are shown in Figure 3 superimposed upon the predicted spectra for the three theoretical models which are normalized to a temperature of 140 K at 20 GHz. We note that Models 1 and 3 predict slopes that are too shallow and too steep, respectively, compared with the data.

Based on a Chi-square analysis for which we assumed nine degrees of freedom (number of data points minus one for the slope), we conclude that Models 1 and 3 can be excluded at the 95% confidence level while Model 2 provides an adequate fit to the data. Since the collision cross section of helium is about a factor of three smaller than that of hydrogen, the total pressure implied by the models depends on the helium-to-hydrogen number mixing ratio. For a pure hydrogen atmosphere, our best estimate of the pressure at a temperature of 130 K is 0.48 atm with a 95% probability that it lies between 0.24 atm and 0.97 atm. Our estimate of the pressure increases by about ten percent if the helium-to-hydrogen number mixing ratio is increased from zero to 0.2.

4. Narrow Band Observations

In July 1972 and May 1973 we attempted to detect narrow band ammonia lines in the Jovian spectrum near 1.25 cm. The observations were made with the 40-m radio telescope at the Owens Valley Radio Observatory operated by the California Institute of Technology. The receiver was a double sideband balanced mixer with a tunable local oscillator. The system temperature was about 2000 K. The intermediate frequency amplifier was followed by a power divider at the input of a six-channel filter bank with filter widths of about 40 MHz. The outputs from the six channels were separately detected and digitally recorded. In addition to the six narrow band channels, the signal from the full IF passband between 20 MHz and 300 MHz was simultaneously detected and recorded. Each individual filter response was normalized to the response of the full IF bandwidth to minimize the adverse effects of antenna tracking errors and differential extinction in the Earth's atmosphere. Spectral features introduced by the receiver and the antenna feed system were removed by dividing all the Jupiter results by measurements of the Moon and other radio sources which were observed for calibration purposes. We obtained confidence that the system was performing properly by observing the water line in the Orion Nebula.

The 1972 observations were made at three different local oscillator frequencies. One of these was repeated and two new frequencies were observed in 1973. The five frequencies were 22.395 GHz, 22.687 GHz, 23.532 GHz, 23.785 GHz, and 24.004 GHz. The observed spectra for each of the five frequencies were compared with the corresponding theoretical spectra which were computed by convolving the model brightness temperatures with the filter response function. The process was repeated for a number

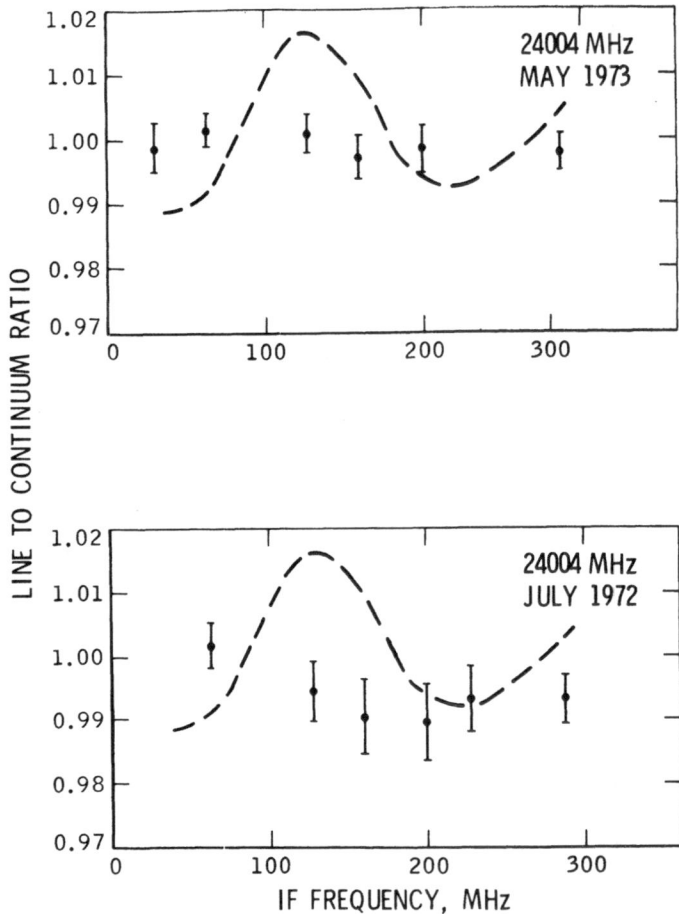

Fig. 4.   Narrow band observational data are shown with one sigma uncertainties indicated. Local oscillator frequency setting used for these data was 24 004 MHz.

of models in which the narrow line features varied from ∼4% emission to ∼4% absorption relative to the broad absorption continuum. A preliminary analysis of our results for the various Chi-square values allows us to conclude that any absorption or emission spectral features that may exist at these five frequencies have amplitudes of less than 2% relative to the continuum. This result is made with 90% confidence factor for line widths between 20 MHz and 100 MHz.

Figure 4 shows our measurements for one representative frequency, 24 004 MHz, along with their one-sigma error bars. These spectra show no evidence of any strong narrow band emission or absorption feature. For comparison, the dashed curve represents a theoretical spectrum for a model in which 4% emission lines are present.

5. Concluding Remarks

The broad band data presented in Figure 3 clearly establishes the existence of a slope

$[\sim 2.5\,K\;GHz^{-1}]$ in Jupiter's microwave spectrum between 20 GHz and 23 GHz. We believe that absorption in the low frequency wing of the pressure-broadened ammonia inversion band is the most plausible explanation of the observed slope. Based on this interpretation the data imply the existence of a saturated ammonia cloud with an equivalent H_2 pressure of 0.48 atm at 130 K.

The absence of narrow ammonia lines in Jupiter's spectrum can most easily be explained by limiting the high altitude ammonia abundance. If ammonia is distributed in the stratosphere according to hydrostatic equilibrium with the reference value of the partial pressure of ammonia chosen equal to its vapor pressure at the minimum temperature reached at the tropopause, then our observations suggest that the minimum temperature is less than 120 K. This result is consistent with the recent measurements by Gillett *et al.* (1969) and Gillett and Westphal (1973) which suggest the presence of a thermal inversion in the stratosphere. However, our results do conflict with a preliminary summary of observations by Wrixon (1969) who also observed the most intense ammonia lines. He reported the detection of narrow emission lines and estimated that they would be explained if 0.06 cm-atm of ammonia were present in a thin layer above the clouds at a temperature and pressure of 145–155 K and 6.5 mb. Using these atmospheric parameters we computed the theoretical spectral response for our filters. The response for one frequency is compared with our observed spectrum in Figure 4. We see no evidence that would confirm the lines reported by Wrixon (1969) in either the 1972 or the 1973 data.

Acknowledgements

We gratefully acknowledge the support provided by the staffs of the Venus Development Station at Goldstone, CA and the Owens Valley Radio Observatory. We are indebted to Dr R. B. Read for collaborating with us in evaluating and calibrating the performance of the 40-m antenna operating at wavelengths near 1.3 cm. We thank Dr Harry Hardebeck, Mr Fred Soltis, and Mr Richard M. Wetzel for their engineering work in support of these observations, and Dr E. T. Olsen for assistance with the narrow band observations. This paper presents the results of one phase of research carried out at the Jet Propulsion Laboratory, California Institute of Technology, under contract number NAS 7-100, sponsored by the National Aeronautics and Space Administration.

References

Barrett, A. H.: 1962, *Coll. Astrophys. Liège* 11, 197.
Gillett, F. C., Low, F. C., and Stein, W. A.: 1969, *Astrophys. J.* 157, 925.
Gillett, F. C. and Westphal, J. A.: 1973, *Astrophys. J.* 179, L153.
Gulkis, S. and Poynter, R.: 1972, *Phys. Earth Planet. Interiors* 6, 36.
Law, S. E. and Staelin, D. H.: 1968, *Astrophys. J.* 154, 1077.
Winter, S.: 1964, 'Expected Microwave Emission from Jupiter at Wavelengths Near 1 cm', Tech. Note, Series 5, Issue 23, Space Sciences Laboratory, University of California, Berkeley.
Wrixon, G. T.: 1969, paper presented at the *3rd Arizona Conf. on Planetary Atmospheres*, held at Tucson, Arizona.
Wrixon, G. T., Welch, W. J., and Thornton, D. D.: 1971, *Astrophys. J.* 169, 171.

DISCUSSION

Owen: Please explain again the determination of the minimum temperature.

Gulkis: If ammonia is distributed in the stratosphere according to hydrostatic equilibrium, with the reference value of the partial pressure at the minimum temperature reached at the tropopause, then the *maximum* value the temperature can reach is 120 K. This limits the high altitude abundance to a sufficiently low value that any narrow lines which might form would be below our detection limit.

Gautier: What is the influence of the assumed thermal model on the level of the peak of the computed weight-function (remembering that the partial pressure of ammonia is, in the zone of saturation, only a function of the temperature)?

Gulkis: We have not computed the brightness temperatures for models in which the lapse rate departs from an adiabatic rate; however I expect that a sub-adiabatic rate will cause the weighting functions to shift to higher altitudes. Consequently the brightness temperature should decrease. It will be interesting to examine how the differential brightness temperature changes as a function of lapse rate.

JUPITER'S RADIATION BELTS AND UPPER ATMOSPHERE

JOSEPH J. DEGIOANNI and JOHN R. DICKEL

University of Illinois Observatory, Urbana, Ill., U.S.A.

Abstract. Models of Jupiter's radiation belts have been constructed to determine the distribution of particles and their energies which will produce the observed decimetric radio emission. Data on the spectrum and the variation of emission with Jovian longitude have been used to show that the relativistic particles have a nearly isotropic distribution with high energies (of order 100 MeV) within 2 Jovian radii and a very flat distribution in the equatorial plane of low energy particles further out in the magnetosphere.

Subtraction of the emission predicted by this model from the total radio emission shows that the thermal contribution in the frequency range between 3000 and 10000 MHz is somewhat less than had been previously expected. (The brightness temperature of the planetary disk is 180 K at 3000 MHz, for example.) This suggests that the ammonia mixing ratio in Jupiter's upper atmosphere may be as high as 0.002.

1. Introduction

The microwave radio emission of Jupiter is composed of 2 parts as seen in Figure 1: non-thermal emission from Jupiter's radiation belts and thermal emission from the upper atmosphere. Detailed study of this emission can therefore be used to improve our knowledge of both regions of the Jovian environment. To this end we have constructed theoretical models of the radiation belts and compared these with the observational data to obtain the physical parameters of the belts. These results are then also used to improve the models of the thermal emission from the atmosphere.

2. The Radiation Belts of Jupiter

2.1. CONSTRUCTION OF THE MODEL

In our simplified model, the Jovian radiation belts are assumed to consist of relativistic electrons, whose motions can be described by the guiding center approximation, trapped in a dipolar magnetic field inclined ten degrees to the axis of the rotation of the planet. Synchrotron emission from these electrons results in the observed non-thermal microwave spectrum. The IBM 360/75 computer at the University of Illinois was used to calculate the Stokes parameters of the synchrotron radiation integrated over a given dipolar shell as a function of the surface magnetic field and the electron energy. The integration takes into account the partial eclipse of the dipolar shells by the planet at different System III longitudes. In addition, the Stokes parameters are integrated separately for two types of electron populations, i.e. one whose pitch angle distribution is isotropic and another whose distribution is sharply confined to the magnetic equator.

2.2. FITTING TO THE OBSERVATIONAL PROPERTIES

2.2.1. *Intensity Profiles*

The magnetosphere is divided into two zones near the approximate distance at which

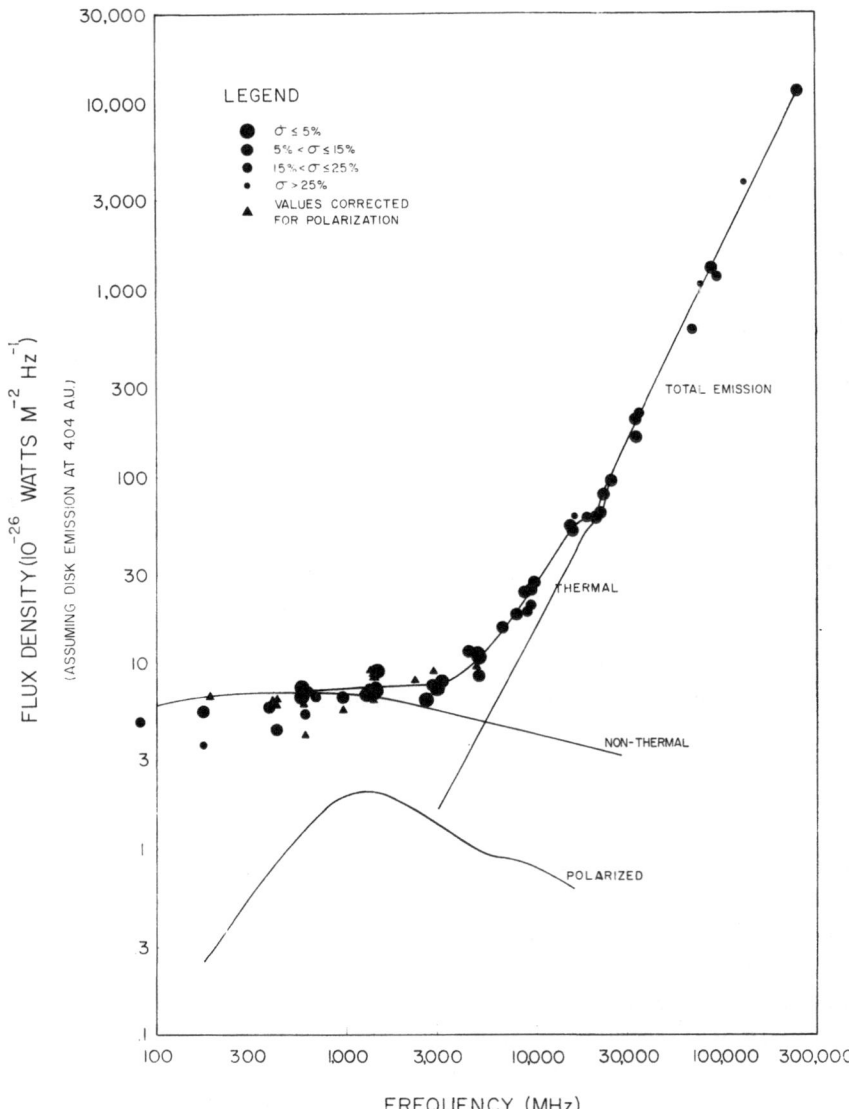

Fig. 1.   The radio spectrum of Jupiter.

the gravitational and the centrifugal forces are equal and oppose each other; this happens at $L=2$ where L is the equatorial distance from the center of the planet in Jovian radii. The intensity of the emission as a function of shell distance for $L>2$ was approximated by a decreasing function of distance from the planet as best fits the observational results of Branson (1968) and Gulkis (1970) at 21 cm. The inner zone $(1<L<2)$ is described by more rapidly varying function whose position of peak intensity is a model parameter which was adjusted to best fit the polarization observations (see Section 3 below). Figure 2 shows how the extent of the emission varies with

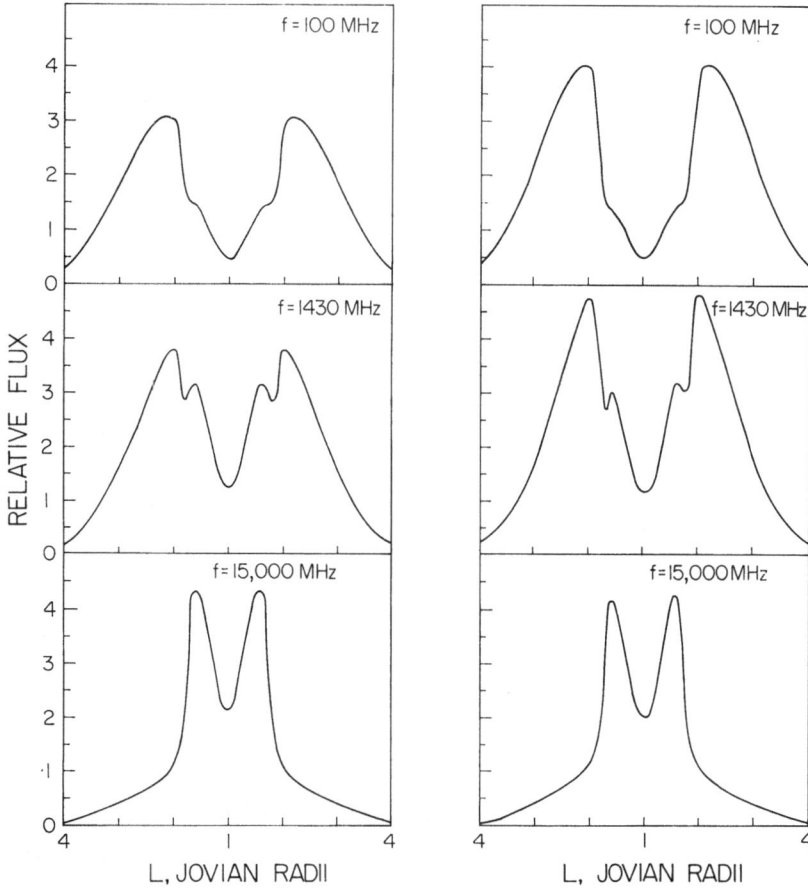

Fig. 2. The intensity profiles at $\lambda_{III} = 198°$ and $\lambda_{III} = 288°$.

frequency. In agreement with the observations by Gulkis (1970), the emission at higher frequencies is mainly produced in the region $1 < L < 2$ (the inner zone), whereas at lower frequencies much of the emission is produced in the outer zone ($L > 2$). The profiles to the left (Figure 2) correspond to System III longitude of 198° and to the right for a longitude of 288°. The latter shows a certain degree of east–west asymmetry caused by the partial eclipse of the planet. This effect has also been observed by Branson (1968).

2.2.2. *Variation of Intensity of Emission with Rotation*

The variations in intensity of emission with rotation of the planet are mainly determined by the outer zone where a major portion of the particles are confined to the magnetic equator. This effect, of course, is a direct consequence of the sharp beaming of synchrotron emission in the plane of oscillation of the relativistic electrons. The amplitude of the variations is a very sensitive function of the degree of confinement as

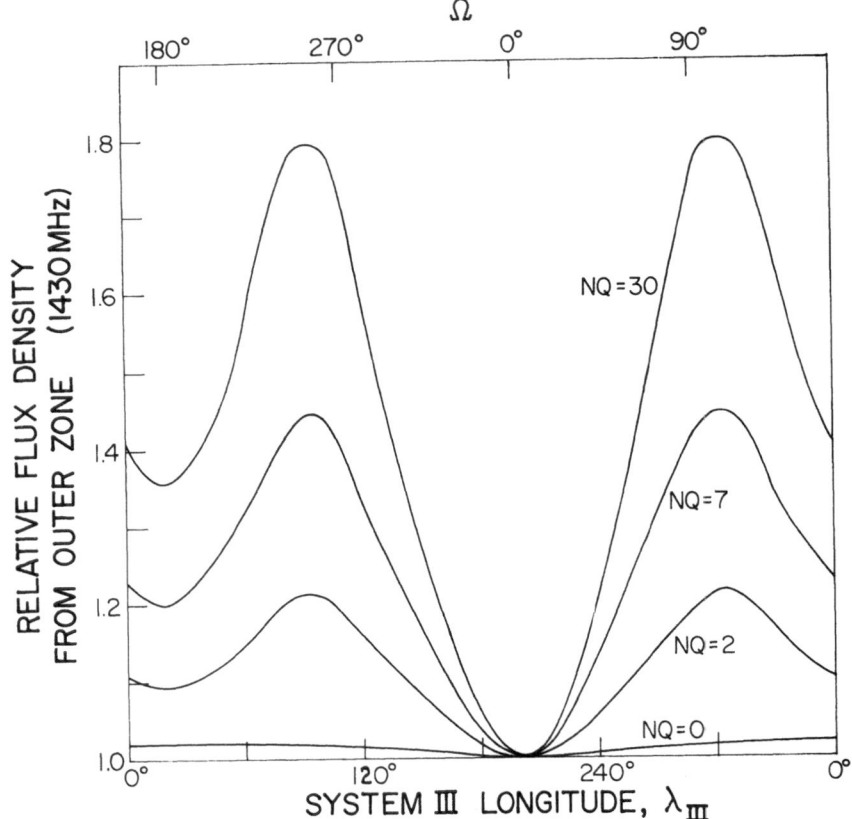

Fig. 3.   Variation of flux density at 1430 MHz (outer zone) as a function of *NQ*.

shown in Figure 3 where NQ is a measure of the fraction of particles whose motion
is limited to the plane of the magnetic equator. Figure 4 shows how the predicted
variations compare with the observations at 2600 MHz and 600 MHz by Roberts and
Ekers (1968). The oscillations at 2600 MHz contain a nonvariable contribution of
20% from the thermal component.

2.2.3. *Spectrum of the Degree of Polarization*

The fractional degree of polarization is a very sensitive function of the equatorial
pitch angle distribution of the particles. Thus, we find that the pitch angle distribution
in the inner zone, where most of the emission at high frequencies is produced, must
be isotropic if we hope to fit the observed polarization spectrum. In addition, for an
isotropic distribution of particles, the degree of polarization is a rapidly varying function of the location within the inner zone, as shown in Figure 5. This fact imposes an
important constraint as to the location and shape of the intensity profile from the
inner zone. The predicted polarization spectrum is shown in Figure 6. In particular,
the rapid decline of the degree of polarization at higher frequencies is in good agree-

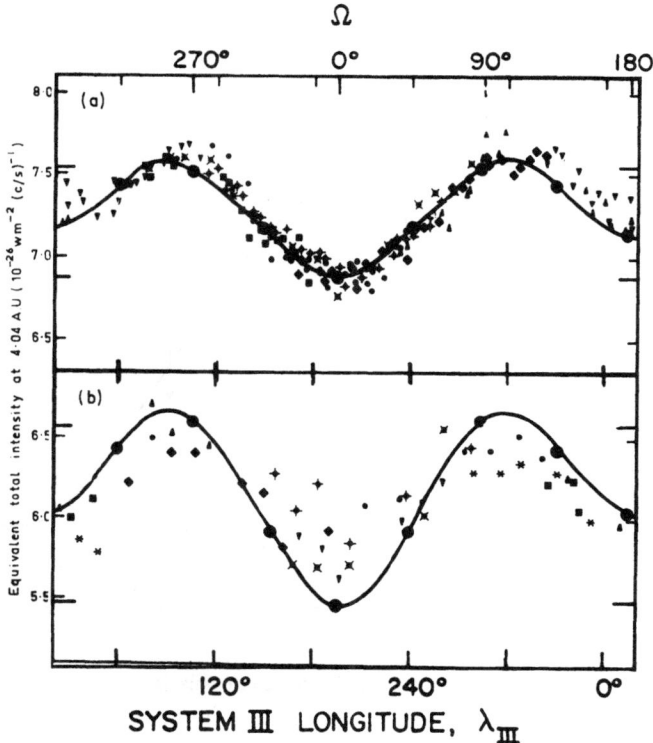

Fig. 4. Variations of flux density at 2600 MHz (upper curve) and 600 MHz (lower curve) as compared with the observations by Roberts and Ekers (1968).

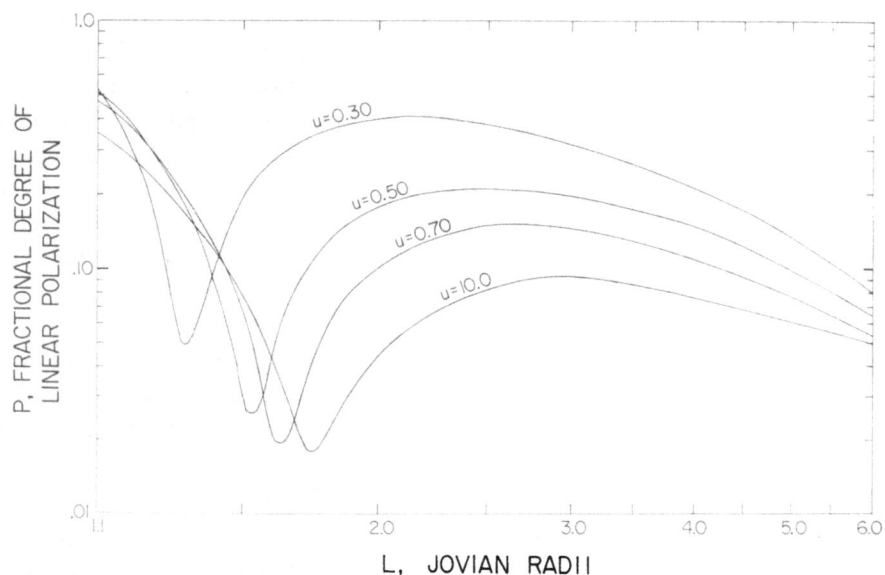

Fig. 5. Fractional degree of polarization as a function of L and u (u is an energy-dependent variable).

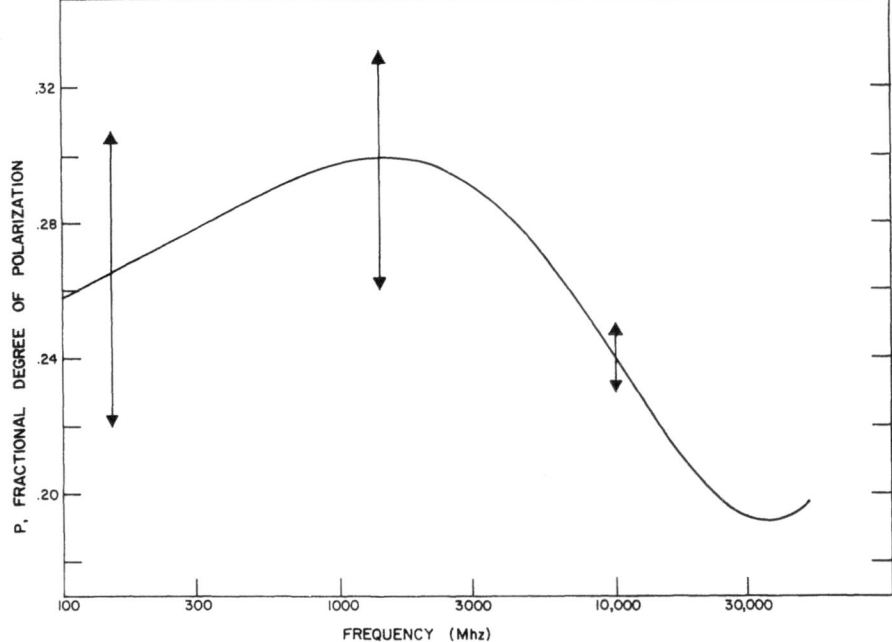

Fig. 6.   Spectrum of the degree of polarization.

ment with the observations (Dickel *et al.*, 1970). The arrows in Figure 6 show the range of variation in the degree of polarization with the rotation of the planet.

2.2.4. *Spectrum of the Non-Thermal Emission*

The shape of the non-thermal spectrum is determined by the energy distribution of the electrons in the outer and inner zones. We find that two distinct energy distributions, rather than a gradient of energy across the dipolar shells, give the best agreement with the observational results (Dickel *et al.*, 1970). Our best model, included in Figure 1, predicts a relatively flat non-thermal spectrum. As a consequence, the thermal contribution to the total emission at higher frequencies (> 3000 MHz) appears to be less than previously anticipated. This result will be elaborated upon in Section 3.

2.3. SUMMARY OF CHARACTERISTICS OF JUPITER'S RADIATION BELTS

The properties of the radiation belts are shown in Table I. The inner zone has a relatively uniform electron density population with a peak intensity of the emission near 1.6 Jovian radii. The electron densities refer to equatorial values and, of course, show a rapid decline with magnetic latitude especially in the outer zone. The equatorial density in the outer zone falls approximately as L^{-4} which is similar to the distribution of the thermal plasma in that region as derived in a model of the plasmasphere by Melrose (1967). The equatorial pitch angle distribution is nearly isotropic for the inner zone ($1 < L < 2$) and becomes 80% confined to a flat equatorial distribution for

TABLE I

Parameters of Jupiter's radiation belts

	Inner zone	Outer zone
Equatorial distance from center of Jupiter	1.6 R	> 2.0 R
Peak relativistic[a] electron density	1 cm⁻³	2 cm⁻³
Equatorial density distribution	uniform	$\sim(R/R)^{-4}$
Mean electron energy[a]	18 MeV	6 MeV
Pitch angle distribution	isotropic	80% of particles confined to equator (with distribution of $\sin^{60}\alpha$)

[a] Values assume B_{surface} of 30 G.

$L > 2$ (i.e., with pitch angles which follow a distribution of the form $\sin^{60}\alpha$). The electron densities and estimated mean electron energies are obtained if we assume a surface magnetic field of 30 G. A plot of the electron flux as function of latitude and planetary distance is shown in Figure 7. There is little fine structure except near $L=2$ where

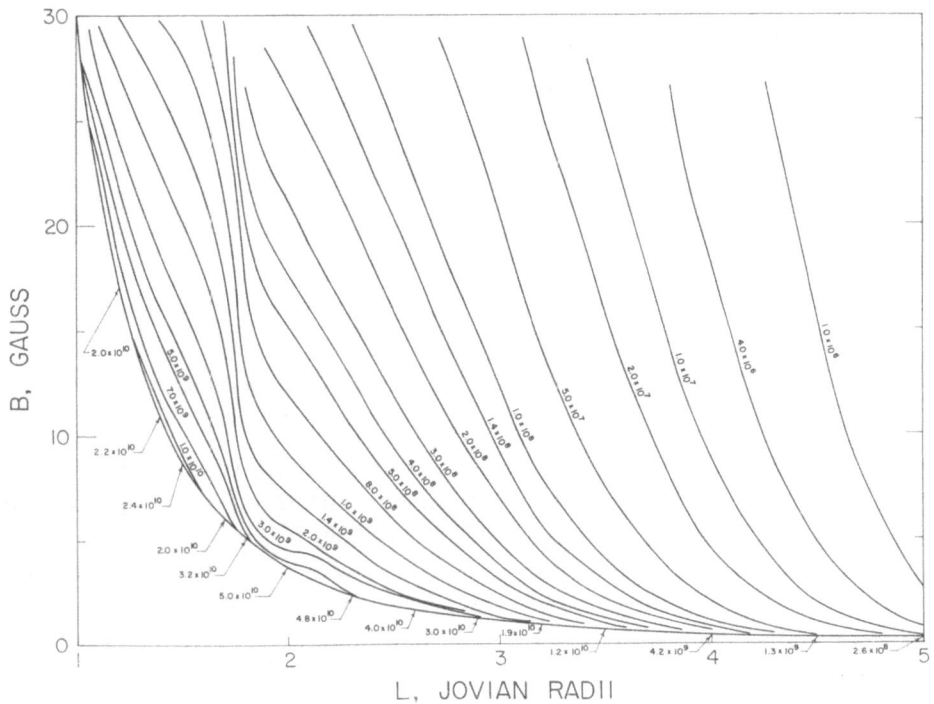

Fig. 7.   *B-L* plot of the electron flux for a surface, equatorial magnetic field of 30 G.

the two zones merge with each other. These values are of the order of 1000 times those which have been observed in the radiation belts of the Earth.

3. Implications of the Non-Thermal Spectrum upon the Jovian Upper Atmosphere

The current more detailed model of the Jovian radiation belts predicts a relatively flat non-thermal spectrum out to fairly high frequencies (3000 MHz to 15000 MHz). Thus, it appears that the thermal contribution to the total emission is less than previously anticipated from the models of Berge (1966) and Branson (1968). In particular, the thermal component at 3000 MHz is reduced to a temperature of approximately 180 K. Figure 8 shows the Jovian thermal spectrum compared with the observational results.

The current model gives a strong indication for the flattening of the Jovian thermal spectrum out to at least 10 cm. Goodman (1969) made several models of the Jovian

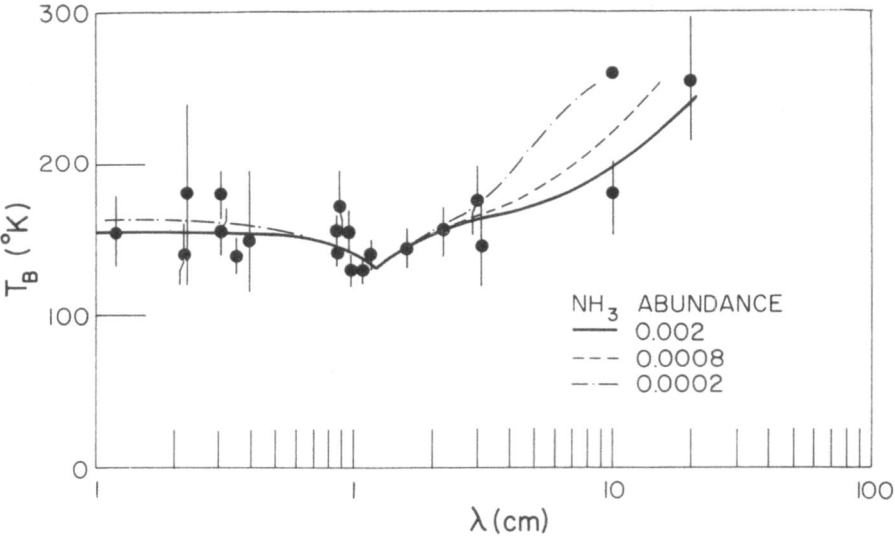

Fig. 8.   Jovian thermal spectrum. The data points shown are from Dickel *et al.* (1970).

atmosphere without a cloud layer and according to his results we would have to raise the ammonia mixing ratio in the upper atmosphere to 0.002 in order to approach the current degree of flattening. Optical observational results (Goody, 1969; Owen, 1969) indicate values closer to 0.0008. Our spectrum, however, is in excellent agreement with models obtained by Hogan *et al.* (1969) where they used cloud-top temperatures near 200 K and an ammonia mixing ratio of 0.001. Sources below the top of the dense cloud layer were not considered in those models. Thus, it is possible that the cloud layer is opaque to microwave radiation because of strong scattering by water ice or other aerosols which may exist in the region or in the lower atmosphere.

4. Conclusion

In summary, the Stokes parameters of synchrotron emission were integrated over a given dipolar shell and obtained as function of surface magnetic field, electron energy and electron pitch angle distributions. The integration takes into account the partial eclipse of the dipolar shells by the planet at different System III longitudes. The solution consists of the belt characteristics which give the best match with the observational data, i.e., the interferometric and occultation data, the variability studies, and the spectral profiles of the intensity and of the degree of polarization. The results indicate a relatively flat non-thermal spectrum to frequencies even higher than 3000 MHz such that the thermal component of emission is less than previously expected.

Acknowledgements

This research was supported by the National Aeronautics and Space Administration Grant NGR 14-005-176.

References

Berge, G. L.: 1966, *Astrophys. J.* **146**, 767.
Branson, N. J.-B. A.: 1968, *Monthly Notices Roy. Astron. Soc.* **139**, 155.
Dickel, J. R., Degioanni, J., and Goodman, G.: 1970, *Radio Sci.* **5**, 517.
Goodman, G.: 1969, *Models of Jupiter's Atmosphere*, Ph.D. Thesis, University of Illinois, Urbana.
Goody, R.: 1969, *J. Atmospheric Sci.* **24**, 997.
Gulkis, S.: 1970, *Radio Sci.* **5**, 505.
Hogan, J. S., Rasool, S. I., and Encrenaz, T.: 1969, *J. Atmospheric Sci.* **24**, 898.
Melrose, D. B.: 1967, *Planetary Space Sci.* **15**, 381.
Owen, T.: 1969, *Icarus* **10**, 355.
Roberts, J. A. and Ekers, R. D.: 1968, *Icarus* **8**, 160.

ANALYSIS OF SPECTROSCOPIC OBSERVATIONS OF JUPITER AND THE VARIABILITY OF THE STRUCTURE OF THE VISIBLE CLOUDS

G. E. HUNT

Meteorological Office, Bracknell, Berkshire, England

and

J. T. BERGSTRALH

Jet Propulsion Laboratory, Pasadena, Calif., U.S.A.

Abstract. Realistic inhomogeneous models of Jupiter's lower atmosphere are used to interpret recent observations of the (3–0) and (4–0) H_2 quadrupole lines at different portions of the visible disc of the planet. The time variations of these observations are analysed in terms of changes in the cloud structure.

1. Introduction

In an earlier publication, Bergstralh (1973) reported a quantitative "mapping" of the strengths of the CH_4 $3\nu_3$ *R*-branch manifolds across the Jovian disk, and concluded from the behaviour of these lines that a layered inhomogeneous scattering model for absorption line formation was the most satisfactory of three considered. We conducted a similar mapping program during the 1972 apparition of Jupiter, this time in the $S(1)$ lines of the H_2 (3–0) and (4–0) quadrupole bands, since an analysis by Hunt (1973a) indicated that the H_2 line strengths were more sensitive than the CH_4 lines to conditions of line formation. To ensure that variations of H_2 absorption with time and/or zenographic longitude did not introduce spurious results, we also instituted a 'patrol' of the H_2 line strengths at the centre of the Jovian disk.

2. Observations

Observations were obtained with the coudé spectrographs of the 208 cm (82 in.) Struve telescope at McDonald Observatory and the 61 cm (24 in.) telescope at Jet Propulsion Laboratory's Table Mountain Observatory. Because of its large image scale, we used the McDonald telescope for spatial mapping of the line strengths. The Table Mountain telescope was used for the central meridian patrol since its image scale was adequate for the purpose, and large blocks of its time could be committed to the program. In addition, a very limited amount of spatial mapping was attempted at Table Mountain.

All spectra were recorded on photographic plates, using a two-stage S-1 image intensifier tube at McDonald and a single-stage S-25 tube at Table Mountain. Spectra were obtained at 1.6 Å mm^{-1} reciprocal dispersion in the (4–0) band and 2 Å mm^{-1} in the

(3–0) band at both observatories. The plates were traced with a digital microphoto-meter at JPL, and the digital data reduced to intensity tracings by a computer routine using conventional techniques of photographic photometry.

All of the spectra reported here were obtained with the image of Jupiter held in a pole-to-pole orientation along the spectrograph slit by means of an image rotator. The scale of the McDonald spectra allowed them to be traced in three strips parallel to the dispersion; a central strip corresponding to the dusky equatorial zone, between about ±15° zenographic latitude, and strips on either side corresponding to the brighter north and south tropical zones, between about ±15° and ±49° latitude. The Table Mountain image scale was such that a rather broad strip, between about ±30° latitude, was the narrowest region that could be traced with signal-to-noise adequate for quantitative work.

To check the photometric calibration, the equivalent widths of several Fraunhofer lines, selected to match approximately the strengths and wavelengths of the Jovian H_2 lines, were measured on each tracing. The measured strengths of the former were systematically low on the tracings of the Table Mountain spectra and systematically a bit high on the McDonald tracings; linear least-squares fits of 'measured' to 'known' line strengths were therefore computed to correct the measurements of the H_2 lines. Error estimates given with individual measurements reported here are standard (1σ) errors computed from the least-squares fits.

When the Table Mountain measurements of (3–0) and (4–0) S(1) were plotted against date of observation, the points appeared to be randomly distributed, with external variances approximately equal to the estimated errors of individual mea-surements. Indeed, the apparent variations of the line strengths were *not* significant at the 5% level by the F-test. However, when the same measurements were plotted against System I longitude, a broad pattern emerged; the strengths of the (3–0) and (4–0) S(1) lines appeared to decrease by about 10 to 15 per cent between 220° and 280° longitude. Evidently, real variations of small amplitude occurred in the hydrogen line strengths. We are attempting now to correlate these apparent variations with visible features of the Jovian cloud deck, photographed at about the time of our ob-servations.

3. Atmospheric Model

The measurements have been interpreted by constructing a realistic model atmosphere of Jupiter in which all the processes of radiative transfer have been accurately taken into account. The details are described in Hunt (1973b). Hydrogen quadrupole lines exhibit the phenomenon of collision narrowing before the more usual pressure broad-ening takes place and it is important that the current line shape is used in an analysis, Hunt and Margolis (1973). We have used the Galatry line shape which accurately represents the quadrupole line shape. The nominal model of the Jovian atmosphere proposed by Divine (1971), which is consistent with our present knowledge, is used as a basis of this analysis.

4. Results

At the equatorial region of Jupiter a structure of two distinct cloud layers is consistent with all observations. Hunt (1937b) showed the upper cloud layer to be optically thin. In the analysis of these observations we find it has negligible effect upon the equivalent widths observed at the centre of the disc. Consequently observations of this type may be interpreted by a reflecting layer technique. Then the changes in the derived effective pressure at the level of line formation of weak lines, such as the (4–0) S(1), will indicate the variability in the top of the dense, lower cloud.

The observations carried out during June 1972 show a *decrease* in all measured line strengths for System I longitude in the range 200 to 280°. For the (4–0) S(1) line this requires the effect pressure at the level of line formation to vary from $\sim 2 \pm 0.5$ to $\sim 1.25 \pm 0.2$ atm and the abundance from $\sim 68 \pm 15$ to 89 ± 13 km amagat. For the (3–0) S(1) line, a saturated and therefore less sensitive line, the pressure varies from $\sim 1.6 \pm 0.5$ to $\sim 1 \pm 0.4$ atm and the abundance from $\sim 54 \pm 16$ to $\sim 33 \pm 13$ km amagat. This requires the lower cloud to *rise* by ~ 1 scale height during the observational period from the ~ 2 atm level.

During August 1972 the observations indicated an *increase* in line strength in all measured lines over the System I longitude range of 180 to 320°. For the (4–0) S(1) line this requires the effective pressure to increase from 1.88 ± 0.37 to 4.43 ± 0.93 atm; the abundance from 61.25 ± 11.75 to 145 ± 30 km amagat. The corresponding change for the (3–0) S(1) line is 1.18 ± 0.52 to 2.26 ± 0.89 atm and 38.25 ± 16.75 to 73.75 ± 29.25 km amagat. This requires the level of the lower cloud to *decrease* by more a scale height during the observational period of two Earth days from the 1.88 atm level.

5. Conclusion

The patrol of H_2 quadrupole line strengths at the centre of the Jovian disk during 1972 apparition indicate great variability in the measured equivalent widths. This has been shown to be correlated with changes in the level of the lower cloud which behaves as a reflecting surface.

References

Bergstralh, J. T.: 1973, *Icarus* **19**, 390.
Divine, N.: 1971, 'The Planet Jupiter' SP, 8069.
Hunt, G. E.: 1973a, *Icarus* **18**, 637.
Hunt, G. E.: 1973b, *Monthly Notices Roy. Astron. Soc.* **161**, 347.
Hunt, G. E. and Margolis, J. S.: 1973, *J. Quant. Spectrosc. Radiat. Transfer* **13**, 417.

DISCUSSION

Gulkis: What is the significance of plotting the data in System I longitude?

Owen: The data were not only restricted to the central meridian longitude but also to the equatorial region.

MOTIONS IN JUPITER'S ATMOSPHERE

R. HIDE

Geophysical Fluid Dynamics Laboratory Meteorological Office, Bracknell, Berkshire, England

Abstract. Recent contributions to the dynamics of Jupiter's atmospheres were reviewed under the following headings: (a) general magnitude of winds and temperature fluctuations; (b) cloud bands and zonal velocity profile (including equatorial jet); (c) transient non-axisymmetric features (spots, and irregular markings); (d) the Great Red Spot.

Some Useful References

Banos, C. J.: 1971, *Icarus* **15**, 58.
Banos, C. J. and Alissandrakis, C. E.: 1971, *Astron. Astrophys.* **15**, 424.
Barcilon, A. and Gierasch, P. J.: 1970, *J. Atmospheric Sci.* **27**, 550.
Gierasch, P. J.: 1970, *Earth Terrest. Sci.* **4**, 171.
Gierasch, P. J.: 1973, *Icarus* **19**, 482.
Gierasch, P. J., Ingersoll, A. P., and Williams, R. T.: 1973, *Icarus* **19**, 473.
Golitsyn, G. S.: 1971, *Atmospheric Oceanic Phys.* **7**, 974.
Goody, R. M.: 1969, *Rev. Astron. Astrophys.* **7**, 303.
Hide, R.: 1969, *J. Atmospheric Sci.* **26**, 841.
Hide, R.: 1974, *Proc. Roy. Soc. London* **A336**, 63.
Hogg, N. G.: 1973, *J. Fluid Mech.* **58**, 517.
Hogg, N. G.: 1973, *Deep Sea Res.* **20**, 449.
Hubbard, W. B.: 1973, *Space Sci. Rev.* **14**, 424.
Inge, J. L.: 1973, *Icarus* **20**, 1.
Ingersoll, A. P. and Cuzzi, J. N.: 1969, *J. Atmospheric Sci.* **26**, 981.
Layton, R. G.: 1971, *Icarus* **15**, 480.
Kuiper, G. P.: 1972, *Sky Telesc.* **43**, 75.
Leovy, G. B. and Pollack, J. B.: 1973, *Icarus* **15**, 195.
Lewis, J. S.: 1973, *Space Sci. Rev.* **14**, 401.
Maxworthy, T.: 1973, *Planetary Space Sci.* **21**, 623.
Newburn, R. L. and Gulkis, S.: 1973, *Space Sci. Rev.* **3**, 179.
Reese, E. J.: 1973, *Icarus* **17**, 57.
Starr, V. P. and Rosen, R. D.: 1972, *Tellus* **24**, 73.
Stone, P. H.: 1973, *Space Sci. Rev.* **14**, 444.
Streett, W. B., Ringmacher, H. I., and Veronis, G.: 1971, *Icarus* **14**, 319.
Williams, G. P. and Robinson, J. B.: 1973, *J. Atmospheric Sci.* **30**, 684.

DISCUSSION

Heard: Are the bands regarded as holes in an upper cloud layer through which we see a lower cloud layer?

Hide: The bands comprise belts and zones. It is generally considered that the zones, which are bright, are regions of rising motion where ammonia cirrus clouds form and that the belts, which are dark, are regions of descending motion, where ammonia cirrus does not form.

Low: Our recent *Astrophys. J.* paper shows a strong correlation between temperature measurements at 5 μ and 'cloud' colour. The blue clouds are hot (T about 250–300 K); the red, brown clouds are cold (T about 200 K) and, therefore, presumably higher.

Gehrels: It seems urgent, and interesting, to use photographs of Jupiter to measure cloud motions.

Baum: Profiles of rotation period as a function of latitude are being measured and analysed by Jay

Inge in my group at Lowell Observatory, particularly seeking short term variations. A paper by Inge concerning sampled time intervals in 1970, 1971, and 1972 will soon appear in *Icarus*. In the hope of detecting more detail (perhaps non-laminar) in the atmospheric flow pattern, I am attempting by image rectification to construct pseudo-time-lapse sequences as they would be seen from a synchronous satellite in an equatorial orbit around the planet; but that is a rather difficult task, it will take time to do, and it may not succeed. Our purpose, however, is to make sure that we detect all of the image information inherently available in the patrol photographs.

RESULTS OF THE BETA SCORPII OCCULTATION BY
JUPITER ON MAY 13, 1971

L. VAPILLON

Groupe Planètes, Observatoire de Paris-Meudon, France

Abstract. The aim of this paper is to present results of temperature and density profiles of the upper atmosphere of Jupiter as deduced from observation of the occultation of β Sco on May 13, 1971.

The occultation light curve of β Sco is shown. The analysis of data is based on the Abelian integral inversion. Jovian refractivity profiles and typical temperature and density profiles are also shown.

The extent of the atmospheric zone where results are significant is discussed.

It is shown that no valid information about the thermosphere has been obtained.

1. Introduction

The interest in the occultation of a star by a planet is that it enables one to investigate the atmosphere of the planet. It is known that the gradual drop in the flux from the star during the occultation is due to differential refraction in progressively denser (and, therefore, deeper) layers of the planet's atmosphere. Clearly, it is desirable to be able to observe the star for as long as possible during the occultation, which requires the capacity of making precise measurements of low stellar fluxes. One must, therefore, have access to a large telescope. Our observations of the occultation of β Sco by Jupiter on May 13, 1971, were made with the 2 m telescope of the Radcliffe Observatory, Pretoria. The observed flux of β Sco fell to 1/10 of its initial value over a period of about a minute, and to 1/100 of its initial value in 3 min. The measurement of these very low values of the stellar flux are all the more difficult close to the light limb of the planet.

The latter difficulty may be overcome by using a photometer equipped with a diaphragm of suitable diameter the outer parts of which reflect the light from the limb of the planet to an auxiliary photomultiplier. From the variations in the flux received by this photomultiplier, one may correct for pointing errors, turbulence etc... even *a posteriori*.

In order to increase the contrast of the star relative to the planet, the observations were made with an interference filter, of 9 Å width, centred on the wavelength of the Ca K line (3934 Å). By this means, we were able to observe β Sco for over 900 s at immersion and 820 s at emersion, (Figures 1a and 1b).

The light curve which we obtained shows that a substantial error is committed if the observations stop too soon or if too small a telescope is used. It is evident from the light curve that only after a period of about 5 min does the flux of the star become truly indistinguishable from the background noise. If the observations stop before this point is reached, the zero of the light curve is placed too high, with the consequences that will be discussed below.

Woszczyk and Iwaniszewska (eds.), Exploration of the Planetary System, 391–399. All Rights Reserved

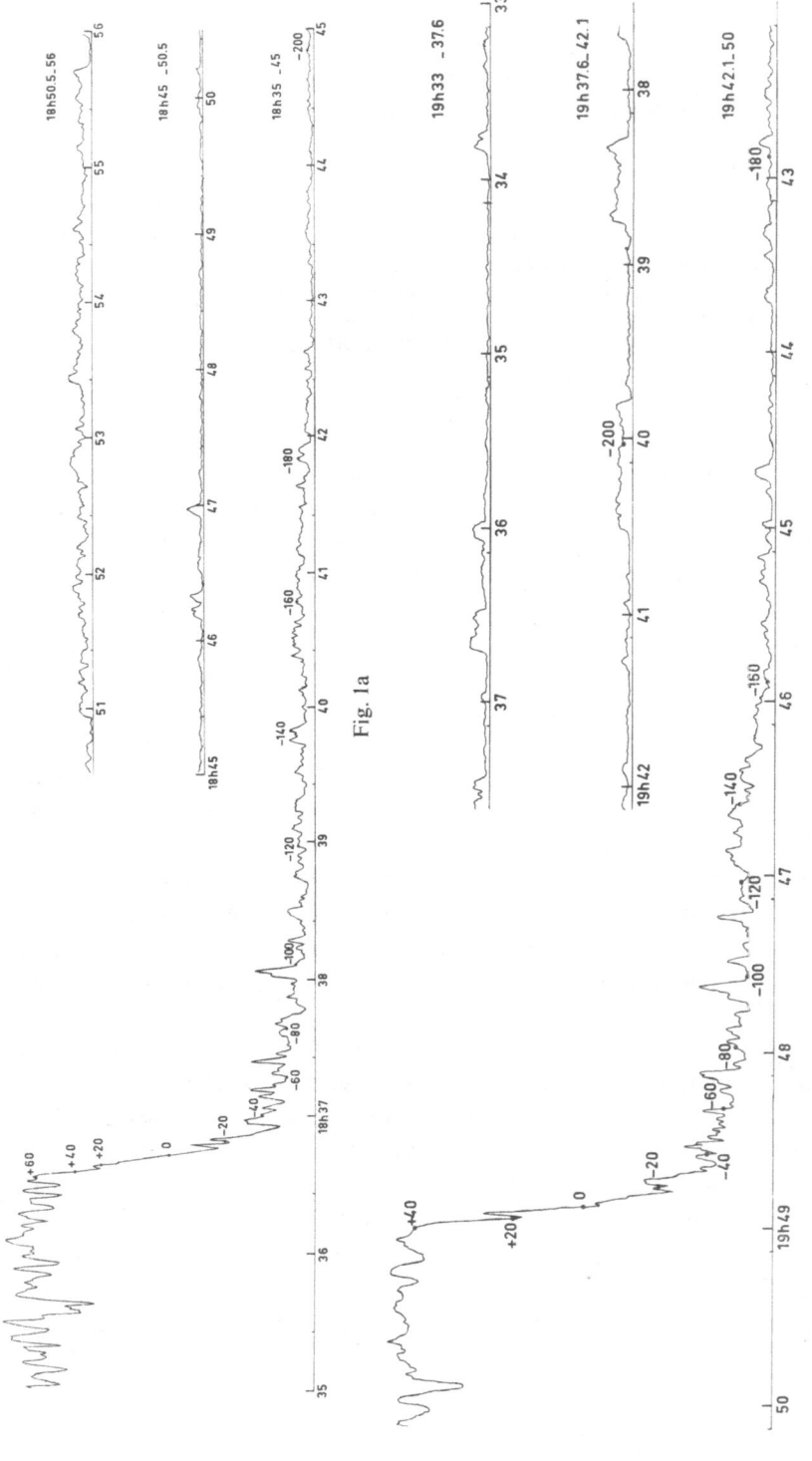

Fig. 1a

Fig. 1b

Fig. 1. (a) The light curve, $\Phi(t)$, of β Sco A at Immersion, from 18^h35^m to 18^h56^m (UT). The altitudes h corresponding to the points of closest approach are marked at 20 km intervals down to $h = -200$ km. (b) The light curve, $\Phi(t)$, of β Sco B at Emersion, from 19^h33^m to 19^h50^m (UT). The time axis is reversed (time increases from right to left). The altitudes h corresponding to the points of closest approach are marked at 20 km intervals down to $h = -200$ km. Note the wide 'plateau', particularly between 19^h36^m and 19^h40^m, and the rapid rise in flux after 19^h46^m.

2. Refractivity Curves

Since the extinction of the stellar flux is due exclusively to differential refraction in progressively denser layers of the planetary atmosphere, only the major constituents of the atmosphere (H_2 and He) are important. In what follows, we consider the atmosphere to consist only of these two gases.

We take, as time origin, the time at which the flux from the star has fallen to one

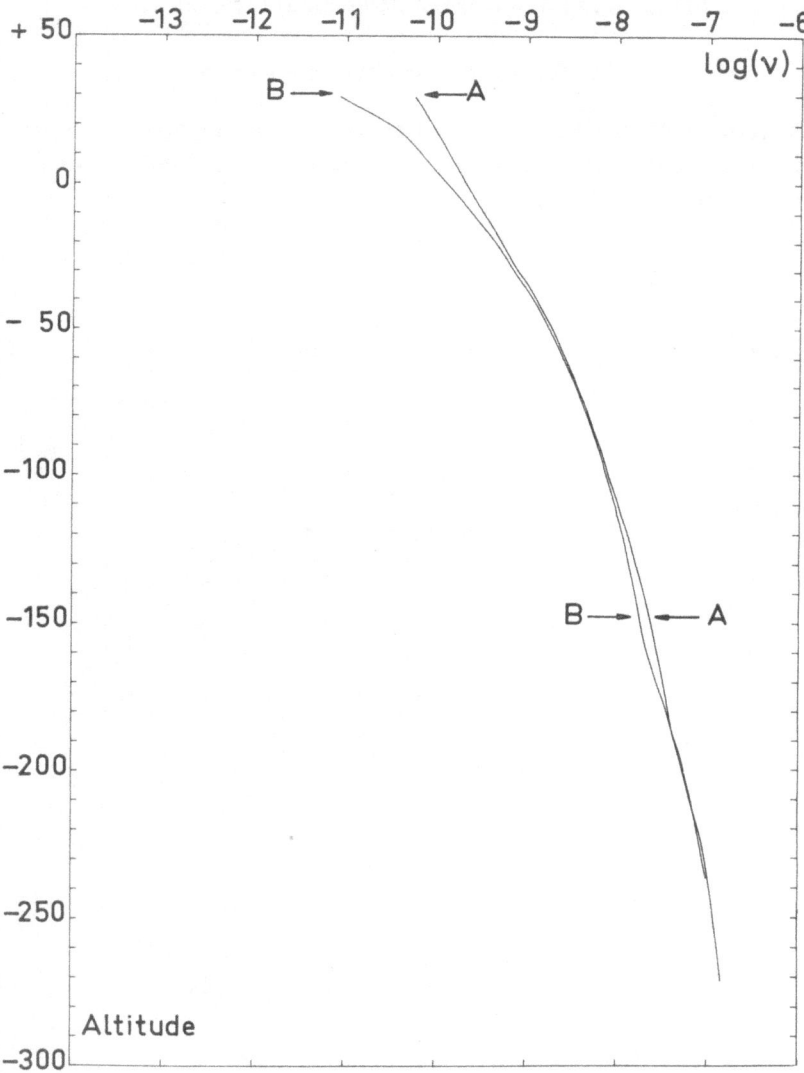

Fig. 2. Refractivity profiles $\log(v(h))$ obtained at Immersion (A) and Emersion (B). Note the divergence between the two profiles above the altitude $h = 0$.

half of its initial value and, as height origin, the point of closest approach of the light beam at this same time. We see that the curves corresponding to immersion and emersion are very similar for all but the largest values of the flux. This is because the beginning and the end of the occultation are badly defined on the light curves, due to turbulence in the Earth's atmosphere, and the exact form of the light curve at these times is essentially unknown. Consequently, we estimate as 100% the error in the refractivity at heights greater than +30 km.

We note in passing that the refractivity curve is not a straight line as the upper atmosphere of Jupiter is not at a constant temperature (Figure 2).

3. Temperature and Pressure Curves

From the refractivity profile, it is possible to deduce the temperature and pressure as functions of altitude. These curves are given in Figures 3a and 3b, subject to the following conditions:

$$q = \frac{H_2}{H_2 + He} = 0.9; \qquad h_0 = -14 \text{ km}; \qquad T(h_0) = 150 \text{ K};$$

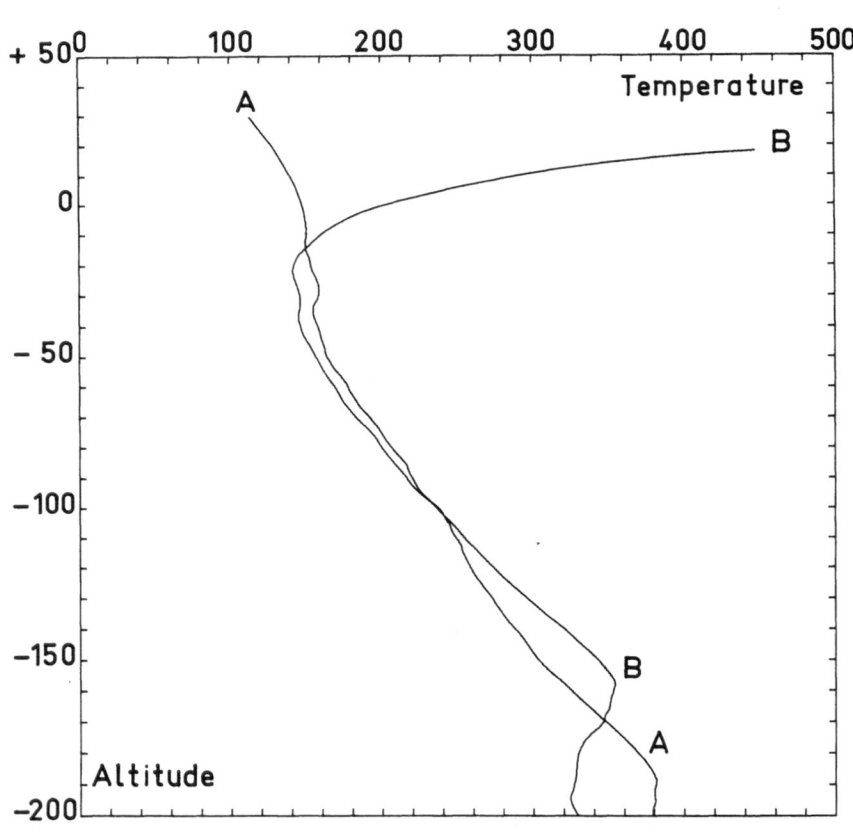

Fig. 3a.

where h_0 is the height at which one begins the numerical integrations and $T(h_0)$ is an integration constant whose value is somewhat arbitrary but which can be fixed to within certain reasonable limits. Curve A is for immersion and curve B for emersion. At intermediate altitudes $(-150 \leqslant h \leqslant -50$ km), the temperature gradient, dT/dh, is constant and equal about -1.5 K km^{-1}.

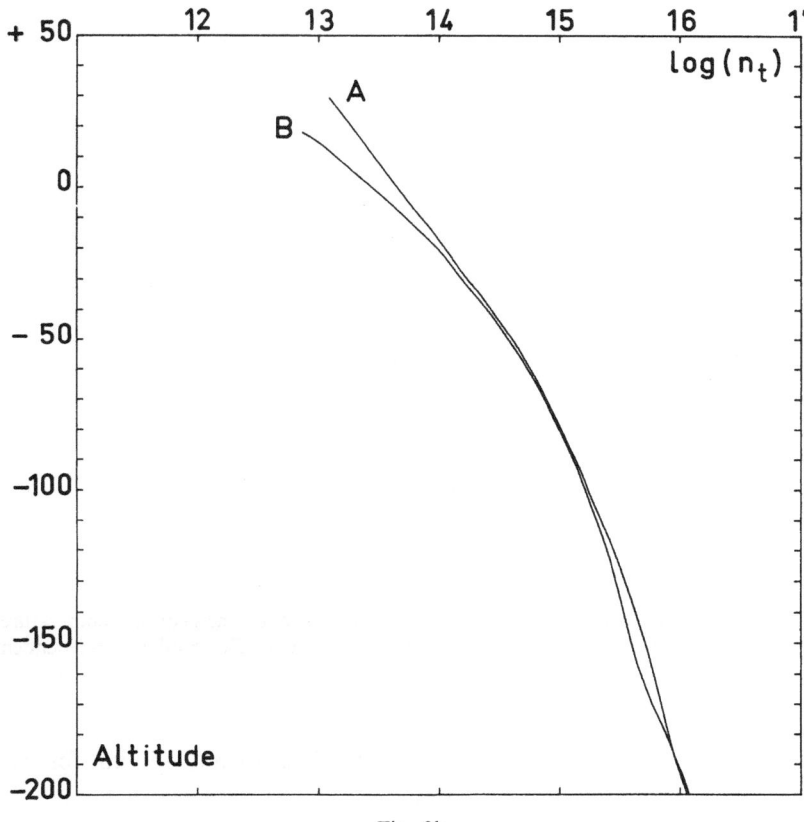

Fig. 3b.

Fig. 3. (a) Typical temperature profiles, $T(h)$, obtained at Immersion (A) and Emersion (B). $q = 0.9$; $h_0 = -14$ km; $T_0 = 150$ K. Note the temperature maximum at $h = -190$ km (A) and -160 km (B). (b) Typical density profiles log$(n_t(h))'$, obtained at Immersion (A) and Emersion (B). $q = 0.9$; $h_0 = -14$ km; $T_0 = 150$ K. Since v and n_t are proportional if the atmosphere is wellmixed, i.e. the mixing ratio is altitude independent, Figure 3b and Figure 2 are similar.

4. Importance of the Boundary Conditions

We now consider the effect on the final results of the choice of $T(h_0)$ (Figure 4).

We cannot choose h_0 too high in the atmosphere, because, as we mentioned above, the corresponding values of the refractivity are very imprecise. Neither can it be too low as the refractivity is then large and this leads to integration problems. We have chosen an optimum value of $h_0 = -14$ km. For this value of h_0, calculations show

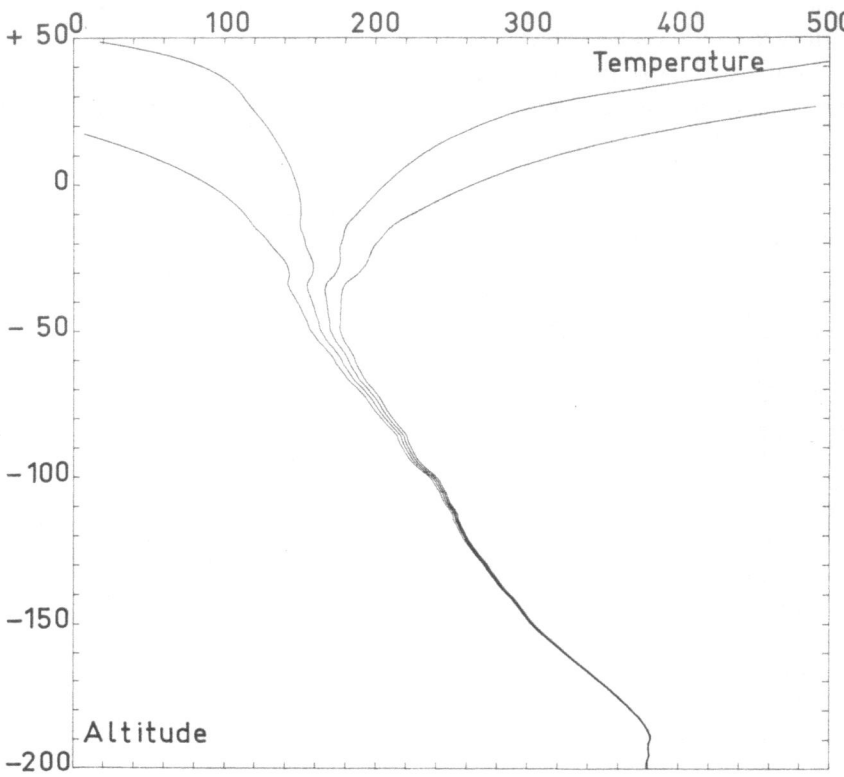

Fig. 4. The influence of the integration parameters T_0 and h_0 upon the deduced temperature profile at Immersion. $q = 0.9$; $h_0 = -14$ km; $T_0 = 120, 150, 180, 210$ K. The profile is independent of T_0 for $h < -100$ km.

that the temperature profiles are practically identical for $h \leqslant -50$ km for values of $T(h_0)$ as high as 180 K or as low as 120 K.

5. Importance of the Determination of the Initial Stellar Flux

As we saw above, the determination of the initial and final values of the stellar flux is affected by turbulence in the Earth's atmosphere and this uncertainty introduces an additional error into the profiles of temperature and pressure. We have tried to estimate the importance of this source of error by looking at the effect of a -1% (curve C) and $+1\%$ (curve A) variation in the value of the unocculted stellar flux (Figure 5).

High in the atmosphere, these variations produce large changes in the profiles: it is not possible to draw any meaningful conclusions about conditions in the atmosphere for $h \geqslant +30$ km. This height, of $+30$ km, is the upper limit of our study. For $h \leqslant +30$ km, however, the effect of the uncertainty in the unocculted stellar flux rapidly becomes negligible.

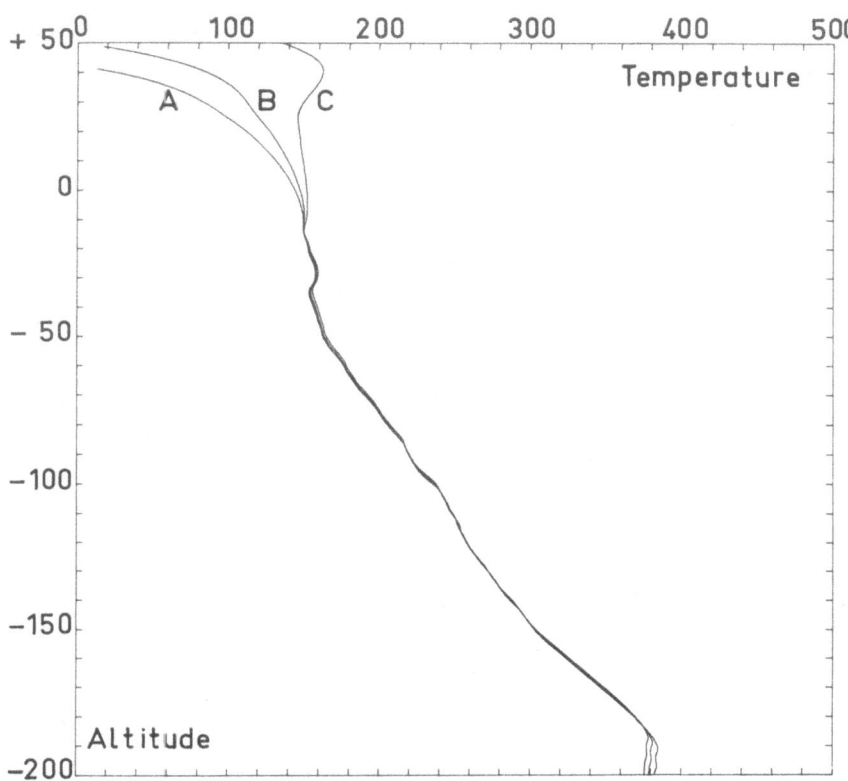

Fig. 5.   The influence of a 1% error in the flux level of β Sco A upon the temperature profile at Immersion. A – Φ_0 increased by 1%; B – reference curve; C – Φ_0 decreased by 1%. $q = 0.9$; $h_0 = = -14$ km; $T_0 = 150$ K. The profile $T(h)$ is significantly affected only high in the atmosphere ($h > 0$ km)

6. Importance of the Determination of the Zero Level of the Stellar Flux

We mentioned in the introduction that the determination of the zero stellar flux level (i.e., the level of background emission from Jupiter) can introduce a significant error. The importance of this error has been estimated by varying the value of this flux level by about $\pm 1\%$. These changes are unimportant for the high atmospheric layers but very important deep in the atmosphere. A calculation which extended down to $h = 280$ km showed that the above variation in the zero flux level results in T being undetermined for $h \leqslant -200$ km, whilst in the region $-150 \leqslant h \leqslant -50$ km the temperature and especially dT/dh are perfectly defined. We have, therefore, taken $h = -200$ km as the lower limit of our investigation (Figure 6).

7. Importance of Helium

Finally, we consider the variation of the calculated profiles with the relative abun-

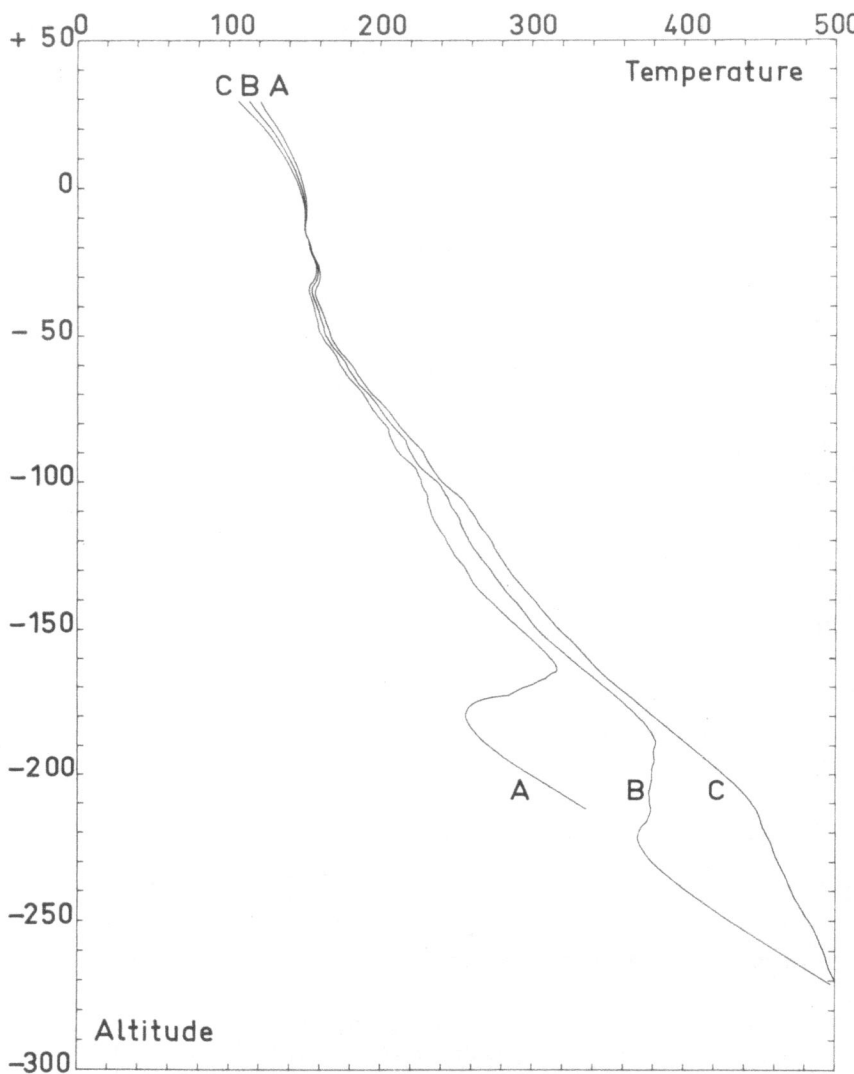

Fig. 6. The influence of an 1 % error in the Jovian background light upon the temperature profile at Immersion. A – background light level increased by 1 %; B – reference curve; C – background light level decreased by 1 %. $q = 0.9$; $h_0 = -14$ km; $T_0 = 150$ K. The profile $T(h)$ is appreciably affected deep in the atmosphere ($h < -100$ km). Note the displacement of the temperature maximum.

dance of helium (Figure 7). Calculations were carried out for $q = 1$, 0.9, 0.8 and 0.7.

Between -50 and -150 km, the gradient of the temperature profile remains independent of h but increases by a factor 1.3 for an increase in the helium abundance from 0% to 30%. Such a result was to be expected as the introduction of helium is known to lead to an increase in temperature for a given value of the refractivity. Our measurements would seem to exclude the possibility that helium is the major con-

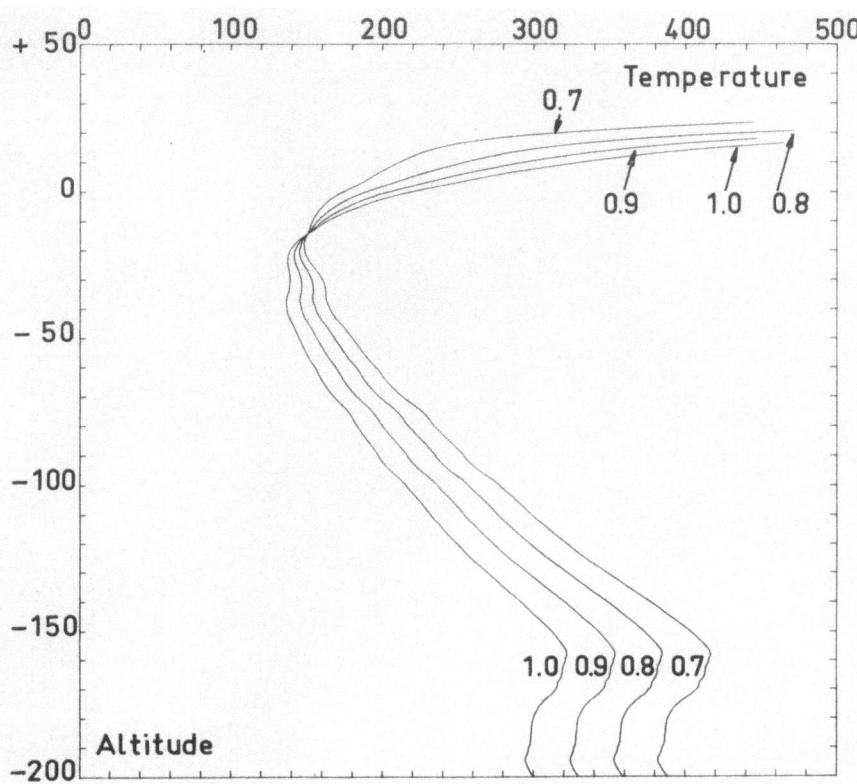

Fig. 7.   Influence of the mixing ratio, $q = H_2/H_2 + He$, at Emersion. $q = 1.00$ (A), 0.90 (B), 0.80 (C), 0.70 (D); $h_0 = -14$ km; $T_0 = 150$ K.

stituent of Jupiter's atmosphere as in this case, the calculated temperatures for $-200 \leqslant h \leqslant -150$ km are unreasonably high.

8. Conclusions

(1) The H_2/He abundance ratio cannot be precisely determined but is probably such that $q > 0.7$. An *in situ* measurement of the temperature in the atmosphere of Jupiter would clarify this point.

(2) The temperature profile is well determined as a function of q for $-150 \leqslant h \leqslant -50$ km and the gradient is $dT/dh \simeq -1.5$ K km^{-1} (slightly dependent on q).

(3) The temperature reaches a maximum value, $T_{max} \geqslant 300$ K, between -160 and -190 km.

(4) No information on the thermosphere can be derived as the light curve for this region is too noise dependent.

Note of the editors. A more extensive description of this observation can be found in: Vapillon, L., Combes, M., and Lecacheux, I.: 1973, *Astron. Astrophys.* **29**, 135.

ATMOSPHERIC PROPERTIES OF JUPITER DETERMINED FROM GALILEAN SATELLITE ECLIPSE LIGHT CURVES

R. W. SHORTHILL*

University of Utah, Salt Lake City, Utah, U.S.A.

and

T. F. GREENE* and D. W. SMITH

University of Washington, Seattle, Wash., U.S.A.

Abstract. Twelve eclipse light curves for the Galilean satellites have been observed at 30 colors. The shape of the curves depend upon Jovian atmospheric properties such as Rayleigh scattering, aerosol distribution, molecular absorption, scale height and cloud top altitude, as well as the satellite diameter. Different zenographic latitudes and longitudes along the sunrise and sunset terminator have been observed. Very long absorption path lengths are obtained compared to normal incidence because of the tangential passage of the Sun's rays. Refractive tails are observed in most cases which allow aerosol distributions to be determined. The other atmospheric properties may also be derived.

1. Introduction

Minor amounts of methane and ammonia in the Jovian atmosphere have been detected spectroscopically along with the molecular hydrogen quadrupole lines. Observation of other species is difficult because of the short absorption path lengths for normal incidence spectroscopy and the small amounts which may be present. As yet, only methane, ammonia and hydrogen have been observed, although there is reason to expect the presences of other gases. The structure of the Jovian atmosphere, however, can be studied in detail by stellar occultation techniques. This method is limited to only the upper most atmospheric layers, however, occultations by bright stars are infrequent.

Eclipses of the satellites, which are frequent, provide another method to study the structure and composition of the Jovian atmosphere. As one of the Galilean satellites pass through the shadow of Jupiter it probes to greater depths in the atmosphere, in some cases, to the cloud deck. Until recently, these observations have been limited to broad band measurements, such as those of Price and Hall (1971). They reported several points on the 'refractive tail' and concluded the existence of an aerosol haze.

Two of the authors (RWS and TFG) started a study of the Galilean satellites in 1970. The observational program with the 200-in. Hale telescope, and the multichannel spectrometer (Oke, 1969) is continuing (Greene *et al.*, 1971; Greene and Shorthill, 1972).

The purpose of this study was to determine the composition and structure of the Jovian atmosphere with the narrow band instrument. Several results of the study will be described. A brief description of satellite eclipse light curves will be given.

---

* Formaly of the Boeing Aerospace Company, Seattle, Wash.

Woszczyk and Iwaniszewska (eds.), Exploration of the Planetary System, 401–405. All Rights Reserved

2. Satellite Eclipse Phenomena

The shape of the Jovian satellite eclipse curve is determined by many physical parameters. The detailed theory has been discussed by Price (1970). Some of these physical parameters will be described briefly.

2.1. SATELLITE RADIUS

If the satellite was a point source the light curve would fall off at a very rapid rate determined by the finite solar diameter only. The fall-off becomes less rapid with increasing satellite diameter. After seven to eight magnitudes of darkening the effect of diameter on the shape of the light curve is insignificant, and the shape depends on the other factors.

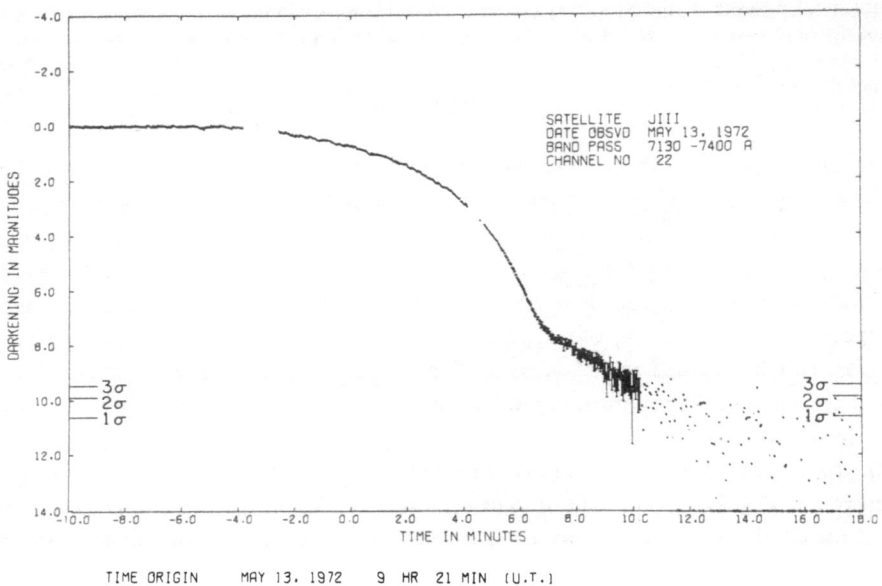

Fig. 1. A typical eclipse light curve for J III, May 13, 1972, in the 7130 to 7400 Å band pass. Data points which are below $m_v = 14$ are plotted at $m_v = 14$. The break in the curve at time $+4.0$ min results when the telescope aperture is opened to 200 in. from 90 in.

2.2. RAYLEIGH SCATTERING

Sun light is refracted and scattered into the geometric shadow resulting in a 'refractive-tail' extension on the eclipse light curves. The major cause of extinction in the Jovian atmosphere is Rayleigh scattering, whose wavelength dependence is easily observable. For example, at 7300 Å the eclipse light curve exhibits a refractive-tail, as shown in Figure 1, lasting for several minutes. At 11000 Å the refractive-tail is present even longer, while at 5000 Å the light curve falls very rapidly and may show no refractive-tail at all.

2.3. SCALE HEIGHT

If there were no atmosphere, the eclipse light curve would be determined only by the finite size of the Sun. As the scale height increases, the effect of the refraction tends to flatten the light curves. As the scale height increases to large values, the light curves fall-off again more rapidly.

2.4. EXTINCTION DUE TO AEROSOLS

If the ratio of aerosols and gaseous particles is not constant throughout the atmosphere, a wavelength dependency different from that due to Rayleigh scattering will be observed.

2.5. EXTINCTION DUE TO MOLECULAR ABSORPTION BANDS

The absorption path is increased greatly compared to normal incidence because the sunlight passes tangentially through the Jovian atmosphere. The presence of an absorbing species may be determined by observing the continuum on either side of the suspected molecular band.

2.6. CLOUD TOP CUT-OFF

At some point $(m_v \geqslant 9)$ a ray passing tangentially through the atmosphere will be cut off by the cloud tops. This should be observed simultaneously at all wavelength channels that are above noise.

2.7. ANOMALOUS REAPPEARANCE BRIGHTENING

If the albedo of a satellite should increase while it is in totality, then upon reappearance the effect will be detectable as an increase in brightness compared to the pre-eclipse brightness.

2.8. ORBITAL PARAMETERS

The relationships between tilts, rotation, etc., allow different zenographic latitudes and longitudes to be investigated by the eclipse method. For example, J IV probes all latitudes, while J I can probe only a limited range in latitude. The ingress light curves probe the sunset terminator while the egress light curves probe the sunrise terminator on Jupiter. Thus, the measurement of Jovian eclipse light curves provide a method to study the composition and structure of the Jovian atmosphere. Different zones, belts, the red spot, aerosol distribution, meteorological conditions, etc., can be systematically investigated.

3. Observations

Measurements were made at 30 different wavelengths on the four Galilean satellites. A total of 12 satellite eclipses have been observed using the 200-in. Hale telescope.

The wavelength channels are distributed from 3160 to 10680 Å. Below the 5720 Å channel, the band pass varied from 60 to 160 Å, above from 40 to 360 Å depending on the band or continuum being observed. Figure 1 shows an example of an eclipse light curve for Ganymede (J III) in the 7130 to 7400 Å channel (Greene *et al.*, 1974). The refractive-tail is clearly evident out to about ten magnitudes of darkening. At time equals + 13 min (see Figure 1), the ray from the Sun to the satellite has passed through approximately 700 km-atm of gas in the atmosphere of Jupiter.

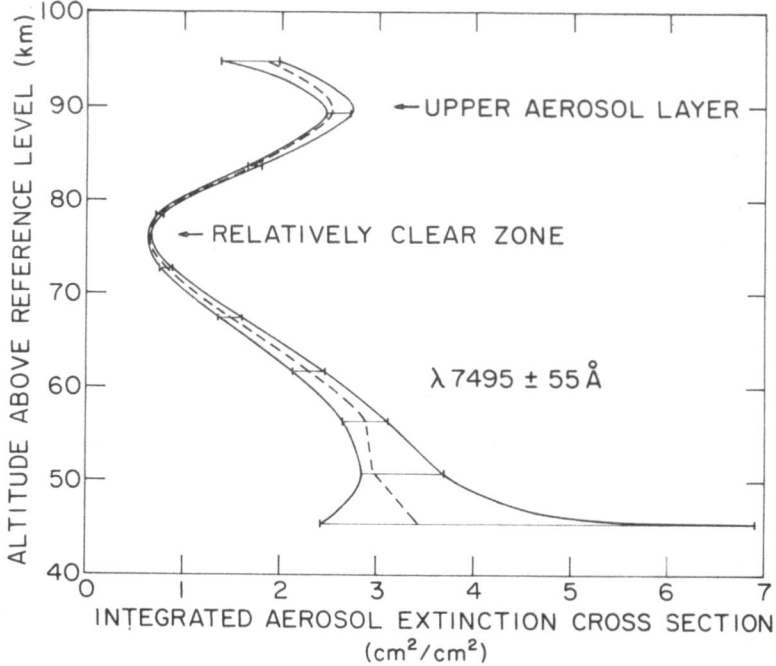

Fig. 2.   Aerosol extinction vs altitude above reference level of 10^{20} cc^{-1}. The two solid lines represent the range of the model. The dashed line is the average fit.

An analysis of six continuum wavelength bands 6325, 6850, 7495, 8260, 9260, 10 500 Å with an average band width of 145 Å was made to determine the aerosol extinction. Figure 2 shows the results of this preliminary analysis for the 7495 ± 55 Å band. The relatively clear zone is at the 10^{18} cc^{-1} level. Further analysis may allow the determination of the aerosol particle size in the layers.

A comparison of the 6420 ± 20 Å band with several continuum channels furnishes evidence for a tentative identification of the pressure induced dipole $S(0)$ line of the (4, 0) band of H_2. This identification is tentative because contamination from nearby ammonia bands is yet to be determined.

Analysis of other absorption bands is continuing. The absolute timing of both ingress and egress provide new information on the figure of Jupiter. In two eclipses, the region above the Red Spot was probed and analysis of this observation is underway.

There are available almost three hundred light curves (12 eclipses at 30 wavelengths) not all, however, of the same quality. With these data a series of parametric fits are made to theoretical eclipse light curves. In this way the physical parameters of the Jovian atmosphere are being studied. In order to cover more zenographic positions, observations of the Galilean satellite eclipses must be continued and extended to other large aperture telescopes at various geographic longitudes.

Acknowledgements

The authors (R.W.S. and T.F.G.) wish to express appreciation to the Hale observatories for the use of the Hale telescope during their stay at Mt. Palomar as Guest Investigators.

References

Greene, T. F., Shorthill, R. W., and Despain, L. G.: 1971, 'Jovian Satellite Eclipse Study I, 1971 Eclipses' (Seattle: The Boeing Co. Doc. D180-14193-1, 1971 Oct.).
Greene, T. F. and Shorthill, R. W.: 1972, 'Jovian Satellite Eclipse Study II, 1972 Eclipse Data' (Seattle: The Boeing Co., Doc. D180-15151-1, 1972 Oct., and Doc. D180-15151-2, 1972 Nov.).
Greene, T. F., Shorthill, R. W., and Smith, D. W.: 1974, 'An Eclipse of Ganymede Observed in Thirty Colors', submitted for publication.
Oke, J. B.: 1969, *Publ. Astron. Soc. Pacific* **81**, 11.
Price, M. J.: 1970, 'On the Inference of the Physical Properties of the Jovian Atmosphere From Photometery of the Galilean Satellites, (Washington, D. C.: NASA TR R-345.)
Price, M. J. and Hall, J. S.: 1971, *Icarus* **14**, 3.

MILLISECOND POLARIZED PULSES IN DECAMETRE-WAVE RADIATION FROM JUPITER AND SUN

C. H. BARROW

Dept. of Physics, University of the West Indies, Jamaica, W.I.

Abstract. A new approach is suggested to the problem of the theory of Jovian decametric radiation and physical conditions at the point of origin. This depends upon a comparison of the characteristics of fast pulse decametre-wave radio emission from both Jupiter and the Sun and invokes the deductive procedure which Sagan (1971) has called 'Propositional Calculus'.

Fast polarized pulses in radiation from Jupiter and the Sun have been studied at fixed frequencies in the range 18 to 26 MHz with time resolutions from one to five msec; a number of similarities between the pulses from both sources have been noted. A comparison of some of the pulse characteristics is being made in order to decide whether or not they are sufficiently alike to be regarded as having a common mechanism of origin at both Jupiter and the Sun. From this 'decision' it is proposed to establish boundary conditions for theoretical study. Fast pulses in the Jupiter radiation are generally supposed to be a source phenomenon although their actual mechanism is not understood. The reasons for this are to some extent inferred rather than proven and so, to check the possible (if unlikely) role of the interplanetary medium, observations are also being made using the large 26 MHz array at the University of Florida to search for possible fast pulses in the radiation from the more distant source Taurus A.

Several different types of burst structure have been recognized in the Jupiter radiation and various classifications have been proposed based upon typical burst durations. In this paper we confine attention to structures and pulses having durations of the order of 100 ms or less and we refer to these as 'millisecond pulses'. These have been studied in some detail by Olsson and Smith (1966), by Torgersen (1969), and by Baart *et al.* (1966). Typical examples are shown in Figures 1 and 2. The pulses can occur singly, in groups, or in prolonged sequences; they have narrow instantaneous bandwidths and they may display a characteristic polarization. Rapid polarization changes

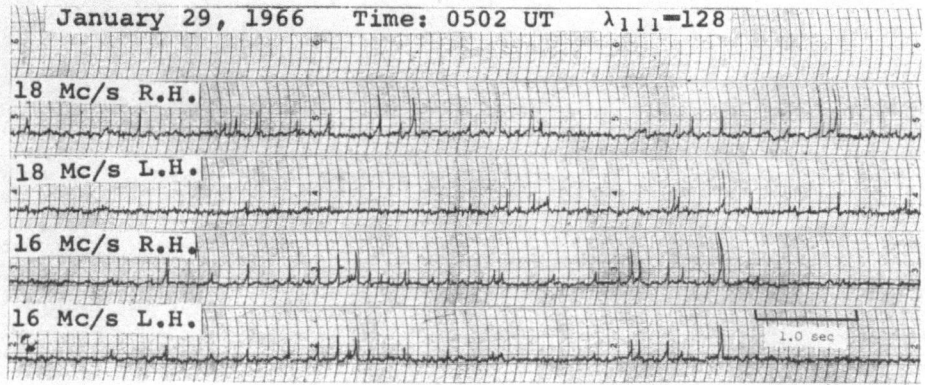

Fig. 1. Left- and right-hand components at 16 and 18 MHz showing isolated millisecond pulses from Jupiter.

Woszczyk and Iwaniszewska (eds.), Exploration of the Planetary System, 407–413. All Rights Reserved
Copyright © 1974 by the IAU

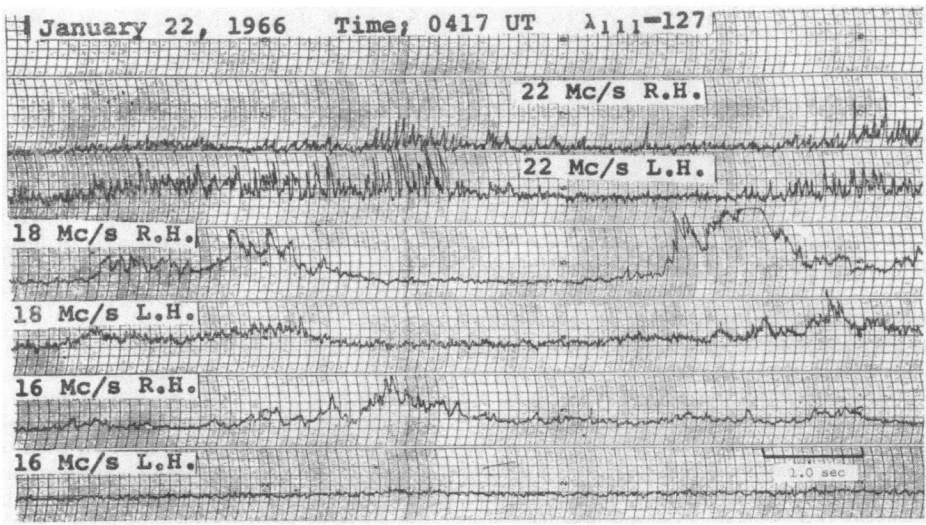

Fig. 2.   High-speed record of groups of millisecond pulses from Jupiter at 22 MHz. In this case the pulses are not present simultaneously on the other channels.

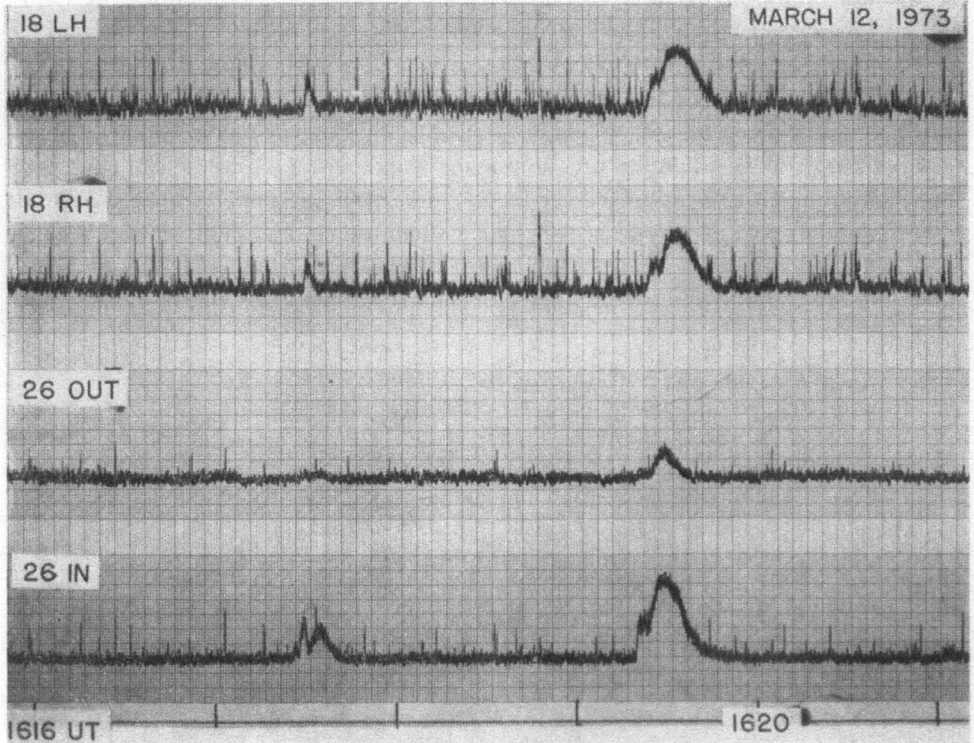

Fig. 3.   Unresolved solar millisecond pulses at 18 and 26 MHz. Some of the pulses recorded on the 26 MHz in-phase channel do not appear simultaneously on the other three channels.

and reversals in sense sometimes occur. The pulses appear to be predominantly associated with the B and C 'sources' on Jupiter.

During the recent minimum in the Jupiter activity inverse sunspot cycle we attempted solar observations using the Jupiter receiving equipment. The radio site at the University of the West Indies, Jamaica (18°N, 77°W) is well shielded by mountains and it is often possible to observe under quite good conditions by day, even at 18 MHz. Solar millisecond pulses were first observed in the summer of 1970 and were reported by Barrow and Saunders (1972). Similar and related observations have been reported by various other workers, notably by Mosier and Fainberg (1972).

The receiving equipment consists of an 18 MHz crossed-Yagi polarimeter and two square corner-reflectors. These latter may be used either for 22 and 26 MHz total-power or as a simple interferometer providing in-phase and out-of-phase channels from a hybrid ring at 26 MHz. It is also possible to receive total-power frequencies separated by 100 to 500 KHz from divided antenna inputs. Magnetic tape and high-speed pen-recording are available for four channels of information. In addition, a swept-frequency receiver, built and operated by Mr A. Achong, covers the band 18–28 MHz with a time resolution of 50 ms.

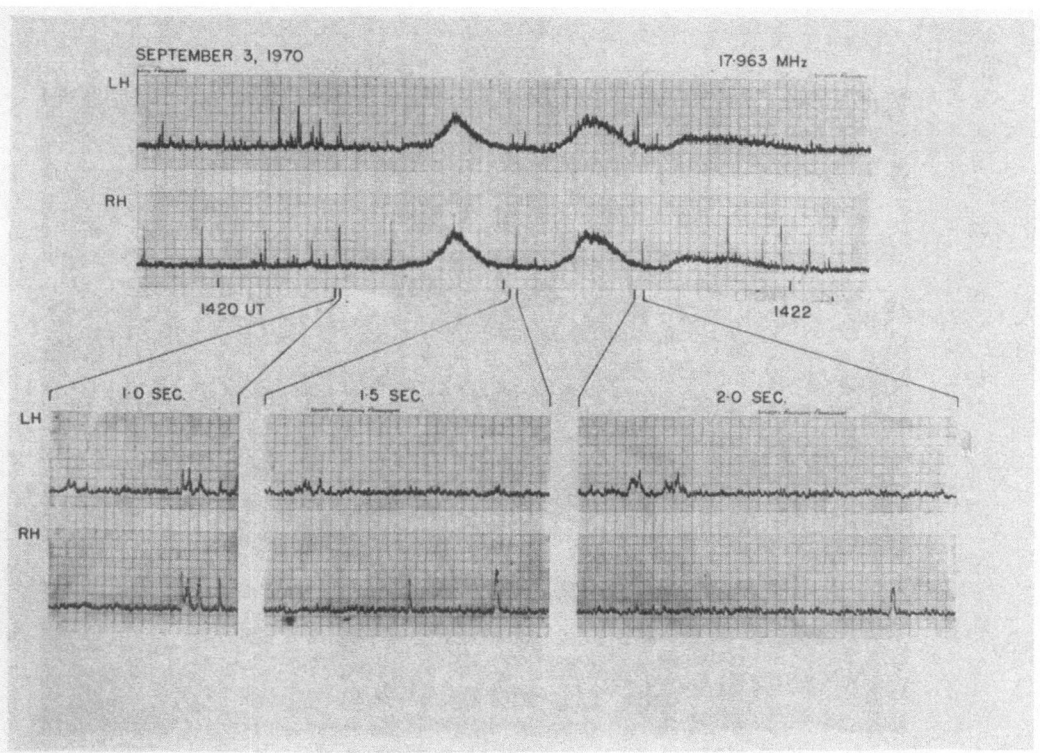

Fig. 4.   Slowed playback sections taken during a type III event. Note the fast changes in polarization sense. Pulse durations at half-height range from about 5 to 30 ms with bandwidths less than 150 KHz.

The problem of receiver bandwidth for the fixed frequencies has been discussed in some detail by Barrow and Saunders (1972). It is necessary, at decametric frequencies, to tune the receivers to operate in gaps between stations and so the bandwidth must be sufficiently narrow to achieve this while still being broad enough to give reasonable time resolution. 2.4 KHz is found to be an acceptable compromise as compared to about 5 KHz often used for Jupiter observations at night.

A general record of each observation is made with a slow-speed chart recorder and at the same time the observation is recorded on magnetic tape following the procedure used by Torgersen (1969) for Jupiter observations. Sections of interest are identified from the general record and then played back either to the high-speed pen recorder or to a CRO where they can be recorded photographically. By using a slower playback speed than recording speed the effective time resolution of the high-speed recorder can be improved. A small post-detection time constant is added to give resolutions of one and five milliseconds for the CRO and the high-speed pen recorder respectively.

Interference discrimination has already been described by Barrow and Saunders (1972). Briefly, the following considerations are involved:

(a) Response of the interferometer system.

(b) Narrow bandwidth and very short duration of the millisecond pulses.

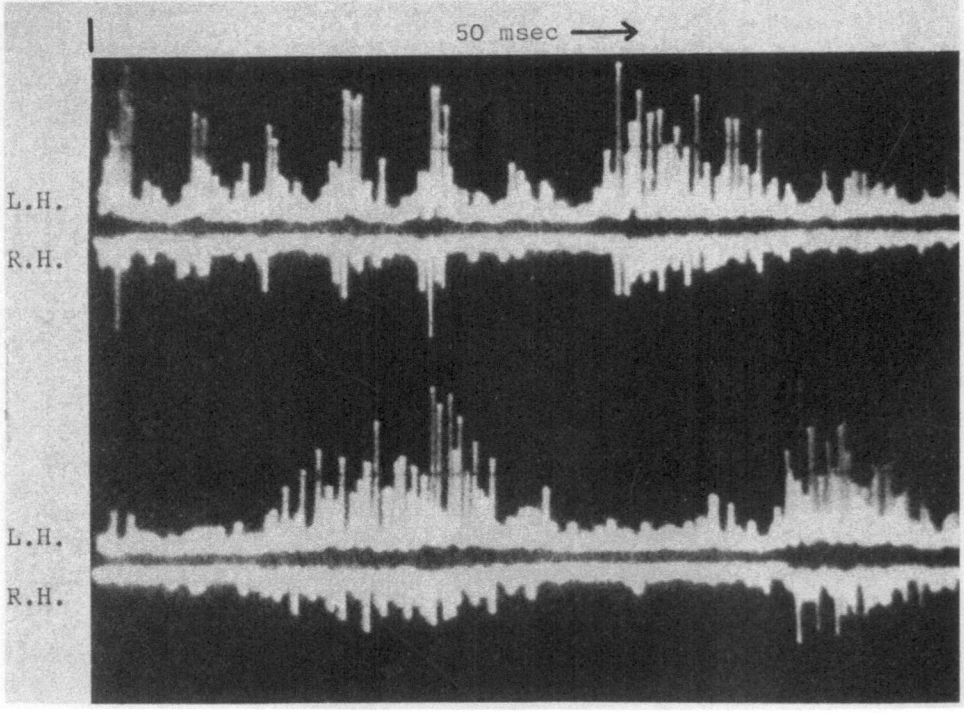

Fig. 5.   CRO playback record showing two 50 ms sweeps taken at 18 MHz. The upper section of each trace is the left-hand polarized channel; the lower section of each trace (displacement downward) is right-hand polarized.

(c) Characteristic polarization of the pulses.

(d) Confirmation of solar radio activity, at the time of the pulses, from *Solar-Geophysical Data* (Prompt Reports), ESSA, U.S. Department of Commerce.

(b) and (c) above distinguish the solar pulses from static pulses which are broadbanded, of longer duration and unpolarized. (d) is necessary as many pulses seem to occur at the onset and the decay of type III bursts as well as during noise storms. Only pulses occurring in association with confirmed solar phenomena are considered.

Some typical examples of solar millisecond pulses are shown in Figures 3, 4, and 5. In Figure 3 the pulses are unresolved but a number can be seen clearly on the 26 MHz in-phase channel that are not shown simultaneously on the other three channels. Figure 4 is a slowed playback record and shows good examples of polarization reversal. Figure 5, in the upper trace, shows a sequence of very fast left-hand polarized pulses each of duration about one millisecond. This sequence is similar to several reported for Jupiter by Torgersen (1969).

Note that in the records shown, we are only recording left- and right-hand components. We can, therefore, assess only the sense of polarization and not the degree of polarization apart from the special case of pure circular. This point has been discussed in detail by Barrow and Morrow (1968).

We see that the solar pulses display a number of similar characteristics to Jovian millisecond pulses, notably with respect to typical durations, intensities, polarization characteristics, bandwidths and short sequences. The solar pulses have not yet been observed in the prolonged, almost continuous sequences that sometimes occur in Jupiter radiation, however.

In Figure 6, half-height pulse durations are compared for some 1600 pulses and the distributions are seen to be similar for both Jupiter and the Sun. The majority of the pulses have durations between 10 and 20 ms. For comparison, Carr and Gulkis (1969) quote 16.0 ± 2.2 ms as a typical average duration for Jovian millisecond pulses.

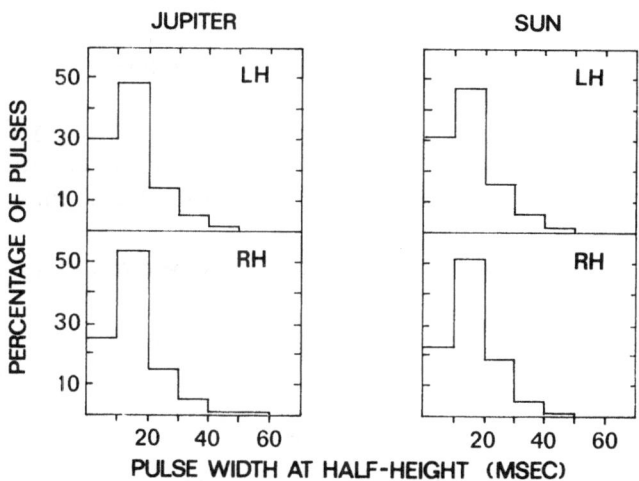

Fig. 6.   Normalized distribution of half-height pulse durations for Jupiter and the Sun.

Rise-and-fall rates and pulse-separation within groups are also being compared but in this respect it is often difficult to find pulses which can be measured unambiguously because of grouping, possible overlapping, and structured peaks. Also a pulse having very fast rise-and-fall rates may split in the manner of an unresolved square pulse [see, for example, Goodyear (1971)]. Thus one must be suspicious of what may appear to be a 'fast double-pulse'.

Millisecond pulses from Jupiter are generally accepted as being a source phenomenon rather than a superimposed effect of the interplanetary medium. The reasons for this are that the pulses do not appear to show drifts in time when observed from separated sites (Slee and Higgins, 1967) and that they are not seen in scintillation studies of radio stars (Warwick, 1967). These latter observations, however, are conducted at higher frequencies and on a longer time scale than Jovian millisecond pulse studies so they can hardly be regarded as conclusive proof. A check can be made (although probably with negative result) by using the same recording-playback technique on large array observations of a more distant source. Such observations are being attempted at the present time by Dr T. D. Carr who is kindly using the large 26 MHz array at the University of Florida to monitor Taurus A. Two sets of observations have been obtained at the time of writing both giving a negative result.

Eventually it is hoped to provide a reasonably objective answer to the question 'Are the millisecond pulses from Jupiter and the Sun sufficiently similar to be regarded as having a common mechanism of origin?' If we can make a decision regarding this question, we can apply the deductive process of propositional calculus to compare all the known characteristics of Jupiter and the Sun that are relative to radio emission and either eliminate or retain properties which might be common to both objects. By this means we may be able to obtain a few more definite starting points than are presently available for the would be theoretician. Conditions of local electron density and magnetic field are two obvious examples. Present estimates of these quantities in theoretical work cover ranges of some two orders of magnitude.

In conclusion four general thoughts are offered for the future:

(a) At the present time the millisecond pulses from Jupiter and the Sun appear to be more similar than dissimilar.

(b) The propositional calculus approach could (as pointed out by Dr S. Gulkis in discussion) obviously be directed towards obtaining solar information rather than Jovian information although at present it appears that solar radio emission is generally better understood than the Jovian emission.

(c) The propositional calculus approach could also be directed at a comparison of Jupiter and Saturn.

(d) Perhaps the time has arrived when we have sufficient grounds for a general study of solar-Jovian relationships.

References

Baart, E. E., Barrow, C. H., and Lee, R. T.: 1966, *Nature* **211**, 808.
Barrow, C. H. and Morrow, D. P.: 1968, *Astrophys. J.* **152**, 593.

Barrow, C. H. and Saunders, H.: 1972, *Astrophys. Letters* **12**, 211.
Carr, T. D. and Gulkis, S.: 1969, *Ann. Rev. Astron. and Astrophys.* **7**, 577.
Goodyear, C. C.: 1971, *Signals and Information*, Butterworth, London.
Mosier, S. R. and Fainberg, J.: 1972, USNC-URSI Meeting, Washington.
Olsson, C. N. and Smith, A. G.: 1966, *Science* **153**, 289.
Sagan, C.: 1971, *Comments Astrophys. Space Phys.* **3**, 65.
Slee, O. B. and Gent, H.: 1967, *Nature* **216**, 235.
Torgersen, H.: 1969, *Physica Norvegica* **3**, 195.
Warwick, J. W.: 1967, *Space Sci. Rev.* **6**, 841.

DISCUSSION

Gulkis: There is a good chance that we will understand the Jupiter emission process prior to the time we understand the solar emission mechanism. Hence a common mechanism should allow us to better understand the environmental conditions on the Sun.

Moore: What is your opinion about the correlation between the position of Io and the Jovian radio emission?

Barrow: The correlation is now well established and is, in fact, sometimes used as a means of Torgersen, H.: 1969, *Physica Norvegica* **3**, 195. predicting major periods of Jupiter activity.

THE ATMOSPHERE OF SATURN

V. G. TEIFEL

Astrophysical Institute, Academy of Sciences Kaz. SSR, Alma-Ata, U.S.S.R.

Abstract. The results of recent studies of the optical properties, temperature, chemical composition and a probable structure of Saturn's atmosphere are reviewed.

1. Introduction

During recent years great attention has been directed to the study of Jupiter, since this planet is the first object of far space which may be studied in the near future by space methods. It seems a quite understandable tendency to obtain more and more observational data about this planet now, since this information is indispensable for the treatment of the aims and methods of the space researches on Jupiter. Nevertheless much new data were obtained about another major planet – Saturn. These data have shown some sufficient discrepancies between these two greatest bodies of our planetary system.

We shall give here a short review of recent studies of Saturn and of the data on optical properties, chemical composition and probable structure of Saturn's atmosphere.

2. Atmospheric Composition

In the atmosphere of Saturn, only methane and molecular hydrogen were discovered by spectral methods. Some indications of the presence of ammonia were derived from spectroscopy but these data are very uncertain and may be rather erroneous. More reliable results were obtained by radioastronomical methods that permit the study of atmospheric layers under the clouds and which have shown the presence of noticeable amounts of ammonia in the lower atmosphere of Saturn. It is necessary to note immediately that the quantitative estimates of the gas abundances, obtained by the spectral measurements of molecular lines and bands, can not be entirely certain because the visible intensity of the absorption features is affected by the multiple scattering in the aerosol medium. Now we shall consider the data on the abundance of some gases derived from the spectral and microwave observations.

2.1. HYDROGEN

According to the measurements of Tanaka (1967), the equivalent width of $S(1)$ line in the (3–0) quadrupole band of H_2 is 0.078–0.089 Å for the center of Saturn's disk. The equivalent path of H_2 (L_{H_2}) may be obtained as about 70 km-atm (neglecting the effect of the line saturation) or about 330 km-atm (if the line is saturated). For this line Owen (1969) has obtained $W = 0.065 \pm 0.010$ Å and $L_{H_2} \approx 190 \pm 40$ km-atm if the line is saturated. There are no other quantitative data on the H_2-abundance derived

Woszczyk and Iwaniszewska (eds.), Exploration of the Planetary System, 415–440. All Rights Reserved

from line intensity measurements. Using the theory of collision narrowing of hydrogen lines developed by Fink and Belton (1969), we can obtain approximately the limits of L_{H_2} from the measurements noted above. Assume for Saturn's atmosphere $T \approx 100 K$ and $P_{eff} \approx 1$ atm. Then we obtain for the S(1) line of H_2(3–0) band the Doppler width $\alpha_D = 0.0373$ cm^{-1} and the parameter

$$y = \frac{\alpha_L}{\alpha_D} = 1.9 \cdot 10^{-2} \lambda P \left(\frac{T}{T_0} \right)^{-1.25} = 0.0543, \tag{1}$$

where α_L – the Lorentz width of the absorption line. For the line S(1) in (4–0) band of H_2 $\alpha_D = 0.0477$ cm^{-1} and $y = 0.0425$. Using the values of y and $W/\alpha_D \sqrt{\pi}$ we have derived from the curve of growth calculated by Fink and Belton that the value

$$u = \frac{N s_0}{\alpha_D \sqrt{\pi}}, \tag{2}$$

(where N – a quantity of absorbing gas, s_0 – the absorption coefficient for line) is about 8–11 and ~ 0.4 for the S(1) lines of each band consequently. If the effective atmospheric mass in a center of Saturn's disk is assumed $\eta = 2$ we have $L_{H_2} \approx 193$–165 km-atm from Tanaka's measurements and $L_{H_2} \approx 97$ km-atm from the Giver and Spinrad (1966) data for the (4–0) H_2 S(1) line in Saturn's spectrum. Thus we do not have a very certain result even if we take into account the collision-narrowing effect: the equivalent path of H_2 varies from ~ 100 km-atm to ~ 260 km-atm for a simple reflecting model of the line formation. The partial pressure of H_2 at the base level of a homogeneous layer with that thickness must be $0.8 < P < 2.2$ atm, and it does not contradict the value of P_{eff} assumed above for our calculations. The values of L_{H_2} noted above were obtained without taking into account the effect of the cloud layer on the formation of absorption lines; these L-values can not be accounted as the true H_2 abundance in the effective zone of atmosphere where the H_2 lines are formed.

There is another way to estimate the H_2 abundance in the atmosphere above the clouds. It uses the Rayleigh scattering effects in the ultraviolet spectrum of Saturn. We shall consider the ultraviolet measurements later, in connection with the question of the aerosol layer structure. From the UV measurements of Saturn's spectral reflectivity at $\lambda\lambda 0.33$–0.45 μ we have obtained the equivalent thickness of H_2, over the aerosol layer, of about 13.5 ± 1.0 km-atm in the equatorial belt and about 19 ± 2 km-atm in the temperate latitudes. It is about one order of magnitude less than that derived from the measurements of quadrupole lines.

2.2. METHANE

From Kuiper's measurements (1947), the equivalent path of CH_4 is about 350 m-atm for Saturn. This is probably an overestimate. As we have found from photographic and photoelectric spectrophotometry (Teifel *et al.*, 1971, 1973) the equivalent width of the 0.62 μ absorption band of CH_4 in the Saturn spectrum is only about 1.3–1.5 times more than in Jovian spectra. This band may be considered as non-saturated. Then if for Jupiter $L_{CH_4} \approx 130$ m-atm according to Owen and Mason (1969) we have for Saturn's equatorial belt $L_{CH_4} \approx 170$ m-atm and for the south temperate belt $L_{CH_4} \approx 200$ m-atm. The same results were obtained from comparison of the intensity of the 0.543 μ CH$_4$ band in the spectra of Saturn and Jupiter scanned by a photoelectric spectrometer (Bugaenko et al., 1971). These values of L_{CH_4} are the upper limits because multiple scattering must increase the intensity of the weak absorption bands. As will be shown later, the most probable upper limit of the CH$_4$ relative abundance on Saturn is about 3.8×10^{-3}, if the scattering effects are taken into account.

2.3. AMMONIA

The question of the presence of ammonia in Saturn's atmosphere is very important in connection with the problem of the cloud composition. The physical conditions in the atmosphere of this planet are favourable for ammonia sublimation and the formation of ammonia clouds.

The spectroscopic observations have not detected a more intensive (than in visible)

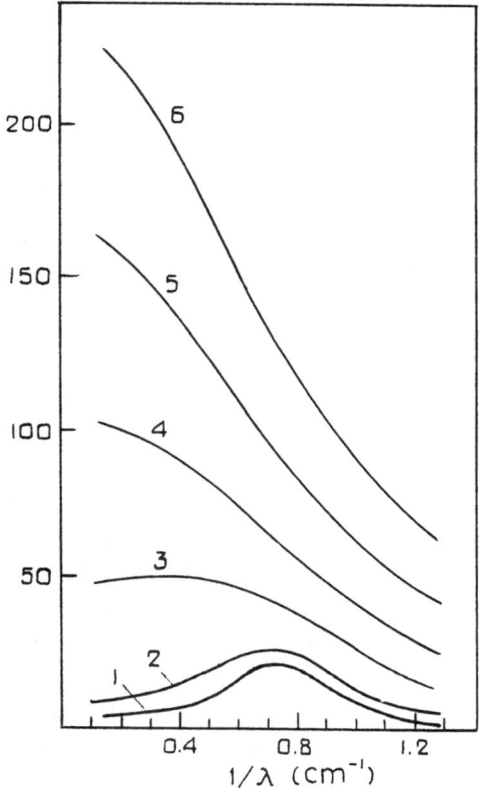

Fig. 1.   The absorption coefficients of NH$_3$, in a hydrogen atmosphere at $T = 293$ K, at the pressures of 7 atm (1), 10 atm (2), 20 atm (3), 30 atm (4), 40 atm (5) and 50 atm (6), after Kuzmin et al. (1972).

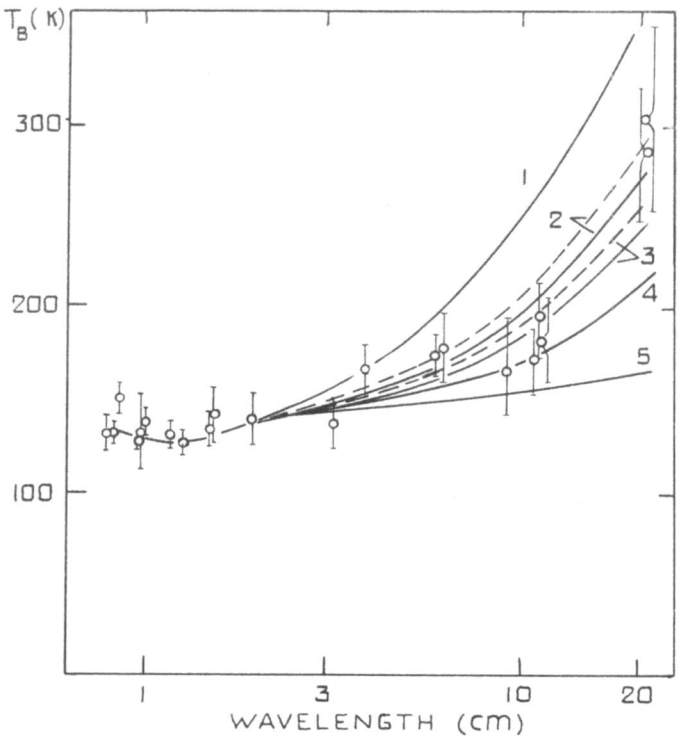

Fig. 2. A comparison of the calculated and observed spectral variations of the brightness temperature of Saturn. Solid lines – a pure hydrogen atmosphere, dotted lines – a hydrogen-helium mixture. The relative abundances of NH_3 in the undercloud atmosphere are 10^{-5} (1), 3×10^{-5} (2), 5×10^{-5} (3), 10^{-4} (4), 10^{-3} (5). After Kuzmin *et al.* (1972).

infrared ammonia band at $\lambda 1.51 \mu$. Moroz (1966) has estimated an upper limit to the ammonia abundance of about 50 cm-atm, from the absence of this band in his scans of Saturn's spectrum. Wrikson and Welch (1970) have concluded from the study of millimeter radio emission from Saturn that the quantity of NH_3 above the level with T 125–130 K can not be more than about 8 cm-atm; hence it is natural that the spectral observations have not detected ammonia absorption in the optical spectrum.

Kuzmin *et al.* (1972) have calculated theoretically the spectral distribution of the brightness temperature in the microwave spectrum of Saturn for different relative abundances of NH_3, taking into consideration the influence of low and high pressures on the absorption coefficients of ammonia. These absorption coefficients were derived for the levels with low pressures ($P < 6.5$ atm) anf for the deep layers of atmosphere ($P > 6.5$ atm) separately. Values of the NH_3 absorption coefficients vs wavelength are shown for some pressures in Figure 1. The best agreement between the calculated dependence of T and λ with the radio observations may be noted at a relative ammonia abundance of about $(3-5) \times 10^{-5}$ in the undercloud atmosphere of Saturn (Figure 2). It is interesting that the radio emission at $\lambda > 30$ cm in this case must be formed in very deep atmospheric layers where the pressure may be more than 1000 atm. In this region multiple molecular collisions must take place and therefore the study of the molecular absorption at very high pressures is necessary.

2.4. HELIUM

The relative abundances of hydrogen and helium are the fundamental characteristics of the atmospheres of major planets, since these gases determine the mean molecular weight and thermodynamic properties of atmospheres. Unfortunately, even for Jupiter the ratio H_2: He is not known with a sufficient accuracy. According to estimates of Owen and Mason (1969), the most probable value of H_2: He may be about 5:1 for the number of molecules. This ratio was derived from measurements of the methane lines halfwidths, which are chiefly determined by collisions with H_2 and He molecules. It is not improbable that the relative abundance of helium may be somewhat less than 0.2 if some part of the line broadening is caused by multiple scattering effect in the aerosol medium where the weak methane lines are formed. There are no halfwidth measurements for the methane lines in Saturn's spectrum and we only speak about the relative helium abundance in Saturn's atmosphere by the analogy with Jupiter.

3. Temperature

After the first radiometric measurements published by Menzel *et al.* (1926), no further investigations of the thermal emission of Saturn were performed for a very long time. The most important results from recent experiments have been obtained by Aumann *et al.* (1969). Their measurements, at a wide range of wavelengths (from 1.5 to 350 μ), have made it possible to estimate the entire flux of the thermal emission of Saturn (about 8.7×10^{-13} W cm^{-2}) and the effective temperature of planet $T_e = 97 \pm 4$ K. This temperature is significantly more than the equilibrium temperature, which is about 77 K if planetary albedo $A_v \approx 0.45$. Thus the heat flux of Saturn is about 2.5 times more than may be due to heating by the solar radiation only. It means that Saturn, like Jupiter, has it's own internal heat sources. The most probable source may be a gravitational contraction of the planet. The heat flux may be carried by convection in the planet interiors.

The temperatures derived from measurements in different parts of the infrared spectrum are not different in principle from T_e but they are somewhat lower than the temperatures obtained by the earliest observations. Now there are many estimates of the brightness temperature of Saturn in the microwave range – from 12 to 50 cm. These estimates give higher means of T_B (Figure 3) than those derived from optical observations. The temperature increases with the wavelength, but not as sharply as is observed for Jupiter, where at $\lambda \geqslant 3$ cm a significant role is played by nonthermal emission. The microwave spectrum of Saturn may be interpreted solely as the thermal radiation output from depths which increase with wavelength: the brightness temperature of Saturn at $\lambda 49.5$ cm is about 390 ± 65 K (Yerbury *et al.*, 1971) and it is about 540 ± 110 K at $\lambda 94.3$ cm (Yerbury *et al.*, 1973). There are no certain data on

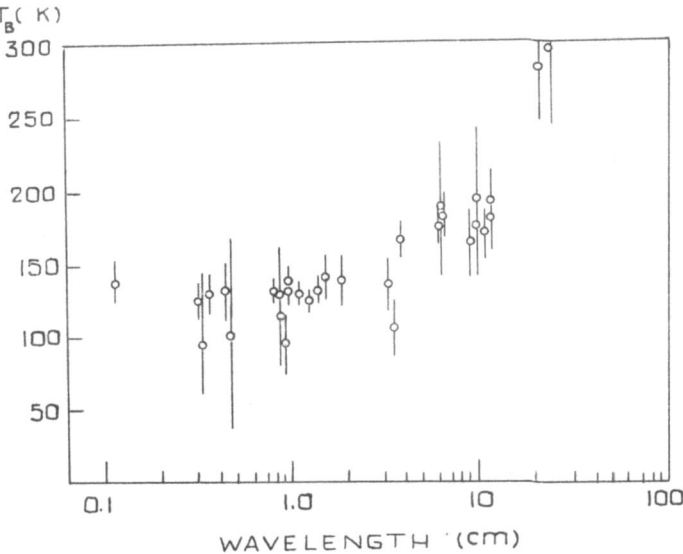

Fig. 3. The spectrum of the microwave emission of Saturn.

the presence of linear polarization of Saturn's radio emission, which might have been considered as evidence of a nonthermal component in the microwave radiation. The interferometric measurements at $\lambda 11$ cm (Kellermann, 1966; Berge and Read, 1968) have shown that practically the entire radio emission is formed in a region whose angular dimensions are not more than the visible disk of planet. This means that there are no radiating belts around Saturn. The presence of significant magnetic field near Saturn is also scarcely probable.

TABLE I

Temperature of Saturn from the optical measurements

Wavelengths, microns	$T(K)$	Author	Notes
8–14	125	Menzel *et al.* (1926)	T_B
8–14	85 ± 2	Low (1964)	T_B
8–14	93 ± 3	Low (1964)	T_B
8–14	100	Murray and Wildey (1963)	T_B
8–14	$107(+5, -10)$	Moroz *et al.* (1968)	T_B
10–14	98.3 ± 3	Allen and Murdock (1971)	T_B
10–14	99.7 ± 3	Allen and Murdock (1971)	T_B
5	120	Low and Davidson (1968)	T_B
20	97 ± 2	Murphy (1973)	T_B
1.5–350	97 ± 4	Aumann *et al.* (1969)	T_e
0.827	88	Spinrad (1964)	T_{rot}
0.643	126 ± 30	Giver and Spinrad (1966)	T_{rot}
0.643	103 ± 20		

The optical measurements of Saturn's temperature are summarized in Table I along with estimates of the rotational temperatures which were obtained by Giver and Spinrad (1966) from the ratio of the hydrogen quadrupole lines $S(1)$ and $S(0)$ intensities in the (4–0) band and by Spinrad (1964) from measurements of the same lines in the (3–0) band.

It is very interesting that there has not been observed on Saturn the anomalously high brightness temperature near $\lambda\,5\mu$ which was discovered on Jupiter. In some regions of the Jovian disk, especially in the North Equatorial Belt, the brightness temperature at $\lambda\,5\,\mu$ was about 230 K or more, according to the observations of Westphal (1969), and Keay *et al.* (1972). This high temperature may be caused by a decrease in the screening of the thermal radiation by the lower clouds whose density is not homogeneous in the different regions of Jupiter. On Saturn, as can be seen from Table I, the temperature at $5\,\mu$ is not much higher than the effective temperature and there are no peculiarities in the temperature distribution on Saturn's disk (Low and Davidson, 1969; Westphal, 1971). It is very hard to suspect the presence of very strong absorption by gases in this spectral range and it is more probable to suggest the presence of a dense lower cloud layer which is opaque for $5\text{-}\mu$ emission. Additional researches are very much needed, especially the study of the spectral variations of Saturn's brightness temperature in the infrared.

4. Optical Properties of the Cloud Layer and Outer Atmosphere

We now have some definite information about the spectral reflectivity of Saturn from the photoelectric measurements by McCord *et al.* (1971), Teifel *et al.* (1971), Irvine and Lane (1971), and Bugaenko (1972). The reflectance of planet in the spectral region outside of the methane absorption bands (at $\lambda < 0.7\,\mu$) is decreased toward the short wavelengths (Figure 4). The most sharp decreasing of reflectivity begins from $\lambda \approx 0.5\,\mu$

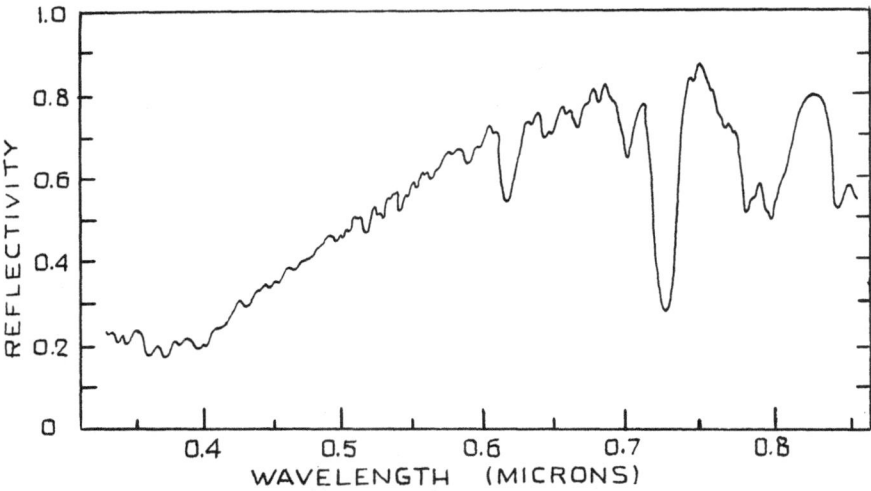

Fig. 4. The absolute spectral reflectivity at the center of Saturn's disk, after Bugaenko (1972).

to $\lambda \approx 0.38 \mu$, but at $\lambda < 0.38 \mu$ the reflectivity is observed to increase and this increase is continued to about 0.25 μ, as was determined by ground measurements (Teifel and Kharitonova, 1972a, 1973; McCord et al., 1971; Krugov, 1972) and by OAO observations (Wallace et al., 1972). It might be supposed that the reflectivity depression near $\lambda 0.38 \mu$ is connected with a wide range absorption band of an aerosol. The short wavelength decrease of the reflectivity is caused undoubtedly by the aerosol continuum absorption but the growth of reflectivity at $\lambda < 0.38 \mu$ is of other nature. If the aerosol absorption is decreased and the single scattering albedo

$$\omega_c = \frac{\sigma_a}{\sigma_a + \kappa_a},\tag{3}$$

(where σ_a is the volume scattering coefficient and κ_a is the volume coefficient of true absorption) is increased accordingly we must observe at $\lambda < 0.38 \mu$ more and more darkening of Saturn's disk toward the limb; this follows from the theory of diffuse reflection of radiation by a scattering medium. However, there is observed a brightening of Saturn's disk near the limb in the ultraviolet (at $\lambda \approx 0.36 \mu$) as seen from the photographs of Saturn obtained by Marin (1968), Reese (1971) and others. Also there

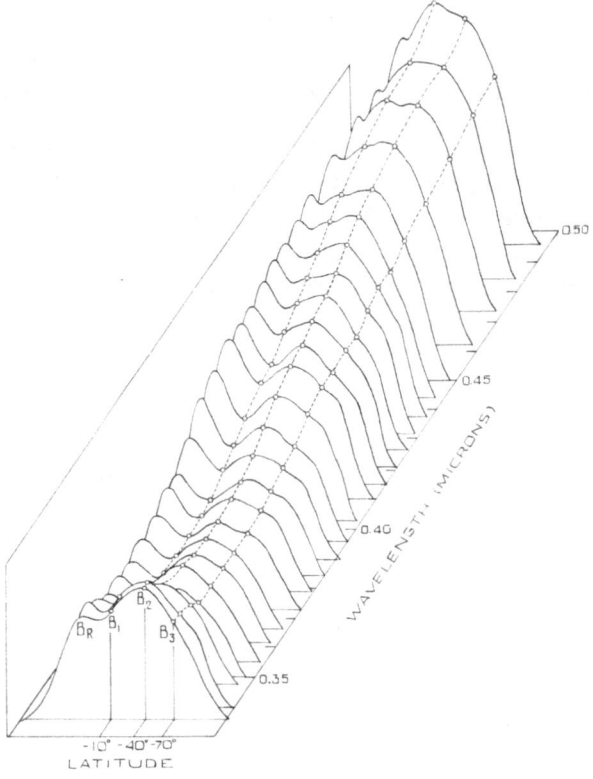

Fig. 5. Spectral variations of photometric monochromatic profiles at the central meridian of Saturn, from photoelectric spectral measurements in 1972 at $\lambda\lambda 0.34$–0.50μ.

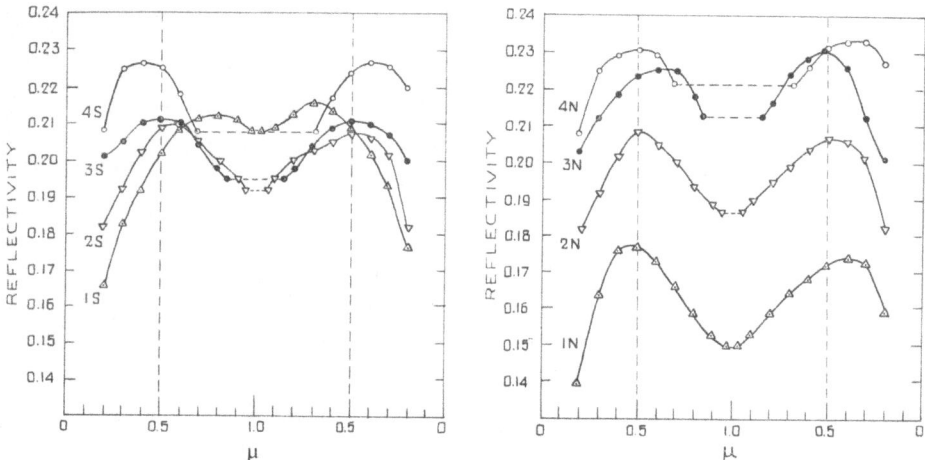

Fig. 6. Variations of reflectivity $r(\mu)$ in some zones of Saturn in 1966, at $\lambda_{\text{eff}} = 0.355\ \mu$, derived from the observations of Marin (1968).

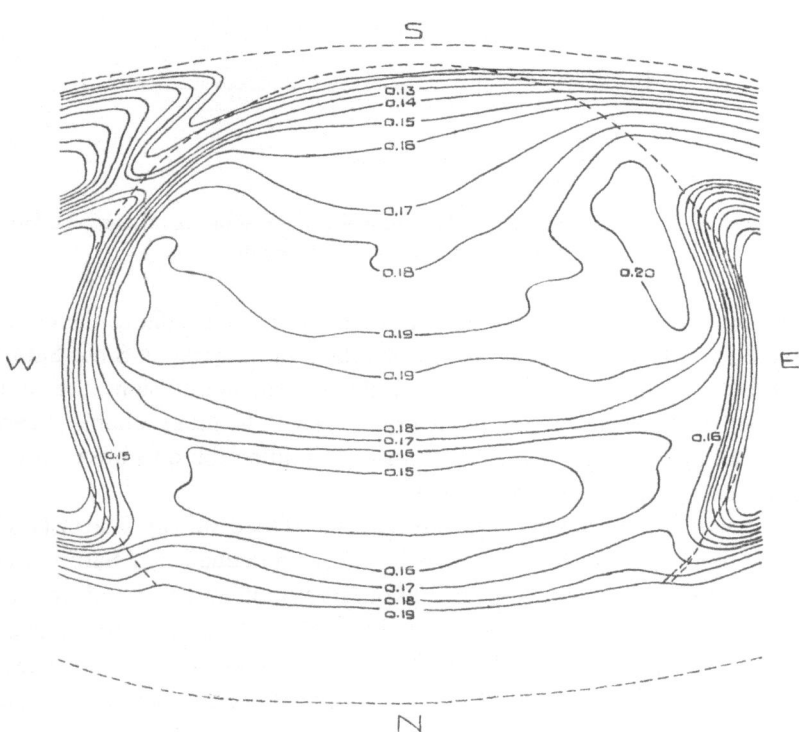

Fig. 7. The isophotes of reflectivity on Saturn's disk in the ultraviolet ($\lambda_{\text{eff}} = 0.36\ \mu$) derived from the measurements of negatives obtained at Mauna Kea Observatory in 1971.

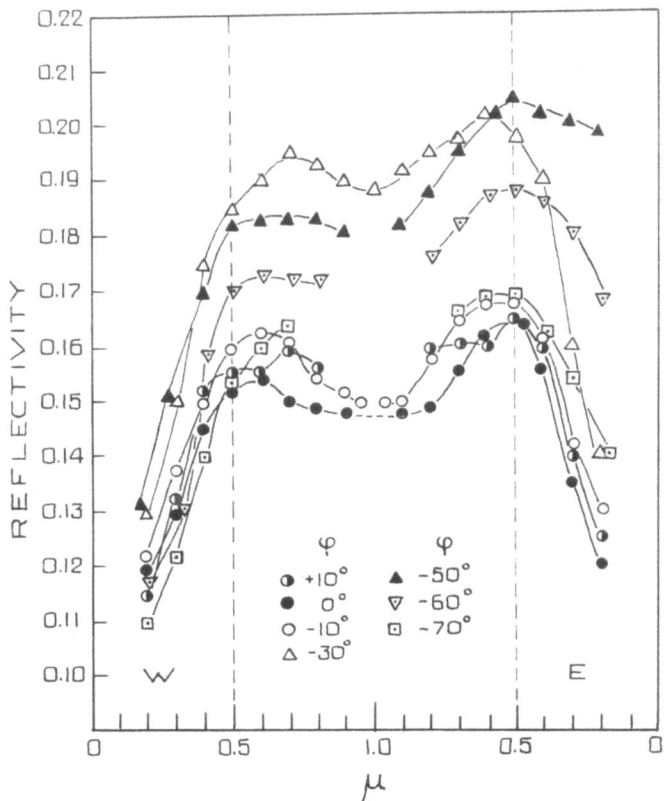

Fig. 8.   Variations of reflectivity $r(\mu)$ in the ultraviolet along some saturnographic latitudes (φ), derived from the preceding figure.

is other peculiarity: the equatorial belt of Saturn, which was light in the visible, turns into most dark belt on the disk in the ultraviolet light (Figure 5). Since Saturn is observed at different angles of the equatorial plane to the line of sight, the comparison of the reflective properties of any planetary zones will be more reliable if carried out for points falling at the identical conditions of illumination and observation, i.e. at the equal means of $\mu_0 = \cos i$ and $\mu = \cos \varepsilon$.

We have used the photometric measurements carried out by Marin (1968) from the photographs obtained by him in 1966, when the equator of Saturn was in the Earth-Saturn-Sun plane, and the film photographs of Saturn obtained in 1971 at Mauna Kea Observatory (Hawaii) and kindly offered to us by Dr Cruikshank. For these images we have calculated the values of μ and planetographic latitudes φ using Schoenberg's formulaes (Schoenberg, 1929) and then we have plotted the visible ultraviolet reflectivity r vs μ for different belts of Saturn. It was assumed from the photoelectric spectrophotometry that $r = 0.15$ at $\lambda 0.36\,\mu$ in the equatorial belt. The results have shown (Figures 6–8) an increase of reflectivity with latitude, excluding the polar regions: the south polar region is more similar to the equatorial belt in ultraviolet. The limb brightening is observed at all latitudes, having maximum of reflectivity near $\mu \approx 0.5$–0.6. The displacement of the maximum from the limb is caused by turbulence of the Earth's atmosphere; however the increasing of brightness toward the limb on Saturn's disk in the ultraviolet may be interpreted as a result of Rayleigh scattering in the overcloud gaseous atmosphere. A theoretical study of Rayleigh scattering in planetary atmospheres (Coulson *et al.*, 1960) has given the brightness of a Rayleigh atmosphere increasing up to limb, but in the presence of atmospheric turbulence the visible intensity distribution is distorted and may be given in the one-dimensional case by the formula

$$ f(x) = \frac{1}{\sigma\sqrt{2\pi}} \int\limits_{-1}^{+1} F(x') \exp\left[-\frac{(x-x')^2}{2\sigma^2} \right] dx', \tag{4} $$

where $F(x')$ is the true distribution of the brightness along the radius (axe x) and σ is a parameter of the Gauss curve which describes the distribution function for the star image oscillation. If $F(x')$ is the theoretical variation of the Rayleigh atmosphere brightness on planetary disk when the ground albedo $A=0$, the visible distribution of intensity at different values of σ (Figure 9) will be qualitatively similar to the observed intensity distribution found for the ultraviolet images of Saturn. However, the ratio of the reflectivities at $\mu=0.5$ and $\mu=1.0$ from the theoretical curve is about 1.4–1.6 but the observed values are less than 1.1 (excluding the north component of equatorial belt where this ratio was about 1.17 in 1966). This discrepancy between the theory and observations can not be explained by the image quality, because the photographs of Saturn were obtained only in conditions of good seeing ($\sigma \leqslant 0.1$). This means it is necessary to take into consideration the reflection of light from the ground (cloud layer of Saturn) in the ultraviolet, assuming $A \neq 0$. The observed limb brightening can not be explained by the outer scattering aerosol layer without also significant Rayleigh scattering in a gaseous layer.

Using the approximate formula

$$ r(\mu) = r_a(\mu) \exp\left(-\frac{2\tau_R}{\mu} \right) + r_R(\mu, \tau_R), \tag{5} $$

where $r_a(\mu) = A\mu^q$ is the reflectivity (visible albedo) of the cloud cover, r_R is the reflectivity of the Rayleigh atmosphere, we can obtain the Rayleigh optical thickness τ_R if $r(\mu)$ is known for different points of Saturn's disk. Passing by the details of the calculations (Teifel, 1974) we shall give here the final results. For Saturn's equatorial belt with a minimal reflectance in ultraviolet we have

$$ \tau_{R1} \approx 0.21 \pm 0.02, $$

and the ground albedo

$$ A_1 \approx 0.10 \pm 0.01. $$

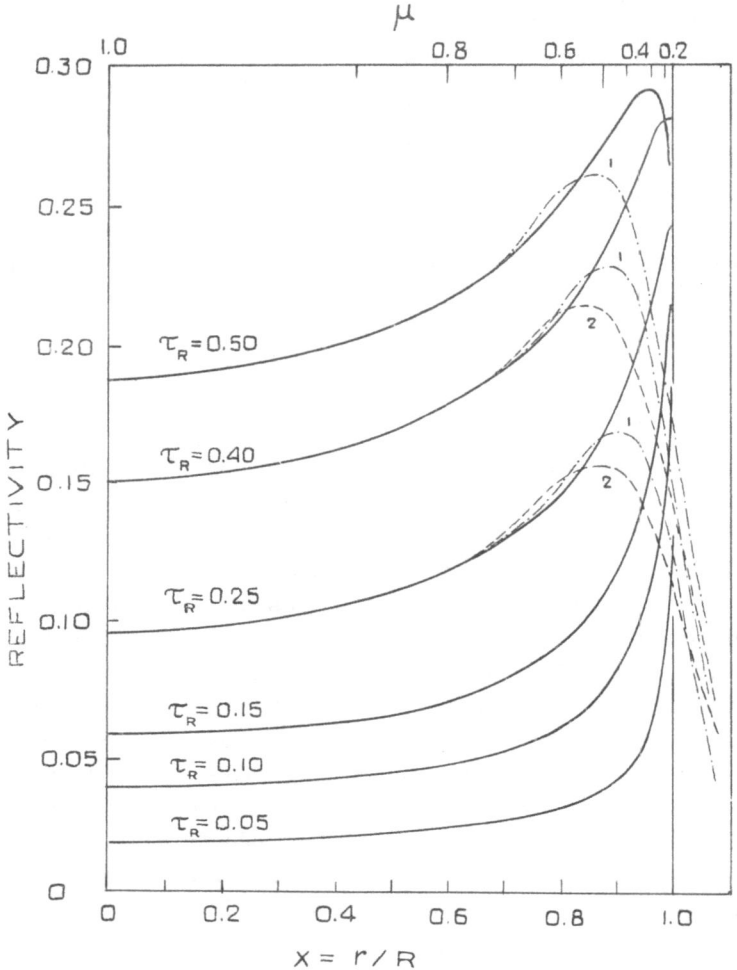

Fig. 9.   Theoretical distribution of reflectivity for a Rayleigh atmosphere along the planetary radius, at some values of τ_R. Solid lines – without the atmospheric turbulence influence ($\sigma = 0$); 1 – at $\sigma = 0.050$, 2 – at $\sigma = 0.075$.

The maximal optical thickness of the Rayleigh atmosphere was obtained for the temperate belt of Saturn:

$$\tau_{R2} \approx 0.30 \pm 0.03$$

and

$$A_2 \approx 0.18 \pm 0.01 .$$

As a Rayleigh scattering in Saturn's atmosphere is mostly caused by hydrogen (because the scattering coefficient of helium molecules is about 15 times less than that of hydrogen) it may be possible to estimate an equivalent thickness of H_2 above the aerosol layer using the values of τ_R noted above. We have obtained

$$U_1(H_2) \approx 13.5 \pm 1.0 \text{ km-atm}$$
$$U_2(H_2) \approx 19.0 \pm 2.0 \text{ km-atm}$$

The light exchange between the atmosphere and ground is neglected in formula (5). If the correction for this exchange would be introduced, we should have somewhat smaller values of τ_R and $U(H_2)$.

According to Trafton's (1967) calculations for a pure hydrogen atmospheric model of Saturn, the equivalent thickness of H_2 above an upper boundary of the convective zone is about 50 km-atm. For models calculated in the 'gray' approximation, with a pressure-induced absorption of the thermal radiation by hydrogen (Teifel, 1974), we have obtained about 30 ± 2 km-atm of H_2 above the convective zone boundary. This value does not depend significantly on the assumed ratio H_2:He. Thus, from the ultraviolet observations we obtain smaller values of $U(H_2)$ than would be observed if the upper boundary of the aerosol layer resembles the convective zone boundary. Probably the aerosol extends to higher levels (Figure 10), especially in the equatorial belt. As will be shown below, this conclusion also follows from the absorption bands measurements on Saturn's disk.

There is another peculiarity of the spectral reflectance of the equatorial and temperate belts on Saturn. As was shown by our photoelectric spectrophotometry, a

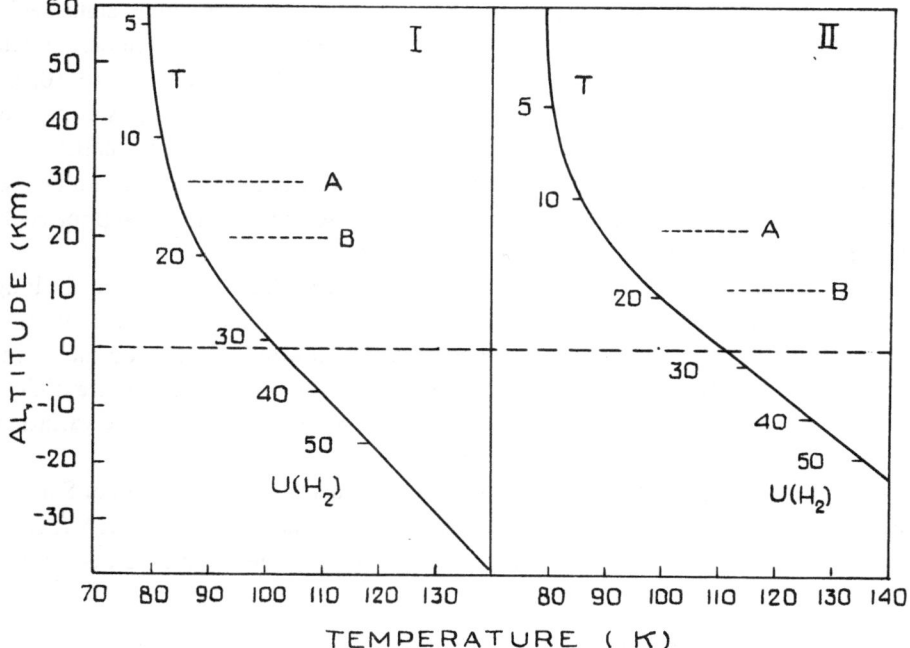

Fig. 10. The location of the upper boundary of the aerosol layer in the equatorial (A) and temperate (B) belts of Saturn, as derived from the ultraviolet measurements for two models of Saturn's atmosphere: I – H_2:He = 5:1, $T_e = 97$ K; II – H_2:He = 1:1, $T_e = 97$ K.

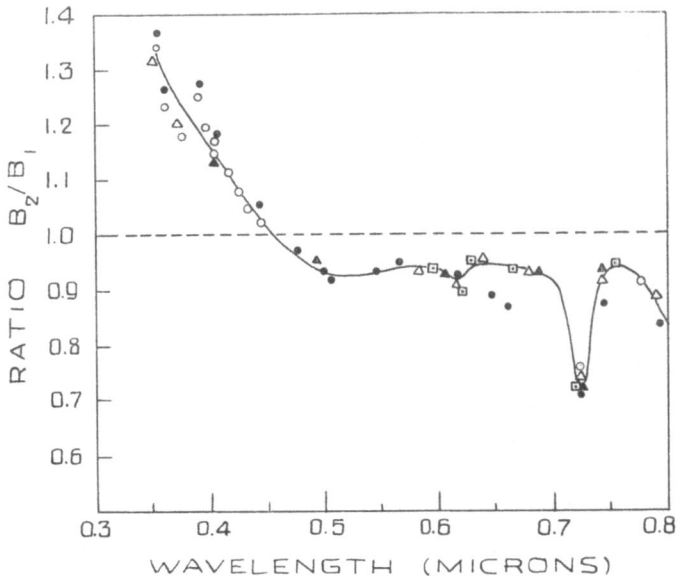

Fig. 11.   The ratio of spectral reflectivities for the temperate (B_2) and equatorial (B_1) belts of Saturn
in 1972, from the photoelectric spectral measurements.

sharp discrepancy in the spectral change of reflectivity between these belts begins
shortward from $\lambda 0.5\ \mu$. The ratio of the brightness of equatorial and temperate belts
increases monotonically towards the longer wavelengths from $\lambda \approx 0.38\ \mu$ to $\lambda 0.50\ \mu$
(Figure 11); but at $\lambda > 0.5\ \mu$ the temperate belt appears more neutral as compared
with the equatorial belt, excluding the methane absorption bands. The methane absorption in the temperate latitudes is more than in the equatorial belt, as can be seen
from a comparison of spectral reflectivity within these bands. This is supported by
measurements of the intensity of the methane bands.

An atmosphere above the clouds has produced some measurable effects at the long-
wavelengths only in the methane bands, but the reflectivity of Saturn in the continuum
is specified by the cloud cover of planet. There are very few reliable data on the brightness distribution over Saturn's disk, since the photometry of the disk is more difficult
when the saturnocentric latitude of the Earth is not equal to zero, because variations
of cloud reflectivity have a zonal character.

We can estimate approximately the parameters for a scattering function of Saturn's
cloud layer, using the data of the photographic photometry done in 1966 (Marin,
1968; Texereau, 1967) and the theoretical calculations of the brightness distribution
for the scattering function

$$X(\cos \gamma) = 1 + X_1 P_1 (\cos \gamma) + X_2 P_2 (\cos \gamma),\tag{6}$$

where $P_1 (\cos \gamma)$ and $P_2 (\cos \gamma)$ are the Legendre polynomials (Loskutov, 1971). The
best fit of the observations and theory (with the addition of the absolute spectrophoto-

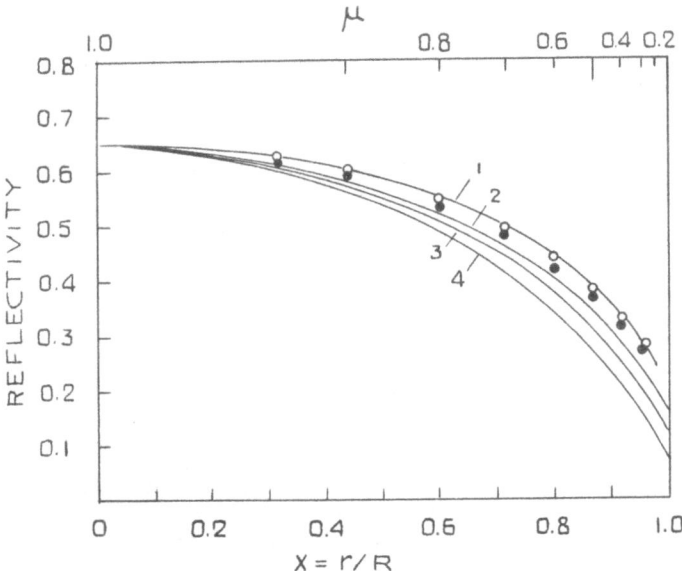

Fig. 12. The distribution of the reflectivity along the equatorial radius of Saturn's disk observed in yellow light and computed theoretically. Open circles – the observations of Texereau (1967), filled circles – the observations of Marin (1968). The parameters of the scattering function are $X_2 = 1.0$ and $X_1 = 1.0$ (1), 1.3 (2), 1.5 (3) and 1.7 (4).

metry data) was obtained for the measurements of Saturn's images in yellow light ($\lambda \approx 0.59 \mu$), if the parameters in (6) are $X_1 \approx 1.10$–1.15 and $X_2 \approx 1.0$ (Figure 12). Some discrepancies between the data of Marin and Texereau, obtained at the same time, may be explained by a not very exact determination of the limb on photographs. At any case, the observed darkening toward the limb of Saturn in yellow light is less than calculated for a scattering function with a great asymmetry. The relation of the parameter $\beta = B(\mu = 0.5)/B(\mu = 1.0)$, for the equator of Saturn in 1966, with the reflectivity of the disk center at different wavelengths is very close to the theoretically calculated relation, as can be seen from Figure 13 for the case of $X_1 \approx 1.1$ and $X_2 \approx 1.0$. Nearly the same result was obtained by Krugov (1973) from measurements in red light. The scattering function for the Jovian clouds in the longwavelength region of the spectrum ($\lambda \approx 0.6 \mu$), is more stretch forward ($X_1 \approx 1.7$, $X_2 \approx 1.0$) as was obtained by Loskutov (1971) and Kartashov (1972).

Some results of Saturn's polarimetry were obtained recently (Hall and Riley, 1969; Bugaenko et al., 1971; Bugaenko and Galkin, 1972). These observations have not been interpreted, but it is interesting to note one strange result: the plane of polarization, near the west and east limbs, does not coincide with the reflection plane, either at zero phase angle or at another angle. The region of this anomalous behaviour of the polarization plane falls into the planetocentric latitudes $\pm 30°$. This fact is noteworthy since the equatorial belt of Saturn is distinguished by its other optical properties.

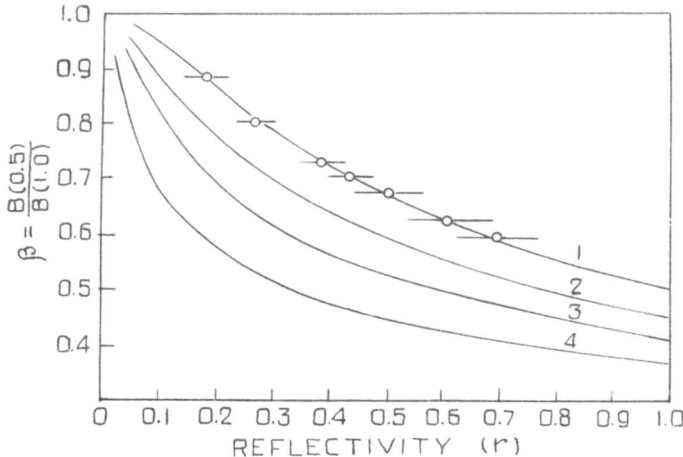

Fig. 13.   The value of $\beta = B(0.5)/B(1.0)$ vs the spectral reflectivity r, from Marin observations of Saturn (dots), and theoretically computed for the parameters of the scattering function $X_2 = 1.0$ and $X_1 = 1.0$ (1), 1.3 (2), 1.5 (3) and 1.7 (4).

5. Morphology of the Methane Absorption on Disk of Saturn

The first measurements of the intensity of the methane absorption band at $\lambda 0.62\ \mu$ were carried out by Hess (1953) in 1950. He found a relation between the increase of absorption and increasing latitude. More detailed studies of latitudinal variations of the intensity of this band on Saturn's disk were carried out in 1965–1970 in Alma-Ata (Teifel, 1969; Teifel and Kharitonova, 1970; Teifel et al., 1971, 1973). As seen from the Figure 14, the least CH_4 absorption was at temperate latitudes. It is important to note that the latitudinal distribution of this absorption is not changed significantly from year to year, independently of the inclination of Saturn's visible equator, which changed from $+3.5°$ to $-22.5°$ during this observational period. The last authors have shown that latitudinal variations in the absorption can not be stipulated by geometrical effects or by variations in planetary heating by solar radiation. The latitudinal variations in the methane absorption have reflected substantially the more or less permanent zonal features of planetary cloud cover. Kozyrev (1968) has also noted the attenuated absorption of CH_4 in the equatorial belt, but he has attributed this effect to the region which fell inside of the ring shadow. However, the attenuation of methane absorption is a property of a wider equatorial belt and it is improbable to assume that the shadowing of Saturn's globe by the rings is the main reason for the decreasing of absorption near the equator.

   The absorption in the $0.62\ \mu$ band of CH_4 is somewhat decreased from the center of disk toward the east and west limbs of Saturn as can be seen from Table II. This is in contradiction with the simple reflecting model for the formation of absorption bands. If this model is assumed the absorption band intensity must increase toward the limbs and the relation between the central depths of band at $\mu = 1\ (R_1)$ and

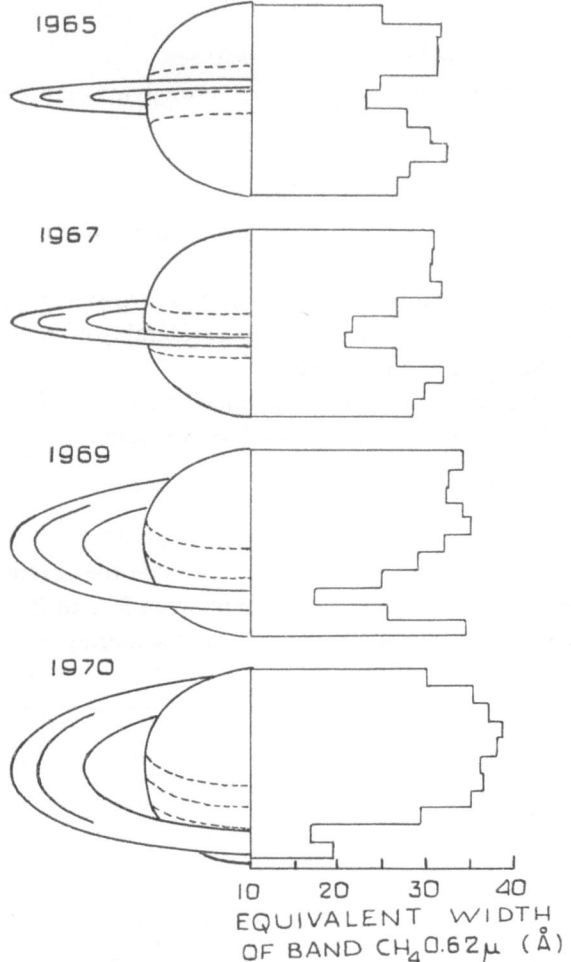

Fig. 14. The meridional distribution of methane absorption on Saturn in 1965–1970.

TABLE II

Equivalent width (W) and central depth (R) of the absorption band of
CH_4 at 0.62μ near the center ($\mu \approx 1$) and near the limb ($\mu \approx 0.5$) of
Saturn's disk in 1969.

Latitudes	$\mu \approx 1.0$		$\mu \approx 0.5$		
	$W_1(\text{Å})$	R_1	$W_2(\text{Å})$	R_2	R^*_2
$0°–14°$	29.2	0.276	26.5	0.255	0.480
$-14 -28$	32.9	0.316	32.8	0.315	0.525
$-28 -45$	34.5	0.346	33.8	0.336	0.572

$\mu = 0.5 (R_2)$ may be presented as a simple formula,

$$R_2 = 1 - (1 - R_1)^2,\tag{7}$$

since for the depth of a band consisting of many strongly overlapped lines

$$R_v = 1 - \exp\left(-\frac{2\tau_v}{\mu}\right),\tag{8}$$

where τ_v is the optical thickness of the atmosphere above the clouds in the absorption band center, $\mu = \mu_0 = \cos \varepsilon$.

The depth of a methane band near Saturn's limb would satisfy relation (8) if the simple reflecting model were true, and would be considerably more than the observed depth, as seen from the last column of Table II.

On the other hand if we assume a scattering model for the formation of the absorption bands, the depth of the band must satisfy the expression

$$R_v = 1 - \frac{\varrho_v(\omega_v, \mu)}{\varrho_c(\omega_c, \mu)},\tag{9}$$

where ϱ_c and ϱ_v are the reflectivities of the cloud layer in the continuum and in the absorption band center respectively. The relation between R_1 and R_2 (the band depths at $\mu = 1$ and $\mu = 0.5$) will be practically the same for any scattering function (Figure 15).

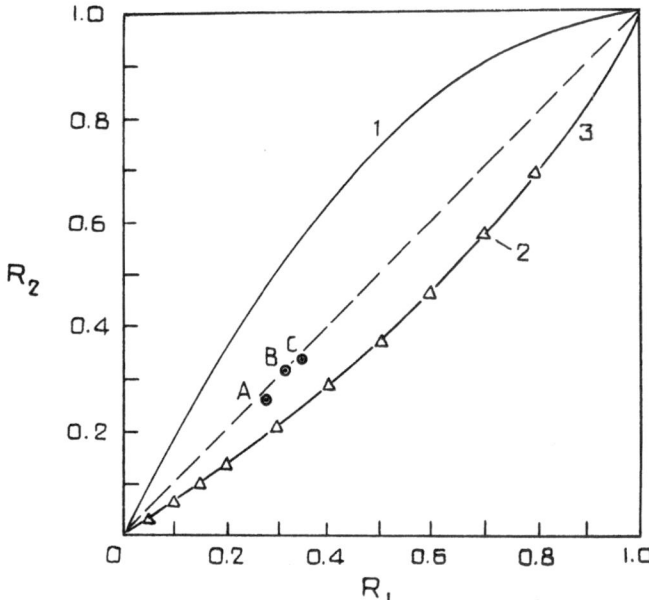

Fig. 15.   Theoretically computed and observed relation between the depths of the 0.62μ CH$_4$ absorption band; R_1 ($\mu \approx 1.0$) and R_2 ($\mu \approx 0.5$). A – for the equatorial belt, B – for the temperate belt, C – for the higher latitudes of Saturn. *Curves*: 1 – for the simple reflecting model, 2 and 3 – for the scattering models with isotropic scattering (2) and with the scattering function characterized by $X_1 = 1.7$ and $X_2 = 1.0$.

The observations do not fit this relationship, as noted on Figure 15 by points. It may be interpreted as a result of the absorption in the overcloud atmosphere of Saturn at the sufficient role of the multiple scattering inside the clouds in the absorption band formation. From the comparison of the observations with the theory we have concluded that the optical thickness of the outer atmosphere in 0.62 μ band of CH_4 is the

TABLE III

Latitudinal variations of the methane absorption in the bands of CH_4 at 0.62 μ and 0.72 μ on Saturn's disk from the photoelectric measurements in 1971

No	Latitudes	$CH_40.62 \mu$			$CH_40.72 \mu$		
		R	$W(Å)$	n	R	$W(Å)$	n
1	$+13° -- 3°$	0.238 ± 0.003	25.7 ± 0.4	88	0.600 ± 0.004	110.7 ± 1.2	90
2	$+ 1 -- 14$	0.263 ± 0.002	27.9 ± 0.4	62	0.666 ± 0.003	122.5 ± 1.2	52
3	$-10 -- 24$	0.287 ± 0.002	29.6 ± 0.4	65	0.712 ± 0.003	132.1 ± 0.9	52
4	$-20 -- 35$	0.295 ± 0.002	31.4 ± 0.3	111	0.760 ± 0.002	138.7 ± 0.9	87
5	$-31 -- 46$	0.292 ± 0.002	29.8 ± 0.4	71	0.770 ± 0.003	141.0 ± 1.4	44
6	$-42 -- 60$	0.282 ± 0.002	30.2 ± 0.4	68	0.755 ± 0.002	135.3 ± 1.3	59
7	$-55 -- 77$	0.278 ± 0.003	29.6 ± 0.6	64	0.742 ± 0.003	136.6 ± 1.7	50
8	$-68 -- 90$	0.288 ± 0.004	32.2 ± 0.7	46	0.735 ± 0.003	132.4 ± 1.8	45

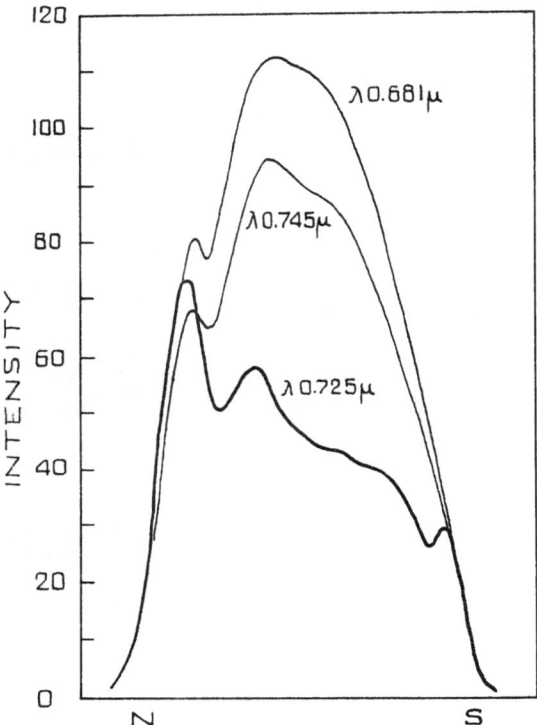

Fig. 16. The photoelectric monochromatic profiles of Saturn's central meridian in 1972, in the center of the 0.72 μ CH_4 absorption band and in the nearest parts of the continuous spectrum.

least in the equatorial belt of Saturn and is increased about 1.5 times in temperate belts of this planet. The volume scattering coefficient is varied very slightly with the latitude and the latitudinal discrepancy of the optical thickness of the outer atmosphere has suggested that the upper boundary of the cloud layer in the equatorial belt is located somewhat higher than in the temperate latitudes. The latitude difference is about 12–13 km at $T \approx 100\,\mathrm{K}$. Let us remember that the same conclusion was reached independently from the observations of Saturn in the ultraviolet.

In 1971 and 1972–1973 measurements of CH_4 absorption bands at $\lambda 0.62\,\mu$ and $0.72\,\mu$ were carried out on Saturn's disk with a photoelectric spectrometer (Teifel and Kharitonova, 1972b, 1974). The latitudinal variations of the equivalent widths and central depths of these bands were studied and the results for 1971 are listed in Table III where n is the number of the measured absorption band profiles.

The change of the methane absorption with latitude is noticeable even on the photometric scans of the central meridian of Saturn: the brightness distribution in the CH_4 $0.72\,\mu$ band is sharply different from the distribution in the nearest parts of continuum (Figure 16). A similar picture is observed in the more intense CH_4 band $0.89\,\mu$ as shown by photographs obtained by Owen (1969) and infrared spectrograms (Figure 17) obtained by Vdovitchenko (1974).

The results of several series of absorption bands measurements near the west and east limbs of Saturn in 1972 are shown in Figure 18. These data have confirmed the

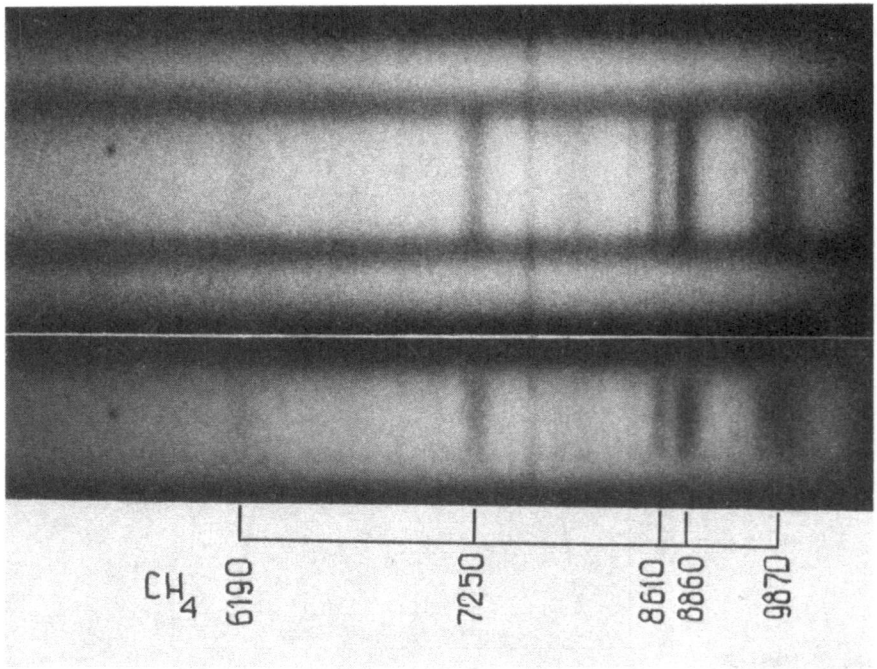

Fig. 17.   The infrared spectrograms of Saturn's equator (1) and of the central meridian (2), at $\lambda\lambda 0.6$–$1.1\,\mu$, photographed by Vdovitchenko in 1972 using the electronic image tube.

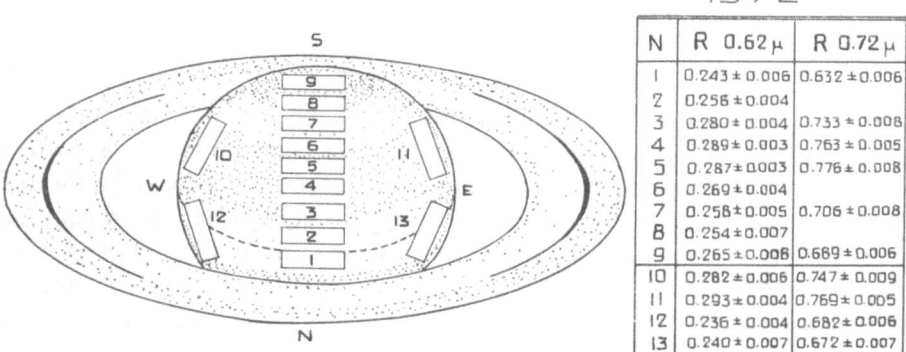

Fig. 18.   Estimates of the central depth of the methane absorption bands at 0.62 μ and 0.72 μ, in some parts of Saturn's disk, obtained from the photoelectric spectral measurements in 1972.

conclusion about the lack of increasing absorption toward the limbs within each belt of Saturn.

An interesting result was obtained both in 1971 and 1972. The intensity of the CH_4 absorption band at 0.62 μ is slightly increased near the south polar limb as compared with the adjoining regions of disk but the CH_4 band at 0.72 μ does not show this effect.

From the measurements of the intensity of the CH_4 band at 0.62 μ in 1971 we have obtained some characteristics of the cloud layer and the atmosphere above the clouds. These are presented in Table IV where ϱ_c are the reflectivities in the continuum of zones near the central meridian ($\lambda \approx 0.62\ \mu$), R_1 and R_2 are the observed values of the absorption band depth at $\mu \approx 1.0$ and $\mu \approx 0.5$, τ_ν^* is the optical thickness of the atmosphere above the clouds in the center of the absorption band for the simple re-

TABLE IV

Some atmospheric parameters derived from the CH_4 0.62 μ absorption band measurements on Saturn's disk in 1971

Parameter	Equatorial belt			Temperature belt		
	I	II	III	I	II	III
ϱ_c	0.775	0.775	0.775	0.736	0.736	0.736
R_1	0.275	0.288	0.262	0.297	0.314	0.290
R_2	0.252	0.239	0.265	0.316	0.299	0.333
τ_ν^*	0.161	0.170	0.152	0.177	0.189	0.172
τ_ν	0.030	0.019	0.040	0.055	0.044	0.067
$\sigma_a H_0 \dfrac{P_o}{P_e}$	2.5	1.3	4.1	5.0	3.5	7.7
R^*_1	0.230	0.261	0.200	0.215	0.240	0.189
$L(CH_4)$, m-atm	218	230	205	239	255	232
$U(CH_4)$, m-atm	40	26	54	74	59	90

flecting model, τ_v is the same but for a model with a multiple aerosol scattering, $b^{-1} = \sigma_a H_0 (P_0/P_e)$ is a value derived from the relation

$$\omega_v = (\omega_c^{-1} + b\tau_v)^{-1} \tag{10}$$

and σ_a is a volume scattering coefficient of the aerosol medium, H_0 is the scale height of the overcloud atmosphere, P_0 is the pressure on the upper boundary of clouds, P_e is the effective pressure in the region of the absorption band formation, R_1^* is the depth of the absorption band formed into the cloud layer only, $L(CH_4)$ is an equivalent path of CH_4 derived from the absorption band depth in the simple reflecting model, $U(CH_4)$ is an equivalent thickness of CH_4 in the overcloud atmosphere for the scattering model. All these parameters were obtained for the mean values of R_1 and R_2 (the column index I) and for the probable limits of R_1 and R_2 in the confidence interval at the probability $\alpha = 0.99$ (indices II and III).

As means we have $U(CH_4) \approx 41$ m-atm in the equatorial belt and $U(CH_4) \approx 72$ m-atm in the temperate belt with the probable error ± 10 m-atm. The upper limit of the CH_4 relative abundance is about 3.8×10^{-3} if we assume that the thickness of the overcloud atmosphere is the same in both the ultraviolet and visible spectral range.

Assuming $H_0 \approx 30$ km and $P_e/P_0 \approx 2$ we obtain $\sigma_a \approx (1-3) \times 10^{-6}$ cm^{-1} in the equatorial belt and $\sigma_a \approx (2-5) \times 10^{-6}$ cm^{-1} in the temperate belt.

6. Some Notes on the Saturn's Atmosphere Structure

Let us consider to what extent the observational data agree with a probable atmospheric model for Saturn. We assume as a working model the 'gray' one with $T_e = 97$ K and the ratio $H_2 : He = 5:1$ for the number of molecules. The altitudinal temperature and pressure distributions in this model are presented on Figure 19, where curves of the changes of saturated NH_3 and CH_4 vapour pressures are also shown. The imimpossibility of methane condensation at the Saturn's atmospheric conditions is evident because the relation

$$P_{CH_4}(z) > E_{CH_4}(T_z) \tag{11}$$

is necessary for the formation of the CH_4 clouds. Here $P_{CH_4}(z)$ is the partial pressure of CH_4 at the level z and $E_{CH_4}(T_z)$ is the pressure of saturated vapours of CH_4 at the temperature of this level. As seen from Figure 19 the partial pressure of CH_4 must be no less than 0.2 of the general pressure for the presence of the methane clouds. In this case, the relative abundance of CH_4 must be at least about 3×10^{-2} of the molecules, but the observations give us a value which is about ten times smaller.

For the ammonia condensation the conditions in the atmosphere of Saturn are more favourable since the pressure of saturated vapours of NH_3 is sharply decreased with the temperature decreasing upward. If the relative abundance of NH_3 is assumed about 4×10^{-5} in the undercloud atmosphere, based on the radio observations, the partial pressure of ammonia under the clouds must be about $3 \times 10^{-4} P$ where P is a general pressure. If so, the sublimation level (or lower boundary of the ammonia

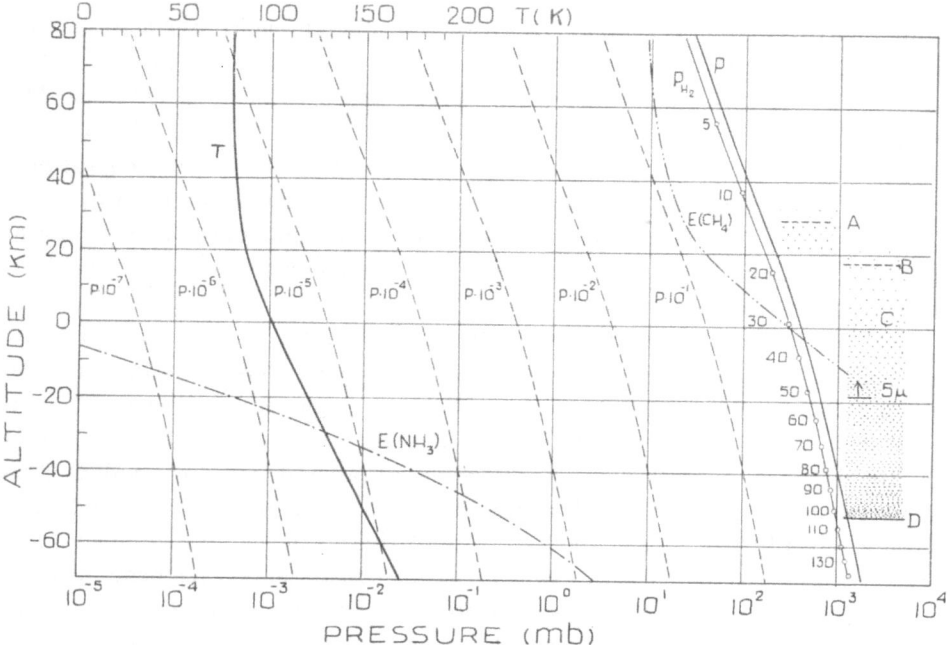

Fig. 19. The working model of the atmosphere of Saturn, A – the upper boundary of the aerosol in the equatorial belt, B – the same in the temperate belt, C – the level of the theoretical boundary of the convective zone, D – the NH_3 sublimation level. For this model $T_e = 97$ K, $H_2 : He = 5 : 1$.

clouds) must be at a depth of about 55 km below the convective zone boundary. The temperature of this level is about 155 K. The probable range on the depth of the lower boundary for ammonia clouds is from 50 to 60 km, if some uncertainty in the relative abundance of NH_3 is taken into consideration.

Thus the vertical extent of the aerosol layer on Saturn is very significant: using the ultraviolet data, it is about 70–80 km, i.e. about 2–3 times more than the probable thickness of the upper cloud layer on Jupiter. The lower part of the cloud cover of Saturn is rather dense and opaque, even for infrared radiation (at least for $\lambda \leqslant 5\,\mu$). As was noted above, the brightness temperature of Saturn at $\lambda 5\,\mu$ is about 120 K. This temperature may be connected with a level which is about 30–40 km above the lower boundary of the ammonia clouds.

Theoretical computations (Teifel, 1970; Teifel et al., 1971) show that the density of the ammonia cloud layer in the convective zone must be comparable with the density of Earth stratus clouds (about 10^{-7} g cm^{-3}). Formation of clouds is also feasible over the upper boundary of the convective zone, because even a very small quantity of NH_3 transported to these layers by mechanical turbulence may produce the oversaturation. The ammonia sublimation process may be facilitated at these small concentrations by the presence of cosmic dust in the upper atmosphere, and by the products of meteor particle evaporation. The density of the aerosol component at these latitudes depends on the vertical mixing intensity, and on the number of particles having a meteor nature which play the role of sublimation nuclei. The aerosol component in the upper cloud zone is a comparatively thin haze, and the particles of this haze may be of a two-layer structure – a solid nucleous with an ice envelope. The optical properties of the haze depend on the particle structure. The problem of the cloud colouring is not solved for Saturn, as for Jupiter. Some chemical compounds as NH_4SH or $(NH_4)_2S$ whose presence in Saturn's atmosphere is supposed by Lewis (1969), Owen and Mason (1969) and Lebofsky (1973) do not contradict the spectral reflectivity of Saturn, but the formation of these substances seems more probable in the deeper layers of the atmosphere.

The cloud cover of Saturn is the only indicator of the atmosphere circulation. Some unstability of the atmosphere is detected from time to time. Thus, for instance in 1966 the south component of Saturn's equatorial belt was a light zone both in visible light and ultraviolet, and the brightness distribution in this zone was significantly different from the distribution in other cloud belts of Saturn (as seen from Figure 6). In 1968 this anomalous component had disappeared. Another interesting feature on Saturn was observed in 1970 (Reese, 1970). It was a little light spot at a latitude of about $-57°$.

The spectral reflectivity differences between the equatorial and temperate belts at $\lambda < 0.5\ \mu$ can not be explained only by differences in the overcloud Rayleigh atmospheric thickness. There is the possibility of an increase in the concentration of a strongly absorbing substance in the equatorial belt. This substance may be formed there at the higher levels.

A few words concerning the great equatorial acceleration on Saturn may be added. It is known that the rotation periods are about 10^h14^m at the equator and about 10^h38^m at latitudes higher than $30°$. If this difference of zonal rotational velocities is attributed to the same equipotential level in the atmosphere, the differential velocity must be more than $330\ \text{m s}^{-1}$ (this velocity on Jupiter is about $120\ \text{m s}^{-1}$). At the boundary between the equatorial and temperate belts of Saturn, there is no turbulence which is so strong as that on Jupiter. On the other hand, we do not detect on Jupiter a significant difference between the intensity of molecular absorption bands in zones with rotation systems I and II. Apparently, the altitude differences of the cloud boundary on Jupiter are less than on Saturn. The visible equatorial acceleration on Saturn, and the higher location of the aerosol boundary are probably connected with one another. Two alternative assumptions may be offered. (1) The equatorial acceleration on Saturn is truly strong and the formation of clouds at higher levels on the equator is promoted by the presence of this acceleration; (2) The true equatorial acceleration is not as strong as it appears but the high clouds on equator are moved more quickly by the stratospheric winds. On both cases, the observed peculiarities of Saturn's atmospheric circulation are more probably caused by inner factors than by solar heating input.

References

Allen, D. A. and Murdock, T. L.: 1971, *Icarus* **14**, 1.
Aumann, H. H., Gillespie, C. M., and Low, F. J.: 1969, *Astrophys. J.* **157**, 69.

Berge, C. L. and Read, R. B.: 1968, *Astrophys. J.* **152**, 755.
Bugaenko, L. A.: 1972, *Astron. Vestnik* **6**, 19 (in Russian).
Bugaenko, L. A., Galkin, L. S., and Morozhenko, A. V.: 1971, *Astron. Zh.* **48**, 602 (in Russian).
Bugaenko, O. I. and Galkin L. S.: 1972, *Astron. Zh.* **49**, 837 (in Russian).
Coulson, K., Dave, J. V., and Sekera, Zd.: 1960, *Tables Related to Radiation Emerging from a Planetary Atmosphere with Rayleigh Scattering, Univ. of Calif. Press.*
Fink, U. and Belton, M. J. S.: 1969, *J. Atmospheric Sci.* **26**, 952.
Giver, L. P. and Spinrad, H.: 1966, *Icarus* **5**, 586.
Hall, J. S. and Riley, L. A.: 1969, *J. Atmospheric Sci.* **26**, 920.
Hess, S. L.: 1953, *Astrophys. J.* **118**, 151.
Irvine, W. M. and Lane, A. P.: 1971, *Icarus* **15**, 18.
Kartashov, V. F.: 1972, *Astron. Tzirkular*, No. 724 (in Russian).
Keay, C. S. L., Low, F. J., and Rieke, G. H.: 1972, *Sky Telesc.* **44**, 296.
Kellermann, K. I.: 1966, *Icarus* **5**, 478.
Kozyrev, N. A.: 1968, *Izv. GAO Akad. Nauk SSSR*, No. 184, 99 (in Russian).
Krugov, V. D.: 1972, *Astron. Vestnik* **6**, 85 (in Russian).
Krugov, V. D.: 1973, Dissertation (in Russian).
Kuiper, G. P.: 1947, *The Atmospheres of the Earth and Planets*, Chicago, p. 304.
Kuzmin, A. D., Naumov, A. P., and Smirnova, T. V.: 1972, *Astron. Vestnik* **6**, 13 (in Russian).
Lebofsky, L. A.: 1973, *Bull. Am. Astron. Soc.* **4**, 362.
Lewis, J. S.: 1969, *Icarus* **10**, 365.
Loskutov, V. M.: 1971, *Astron. Vestnik* **5**, 153 (in Russian).
Low, F. J.: 1964, *Astron. J.* **69**, 550.
Low, F. J. and Davidson, A. W.: 1969, *Bull. Am. Astron. Soc.* **1**, 200.
Marin, M.: 1968, *J. Obs.* **51**, 179.
McCord, T. B., Johnson, T. V., and Elias, J. H.: 1971, *Astrophys. J.* **165**, 413.
Menzel, D., Coblentz, W., and Lampland, C.: 1926, *Astrophys. J.* **63**, 177.
Moroz, V. I.: 1966, *Astron. Zh.* **38**, 1080 (in Russian).
Moroz, V. I., Vasiltchenko, N. V., Danilanz, L. B., and Kaufman, S. A.: 1968, *Astron. Zh.* **45**, 189 (in Russian).
Murphy, R. E.: 1973, preprint.
Murray, B. C. and Wildey, R. L.: 1963, *Astrophys. J.* **137**, 692.
Owen, T.: 1969, *Icarus* **10**, 355.
Owen, T. and Mason, H. P.: 1969, *J. Atmospheric Sci.* **26**, 870.
Reese, E. J.: 1971, *Icarus* **15**, 466.
Schoenberg, E.: 1929, *Handbuch der Astrophysik*, Band II, Berlin.
Spinrad, H.: 1964, *Appl. Opt.* **3**, 181.
Tanaka, W.: 1967, *Proc. Japan. Acad.* **43**, 971.
Teifel, V. G.: 1969, *J. Atmospheric Sci.* **26**, 854.
Teifel, V. G.: 1970, *Astron. Vestnik* **4**, 81 (in Russian).
Teifel, V. G.: 1974, *Astron. Vestnik* **8**, 3 (in Russian).
Teifel, V. G., Usoltzeva, L. A., and Kharitonova, G. A.: 1971, in C. Sagan, T. C. Owen, and H. J. Smith (eds.), 'Planetary Atmospheres', *IAU Symp.* **40**, 375.
Teifel, V. G., Usoltzeva, L. A., and Kharitonova, G. A.: 1971, *Astron. Zh.* **48**, 380 (in Russian).
Teifel, V. G., Usoltzeva, L. A., and Kharitonova, G. A.: 1973, *Astron. Zh.* **50**, 167 (in Russian).
Teifel, V. G. and Kharitonova, G. A.: 1970, *Astron. Tzirkular*, No. 549 (in Russian).
Teifel, V. G. and Kharitonova, G. A.: 1972a, *Astron. Tzirkular*, No. 683 (in Russian).
Teifel, V. G. and Kharitonova, G. A.: 1972b, *Astron. Tzirkular*, No. 735 (in Russian).
Teifel, V. G. and Kharitonova, G. A.: 1974, *Astron. Zh.* **51**, 167 (in Russian).
Texereau, J.: 1967, *Sky Telesc.* **33**, 226.
Trafton, L. M.: 1967, *Astrophys. J.* **147**, 765.
Vdovitchenko, V. D.: 1974, *Astron. Vestnik*, in press.
Wallace, L., Caldwell, J. J., and Savage, B. D.: 1972, *Astrophys. J.* **172**, 755.
Westphal, J. A.: 1969, *Astrophys. J.* **157**, 63.
Westphal, J. A.: 1971, private communication.
Wrixon, G. T. and Welch, W. J.: 1970, *Icarus* **13**, 163.
Yerbury, M. J., Condon, J. J., and Jauncey, D. L.: 1971, *Icarus* **15**, 459.

Yerbury, M. J., Condon, J. J., and Jauncey, D. L.: 1973, *Icarus* **18**, 177.

DISCUSSION

Gulkis: I have also analysed the microwave spectrum of Saturn and found a significantly larger ammonia abundance than reported in this paper. I believe the reason for the discrepancy is the fact that my analysis used the recent Cornell University data at 50 cm and 70 cm whereas the longest wavelength data used in your analysis was 21 cm.

MODELS OF SATURN'S RINGS WHICH SATISFY THE OPTICAL OBSERVATIONS*

Y. KAWATA and W. M. IRVINE

Dept. of Physics and Astronomy, University of Massachusetts, Amherst, Mass., U.S.A.

Abstract. A theoretical model of Saturn's rings is investigated which includes the shadowing effect and realistic anisotropic phase functions for the ring particles. The effects of multiple scattering and the finite size of the Sun, including the penumbra, are rigorously included. The permissible range of the relevant parameters, including optical thickness, single scattering albedo, volume density, and phase function are investigated by comparing the theoretical results to observations of the ring brightness vs phase angle, wavelength, and elevation of the Sun and Earth. Anisotropic scattering by the ring particles is necessary in order to match the observations. The colour dependence of the opposition effect is interpreted in terms of the albedo spectrum of the ring particles.

1. Introduction

The data provided by optical observations has traditionally been the core material for attempts to understand the nature of Saturn's rings. Although critical observations are now becoming available in the infrared and microwave regions of the spectrum, models of Saturn's rings must continue to satisfy the constraints provided by the optical data.

By optical in the present context we mean the extended visible portion of the spectrum (roughly 0.3–1.0 μ) in which the radiation received at the Earth from Saturn's rings is reflected sunlight, and hence does not include thermal emission by the ring particles. Previous analyses of the optical observations have suffered from limitations which are no longer necessary in view of improved computational and theoretical methods. We have accordingly endeavored to apply the best procedures currently available in an effort to see what limitations are imposed upon the physical parameters of the ring system and the particles which it contains.

The procedure which we shall use is a refinement of that originally proposed by Seeliger (1887) and subsequently employed in the fundamentally important work of Bobrov (e.g., 1970) and Franklin and Cook (1965).

2. Available Observations and Outline of the Procedure

Since the most complete and reliable photometric data are available only for the bright ring B, we shall concentrate our attention on this ring. The fundamental observation which must be matched by any theoretical model is the phase curve of the rings; that is, the surface brightness normalized at phase angle 0 vs phase angle α. This curve is normally plotted in stellar magnitudes per square arc-second of the rings, and has the following three characteristic features:

---

* Contribution Number 172 from the Five College Observatories.

Woszczyk and Iwaniszewska (eds.), Exploration of the Planetary System, 441–464. All Rights Reserved

(a) A very sharp surge in brightness near $\alpha = 0$ which is known as the opposition effect;

(b) A linearly decreasing brightness as α increases for $\alpha \gtrsim 2°$;

(c) A dependence upon wavelength.

The most reliable photometric phase curves appear to be those of Franklin and Cook (1965) which were obtained in the B and V wavelength bands. This data is illustrated in Figure 1. Since Saturn did not reach exact opposition ($\alpha = 0$) during their observations and the extrapolation to $\alpha = 0$ is somewhat arbitrary, we have normalized all data at the minimum phase angle observed ($\alpha = 0.094°$).

Another observation of critical importance to the understanding of the rings is the absolute surface brightness as a function of wavelength. Data for B and V were obtained by Franklin and Cook (1965). Corresponding data for other wavelengths may be obtained from the relative spectral photometry of Lebofsky *et al.* (1970), Irvine and Lane (1973), and Kharitonova and Teifel (1973), although care must be taken to insure that observations made under corresponding conditions are compared. This is important because the brightness of the rings may depend upon a number of geometric factors, including α and declination of the Sun and Earth relative to the ring plane, as well as possibly on distance of Saturn from the Sun or the position of the ring particles relative to their eclipse by Saturn.

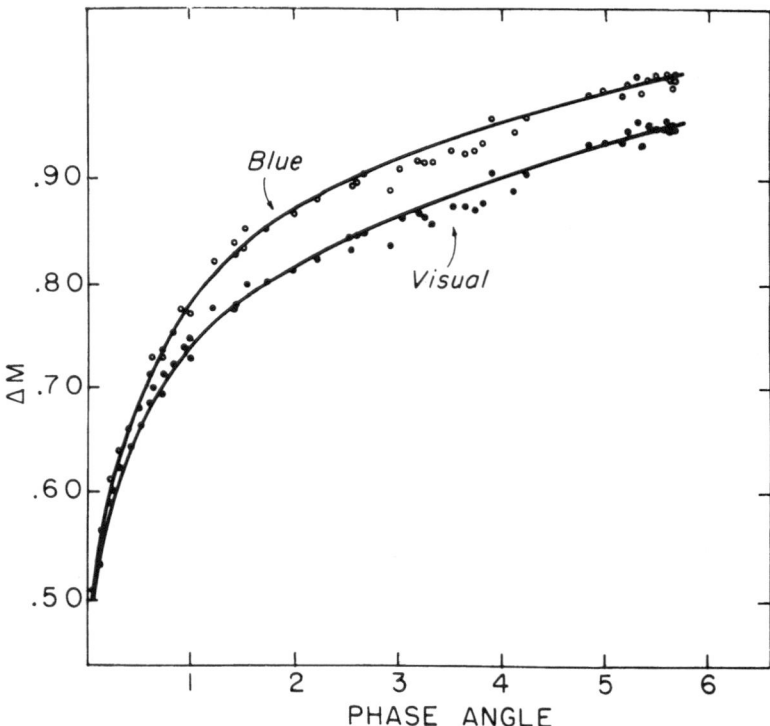

Fig. 1.   Phase curves for ring B at two wavelengths normalized
at $\alpha = 0°.094$ (from Franklin and Cook, 1965).

Additional photometric observations of potential importance are the variation in ring brightness with declination of the Sun and Earth. Observations have been made by Camichel (1958) and Price (1973). Complete phase curves of the rings at wavelengths in the red and near ultraviolet would be most desirable in the future.

The principle diagnostic characteristic of the phase curve is the opposition effect. We shall procede on the assumption that this effect is produced by the mutual shadowing of the ring particles, an idea originally proposed by Seeliger (1887). In more detail, we assume that the rings consist of a plane layer containing many independent particles which are illuminated by the Sun and observed from the Earth. Those particles nearer the Sun cast shadows upon the particles behind. At exact opposition an observer on Earth will see only sunlit particles and so will observe a maximum surface brightness. As the phase angle increases, the shadowed particles which were formerly shielded from view by the sunlit particles may now be observed from the Earth, so that the surface brightness falls off. This initial decrease in brightness takes place very rapidly as a function of phase angle.

Observations of stellar occultations by Saturn's rings indicate that the optical thickness of ring B is near unity. Because of the relatively high albedo of the ring particles (see below), it then follows that multiple scattering will play an important role in determining the photometric properties of the ring system. In previous computations of the shadowing mechanism multiple scattering has been included in only an approximate manner, by assuming that higher order scattering will be isotropic. In the present paper we treat rigorously the multiple scattering problem for more realistic, anisotropic particle phase functions.

The close relationship between the shadowing mechanism described above and the usual multiple scattering theory of radiative transfer has been discussed by Irvine (1966). Radiative transfer theory can be applied when the interparticle distance in the layer is sufficiently large that each particle is effectively in the far field for scattering by the other particles, so that shadows may be neglected. When the particles are large enough and their number density is great enough, they will cast shadows upon each other, and the usual multiple scattering theory must be modified to include the effect of shadowing. Fortunately, this can be done in a straight-forward manner.

3. Theoretical Procedure

We shall assume in our model that the rings are plane-parallel and homogeneous with respect to optical depth. We thus neglect the possibility that such properties as mean particle size or composition depend on altitude with respect to the center of the ring plane. We shall furthermore assume for the present that the ring particles may be characterized by a single effective radius ϱ and for the purposes of the shadowing computation may be treated as spheres. We shall return in Section 5 below to the possibility of a distribution of particle sizes, which introduces into the theory such quantities as $\langle \varrho^2 \rangle^{1/2}$ and $\langle \varrho^3 \rangle^{1/3}$ in addition to the mean radius ϱ. The assumption of sphericity will not significantly effect the applicability of our results, since it can be shown that the magnitude of the shadowing effect at opposition is independent of the particle shape (Seeliger, 1895), and we do not require that the individual particle phase function be given by Mie theory.

Following the procedure of Irvine (1966), we may express the specific intensity I of the radiation reflected by the rings as a sum of successive orders of scattering:

$$I = I_1^s + \sum_{n=2}^{\infty} I_n a^n, \tag{1}$$

where I_1^s is the contribution from once-scattered radiation including the necessary shadowing correction, and $I_n a^n$ is the contribution from radiation scattered n times. Fortunately, the effect of mutual shadowing is important only in the calculation of the primary scattered intensity and may be neglected in the computation of I_n for $n \geqslant 2$.

3.1. SHADOWING MECHANISM

Let us discuss first the computation of I_1^s. If the wavelength of light λ is such that

$$\lambda \ll \varrho^2 / \varDelta,$$

where $\varDelta$ is the lesser of the mean free-path of a photon in the layer and the thickness of the layer, a shadow will be formed behind each particle which will be described by geometric optics (van de Hulst, 1957). Let us introduce coordinates such that $\theta = = \arccos \mu$ is the polar angle with respect to the outward normal to the ring layer and ϕ is the corresponding azimuthal angle measured from the plane of incidence. We shall use the notation $\Omega = (\theta, \phi)$ to specify a particular direction, and shall initially assume that solar radiation is incident only in the direction $\Omega_0 = (\theta_0, \phi_0)$. For convenience we take $\mu_0 \equiv |\cos \theta_0|$. Let the thickness of the ring layer be t, and let the fraction of the ring volume occupied by particles be D, so that

$$D = \tfrac{4}{3} \pi \varrho^3 n, \tag{2}$$

where n is the number density of particles in the rings.

The physical mechanism operative in the shadowing effect can be understood by referring to Figure 2 which shows a particle of radius ϱ at a depth h within the ring layer. The volumes V_1 and V_2 are (except for a small correction near the surface) cylinders of base area $\pi \varrho^2$ and height h/μ_0 and h/μ, respectively. A small element ε of the projected area of the particle will be both sunlit and observable from the Earth provided that the centers of all other particles in the ring layer are outside of the volumes V_1 and V_2. If, as implied by our postulate of homogeneity, the ring particles are randomly distributed through the ring volume, the probability of the above situation occurring may be easily computed upon the assumption that the fractional volume occupied by particles is sufficiently small ($8D \ll 1$). This probabilistic approach yields the familiar exponential attenuation for both the average radiation field at a depth h in the layer and for the primary scattered radiation emerging from the layer, provided that the volumes V_1 and V_2 do not significantly overlap. If they do so overlap,

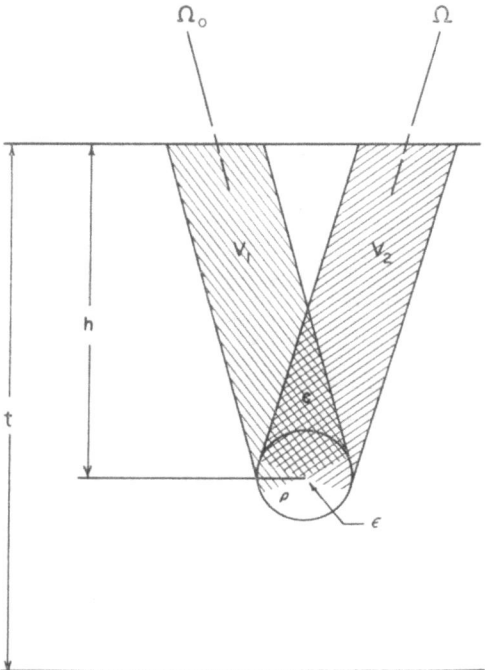

Fig. 2. Geometry of the shadowing effect. Direction towards
the Sun is Ω_0, towards the Earth is Ω.

an anomalously high intensity is produced because the probability for photon escape
from a depth h becomes highly correlated with the probability of photon penetration
to the same point. This is the shadowing effect.

The analysis shows that, if the incident solar flux through the upper surface is π,
the single-scattered intensity at an optical depth $\zeta = n\pi \varrho^2 h$ is given by

$$I_1^s = \frac{a\Phi(\Omega, \Omega_0)}{4\mu\,\mu_0}\, e^{\zeta/\mu} \int\limits_\zeta^\tau d\zeta' \exp\left[-\zeta'(\mu^{-1} + \mu_0^{-1}) + nC\right], \qquad (3)$$

where n is the number density of particles, C is the overlap volume shown in Figure 2,
a is the single-scattering albedo of the ring particles, the particle phase function is Φ,
and the polar angles of incident and scattered light arccos μ_0 and arccos μ are measured from the outward normal. Clearly, when $C = 0$, I_1^s is the primary scattering obtained from the usual multiple scattering theory. Since we have assumed that the
particles are large and diffraction may be neglected (that is, the efficiency factor for
extinction is unity), the optical thickness of the rings is related to the parameters
previously introduced by

$$\tau = n\pi\varrho^2 t. \qquad (4)$$

The quantity nC has the form (Irvine, 1966)

$$nC = \begin{cases} v_1 & (\zeta' - \zeta) \geqslant Z, \quad \alpha \neq 0 \\ v_1 - \Sigma & (\zeta' - \zeta) \leqslant Z, \quad \alpha = 0 \\ (\zeta' - \zeta)/\mu_0 & (\alpha = 0), \end{cases} \tag{5}$$

where

$$v_1 = \frac{(1 + \cos\alpha)\, D}{\pi \sin\alpha}$$

$$\Sigma = \frac{3\,(\cos\theta + \cos\theta_0)^2\, D}{4\pi \sin\alpha \cos\theta \cos\theta_0 \cos v} \left[\cos\psi - \frac{\cos^3\psi}{3} - (\pi/2 - \psi)\sin\psi\right]$$

$$\cos\alpha = \cos\theta \cos\theta_0 + \sin\theta \sin\theta_0 \cos(\phi - \phi_0)$$

$$\cos\delta = (\cos\theta_0 - \cos\theta \cos\alpha)/(\sin\theta \sin\alpha)$$

$$\tan v = \sin\theta \sin\delta \sin\alpha/(\cos\theta + \cos\theta_0)$$

$$\sin\psi = \frac{4\,(\zeta' - \zeta)\sin\alpha \cos v}{3\,(\cos\theta + \cos\theta_0)\, D}$$

$$Z = \frac{3\,(\cos\theta + \cos\theta_0)}{4} \frac{}{\sin\alpha \cos v}.$$

When $\zeta = 0$, we find the reflected intensity as

$$R_1^s \equiv I_1^s(\zeta = 0) \qquad (\mu > 0). \tag{6}$$

The above approach is essentially that used by Seeliger (1887). Because of the geometry in Figure 2, it is referred to as the cylinder-cylinder model. Bobrov (cf. 1970) has pointed out the important effect introduced by the finite angular diameter of the Sun at the distance of Saturn. He modified the previous theory by replacing the volume V_1 in Figure 2 by a conical volume, producing a 'cone-cylinder model'. This procedure, however, ignores the penumbra of the shadows cast.

Franklin and Cook (1965) observed that the opposition effect appears to be wavelength dependent, and proposed a model in which this dependence was produced by the wavelength variation of diffraction into the shadow zone. They treated this situation by using a 'cone-cone' model for the shadowed volumes, with the dimensions of the 'diffraction cone' being wavelength dependent. This model led to an unreasonably small physical thickness of the ring, however, and it is desirable to search for an alternative mechanism for producing the wavelength dependence. Franklin and Cook also considered the possibility that the wavelength effect might be due to differences in the glory produced by small Mie-scattering spheres forming a surface structure on larger particles. This seems extremely unlikely, however, because of the high degree of symmetry of the scattering centers needed to produce the glory phenomenon.

In the present model we propose to take into account the effect of the Sun's finite size by performing a numerical average of the intensity obtained under the assumption of a point (infinitely distant) Sun. This will rigorously include the effect of the penumbra and also any influence of solar limb darkening. The computation can be carried out quite rigorously, because the necessary arithmetic in Equations (3) and (5) is efficiently and rapidly performed on an electronic computer. The resulting model will have a reduced opposition effect relative to the point Sun model, because there is no longer an exact opposition for the total solar flux.

If the angular diameter of the Sun at Saturn's distance is β (about 3'), and the Sun is assumed symmetric about the angular direction $\Omega_{00} = (\theta_{00}, 0)$ of its midpoint, the average reflected intensity in the direction (θ, ϕ) will be

$$\langle R_1^s (\theta, \phi) \rangle \equiv \frac{\int\limits_0^{\beta/2} d\theta' \int\limits_0^{2\pi} d\phi' \sin \theta' \, W(\cos \theta') \, R_1^s [\theta, \phi; \theta_0(\Omega'), \phi_0(\Omega')] \cos \theta_0(\Omega')}{\int\limits_0^{\beta/2} d\theta' \int\limits_0^{2\pi} d\phi' \sin \theta' \, W(\cos \theta') \cos \theta_0(\Omega')},$$

(7)

where the solar limb darkening

$$W(\mu') = a_\lambda + b_\lambda \mu' + c_\lambda [1 - \mu' \ln(1 + (\mu')^{-1})]$$
$$\mu' = \cos \theta'$$

is taken from Pierce and Waddell (1961). The primed coordinates are measured with respect to the direction Ω_{00} as polar axis. The relevant angles may be obtained from spherical trigonometry as

$$\cos \theta_0 = \cos \theta' \cos \theta_{00} - \sin \theta' \sin \theta_{00} \cos \phi'$$

$$\sin \phi_0 = \sin \theta' \frac{\sin \phi'}{\sin \theta_0}.$$

The integrations were carried out by a Gaussian procedure using as many as 14 points in both θ and ϕ. The results are quite insensitive to the particular limb darkening law chosen.

One measure of the theoretical amplitude of the mutual shadowing effect which provides some insight into the differences between the present model and previous ones is the magnitude difference $\delta M(D)$ between the primary scattered reflection at $\alpha = 0$ (where the shadowing effect is maximum) and at $\alpha = 6°$ (where its effect is small):

$$\delta M(D) = -2.5 \log \frac{[R_1^s (\alpha = 6°)]}{[R_1^s (\alpha = 0°)]}.$$

(8)

We have written $\delta M(D)$, since the volume density D is the principal parameter determining the strength of the shadowing effect. Figure 3 presents results of computations for $\delta M(D)$ for the case $\tau = 1$, $\theta = \theta_0 = 64°$ (which approximates the Saturnocentric declination of the Sun and Earth during the observations of Franklin and Cook), and $\Phi = 1$. The notation 'point $\odot$' refers to results based on the cylinder-cylinder model described above; 'finite $\odot$' represents the refined cylindrical model which we shall employ and which is described by Equation (7); and 'Bobrov' refers to the cone-cylinder model of that author (see Bobrov, 1970).

For both the point $\odot$ and the finite $\odot$ models the shadowing effect (neglecting

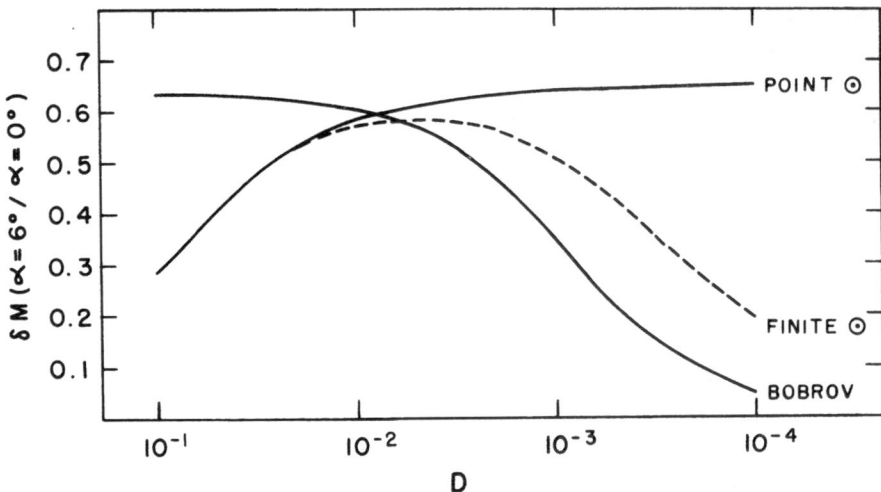

Fig. 3.   Magnitude of the shadowing effect $\delta M(D)$ for first order scattering as a function of fractional volume D occupied by ring particles. Geometry appropriate to Franklin and Cook (1965) data, and $\tau = 1$. See Equation (8).

multiple scattering) decreases for large volume densities because the surface of the rings appears to become smoother (uniformly filled). Because of certain simplifying assumptions in the mathematics, the Bobrov model does not exhibit this behavior. For very small values of D, the point ⊙ model approaches an asymptotic value for δM; this behavior is a result of the infinite extent of the cylindrical shadows, and the fact that for a given optical depth τ, as D decreases, the thickness of the layer t must increase (see Equation (4)). In contrast, both the finite ⊙ and Bobrov models produce shadows of finite length, so that for sufficiently small D (sufficiently large inter-particle distance) the shadowing effect vanishes. The Bobrov model produces a smaller shadowing effect for small D because of inaccuracies in its treatment of the penumbra.

A more detailed view of the relation between δM and D may be obtained by considering the entire phase curve $M_1(\alpha)$ as a function of D. Sample phase curves for various D computed from our shadowing theory are shown in Figure 4 for the same parameters used in Figure 3. We note the following points:

(a) The smaller the volume density D, the steeper the initial decrease in brightness with increasing phase angle (that is, the more peaked is the opposition effect).

(b) For $D \lesssim 0.01$, the total shadowing effect δM over the phase angle range 0–6° increases as D increases.

(c) For $D \gtrsim 0.01$, the opposition effect has become so broad that δM begins to decrease although the phase curve is still falling off at 6°. We should point out that the values of δM in Figure 4 differ from those in Figure 3 for the same D because in the former case they have been normalized at $\alpha = 0.094°$ for comparison with the data of Franklin and Cook, while in the latter case the normalization was at exact opposition ($\alpha = 0$).

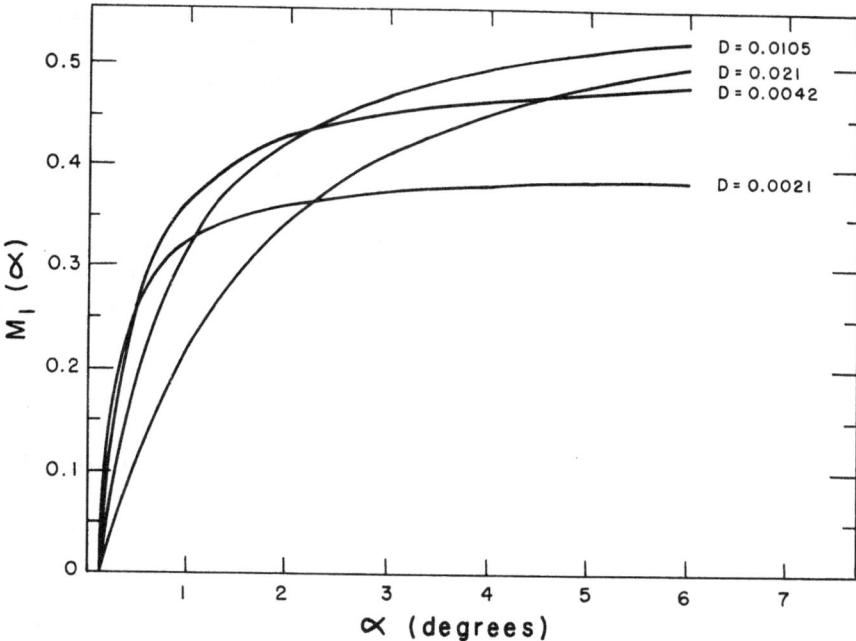

Fig. 4.   Theoretical phase curves for first order scattering with $\tau = 1$, $\theta = \theta_0 = 64°$. Curves labeled with the density parameter D.

In fact none of the curves presented in Figure 4 agrees with the data (cf. Figure 1). If the theoretical curves are sufficiently steep for small phase angles, they are too flat for $\alpha \gtrsim 2°$. We therefore must consider the additional effects of the particle phase function Φ and of higher order scattering (next section). Note, however, that since the sharp opposition peak is primarily the result of the shadowing mechanism and not these other effects, we may say from a comparison of Figures 1 and 4 that $D \lesssim 0.02$ if we are to produce a sufficiently sharp peak.

3.2. Multiple scattering

It seems to us reasonable to suppose that the wavelength dependence of the phase curve may be due largely to variation of the single-scattering albedo a of the particles. The spectra obtained by Lebofsky *et al.* (1970) show that the ring reflectivity varies significantly between the ultraviolet and the infrared. Increasing the value of a will significantly change the multiply scattered contribution to the reflected intensity, with a resulting dilution of the opposition effect and change in shape of the phase curve. Let us compute the magnitude of this effect.

Apart from the shadowing effect, the intensity reflected by Saturn's rings will satisfy the equation of radiative transfer

$$\mu \frac{dI}{d\zeta} = -I + \frac{a}{4\pi} \int_{4\pi} d\Omega' \Phi(\Omega, \Omega') I(\Omega'), \tag{9}$$

where a and Φ have been defined above. Since the optical thickness of the rings is not large, it is convenient to express the solution to Equation (9) as the sum of successive orders of scattering (van de Hulst, 1948; van de Hulst and Irvine, 1963; Irvine, 1964), so that

$$I(\Omega, \zeta) = \sum_{n=0}^{\infty} a^n I_n(\Omega, \zeta), \tag{10}$$

where I_n is the nth order intensity for $a=0$. Since we take the optical thickness of the rings as known in the computation of a particular model, Equation (10) provides an efficient means for determining the effect of a change in particle albedo. The desired ring reflectivity including the opposition effect is thus, from Equations (1), (6), and (10),

$$R(\Omega) = R_1^s(\Omega) + I(\Omega, 0) - a I_1(\Omega, 0) \qquad (\mu > 0). \tag{11}$$

The ring reflectivity will be a function of the parameters a, τ, Ω_0, and D, as well as direction Ω and the properties of the phase function Φ.

The successive terms in Equation (10) are found, for incident flux π through the horizontal upper boundary of the rings, from the relations

$$I_0(\Omega, \zeta) = \frac{\pi \delta(\mu - \mu_0)\, \delta(\phi - \phi_0)\, e^{-\zeta/\mu_0}}{\mu_0}$$

$$B_n(\Omega, \zeta) = \frac{1}{4\pi} \int_0^{2\pi} d\phi' \int_{-1}^{1} d\mu' \Phi(\Omega, \Omega')\, I_{n-1}(\Omega', \zeta)$$

$$I_n^-(\Omega, \zeta) = \int_0^{\zeta} d\zeta' e^{-(\zeta'-\zeta)/\mu}\, \frac{B_n(\Omega, \zeta')}{-\mu} \tag{12}$$

$$I_n^+(\Omega, \zeta) = \int_{\zeta}^{\tau} d\zeta' e^{-(\zeta'-\zeta)/\mu}\, \frac{B_n(\Omega, \zeta')}{\mu}$$

$$I_n(\theta = \pi/2, \phi, \zeta) = B_n(\theta = \pi/2, \phi, \zeta)$$

where B_n is the so-called source function for nth order scattering and the superscripts $+$ and $-$ refer to the cases $\theta < \pi/2$ and $\theta > \pi/2$, respectively. As has been emphasized in the above references, the ratio of successive terms I_n/I_{n-1} approaches a constant value as n increases, so that the series (10) may be truncated and the remainder replaced by a geometric series.

The double integrations over θ and ϕ in the above equations may be eliminated by expanding the phase function in a cosine series in $(\phi - \phi_0)$ (e.g., Hansen, 1969). Setting

$$\Phi(\mu, \phi; \mu, \phi_0) = \sum_{m=0}^{\infty} \Phi^m(\mu, \mu_0) \cos m(\phi - \phi_0), \tag{13}$$

we have

$$B_n(\mu, \phi, \zeta; \mu_0, \phi_0) = \sum_{m=0}^{\infty} B_n^m(\mu, \zeta; \mu_0) \cos m(\phi - \phi_0)$$

$$I_n^+(\mu, \phi, \zeta; \mu_0, \phi_0) = \sum_{m=0}^{\infty} I_n^{m+}(\mu, \zeta; \mu_0) \cos m(\phi - \phi_0) \qquad (14)$$

$$I_n^-(\mu, \phi, \zeta; \mu_0, \phi_0) = \sum_{m=0}^{\infty} I_n^{m-}(\mu, \zeta; \mu_0) \cos m(\phi - \phi_0),$$

where

$$B_1^m = \frac{1}{4\mu_0} e^{-\zeta/\mu_0} \Phi^m(\mu, \mu_0)$$

$$B_n^m = \frac{1}{(4 - 2\delta_{0,m})} \int_{-1}^{1} d\mu' \Phi^m(\mu, \mu') I_{n-1}^m(\zeta, \mu', \mu_0) \qquad (15)$$

$$I_n^{m-}(\mu, \zeta, \mu_0) = \int_{0}^{\zeta} \frac{d\zeta'}{-\mu} e^{-(\zeta'-\zeta)/\mu} B_n^m(\mu, \zeta', \mu_0)$$

$$I_n^{m+}(\mu, \zeta, \mu_0) \int_{\zeta_0}^{\tau} \frac{d\zeta'}{\mu} e^{-(\zeta'-\zeta)/\mu} B_n^m(\mu, \zeta', \mu_0)$$

Since the ring particle phase function is not known *a priori*, we shall choose a simple analytic expression which may be parameterized to conveniently describe a wide variety of phase functions. Such a function, which is also easily expressed in the form of Equation (13), is the Henyey-Greenstein function

$$\Phi_{HG}(\gamma, g) = \frac{(1 - g^2)}{(1 + g^2 - 2g \cos \gamma)^{3/2}}, \qquad (16)$$

where

$$g = \frac{1}{2} \int_{-1}^{1} d\mu \, \mu \Phi(\mu). \qquad (17)$$

We note that

$$\Phi_{HG}(\gamma, g) = \sum_{n} (2n + 1) g^n P_n(\cos \gamma), \qquad (18)$$

where

$$\cos \gamma = \mu \mu_0 + \sqrt{1 - \mu^2} \sqrt{1 - \mu_0^2} \cos(\phi - \phi_0)$$

and P_n and P_n^m are the Legendre and the associated Legendre polynomials, respectively. Using the addition theorem we thus obtain

$$\Phi_{HG}(\gamma, g) = 1 + \sum_{n=1}^{\infty} (2n + 1) g^n \left[P_n(\mu) P_n(\mu_0) + 2 \sum_{m=1}^{n} \frac{(n-m)!}{(n+m)!} \times \right.$$

$$\left. \times P_n^m(\mu) P_n^m(\mu_0) \cos m(\phi - \phi_0) \right] \qquad (19)$$

which may be written

$$\Phi_{HG}(\gamma, g) = \sum_{\kappa=0}^{\infty} \Phi_{HG}^{\kappa}(\mu, \mu_0, g) \cos \kappa (\phi - \phi_0), \tag{20}$$

where

$$\Phi_{HG}^0(\mu, \mu_0, g) = 1 + \sum_{n=1}^{\infty} (2n + 1) g^n P_n(\mu) P_n(\mu_0) \tag{21}$$

$$\Phi_{HG}^{\kappa}(\mu, \mu_0, g) = 2 \sum_{n=\kappa}^{\infty} (2n + 1) g^n \frac{(n - \kappa)!}{(n + \kappa)!} P_n^{\kappa}(\mu) P_n^{\kappa}(\mu_0), \qquad \kappa \neq 0.$$

In our models for this paper we shall use the phase function

$$\Phi(\gamma) = b\Phi_{HG}(\gamma, g_1) + (1 - b) \Phi_{HG}(\gamma, g_2) \tag{22}$$

which is normalized such that

$$\frac{1}{2} \int_0^{\pi} d\gamma \sin \gamma \Phi(\gamma) = 1. \tag{23}$$

Equation (22) allows us to investigate particle phase functions which are isotropic, principally forward directed, principally backward directed, or which contain both a forward and backward peak.

4. Comparison with the Observations

Using the results of the previous section we may write for the theoretically predicted phase curve $M(\alpha)$

$$M(\alpha) = -2.5 \log \left[\frac{\langle R_1^s(\alpha) \rangle + \sum_{n=2}^{\infty} R_n(\alpha)}{\langle R_1^s(0) \rangle + \sum_{n=2}^{\infty} R_n(0)} \right], \tag{24}$$

where we write $R_n = a^n I_n^-(\zeta = 0)$ and the angular brackets denote an integration of the incident radiation over the disk of the Sun. In order to illustrate more clearly the role of the parameters involved, we may rewrite (24) as

$$M(\alpha) = -2.5 \log \left[\frac{a\Phi(\pi - \alpha) \langle S(\alpha) \rangle + \sum_{n=2}^{\infty} R_n(\alpha)}{a\Phi(\pi) \langle S(0) \rangle + \sum_{n=2}^{\infty} R_n(0)} \right], \tag{25}$$

where

$$\langle S(\alpha) \rangle \equiv \left\langle \frac{1}{4\mu \mu_0} \int_0^{\tau} d\zeta' \exp\left[-\zeta'\left(\frac{1}{\mu} + \frac{1}{\mu_0}\right) + nC(\alpha) \right] \right\rangle \tag{26}$$

is the primary scattered intensity including the shadowing effect for the case of conservative, isotropic scattering.

Equation (25) may be further transformed to simplify the comparison with the observations. Because the higher order scattering component of the intensity does not change rapidly with angle and because the maximum phase angle observable for Saturn is 6°, we will have to a good approximation that

$$\sum_{n=2}^{\infty} R_n(0) \simeq \sum_{n=2}^{\infty} R_n(\alpha). \tag{27}$$

In fact, for the cases investigated below, Equation (27) holds to better than 1%. It is then convenient to rewrite Equation (25) as

$$M(\alpha) = -2.5 \log \left[\frac{\dfrac{\Phi(\pi-\alpha)}{\Phi(\pi)} \dfrac{\langle S(\alpha)\rangle}{\langle S(0)\rangle} + x}{1+x} \right], \tag{28}$$

where

$$x \equiv \frac{\sum\limits_{n=2}^{\infty} R_n(0)}{a\Phi(\pi)\langle S(0)\rangle} \tag{29}$$

is the fraction of the intensity observed at exact opposition which is due to multiple scattering, while the other ratio in the numerator of Equation (28) is the intensity ratio which appeared in the definition of $\delta M(D)$. We thus see that the shape of the phase curve will depend on the quantities x, the phase function Φ, and the optical thickness τ and volume density D through $\langle S\rangle$. The single scattering albedo a enters indirectly through x. It is these parameters D, τ, a, and the quantities characterizing the phase function which we wish to determine.

In addition to the phase curve, the absolute surface brightness of ring B at opposition is a critical measurement for defining the ring parameters. Noting that the incident flux on the rings will be $\pi F \mu_0$, where πF is the solar flux at the distance of Saturn through an area normal to the direction to the Sun, we may relate the observed absolute brightness at opposition to that intensity $R^\circ \equiv R(\alpha=0)$ calculated from Equation (11) by

$$R^0 = \frac{\pi I_R}{\pi F \mu_0} \frac{I_D}{I_D} = \frac{1}{\mu_0} \frac{I_R}{I_D} \frac{I_D/\pi F}{(1/\pi)}$$

$$R^0 = \frac{1}{\mu_0} p_S \frac{I_R}{I_D}, \tag{30}$$

where I_D is the mean specific intensity averaged over Saturn's disk, I_R is the mean specific intensity of the ring B, and p_S is the geometric albedo of Saturn's disk. Using the data of Cook et al. (1973) for I_R/I_D and the value $p_S = 0.429$ for V from Irvine and Lane (1973), we find a value in the visual of $R_V^0 \simeq 1.2$. Because the disk was partly shielded by the rings during the Franklin-Cook observations, this value of R_V^0 is based on an I_D which will be biased towards Saturn's equatorial regions. We may obtain an independent estimate of the brightness of ring B by multiplying the corresponding data of Price (1973), which apply to the total ring system, by a factor of 1.2, which is the correction determined from Franklin and Cook necessary to transform to ring B alone. The Price data give $R_V^0 = 1.1 \pm 0.1$ at a ring inclination of 26°, which is appropriate to the present discussion.

The wavelength dependence of the surface brightness is of critical importance in determining the ring parameters. The principal data relevant to this problem are the observations in B and V of Franklin and Cook (1965), the relative spectral reflectivity measurements of ring B by Lebofsky et al. (1970), the similar data from Irvine and Lane (1973) which were deduced from observations of the combined light of the Saturn system, and some recent spectral scans by Kharitonova and Teifel (1973). The data are in reasonable agreement for $\lambda \lesssim 6000$ Å if we bear in mind the color dependence of the opposition effect as reported by Franklin and Cook and Irvine and Lane. At longer wavelengths, however, there are some serious disagreements, which might reflect differences in inclination angle of the rings during the observations or possibly a temporal variation. We shall limit ourselves to the observations by Lebofsky et al. as the most direct and completely reported results at this time. Using their data to scale the visual reflectivity, we obtain in the B band $R_B^0 \simeq 0.83$ and a maximum value near 1 μ of $R_R^0 \simeq 1.3$. These data do not include a differential opposition effect. Since the differential opposition effect between the blue and the visual is approximately 5% (Franklin and Cook, 1965; Irvine and Lane, 1973), we shall take $R_B^0 = 0.87$. We shall for the present neglect any differential opposition effect between the visual and the red, in spite of the indication for such an effect from Irvine and Lane. We note that the peak reflectivity in the red is much higher in the results of Irvine and Lane and Kharitonova and Teifel than that given by Lebofsky et al.

We wish our theoretical model to match both the absolute brightness measurements and the shape of the phase curves in B and V. We may facilitate this comparison by considering the diagram in Figure 5. The vertical axis represents the primary scattered radiation, including the shadowing effect, computed at $\alpha = 0.094°$. This will be given theoretically by $a \, \Phi(\alpha = 0°\!.094) \langle S(\alpha = 0°\!.094) \rangle$, where we recall that this value of α is the minimum obtained during the observations of Franklin and Cook. The horizontal axis in Figure 5 represents the sum of the higher order scattering, which according to the model is $\sum_{n=2}^{\infty} R_n$. The dashed curves designated R, V and B are the *loci* of points which satisfy the observed absolute brightness in the red, visual, and blue, respectively. For agreement with the model the absolute brightness must be

$$R(\alpha = 0°\!.094) = a\Phi(\alpha = 0°\!.094) \langle S(\alpha = 0°\!.094) \rangle + \sum_{n=2}^{\infty} R_n(a, \Phi), \qquad (31)$$

where the indicated arguments draw attention to the dependence of the quantities on phase angle α and phase function Φ. We have plotted as an example the lower observational limits of the absolute surface brightness in V and R, and have shown an uncertainty of ± 0.05 for B as an example of the possible uncertainty in these measurements. In fact, these lower limits on the observed brightness allow the largest possible range of particle albedo a in the comparison with theory, and also lead to a lower limit on the volume density D.

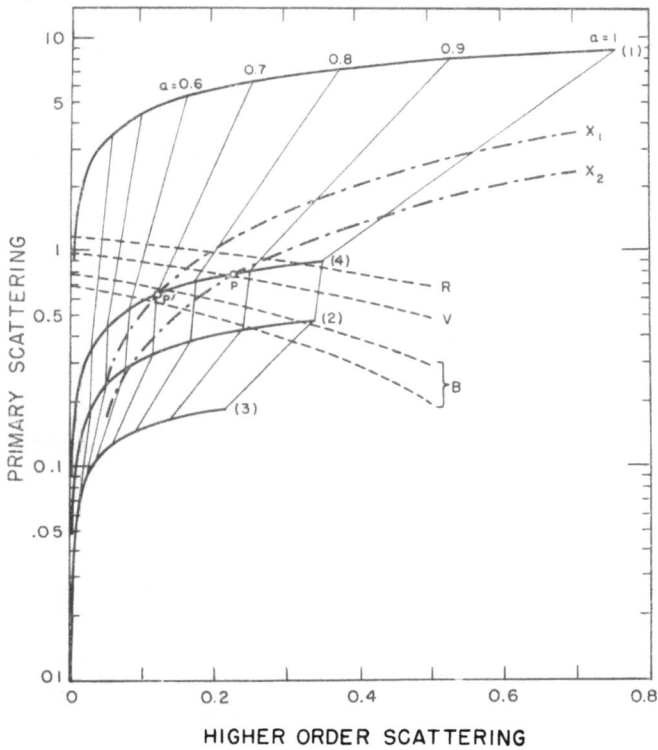

Fig. 5. Diagram for comparison of observation and theory. Vertical axis gives primary scattering contribution to total brightness, horizontal axis gives multiple scattering contribution. See text.

TABLE I

Multiple scattering contribution x for $\tau = 1$

$D = 0.012$

B	0.14 ± 0.02
V	0.26 ± 0.02

$D = 0.010$

B	0.17 ± 0.02
V	0.29 ± 0.02

$D = 0.008$

B	0.22 ± 0.22
V	0.34 ± 0.02

For given D and τ, the overall shape of the phase curve $M(\alpha)$ depends principally upon the fraction of multiple scattering x. By experimenting with a wide choice of values for these parameters and also for the phase function Φ and the single scattering albedo a, we find that the sharp peak in the opposition effect depends primarily upon the value of D, and that the observations restrict D to a narrow range around the value

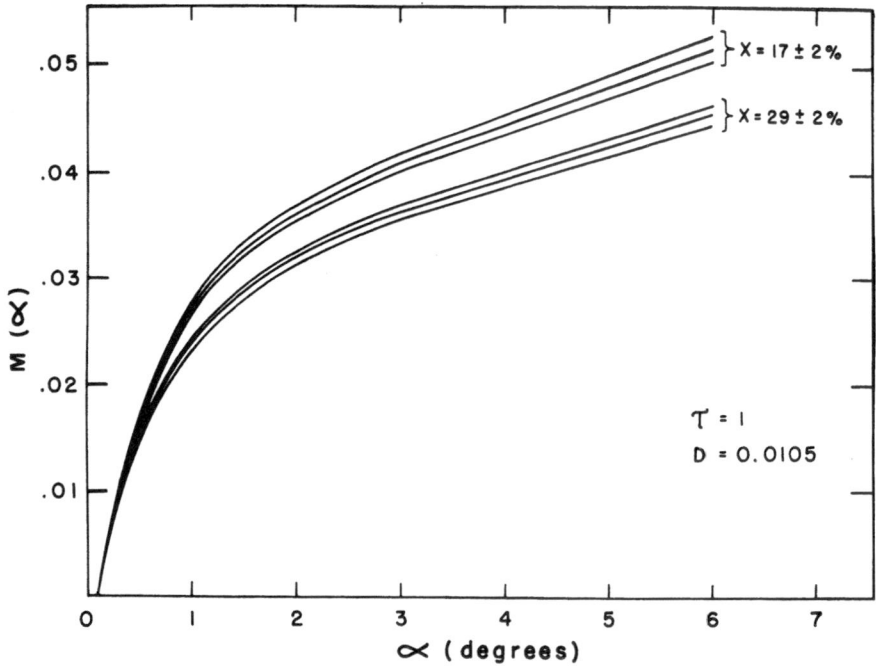

Fig. 6. Theoretical phase curves which match the observations of Franklin and Cook (1965). Fraction of multiple scattering is x, other parameters are shown. Note that $\theta = \pi - \alpha$.

0.01. Let us for the present take $\tau = 1$ on the basis of the observations of stellar occultations by the rings discussed by Cook *et al.* (1973). We then list in Table I values x for several choices of D which produce theoretical phase curves according to Equation (11) which agree with the observations in B and V. For example, for $D = 0.010$, the appropriate values of x are 0.29 and 0.17 in V and B, respectively. The dash-dot curves plotted in Figure 5 are now the *loci* of points for which the fraction of multiple scattering is $x_1 = 0.17$ and $x_2 = 0.29$, respectively. The corresponding phase curves are shown in Figure 6.

We may now use Figure 5 to determine the single scattering albedo a and properties of the phase function Φ for the ring particles if we assume that the differences in brightness and phase curve for B and V result only from a change in particle albedo a. In other words, we assume that as the albedo of the particles changes with wavelength, the relative angular distribution of scattered light remains unchanged. This is a reasonable approximation for large, bright particles (which are necessary to produce the shadowing effect and the observed high ring brightness) for which geometrical optics is valid. Our whole approach to the shadowing effect through geometrical optics also requires that τ be independent of wavelength. If we now call the intersection between the curves V and x_2 in Figure 5 a point P, and the intersection between the middle of the range B and the curve x_1 a point P', then the theoretically computed brightness curve for the rings which passes through both the points P and P' will match the observed absolute brightness and also the observed phase curves. We have plotted in Figure 5 such theoretical brightness curves for four different phase functions obtained from Equation (22). We have taken the center of the Sun and Earth in directions corresponding to the Franklin and Cook observations. The solid lines represent such brightness curves for the phase functions (1) $b=0$, $g_1=0$, $g_2=-0.7$ (a very strong backward peak with no forward scattering); (2) $b=1$, $g_1=0$, $g_2=0$ (iso-

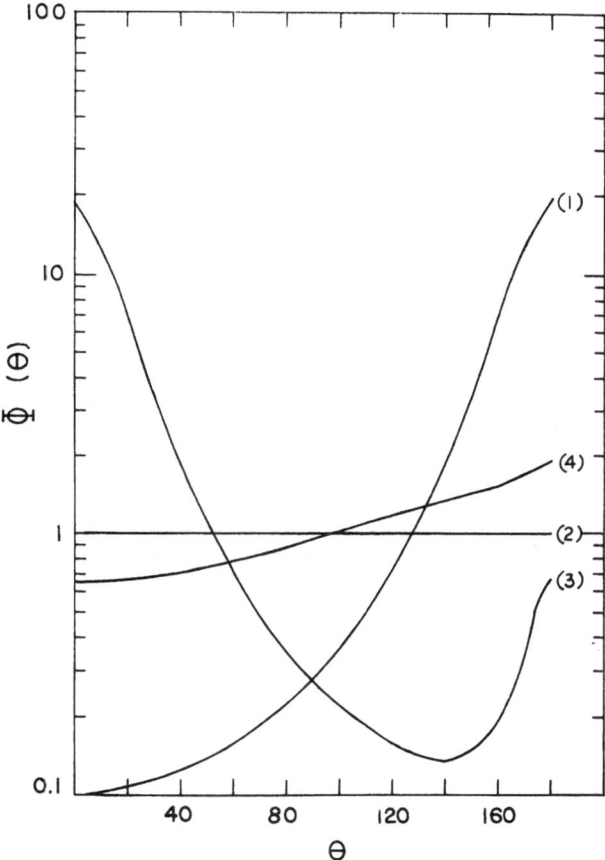

Fig. 7. Four sample phase functions Φ (see Equation (22)).
Parameters described in the text.

tropic scattering); (3) $b=0.988$, $g_1=0.7$, $g_2=-0.805$ (a very strong forward peak with a small backward peak, reminiscent of the phase function for terrestrial clouds); (4) $b=0.995$, $g_1=-0.14$, $g_2=-0.84$ (a more slowly varying backward scattering phase function with a slight peak near 180°). These phase functions are illustrated in Figure 7. All of them except (2), isotropic scattering, have a similar slope near 180° which produces a roughly satisfactory shape to the phase curve in the linearly varying region. The albedo a is the only variable unspecified in the theoretical brightness curves in Figure 5, and it thus serves as a parameter whose variation along the curves is indicated.

The power of this procedure is illustrated by the large differences between the curves (1–4) in Figure 5. The requirement that the model match both the shape of the phase curve and the absolute brightness clearly puts significant restrictions on the form of the phase function. In particular, it is quite evident that neither the phase function with a very strong backward peak nor that with a very strong forward peak can match the observations. Some degree of backscatter is required to match the phase curve, so that the phase curve must be similar in shape to the curve (4). Although the shape of the phase curve (apart from the opposition peak) depends principally on the values of Φ near 180° (corresponding to the small phase angles observable for Saturn), an appropriate phase function cannot be very different from the curve (4). If it decreased much more sharply with decreasing $\theta = \pi - \alpha$, it will not satisfy the normalization condition. The addition of a shallow forward peak to the phase function would be possible and would require a lower backward peak; that is, the phase function would become more isotropic.

We may now determine the single scattering albedo from the position of the points P and P' on the curve (4) in Figure 5. We find $a_V = 0.87$ and $a_B = 0.70$. By normalizing the phase function (4) to unity at $\alpha = 180°$ and integrating, we may obtain the phase integral q for the ring particles as $q = 2.1$. The resulting geometric albedos in the visual and blue for the ring particles are then $p_V = a_V/q = 0.41$ and $p_B = 0.33$. The shape of the particle phase function is compared to those for the Moon and for a Lambert surface in Figure 8. The shape is quite similar to the Moon near $\alpha = 0$, but the ring particle brightness falls off less rapidly with increasing α than does that of the

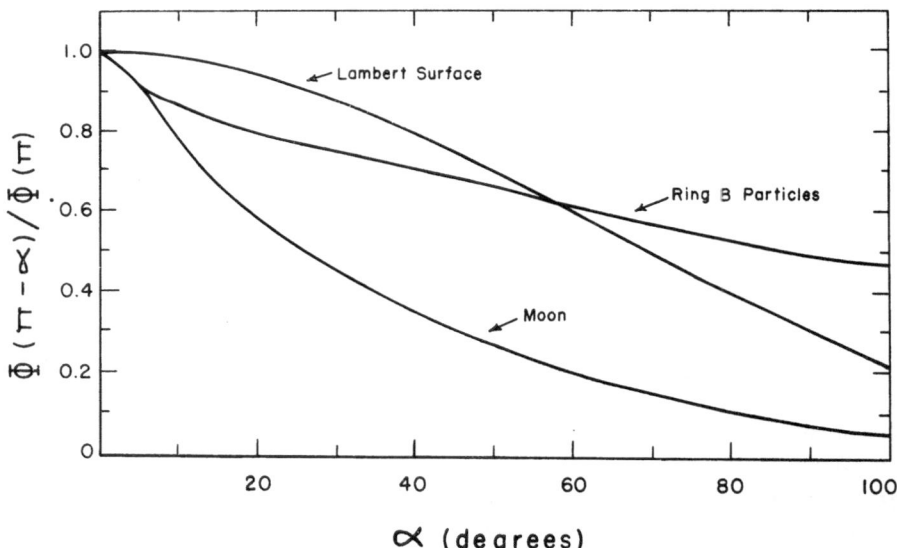

Fig. 8.   Ring particle phase curve $\Phi(\pi - \alpha)/\Phi(\pi)$ compared with lunar phase curve from Rougier (1934) and with phase curve for Lambert sphere.

Moon. This is in agreement with the results of Veverka (1973) for snow covered objects.

5. Allowable Range of Parameters

As a result, we may say that a satisfactory model of the B ring which matches the observed phase curves in B and V and also the corresponding absolute brightnesses has optical thickness $\tau = 1$, a volume density $D = 0.010$, and ring particles with a phase function given by (4) in Figure 7 and Bond albedos of $a_V = 0.87$ and $a_B = 0.70$.

It is of course important to see how much each of these parameters can be varied without disrupting the fit of the model to the data. Because the parameters cannot be varied independently if the model is going to continue to match the observations, the problem is difficult. Some possible directions in which changes may occur are sketched below.

If the optical thickness τ of the layer is kept constant at a value of unity, the value of D can be reduced only slightly from that discussed above. When $D = 0.008$, computations (whose results are shown in Table I) indicate that the multiple scattering contributions x needed to match the phase curves are $x_1 = 0.22$ (B) and $x_2 = 0.34$ (V). Extrapolation of the curve connecting the new points P' and P in this case shows that the curve will cross the dashed curve R when the particle albedo in the red $a_R \simeq 1.0$. In other words, lowering of D must be compensated by an increase in multiple scattering which requires an increase in the particle albedo in V and B, and a resulting increase in the particle albedo for R also. But $a_R \leqslant 1$ on physical grounds, so that the brightness of the rings in the red could not be matched by the model if $D \leqslant 0.008$. It might be expected from Figure 3 that reduction of D to a value less than 0.005 would begin to reduce the shadowing effect. This is in fact true, but the resulting shape of the opposition effect is too steep to agree with the observations.

As has been pointed out in a previous section, if the value of D is too large, the opposition peak will be too broad. For the present choice of the other parameters the upper limit on D is approximately 0.012. With this value we obtain multiple scattering contributions x of 0.26 in V and 0.14 in B. The corresponding values of albedo parameterizing the theoretical brightness curve are $a_V = 0.82$ and $a_B = 0.65$. The phase function in this case will be slightly more backward scattering than the phase function (4).

We may summarize these results by saying that, for $\tau = 1$ and the minimum surface brightness allowed by the observations

$$0.008 < D < 0.012$$
$$0.65 < a_B < 0.75$$
$$0.82 < a_V < 0.9$$
$$0.93 < a_R < 1.0.$$

We have also examined the case $\tau = 0.7$ as an estimate of the effects of this possible lower limit for ring B. We find that a somewhat smaller value of $D \simeq 0.006$ is required to match the shape of the opposition peak. The magnitude of the shadowing effect due to single scattering decreases with decreasing τ, but this may be compensated by decreasing the dilution of the shadowing effect due to multiple scattering. In the present case we find it necessary to take approximately $x(V)=0.21$ and $x(B)=0.10$. Constructing the figure analogous to Figure 5 for the present case we find that a somewhat more backscattering phase function is required, although in general its shape will be similar to that for the case $\tau=1$. The corresponding albedos are found to be $a_V=0.85$ and $a_B=63$, which are not drastically different from their values for $\tau=1$. The value of τ cannot be much less than 0.7, since lower values do not produce a sufficiently large opposition effect in the ultraviolet (cf. Irvine and Lane, 1973).

To investigate the effects of choosing a larger optical thickness τ_0, we have also carried out computations for $\tau=2$. In this case we find $D \approx 0.013$ and multiple scattering contributions of $x(V)=0.40$ and $x(B)=0.28$. These larger fractions of multiple scattering are necessary to dilute the larger shadowing effect produced by primary scattering as τ increases. The albedos are slightly increased relative to $\tau=1$, the new values being $a_V \approx 0.90$ and $a_B \approx 0.78$, while the corresponding phase function has less of a backward peak than for $\tau=1$.

In addition to the lower limit for the absolute surface brightness R_V^0, we must investigate the effect on the model of choosing the apparent mean observational value $R_V^0 \simeq 1.1$. The range of D and x which match the phase curves remains unchanged, since the phase curves measure only relative brightness. The particle albedos are increased, but the requirement that $a_R \leqslant 1$ in the red provides also an upper limit on a_V and a lower limit on D (through interaction with x). As a result, the allowable range of particle albedo is reduced relative to that for a lower absolute surface brightness. We find $0.85 \leqslant a_V \leqslant 0.90$ and $0.66 \leqslant a_B \leqslant 0.72$ for $\tau=1$. The corresponding phase function has approximately a 10% larger backward peak than in the previous case, so that the phase integral becomes $g \simeq 1.95$ instead of 2.1.

5.1. DISTRIBUTION OF PARTICLE SIZES

Up to this point we have assumed that the ring particles can be characterized by a single effective radius ϱ. Within the framework of this assumption we can determine that size if we know the geometric thickness of the rings t. From the definitions of the optical depth $\tau=\pi \varrho^2 n t$ and $D=(4/3)\,\pi \varrho^3 n$ we find that

$$\varrho = \frac{3}{4}\frac{D}{\tau}\,t. \tag{32}$$

The ring thickness has been measured by Kiladze (1969) and Focas and Dollfus (1969) to be approximately 2 km. Taking $\tau \simeq 1$, $D \approx 0.01$ and $t \simeq 2$ km, we find that $\varrho \simeq 15$ m. This size is consistent with the recent radar results obtained by Goldstein and Morris (1973).

The above result is quite deceptive, however. To see this we must investigate the possible influence on our results of allowing for a distribution in particle sizes. Let us assume that the number of particles with radii between ϱ and $\varrho+d\varrho$ is given by

$$dn = f(\varrho)\,d\varrho$$

per unit volume. If we assume that the volume density D remains small enough so that the small element of area ε on a test particle in Figure 2 shielded from the Sun and Earth by a particle in the range $\varrho \to \varrho + d\varrho$ is independent of the probability that it is shielded by a particle of any other radius, we may write for the first order reflected intensity including the shadowing effect

$$
R_1^s = \left(\frac{a\Phi}{4\mu\,\mu_0}\right)\left(\pi \int_{\varrho_1}^{\varrho_2} f(\varrho)\,\varrho^2\,d\varrho\right) \times \int_0^t \exp\left[- h'\left(\frac{1}{\mu} + \frac{1}{\mu_0}\right) \times \right.
$$

$$
\left. \times \left(\pi \int_{\varrho_1}^{\varrho_2} f(\varrho)\,\varrho^2\,d\varrho\right) + \int_{\varrho_1}^{\varrho_2} f(\varrho)\,C(\varrho, h')\,d\varrho\right]dh', \qquad (33)
$$

where the upper and lower limits on particle size have been labeled ϱ_2 and ϱ_1. The rest of the theory remains the same.

The choice of possible forms for the particle distribution function is of course infinite. Bobrov has investigated the relation

$$
f(\varrho) = K\varrho^{-s}, \qquad (34)
$$

where K is a constant and s is a parameter describing the shape of the distribution. This distribution law is common in meteor astronomy, and of course can lead to a predominant number of quite small particles.

We have investigated the effect of a uniform particle distribution ($s=0$) and also the cases $s=2$ and 3. In all cases we find that the results are quite insensitive to the lower limit ϱ_1 of the particle size distribution provided that $\varrho_1 \lesssim 1$ cm, but do remain quite dependent upon the value of D. This shows that the shadowing effect in the case when a dispersion of particle sizes is present continues to determine the volume density D quite precisely, but that the mean particle size remains uncertain provided that it is large enough to produce the required geometrical optics shadowing.

On the other hand, the size distribution cannot be too sharply varying. When the parameter $s=3$ in Equation (34), it is not possible to obtain agreement with the observations. For a value $s=2$, the range of permissible values for the volume density D is somewhat increased ($0.005 \lesssim D \lesssim 0.013$).

For $s=0$ the results are quite similar to the monodisperse case. If we fix the optical depth $\tau=1$ and the geometric thickness $t=2$ km, the value of D will be a function of the upper limit ϱ_2 of the size distribution. To obtain the necessary value of $D \approx 0.012$, we must have $\varrho_2 \lesssim 25$ m. The mean particle size is then $\langle\varrho\rangle \approx 12$ m.

The corresponding upper limit of the particle size distribution when $s=2$ is 15 m $\leqslant$ $\leqslant \varrho_2 \leqslant 50$ m. The range of permissible values of the multiple scattering contribution x is also increased, so that $0.17 \leqslant x_V \leqslant 0.30$ and $0.05 \leqslant x_B \leqslant 0.18$. Using the same method as in the monodisperse case (cf. Figure 5), we find the allowable range for the particle albedo to be $0.72 \leqslant a_V \leqslant 0.82$ and $0.42 \leqslant a_B \leqslant 0.6$ for $R_V^0 = 1.0$. The upper limit on the mean particle size is estimated to be less than 2 m.

5.2. Tilt effect

We may test the validity of the type of model chosen here by comparison with other types of observation. Particularly important are the data on the surface brightness as a function of solar illumination angle θ_0 (tilt angle). Observations over a limited range of θ_0 have been made by Camichel (1958) and Focas and Dollfus (1969). More accurate and homogeneous measurements have been published by Price (1973), although this data includes both rings together. At $\theta = \theta_0 = 64°$, the mean surface brightness of ring B is 20% greater than the mean of A and B (Franklin and Cook, 1965; Camichel, 1958). As a first approximation towards removing the effect of ring A, we may assume that this ratio applies also at other θ_0. Price's data so corrected are plotted in Figure 9, together with Camichel's results scaled to agree with Price at $\theta_0 = 26°$. The corresponding theoretical curve was computed for $\tau = 1$, $a_V = 0.9$, and the phase function (22)

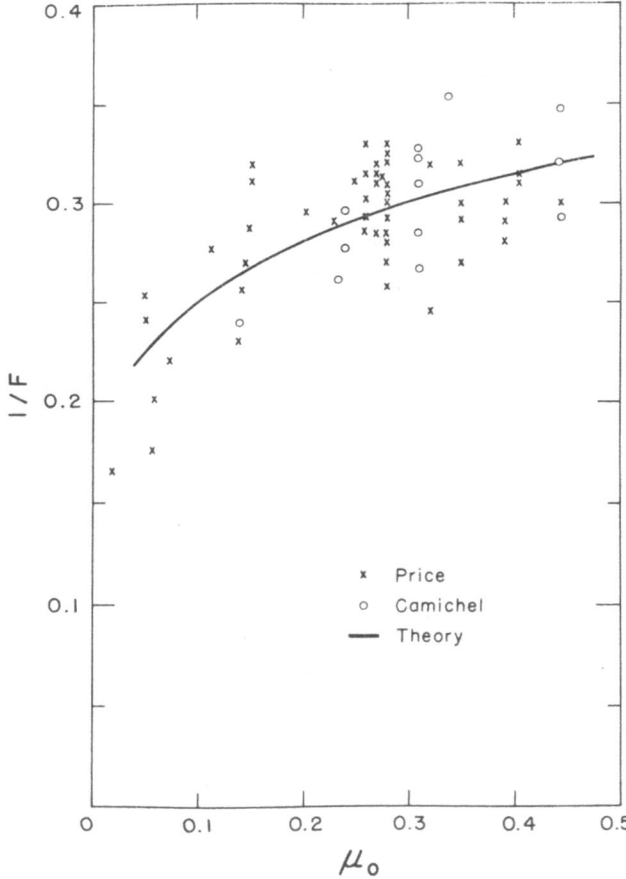

Fig. 9. Surface brightness $\mu_0\, R(\alpha = 6°)$ for ring B as a function of solar illumination angle arccos μ_0 for the V band. Theoretical curve computed for $\tau = 1$, $a_V = 0.9$, and the phase function noted in the text.

with $b = 0.995$, $g_1 = -0.16$, and $g_2 = -0.85$. This form of Φ has $q = 1.95$ and fits the observed phase curves and their mean absolute surface brightness $R_V^0 = 1.1$.

The agreement between theory and observation is quite good. This result clearly illustrates the importance of multiple scattering in the rings, since the surface brightness for primary scattering alone would decrease with increasing θ_0, in opposition to the observations and to the multiple scattering model.

6. Conclusion

The models of the rings described here match the observed phase curves, absolute surface brightness, and tilt effect for ring B. They prescribe a value of the volume density D close to 0.01 and a phase function which is somewhat backscattering (of the general form shown by curve 4 in Figure 7). The maximum range for the particle Bond albedos in the visual and blue are

$$0.72 \leqslant a_V \leqslant 0.90$$
$$0.42 \leqslant a_B \leqslant 0.75.$$

It is not possible to obtain a reliable estimate of the mean particle size without a knowledge of the particle size distribution, but it appears that $\langle \varrho^2 \rangle^{1/2} \lesssim 15$ m.

Acknowledgements

We are grateful to Drs J. Cuzzi, D. Van Blerkom, F. Franklin, and R. Murphy for several stimulating discussions. This research was supported in part by NASA Grant NGL 22-010-023.

References

Bobrov, M. S.: 1970, *Solar System Res.* **4**, 127 (English transl.).
Camichel, H.: 1958, *Ann. Astrophys. Paris* **21**, 231.
Cook, A. F., Franklin, F. A., and Palluconi, F. D.: 1973, *Icarus* **18**, 317.
Focas, J. and Dollfus, A.: 1969, *Compt. Rend. Acad. Sci. Paris* **268**, 100.
Franklin, F. A. and Cook, A. F.: 1965, *Astron. J.* **70**, 704.
Goldstein, R. M. and Morris, G. A.: 1973, *Icarus* **20**, 260.
Hansen, J. E.: 1969, *Astrophys. J.* **155**, 565.
Irvine, W. M.: 1964, *Bull. Astron. Inst. Neth.* **17**, 266.
Irvine, W. M., 1966, *J. Geophys. Res.* **71**, 2931.
Irvine, W. M. and Lane, A. P.: 1973, *Icarus* **18**, 171.
Kharitonova, G. A. and Teifel, V. G.: 1973, *Astron. Tsirk.*, No. 747.
Kiladze, R. I.: 1969, *Abastumani Astrophys. Obs. Bull.*, No. 37, 151.
Lebofsky, L. A., Johnson, T. V., and McCord, T. B.: 1970, *Icarus* **13**, 226.
Pierce, A. K. and Waddell, J. H.: 1961, *Mem. Roy. Astron. Soc.* **68**, 89.
Price, M. J.: 1973, *Astron. J.* **78**, 113.
Rougier, G.: 1934, *Astronomie* **48**, 220; **48**, 281.
Seeliger, H.: 1887, *Abhandl. Bayer. Akad. Wiss., Math.-Naturn., Kl. II* **16**, 405–516.
Seeliger, H.: 1895, *Abhandl. Bayer. Akad. Wiss., Math.-Naturn., Kl. II* **18**, 1.
Van de Hulst, H. C.: 1948, *Astrophys. J.* **107**, 220.
Van de Hulst, H. C.: 1957, *Light Scattering by Small Particles*, John Wiley & Sons, Inc., New York, p. 106.

Van de Hulst, H. C. and Irvine, W. M.: 1963, *La physique des planètes*, 11th Liège Astrophys. Symp., p. 78.
Veverka, J.: 1973, *Icarus* **20**, 304.

DISCUSSION

Smith: Does your model take into account the recent radar results from Goldstein, which appear to require very large (preferably meters if not even tens of meters) particles?

Irvine: It is reassuring that our model results in particle sizes which are consistent with the radar results. In my mind the most puzzling question at the moment is the lack of a definitive passive radio detection of the rings.

Feigelson: I doubt that it is possible to apply radiative transfer equations to Saturn ring particles, which are relatively large, if the optical density is low.

Irvine: The optical thickness of the rings is of the order of unity. If the geometric thickness is a couple of kilometers, and the particle sizes are in the meter range, I think a multiple scattering procedure is reasonable. Of course if the particles have radii in the kilometer range, or the rings are a monolayer, the situation is different.

Bobrov: I have some comments. (1) Several important parameters of Saturn's rings are poorly known; for instance, optical thickness of B-ring, phase curve of A-ring, etc. The C-ring is practically unexplored. This means that, for progress in research, further observations are needed.

(2) The mutual shadowing effect theory supposes the rings to be a system many particles thick, which faces very serious dynamical difficulties. Possibly the extremely sharp peak of the rings' phase curve results from the scattering by single particles. To clear up this question we need laboratory photometry down to 1° phase.

(3) One must also not forget that up to now we know only a small portion of the rings' phase curve 0–6°, and therefore the observation of the phase variation of the rings from space vehicles may add sufficient information to our knowledge.

A STUDY OF THE OUTERMOST RING OF SATURN

M. S. BOBROV

Astronomical Council of the U.S.S.R. Academy of Sciences, Moscow, U.S.S.R.

Abstract. The attention is called to the fact that the discovery by Feibelman (1967) of the rarefied outer ring of Saturn is confirmed by the observations of Kuiper (1972). It is proposed to designate this object as E-ring (exterior) in order to avoid confusion with the innermost, also rarefied, D-ring observed by Guérin (1970) and earlier by Barabashov and Semejkin (1933). The effects of the interaction of E-ring with inner Saturn's satellites are briefly discussed. The conclusion is drawn that in cosmogonic time scale these effects are small. It is also shown that the optical thickness of E-ring is lower than 1/20000; the available photometric estimations of the geometric thickness of A- and B-rings need not be corrected for the light scattering and absorption by E-ring.

1. The Existence of the Ring

Now it appears unquestionable that outside the well-known ring system of Saturn there is one more ring, broader than others but extremely rarefied. It is very faint and may be detected only near the edgewise position of the ring system.

I should like to remind that the first photographs showing this object were taken by Feibelman (1967). Figure 1 displays his drawing from the negative taken on November 14, 1966, and the microdensitometer trace along the path crossing the new ring.

For several years Feibelman's result was not confirmed by any other observer. Recently Kuiper (1972) reported that he has also found the outermost ring of Saturn

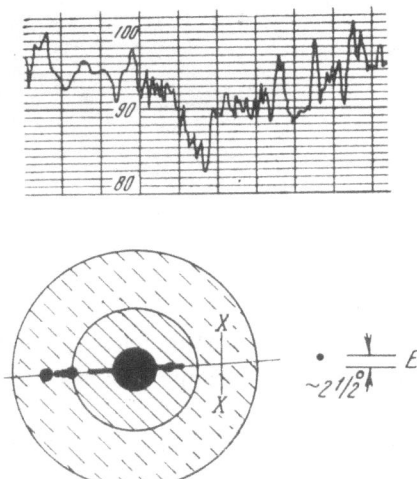

Fig. 1.   Below: Aspect of Saturn, 14 November, 1966. Top: Microdensitometer trace along path $X-X$. The central dip corresponds to the location of the visible thin line of the outermost ring (Feibelman, 1967).

Woszczyk and Iwaniszewska (eds.), Exploration of the Planetary System, 465–470. All Rights Reserved
Copyright © 1974 by the IAU

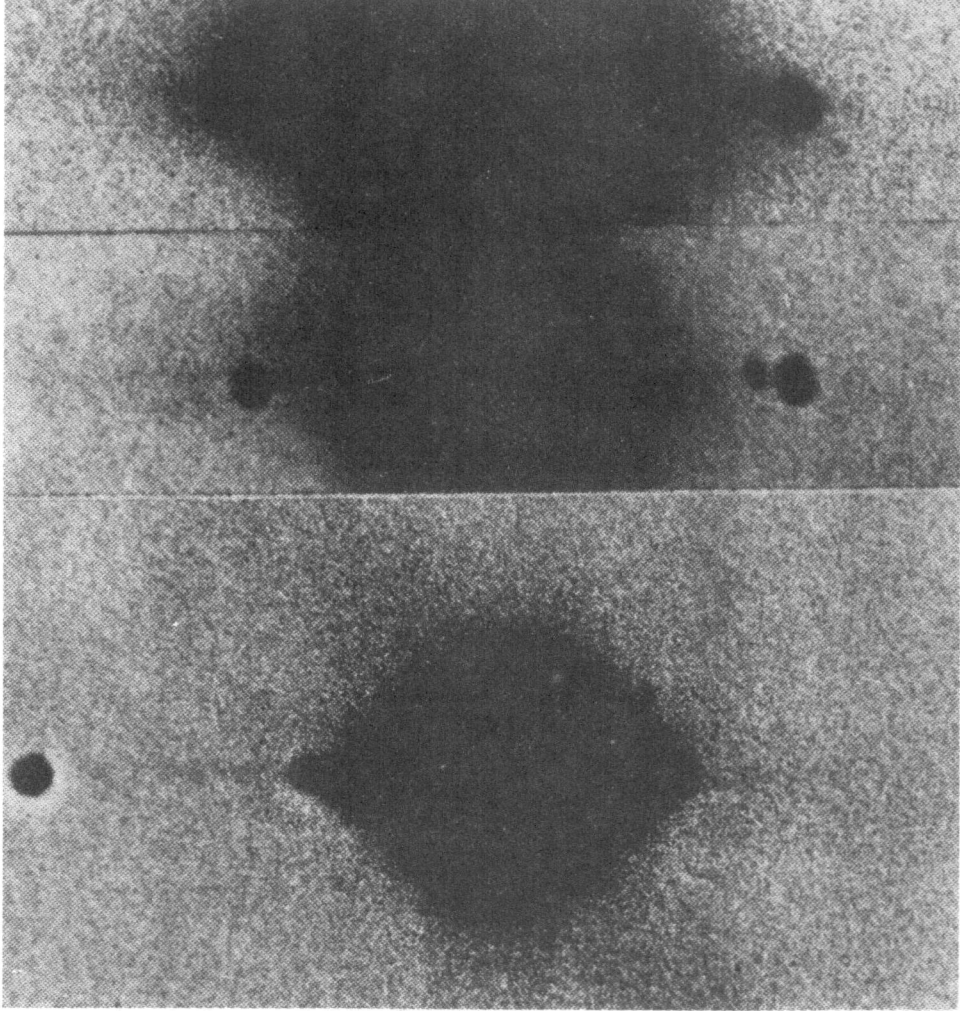

Fig. 2. Some negatives of Saturn system taken at Mt. Catalina Observatory in December 1966 –
January 1967 and showing faint line of the outermost ring (Kuiper, 1972).

when examining his negatives taken with 61-in. telescope of Mt Catalina Observatory
nearly simultaneously with Feibelman. Figure 2 shows one of the negatives taken by
Kuiper.

So the existence of this ring is now confirmed.

2. Designation of the Ring

In old literature and in Feibelman's paper it is called D-ring. Some years later Guérin
(1970) reported on the discovery of another ring of Saturn, inside of the known ring

system, and called it also 'ring D'. Since that time the majority of authors designate Feibelman's ring as 'D'-ring'.

This is not convenient and may lead to misunderstanding. Possibly it would be more reasonable to keep the designation 'D' for the innermost ring, as the ring immediately following A-, B- and C-rings, and for the outermost ring to introduce the new designation 'E-ring' (exterior).

Moreover, D-ring cannot be considered as a new object. About forty years ago Barabashov and Semejkin (1933) revealed the presence of fine rarefied matter in space between C-ring and Saturn's ball. Figure 3 displays the photometric cross-section

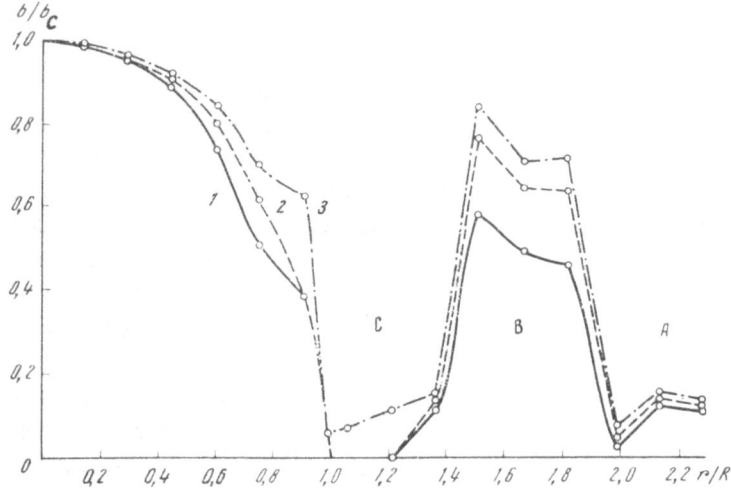

Fig. 3. Photometric cross-section through the disk and rings of Saturn along the major axis of the rings; r – the distance from the center of the system, R – radius of Saturn's disk, b – brightness in a given point of cross-section, b_c – brightness in the center of the disk. Curves 1, 2, 3 correspond to the data obtained with red, yellow, and blue filters, respectively. Blue curve shows the presence of rarefied matter between C-ring and limb of the planet. All data are corrected for seeing and diffraction effects using the method of artificial planet (Barabashev and Semejkin, 1933).

through the rings and disc of Saturn according to observations of these authors. It may be seen that in blue light the intensity inside C-ring is not zero. This fact was formulated by the quoted authors as "the continuation of the ring system down to Saturn's atmosphere."

3. E-Ring and Saturn's Satellites

E-ring is so wide that the orbits of some inner satellites of Saturn lie inside it. According to Feibelman, E-ring spreads out to the orbit of Enceladus, according to Kuiper – even to Dione. An interesting problem is arising from the effects due to interaction of the ring matter with the satellites.

Evidently the interaction must lead to secular diminution of the excentricities and inclinations of the satellites' orbits. But actually Mimas and Tethys have inclinations $1°5$ and $1°1$, respectively; Mimas has also noticeable eccentricity (0.02). These facts indicate that even in cosmogonical time scale the interaction effects are small.

4. E-Ring and Photometric Estimations of the Ring System Thickness

Due to observations of Focas and Dollfus (1969) and Kiladze (1969) we have now for the first time photometric estimations of physical thickness Z_0 of Saturn's rings (1–3 km). The estimations are based on measurements of light flux Φ per unit length of the image of nearly edgewise rings.

In Figure 4 the observed values of Φ are plotted as a function of Earth's elevation angle B above the ring plane. One may see practically linear course for both branches of photometric curve (for bright and dark sides of the rings, respectively). This is consistent with the simple ring model – a plane-parallel homogeneous sheet of diffusely reflecting particles (for details see Bobrov, 1972).

But using the plane-parallel model we neglect the fact that the bright rings are surrounded by ring E. Let us show that such neglection does not introduce significant errors into photometric estimations of physical thickness of the bright rings.

The observations of E-ring in 1966 were carried out during the period of very small B when the rings' structure along the normal to their major axis could not be resolved by telescope, i.e. the rings were the linear light source. For such source a good photometric characteristic is the above-mentioned magnitude Φ – the light flux per unit length (in the case under consideration Φ is a function of the distance of a given photometric cross-section from the center of Saturn's disk). All the observations show that for sections crossing only E-ring the values of Φ are much lower than for sections crossing both E- and A-ring:

$$\Phi_E \ll \Phi_{AE}. \tag{1}$$

This result follows from the examination of Figure 1 and 2 ,as well as from the fact that Focas and Dollfus (1969) and Kiladze (1969) did not obtain the image of E-ring even at extremely small values of B. In particular, on December 17, 1966, only a few hours before the ring-plane crossing by the Earth, no trace of E-ring was detected. The ratio of light fluxes from the dark side of B-, A- and E-rings and from the edge of A-ring was

$$\Phi_{dark\ BAE}/\Phi_{edge\ A} = 13/87 \tag{2}$$

(Bobrov, 1970). Hence taking into account that Equation (1) is also valid for the periods of dark side visibility, and that

$$\Phi_{dark\ AE} \approx \Phi_{dark\ BAE} \tag{3}$$

we may write

$$\Phi_{dark\ E} \ll \Phi_{edge\ A}. \tag{4}$$

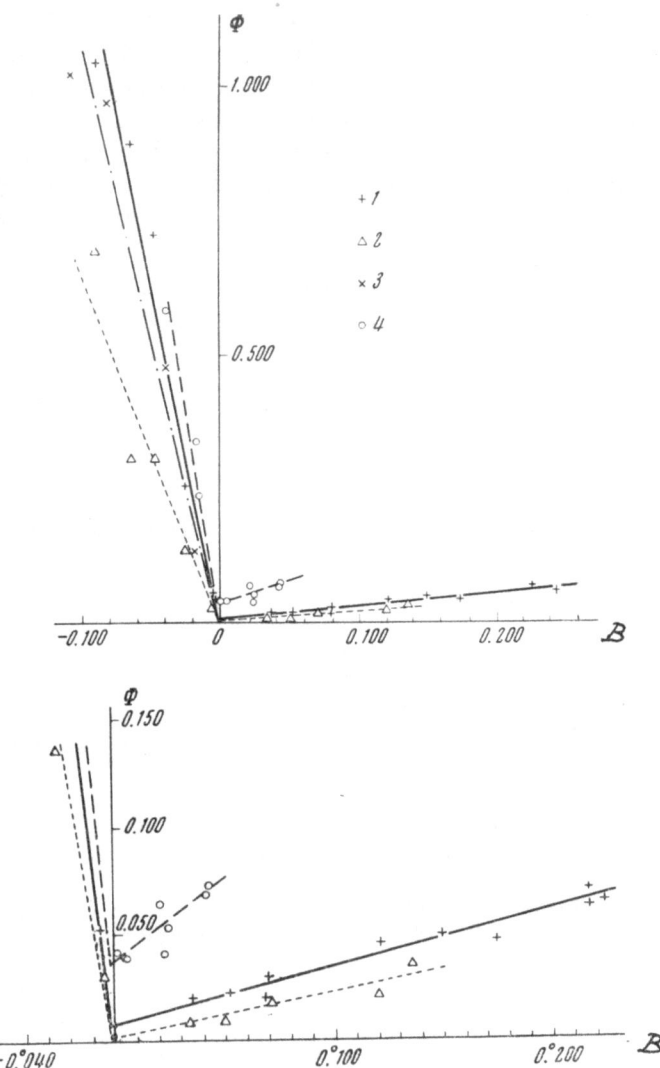

Fig. 4. Light flux Φ, per unit length of the rings, plotted as a function of the elevation angle B of the Earth above the ring plane. Data of observations near the moments $B = 0$. 1 and 2 – October–November, 1966, blue and red light, respectively (Kiladze, 1969); 3 – December 1966, blue light (*ibid.*), 4 – December, 1966, yellow light (Focas and Dollfus, 1969). Within the accuracy of measurements the course of $\Phi(B)$ is linear.

Since the Z_0 estimation was, in the end, based on the measured values of $\Phi_{\text{edge A}}$, Equation (4) means that the influence of the light flux from E-ring on the ring thickness estimation was small.

E-ring may also cause another effect: absorption of the light of the bright rings. In assumption that B-, A- and E-rings have equal or comparable physical thickness, there will be absorption of the light falling on the edge of A-ring and being reflected by this edge to the Earth. If the thickness of E-ring is much greater than that of the bright rings there is also absorption of the light from the northern or southern side of the bright rings. However, the observations made in 1966 did not reveal such effects. Indeed, the linear course of $\Phi(B)$ in cross-sections 0.66 R_A and 0.85 R_A (where R_A is the outer radius of A-ring) took place down to very small B (see Figure 4). In particular, Focas and Dollfus (1969) observed down to $B_{min} = 0°003$; the path of the light ray inside E-ring in that moment was $Z_0 \operatorname{cosec} 10''8 = 2 \times 10^4 Z_0$, but it did not cause any appreciable absorption.

Furthermore these data permit to estimate the upper limit of the E-ring optical thickness, τ_{OE}. As it follows from the absence of absorption, $\tau_{OE} \operatorname{cosec} B_{min} < 1$. Substituting the numerical data, we obtain

$$\tau_{OE} < \frac{1}{20\,000}. \tag{5}$$

Thus the optical thickness of E-ring is extremely small.

The final conclusion is that within the accuracy of 1966 photometric measurements the influence of the reflecting and absorbing properties of E-ring on the estimations of rings' physical thickness is negligible.

References

Barabashov, N. P. and Semejkin, B. E.: 1933, *Astron. Zh.* **10**, 381.
Bobrov, M. S.: 1970, in A. Dollfus (ed.), *Surfaces and Interiors of Planets and Satellites*, Section 7, Academic Press, London, p. 429
Bobrov, M. S.: 1972, *Astron. Zh.* **49**, 424.
Feibelman, W. A.: 1967, *Nature* **214**, 793.
Focas, J. H. and Dollfus, A.: 1969, *Astron. Astrophys.* **2**, 251.
Guérin, P.: 1970, *Sky Telesc.* **40**, 88.
Kiladze, R. I.: 1969, *Abastumansk. Obs. Bull.*, No. 37.
Kuiper, G. P.: 1972, AAS Meeting, invited paper on 'Origin of the Solar System'.

THEORETICAL STUDIES OF AN ATMOSPHERE
AROUND SATURN'S RINGS

M. DENNEFELD

Service d'Aéronomie, Verrières-le-Buisson, France

Abstract. The subject of this paper is an attempt to predict what can be seen in the vicinity of Saturn through a UV photometer on board the Mariner-Jupiter-Saturn mission. We will restrict ourselves to the study of atomic hydrogen and OH and to H_2O which can be the parent molecule and will consider only the particles ejected from the rings and from Titan.

Until recently, only a few experiments have been conducted to detect an atmosphere around the rings of Saturn and the results are controversial (e.g. Franklin and Cook, 1969; Feibelman, 1967).

The subject of this paper is an attempt to predict what can be seen in the vicinity of Saturn through a UV photometer on board the Mariner-Jupiter-Saturn mission. Therefore we will restrict ourselves to the study of atomic hydrogen and OH and to H_2O which can be the parent molecule.

We can, *a priori*, expect three sources:
– the atmosphere of Saturn,
– the rings,
– the atmosphere of Titan.

In fact, the atmosphere of Saturn can be neglected owing to the great mass of the planet.

It appears very rapidly that the atoms or molecules ejected from the rings form an 'atmosphere' totally uncoupled from the 'atmosphere' formed by the atoms ejected from Titan. Therefore, the two problems will be treated separately.

First, we neglect the hypothetical existence of a magnetic field on Saturn and all the dynamical problems will be treated as two-bodies problems.

1. Particles Ejected from the Rings

1.1. SOURCES

The only parameters of the rings we need to know are the optical thickness, the temperature and composition of the surface, which are directly obtained from observations. We use $\tau = 1$ for ring B and $\tau = 0.5$ for ring A, surface temperature $T \simeq 80 \, \mathrm{K}$ and we assume that the ring's particles are coated with pure water ice. We neglect rings other than A or B.

Under these conditions, the sources are:

1.1.1. *Evaporation of Ice*

At $T = 80 \, \mathrm{K}$, the vapor pressure is $P = 1.1 \times 10^{-22}$ torr which gives a yield S of $1.0 \times \times 10^3 \, H_2O$ molecules $m^{-2} \, s^{-1}$, with a sticking coefficient equal to 1.

Woszczyk and Iwaniszewska (eds.), Exploration of the Planetary System, 471–481. All Rights Reserved
Copyright © 1974 by the IAU

The total production is then 1.36×10^{20} H_2O s^{-1}, with a mean square velocity $\bar{v}$ of 330 ms^{-1}. But it must be pointed out that this production rate can be increased by a factor of 4×10^6 if the surface temperature is 100 K.

1.1.2. *Bombardment by Meteorites*

The density of meteorites in the vicinity of the rings is taken from Cook and Franklin (1970) and is equal to 1.2×10^{-23} g cm^{-3} with a mean velocity perpendicular to the plane of the rings $\bar{v} = 15$ km s^{-1}.

The effect of the bombardment depends on the structure of the ring's particles.

(a) If the surface is coated with ice-crystals, solid particles are ejected the mass of which can be evaluated from the results of Gault *et al.* (1963) on lunar material. The ejected mass is about 3700 times the incoming mass for the range of velocities we consider. The ejection is then of about 2×10^4 kg s^{-1}. This production is insufficient by several orders of magnitude to supply the density of small particles required by the model of Franklin and Cook (1965).

(b) If the surface is coated with snow, we can assume that a fraction η (here taken equal to $\frac{1}{3}$, as in Cook and Franklin (1970)) of the incoming energy is used to evaporate water. This gives a mass of water vapor equal to 70 times the incoming mass but only 2.7% of this vapor can escape the surface (owing to the depth of the hole produced by the impact).

This gives a total production of 3.7×10^{26} H_2O s^{-1}. The velocity of the ejected molecules depends on the heating but is unlikely to be greater than 3 km s^{-1}.

1.1.3. *Bombardment by the Solar Wind*

As pointed out by Harrison and Schoen (1967), the bombardment by the solar wind can be an important source of mass loss for the rings. To evaluate the ejection rate more precisely, we extrapolate the results used for metals by, e.g., Dienes and Vineyard (1957) or Pease (1960) to the water ice case. It is not possible to predict with certainty whether the ejected particles are H, OH or H_2O. We assume that the actual production rate is directly proportional to the production rate calculated by assuming that H or OH alone are present in the crystal. More detailed calculations, taking into account the actual structure of ice, will be achieved very soon. Assuming that the displacement energy $E_d = 36$ eV is very high (by the experimental law $E_d = 7$ E_s where E_s is 5.16 eV, binding energy H–OH), we obtain a lower limit for the yield if the mechanism proposed is operating. The results are given in Table I, for impact of a proton or an alpha-particle on H or OH, assuming a velocity of 400 km s^{-1} for the solar wind. The same calculation was made for ejection of H_2O assuming that in this case $E_s = 0.52$ eV, the energy of sublimation of water ice.

These values have to be considered carefully because the theory is very elementary. Laboratory measurements are planed, in collaboration with the group of M. Maurette, Laboratory R. Bernas, Orsay, France. The calculations don't allow us to predict the velocity at ejection. We can only say that the kinetic energy is at most of the order of one or two times the energy of sublimation. We then obtain the following produc-

TABLE I

Sputtering yields

Target	Incoming particle	
	H^+	He^{++}
H	7.2×10^{-3}	4×10^{-2}
OH	1.4×10^{-2}	9.6×10^{-2}
H_2O	0.18	1.13

tion, assuming the flux of solar wind particles is 2.2×10^6 H^+ cm^{-2} s^{-1} and 8.8×10^4 He^{++} cm^{-2} s^{-1} at 9.54 AU:

$$7.5 \times 10^{23} \text{ H s}^{-1} \text{ and } 3 \times 10^{24} \text{ OH s}^{-1} \text{ if there is no water ejection,}$$

or

$$8.7 \times 10^{22} \text{ H s}^{-1}, 3 \times 10^{23} \text{ OH s}^{-1}, 5 \times 10^{25} \text{ H}_2\text{O s}^{-1}.$$

More details can be found in Dennefeld (1973).

1.1.4. *Optical Erosion*

This mechanism is probably not important enough because of the very low momentum of the photons with respect to the thermal momentum of the atoms in the crystal (Dexter, 1964).

1.1.5. *Effect of the Interstellar Wind*

The neutral hydrogen atoms flowing in the heliosphere have a kinetic energy at impact on the rings of Saturn which is at most equal to 16.4 eV, considering a velocity of 8 km s^{-1} at infinity (with respect to the Sun) and a balance between radiation pressure and gravitational attraction of the Sun. The threshold for ejection is about 20 eV and therefore the hydrogen atoms of the interstellar wind have no effect.

On the other hand, if He atoms are present in the interstellar wind with the same velocity at infinity of 8 km s^{-1}, their energy at impact will be sufficient to eject H_2O molecules, but not hydrogen or OH.

Using the model of interstellar wind calculated by Ammar (1973) (density of 0.01 He cm^{-3} at infinity), we obtain a production rate of water between 4.4×10^{24} H_2O s^{-1} and 1.3×10^{24} H_2O s^{-1}, depending on the position of Saturn on his orbit. This production rate is ten times lower than the production rate by solar wind particles and will be preponderant only when the rings are seen edgewise from the Sun.

The production rates by all these mechanisms are summarized in Table II. The uncertainty can be as great as a factor of 2 or 3.

1.2. DISTRIBUTION OF EJECTED PARTICLES

The orbit of each individual particle has been calculated in a two-bodies problem, assuming that all ejected particles have the same velocity and assuming a random distribution of the direction of ejection, in a reference frame bounded to the rings.

TABLE II

Production of the particles

Molecules	Mechanisms	Production rates	Velocity of ejected particles
H_2O	Evaporation of ice	1.4×10^{20} s^{-1}	330 ms^{-1}
	Meteoroidal bombardment	3.7×10^{26} s^{-1}	1 kms^{-1}
	Solar wind bombardment	5.1×10^{25} s^{-1}	<3 kms^{-1}
	Interstellar wind (He) bombardment	4.4×10^{24} s^{-1}	<3 kms^{-1}
H	Solar wind bombardment	7.5×10^{23} s^{-1}	<44 kms^{-1}
OH	Solar wind bombardment	3×10^{24} s^{-1}	<10 kms^{-1}

We have the following results:

– As long as the ejection velocity V_i is much lower than the circular velocity, the dependency of the characteristics of the orbit upon the distance of ejection is small.

– The loss rate by escape or falling into the planet is less than 5% if $V_i < 3$ km s^{-1} (Figure 1).

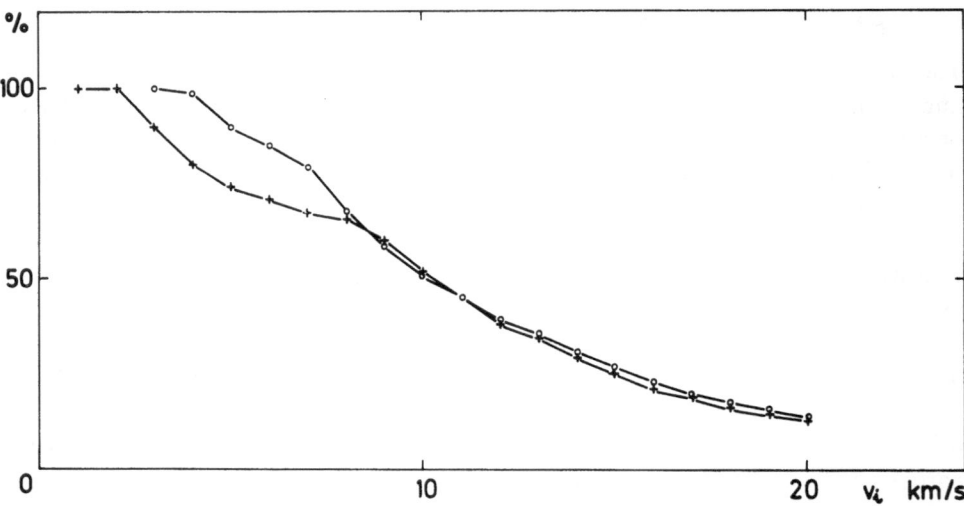

Fig. 1.   Particles in 'stable' orbits around the rings. *Abcissa:* velocity of ejection with respect to the ring's particles. We assume that all particles have the same velocity.
Ordinates: percentage of ejected particles.

If $V_i < 3$ km s^{-1}, the volume occupied by the particles depends only on the ejection velocity and on the greatest distance of ejection. It has a toroïdal form, the section of which can be seen in Figure 2.

After one orbit, the atom or molecule collides with the rings and the probability of flowing through without capture is $\varepsilon = 1 - \tau S_t$ where τ is the optical thickness and S_t the sticking coefficient. The total number of atoms in the volume is then $QT/(1-\varepsilon)$, where Q is the production rate, T the mean orbital period and $(\varepsilon < 1$.

The lifetime against capture is then $T/(1-\varepsilon)$, to be compared with the lifetime against photodissociation by solar EUV photons.

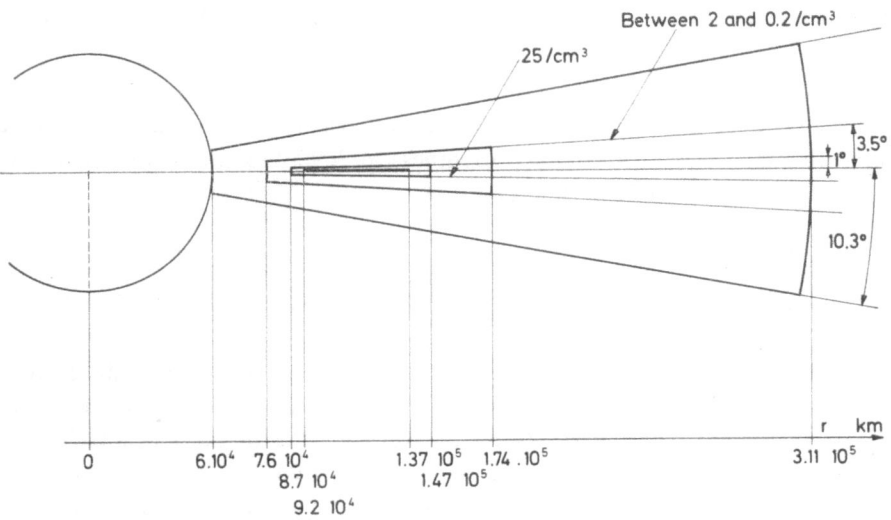

Fig. 2. Volume occupied by water vapor around the rings: section perpendicular to the plane of the rings. We distinguish between different origins of water.

1.2.1. Case of H_2O Molecules

We take $S_t = 0.5$ (e.g. Bruce, 1954) and $\tau = 1$ everywhere. Then $\varepsilon = 0.5$, $T/(1-\varepsilon) = 2\,T \simeq 8 \times 10^4$ s. The lifetime against photodissociation is 9×10^6 s at 9.54 AU for H_2O (Massey et al., 1969; Hinteregger, 1970). We therefore deduce that water is not dissociated before capture by the rings. This means that the total loss of matter by the rings due to the ejection of water is very low, contrary to what has been claimed previously (e.g. Harrison and Schoen, 1967).

The densities obtained are the following, depending on the ejection mechanism which governs the velocity after ejection:
  - Evaporation 5×10^{-5} H_2O cm^{-3}
  - Solar wind bombardment   0.18 H_2O cm^{-3}
  - Meteoroidal bombardment   24 H_2O cm^{-3}
  - Interstellar wind bombardment   0.02 H_2O cm^{-3}.

All these volumes are shown in Figure 2.

Detection by H Lα absorption is impossible: the optical depth is about 10^{-5} and detection in the infrared seems to be very difficult.

1.2.2. *Case of H Atoms*

The lifetime is limited by photoionization and charge exchange: 1.5×10^8 s at 9.54 AU because the sticking coefficient in ice is very low, and therefore an H atom suffers many collisions before destruction. The volume occupied must then be spherical, with a radius of 3.1×10^5 km if the velocity at ejection is 3 km s^{-1}. The density is then about 1 H cm^{-3}, giving a brightness of 1.3 rayleigh, detectable by a Lα photometer. This density can be considerably reduced if the velocity at ejection is higher than 3 km s^{-1}.

1.2.3. *Case of OH Radicals*

If the velocity at ejection is 3 km s^{-1}, density becomes 9 OH cm^{-3} in the same spherical volume than before. The brightness at 3090 Å is then about 6.5 rayleigh. The same remark as for H is valid here.

1.3. CONCLUSION

We have pointed out the existence of toroïdal or spherical atmospheres around the rings of Saturn.

The observation of these atmospheres seems to give no very important results concerning the size of particles in the rings. The only directly observable parameter is the velocity at ejection which depends on the surface-state of the rings-particles in a manner which can be well known only after laboratory-experiments.

On the other hand, water density is very sensitive to surface temperature. The mass of the rings can be attained through perturbation of individual orbits only if it is higher than about 10^{-3} times the mass of Saturn, and this is very unlikely.

However, if there are Van Allen Belts around Saturn, the flux of particles colliding with the rings can be much higher. In this case, the brightness of the 'atmosphere' will be increased too and the flux of trapped particles can be deduced from the study of H Lα emission around the rings.

2. Atoms Ejected from Titan

We can distinguish between escape of atomic hydrogen and escape of molecular hydrogen.

The escape flux depends strongly on the model used.

2.1. ESCAPE OF ATOMIC HYDROGEN

We shall try to set an upper limit to the escape flux of atomic hydrogen in the following manner: the parent molecules of atomic hydrogen can be H_2 or CH_4 in the atmosphere of Titan (we shall speak later about NH_3). We shall assume that all the solar EUV flux able to destruct these parent molecules is absorbed with an efficiency of 1. In this case, the flux to be used is that below 1450 Å. The input at 9.54 AU is then 5.5×10^9 photons cm^{-2} s^{-1} (Hinteregger, 1970). Assuming that once a photon of

$\lambda < 1450$ Å penetrates below a radial distance of, say, 10 000 km, it produces a H atom, the total number of H atoms produced is then 1.5×10^{28} s^{-1}. We further assume that, in a steady state, each atom produced will escape. These atoms will have a Maxwellian distribution of velocities at an exospheric temperature of about 80 K. To obtain an upper limit, we assume that the mean square velocity is equal to the escape velocity (limit of blow-off) at the critical level. Table III shows then that only a few percent of the atoms can escape from the Saturnian system, as long as the exospheric temperature is lower than 120 K. See also Figure 3.

The volume occupied has a toroïdal form, the section of which is shown in Figure 4.

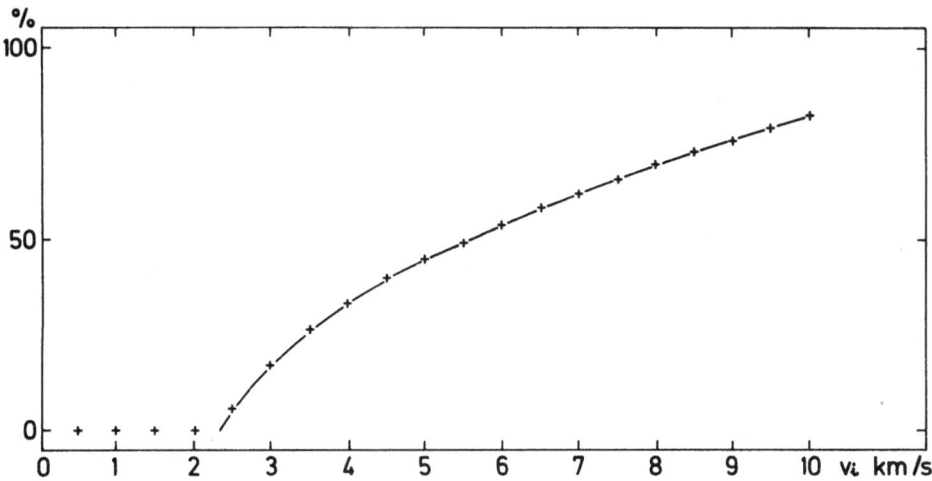

Fig. 3.   Particles ejected from Titan which also escape from the system of Saturn. *Abcissa:* ejection velocity with respect to Titan. We assume here that all particles have the same velocity. *Ordinates:* percentage of ejected particles.

TABLE III

Escape of particles from the Saturnian system

Exospheric temperature	% of produced particles with escape velocities from Titan	% of produced particles with velocities > 2.3 km s^{-1}	Rates of maximum escape	Escape from the Saturn system (in % of ejected particles)	Escape from the Saturn system (in % of particles escaped from Titan)
70	39	2.8	6.0	0.17	0.44
80	39	4.7	11.0	0.52	1.3
90	39	7.0	16.0	1.1	2.9
100	39	9.5	19.5	1.9	4.9
110	39	12.3	22.0	2.7	6.9
120	39	15.1	23.0	3.5	9.0

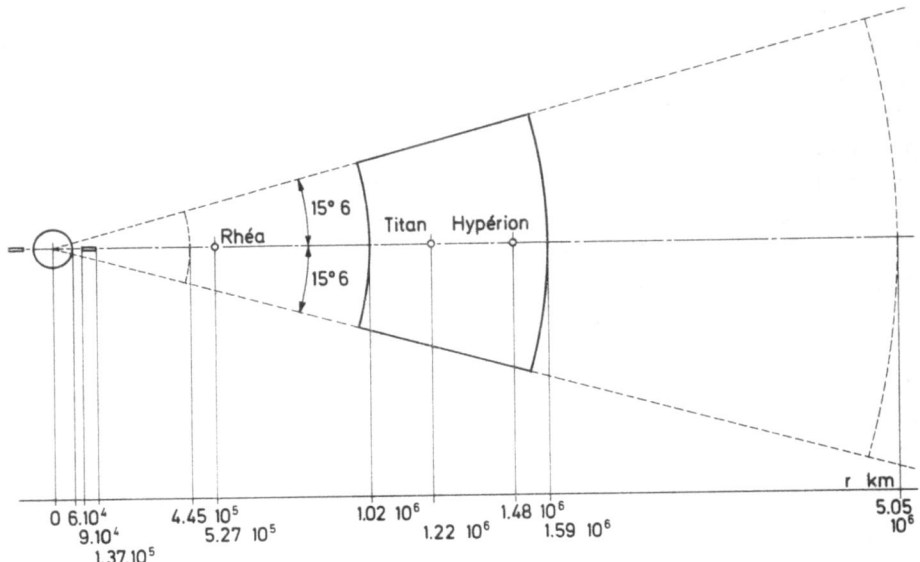

Fig. 4.   Volume occupied by atomic hydrogen around the orbit of Titan: section perpendicular to the plane of orbit. The section in full lines has density 3 times greater than the mean density.

The mean density is given by:

$$\bar{n} = \frac{Q}{\dfrac{V}{\tau} + S_e v_T},$$

where Q is the production rate, τ the lifetime against ionization and charge exchange, v_T the orbital velocity of Titan and S_e the surface of capture (taken as before equal to a sphere of radius 10000 km).

With $Q = 1.5 \times 10^{28}$ s^{-1}, $V = 1.5 \times 10^{29}$ m^3, $\tau = 1.5 \times 10^8$ s, $S_e = 3.14 \times 10^{14}$ m^2, $v_T = 5.57$ km s^{-1}, the mean density becomes:

$$\bar{n} = 15 \text{ H cm}^{-3}.$$

The loss rate by capture by Titan (term $S_e v_T$) is negligible with respect to the loss rate by photoionization or charge exchange.

In the volume shown in full lines in Figure 4 the density is about 3 times greater than the mean density. This atmosphere is optically thick in Lα 1216 Å. The optical depth is about 4 along a line of sight lying in the plane of the rings. Assuming that we can see up to an optical depth of $\tau = 1$, the brightness is then

$$4\,\pi B = 95 \text{ rayleigh}.$$

2.2. Escape of molecular hydrogen

We use the model of Trafton (1972). The escape flux is 2×10^9 cm^{-2} s^{-1} for an exospheric temperature of 74 K. This gives a production rate equal to the production rate of atomic hydrogen calculated before. Once ejected from Titan, the principal destruction process of molecular hydrogen is photoionization ($\sigma \sim 10^{-17}$ cm^2, $\lambda < 804$ Å) rather than photo-dissociation ($\sigma \sim 10^{-18}$ cm^2, $\lambda < 840$ Å).

The contribution to the density of atomic hydrogen is therefore only 20% of that calculated previously.

But if this escape of molecular hydrogen is the dominant one (the escape of atomic hydrogen can be much lower than 1.5×10^{28} s^{-1} if recombination in the atmosphere takes place) we have a mean density 3 H cm^{-3} and a brightness of 72 R. In this case, the 'atmosphere' is optically thin.

In all cases, we can conclude with the existence of an 'atmosphere' of atomic hydrogen around Titan.

More details can be found in Dennefeld (1973). Recently, Hunten (1973) published a model of Titan with an escape flux of the order of 10^{11} cm^{-2} s^{-1} at a critical level of 11 000 km. The total escape is then 1.5×10^{30} H$_2$ s^{-1} which is 300 times greater than the value used by Trafton. With the same escape velocity as before, 1.4 km s^{-1}, this gives a mean density of 1150 cm^{-3} (only 10% of the molecules are dissociated into H + H).

In this case, we have still a mean brightness of the order of 95 R, but the optical thickness at center is very high, of the order of 100. A model using multiple diffusion of radiation must then be used to determine the actual brightness. With such a high density, each atom will suffer many collisions before destruction and we can infer from this that the volume will not be inevitably toroïdal, but can be spherical in shape, or flat, depending on the momentum-transfer in collisions.

The observation of the Lα airglow in the vicinity of the orbit of Titan can therefore give us the escape flux and an indication about the existence of NH$_3$ in the atmosphere of Titan which seems to be the only candidate to supply such a high escape flux.

It must be pointed out that in the case of the presence of NH$_3$ the upper limit of escape of atomic hydrogen will be increased by more than a factor of 100 ($\lambda < 2400$ Å, instead of $\lambda < 1450$ Å).

An other point is the existence in the vicinity of Titan of a flux of charged particles in the magnetosphere of Saturn. This has been studied by Brice and McDonough (1973). The density of atomic hydrogen can be reduced if the flux of charged particles is as high as the flux of the solar wind at 9.54 AU, 2×10^6 H$^+$ cm^{-2} s^{-1} which seems quite possible. In this case, the distribution in the toroïdal volume will be asymmetric owing to the fact that the flux of charged particles decreases with radial distance to Saturn. Therefore it is possible to distinguish between a loss due to magnetospheric particles and a lower escape flux from Titan.

2.3. CONCLUSION

There are two independent 'atmospheres' in the system of Saturn: one around the rings, the other around the orbit of Titan.

– The 'atmosphere' around the rings contains water, the density of which can give us the surface state of the ring's particles: ice or snow. Predictions about H or OH are not easy as long as laboratory experiments are not made.

The existence and the flux of trapped particles in hypothetical Van Allen Belts are directly related to the brightness of this atmosphere.

– The 'atmosphere' around Titan contains H and H_2, his size and extent depends on the escape flux from Titan.

If the form is spherical, the magnetic field of Saturn has no effect on the lifetime of H atoms at the distance of Titan and the escape flux is at least of the order of 5×10^{10} cm^{-2} s^{-1}. This possibly indicates the presence of NH_3. A toroïdal form indicates a mean density low enough to make collisions negligible or an exchange of momentum in collisions which takes place preferentially perpendicularly to the plane of the orbit of Titan. The observation of the Lα glow gives us the escape flux and the exospheric temperature of Titan.

– The same calculations as before can be applied to the Galilean satellites of Jupiter, except Io, if we assume as a first step that they have no atmosphere. Assuming a surface temperature of the order of 130 K, the water molecules evaporated (this is the dominant mechanism) from the surface (which is partially coated with ice, (Pilcher et al., 1972)) cannot escape the gravitational attraction of the body. They follow ballistic orbits before sticking on the surface. We have then a layer of water vapor on the surface, 75 km thick, with a density of the order of 5×10^7 cm^{-3}, detectable by IR techniques. This number must be multiplied by a weighting factor less than one, which gives the proportion of the area coated with ice (Pilcher et al., 1972). This density is low enough to make flight collisions negligible. But, as suggested by Lewis, if ammonia-hydrates are present, the density can rise by a factor as high as 10^5, leading to the existence of a real atmosphere. Full calculations will be published later.

Acknowledgements

I thank Drs J. Lewis and M. Maurette for helpful discussions. I thank Drs Brice and McDonough for kindly sending a preprint.

References

Ammar, A.: 1973, 3ème Cycle Thesis, to appear at University of Paris VI, France.
Brice, N. M. and McDonough, T. R.: 1973, *Icarus* **20**, 136.
Bruce, E. W.: 1954, *Trans. Instr. Chem. Engrs. London* **32**, 192.
Cook, A. F. and Franklin, F. A.: 1970, *Icarus* **75**, 195.
Dennefeld, M.: 1973, 3ème Cycle Thesis University of Paris VI, France.
Dexter, D. L.: 1964, *Nuovo Cimento* **32**, 90.
Dienes, G. J. and Vineyard, G. A.: 1957, in *Radiation Effects in Solids*, Interscience, New York, Ch. 3.
Feibelman, W. A.: 1967, *Nature* **214**, 793.
Franklin, F. A. and Cook, A. F.: 1965, *Astron. J.* **70**, 704.
Franklin, F. A. and Cook, A. F.: 1969, *Icarus* **10**, 417.
Gault, D. E., Shoemaker, E. M., and Moore, H. J.: 1963, NASA Techn. Note D-176.
Harrison, H. and Schoen, R. I.: 1967, *Science* **157**, 1175.
Hinteregger, H. E.: 1970, *Ann. Geophys.* **26**, 547.

Hunten, D. M.: 1973, *J. Atmospheric Sci.* **30**, 726.

Lewis, J. S.: 1971, *Icarus* **15**, 174.

Massey, H. S. W., Burhop, E. H. S., and Gilbody, H. S.: 1969, in *Electronic and Ionic Impact Phenomena*, Clarendon Press, Oxford, Vol. II.

Pease, R. S.: 1960, *Rendiconti S.I.F.* **13**, 158.

Pilcher, C. B., Ridgway, S. T., and McCord, T. R.: 1972, *Science* **178**, 1087.

Trafton, L.: 1972, *Astrophys. J.* **175**, 285.

THE SATURNIAN 'GAS-DOUGHNUT' HYPOTHESIS

THOMAS R. McDONOUGH and NEIL BRICE

School of Electrical Engineering, Cornell University, Ithaca, N.Y., U.S.A.

Abstract. It is hypothesized that constituents escaping the atmosphere of Titan are unlikely to have sufficient velocity to escape from Saturn. Thus most of the material will enter into elliptical orbits which, at some point, cross Titan's orbit at a small angle. In total, this matter will form a 'doughnut' shaped feature, roughly centered on Titan's orbit and having dimensions of about 20 R_S (Saturn radii) in the Saturnian-equatorial plane and 10 R_S normal to this plane. It is assumed that the material is lost from the doughnut when it is ionized or is recaptured by Titan's atmosphere. The upper limit on the density of hydrogen in the ring is estimated to be about 10^3 cm^{-3}, with up to 99% of the particles in the doughnut being recaptured by Titan.

A brief summary of the work which has been done on this problem is given, and areas needing further research are suggested.

It has been calculated that most atoms or molecules which escape from the atmosphere of Titan probably do not have enough speed to escape from Saturn (McDonough and Brice, 1973a). If this is true, then the Titanian constituents will orbit Saturn, forming a gaseous ring around the planet which is much larger than the visible rings of Saturn. The purpose of this note is to summarize the current theoretical studies of this ring, which have been given by McDonough and Brice (1973a, b), McDonough (1974), Dennefeld (1973, 1974), Tabarié (1974), and Sullivan (1973), and to point out areas needing further research.

If the particles which escape Titan but not Saturn are unperturbed, they will describe elliptical orbits which intersect Titan's orbit. Because they intersect Titan's orbit, the interesting possibility exists that some of them will be recaptured by Titan. If Titan has an exospheric radius comparable to its visible radius, the fraction of particles which is recaptured is apt to be small, but if the exospheric radius is of the order of 10 Titan radii, the recaptured fraction could be as much as ~99% (McDonough and Brice, 1973a, b). Hunten (1974) has suggested that recapture may increase the gross escape rate from Titan in such a way that the net loss rate is the same as if there were no capture.

There are several potential sources of perturbations of the particle orbits, each of which may decrease the recapture probability: particle-particle collisions, solar radiation pressure, gravitational attraction by the satellites and by the aspherical component of the Saturnian gravitational field, interaction with the visible rings, interaction with the solar wind, and, if Saturn possesses a magnetic field, interaction with corotating magnetospheric plasma and energetic particles.

The importance of particle-particle collisions has been emphasized by Hunten (1974) and by Dennefeld (private communication), and has been studied under certain approximations by Sullivan (1973). The effect of such collisions is unclear and should be investigated. Solar radiation pressure, particularly that of the Lα line of atomic hydrogen, may be significant despite the fact that the pressure is approximately

Woszczyk and Iwaniszewska (eds.), Exploration of the Planetary System, 483–485. All Rights Reserved
Copyright © 1974 by the IAU

balanced by the solar gravitational force, because (1) in Saturn's frame of reference, i.e., in free fall, the solar gravitational force is balanced (neglecting field gradients) by the centrifugal force due to motion around the Sun, while the atoms still experience radiation pressure, and (2) the quantization of the solar photons implies discontinuities in particle velocities. The principal effect of corotating plasma, if Saturn has a strong magnetic field, and of the solar wind, if it has not, is expected to be charge-exchange ionization of the orbiting atoms and molecules. Because the principal constituent of the doughnut is probably hydrogen, if Trafton's (1972) detection of Titanian H_2 is correct, charge-exchange will probably control the ratio of atomic to molecular hydrogen in the doughnut (Blamont and Dennefeld, private communications). All of the above effects on particle orbits deserve further study.

For a Titanian exospheric temperature of $\sim 100 \text{K}$, celestial mechanics implies a hydrogen torus with a radius of the order of 20 Saturn's radii and with a thickness of the order of 10 Saturnian radii normal to Titan's orbital plane. Using the Titanian escape fluxes of Trafton (1972) and Hunten (1973), one can compute an upper limit to the hydrogen density in the doughnut of $\sim 10^3 \text{ cm}^{-3}$, if recapture by Titan is efficient (McDonough and Brice, 1973a, b). A smaller recapture efficiency could increase this density.

The Saturnian 'doughnut' hypothesis appears to be viable, and is worth pursuing theoretically and experimentally as a probe of the Saturn–Titan system.

Acknowledgements

We are indebted to Drs J. E. Blamont and M. Dennefeld of the University of Paris, and to Dr D. M. Hunten of Kitt Peak National Observatory, for stimulating conversations. This research was sponsored in part by the NASA Physics and Astronomy Program under Grant number 33-010-161, and the National Science Foundation Atmospheric Sciences Section under Grant numbers GA-11415 and GA-36916.

References

Dennefeld, M.: 1973, Ph. D. Thesis, University of Paris.
Dennefeld, M.,: 1974, this volume, p. 471.
Hunten, D. M.: 1973, *J. Atmospheric Sci.* **30**, 726.
Hunten, D. M. (ed.): 1974, *The Atmosphere of Titan* (NASA SP-340), p. 110.
McDonough, T. R.: 1974, in D. M. Hunten (ed.), *The Atmosphere of Titan* (NASA SP-340), p. 118.
McDonough, T. R. and Brice, N. M.: 1973a, *Nature* **242**, 513.
McDonough, T. R. and Brice, N. M.: 1973b, *Icarus* **20**, 136.
Sullivan, R.: 1973, to be published.
Tabarié, N.: 1974, in D. M. Hunten (ed.), *The Atmosphere of Titan* (NASA SP-340), p. 123.
Trafton, L.: 1972, *Astrophys. J.* **175**, 285.

DISCUSSION

Smith: Will the solar wind not tend to sweep out the proposed exospheric cloud of hydrogen?

McDonough: In deriving the density of the gaseous torus, we first assumed that charge-exchange of the hydrogen with solar wind was the dominant loss process. For H, this lifetime is about 6 yr at Saturn's distance, and for H_2, about 50 yr. Thus the lifetimes are quite long, and you can get densities up to about 10^5 cm^{-3} even with the solar wind unshielded by a Saturnian magnetosphere. Our second model included recapture by Titan. If there is recapture, the density could still be as much as about 10^3 cm^{-3}.

Baum: What observations would best set a limit on the amount of gas in the postulated 'Titan' ring? Can we already put a limit on it from existing data? Might spectra of Titan be detectably different when at different positions in its orbit and therefore seen through different amounts of the ring?

McDonough: Lα and Hα (Balmer alpha) observations appear to be the most promising method of detecting or setting limits to the proposed Saturnian gas-torus. Existing Lα and Hα data do not appear to be sufficient to detect even the densest hydrogen torus. A previous search for an atmosphere around the visible rings of Saturn did not appear to set useful limits on the probable constituents of the Saturnian gas-torus. It is possible that Titanian spectra may be different at different positions, due to the torus, but the torus would be essentially transparent in most wavelengths, Lα being a possible exception.

Gorgolewski: Is your model applicable to Jupiter satellites?

McDonough: Yes, it may apply to all of the giant planets. Jupiter's magnetosphere protects such a toroidal ring from the solar wind, although corotating plasma may reduce a ring's density somewhat. Two of Jupiter's Galilean satellites have been proposed as sites of atmospheres, and all of the Galilean satellites are deep enough in the gravitational potential well of Jupiter for them to trap escaped atmosphere in toroidal rings around Jupiter.

OBSERVATIONAL CONSTRAINTS ON MODEL ATMOSPHERES FOR URANUS AND NEPTUNE

THÉRÈSE ENCRENAZ*, JOHANNES HARDORP, and TOBIAS OWEN**

Dept. of Earth and Space Sciences, State University of New York at Stony Brook, Stony Brook, N.Y., U.S.A.

and

JERRY H. WOODMAN

McDonald Observatory, University of Texas, Fort Davis, Tex., U.S.A.

Abstract. We have re-examined the visible and near-IR regions of the spectra of Uranus and Neptune to provide additional data for constructing atmospheric models. We find that a true continuum exists only at wavelengths below 4700 Å, that the 6420 Å absorption previously attributed to hydrogen is probably caused by methane, that there is no evidence for ammonia, ethylene or ethane absorptions in our spectra, and that the abundance of methane is probably much higher than previous estimates suggest. This last finding implies that the value of H/C in the atmospheres of both planets is much less than 1/10 the solar (or Jovian) value. Clear, Rayleigh-scattering model atmospheres are not compatible with the observations, but more work is needed to establish viable alternatives.

1. Introduction

There has been considerable interest in recent years in the structure and composition of the atmospheres of Uranus and Neptune. These distant planets are not easy objects to study at the telescope, so efforts to understand their atmospheres have been severely hampered by a lack of useful observational data. In the present paper, we first review some current ideas about both planets and then offer some new data which should be incorporated in future atmospheric models.

Belton, McElroy, and Price (1971 – hereinafter referred to as BMP) have suggested that the atmospheres of both of these planets are devoid of clouds in the regions accessible to optical observations. This means that the transfer of radiation outside molecular absorption bands proceeds by Rayleigh and Raman scattering in the semi-infinite, pure molecular atmospheres. A further refinement to the original model was offered by Wallace (1972) who considered the effects of Raman scattering in greater detail.

This model has been challenged by several authors. Sinton (1972) and Westphal (1972) both reported observations of limb brightening on the disk of the planet when it was viewed in the region of a strong methane absorption band. Sinton (1972) also found brightening over the south pole of the planet. By analogy with similar observations of Jupiter (Owen, 1969), both Sinton and Westphal concluded that their results supported a model in which a high cloud layer was overlain by a layer of absorbing and scattering gas. Belton and Price (1973) have responded that limb brightening could be expected in strong methane bands in a pure molecular model as well.

---

* Present address: Groupe Planètes, Observatoire de Meudon, Meudon, France.
** Guest investigator, Hale Observatories, 1972.

Woszczyk and Iwaniszewska (eds.), Exploration of the Planetary System, 487–496. All Rights Reserved
Copyright © 1974 by the IAU

However, their discussion cannot account for the bright features on the disk itself that are apparent in Sinton's (1972) photograph.

Prinn and Lewis (1973) have presented thermodynamic arguments for the presence of a high methane haze in the atmosphere of Uranus. Danielson *et al.* (1972) found that limb darkening of Uranus in the visible and near UV could not be matched by a semi-infinite atmosphere, but could be explained by a model incorporating a cloud layer beneath a layer of gas. Belton *et al.* (1973) have found evidence for Raman scattering in the atmosphere of Uranus, but they argue that this evidence does not really permit a conclusive discrimination among various models. In a new discussion of the pressure-induced absorptions of hydrogen, Belton and Spinrad (1973) concluded that the semi-infinite model does not fit the observations as well as a model with a cloud layer. Finally, Savage and Caldwell (1973) have presented the results of OAO short wavelength photometry which again do not agree with the predictions for a semi-infinite atmosphere: the planets are not as bright at short wavelengths as would be expected in the absence of clouds or haze.

In short, the weight of presently available evidence appears to be against the model originally proposed by BMP. We now wish to present some additional new data, including our own, which should be incorporated in attempts to develop more sophisticated models in the future.

2. Density and Bulk Composition

New determinations of the radii of both planets have suggested that they are larger than had been thought. This means the densities are lower, as shown below:

	M	R	ϱ
Uranus	15	4.04	1.2
Neptune	17	3.87	1.7

Masses and radii are given in terms of terrestrial values; the new radii are taken from Danielson *et al.* (1972) – Uranus, and Kovalevsky and Link (1969) – Neptune.

While these densities are lower than the 'textbook' values of 1.60 and 2.25 (Allen, 1963), they still do not permit the planets to be composed primarily of hydrogen, as is the case for Jupiter and Saturn (Smoluchowski, 1973). The precision of the present data is probably still too poor to exclude the possibility that the bulk composition of the two planets is identical.

3. The Radio and Thermal IR Spectra

A useful summary of available observations from $10\,\mu$ to 10 cm has been given by Morrison and Cruikshank (1973), who pointed out the essential similarity of the thermal spectrum from each planet. These authors used this similarity to suggest that if the chemical composition of the two planets is also the same, 'it appears that they have essentially the same atmospheric structure in spite of their differences in distances from the Sun and in inclination of their axes of rotation'. We shall return to this point since it has an obvious bearing on any attempts to develop model atmospheres.

A recent observation of Uranus at 21 cm by Briggs (1973) revealed a temperature of 280 ± 60 K at this wavelength, suggesting that these planets, like Jupiter and Saturn, probably have regions in their lower atmospheres that are well above the freezing point of water.

4. The Near-IR Continuum

BMP suggested that the depressed near-IR continuum of the spectra of both planets could be explained by the far wing of the pressure-induced fundamental of H_2. This discussion ignored the effects of weak methane bands. In reality, it is virtually impossible to find a region in this part of the spectrum of either planet where weak methane lines are not present. The contrary impression exists simply because of observations made at low spectral resolution.

An excellent example of this is the 7500 Å region. At very low resolution, this appears to be a 'window' in the planetary spectra, but as the resolution is increased, more and more lines appear (Kuiper, 1949; Owen, 1967; Lutz and Ramsay, 1972). These lines are all methane, and one may predict with reasonable certainty that still higher resolution would probably add to their number.

Another place where this effect occurs is at 6400 Å. Scanner observations unaffected by the vagaries of photographic emulsions clearly show that the continuum in this region is badly distorted. Once again, high resolution coupled with appropriate laboratory data indicates that weak methane lines are present. It is our current assessment that an undistorted continuum exists only at wavelengths below 4700 Å (see below).

5. The 6420 Å Region and the 4–0 H₂ Dipole Absorptions

Spinrad (1963) felt that he had identified the $S_4(0)$ pressure-induced absorption from H_2 in the spectra of both Uranus and Neptune at 6420 Å. A careful study of the problem by Belton and Spinrad (1973) has led to the conclusion that this identification was probably incorrect because of the absence of the $S_4(1)$ line in the planetary spectra. It would occur at 6370 Å, in a region that seems relatively free of methane absorption.

We have confirmed the absence of the $S_4(1)$ line and we feel we have been able to show that the 6420 Å absorption is in fact caused by methane (Figure 1 – Owen et al., 1974). This gas also appears to be responsible for an absorption at 6490 Å reported by Spinrad (1963) and confirmed by Teifel and Kharitonova (1970). This at present there seems to be no reason to assume that the 6420 Å feature is caused by H_2, although this gas make some contribution to the observed absorption. Abundances of H_2 based interpretations of this feature are therefore unlikely to be correct.

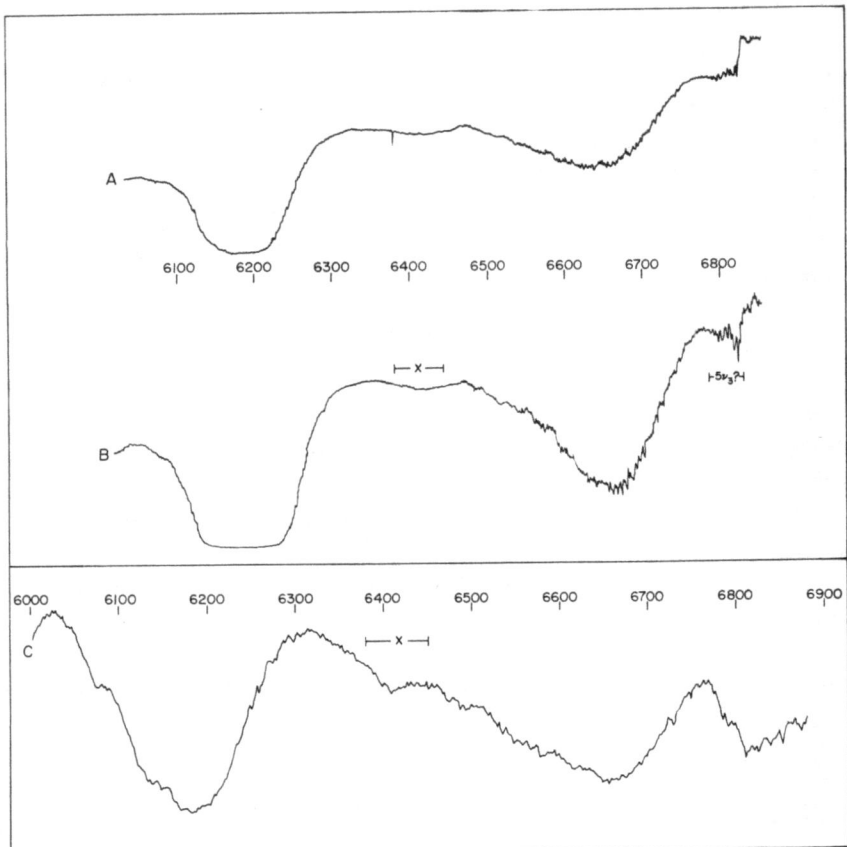

Fig. 1.   Density tracings of spectrograms of methane – (A) 5.28 km amagat at 1.0 atm, (B) 10.1 km amagat at 1.92 atm. Original dispersion 4.56 Å mm⁻¹, NRC, Ottawa, Canada. The sharp spike at 6420 Å in (A) is a plate defect. (C) The spectrum of Uranus divided by that of the Moon. Original resolution, 2 Å, Tull coudé scanner at 2.7-m telescope, McDonald Observatory, U.S.A. (July, 1973).

6. Abundances of Atmospheric Constituents

6.1. HYDROGEN

A summary of recent abundance determinations from the H_2 quadrupole lines in the spectrum of Uranus is given in Table IV of Encrenaz and Owen (1973). While there is still a considerable scatter in these results, convergence toward a value of 450 ± 100 km am (for simple reflecting models) seems to be occurring. The first observations of quadrupole lines in the spectrum of Neptune have been reported at this conference by Trafton (1973), who found the equivalent widths to be similar to those for Uranus.

6.2. METHANE

The original estimates of abundances by Kuiper (1952) and the revisions suggested by

Owen (1967) are given below:

	Kuiper (1952)	Owen (1967)
Uranus	2.2 km am	3.5 ± 1.2 km am
Neptune	3.7	6

The revisions seemed necessary to provide enough methane to account for the weak 7500 Å lines referred to above. However, both sets of estimates were simply based on attempts to match observed intensities of the planetary absorptions with intensities in laboratory spectra of the gas. This a crude approach at best, since it ignores possible differences in pressure and temperature between the laboratory and the atmospheres of the planets.

If the hydrogen abundances mentioned above are correct, the effective pressure at the level of line formation will be of the order of 4 atm (Encrenaz and Owen, 1973). This means that weak methane bands in the planetary atmospheres will not be saturated and hence only the temperature difference is significant. By integrating the absorption over an entire band, we should be able to allow for this effect. We have therefore looked for weak bands in the planetary spectra at short wavelengths where we have some hope of identifying the true continuum.

In the region from 4500 to 5300 Å, we have found the following bands in spectra of both planets (Figure 2):

5210 Å	N	GBBM
5090		GBBM
4950		GBBM
4860		GBBM
4590	N	

This region was studied previously by Adel and Slipher (1934) who found all of the bands listed above except the one at 4950 Å in the spectrum of Neptune. The two bands marked N were not found by them in the spectrum of Uranus, although they are definitely present (Figure 1). The bands marked GBBM were reported by Galkin *et al.* (1971) who only observed Uranus, and then only for $\lambda > 4800$ Å. We confirm their identification of the band at 4950 Å, the weakest member of this set.

In ratio spectra, we find that the 4860 and 4590 Å bands are especially enhanced on Neptune compared with Uranus, while those at longer wavelengths (on our list) are more equal (Figure 2). But while Kuiper (1952) reported a ratio of 1.7 for the methane abundances in the atmospheres of the two planets, we find it to be closer to 1.25, assuming that a ratio of the equivalent widths of these weak bands is in fact a measure of the abundance ratio. This result was obtained both from the McDonald observations shown in Figure 2 and from observations made by one of us (J.H.) at much lower resolution at the Smithsonian Astrophysical Observatory at Mt. Hopkins. The next step in this preliminary approach to the problem is to try to match the

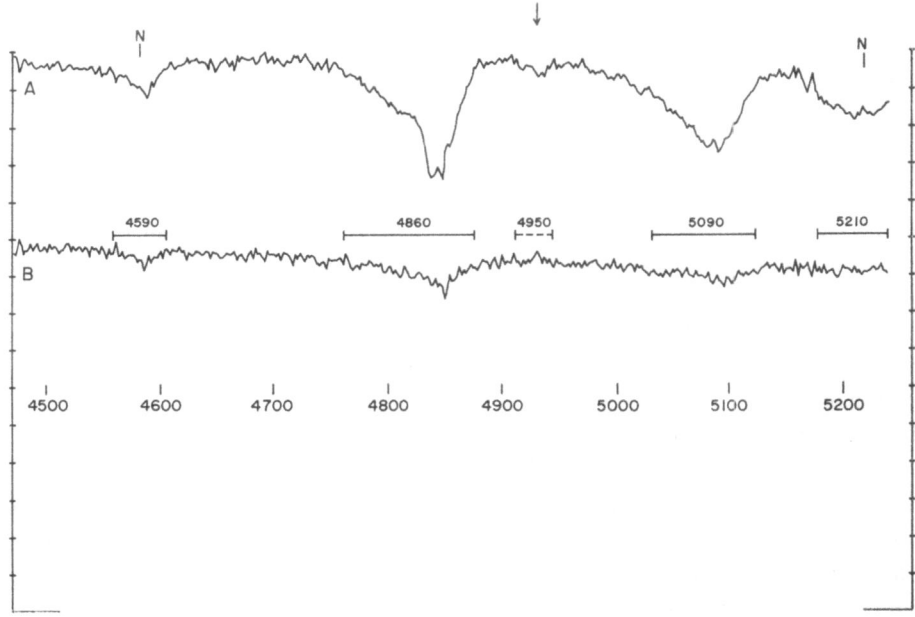

Fig. 2.   (A) The spectrum of Uranus divided by that of the Moon. Arrow denotes location of 4950 Å band, N refers to bands seen by Adel and Slipher (1934) in spectrum of Neptune only (see text). (B) The spectrum of Neptune divided by the spectrum of Uranus. Note residual absorptions. Both (A) and (B) obtained at 2 Å resolution with Tull coudé scanner at 2.7-m telescope, McDonald Observatory U.S.A. (July, 1973).

intensity of one of these weak bands in the planetary spectra with a suitable laboratory path length of methane.

We have examined spectra of 5.28 and 10.1 km of methane kindly obtained for us by Dr D. A. Ramsay at the NRC in Ottawa. The respective pressures were 1 and 1.9 atm, always at room temperature (cf. Figure 1). We found traces of the 5970 Å and 5760 Å bands, but no hint of 5210 Å, which is considerably stronger than 4590 Å or 4950 Å. One is thus led to the conclusion that the effective optical paths of methane in the atmospheres of these planets (including the air mass factor η) are much larger than 10 km am and may be in the range 50 to 100 km am. This in turn implies a value of $H/C \approx 40$, almost two orders of magnitude below the cosmic value exhibited by Jupiter and Saturn.

Clearly this estimate requires quantitative verification and we hope to be able to provide that in the near future through a study of the 6800 Å band, tentatively identified as $5v_3$ (cf. Figure 1), which seems to have a structure that is sufficiently simple to permit precise analysis (Owen, 1966). A recently obtained spectrogram of this band that illustrates its potential is reproduced in Figure 3 (see also Teifel and Kharitonova, 1970).

One other point should be mentioned. The fact that the weak methane absorptions in regions where the continuum is well-defined are definitely stronger on Neptune

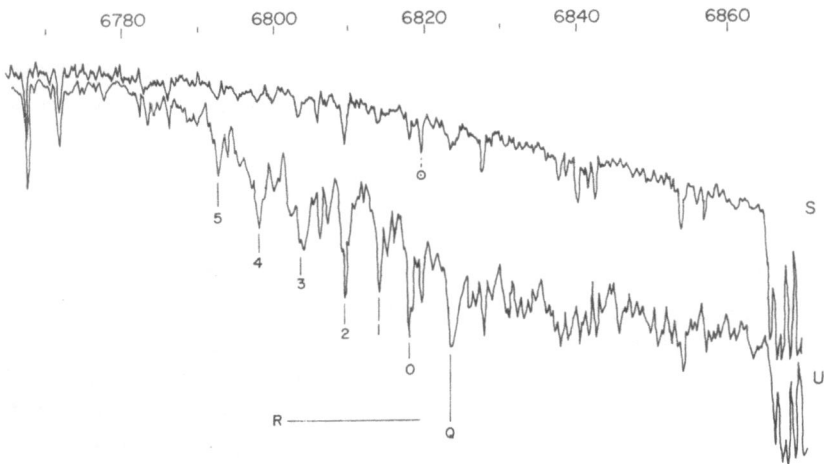

Fig. 3. Reproductions of spectrograms of Saturn (top, labeled 'S') and Uranus (bottom, labeled 'U') in the region of the putative $5\nu_3$ band. The J assignments are very tentative. Original dispersion 7 Å mm^{-1}, coudé spectrograph, 5-m telescope, Hale Observatories (March, 1972).

than on Uranus is an argument against the BMP model of a pure molecular semi-infinite atmosphere. It also suggests, counter to Morrison and Cruikshank (1973) and Belton and Spinrad (1973) that the atmospheric structures of these two planets are not identical.

6.3. AMMONIA

Danielson (private communication) has pointed out that in the case of a semi-infinite atmosphere on Uranus, a substantial number of photons should reach levels at which ammonia absorption will occur in the 6450 Å region. Accordingly, we have examined the spectrum of Uranus in this region to see if any ammonia lines were present. We have not been able to find any evidence for ammonia, with a lower limit of 10–15 mÅ on the equivalent widths of the lines. Comparison with observations of this band in the spectrum of Jupiter (Mason, 1970) indicates that the total amount of ammonia in the line of sight must be less than 20% of the amount on Jupiter. This suggests $NH_3 \leqslant 2.5$ m am in a one-way path in the atmosphere of Uranus. We have not carried out this kind of analysis for Neptune.

6.4. OTHER GASES

In their original identification of methane features in the spectra of these two planets, Adel and Slipher (1934) also examined the absorption spectra of long paths of ethylene (C_2H_4) and ethane (C_2H_6). They found no indication that either of these gases was present.

In the region we have investigated, the relevant absorptions Adel and Slipher reported are:

$$C_2H_6 \quad 6320 \text{ Å}$$
$$C_2H_4 \quad 6040 \text{ Å}.$$

We must also report that we find no compelling evidence to suggest that either one of these gases is in the atmosphere of either planet. However, because of its interest in atmospheric photochemistry, we are undertaking a new investigation of the C_2H_6 spectrum in the laboratory and will reexamine the question of its existence in these atmospheres at a later date.

7. Raman Effect

Belton *et al.* (1973) have reported detection of the Raman effect in the spectrum of Uranus. Specifically, they have found an absorption near 4025 Å which they attributed to the overlapping H and K lines shifted appropriately for the $S(0)$ and $S(1)$ rotational transitions in H_2. The depression corresponds to an absorption of about 10%, in accord with the prediction by Wallace (1972). They found the relative $S(0):S(1)$ Raman contributions to be of the order of $0.3:0.7$. The published observations show no indication of the K_0 line, and only weak evidence for H_1.

We have also looked for this effect in observations carried out with the Tull scanner at the McDonald Observatory in March and July of this year (1973). We studied both Uranus and Neptune, using Saturn's rings and the Moon to provide solar comparison spectra. Details will be presented elsewhere (Woodman *et al.*, 1974), but we can report at this time that we find only weak evidence of an absorption at 4025 Å, with a central intensity no greater than 5%. Additional study of the problem would seem to be advisable.

Even if Belton *et al.* (1973) are correct, however, the reader should not conclude that this observation supports the clear atmosphere model of BMP. In their discussion, Belton *et al.* (1973) point out that it is possible to have atmospheres with a tenuous haze or with a cloud layer that would also satisfy their observations.

8. Conclusions

The main results of this survey may be summarized as follows:

(1) The weight of present observational evidence is against the semi-infinite, pure-molecular atmosphere originally proposed by BMP. It is not yet clear what the best alternative model is.

(2) The continuum in spectra of both planets is distorted by methane absorptions at wavelengths greater than 4700 Å.

(3) The absorption at 6420 Å in spectra of both planets is mainly caused by methane, rather than the $S_4(0)$ dipole absorption of H_2.

(4) The abundance of methane is higher than had been thought, perhaps by as much as a factor 10, although this conclusion requires quantitative verification by studies of individual bands.

(5) The mixing ratio of methane to hydrogen is much higher in these atmospheres than in the atmospheres of Jupiter and Saturn, with values of H/C below 100 a real possibility. Aside from its cosmogonical implications, this low ratio increases the

chances of methane condensation in the upper atmospheres (Prinn and Lewis, 1973).

(6) The atmospheres of the two planets do not appear to be identical, despite many similarities.

Acknowledgements

Observations at the McDonald and Hale Observatories were made possible through the courtesy of Dr H. J. Smith and Dr H. W. Babcock, respectively. We are grateful for assistance in plate reductions from Ms Carolyn Porco and for many helpful discussions with Dr Barry Lutz. The laboratory spectra which were critical to this work were generously obtained and made available by Dr D. A. Ramsay. The research was supported by NASA grants NGR-33-015-141, NGR-33-015-169, and NGR-33-015-165.

References

Adel, A. and Slipher, V. M.: 1934, *Phys. Rev.* **46**, 902.
Allen, C. W.: 1963, *Astrophysical Quantities*, Athlone Press, London, p. 143.
Belton, M. J. S., McElroy, M. B., and Price, M. J.: 1971, *Astrophys. J.* **164**, 191.
Belton, M. J. S. and Price, M. J.: 1973, *Astrophys. J.* **179**, 965.
Belton, M. J. S. and Spinrad, H.: 1973, *Astrophys. J.* **185**, 363.
Belton, M. J. S., Wallace, L., and Price, M. J.: 1973, *Astrophys. J.* **184**, L143.
Briggs, F. H.: 1973, *Astrophys. J.* **182**, 999.
Danielson, R. E., Tomasko, M. G., and Savage, B. D.: 1972, *Astrophys. J.* **178**, 887.
Encrenaz, T. and Owen, T.: 1973, *Astron. Astrophys.* **28**, 119.
Galkin, L. S., Bugaenko, L. A., Bugaenko, O. I., and Morozhenko, A. V.: 1971, in C. Sagan, T. C. Owen, and H. J. Smith (eds.), 'Planetary Amospheres', *IAU Symp.* **40**, 392.
Kovalevsky, J. and Link, F.: 1969, *Astron. Astrophys.* **2**, 398.
Kuiper, G. P.: 1949, *Astrophys. J.* **109**, 540.
Kuiper, G. P. (ed.): 1952, *The Atmospheres of the Earth and Planets*, Univ. of Chicago Press, Chicago, p.374.
Lutz, B. L. and Ramsay, D. A.: 1972, *Astrophys. J.* **176**, 521.
Mason, H. P.: 1970, *Astrophys. Space Sci.* **7**, 424.
Morrison, D. and Cruikshank, D. P.: 1973, *Astrophys. J.* **179**, 333.
Owen, T.: 1966, *Astrophys. J.* **146**, 611.
Owen, T.: 1967, *Icarus* **6**, 108.
Owen, T.: 1969, *Icarus* **10**, 355.
Owen, T., Lutz, B. L., Porco, C. S., and Woodman, J. H.: 1974, *Astrophys. J.* **189**, 379.
Prinn, R. G. and Lewis, J. S.: 1973, *Astrophys. J.* **179**, 333.
Savage, B. D. and Caldwell, J. C.: 1974, *Astrophys. J.* **187**, 197.
Sinton, W. M.: 1972, *Astrophys. J.* **176**, L131.
Smoluchowski, R.: 1973 in D. Gautier (ed.), *Physics of Planets*, Meudon: Groupe Planètes, p. 105.
Spinrad, H.: 1963, *Astrophys. J.* **138**, 1242.
Teifel, V. G. and Kharitonova, G. A.: 1970, *Astron. Zh.* **13**, 865.
Trafton, L.: 1974, this volume, p. 497.
Wallace, L.: 1972, *Astrophys. J.* **176**, 249.
Westphal, J. A.: 1972, *Bull. Am. Astron. Soc.* **4**, 361.
Woodman, J. H., Owen, T., and Encrenaz, T.: 1974, in preparation.

DISCUSSION

Fox: Is it possible that the 6420 Å feature is a combination of H_2 and CH_4 absorption?

Owen: It may be, but there are other arguments against the H_2 interpretation; first, the absence of the S(1) feature at 6470 Å, despite the measured continuum level at this wavelength, and second, the presence of structure at high resolution in the 6420 Å region.

Williams: Have there been supporting laboratory studies at Uranus/Neptune temperatures?

Owen: There have only been a few studies at relatively short pathlengths at the appropriate temperatures. We are hoping to carry out better studies when a new, temperature-controlled cell is completed in Canada next year.

Trafton: What is the spectral resolution of the laboratory CH_4 data at 6400 Å?

Owen: The dispersion is 4.5 Å mm^{-1}. The planetary spectrum has a resolution element of 4 Å.

NEPTUNE: OBSERVATIONS OF THE H₂ QUADRUPOLE LINES IN THE (4–0) BAND

L. TRAFTON

*McDonald Observatory and Dept. of Astronomy, University of
Texas at Austin, Tex., U.S.A.*

Abstract. The first measurement of Neptune's quadrupole H₂ lines is reported. The equivalent widths of the S(0) and S(1) lines of the (4–0) band are given along with the corresponding widths measured from comparison spectra of Uranus taken on the same nights. These are interpreted in terms of both an inhomogeneous atmosphere overlying a reflecting layer and a homogeneous, semi-infinite, scattering atmosphere. Only the scattering model proves to be consistent with Neptune's spectrum in this wavelength region. The H₂ abundance along the scattering mean free path is found to be less than the value for Rayleigh scattering in pure H₂. This result is interpreted in terms of the presence of H₂, CH₄ and at least one other gas, instead of the more conventional interpretation in terms of the presence of an aerosol mixed with H₂. Weak features in the continuum were observed. Their widths and the strength of the H₂ features indicate that H₂ is more abundant than the sum of the remaining gases in these atmospheres.

1. Introduction

The only previous evidence that H_2 is a constituent of Neptune's atmosphere is the presence of the pressure-induced absorption feature of H_2 at $\lambda 8270$, first identified by Herzberg (1952). This feature is also present in the spectrum of Uranus but not in that of Jupiter or Saturn owing to greater aerosol scattering in their atmospheres. Quadrupole lines from both the (3–0) and (4–0) bands have been measured in the spectra of Jupiter and Saturn (see McElroy, 1969 for a review). Likewise, Uranus' quadrupole lines of the (4–0) band have been measured (Giver and Spinrad, 1966; Trafton, 1973; Price, 1973) as well as the quadrupole lines of the (3–0) band (Lutz, 1973; Trafton, 1973). Quadrupole H_2 observations were not accomplished for Neptune owing to the faintness of Neptune's spectrum.

This difficulty has resulted in very little data being obtained by means of which Uranus and Neptune might be differentiated. Consequently, it is commonly assumed that the atmospheres of these two planets are essentially alike. Nevertheless, there are indications that their spectra differ in some respects. For example, Wamstecker (1973) recently has found that Neptune's spectrum exhibits CH_4 bands which absorb more weakly in their central regions than do Uranus' bands. Neptune's geometric albedo also appears to be somewhat lower in the continuum than Uranus' is. Neptune's seasonal changes certainly cannot be as extreme as Uranus' since Neptune's equator is inclined only 29 deg to its orbital plane. As Uranus' axis moves towards the Sun during the next decade, the spectra of the two planets may become entirely different. Only observations of Neptune can establish Neptune's character; reliance should not be placed on any assumed similarity between these two planets.

I report herein the first detection of the quadrupole lines of H_2 in the spectrum of Neptune. The equivalent widths of the S(0) and S(1) lines of the (4–0) overtone band

Woszczyk and Iwaniszewska (eds.), Exploration of the Planetary System, 497–511. All Rights Reserved

of H_2 are presented along with their implications for the state of H_2 in Neptune's atmosphere. Particular attention is paid to differences between the H_2 spectra of Neptune and Uranus.

2. The Observations and Their Reduction

The observations of Neptune occurred during the months of May, June and July, 1973. Observations of Uranus were also acquired to serve as comparisons for Neptune and to facilitate setting up the spectrograph for the observations of Neptune. The dates and other observational conditions are given in Tables I and II for Neptune and Uranus, respectively. I employed the coudé scanner (Tull, 1972) of the 272 cm (107-in.) telescope of the McDonald Observatory, using an RCA 31034 photomultiplier having a GaAs photocathode and cooled by dry ice to yield a dark count of 1.5 s^{-1}. The spectrograph employed an echelle grating in a double-pass Littrow mode with a cross-dispersing grating to separate the overlapping orders. The spectrographic parameters are given in Table III.

The entrance slit width projected to 3.5″ on the sky so that it included practically all of Neptune's light. The signal was typically 11 s^{-1} for the S(0) line and 8 s^{-1} for the S(1) line. This difference arose largely because of the different settings with respect to the blaze of the echelle. Observational sessions of Neptune ranged from 2 to 6 h each. During an observation, 40 channels, each of width 50 mÅ or 46 mÅ for the S(0) or S(1) line, respectively, were sequentially scanned in each direction of the

TABLE I

Observations of Neptune

PLN #	Date	Scan cycles	Final airmass	Seeing (arc-sec)	Temp/ humidity (C/%)	Maximum counts/ch		Doppler shift	
						Net	Dark	(Å)	(ch)
S(0) $\lambda 6435.0$									
3616	17 May 73	5654	2.2	1	16/31	2650	429	− 0.12	− 2.3
3622	18 May 73	6722	1.7	1 +	16/22	3225	510	− 0.11	− 2.1
3694 [a]	10 Jun 73	3683	1.7	3	14/51	1288	369	+ 0.14	+ 2.9
3805	17 Jun 73	10767	4.1	1	23/15	5419	525	+ 0.22	+ 4.3
3832	18 Jun 73	9345	3.2	1–2	22/10	4574	456	+ 0.23	+ 4.5
S(1) $\lambda 6367.7$									
3563	10 May 73	5042	1.8	1	18/34	2330	262	− 0.19	− 3.8
3730	13 Jun 73	5288	3.5	2	16/45	1487	264	+ 0.18	+ 3.5
3749	15 Jun 73	7551	2.2	$1\frac{1}{2}$	21/30	3130	410	+ 0.20	+ 3.9
3767	16 Jun 73	5042	3.0	2–3	19/46	1700	260	+ 0.21	+ 4.1
3866	7 Jul 73	5648	2.4	1–3	21/21	1866	406	+ 0.40	+ 8.0
3943	8 Jul 73	4475	3.0	2–3	18/57	1487	250	+ 0.41	+ 8.2
3984	9 Jul 73	5061	3.0	2–3	18/52	1422	293	+ 0.42	+ 8.4

[a] Omitted from sums owing to cirrus and poor seeing.

TABLE II

Comparison observations of Uranus

S(0) λ6435.0

PLN #	Date	Scan cycles	Final hour angle	Seeing (arc-sec)	Temp/ humidity (C/%)	Maximum counts/ch Net	Maximum counts/ch Dark	Doppler shift (Å)	Doppler shift (ch)
3601	12 May 73	856	–	1	23/12	2507	75	+ 0.33	+ 7.1
3615	17 May 73	423	– 0:16	1	16/24	1863	32	+ 0.36	+ 7.9
3620	18 May 73	2042	– 1:46	1 +	17/22	10305	153	+ 0.38	+ 8.2
3691	10 Jun 73	285	0:34	2	14/51	1142	40	+ 0.54	+ 11.8
3692	10 Jun 73	707	1:00	2	14/51	2719	100	+ 0.54	+ 11.8
3804	17 Jun 73	456	0:31	1	23/15	2399	27	+ 0.57	+ 12.4
3831	18 Jun 73	579	0:25	1	27/11	2961	39	+ 0.58	+ 12.6

S(1) λ6367.7

PLN #	Date	Scan cycles	Final hour angle	Seeing (arc-sec)	Temp/ humidity (C/%)	Maximum counts/ch Net	Maximum counts/ch Dark	Doppler shift (Å)	Doppler shift (ch)
3729	13 Jun 73	521	2:26	2	16/45	1765	29	+ 0.55	+ 12.1
3748	15 Jun 73	323	0:40	1½	21/30	1250	10	+ 0.56	+ 12.3
3765	16 Jun 73	982	0:50	2 –	22/40	2730	50	+ 0.57	+ 12.4
3865	7 Jul 73	354	2:42	1	21/28	1190	20	+ 0.62	+ 13.5
3942	8 Jul 73	387	3:49	2-3	19/55	1068	31	+ 0.62	+ 13.6
3983	9 Jul 73	452	3:12	2 +	19/50	1218	32	+ 0.62	+ 13.6

TABLE III

Spectrographic parameters

Detector	RCA 31034 photomultiplier with GaAs photocathode
Dark count	1.5 s⁻¹
Gratings	Echelle, with 79 grooves mm⁻¹ Cross disperser, with 300 grooves mm⁻¹ blazed at 1 μ
Orders	
S(0)	35
S(1)	36
Filter	OG-515
Slit width	1500 μ
Resolution element	0.15 Å
Channel width	10 steps per channel
Step duration	2 ms

spectrum with a dwell time of 20 ms per channel. The scans were coadded and displayed in real time.

I divided the observed spectra by a white light spectrum in order to reduce the effect of vignetting in the spectrograph. A white light spectrum was obtained immediately after each observation of Neptune, without altering the spectrographic parameters, by scanning a ground glass screen placed before the entrance slit and illuminated by an incandescent source. This procedure was only partially successful in eliminating the vignetting because the ground glass screen illuminated more of the collimator than did the $f/33$ beam from the telescope. However, the residual slopes of the continuum for the separate observations are in mutual agreement, including those of the comparison spectra taken of Uranus. These covered a wider bandpass, typically 60 channels, so they served the useful purpose of reducing the ambiguity in the residual slope of Neptune's continuum caused by weak features.

Because of the low signal, the individual observations of Neptune contained unacceptably high photon noise (as much as $\pm 3\%$ s.d.) for equivalent width measurements. To reduce this to acceptable values, I later summed the separate spectra obtained for a given line, shifting each in wavelength to account for the varying scan bandpass and Doppler shift. Summations of Neptune's spectra are shown in Figures 1 through 4. Figure 1 shows two partial summations of the S(0) line at

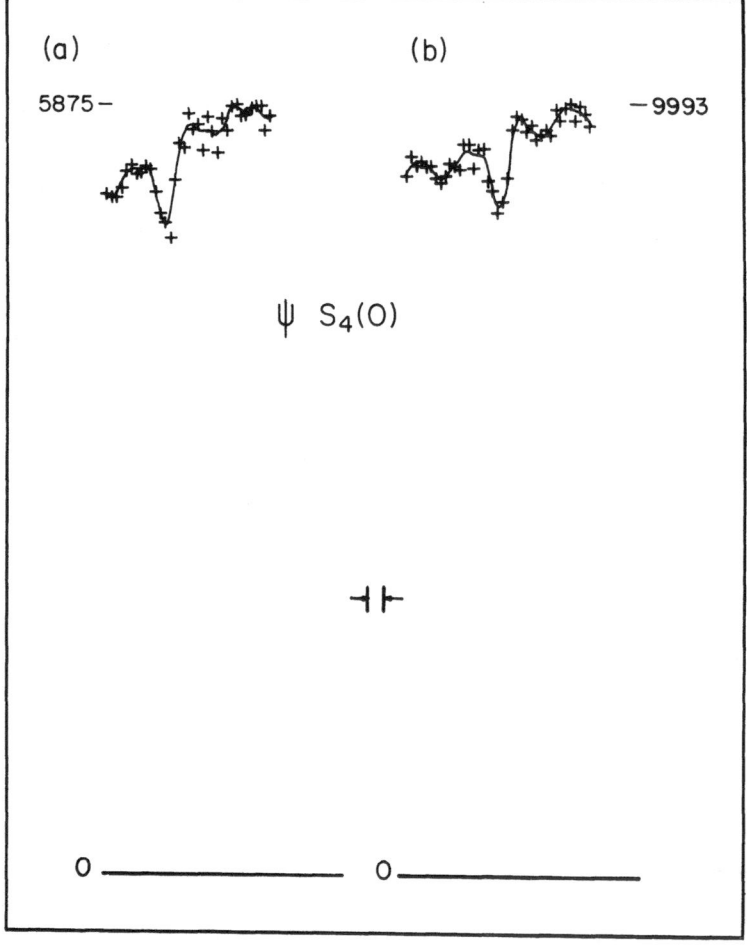

Fig. 1.   Partial summations of Neptune's $S_4(0)$ H_2 line. The pluses denote the data and the continuous line is the optimally smoothed spectrum. The numbers indicate the maximum number of counts acquired per channel for each summation, respectively. (a) Summation of the May observations. (b) Summations of the June observations minus PLN $\#3694$.

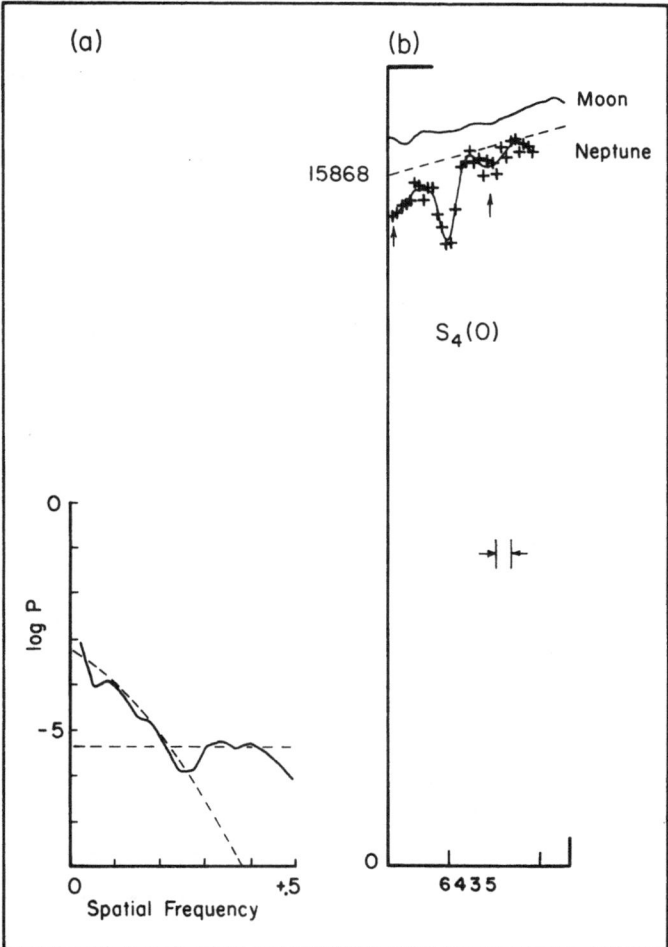

Fig. 2.   (a) Power spectrum for the summation of Neptune's $S_4(0)$ H₂ line. The ordinate is the logarithm to base 10 of the spectral power and the abscissa runs from 0 to the Nyquist frequency. The dashed lines indicate a parabolic modelling of the signal part of the power spectrum and the whitenoise level. This model of the power spectrum determines the optimum smoothing. (b) Summations of Neptune's $S_4(0)$ H₂ line with Lunar comparison. The dashed line is the adopted continuum level; its slope was determined from wider spectra of Uranus. The pluses denote the data and the curve depicts the optimally smoothed spectrum. The vertical arrows mark the positions of weak features distinctly visible in the spectra of both Uranus and Neptune. The spectrographic resolution element is depicted schematically. The number is the maximum count acquired per channel and the scale is 20.0 ch Å⁻¹.

$\lambda 6435.0$ for comparison and Figure 2 shows the summation of all the observations for S(0) except PLN # 3694, which was omitted owing to poor observing conditions. Figure 3 presents two partial summations for Neptune's S(1) line at $\lambda 6367.7$ and Figure 4 gives the summation of all the observations of this line. Summations of the S(0) and S(1) lines in the spectra of Uranus are shown in Figure 5 for comparison with Neptune. The partial sums are entirely independent and so illustrate the effect

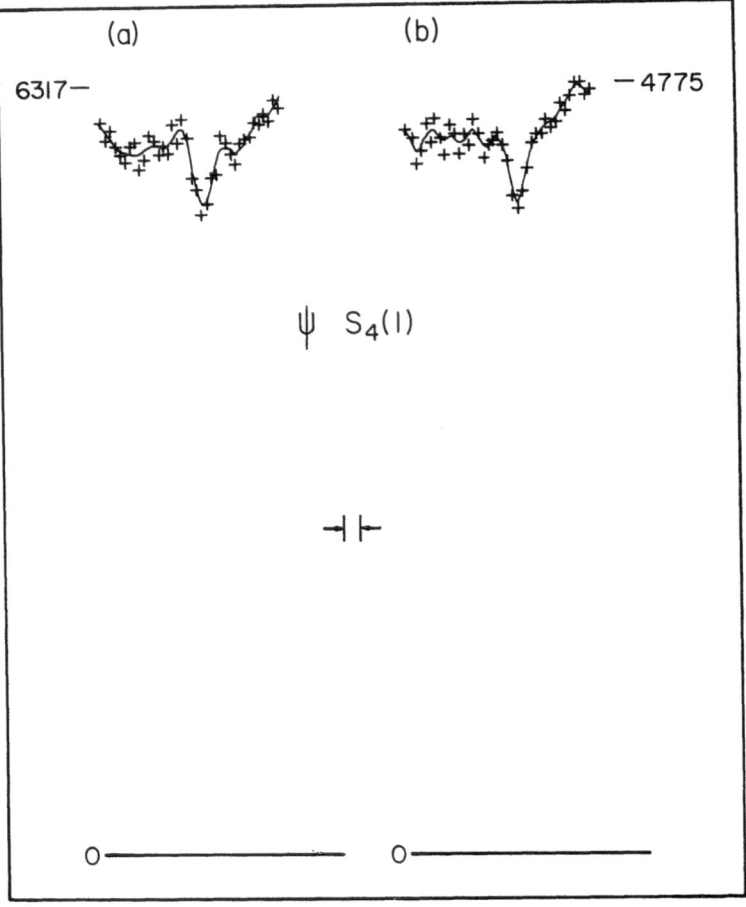

Fig. 3.   Partial summations of Neptune's $S_4(1)$ H₂ line. The pluses denote the data and the continuous line is the optimally smoothed spectrum. The numbers indicate the maximum number of counts acquired per channe for each summation, respectively. (a) Summation of the June observations. (b) Summation of the July observations.

of photon noise and scintillation on the observations. The photon noise in the total sums is less than 1% since the total count is greater than 10^4 per channel.

Figures 3 and 4 also show lunar comparison spectra and the power spectra for the observations of Neptune. The power spectrum is a measure of the contribution that each frequency makes to the observed spectrum (cf. Bracewell, 1965) and is used to determine the optimum smoothing of the data. Following Brault and White (1971), I modelled the signal part of the power spectrum by a parabola and the noise part by a horizontal line corresponding to a white noise level. Each spectral frequency was weighted proportionally to the ratio of the signal power spectrum to the total power spectrum. The inverse transform then gives the optimally smoothed spectrum (Brault and White, 1971). The only subjective element in this smoothing arises from fitting

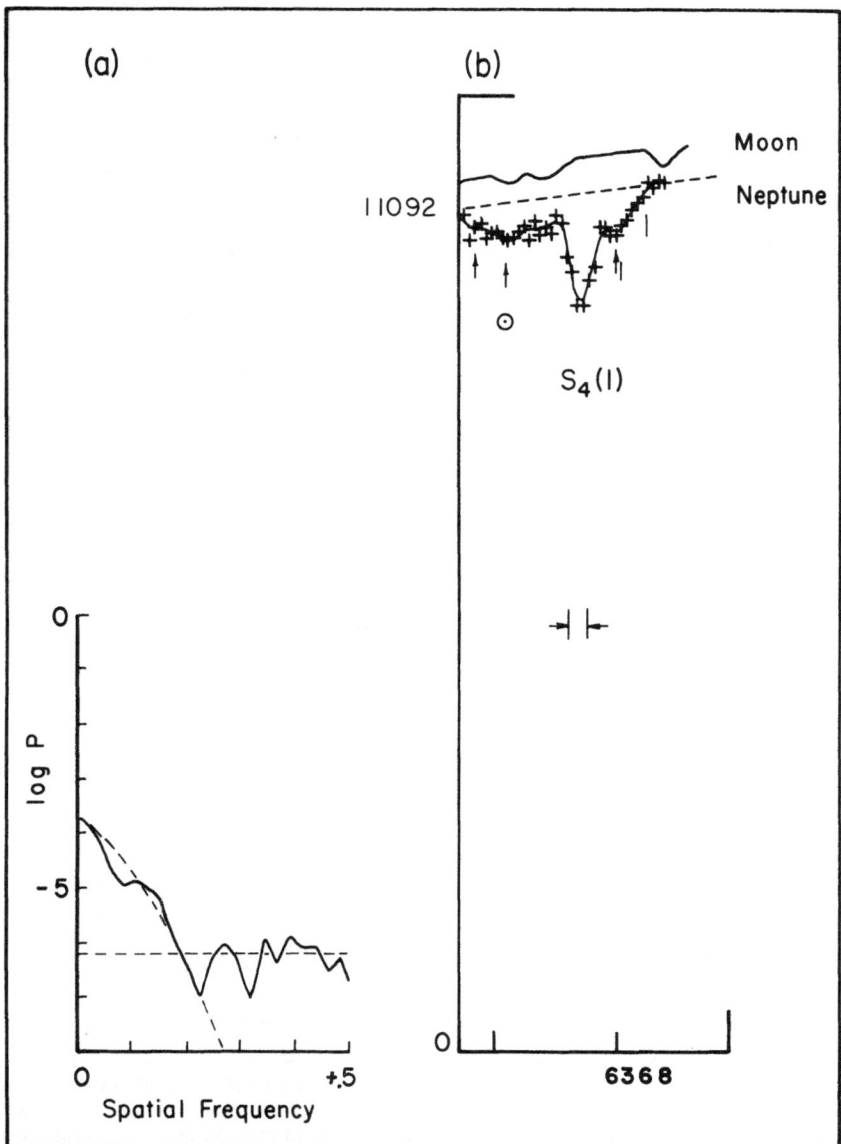

Fig. 4.   (a) Power spectrum for the summation of Neptune's S$_4$(1) H$_2$ line. The ordinate is the loga-rithm to base 10 of the spectral power and the abscissa runs from 0 to the Nyquist frequency. The dashed lines indicate a parabolic modelling of the signal part of the power spectrum and the white noise level. This model of the power spectrum determines the optimum smoothing. (b) Summations of Neptune's S$_4$(1) H$_2$ line with lunar comparison. The dashed line is the adopted continuum level; its slope was determined from wider spectra of Uranus. The pluses denote the data and the curve depicts the optimally smoothed spectrum. The vertical arrows mark the positions of weak features distinctly visible in the spectrum of Uranus. The lines denote the location of a telluric H$_2$O feature occurring in Doppler shifted positions in Neptune's summation spectrum. This corresponds to the rightmost feature in the lunar spectrum. The symbol ⊙ indicates that a solar line may explain the corresponding mean absorption feature. The spectrographic resolution element is depicted sche-matically. The number is the maximum count acquired per channel and the scale is 21.9 ch Å^{-1}.

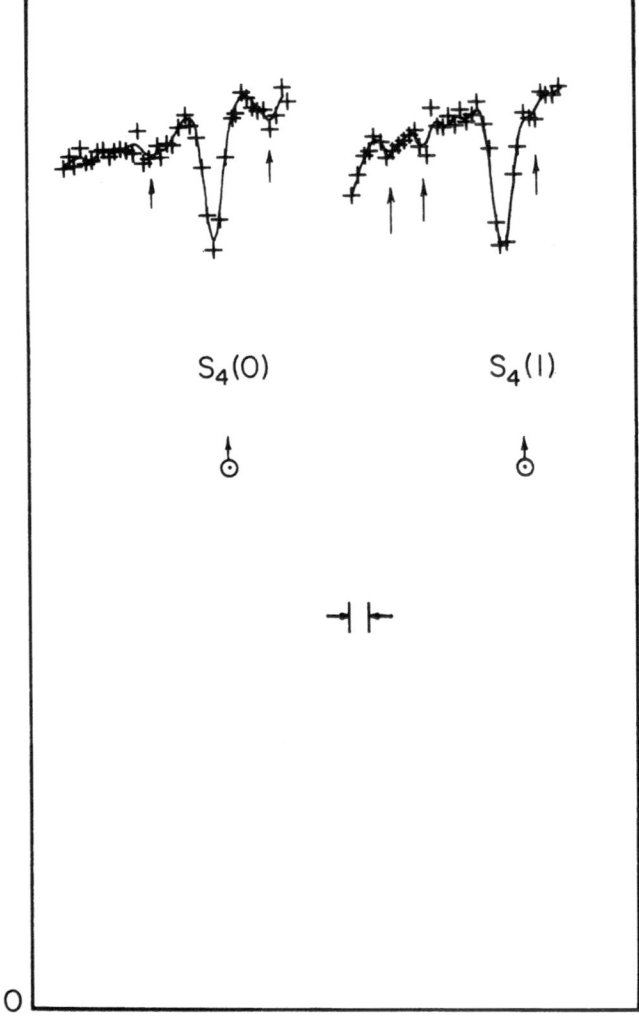

Fig. 5.   Observations of Uranus' $S_4(0)$ and $S_4(1)$ H$_2$ quadrupole lines obtained for comparison with Neptune's spectra. The pluses denote the data and the curves are the optimally smoothed spectra. The arrows mark the positions of weak absorption features. The $S_4(0)$ line is PLN $\#$3620 and has 10305 counts in its most highly filled channel. The $S_4(1)$ line is the summation of all the spectra obtained during the same nights as Neptune's $S_4(1)$ observations plus one more taken earlier. This summation has 10488 counts in its most highly filled channel. The standard deviation of the photon noise should be less than 1 % of the continuum.

this model of the power spectrum to the actual one. Four points in the power spectrum must be chosen to effect the smoothing. The modelled power spectra are depicted in Figures 2a and 4a. The corresponding smoothed spectra are presented in Figures 2b and 4b. Note that the dispersion of the data points about their smoothed value is significantly reduced in the total summation of the observations for each line.

The greatest uncertainty in the measurement of the equivalent widths of the H$_2$ lines arises from the uncertainty in the location of the continuum rather than from the shape of the lines. Although the spectroscopic resolution element is 0.15 Å, there appears to be no region within the spectral bandpass of our observations which is sufficiently free of absorption features to permit establishing the continuum level accurately. The neighborhood of the $S(0)$ line shows weak planetary features in both the spectra of Uranus and Neptune at $+0.50$ Å and -0.60 Å relative to the H_2 line. The $S(1)$ line neighborhood shows features at -0.62 Å and -0.87 Å relative to the H_2 line and a blend at $+0.27$ Å in the wing of the Neptune H_2 line. The feature at -0.62 Å is probably the 2.5 mÅ solar feature at $\lambda 6367.13$ and the blend is partly a telluric H_2O line at $\lambda 6368.46$ and a weak planetary feature visible in the partial summation in Figure 3. Here, the Doppler shift clears the H_2O feature, revealing a weak absorption feature in the wing of the $S(1)$ line. These features may be slightly stronger in the spectrum of Neptune.

The adopted continuum levels are depicted in Figures 2b and 4b where the slopes have been determined using the wider spectra of Uranus as discussed above. The equivalent widths were measured by approximating the shape of the H_2 lines by a triangle. The legs of the triangle were placed tangent to the line and the continuum

TABLE IV

Equivalent widths for
H_2 lines in the (4-0) band

Neptune		Uranus comparisons	
S(0)	28 ± 4 mÅ	S(0)	29 ± 4 mÅ
S(1)	31 ± 4 mÅ	S(1)	33 ± 4 mÅ

level. Table IV gives the measured equivalent widths of Neptune's (4-0) H_2 features, as well as Uranus' for comparison. The uncertainties are estimated errors. Any systematic error resulting from the continuum placement would affect the spectra of both planets equally so that the relative measurements would still be meaningful.

3. Analysis

Upon comparing the spectra of Uranus and Neptune in Figures 2, 4 and 5, one notices that the H_2 lines in the spectra of Neptune are partially washed out with respect to their counterparts in the spectrum of Uranus. The lines are wider and their central intensities are greater. Their equivalent widths are only slightly less than the values measured for Uranus' features (refer to Table IV). Furthermore, the resolution of the weak features in the continuum appears to be similarly degraded in Neptune's spectrum. Since the pressure broadening coefficient of H_2 is small (0.0015 cm^{-1} atm^{-1} at room temperature) compared to that for CH_4, it appears unlikely that this could be simply a pressure effect. Pressures high enough to broaden significantly an H_2 spectral line of width 0.15 Å would wash out any weak CH_4 features.

Because of Doppler shifts arising from planetary rotation, one would expect the observed spectral resolution element to be larger than that corresponding to the slit width. Neptune's rotation period is $\frac{3}{2}$ of that for Uranus but Uranus' poles are inclined 57 deg to the line of sight while Neptune's is inclined much more. Furthermore, the observations of Uranus were made near the central meridian where at the coudé focus, only about 88% of Uranus' projected equatorial diameter entered the slit. These effects would reduce the Doppler broadening somewhat for Uranus.

The spectrograph itself can increase or decrease the resolution of spectra obtained from a rotating planet as the angle of the rotation axis with the slit is varied (Deeming and Trafton, 1970). For the June scans, Uranus' orientation favored an enhancement in the resolution. On the other hand, Neptune's orientation seldom was favorable because its axis rotated considerably with respect to the spectrographic slit during the long exposures. From these considerations, I estimate that the Doppler smearing of Uranus' features should be no more than 70% of the smearing for Neptune. The resulting Doppler shifts of the equatorial limbs of the two planets are 1.4 and 2.0 channels, respectively. Limb darkening and the circular geometry will weight the less Doppler shifted components more. In any case, the residual Doppler shift is less than 0.6 channel between the two planets. This appears to be less than the observed difference so that Neptune's spectral features may be intrinsically broader than Uranus' are, depending on the degree of possible enhancement of Uranus' spectral resolution.

The scan summations for both planets are slightly degraded in resolution owing to the change in Doppler shift with time and the constraint of allowing shifts of only integral channels in the summations. This should essentially affect the spectra of both planets equally since the number of spectra for each planet is about the same and a spectrum of each planet was obtained during a given night. This should not result in a spectral widening of more than one channel.

Since the entire disk of Neptune was included in the observations, it was necessary to average over the disk the equivalent width $W(\mu)$ calculated from models in order that theory may be compared with the observations. Here, μ is the direction cosine of the emergent radiation. The appropriate average is

$$\bar{W} = \frac{\int\limits_0^\infty W(\mu) I(\mu) \mu \, d\mu}{\int\limits_0^\infty I(\mu) \mu \, d\mu}.$$

For $I(\mu)$, I assumed the result for Uranus in the range $\lambda\lambda 3800-5800$ obtained from Stratoscope II by Danielson et al. (1972).

I considered two extreme models for $W(\mu)$. The first is an inhomogeneous model consisting of a nonscattering gas overlying a reflecting layer. The structure of this model follows the calculations of Trafton (1967) and neglects all scattering. I have computed results for the inhomogeneous case for effective temperature 60 K since measurements of Neptune's $20\,\mu$ flux by Morrison and Cruikshank (1973) indicate that Neptune's effective temperature may be as high as Uranus' is.

This model shows a curve of growth for the $S(1)$ line which increases much more steeply than that for the $S(0)$ line. The reason for this is that the state $J=1$ is sparsely populated in the upper layers of Neptune's atmosphere but becomes considerably more populated in the deeper layers. The population of the $J=0$ state, on the other hand, varies slowly over this range. This line is consequently less sensitive to the temperature structure. It implies an equilibrium H_2 abundance above the reflecting layer of 500 to 800 km-amagats, depending on the pressure. Other gases in the atmosphere would result in a lower H_2 abundance because they would increase the pressure.

The $S(1)$ line is sensitive to the effective temperature of the model. For my $T_e = =60\,K$ model, it implies an H_2 abundance of only 300 to 400 km-amagats for the same pressure range. Lowering the value of T_e would bring the ortho-para H_2 ratio closer to that for an equilibrium mixture, as would increasing the pressure by adding other gases.

The reflecting layer model is not appropriate for Neptune's atmosphere at these wavelengths because it implies quantities of gas so large that Rayleigh scattering cannot be ignored. The mean free path for light at these wavelengths is about 715 km-amagats H_2. My observations include the whole disk and would correspond to passage twice through the atmosphere. The observed strength of the $S(0)$ line then implies that most of the light would have been scattered. This result follows even for non-equilibrium ortho-para H_2 mixtures in Neptune's atmosphere. That scattering cannot be neglected in the formation of the (4–0) H_2 quadrupole lines is a conclusion Belton *et al.* (1971) came to regarding Uranus' spectrum.

Therefore, I considered the opposite extreme; namely, a homogeneous model consisting of a semi-infinite, isotropically scattering atmosphere (see, for example, Chamberlain, 1970). The results are analyzed in terms of the abundance of H_2 along a scattering mean free path (unit optical depth for scattering). I determined the continuum geometric albedo at this wavelength by measuring Neptune's spectral reflectivity with sufficient resolution to reveal the large-scale structure of the CH_4 bands and the interspersed continuum. The spectral reflectivity I derived by ratioing scans of Neptune to scans of the Moon and then correcting the result by the lunar reflectivity values obtained by McCord and Johnson (1970). The geometric albedos of Wamstecker (1973) at $\lambda 5000$ provided the necessary normalization to the spectral reflectivity scans. The continuum reflectivities between the CH_4 bands near $\lambda 6500$ fall nearly on a smooth curve so I used the value of this curve at $\lambda 6400$ to derive Neptune's geometric albedo in the continuum for the (4–0) H_2 lines. Assuming isotropic scattering, this leads to a single scattering particle albedo of 0.91. This value indicates that overlap from CH_4 lines is probably significant in the continuum.

Values of $\bar{W}$ were calculated from this model for a variety of temperatures and pressures. Figure 6 shows a self-consistent result giving 320 km-amagats of H_2 along a scattering mean free path for $W=28$ mÅ in the $S(0)$ line and $W=32$ mÅ in the $S(1)$ line. The corresponding temperature is 95 K and the effective pressure, P_e, is

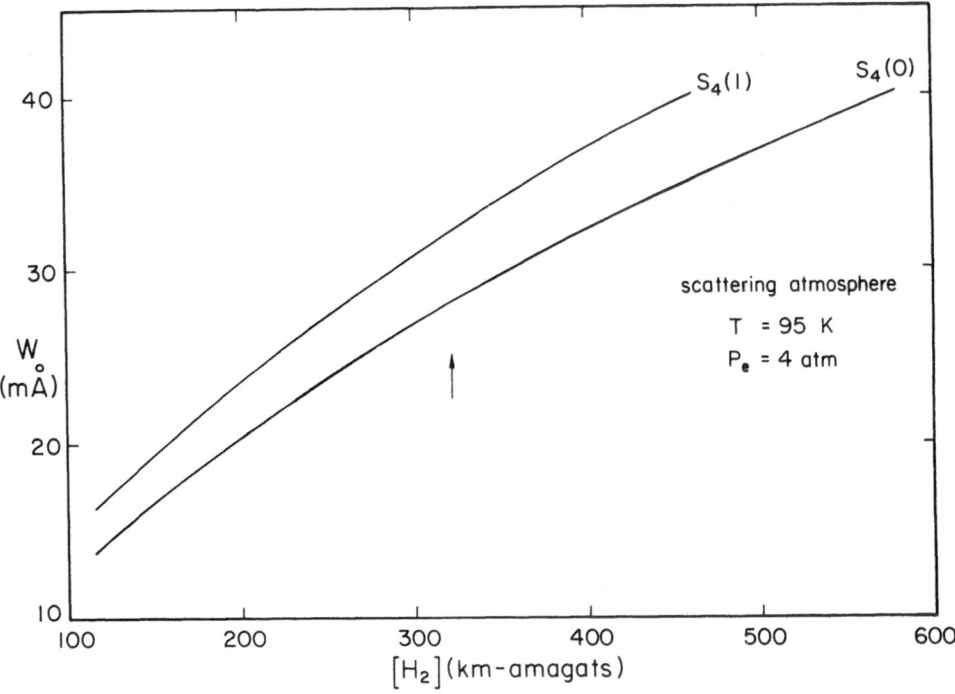

Fig. 6. Curves of growth for Neptune's (4–0) H_2 quadrupole lines. A semi-infinite, isotropically scattering atmosphere is employed, so the abscissa denotes the amount of H_2 along a scattering mean free path. To make the model consistent with the observations, a temperature of 95 K was chosen and an equilibrium distribution of ortho- and para H_2 was assumed. The results are for an effective pressure of 4 atm. The arrow indicates that a specific abundance of 320 km-amagats of H_2 is consistent with 28 mÅ and 32 mÅ for S(0) and S(1), respectively. The actual specific abundance of H_2 depends on the relative abundance and weight of the other gases in Neptune's atmosphere (see text).

4 atm. The value of the effective temperature is rather insensitive to P_e. This pressure is close to that which this quantity of H_2 would have at its base (3.2 atm) if arranged in a vertical column. Because of the inhomogeneous structure of the atmosphere and the random orientation of the scattering paths, the actual effective pressure may be less than this value, but this inference is uncertain because other gases may be present. For the physical conditions in this scattering model of Neptune's atmosphere, an increase in effective pressure results in an increase in $\bar{W}$. Therefore, the pressure broadening term dominates the collisionally narrowing term (cf. James, 1969) in the expression for the equivalent width. Reducing P_e by a factor of four increases the H_2 abundance along the scattering mean free path by only a factor of 1.39 when the absorption is held fixed. It is unlikely that the effective pressure is less than 1 atm because of the weight of the gas inferred from the observations and the isotropic character of the scattering. This minimum pressure sets a conservative upper bound of 450 km-amagats on the amount of H_2 in a mean free path, regardless of the model parameters.

Since the mean free path for Rayleigh scattering is 715 km-amagats H_2, the observations imply the existence of an aerosol or another gas in Neptune's atmosphere which reduces the scattering mean free path well below the Rayleigh scattering value for H_2.

It is unlikely that the effective pressure is much larger than about 3 atm because weak features in Neptune's spectrum are visible giving half-widths of about 6 channels, or a third of an ångström. If these arise from CH_4, their Lorentz width at 100 K can be estimated from the result Varanasi *et al.* (1973) obtain for the v_3 band for CH_4 broadened by H_2, $\alpha_0 = 0.13 \pm 0.01$ cm^{-1} atm^{-1}. Using the relation $\alpha = \alpha_0 P_e$, where α is the half-width at half maximum, I estimate that $P_e \lesssim 2.8$ atm. For Uranus, $P_e \lesssim 1.9$ atm. If the continuum is higher, the width of the weak features would be greater and so also would the value of P_e. It is quite unlikely, however, that P_e exceeds twice this estimate.

4. Conclusions

Comparing Neptune with Uranus, the equivalent widths of the ortho- and para H_2 lines have about the same ratio in the spectra of the two planets. This is consistent with the notion of an equilibrium mixture of ortho- and para H_2 having the same temperature at the levels for which absorption effectively occurs in the scattering atmospheres of the two planets. The effective pressure may well be greater in Neptune's atmosphere since Neptune's surface gravity is 1.35 times greater. This would be consistent with the residual broadening of the weak features observed in Neptune's spectrum. In this case, the amount of H_2 in a scattering mean free path would be less in Neptune's atmosphere since the equivalent widths are comparable.

Since the geometric albedo of Neptune is less than Uranus' in this region of the spectrum (Wamstecker, 1973), the effective level of line formation may be higher in Neptune's atmosphere. If this is the case, the near equality in the rotational temperatures derived for these two planets would support the high effective temperature inferred from the measurements of Morrison and Cruikshank (1973). This reasoning depends on the steep scattering model for the atmospheres of these planets and the assumption that the relative strengths of the ortho- and para H_2 lines are determined by the local temperature.

Current thinking attaches any discrepancy between the equivalent widths of Uranus and those corresponding to a purely Rayleigh scattering atmosphere to the presence of an aerosol in the atmosphere (Belton *et al.*, 1971; Price, 1973). I find it odd that the density of any such aerosol would be just that giving a scattering mean free path within a factor of two or three of the Rayleigh value for pure H_2, particularly for both Uranus and Neptune. Aerosol densities several orders of magnitude higher or lower than this value might just as well occur. It seems much more likely that the observed H_2 abundance along the scattering mean free path results from the Rayleigh scattering in a *mixture* of H_2 with other gases in the atmospheres of these planets. In this case, no aerosol at all is required since the amount of H_2 along a mean free path is reduced when other gases are added. The mean free path itself is reduced since a photon may then be scattered by molecules other than H_2. Furthermore, since the ratio of H_2 to the remaining gases in these atmospheres is not likely to deviate more than an order of magnitude from unity, a mixture would explain the narrow disparity between the observed H_2 specific abundances and that corresponding to Rayleigh scattering in pure H_2.

These two physical effects can be distinguished by observation since Rayleigh scattering varies with wavelength as λ^{-4} while aerosol scattering varies much more slowly. Furthermore, if other gases are responsible for the increased scattering, their weight would increase the effective pressure considerably while the aerosols have hardly any effect on the total pressure. The presence of other gases may alter the line broadening coefficient of H_2 and so affect the equivalent widths. Certainly, the increased pressure would reduce the estimates for the H_2 specific abundance since less H_2 would be required to produce the same absorption. This effect raises the estimate for the quantity of other gases along the scattering mean free path, and the resulting increased pressure feeds back to amplify this process. Convergence quickly occurs, though, since the optical depth from Rayleigh scattering along a mean free path is constrained to unity.

I have considered the effect of a number of gases in Neptune's atmosphere by neglecting their influence on the line broadening coefficient of H_2 but including their effect in broadening the H_2 lines as a result of increasing the total pressure. The amount of unknown gas required to give the observed H_2 equivalent widths depends on its mean molecular weight as well as on its refractive index and the amount of H_2.

This method is not effective for inferring the He/H_2 ratio because the Rayleigh scattering of He is 1/15 that of H_2. For example, if Neptune's equivalent widths are interpreted in terms of a H_2–He atmosphere having no aerosols, about 150 km-amagats of H_2 and 8800 km-amagats of He are required to give unit optical depth in Rayleigh scattering at $\lambda 6400$ and also to give the observed equivalent widths for the homogeneous scattering model at 95 K. The effective pressure is roughly approximated by half the weight of the gas in a mean free path when stacked in a vertical column. This yields an effective pressure of about 90 atm, a result clearly excluded by the observed widths of the weak features in the continuum. This means that any He in Neptune's atmosphere has a negligible effect on the scattering mean free path and can be ignored except for its effect on the total pressure.

A number of gases, such as A, N_2, O_2 and H_2O have very similar refractive indices; namely twice that of H_2. If these are lumped together with a designated mean molecular weight of 30, then they are compatible with the H_2 observations if there is 280 km-amagats of H_2 along the mean free path and 102 km-amagats of this unknown mixture. The resulting effective pressure would be about 8 atm. This pressure also appears to be higher than permitted by the widths of the weak features in Neptune's continuum. The conclusion must be that gases of these refractivities and molecular weights are not primarily responsible for the reduction in the mean free path from the Rayleigh scattering value in pure H_2.

The final category of other gases contains those gases having large refractive indices and low molecular weights. Prime examples are CH_4 and NH_3. To illustrate, a H_2–CH_4 atmosphere for Neptune would contain about 320 km-amagats of H_2 and

38 km-amagats of CH_4 along the scattering mean free path for an effective pressure of about 3 atm. This would correspond to a CH_4/H_2 ratio of about 0.12. This result is consistent with the observations and implies a rather high relative abundance of CH_4 compared to that in Jupiter or Saturn. However, the laboratory comparisons of Owen (1967) suggest smaller specific CH_4 abundances. Of course, other gases of high refractivity and low molecular weight may also be contributing to the reduction of Neptune's mean free path, in which case this abundance is characteristic of the sum of CH_4 and these unknown gases.

The major conclusion I draw from this analysis is that merely the visibility of weak features in the spectra of Uranus and Neptune indicates that H_2 must be more abundant than the sum of the other atmospheric constituents in these atmospheres because the spectra of these planets also imply large amounts of H_2 along the mean scattering paths. This result was previously suspected but until now lacked an observational basis. It contradicts an early deduction of Herzberg (1952), the basis of which has been recently questioned (cf. McElroy, 1969), that He should be more abundant than H_2 in these atmospheres. The second most abundant gas is likely to have a mixing ratio significantly smaller than unity. If an aerosol were responsible for the increased scattering, there would be no possibility of having gases other than H_2 and CH_4 present in the visible atmospheric regions in more than trace amounts.

Acknowledgements

It is a pleasure to acknowledge the assistance of J. Woodman in obtaining these observations. This research was supported as one phase of NASA Grant NGR 44-012-152.

References

Belton, M. J. S., McElroy, M. B., and Price, M. J.: 1971, *Astrophys. J.* **164**, 191.
Bracewell, R.: 1965, in *The Fourier Transform and Its Applications*, McGraw-Hill Book Co., New York, Chapter 6.
Brault, J. W. and White, O. R.: 1971, *Astron. Astrophys* **13**, 169.
Chamberlain, J. W.: 1970, *Astrophys. J.* **159**, 127.
Danielson, R., Tomasko, M., and Savage, B.: 1972, *Astrophys. J.* **178**, 887.
Deeming, T. J. and Trafton, L. M.: 1970, *Appl. Opt.* **10**, 382.
Giver, L. P. and Spinrad, H.: 1966, *Icarus* **5**, 586.
Herzberg, G.: 1952, *Astrophys. J.* **115**, 337.
James, T. C.: 1969, *J. Opt. Soc. Am.* **59**, 1602.
Lutz, B. L.: 1973, *Astrophys. J.* **182**, 989.
McCord, T. B. and Johnson, T. V.: 1970, *Science* **169**, 855.
McElroy, M. B.: 1969, *J. Atmospheric Sci.* **26**, 798.
Morrison, D. and Cruikshank, D. P.: 1973, *Astrophys. J.* **179**, 329.
Owen, T. C.: 1967, *Icarus* **6**, 108.
Price, M. J.: 1973, *Bull. Am. Astron. Soc.* **5**, 291.
Trafton, L. M.: 1967, *Astrophys. J.* **147**, 765.
Trafton, L. M.: 1973, *Bull. Am. Astron. Soc.* **5**, 290.
Tull, T. G.: 1972, in S. Laustsen and A. Reiz (eds.), *Proc. ESO/CERN Conference on Auxiliary Instrumentation for Large Telescopes*, Geneva, p. 259.
Varanasi, P., Sarangi, S., and Pugh, L.: 1973, *Astrophys. J.* **179**, 977.
Wamstecker, W.: 1973, *Astrophys. J.* **184**, 1007.

LA ROTATION, LA CARTOGRAPHIE ET LA PHOTOMETRIE
DES SATELLITES DE JUPITER

AUDOUIN DOLLFUS et JOHN B. MURRAY

Observatoire de Meudon, France

Résumé. Les 4 satellites galiléens de Jupiter ont été observés fréquemment depuis 20 ans avec les télescopes à haute résolution du Pic-du-Midi. Des configurations permanentes ont été vues et photo-graphiées à leur surface. Elles ont fait l'objet de cartes-planisphères. Leur identification prouve que les périodes de rotation sont synchrones avec les périodes de révolution pour chacun des 4 satellites, avec un axe de rotation perpendiculaire au plan des orbites dans la limite de la précision des mesures. Les passages des satellites devant le disque de Jupiter donnent une méthode simple pour déterminer les contrastes et les albedos des différentes taches observées sur les satellites. Pour Callisto, les albedos pour l'angle de phase 5° couvrent le domaine de 0,19 à 0,09, le contraste est entre les différentes régions valant 0,5; ces valeurs indiquent une surface de type lunaire, bien que la faible densité entraîne une structure interne différente. Pour Europe, les albedos très élevés sont compris entre 0,73 et 0,52 avec un faible contraste de 0,3, tout à fait compatible avec un dépôt de neige ou de givre, l'observation spectroscopique indiquant la présence de l'eau. Pour Ganymède, les contrastes très élevés peuvent atteindre 0,7 et les albedos s'étendent de 0,5 à 0,15; des grandes étendues de neige sur un sol du type de Callisto rendraient compte de ces valeurs. Io a un albedo très élevé de 0,83 dans la région équatoriale, mais de 0,46 au voisinage des deux pôles assombris; la couleur jaune et l'absence des bandes de H_2O dans le spectre indiquent une surface de composition différente.

Abstract. The four Galilean satellites of Jupiter have been observed frequently during the past 20 years with the high resolution telescopes of Pic-du-Midi. Reliable features are seen and photographed at their surface. They have been mapped on planispheres. Their identification proves that the rotation period is synchronous with the revolution for each satellite, with an axis not significantly departing from the normal to the orbital plane. The transits of satellites in front of the limb-darkened Jupiter disk provide a technique for accurate contrast and albedo determinations of the features observed. For Callisto, the albedos at 5° phase angle range from 0.19 to 0.09 with a contrast 0.5, indicating a lunar type surface, despite the low density involving a different internal structure. For Europa, the albedos are very high, between 0.73 and 0.52, with low contrasts of 0.3, readily explainable by a snow or frost deposit, the spectroscopic evidence indicating water. For Ganymedes, the very high contrasts reach 0.7, and albedos span from 0.50 to 0.15; patches of snow on Callisto-type terrains are suggested. Io has very high albedos of 0.83 for the equatorial zone to 0.46 for the two darkened poles; the yellow color and absence of H_2O bands in the spectrum indicate a special surface composition.

1. Introduction

Les quatre satellites galiléens de Jupiter, Io, Europe, Ganymède et Callisto mesurent entre 3100 km pour Europe et 5500 km pour Ganymède et ces corps ont des volumes comparables à ceux de la Lune, Mercure ou Mars. Ce sont donc des objets célestes d'une importance évidente. De plus, leur situation en orbite autour de Jupiter et, par conséquent, leur origine cosmogonique particulière, contribuent encore à leur conférer une valeur spéciale pour la compréhension des grands problèmes sur l'origine et l'évolution du système solaire.

Vu de la Terre, leur diamètre apparent est compris entre 0,8″ et 1,2″, tandis que le pouvoir séparateur angulaire de nos meilleurs télescopes peut atteindre, lorsque la turbulence atmosphérique l'autorise exceptionnellement, 0,15″. C'est dire que la

Woszczyk and Iwaniszewska (eds.), Exploration of the Planetary System, 513–525. All Rights Reserved

surface de leur disque apparent comprend 20 à 60 données d'information indépendantes et la couverture topologique est suffisante pour permettre l'observation des
configurations à leur surface. Ainsi, l'observateur discerne au télescope des taches
claires ou sombres, il peut les comparer et suivre leurs transformations selon les
paramètres orbitaux.

L'analyse de ces propriétés de surface n'est pas aisée; elle a nécessité, pour nous,
l'emploi des télescopes de haute montagne du Pic-du-Midi. Il faut alors que Jupiter
soit assez haut dans le ciel, que l'observateur se trouve sur place, que le télescope soit
disponible, que le ciel soit dégagé de nuages, que les images télescopiques soient
exceptionnelles, et ces contraintes étalèrent le programme sur de nombreuses années
avant que ne puisse être assemblé le nombre d'observations nécessaire.

De 1941 à 1945, B. Lyot, ainsi que H. Camichel et M. Gentili ont entrepris une
première reconnaissance des configurations de surfaces des satellites, avec les réfrac-
teurs de 38 à 60 cm alors disponibles au Pic-du-Midi (Camichel *et al.*, 1943; Lyot,
1953). Notre programme de surveillance a débuté en 1958 avec le réfracteur de 60 cm,
puis le réflecteur de 107 cm du Pic-du-Midi, ainsi que la grande lunette de 83 cm de
Meudon.

2. Observations recueillies

La figure 1 montre l'aspect télescopique des quatre satellites, pour différentes longitudes
orbitales comptées de 0 à 360° à partir de la conjonction supérieure géocentrique. Les
grossissements employés sont généralement voisins de 1000.

Io, dont la couleur est jaune, montre des régions polaires fortement assombries,
tandis que les contrées équatoriales, très claires, sont marbrées de faibles traînées
souvent méridiennes.

Europe, très blanc, révèle au contraire des régions polaires toujours bien claires,
et quelques marbrures très pâles, de préférence dans la zone équatoriale.

Ganymède, beaucoup plus facile à observer, montre aisément des régions blanches
et des taches sombres formant de forts contrastes, la région polaire nord est volontiers
claire et très blanche.

Callisto est sombre; l'observation, un peu gênée par le manque de lumière, laisse
percevoir des taches permanentes de contrastes plus modérés.

Nos séries visuelles comportent maintenant près de 50 observations. A cette
documentation, s'ajoutent les séquences d'observation de Lyot, Camichel et Gentili
recueillies de 1941 à 1945 mentionnées ci-dessus, 7 dessins de Ganymède et 3 de Callisto
obtenus par A. Danjon à Strasbourg 1934 avec un instrument plus petit de 48,6 cm
(Danjon, 1944), 2 observations isolées de Ganymède par E. Whitaker à Catalina
(Arizona).

Le degré de confiance que l'on peut prêter à ces explorations est illustré par la
figure 2 sur laquelle ont été groupées six observations indépendantes de Ganymède à
peu près sous la même présentation orbitale. Le dessin de Danjon relève d'un télescope
moins puissant. Tous les observateurs s'accordent à voir une région Nord très blanche,
limitée par un rebord sombre, avec une tache sombre au nord-est, des marbrures en

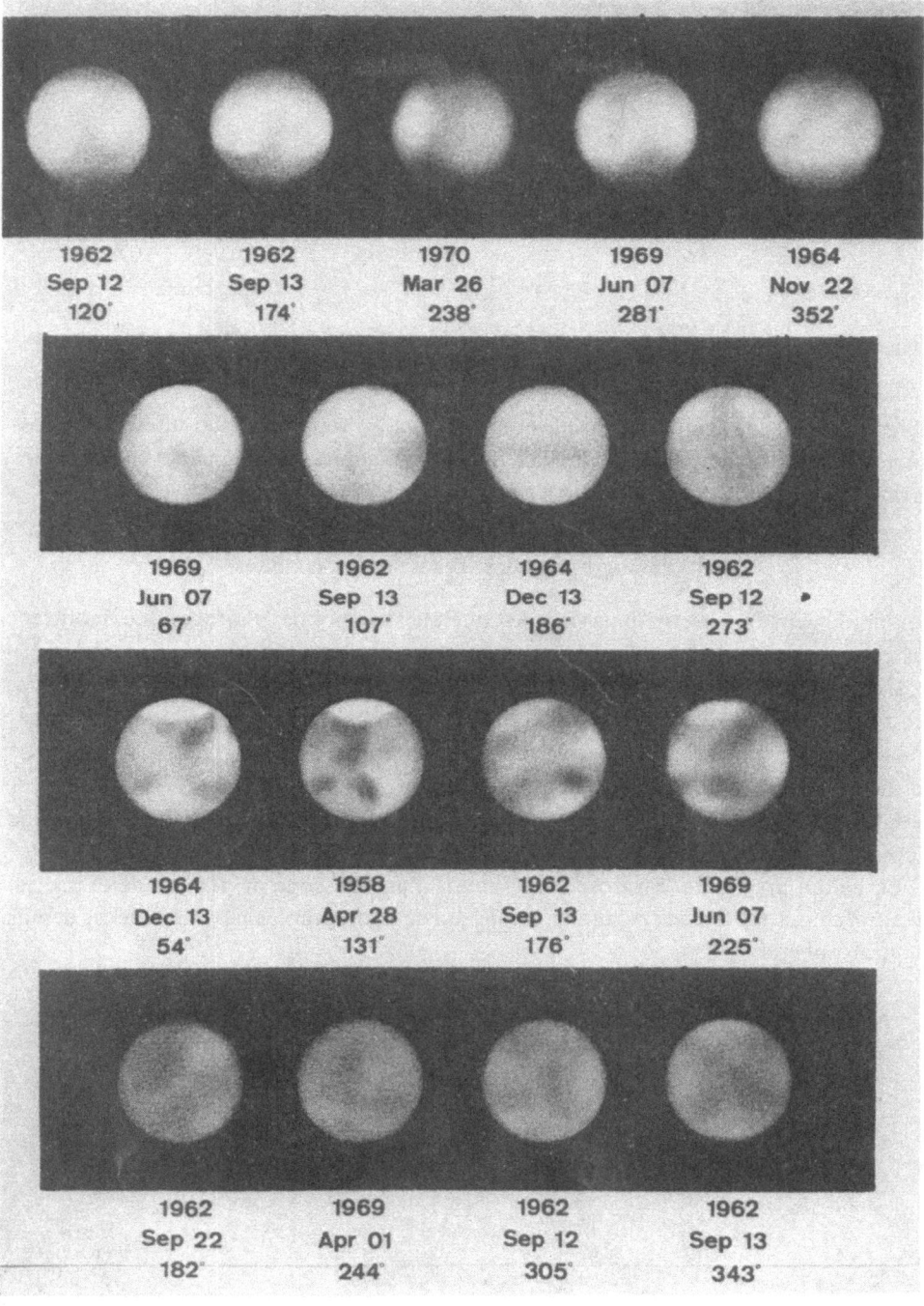

Fig. 1.   Observation télescopique des satellites de Jupiter, par A. Dollfus. Réflecteurs de 107 cm du Pic-du-Midi. Grossissement 790 à 1200. *De haut en bas:* Io, Europe, Ganymède et Callisto. Les dessins sont présentés le nord en haut. Les angles indiqués donnent les longitudes planétocentriques comptées à partir de la conjonction supérieure.

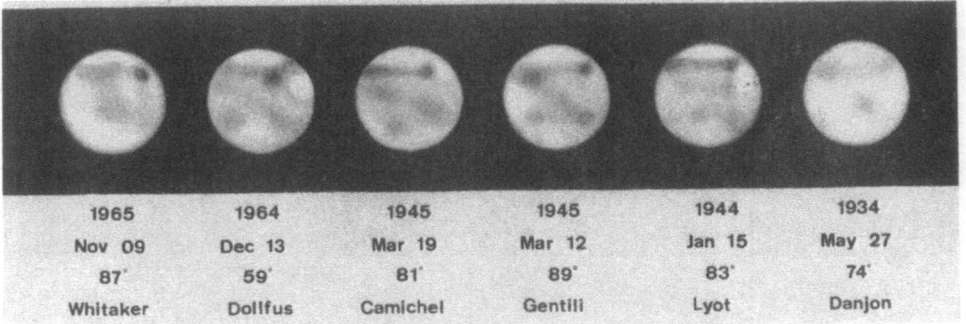

Fig. 2.   Six observations télescopiques de Ganymède vu sous des longitudes planétocentriques voisines, par six observateurs différents. Le nord est en haut.

diagonale dans la région équatoriale et une formation sombre au sud-ouest. La reproductibilité des taches montre qu'elles sont attribuables au sol même et non à des formations atmosphériques temporaires.

3. Photographies des taches sur les satellites

En supplément des observations visuelles, quelques très bonnes photographies montrent les configurations de la surface avec une certaine évidence.

La figure 3 montre Io, obtenu au Pic-du-Midi le 22 Novembre 1964 par P. Guerin. Le tirage, effectué avec deux temps de pose différents, montre sur l'une des images l'assombrissement polaire et, sur l'autre, les marbrures équatoriales. Un dessin de la même région est donné pour comparaison.

La figure 4 se rapporte à Ganymède, localisé exactement au bord du disque de Jupiter, et montrant les taches de sa surface.

Le ballon américain 'Stratoscope 2', muni d'un télescope de 100 cm de diamètre, avait permis à M. Schwarzschild et Danielson de photographier Io et ses taches depuis la stratosphère.

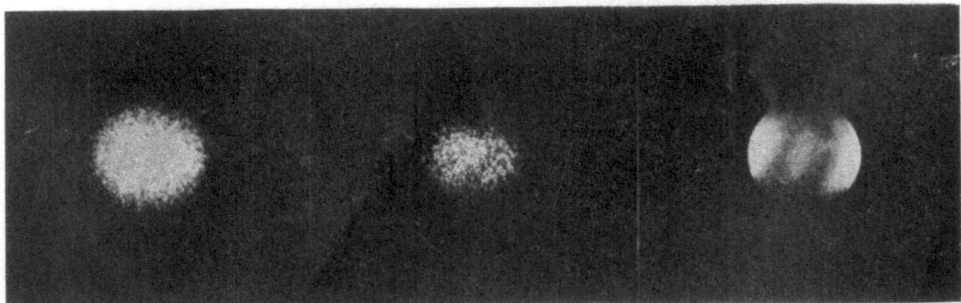

Fig. 3.   Photographie de Io le 22 Novembre 1964 à 21^{h}50^m TU. Réflecteur de 107 cm du Pic-du-Midi. Observateur P. Guerin. *A gauche et au centre:* deux tirages de la même image. *A droite:* dessin par Lyot en 1943 sous la même présentation. Le nord est en bas.

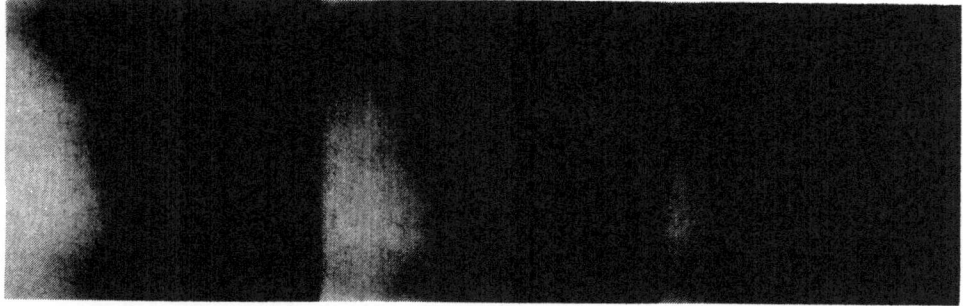

Fig. 4. Photographie de Ganymède au moment du début du passage devant le disque de Jupiter le 13 Septembre 1962 vers 22ʰ07ᵐ TU. Réflecteur de 107 cm du Pic-du-Midi. Observateur A Dollfus. Le nord est en bas.

D'autres clichés du Pic-du-Midi, beaucoup plus nombreux, se rapportent aux taches des satellites en projection devant le disque de Jupiter, et quelques-unes sont reproduites et commentées ci-après, figures 6 et 7.

4. Périodes de rotation des satellites

L'analyse détaillée des récurrences des mêmes configurations de surface, et la réalisation des planisphères qui suivent, conduisent à la conclusion que chacun des quatre satellites tourne sur lui-même autour d'un axe sensiblement perpendiculaire au plan de leur orbite. Les périodes de rotation sont synchrones avec celle des révolutions autour de la planète, et ces astres présentent donc toujours la même face à l'astre central, comme dans le cas de la Lune pour la Terre.

Les mesures des positions des taches liées à l'exécution des planisphères décrits ci-dessous conduisent à des valeurs plus exactes des périodes de rotation et à des estimations de la précision. Le tableau I ci-dessous donne les résultats.

Dans les quatre cas, les limites d'erreur recouvrent la période de révolution sidérale. Les phénomènes de freinages par marées ont donc exactement figé le synchronisme des rotations et révolutions, à la précision des mesures.

TABLEAU I

Satellites de Jupiter

(Période de rotation et révolution synodique)

Satellite		Période	Jours	Heures	Minutes	Secondes
Io	(J I)	rotation	1	18	28	$35{,}9 \pm 0{,}2$
		révolution	1	18	28	35,9
Europe	(J II)	rotation	3	13	17	56 ± 6
		révolution	3	13	17	53,7
Ganymède	(J III)	rotation	7	03	59	36 ± 1
		révolution	7	03	59	35,9
Callisto	(J IV)	rotation	16	18	05 ± 2	
		révolution	16	18	05	06,9

5. Cartographie des configurations de surface

L'analyse de l'ensemble de la documentation conduit à la mise en planisphère des taches observées sur les sols de ces astres. Le travail cartographique, entrepris à l'Observatoire de Meudon, n'est pas encore achevé car de nouveaux documents viennent d'être acquis. Les mesures de coordonnées ont déjà porté sur 7 détails caractéristiques de la surface de Io, 4 de la surface de Europe, 4 pour Ganymède et 15 pour Callisto. Le système de longitude et latitude est celui défini par l'Union Astronomique Internationale à l'Assemblée Générale de Sydney en Août 1973. L'axe de rotation est supposé exactement perpendiculaire au plan orbital. Le méridien origine contient le point sub-terrestre lors de la conjonction supérieure exacte à la phase nulle. La longitude du méridien central augmente avec le temps.

La figure 5 reproduit les quatre planisphères provisoires. Le nord est en haut; les coordonnées sont indiquées pour les latitudes $0°$, $\pm 30°$ et $\pm 60°$, et pour les longitudes de 30° en 30°. Le méridien origine (longitude 0°) est à droite; le centre de chaque carte correspond à la longitude 180° et le dernier méridien à gauche donne 360°.

On remarque les caractères très différents des configurations de surface pour les quatre satellites.

6. Passage des satellites devant le disque de Jupiter

Le satellite Io se projette devant le disque de Jupiter toutes les 42 heures; Europe passe tous les trois jours et demi; le cas se présente pour Ganymède une fois toutes les semaines. Dans le cas de Callisto, le phénomène, beaucoup plus rare, se produit tous les 17 jours, mais seulement pendant deux années consécutives tous les onze ans, lorsque la Terre est suffisamment proche du plan orbital des satellites.

Les satellites se projettent alors sur le fond brillant de la planète, dont la luminance correspond au point subsolaire à un albedo A_0 de l'ordre de 0,7 et s'assombrit vers le bord du disque en fonction de la distance planétocentrique θ à peu près comme la loi de Lambert $A = A_0 \cos \theta$.

Les régions du satellite apparaissent plus brillantes ou plus sombres que le fond sur lequel ils se projettent, selon que leurs albedos sont plus forts ou plus faibles que celui de la planète à la distance θ du point subsolaire. Ces régions disparaissent totalement lorsque leur albedo vaut exactement celui du fond. Comme le satellite se déplace sur le disque en raison du mouvement orbital, $\cos \theta$ varie et l'on observe différentes configurations de taches claires ou sombres. Ces apparences donnent les luminances des différentes régions de la surface du satellite.

La figure 6 montre une photographie du satellite Io près du bord du disque de Jupiter, ainsi que son ombre projetée sur le disque. Les régions équatoriales sont beaucoup plus brillantes que le limbe de Jupiter, ce qui n'est pas le cas des régions polaires, et on observe une sorte d'ellipse lumineuse.

La figure 7 montre Ganymède au centre du disque et son ombre projetée sur la planète. Le diamètre de l'ombre est augmenté par la pénombre et les effets photo-

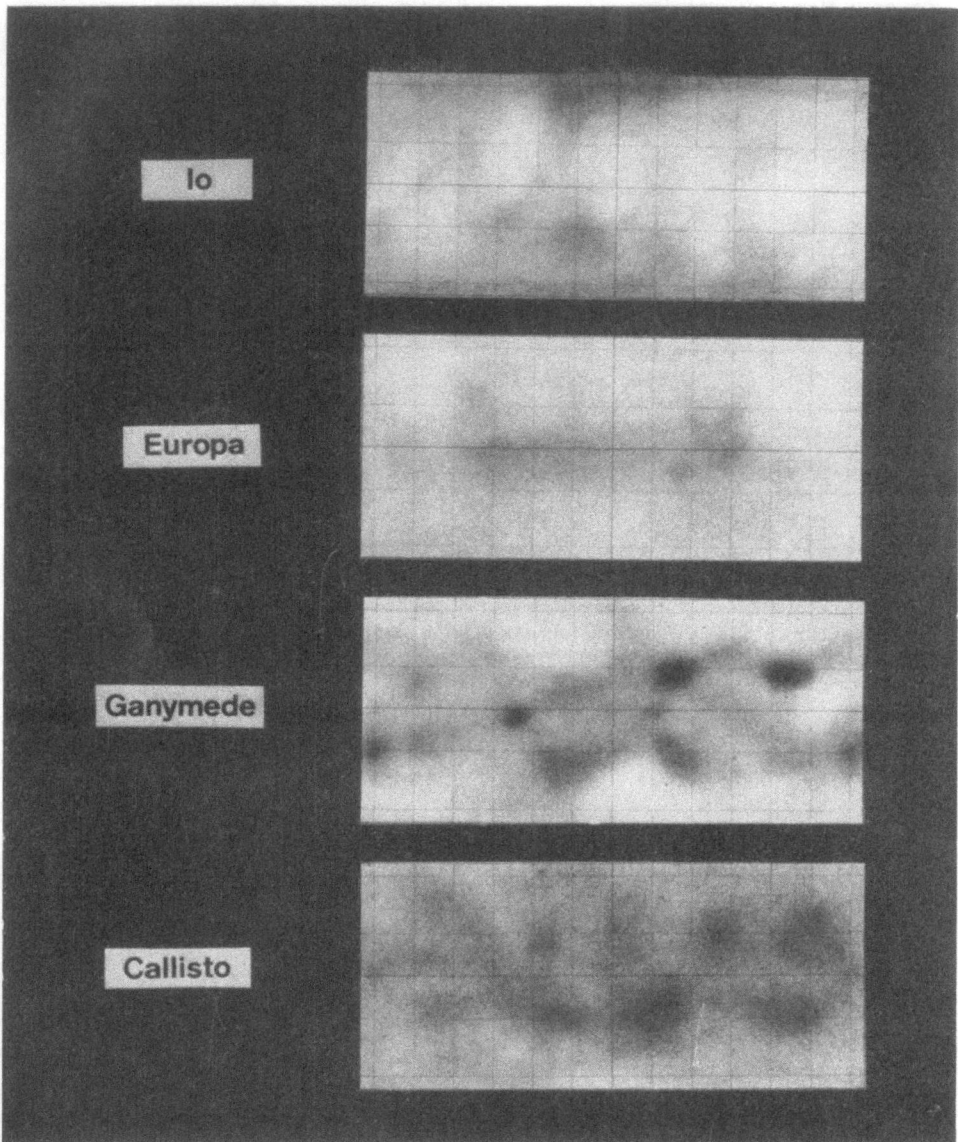

Fig. 5. Planisphères des satellites de Jupiter d'après l'ensemble des observations disponibles en 1972, par J. B. Murray. Le nord est en haut. Les longitudes du centre des cartes sont 180° et correspondent à la conjonction inférieure. Les longitudes croissent de droite à gauche. Les coordonnées sont données de 30° en 30°.

graphiques; mais malgré cela, l'image du satellite est vue plus petite que l'ombre car ce n'est pas le disque entier que l'on observe, mais seulement une tache sombre de la surface ne couvrant qu'une partie du disque. Les autres régions du disque apparent, beaucoup plus claires, ont un éclat comparable à celui du fond et disparaissent presque

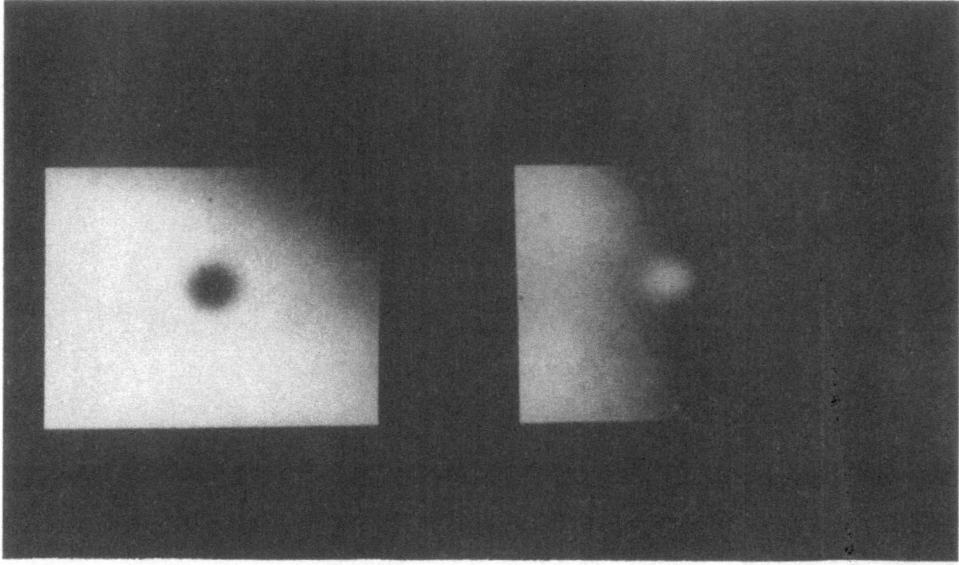

Fig. 6. *A droite:* Photographie de Io au bord du disque de Jupiter le 14 Septembre 1962 à 00ʰ42ᵐ TU. Observateur A. Dollfus. Noter les régions polaires assombries et une marbrure méridienne à gauche. *A gauche:* Photographie de l'ombre de Io projetée sur le disque de Jupiter, le 26 Novembre 1964 à 22ʰ23ᵐ TU. Observateur P. Guerin. Réflecteur de 107 cm du Pic-du-Midi. Le nord est en bas.

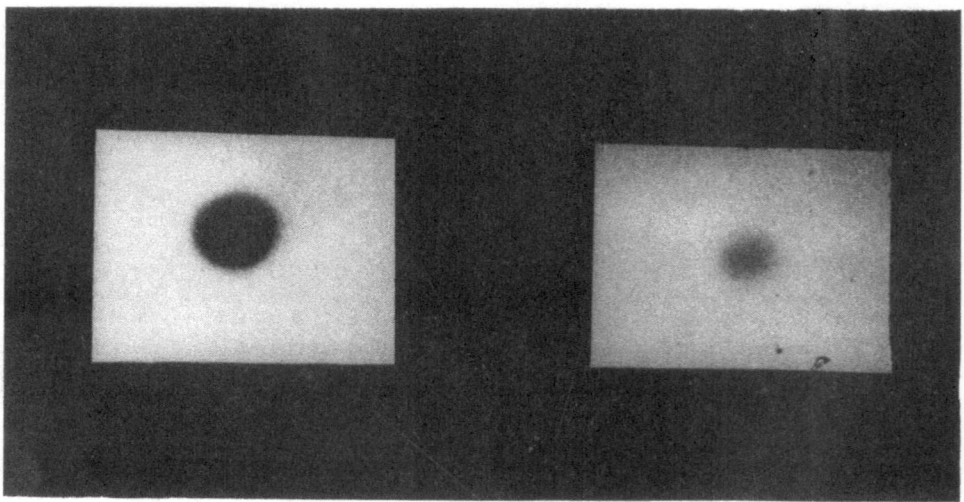

Fig. 7. Photographie de Ganymède en projection sur le disque de Jupiter (à droite) le 13 Septembre 1962 à 23ʰ00ᵐ TU et de son ombre (à gauche). Réflecteur de 107 cm du Pic-du-Midi. Observateur A. Dollfus. Le nord est en bas.

complètement; l'une d'entre elles, en haut à gauche, légèrement plus brillante, se distingue encore faiblement en plus clair. Si Ganymède avait un assombrissement près du limbe, on observerait un anneau sombre qui n'est past décelé; la surface de ce satellite, comme celle de la Lune, n'est donc pas uniformement lisse, mais tourmentée.

Les observations visuelles révèlent tous ces caractères avec encore beaucoup plus de précision.

La figure 8 montre les étapes successives d'un passage de Ganymède, observé le 13–14 Septembre 1962 au Pic-du-Midi. On voit, sur l'image centrale, les régions les plus brillantes avec un albedo encore légèrement supérieur à celui des nuages de Jupiter. Les taches les plus sombres apparaissent sur l'image de gauche encore plus assombries que le bord du limbe de Jupiter pour un angle de phase de 2°5.

La figure 9 concerne un passage de Callisto. L'astre, beaucoup plus sombre que la surface nuageuse de Jupiter, apparaît comme un disque noir (image de droite), avec une région moins sombre en bas à gauche. Cette région devient un peu plus claire que le bord du limbe sur le dessin de 21^{h}42^m. L'observation de 21^{h}55^m montre l'égalité d'éclat des régions les plus sombres et du bord extrême du limbe.

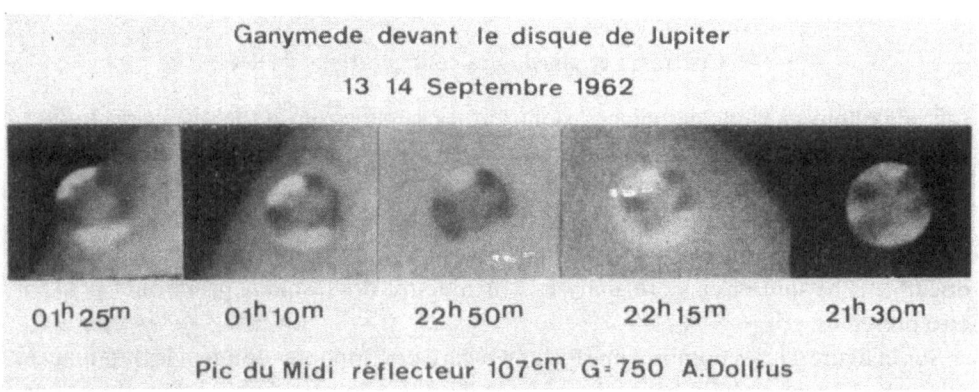

Fig. 8.   Passage de Ganymède devant le disque de Jupiter.

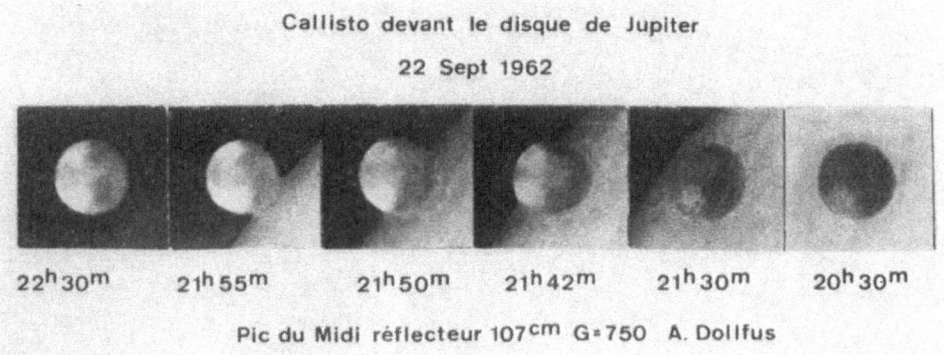

Fig. 9.   Passage de Callisto devant le disque de Jupiter.

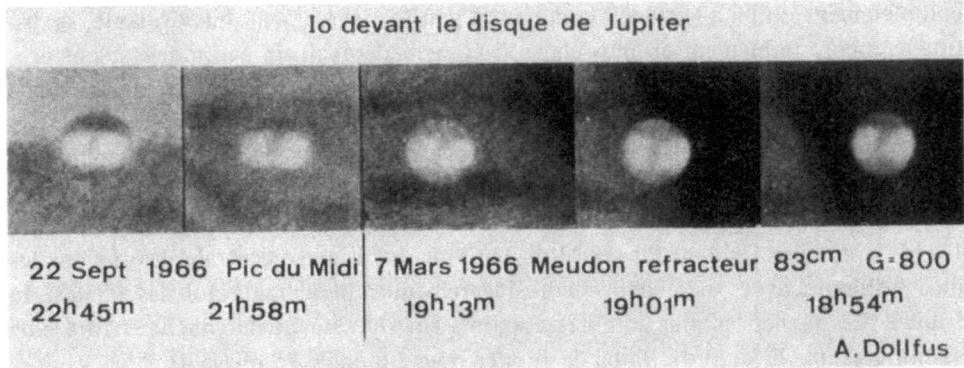

Fig. 10.   Passage de Io devant le disque de Jupiter.

Deux passages d'Io sont combinés dans la figure 10, l'un observé au Pic-du-Midi le 22 Septembre 1966, l'autre à Meudon, avec la grande lunette de 83 cm, le 7 Mars 1966. On peut encore étudier les jeux de contrastes et d'albedo.

7. Contraste et albedo des configurations du sol

L'étude photométrique des clichés montrant les satellites devant le disque de Jupiter, et l'analyse des observations visuelles de passages du type de celles reproduites figures 8, 9 et 10 permettent de construire les isophotes à la surface des satellites.

Le travail n'est pas achevé; de nouvelles observations de passages ont été recueillies récemment, ainsi que des observations d'occultations mutuelles de satellites. Ces documents ne sont pas encore analysés. Néanmoins, des résultats provisoires peuvent être présentés.

Sur la figure 11, les nombres marqués au regard des isophotes donnent les luminances relatives, L, rapportées à celles des régions les plus sombres de la surface du disque. Le contraste maximum entre les taches du sol est défini par $(L_{max} - L_{min})/L_{max}$. Sur Europe, figure 11, les contrastes restent faibles, toujours inférieurs à 0,3. Sur Ganymède,

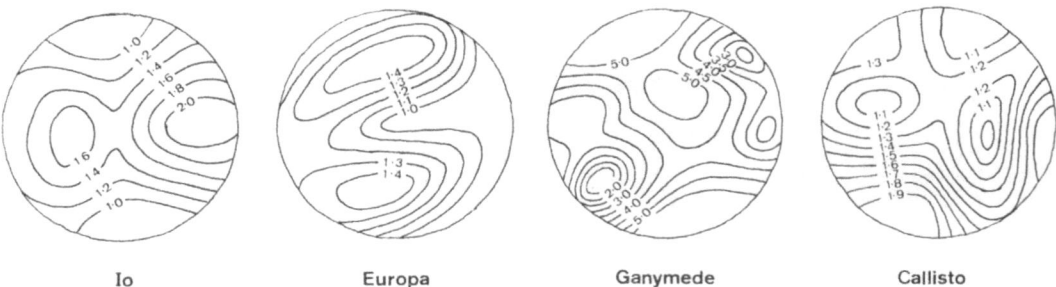

Fig. 11.   Isophotes sur les disques des satellites de Jupiter en projection devant la planète. Les nombres donnent les luminances relatives rapportées aux régions les plus sombres.

TABLEAU II

Contrastes et albedos des taches sur les satellites

	Albedo géométrique pour l'ensemble du disque		Contraste maximum	Albedo géométrique pour la phase 5°	
	Phase 0°	Phase 5°		Régions les plus claires	Régions les plus sombres
Io	0,82	0,68	0,5	0,83	0,46
Europe	0,75	0,65	0,3	0,73	0,52
Ganymède	0,53	0,46	0,7	0,50	0,15
Callisto	0,20	0,14	0,5	0,19	0,09
Lune	–	0,12	0,45	0,14	0,08
Mars (sol)	–	0,16	0,40	0,18	0,10
Mars (calottes polaires vues sous 70°)			0,7	–	–

ces contrastes sont au contraire extraordinairement élevés et atteignent 0,7 (le noir et blanc donne 1,0). Io et Callisto donnent 0,5. Ces valeurs sont reportées dans le tableau II ci-dessous.

Pour déterminer maintenant les albedos géométriques réels de ces taches claires et sombres, il faut connaître les albedos moyens des disques. Ces valeurs ont été discutées récemment par Morrison (1973), sauf dans le cas de Europe, mais la comparaison avec les déterminations antérieures de Harris (1961) et Wamstecker (1972) permettent de les reconstituer. Le tableau II donne les valeurs (première colonne de chiffres). Ces déterminations se rapportent à peu près à l'angle de phase nulle, tandis que toutes les surfaces de comparaison mesurées au laboratoire, dans la nature ou sur les autres planètes ne sont pas facilement observables sous cet éclairement. De plus, des effets de halo se produisent par rétro-diffusion pour les angles de phase inférieurs à 5°, qui compliquent les comparaisons. Il convient donc, pour les comparaisons, de normaliser les albedos à l'angle de phase 5°. Nous utilisons pour cela les courbes photométriques des satellites de Jupiter de Stebbins et Jacobsen (1962).

Les valeurs d'albedos réduites à la phase 5° sont données tableau II, pour l'ensemble du disque et finalement pour les régions les plus claires et les plus sombres observées sur chacun des satellites.

Bien que ces valeurs soient encore provisoires, il est suggestif de les comparer avec celles que donnent, dans les mêmes conditions d'observation, d'autres corps célestes tels que la Lune et Mars.

Sur la Lune, les 'mers' les plus sombres ont des albedos géométriques sous l'angle de phase 5° compris entre 0,074 et 0,096; vus sous une résolution angulaire relative comparable à celle des observations des satellites, elles donneraient des valeurs de l'ordre de 0,08. Les 'continents' s'échelonnent entre 0,096 et 0,16 et donneraient 0,14. Le contraste maximum observé vaudrait 0,45.

Sur Mars, les mesures photométriques de Dollfus (1957a, b) et de Vaucouleurs (1970) donnent un albedo de 0,18 pour les grandes régions claires du disque. Les taches les plus sombres (Syrtis Major) donnent les contrastes de 0,40 et un albedo 0,10.

Les calottes polaires de Mars, que l'on ne peut observer que très obliquement donnent, sous une inclinaison de 70°, une luminance 1,7 fois celle des régions claires du sol, selon Focas (1958) et cette valeur s'abaisse à 1,45 à la fin de l'été martien lorsque les poussières se sont accumulées à la surface. Sous une incidence moins rasante, le contraste serait nettement plus élevé.

Toutes ces valeurs sont reportées dans le tableau II et vont être discutées ci-dessous.

8. Nature de la surface des satellites

Le tableau précédent montre que Callisto a un albedo visuel moyen comparable à celui de la Lune. De plus, les contrastes entre les régions les plus claires et les plus sombres, et leurs albedos correspondants, sont semblables à ceux que montrent, sur la Lune, les mers et les continents dans les mêmes conditions d'observation. Du point de vue optique, la surface de Callisto apparaît donc du type lunaire. Cette conclusion n'était pas évidente si l'on compare les densités respectives de 1,48 et 3,34 indiquant pour Callisto une constitution interne tout à fait différente de la composition lithophile de la Lune.

Europe, au contraire, a un albedo moyen aussi élevé que 0,75, que seuls des corps très blancs tels que la magnésie, la craie fraîchement cassée ou un dépôt de neige fraîche, de givre propre ou de cristaux blancs, peuvent reproduire. Le spectre infra-rouge de Europe montre les bandes caractéristiques du givre d'eau (Kuiper, 1972). Il y a donc tout lieu de penser que la surface de Europe est presque entièrement recouverte de cristaux de neige, ou givre, dont un constituant majeur doit être l'eau solidifiée. Les quelques marbrures que l'on observe à sa surface, et qui conservent un albedo assez élevé de 0,5, correspondent sans doute à des silicates répartis sur la neige, à des nappes de cendres volcaniques, ou à des roches d'origine tectonique non recouvertes.

Il reste à expliquer comment une structure superficielle d'eau cristallisée parfaitement blanche (albedo = 0,73) peut rester vierge au cours des époques géologiques malgré le bombardement météoritique, l'accumulation lente de matière cosmique, etc. ..., le dépôt doit pouvoir se régénérer par exhalation d'eau, ou être assez épais pour absorber dans sa masse toute l'énergie des impacts. Une légère atmosphère animée de vent très rapide pourrait aussi contribuer à transporter les cristaux.

Le satellite Ganymède a des propriétés optiques intermédiaires entre celles de Europe et Callisto; le contraste très élevé entre les régions claires et sombres de sa surface indique deux natures. Les contrées sombres ont un albedo de 0,15, rappelant celui du sol lunaire continental. On peut penser à un sol du type de Callisto, peut-être parsemé de quelques plaques de givre éparses. Les régions claires, bien plus blanches, ont au contraire un vif albedo de 0,50; les spectres de Kuiper (1972) décèlent les bandes de la glace d'eau, de sorte que l'on peut évoquer de grandes étendues givrées comme sur Europe. L'éclat n'atteint pas tout à fait celui de Europe, mais les calottes polaires de Mars sont ternies par une longue exposition à l'air à la fin de l'été martien, et une atténuation semblable pourrait provenir sur Ganymède, de la poussière de roche projetée depuis les régions sombres par les impacts, ou transportée par le vent.

Le satellite Io est un cas différent. L'albedo des régions équatoriales est très élevé mais la couleur est jaune et le spectre ne montre pas de trace de givre d'eau.

Le satellite gravite dans la ceinture de radiation de Jupiter et T. Owen a suggéré, lors de la discussion du présent travail, que le bombardement des particules sur les composés simples type CH_4, NH_3, etc.... pourrait produire une substance visqueuse jaune formée sur le sol.

Bibliographie

Camichel, H., Gentili, M., et Lyot, B.: 1943, *Astronomie* **57**, 49.

Danjon, A.: 1944, *Astronomie* **58**, 33.

De Vaucouleurs, G.: 1970, dans A. Dollfus (éd.), *Surfaces and Interiors of Planets and Satellites*, Academic Press, Chap. 5, p. 225.

Dollfus, A.: 1957a, *Compt. Rend. Acad. Sci. Paris* **244**, 162.

Dollfus, A.: 1957b, *Compt. Rend. Acad. Sci. Paris* **244**, 1458.

Dollfus, A.: 1965, *Ann. Astrophys.* **28**, 722.

Focas, J. H.: 1958, *Compt. Rend. Acad. Sci. Paris* **246**, 1665.

Harris, D. C.: 1961, dans G. P. Kuiper et B. H. Middlehurst (eds.), *Planets and Satellites*, Univ. Chicago Press, p. 272.

Kuiper, G. P.: 1972, *Comm. Lunar Planet. Lab.* **9**, No. 172, 218.

Lyot, B.: 1953, *Astronomie* **67**, 3.

Morrison, D.: 1973, *Icarus* **19**, 1.

Stebbins, J. et Jacobsen: 1962, *Lick Obs. Bull.*, No. 401; cf. A. Dollfus dans *Handbuch der Physik* **54**, p. 218.

Wamsteker, W.: 1972, *Comm. Lunar Planet. Lab.*, No. 167.

DISCUSSION

Icke: Is it possible to observe the surface markings of these satellites by optical interferometry? If so, why hasn't it been done? One could for example use speckle interferometry with laser phase referencing.

Dollfus: Such interferometers are not yet in operation.

Owen: Infrared reflectivity studies of the Galilean satellites confirm the presence of water ice on Europa and Ganymede but not on Io. This is consistent with reflectivity at 5 μ and the reddest colour of Io. This satellite must be covered by something other than pure water ice.

Vsehsvyatsky: When investigating comets about 20 years ago in Kiev, we found a high activity of satellites, which could be a source of short-period comets and other small bodies. The presence of ice and snow on the surface of satellites can be treated as a strong evidence confirming this consideration.

Gorgolewski: The Bonn 100 m radio-telescope was used by J. Pauliny-Toth and A. Witzel from Bonn and S. Gorgolewski from Toruń Observatory, to observe Ganymede and Callisto at 2.8 cm wavelength during the 1973 opposition of Jupiter. All observations were made between July 22nd and August 3rd of 1973 with a total integration time of about 9 h for each satellite. The system noise temperature was about 200 K, bandwidth – 40 MHz and beam width – 80″. The observed brightness temperatures are: 98 $\pm$ 12 K for Callisto and 68 $\pm$ 16 K for Ganymede. Both satellites are significantly cooler than Jupiter, whose brightness temperature at 2.8 cm was found to be 176 $\pm$ 12 K.

OPTICAL LINE EMISSION FROM IO

ROBERT A. BROWN

Center for Earth and Planetary Physics, Harvard University, Cambridge, Mass., U.S.A.

Abstract. In early summer of 1972, we discovered an anomalous brightness in the spectrum of Io near the sodium D-lines. This effect is revealed to be time-varying emission by free sodium atoms on Io.

A year and a half ago there was no reason to expect the new and the unusual in the visible spectra of the Galilean satellites of Jupiter. They were, after all, discovered in January, 1610, and they have been easy objects of spectroscopic studies for more than 50 years. Devoid as those moons are of any yet sensible atmosphere, one would expect to find reflected from them the detailed spectrum of the Sun, variously tempered by broad color differences between the satellites (Johnson and McCord, 1970; Johnson, 1971).

Io is the innermost Galilean satellite. It is known to demonstrate two unusual behaviors: a possible post-eclipse brightening (Binder and Cruikshank, 1964) and a modulation of Jovian decametric radiation (Bigg, 1964). In the early summer of 1972 we discovered an anomalous brightness in the spectrum of Io near the sodium D-lines, which in the solar spectrum are marked by strong absorptions. A series of subsequent observations have revealed this effect to be time-varying emission by free sodium atoms on Io. This is a brief review of our current knowledge of this phenomenon.

Figure 1 shows interferometric spectra of two satellites, Io and Ganymede. The bold

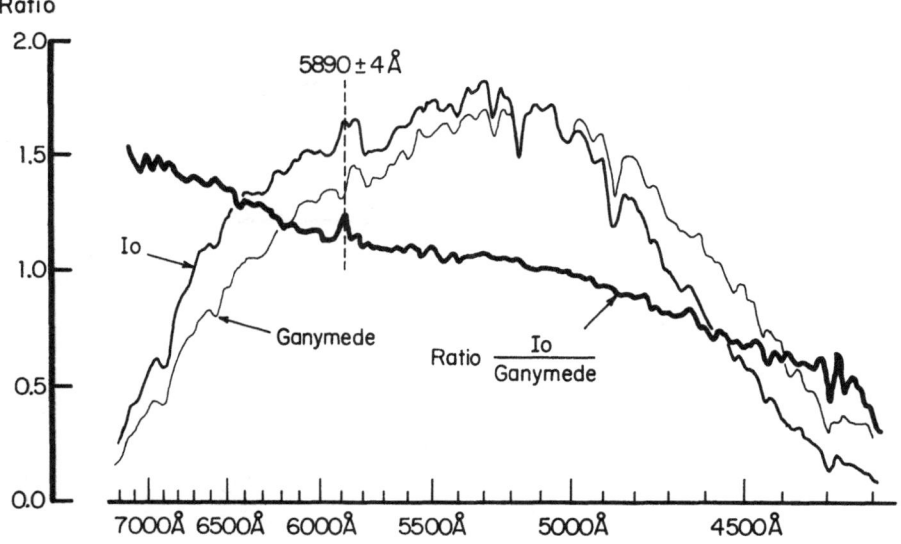

Fig. 1. The emission feature shown in the spectrum of Io when it is normalized to the spectrum of Ganymede. Data taken with interferometer.

Woszczyk and Iwaniszewska (eds.), Exploration of the Planetary System, 527–531. All Rights Reserved
Copyright © 1974 by the IAU

line shows the (arbitrarily normalized) ratio between those two spectra and shows
the sharp brightening of Io at the D-lines, which are not separated at this resolution.
Note that other deep Fraunhofer features in the original spectra are neatly cancelled
in the ratio.

Ratio spectra in each combination for all four of the Galilean satellites are shown
in Figure 2. These spectra were taken on a different night and were truncated by a
broad optical filter which preceded the spectrometer. Note that the anomaly is present
only in the spectrum of Io. A level ratio would indicate the normal solar D strength
of -1.2 Å equivalent width in absorption. In the two similar examples of Figures 1 and
2 Io had a net D-line strength of about 2.5 Å *in emission*.

Figure 3 shows a closer look with a high resolution spectrograph on a different
night. A sharp and slightly shifted emission line is seen in each of the broad solar
D absorptions. A Doppler shift analysis on the emission features proceeded as follows.
A wavelength calibration was obtained by fitting the positions of the 10 indicated
telluric water vapor lines. Figure 4 shows the solar absorption lines interpolated and
subtracted from the observed spectrum to isolate the emission lines. In Figure 5

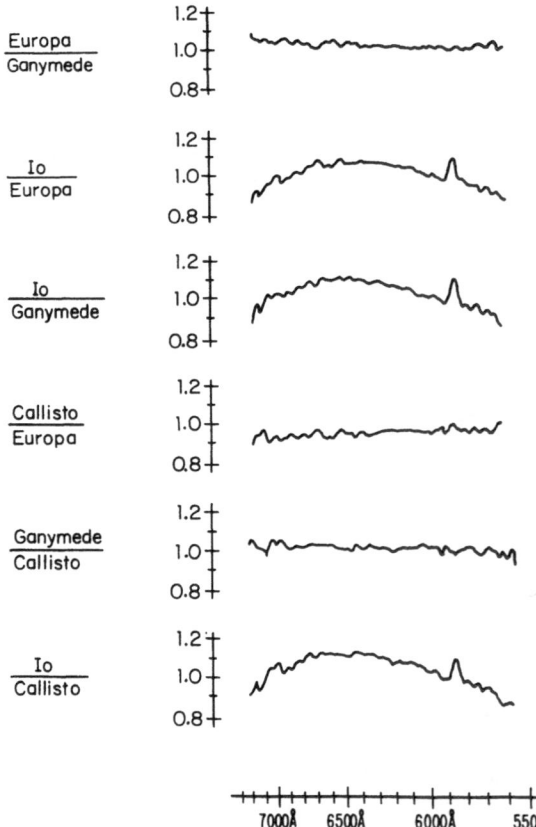

Fig. 2.   Ratio spectra of Galilean satellites. May 25, 1972, 04:38–08:06 UT.

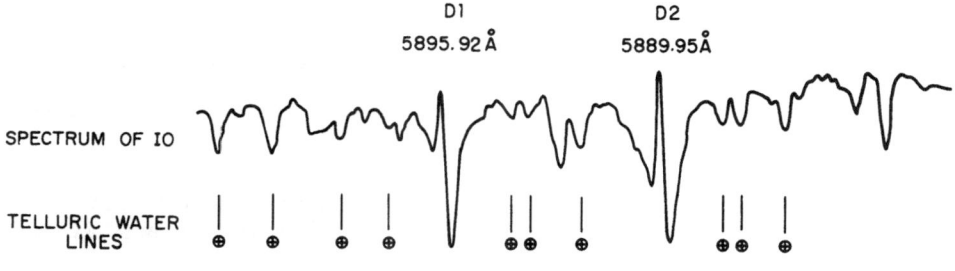

Fig. 3.   High resolution spectrum of Io taken with echelle spectrograph.
July 6, 1973, 10:19–10:49 UT.

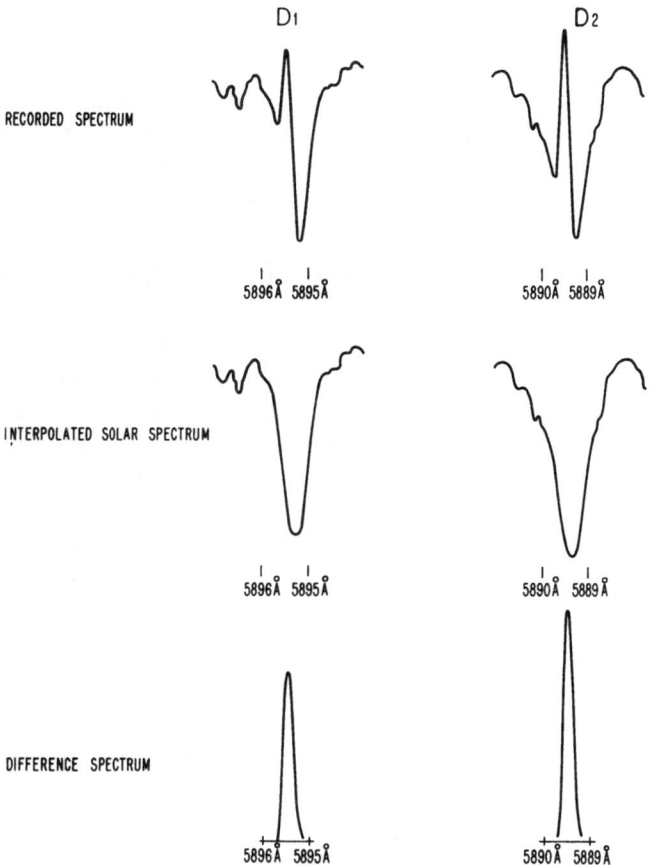

Fig. 4.   Solar absorption lines interpolated and substracted from the observed spectrum to isolate the emission lines.

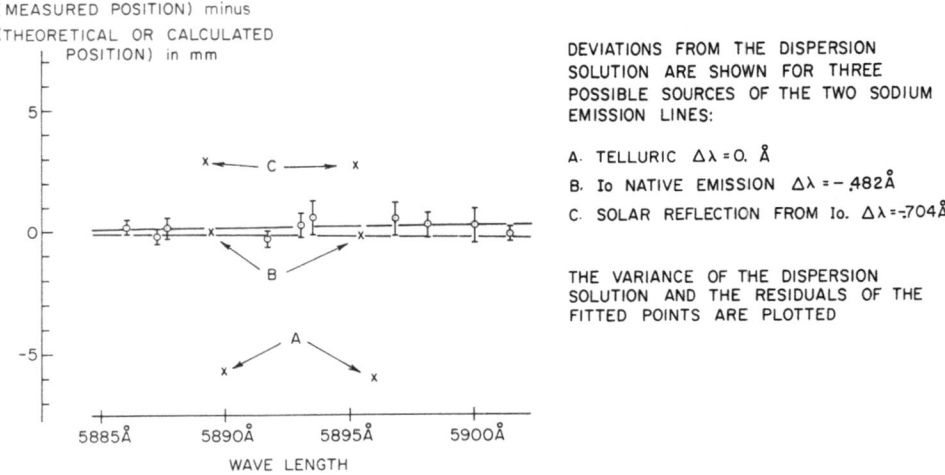

Fig. 5.   Doppler shift identification of Io as the emission source.

the variance of the wavelength calibration is shown by the lines, and the residuals at the fitted points are plotted. The relative velocities of all the bodies in question are known for the observation period. The x's show the residuals when the emission lines are variously interpreted as originating on the Earth, the Sun or Io. Io is confirmed as the source of the emission.

The ratio of the equivalent widths of the two D-line components in the high resolution spectrum is $D_2/D_1 = 1.23 \pm 0.08$. A value of 2 is expected for a source in which fine structure sublevels populated according to their multiplicity and in which self-absorption may be neglected. On the premise that there is indeed considerable self-absorption, the observed ratio and line width can be used to calculate the sodium abundance, but such an estimate is model dependent. A simple view is that the absorption and emission line profiles are the same, and that the two processes are occurring in the same uniform layer, with resonant re-emission neglected. The derived optical thickness is 3.5^{+3}_{-2} in the center of the D_1 line, and the sodium column abundance is $(1.8^{+1.8}_{-0.6}) \times 10^{12}$ cm^{-2}.

The strength of this emission from Io is varying in time. Figure 6 shows 17 independent measurements with 4 different spectrometers of the net D-line strength vs the departure of Io from superior geocentric conjunction. These measurements show a roughly uniform distribution when they are plotted against sub-Io Jovian longitude.

We are initiating two observational programs to study this interesting effect on Io. High resolution spectroscopy will be used to examine other sodium lines and to search for line emission by other species. Secondly, we will monitor the strength of the D-line emission and seek correlations with other phenomena. In all probability this discovery will prove a useful tool in probing both the Jovian environment and the conditions on this fascinating satellite.

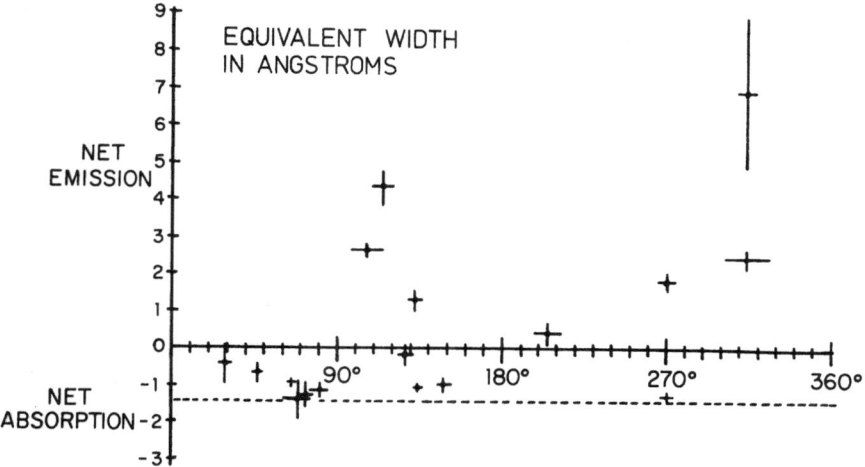

Fig. 6. Net strength of sodium D-lines vs departure from superior geocentric conjunction. Measurements with four spectrometers with different dispersions and spacial apertures are shown.

Acknowledgements

Richard Goody suggested our first work on the Galilean satellites, which resulted in this discovery. Lawrence Mertz constructed the interferometer which was used, and he has been tireless in encouraging and assisting this work. Roger Schatel, Frank Murcray and David Latham have assisted with observations and data reduction. Frederic Chaffee is collaborating in the high resolution spectroscopic observations. A spectrum kindly taken by Helmut Abt is included in Figure 6.

This research was funded by the Atmospheric Sciences Section of the National Science Foundation under Grant No. GA-33990X.

References

Bigg, E. K.: 1964, *Nature* **203**, 1008.
Binder, A. B. and Cruikshank, D. P.: 1964, *Icarus* **3**, 299.
Johnson, T. V.: 1971, *Icarus* **14**, 94.
Johnson, T. V. and McCord, T. B.: 1970, *Icarus* **13**, 37.

RELATION BETWEEN LIGHT VARIATIONS OF SOLAR SYSTEM SATELLITES AND THEIR INTERACTION WITH INTERPLANETARY MEDIUM

C. BLANCO and S. CATALANO

Osservatorio Astrofisico di Catania, Istituto di Astronomia dell'Università di Catania, Italy

Abstract. The behaviour of the available light curves of Jupiter and Saturn satellites is shown to change steadily from inner to outer satellites.

The change in reflectivity of satellites with orbital motion could be explained assuming that dark meteoroidal material accumulates on a given portion of the satellite's surface, which was once mainly covered with ice or snow. The preferential impact location on the satellite's surface could be a consequence of the perturbing action of the planet on the motion of interplanetary material.

1. Introduction

Periodic intrinsic light variations are known for many satellites in the solar system (Harris, 1961). These variations give information about the diffuse reflection properties of satellite surfaces or atmospheres, about their interaction with the interplanetary medium, and the rotation about their own axes. The light variation periods of satellites are equal to their orbital periods. This fact suggests that the rotation and revolution periods are the same. In the Jupiter and Saturn systems, the maximum of the light curve shifts from near eastern elongation, for the inner satellites, to western elongation for the outer ones.

Observations of the Saturn satellites Rhea and Titan and of the Galilean satellites of Jupiter were carried out at the Catania Observatory (Blanco and Catalano, 1971, 1974) to investigate a possible mechanism responsible for the different reflecting properties of the leading and trailing hemispheres for the inner and outer satellites.

Spectral reflectivity measurements have led to the suggestion that Saturnian satellite surfaces are composed primarily of ice (Kuiper, 1952). The four large, bright Jupiter satellites are also partially covered with ice or water frost (Kuiper, 1952; *Sky Telesc.*, 1973). However, the reddish colour observed for many satellites indicates that there would have to be impurities in the frost or ice.

In order to explain these results we make the hypothesis that the light variations of satellites are due to erosion or accumulation of dark meteoroidal material on a portion of their surfaces once covered with ice, taking into account the perturbing action of the planet on the orbits of interplanetary material.

2. Discussion

The best available light-curves in the *V* band of the main Jupiter and Saturn satellites are displayed in Figure 1. In this figure, data are corrected for the solar phase angle effect, and orbital phase angles θ are given according to synodical periods, starting

Woszczyk and Iwaniszewska (eds.), Exploration of the Planetary System, 533–538. All Rights Reserved

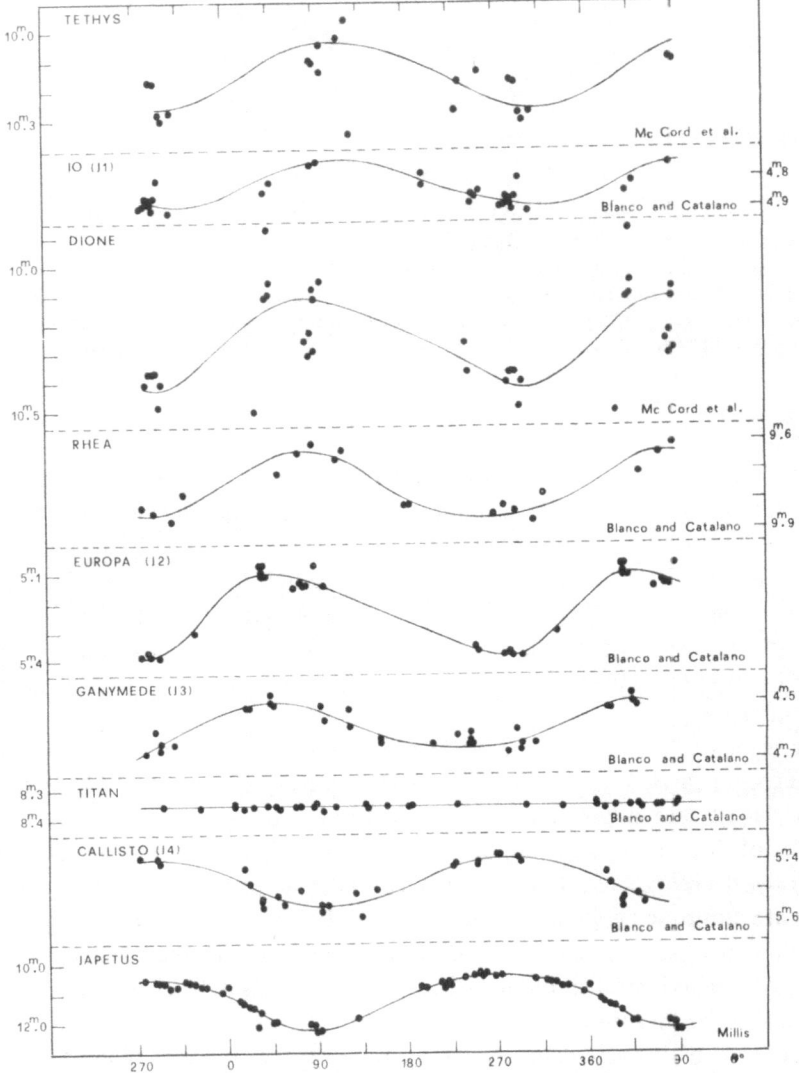

Fig. 1. Light curves in the V band for Jupiter and Saturn satellites.

from the geocentric superior conjunction. Since the satellites are synchronous, as indicated by their light curves, there will be a one-to-one correspondence between θ and the longitude of the sub-Earth point on the satellite's surface. The rotational phase angle will hereafter be indicated by the corresponding orbital phase angle. In Figure 1 the light-curves of satellites are given from top to bottom in order of increasing distance from the planet; a shift of the orbital phase angle of minimum (or maximum) light is clearly visible. Among all the well-observed satellites in Figure 1, only Titan shows no variation of brightness with orbital phase. This is not surprising,

TABLE I

Distance from the primary expressed in units of the planet's radius, average rotational phase angle of maximum and minimum light, and number of estimations for Jupiter and Saturn satellites

Satellite	d	Average rotational phase angle		n	References
		θ_{max}	θ_{min}		
Tethys	4.9	120°	310°	1	McCord *et al.* (1971)
Io ($J1$)	5.9	143	328	3	Blanco and Catalano (1974) Harris (1961) Johnson (1971)
Dione	6.25	60	290	1	McCord *et al.* (1971)
Rhea	8.8	60	250	3	Blanco and Catalano (1971) Harris (1961) McCord *et al.* (1971)
Europa ($J2$)	9.4	78	287	3	Blanco and Catalano (1974) Harris (1961) Johnson (1971)
Ganimede ($J3$)	15	50	223	3	Blanco and Catalano (1974) Harris (1961) Johnson (1971)
Callisto ($J4$)	26.4	280	143	3	Blanco and Catalano (1974) Harris (1961) Johnson (1971)
Japetus	59.7	270	90	3	Harris (1961) McCord *et al.* (1971) *Sky Telesc.* (1973)

since Titan has an atmosphere. The rotational phase angles of minimum and maximum light, averaged from estimations made on the available light curves, are given in Table I for each satellite, together with mean distance from the planet, expressed in units of the planet's radius. The averaged values of θ_{min} and θ_{max} vs the distance are plotted in Figure 2. From this figure it appears that the phase shift of the maximum and minimum light with distance from the planet is quite regular. The data are fitted quite well by the following simple relation

$$\theta = K \cdot d^{-1/2} + K' \tag{1}$$

represented in Figure 2 by the line obtained for $K=833.3$ and $K'(\theta_{min})= -16°7$, or $K'(\theta_{max})=163°3$.

The values of θ corresponding to the minimum light fit the above relation better than the maximum ones. We note that the relation (1) seems to be independent of the particular planet-satellite system. In fact the data for the Jupiter and Saturn satellites, all represented in the Figure 2, agree quite well with one another.

Light variations of Japetus, which appears darker in the leading hemisphere, have been explained by an erosion process resulting from interplanetary impacts of meteoroids on the leading hemisphere once covered with ice or water frost (Cook and Franklin,

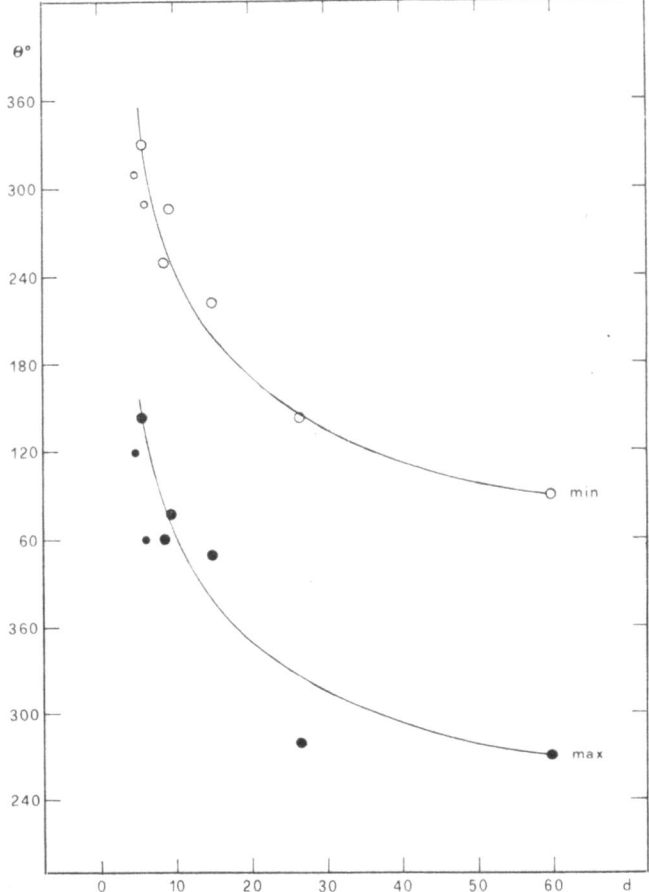

Fig. 2. Average rotational phase angle of maximum and minimum light for Jupiter and Saturn satellites vs. distance from the primary, expressed in units of the planet's radius. The size of dots indicates the number of estimations on different authors' light curves. The lines were computed according to formula (1).

1970). Such an interpretation cannot explain the behaviour of light variations for inner satellites which, as they appear darker in the trailing hemisphere, would suffer more impacts in this hemisphere, in spite of their higher orbital velocity.

Taking into account planetary perturbations on the interplanetary material, the regular shift of θ_{min} (θ_{max}) with distances in Figure 2 might be better explained by meteoroidal impact mechanism, in which the distance from the planet plays the main role. Meteoroidal material entering the planet's sphere of influence will describe open hyperbolic orbits, may be captured in elliptic unstable orbits, or may become, if they loose energy by impact or drag, permanent stable satellites of the planet, thus increasing the dust density near the planet. Figure 3 shows a schematic representation of trapped and non-trapped particles near the planet, in direct and retrograde orbits. The regular

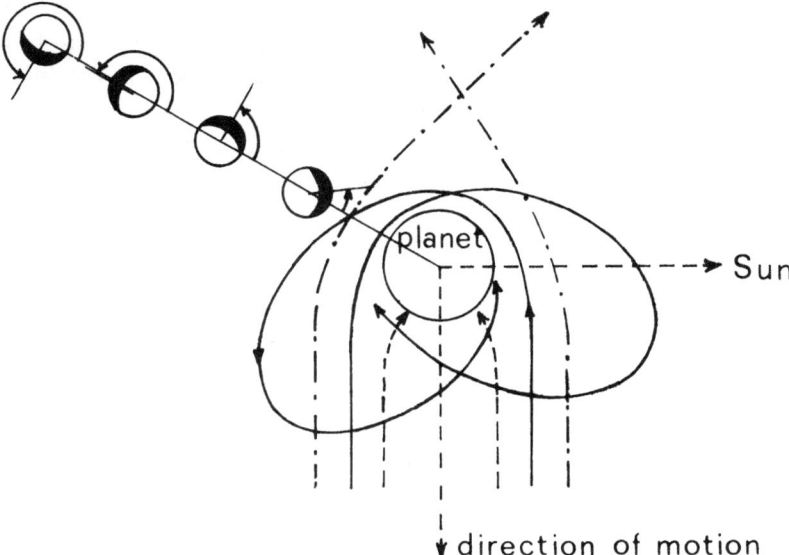

Fig. 3. Schematic representation of orbits of trapped and non-trapped particles in the sphere of influence of a planet. The darked part of satellites' surfaces represents the location of low reflectivity. θ is the rotational phase angle of minimum light according to the light curves.

shift of θ_{min} (θ_{max}) with the distance may be better understood in terms of impacts on the satellite surface by particles mainly moving in direct, high-eccentricity orbits. Some interplanetary particles, when within the planet's sphere of influence, will impinge on the satellite surfaces. The correlation between rotational phase angle of minimum light of the satellite and distance from the planet might be explained, bearing in mind that the location of impacting material on the satellite surface depends on the particle-satellite relative vectorial velocity, which in turn depends on the V_∞ velocity vector and on the distance of the incoming asymptote. We note that the orbital velocities of both satellites and dust particles depend on the inverse square root of distance from the planet, so the relative velocity would show the same dependence on d. If so, a correlation follows between the impact velocity of the dust particles and the rotational phase of minimum reflectivity, that is the satellite rotational angle of the greatest meteoroid accumulation.

3. Conclusion

The light variation of Jupiter and Saturn satellites could be interpreted in terms of meteoroidal material accumulation on a given part of their surface.

The constancy of Titan could be explained by taking into account its atmosphere. The light is mainly diffused by the atmosphere, which has a more uniform reflectance than we can expect from solid surfaces. In addition, the atmosphere could also brake the interplanetary impacts so that the surface features of the leading and trailing hemispheres are not greatly different.

However, a more detailed analysis is needed. In a model of capture of interplanetary medium by the planets, the orbital dynamics of particles affected has to be carefully considered to explain the correlation between the impact velocity of the dust particles and the rotational angle of minimum reflectivity.

In this context, detailed analysis of data on particle orbits obtained by means of space probes in the Earth-Moon system could be very useful.

References

Blanco, C. and Catalano, S.: 1971, *Astron. Astrophys.* **14**, 43.
Blanco, C. and Catalano, S.: 1974, *Astron. Astrophys.* (in press).
Cook, A. F. and Franklin, F. A.: 1970, *Icarus* **13**, 282.
Harris, D. L.: 1961, in G. P. Kuiper and B. M. Middlehurst (eds.), *Planets and Satellites*, The University of Chicago Press, Chicago, p. 272.
Johnson, T. V.: 1971, *Icarus* **14**, 94.
Kuiper, G. P. (ed.): 1952, *The Atmospheres of Earth and Planets*, The University of Chicago Press, Chicago, p. 306.
McCord, T. B., Johnson, T. V., and Elias, J. H.: 1971, *Astrophys. J.* **165**, 413.
Sky Telesc.: 1973, **45**, 22.

DISCUSSION

Dennefeld: If the Galilean satellites are covered with ice, the thermal velocity of evaporated water molecules is not sufficient for allowing escape from the satellite. Then, the water molecules follow ballistic orbits and then stick onto surface. This gives evidence for the existence of a layer of water vapor, with thickness of the order of 70 km and density of the order of 10^7 H_2O cm, if the temperature at the surface is 130 K. The density can be increased by 10^4 if hydrated ammonia is present as suggested by J. Lewis. Furthermore, these molecules can be destroyed by charged particles, and this can explain the absence of ice on Io.

Baum: Astronomers who have an opportunity to obtain light curves of the Jovian satellite events and who would like their results to be included in a combined analysis are invited to send a copy of their light curves to Dr Robert Milis at Lowell Observatory. Timing accuracy, to one second or better, is needed.

PART IV

FUTURE EXPLORATIONS OF THE SOLAR SYSTEM

SPECTROSCOPIC OBSERVATION OF VENUS

A. MONFILS and J. C. GÉRARD

Astrophysical Institute, University of Liège, Belgium

Abstract. The interest of a spectroscopic investigation of the upper atmosphere of Venus in the ultra-violet is briefly shown. The conditions for the observations to be made are defined and the main properties of the necessary instrument are deduced and described. The proposed package is a quadruple miniature spectrometer covering the entire spectral range extending from 1200 Å to 8000 Å with a resolving power of 500.

1. Introduction

Venus is the third brightest object in the sky, but not much is known about it, especially if we compare it with the two brightest objects and with Mars. It can be said, in particular, that the structure and mechanisms of Venus' atmosphere are very little known.

Spectroscopy is a powerful tool which has provided, in the past, most of the information we now possess concerning celestial bodies. The spectra of Venus may permit:
- the identification, evaluation and localization of the concentration of minor components of the atmosphere
- the measurement of scale heights and their variations with solar illumination, cycles and activity
- the determination of the temperature of the upper layers; this requires a higher spectral resolution, but limited spectral ranges
- the detection, identification of any airglow and its correlation with solar activity.
A comparison with the spectra obtained by Barth *et al.* (1971) in the case of Mars and his colleagues shows that at least one may expect to record:
- the Fox-Duffendack-Barker bands of CO_2^+
- the 2900 Å system of CO_2^+
- the Cameron bands of CO
- the comet-tail bands of CO^+
- $L\alpha$
- O I lines at 1305, 1356 and 2972 Å
- C I line at 1657 Å
One can furthermore expect to detect other systems and lines such as the 5577 and 6300 Å [O I] lines and the N_2 1st and 2nd positive systems (or even rare gas emissions). On the other hand, O_3 (2000–3000 Å), CO_2 (1250–2000 Å) and CO_2^+ (3000–3500 Å) may be observed in absorption.

2. Conditions of Observation

Numerous conditions of observation are possible. We have selected here what would appear to be the most simple and economic case: a satellite with an elliptical orbit perpendicular to the ecliptic plane, 24 h period, 300 or 1000 km perapsis.

Woszczyk and Iwaniszewska (eds.), Exploration of the Planetary System, 541–545. All Rights Reserved
Copyright © 1974 by the IAU

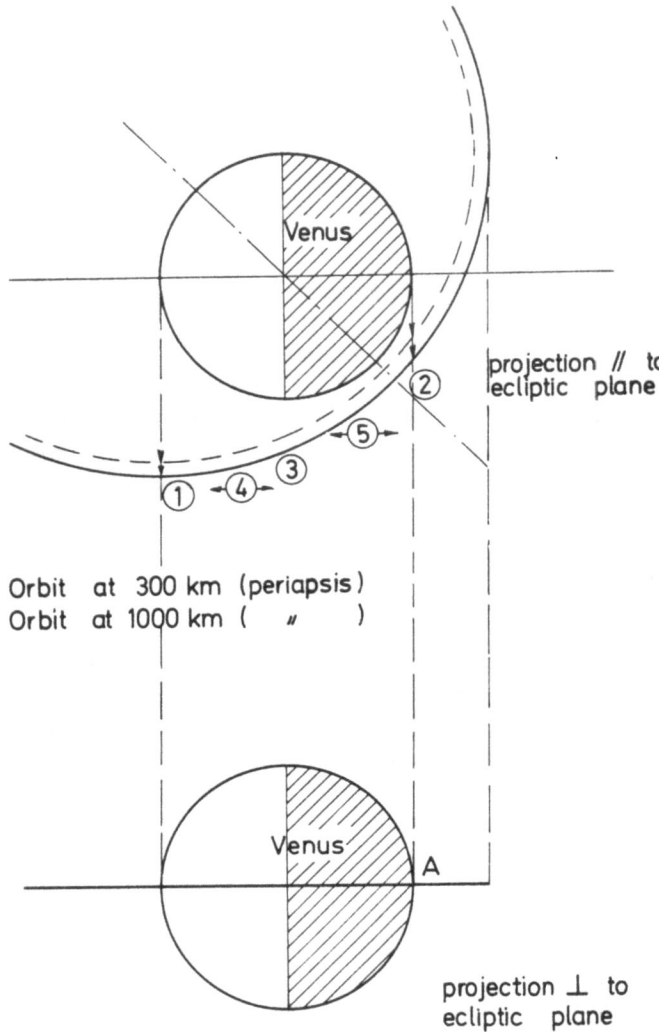

Fig. 1.   Observational configurations in the case of an eccentric Venus Orbiter probe whose orbital
plane is perpendicular to the ecliptic.

The observational configurations may be seen on Figure 1. In particular, the detection of some nightglow would be valuable in the perspective of the controversial ashenlight problem.

Four different satellite positions are of interest. The corresponding distances can be seen in Table I.

It can easily be shown that the observing times are of the order of 30 to 100 s for one profile.

TABLE I

Conditions of observation

Inclination	Periapsis	Approx. distances (km)				
		(1)	(2)	(3)	(4)	(5)
90°	300 km	8000	2600	1400	4500	700
	1000 km	9000	4000	2200	5200	1700

3. Spectroscopic Device

A spectroscopic device designed for operation aboard a spinning probe orbiting around Venus should be compatible with:
– the very low weight available
– the fact that the image of the target is very probably spinning around the line of sight (in the best case).
– the very short observation periods (minutes of time)
– the low power and low transmission rate available
– the advantage of being able to modify the resolution in orbit.
Advantage should also be taken of the long overall lifetime of the mission.

4. Angular Resolution of the Telescope

This depends on the scale heights which may be readily evaluated. Taking the observing distances into account, it may be shown that an angular resolution of 1/200 rad is necessary. This settles the specifications of the telescope and simultaneously the diameter of the entrance hole of the spectrometer. The latter must be circular in order to maintain the spatial resolution independent of the attitude of the satellite. With a focal distance (F.D.) of 14 cm for the telescope, a circular hole of 0.7 mm diam maintains the desired spatial resolution.

5. Optical Principles

The luminosity of a spectrometer is proportional to the square of the F.D. (when the slit shape is constant). The weight is proportional to the cube of the F.D.

Consequently, the specific light gathering power per unit of weight is inversely proportional to the focal distance at constant weight, and is the main parameter. The total gathering power may then be recovered by multiplying the number of spectrographs; the resolving power may, in turn, be recovered by adding Fabry-Pérot type dispersing devices, and by using the gratings at the maximum number of lines per millimeter. This is made possible by the multiplication of the number of spectrometers, which helps here to keep each grating to a limited spectral range. As the latter are chosen so as to match the properties of the photocathodes, the whole system is optimized.

6. Description of the Experimental Package

The package is typically composed of four parallel 9 cm F.D. Ebert-Fastie monochromators, each one following a 14 cm F.D. telescope (Figure 2). The four gratings
are mounted on the same axis, and are simultaneously scanning their spectral range,
which are given in Table II.

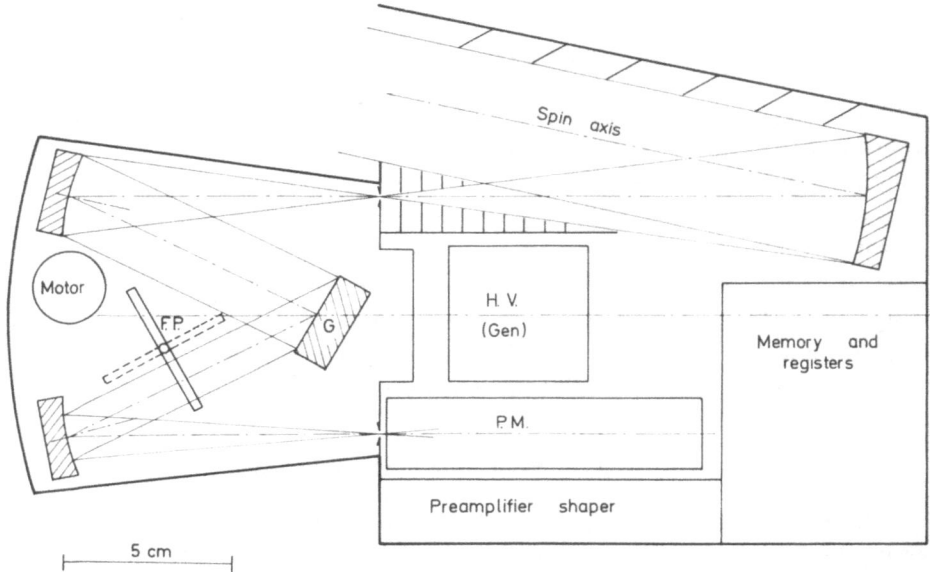

Fig. 2.   View of the experimental package showing how the various optical and electronic constituents
may be disposed. The package is typically composed of four such assemblies.

TABLE II

Gratings and their spectral range

Grating constants (mm^{-1})	Wavelength ranges (Å)
6.000	1000–2000
3.600	1650–3300
2.400	2500–5000
1.500	4000–8000

In the parallel beams, thin Fabry-Pérot interferometers, all mounted on the same axis,
permit a further separation of the bandpass obtained by the Ebert-Fastie. It may be
shown that the resolving power remains at the level of 500 from 1200 Å to 8000 Å.

The electronics must be quite sophisticated as different observational programmes
must be foreseen and as spectra must be added channel per channel prior to tele-
metering. Three operating modes are considered

– low resolution (100) monochromators scanning mode, interferometers off
– medium resolution (500) monochromators and interferometer alternate scanning mode
– surface mapping in four spectral ranges (all spectral devices stopped at a chosen position).

The total power required may be kept between the limits of 5 and 7 W.

The total weight may drop as low as 4.2 kg, if magnesium is used for the casing.

If the total number of spectrographs is cut down to 3 or 2, the total weight drops down to 3.2 and 2.4 kg respectively, but the interest of the concept is progressively decreasing.

Reference

Barth, C. A., Hord, C. W., Pearce, J. B., Kelly, K. K., Anderson, G. P., and Stewart, A. I.: 1971, *J. Geophys. Res.* **76**, 2213.

PHOTOPOLARIMETRY OF PLANETS

T. GEHRELS

The University of Arizona, Tucson, Ariz., U.S.A.

Abstract. Polarimetry is reviewed in a new book of which I am the editor (Gehrels 1974) while detailed papers of our group in Arizona continue to appear in a series 'The Wavelength Dependence of Polarization' in the *Astronomical Journal*.

The PIONEER missions provide the first encounter of Jupiter on 4 December 1973. Our 2.5-cm Maksutov $F/3.4$ telescope has three square focal-plane diaphragms: $2°3$ for zodiacal light, $0°5$ for photopolarimetry and $0°03$ for imaging of Jupiter and satellites. There are two spectral bands, namely 390–500 nm and 595–720 nm. Weinberg and Hanner (1973) are the investigators for the zodiacal light. The instrument was built by Santa Barbara Research Center (Pellicori *et al.*, 1973); specifications have also been published by KenKnight (1971). Maps of brightness and linear polarization of Jupiter are to be made during encounter at about 40°, 100°, and 140° phase angle. We expect to determine aerosol characteristics and the amount of gas above the clouds. We also plan to observe the Galilean satellites at about 20°, 60° and 120° phase. The Pioneer spacecraft spins at 5 rpm and the imaging is done by compositing consecutive scans on the planetary disk (Gehrels *et al.*, 1972). The images will be rectified for changing geometry, during the hour needed for a picture, and other image enhancement techniques will be applied. The images will have resolution somewhat higher than those obtained from Earth, but the precise increase depends on uncertain flight hazards, e.g. the radiation belts. Groundbased observational programs of Jupiter during 1973 and 1974 are needed for tie-in with the Pioneer measurements. The Pioneer program is conducted by the Ames Research Center (see NASA SP-268). My co-investigators at the University of Arizona are D. L. Coffeen, C. E. KenKnight, W. Swindell, and M. G. Tomasko.

References

Gehrels, T. (ed.): 1974, *Planets, Stars and Nebulae Studied with Photopolarimetry*, The University of Arizona Press, Tucson, Ariz.

Gehrels, T., Suomi, V. E., and Krauss, R. J.: 1972, 'The Capabilities of the Spin-Scan Imaging Techniques', in *Space Research XII*, Akademie-Verlag, Berlin, p. 1765.

Hanner, M. S. and Weinberg, J. L.: 1973, *Sky Telesc.* **45**, 217.

KenKnight, C. E.: 1971, 'Physical Studies of Minor Planets', in *NASA SP-267*, p. 633.

NASA SP-268, 'The Pioneer Mission to Jupiter', National Aeronautics and Space Administration, Washington.

Pellicori, S. F., Russell, E. E., and Watts, L. A.: 1973, *Appl. Opt.* **12**, 1246.

DISCUSSION

Irvine: Are there new data on the center-to-limb variation of polarization for Jupiter? This is important for comparison with the center-to-limb variation of equivalent widths of absorption lines, such as those of methane and ammonia.

Gehrels: Yes, J. S. Hall and L. Riley have recently published, in the *Lowell Obs. Bull.*, detailed polarization measurements made with an area scanner.

Woszczyk and Iwaniszewska (eds.), Exploration of the Planetary System, 547. All Rights Reserved
Copyright © 1974 by the IAU

RATIONALE FOR NASA PLANETARY
EXPLORATION PROGRAM

I. RASOOL, D. HERMAN, D. KERRISK, and W. BRUNK

NASA Headquarters, Washington, D.C., U.S.A.

Abstract. NASA program of future solar system exploration in the years 1973–1990 is presented.

1. Background

The guiding philosophy in the formulation of the NASA Planetary Program is the concept of balanced exploration. This concept involves a strategy of broad exploration of the entire solar system, rather than concentration on the detailed exploration of one or two planets. Thus, within appropriate levels of resources and technology, the NASA program seeks a balance between the expansion of knowledge by the inclusion of new targets, and the refinement of knowledge obtained on previously investigated targets. It also favors a reasoned balance between expenditures on immediate missions and investment in the development of the technology which make future missions possible.

In formulating its solar system exploration program, a major consideration has been the development of a progressive understanding of each target, so that proposed missions may build on the results and understanding generated by their predecessors. Generally, the sequence of proposed exploration goes as shown in Figure 1, with Earth-based observations being first augmented by flyby missions, then considerably improved by orbiter missions. Details of planetary atmospheres, composition, structure, and dynamics are then investigated by atmospheric probes, followed by landers to study geology, and, where deemed appropriate, biology. Finally, sample return missions to allow Earth-based analysis are considered.

A second consideration in formulating the Planetary Program has been the availability of spacecraft technology. The program has been geared to an evolutionary spacecraft development, with each mission building on the technology of its predecessors. This minimizes step-function changes in the level of technology required for any spacecraft. Where such step-function changes are required (as, for example, in the introduction of RTG power sources) research and technology development programs have been formulated to bring the new technology items to a state of flight readiness, prior to committing the mission.

A third consideration in formulating the program has been the availability of the propulsive capability required for each mission. Requirements are set by the laws of celestial mechanics, mission objectives, target and year of opportunity. Against these requirements we have the finite capabilities of vehicles currently being developed. The phasing of the program must be consistent with matching the mission requirements to launch vehicle availability.

Woszczyk and Iwaniszewska (eds.), Exploration of the Planetary System, 549–561. All Rights Reserved

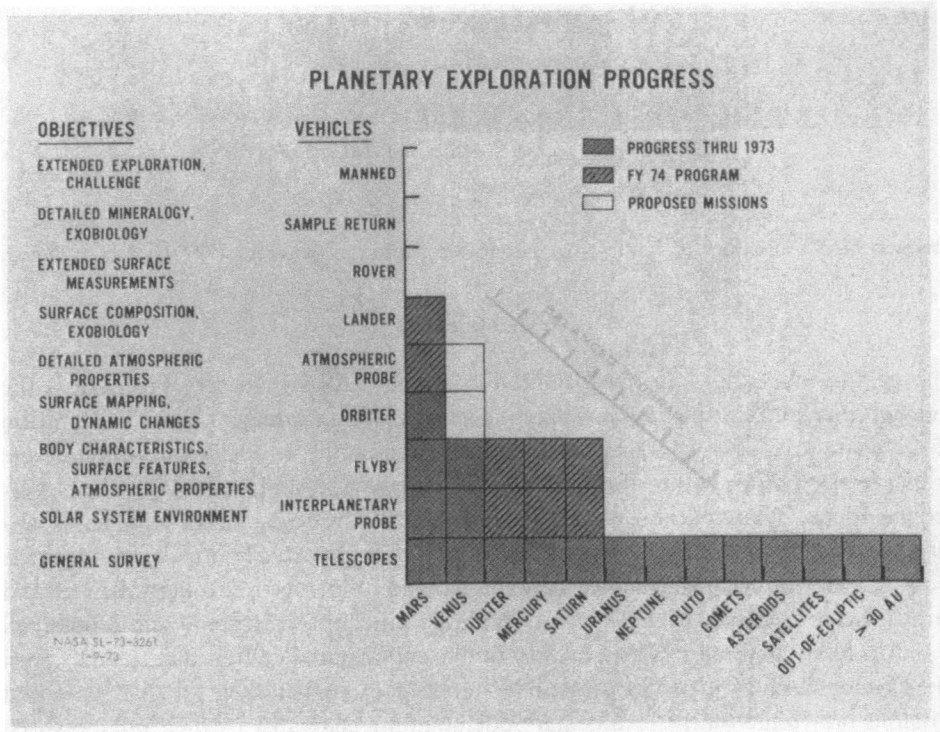

Fig. 1.   Planetary exploration progress.

Finally, the program must be formulated within the bounds of realistic fiscal constraints, taking into account not only the mission costs, but also the costs and cost phasing of new technology and enhanced launch vehicle capability development.

The operation of these constraints can be observed in the development of the Planetary Program to date. Planetary exploration started with the mission of Mariner 2 to Venus in 1962, with spacecraft and launch vehicle based on Ranger technology. The introduction of the Atlas/Centaur launch vehicle allowed the Mariner 4, 5, 6, and 7 missions to Mars and Venus, with increasingly sophisticated spacecraft subsystems. Development of liquid retro-propulsion capability preceded the Mariner 9 Mars-Orbit mission and the development of RTG power sources allowed the extension of our exploration to the outer planets with Pioneer 10 and 11. Increasingly sophisticated trajectory, guidance, navigation, and orbit determination software programs have made the use of planetary gravity assisted swingbys a feasible means of extending launch vehicle capability to targets that are outside their capability for direct flights; this will be used for the first time later this year in the Mariner Venus/Mercury Program to extend the Atlas/Centaur capability to provide a flyby of Mercury.

The introduction of the Titan/Centaur launch vehicle in 1974 will provide another step function increase in the NASA exploration capability. This will be used for the

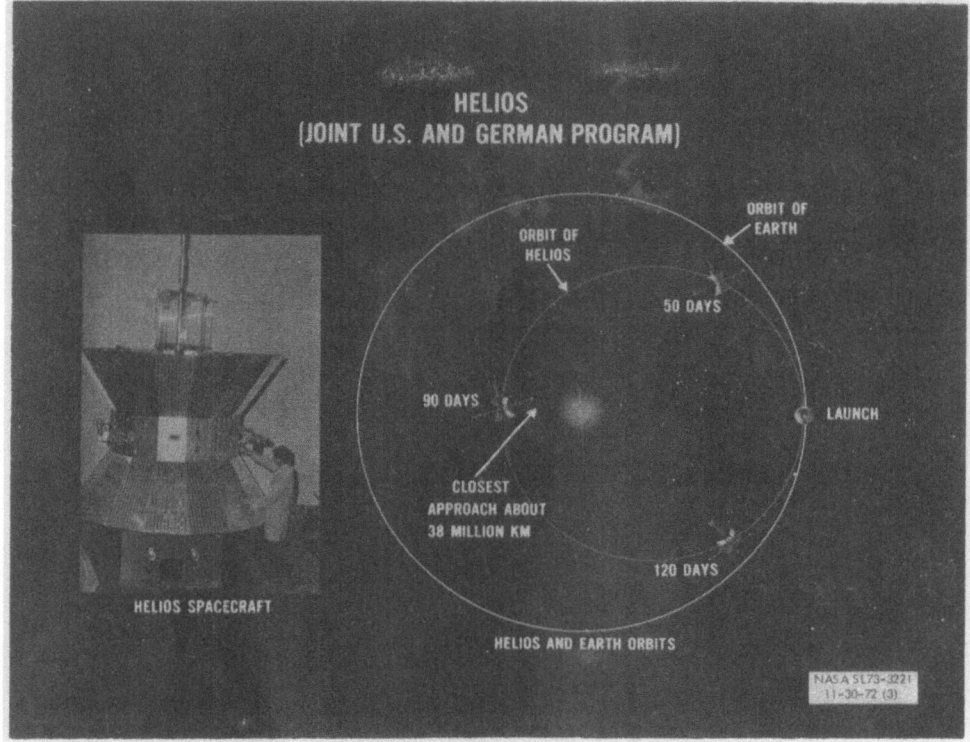

Fig. 2.   Helios mission.

first time to obtain a close solar probe in the joint U.S./West German Helios Program
(Figure 2) and then will in 1975 send Viking to Mars to perform NASA's first planetary soft landing. In 1977, again taking advantage of planetary gravity assist to enhance
the launch vehicle capability, NASA will expand its exploration with the MJS flyby
of Jupiter and Saturn providing the first close pictures of Jupiter, Saturn, Saturn's
Rings, and the satellite Titan.

It is in the context of this background of completed and ongoing programs that
the Planetary Mission Model, defining potential future programs has been constructed.

2. The Planetary Mission Model

The present Planetary Mission Model (Figure 3) is based on a continuation of the
balanced exploration philosophy, tempered by a conception of the realities of technology availability and fiscal constraints. It must be emphasized that the mission
model contains projections of typical missions that might occur in a future time frame
and, as such, is subject to continual modification and updating. Iterations and
modifications will be made to reflect changes necessitated by budgetary or programmatic considerations; to incorporate results from advanced mission studies; and to

NASA MISSION MODEL - 1973　　　　　　　15 February 1973

PLANETARY EXPLORATION

	73	74	75	76	77	78	79	80	81	82	83	84	85	86	87	88	89	90	91
APPROVED PROGRAMS																			
Mariner Venus/Mercury	x																		
Pioneer Jupiter Flyby	x																		
Helios		x		x															
Viking '75			x																
Mariner Jup/Sat '77						x													
INNER PLANETS																			
Viking Orbiter/Lander							x												
Surface Sample Return												x							
Satellite Sample Return																		x	x
Venus Pioneer						x													
Inner Pl. Follow-On							x	x	x		x								
Venus Radar Mapper									x										
Venus Buoyant Station											x								
Mercury Orbiter															x				
Venus Large Lander																			
OUTER PLANETS																			
Mariner Jup/Uranus Flyby							x												
Pioneer Saturn Probe							x												
Pioneer Sat/Uranus Flyby (U Probe)								x											
Mariner Jupiter Orbiter									x	x									
Pioneer Jupiter Probe											x								
Mariner Saturn Orbiter												x	x						
Mariner Uranus/Nep Flyby																		x	x
Jupiter Satellite Orb/Lander																			
COMETS AND ASTEROIDS																			
Comet "X" Slow flyby							x		x										
Encke Rendezvous														x					
Halley Flyby															x				
Asteroid Rendezvous																			

Fig. 3.　Planetary mission model.

incorporate results from previous missions. The mission model envisions the continued use of existing spacecraft technology (i.e. Mariner, Viking, and Pioneer) with only evolutionary changes, but with the introduction of atmospheric probe technology in the late 70's. It envisions the continued use of the Titan/Centaur vehicle through 1980, with the introduction of solar-electric propulsion (SEP) capability about 1979. Beyond 1980, shuttle capability is assumed, although all of the missions through 1985 are also compatible with Titan/Centaur or Titan/Centaur/SEP.

The sequencing and phasing of the missions incorporates the recommendations of the scientific community. A brief explanation of the rationale behind the sequences at each target will be presented.

2.1. MERCURY

The very high energy requirements for reaching Mercury were beyond the capability of the Atlas/Centaur until the gravity assist mode became practical. Using a gravity assist at Venus, the MVM'73 spacecraft will obtain our first exploratory information on Mercury early in 1974. This mission is depicted in Figure 4. The spacecraft has a complement of seven experiments including a television camera to obtain images of the planetary surface. Other experiments will provide data on Mercury's physical properties, surface temperature, atmospheric composition and particle and fields environment. Orbiting the planet however will require the use of multiple Venus

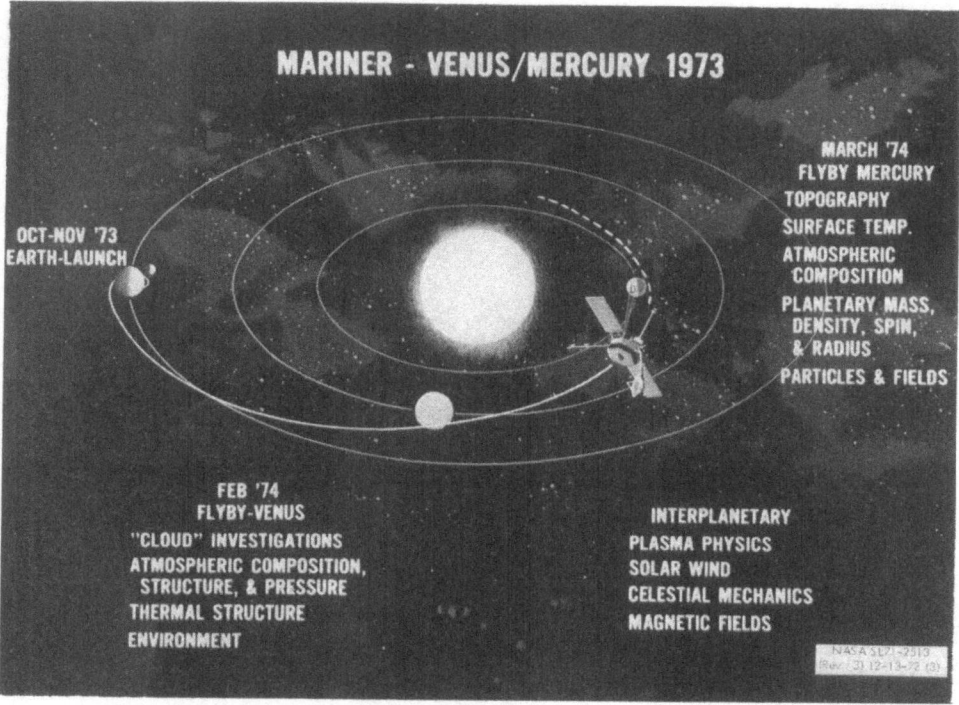

Fig. 4. Mariner-Venus/Mercury mission.

swingbys with very restricted launch opportunities or an advanced SEP system; currently it is estimated this capability will be available in the mid 80's. Before such a mission can be planned, the data from MVM'73 must be thoroughly assessed. Pending this assessment and based on estimated technology availability and funding constraints, a Mercury Orbiter mission is tentatively scheduled for 1987.

2.2. VENUS

Venus, as Earth's nearest neighbour and the planet most closely resembling Earth in gross physical characteristics, is an object of considerable interest. The MVM'73 spacecraft will obtain data on the planet's atmospheric properties including composition and structure as well as 'cloud' investigations. It will image the planet, and imaging will considerably enhance present Earth-based resolution to the level shown in Figure 5. Because of the dense cloud cover, however, it cannot be visually mapped. Also the apparent extreme hostility of its surface environment, as deduced by the Mariner 2 and 5 flyby experiments and the Russian Venera spacecraft, means that survivable landing vehicles will require considerable advance in packaging, thermal control, and corrosion resistance technologies.

From one aspect, however, the hostility of the Venusian atmosphere is an aid to scientific investigation. There are strong cosmological arguments that the very dense

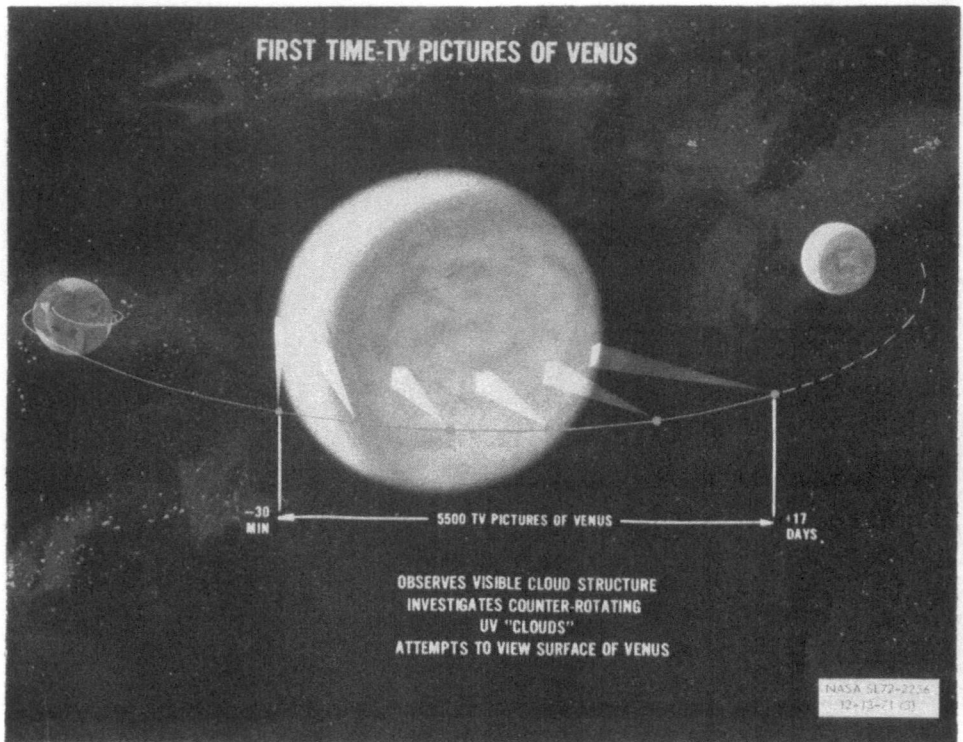

Fig. 5.   First time – TV pictures of Venus.

atmosphere of Venus, studied in conjunction with the very tenuous atmosphere of
Mars, would yield invaluable insight into the processes and dynamics of Earth's own
atmosphere, and aid greatly in our ability to predict the reaction of Earth's atmosphere
to the changing conditions on its surface. Exploration of the Venus atmosphere is
within our current technological capability, and, in addition to the noted benefits,
will provide the data base needed for future surface vehicle design. The program
rationale therefore calls for initial lower atmospheric investigations with the Pioneer
Venus Probe mission (Figure 6) and more extensive environmental and upper atmosphere studies with Pioneer Venus Orbiter. These will be followed by more detailed
atmospheric studies, including a buoyant station to obtain data on atmospheric
dynamics.

Mapping the Venusian surface will require the use of radar. Earth-based observations from the large radar telescope near Arecibo, Puerto Rico, will provide surface
resolution to about 2 km over a portion of the planet in the 1975–1980 time period;
to obtain resolution to the level of the Mariner 9 Mars maps, will require the use of
an orbiting spacecraft with a fully focussed synthetic aperture radar. This is planned
for the 1983 time frame.

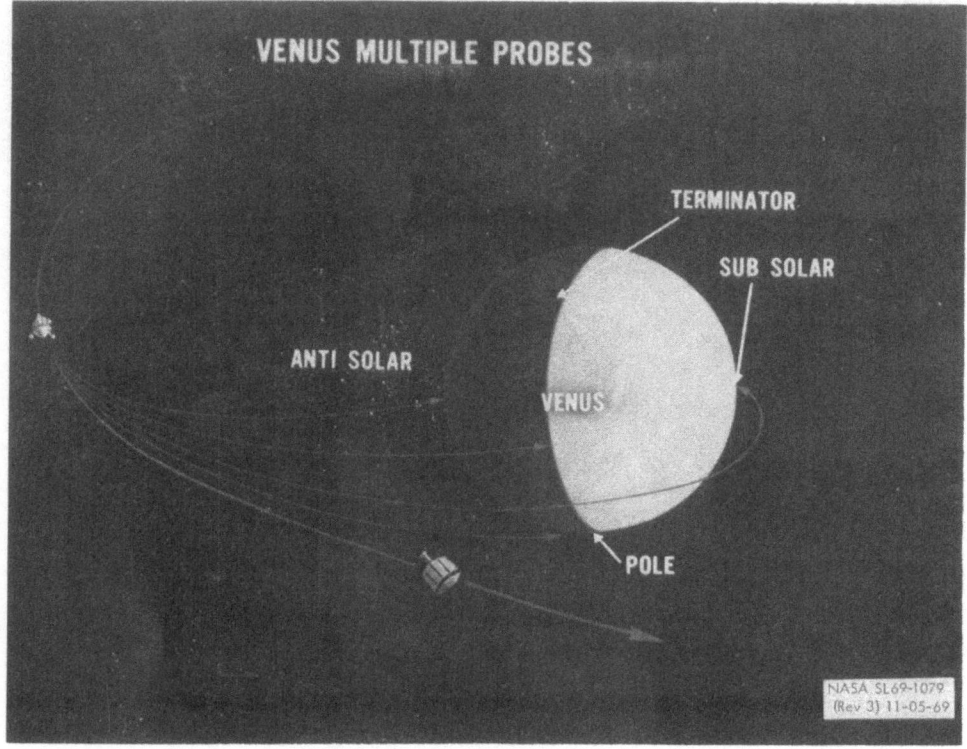

Fig. 6.   Pioneer Venus probe mission.

The combination of high resolution surface maps and detailed environmental data are necessary before rational planning for a soft landing mission can commence. Hence no landing on Venus is comtemplated prior to 1989.

2.3. MARS

The combination of the low transfer energies required and relatively benign environment allows us to push the exploration of Mars at a much greater pace than any other target. The Mariner 9 mission (Figure 7) has provided us with excellent photographs of the entire Martian surface, as dramatically depicted in Figures 8 and 9, and the Viking mission in 1975 (Figure 10) will provide *in situ* surface sample analysis and preliminary biological information.

The continuing exploration of Mars now planned includes further Viking Landings in 1979, perhaps with the addition of some roving capability. Whether these missions would be biologically or geologically oriented will depend on the results of the 1975 mission.

Beyond Viking, the next logical step, based on our lunar experience would be the return to Earth of a sample of the Martian surface. This is within the capability of our existing technology and is planned for 1984.

Fig. 7.   Mariner 9 mission to Mars.

Also within the scope of existing technology is the landing on and sample return from one of the Martian satellites. However, in the context of the balanced program, this mission is given relatively low priority at the present time and is not contemplated until the 1990 time frame.

2.4. THE OUTER PLANETS

There is keen scientific interest in the outer planets and their satellites. The major planets, totally different from the terrestrial planets in size, composition, rotational speed, and other physical characteristics display many anomalous properties which could contain clues to solar system formation processes. The satellites on the other hand have many properties similar to terrestrial planets, and atmospheres have been detected on at least two (Titan and Ganymede). The investigation and understanding of the phenomena of the outer planets ranks high on the priority list of the scientific community.

Until recently, however, the outer planets were beyond our technological capability. Only Jupiter falls within the capability of the Atlas/Centaur, and then only marginally. Further, the distance from the Sun makes the use of solar power impractical, and no other energy source was available. The latter constraint was removed with the develop-

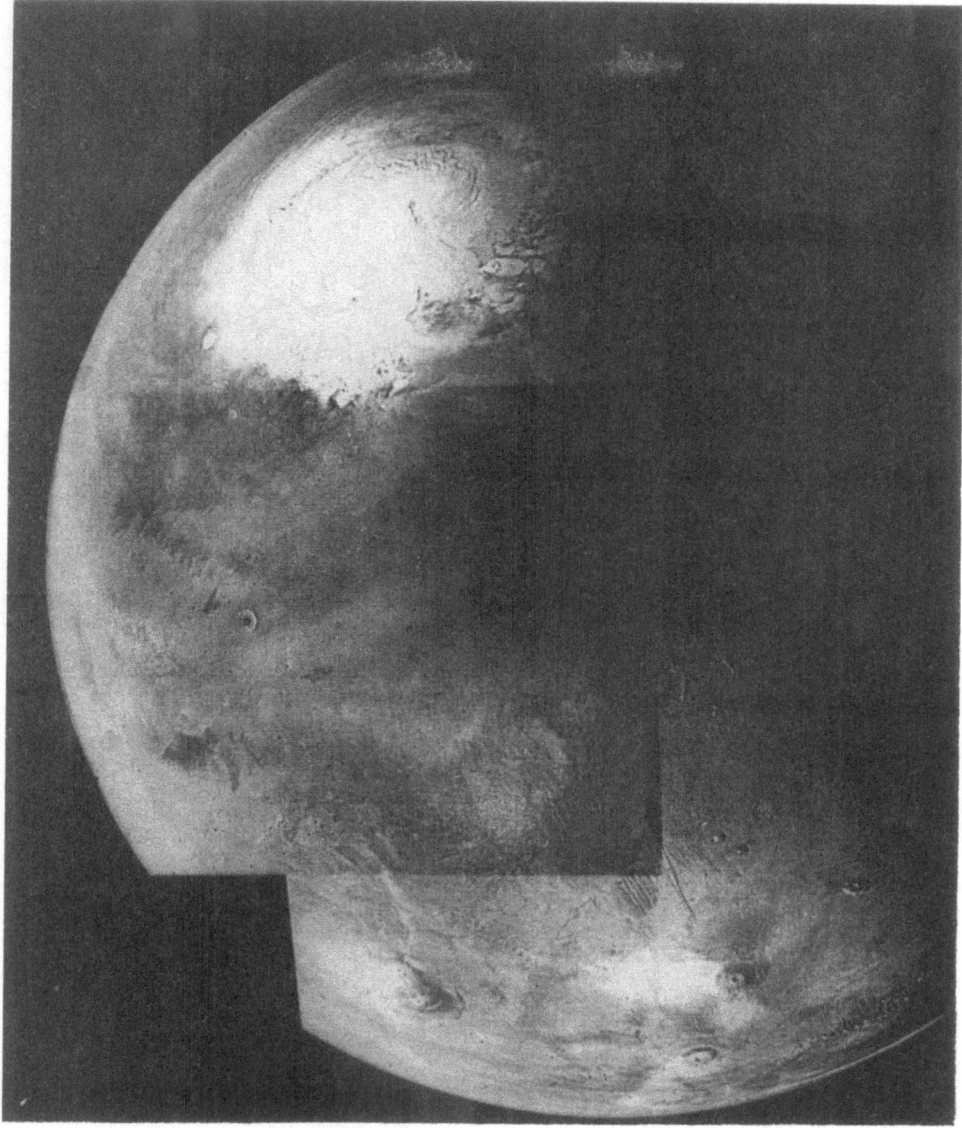

Fig. 8. Mariner 9 – global view of Mars.

ment of RTG power supplies, and initial exploration of the outer planets was started with the Pioneer 10 and 11 missions launched in 1972 and 1973. The primary purpose of these spacecraft is to characterize the Jovian magnetosphere by completing the missions shown in Figure 11. This information will be obtained when the spacecraft encounter Jupiter in December 1973 and December 1974, and will provide the data base for future trajectory and spacecraft design constraints.

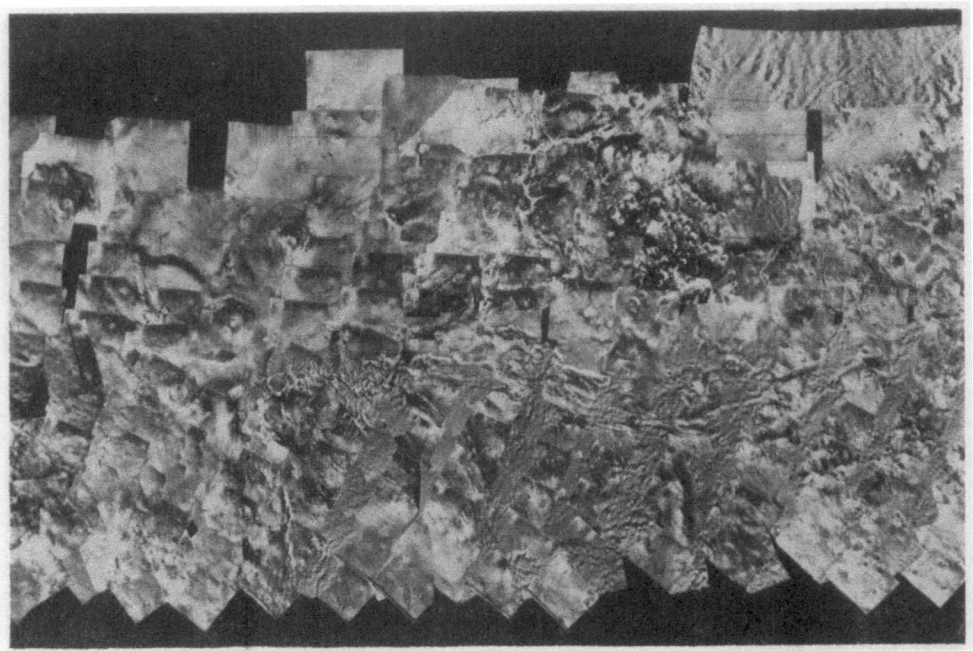

Fig. 9.   Mariner 9 – photomosaic of portion of Mars equatorial region.

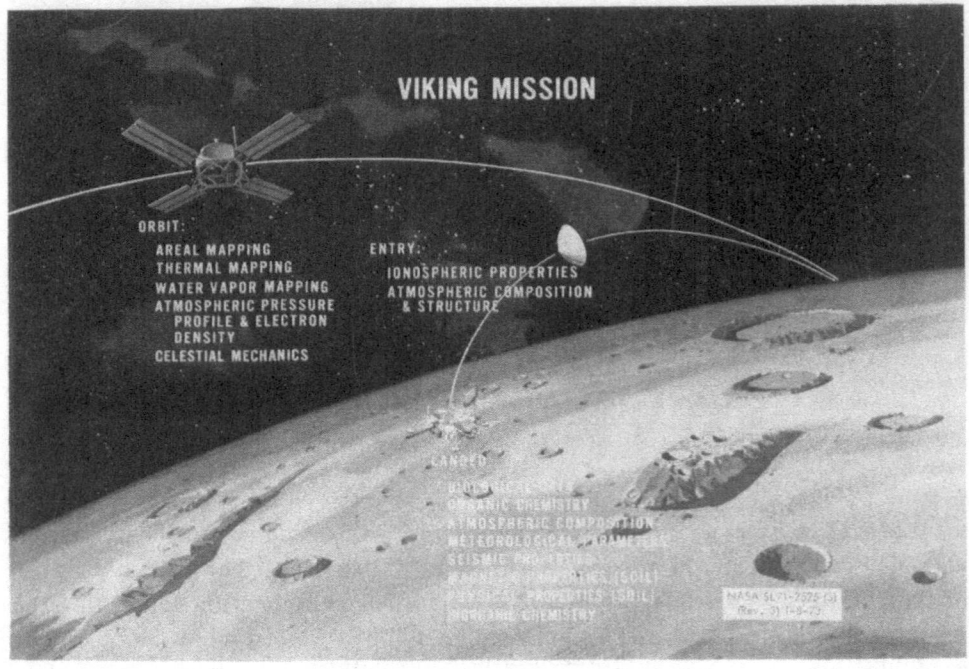

Fig. 10.   Viking mission to Mars.

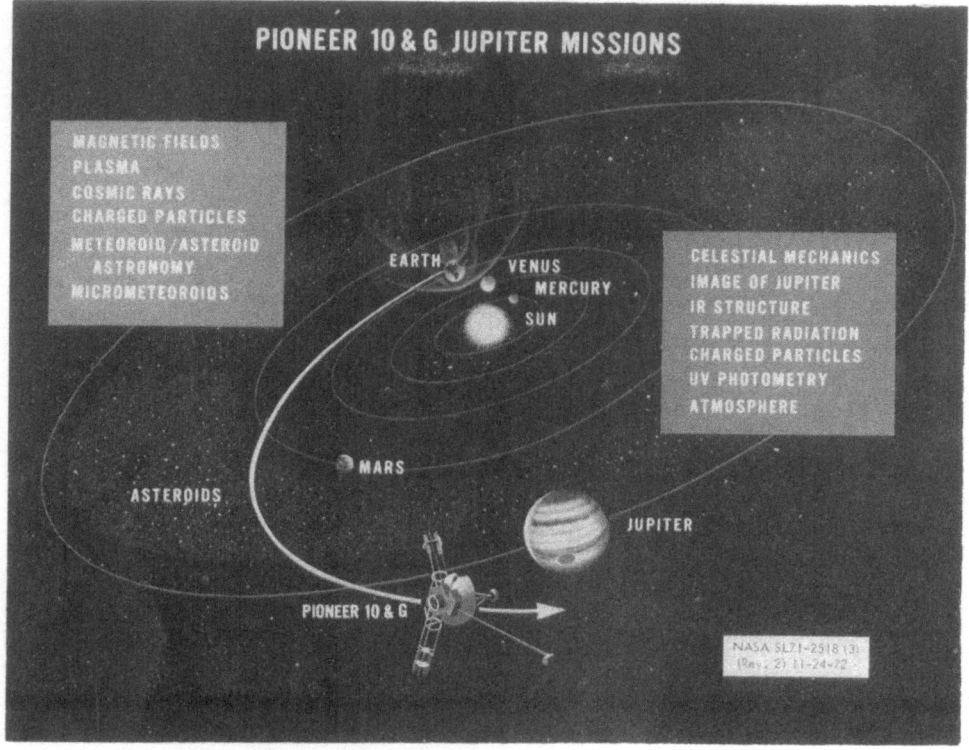

Fig. 11. Pioneer 10 and 11 Jupiter missions.

The launch vehicle capability constraint will be considerably alleviated with the introduction of the Titan/Centaur, and by the unique alignment of the outer planets in the late 70's and early 80's which allow Jupiter and then Saturn to be used as stepping stones to the more distant targets.

Detailed investigations of the outer planets will be pursued along two parallel paths, one emphasizing planetological studies with sophisticated flyby and orbiting spacecraft, and the other emphasizing atmospheric investigations with simpler flyby spacecraft delivering atmospheric probes.

The planetological program will commence with the Mariner Jupiter/Saturn mission in 1977. The objectives of this mission are depicted in Figure 12. The mission will provide close-up imaging in all spectral bands of Jupiter, Saturn, Saturn's Rings, and Titan, and will perform a variety of other science experiments.

In 1979, essentially the same spacecraft will use a Jupiter gravity assist to obtain similar data at Uranus. (Because of spacecraft life time limitations, it is not felt that Neptune could be reached on this mission, although energy-wise, it could be reached using a gravity assist at Uranus.) Following this mission, more detailed investigation of Jupiter is planned with orbiter missions in 1981 and 1982. These missions would provide detailed maps of the planet and its environment and multiple flyby encounters

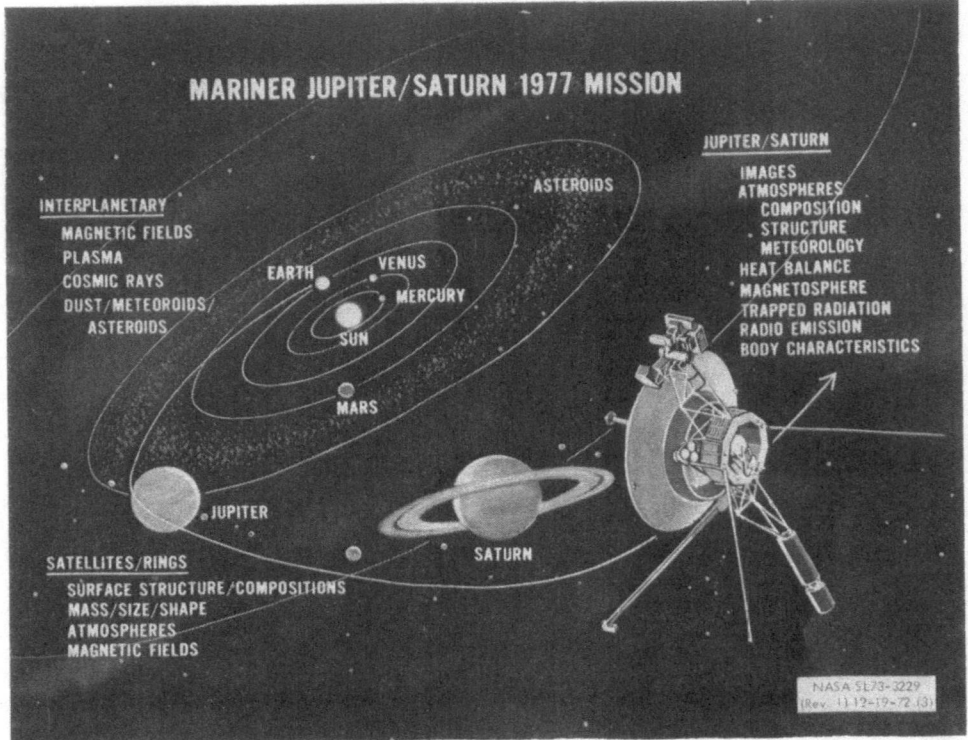

Fig. 12.   Mariner Jupiter/Saturn 1977 mission.

with each of the Galilean satellites. By 1985, the introduction of solar-electric capability should permit the orbiting of Saturn, and bring Neptune within reach using a Uranus gravity assist. The latter is planned by 1986. Finally, it is contemplated that the exploration of the outer planets would enter the lander stage, but this requires propulsion technology beyond anything currently planned. A Ganymede Lander is indicated in the 1990–1991 time frame primarily to serve as a technology driver.

Atmospheric investigation of the outer planets is less demanding, energy-wise, than the planetological exploration because of the generally lighter spacecraft required, and can proceed at a more accelerated pace. Atmospheric probes, derived from the Pioneer Venus technology, would be sent to Saturn in 1979 and to Uranus in 1980. Saturn and Uranus are selected for initial investigation rather than Jupiter because the probe technology requirements are considerably less severe. The third probe of this set will be launched with target selection to be performed en route. This probe could be targeted to Saturn, to Uranus, or possibly to Titan, depending on the status of the earlier probe missions. A Jupiter probe is then contemplated in 1984.

2.5. COMETS AND ASTEROIDS

There is growing interest in comet and asteroid missions, based on the supposition that they are composed of primordial matter essentially unchanged from the time of their composition. The small bodies thus may hold the best clues on solar system formation processes, and there is growing interest in their early investigation. This is hampered by two considerations. First, only limited information can be obtained from flybys of these objects, particularly asteroids. The very short encounter times severely restrict data acquisition, and their size and the uncertainty in their ephemerides make the problem of navigation for close encounters a difficult one. Second, rendez-vous using chemical propulsion systems is prohibitively expensive energy-wise because of the absence of any significant assistance from local gravity fields.

For many comets and asteroids these limitations can be removed with the introduction of solar-electric propulsion capability, and the mission model envisions a rendez-vous with the comet Encke during its 1984 perihelion passage. This would be preceded by a precursor mission during its 1980 perihelion passage which would provide a test bed for cometary science. This precursor could also provide the test flight for SEP, in which case both the coma and tail would be traversed; however, it is possible that fiscal constraints will not permit this. Consequently, a ballistic precursor mission to Encke in 1980 is being studied, with either a single or a dual spacecraft launch. In the single launch mode, the spacecraft would be targeted for the coma; in the dual launch case, the second spacecraft would traverse the tail region. On both the rendez-vous and flyby missions initial flybys of asteroids would also be accomplished, leading to an asteroid rendez-vous mission in 1986, and docking and sample return sometime in the 1990's.

Halley's comet will make a perihelion passage in 1986, and there is great interest in visiting this most famous body. Halley's orbit is such that no existing or contemplated propulsion system is capable of providing rendez-vous; however, interest is such that even a very fast ballistic fly through (55 km s^{-1}) has strong support. Therefore, a Halley's flyby mission is also being planned to be launched in 1985.

It is considered that the mission model presented here represents a realistic and well balanced program for solar system exploration, based on the fiscal and technological constraints as we presently understand them. The model, however, is readily subject to change as our understanding of the programmatic environment changes.

DISCUSSION

Gorgolewski: Do you intend to fly radio astronomy experiments to the outer planets?

Brunk: Yes, there is a radio astronomy experiment being considered for the Mariner-Jupiter-Saturn mission. Such experiments may also be flown on later missions. All programs will also make maximum scientific use of the spacecraft transmitter for radio occultation experiments, etc.

Dollfus: What are the NASA programs for asteroids and comets?

Brunk: There are several missions being studied, such as a Comet Encke mission and one to Halley's Comet. Also under study are multiple flyby missions which would include one or more asteroids and one or more comets.

CONCLUDING REMARKS

T. OWEN

New York State University, Stony Brook, N.Y., U.S.A.

I have been asked to make some closing remarks but I shall try to be very brief because we have had a full program and I think we are all rather tired.

We began with some general considerations about the origin and evolution of the solar system and we have proceeded from that point through many discussions of new observations and new ideas about the atmospheres, surfaces, interiors and magnetic fields of our neighbouring planets. It has been clear from these discussions that we are living in an age during which the exploration of the solar system is being pursued with unusual vigor – helped by much sophisticated equipment, and especially by the results given to us by spacecraft which are actually visiting these planets.

I think it is a special honour for all of us to be having these discussions in Toruń, on the occasion of the 500th anniversary of the birth of Copernicus. Throughout this symposium, I have been fascinated with the picture of Copernicus that is hanging on the wall. In this particular portrait he appears to be almost smiling, and I have been asking myself: what is the meaning of this smile? It seems to me it is the kindly expression of a very wise man whose vision was so profound that his existence divided astronomy into two eras – pre-Copernican and post-Copernican. Perhaps he is slightly amused at the slow rate of progress in the understanding of the solar system that has been achieved in the 500 years since he was born.

But I would like to offer the following perspective. It seems that perhaps Copernicus represents the pure simple childhood of astronomy. Indeed his vision is similar to the difficult moment in the life of a child when he discovers that he is not the center of the universe as he thought when he was a baby. It is a wise child who happily survives the shock! In contrast, I would suggest that we are living in a stormy adolescent phase of solar system science, rich in passion and new discoveries. Nevertheless, we continue to carry vestiges of pre-Copernican thinking with us – the very designation 'terrestrial planets' is one example, and this Earth-centred view undoubtedly contributed to the mistaken notion that Venus and Mars must have atmospheres rich me nitrogen like our own, a notion that has died out only in the last decade.

Must we wait another 500 years for maturity? I think not. It seems possible to us that 70 years from now, when perhaps in this same auditorium our grandchildren will be commemorating the 500th anniversary of the death of this great man, they will have such an immense variety of information about the planets at their disposal that their perspective may once again achieve the unifying simplicity that distinguishes the Copernican world view.

Meanwhile, it is up to us to do our best to help this come about, and surely international meetings such as this one are an excellent way to stimulate research in this exciting field.

Woszczyk and Iwaniszewska (eds.), Exploration of the Planetary System, 563–564. All Rights Reserved.
Copyright © 1974 by thr IAU

As one of the last speakers in this symposium, I should like once again to express our deep appreciation to Dr Woszczyk, to Prof. Iwanowska, to the Rector of the University, to the Chairman of the City Council and to their many associates for sharing their beautiful city and its fine university so graciously with us on this occasion.

Dziękuję bardzo. Thank you.

CHAIRMAN'S CLOSING REMARKS

P. SWINGS

Astrophysical Institute, University of Liège, Liège, Belgium

Dear Colleagues,

The Copernicus Symposium No. IV or IAU Symposium 65 which comes to an end has been a great success. I am tempted to be selfish and to suggest that our Polish colleagues repeat such a Copernican Symposium on the planetary system (or something related, such as cometary or asteroidal symposium) periodically, say every 5 years, always carrying the name of the great genius of Toruń, Nicholas Copernicus.

The success of the symposium is due to the tremendous amount of work performed by Prof. Iwanowska and her associates, especially Drs Woszczyk, Iwaniszewska, Stawikowski and others, among them, a special emphasis being placed on the role of Andrzej Woszczyk.

Let us not forget the great encouragement given by the academic authorities, especially the Rector Magnificus of the Copernicus University, and also the encouragement given by the national and local authorities. I want to tell these authorities that all the foreign scientists who are here appreciate this help given to their colleagues from Toruń, who deserve it wonderfully.

We shall be happy to have the communications published as speedily as possible. I wish to tell all the participants how happy the organizers have been to have such a large and interested audience.

To all these colleagues the organizers wish a pleasant journey back home. I also thank the COSPAR for having co-sponsored our symposium.

To the local organizers we say again 'Dziękuję bardzo', 'thank you very much'.

INDEX OF AUTHORS